Molecular Mechanisms and Physiology of Disease

Nilanjana Maulik • Tom Karagiannis
Editors

Molecular Mechanisms and Physiology of Disease

Implications for Epigenetics and Health

Springer

Editors
Nilanjana Maulik
Department of Surgery
Molecular Cardiology and
Angiogenesis Laboratory
University of Connecticut
School of Medicine
Farmington, CT, USA

Tom Karagiannis
Epigenomic Medicine
Baker IDI Heart and Diabetes Institute
Melbourne, VIC, Australia

ISBN 978-1-4939-0705-2 ISBN 978-1-4939-0706-9 (eBook)
DOI 10.1007/978-1-4939-0706-9
Springer New York Heidelberg Dordrecht London

Library of Congress Control Number: 2014938027

Printed on acid-free paper

Springer is part of Springer Science+Business Media (www.springer.com)

This book is dedicated to our beloved grandmother (didibhai) for being there always for us

Nilanjana Maulik, Ph.D., F.A.H.A.

Preface

Although coined in the 1940s by Conrad Waddington and now representing an intense field of biomedical research, the precise definition of Epigenetics remains controversial. In its simplest form epigenetics refers to heritable changes that are not due to changes in the underlying DNA sequence. Whereas DNA methylation is a well-known and relatively well-characterized epigenetic mechanism, it is still debatable as to whether other processes such as histone posttranslational modifications represent epigenetics per se, given their transient nature. In this volume, we discuss DNA methylation, histone posttranslational modifications and their alteration by chromatin-modifying compounds, and regulation of gene expression by noncoding RNA and mi-RNA as part of the epigenetic umbrella. Indeed, this volume is quite broad consisting of an array of topics including epigenetic effects in various diseases such as autoimmune conditions, cardiovascular diseases, and asthma. The second part of this volume is dedicated to cancer and highlights epigenetic dysregulation in malignancy as well as a number of chapters related to emerging cancer therapeutics.

In Chap. 1, Hussain discusses epigenetic mechanisms associated with childhood diseases. Hussain provides an excellent overview of epigenetic processes in health and disease. This chapter provides a detailed overview of various childhood conditions with an epigenetic association, including various imprinting disorders, childhood malignancy, and diabetes. This is a very extensive chapter, and it highlights the broad spectrum of childhood diseases that can be linked with epigenetic perturbations. Detailed molecular and epigenetic aspects are further explored in Chap. 2 by Torano et al. in the context of neuronal differentiation. In this chapter, Torano et al. describe epigenetic processes including DNA methylation, histone posttranslational modifications, chromatin remodeling and regulation by noncoding RNA in great detail in the context of maintenance of pluripotency. They further explore how aberrant epigenetic mechanisms are associated with a myriad of neurological diseases and malignancy. Further, a detailed overview of epigenetics is provided by Westerland in Chap. 3. This chapter provides a thorough outline of the major epigenetic mechanisms and potential interventions using DNA methyltransferase inhibitors (DNMTs) and histone deacetylase inhibitors (HDACi), compounds which are further explored in the following chapters. In a different light, regulation by modifying the histone tails by proteolytic processing is described in Chap. 4 by Mandal et al. Although not as well characterized as the other processes such as histone acetylation or methylation, as described by Mandal et al., proteolytic processing of the core histones is emerging as an important form of regulation of chromatin organization.

In Chap. 5, Turker et al. describe health implications associated with probiotics and their metabolites. This chapter focuses on anti-inflammatory effects of major probiotic metabolites which include the well-known short chain fatty acid HDACi, butyrate. In Chap. 6, Coppede and Migliore provide an excellent outline and overview of the epigenetic mechanisms associated with autoimmune diseases. Indeed, in Chap. 6 epigenetic phenomena, particularly aberrant mi-RNA processes and histone posttranslational modifications, related to major autoimmune conditions including systemic lupus erythematosus and rheumatoid arthritis are identified and described. Emerging evidence for deregulated epigenetic mechanisms associated with other autoimmune diseases such as Sjrogren's disease, psoriasis, and multiple sclerosis is also discussed in Chap. 6. Chapter 7 continues the disease-specific consideration with a discussion by Whayne, of epigenetic processes associated with cardiovascular disease. This also describes aspects of nutrition, including the well-known methyl-donor folate and tobacco use that are associated with disease. Nutritional aspects are further expanded in Chap. 8 by Burgio and Migliore, in the context of obesity and diabetes. This chapter details genetic and epigenetic mechanisms associated with metabolic syndrome, diabetes, and cardiovascular disease.

Both Chaps. 9 and 10 by Tortorella describe genetic and epigenetic mechanisms and emerging therapies associated with asthma. Traditionally, asthma has been viewed as a disease with a heritable genetic component which is exacerbated by various environmental exposures in early childhood or later in life. It is also becoming apparent that prenatal exposure can influence the risk of developing asthma in accordance with the Barker hypothesis of fetal programming (i.e., perturbations in nutritional or environmental conditions in utero lead to altered developmental programming of organs, influencing the propensity to develop the disease later in life). This idea has been investigated predominantly by exploring the use of tobacco and increased risk of developing various lung pathologies, including asthma later in life. In Chap. 9, Tortorella overviews epigenetic effects associated with prenatal and postnatal environmental exposures. Potential therapeutic avenues in the form of trefoil factor 2 (Chap. 10) and emerging nanotechnologies (Chap. 10) for managing asthma are also explored.

Although mechanisms in cancer formed parts of some of the preceding chapters, the remaining chapters are focussed entirely on the aspects of malignancy including carcinogenesis, cancer metabolism, and emerging cancer therapeutics. For example, in Chap. 11, Masih et al. describe the epigenetic effects of one-carbon metabolism and aberrant DNA methylation in cancer. In this chapter, Masih et al. provide a thorough overview of nutrients involved in one-carbon metabolism including folate, vitamins B6 and B12, choline, and betaine. This chapter also provides a comprehensive review of the effects of these nutrients in ten common malignancies of the gastrointestinal and reproductive systems. A different aspect of aberrant metabolism, namely the Warburg effect, in cancer is the subject of Chaps. 12 and 13 by Molino et al. and Balding et al., respectively. The Warburg effect which postulates that cancer cells predominantly utilize aerobic glycolysis rather than mitochondrial respiration was first described in the 1920s. Although various small groups continued the work sporadically, it was not until the past few years that this topic is reemerging and has been recognized as a critical component of cancer biology. Although the epigenetic component of the Warburg effect is still not well characterized, there is emerging evidence for important links. Indeed, the epigenetic component of all aspects of carcinogenesis is now widely recognized, and epigenetic lesions are now not only being considered as the hallmarks of disease but also being incorporated into the multi-hit model. This is described and characterized by Migheli and Migliore in Chap. 14. Keeping on the topic of epigenetics, nutrition, and metabolism in Chap. 15, Pan et al. describe the effects of nutrition and energy intake in colon cancer. In this chapter, the key colon cancer-associated oncogenes and tumor suppressor genes are discussed in the context of tumor progression, and the cancer cell growth inhibitory effects of nutritional factors are described. Importantly, the potential prophylactic role of an anti-inflammatory diet in colon cancer is outlined.

The final three chapters in this volume deal with potential cancer therapeutics. In Chap. 16, Mazarakis describes the potential protective and therapeutic effects of dietary antioxidants and chromatin-modifying compounds in cancer. This chapter focuses on phenolic compounds from olive and HDACi from a variety of foods. Similarly, the potential of HDACi in combinatorial therapies, in this case phototherapy, is discussed in Chap. 17 by Sung. HDACi have emerged as an important new class of anticancer therapeutics with two compounds, suberoylanilide hydroxamic acid (Zolinza) and depsipeptide (Romidepsin), being approved by the FDA for the treatment of cutaneous T-cell lymphoma and more recently depsipeptide for peripheral T-cell lymphoma. It is widely accepted that the clinical utility of HDACi will predominantly involve combination with other anticancer therapeutics. In Chap. 17, the potential anticancer effects of combinations of HDACi with ultraviolet phototherapy are considered. Anticancer therapy is also the subject of Chap. 18 by Mah et al. in which, nanoparticle formulations for targeted drug delivery are described. There is much excitement regarding the potential of nanoparticles to deliver cytotoxic agents including epigenome-modifying siRNA to selectively induce apoptosis and cell-death in cancer cells. This chapter outlines varying approaches for appropriate nanoparticle preparations with potential clinical applicability.

Overall, this volume encompasses a wide range of topics related to epigenetic mechanisms in health and disease. The scope of the volume spans from descriptions of fundamental epigenetic processes to potential epigenetic interventions for preventing or treating various diseases. Epigenetic phenomena associated with numerous conditions including autoimmune diseases, cardiovascular disease, asthma, and a variety of malignancies are detailed. Given the scope, this volume would be of appeal to a wide readership including those with interests in epigenetics and chromatin biology, disease-specific epigenetic aberrations, and emerging epigenetic-based cancer therapies.

Farmington, CT, USA — Nilanjana Maulik
Melbourne, VIC, Australia — Tom C. Karagiannis

Acknowledgments

My deepest thanks to all the authors for their contributions. I would also like to thank my colleagues, friends, and students without whom this book would have been a distant reality. I also extend my heartfelt thanks to my husband who always encouraged and helped me to complete the book. I thank you all.

Nilanjana Maulik, Ph.D., F.A.H.A.

I am eternally grateful to my friends, students, and colleagues in the Epigenomic Medicine Laboratory at the Baker IDI Heart and Diabetes Institute for their help and support. You have all made my life a bit easier. This volume is dedicated to all of you.

Tom Karagiannis, Ph.D.

Contents

Chapter 1
Epigenetics in Childhood Health and Disease

Naveed Hussain

Abstract The part that epigenetic modifications play in the development of childhood health and disease is being established by ongoing research and discoveries in this field. Right after the establishment of the genetic blueprint at the time of fertilization and zygote formation, the human organism is subject to complex and necessary series of epigenetic modifications of this genetic code to bring about differentiation and development. There are well-recognized stages during this process where the epigenetic changes have the most lasting and profound effects and these are considered critical periods of vulnerability. Depending on the timing of insult within the critical time periods in the human life cycle where epigenetic modifications occur, the effect on health and disease could be transient or may persist across many generations. In this chapter classification of human conditions based on the timing and etiology of epigenetic change has been attempted. Beginning with the time of fertilization of the egg with the sperm and subsequent fetal development and continuing from birth to the attainment of puberty, adulthood, and the generation of gametes for the next generation, the list of conditions where epigenetics has been found to play a key role have been listed and described. The role of epigenetics in certain special circumstances such as assisted reproductive technologies, developmental origins of adult disease, and in the brain and behavioral disorders are also discussed. Understanding the critical period of causation of epigenetic effects may yield important clues in prognostication and in designing therapeutic approaches for these conditions.

Keywords Infant • Newborn • Epigenetics • Imprinted genes • Development • DNA methylation • Histone modification • microRNA • X-inactivation

---

N. Hussain, M.B.B.S., M.D., D.C.H. (✉)
Department of Pediatrics, University of Connecticut School of Medicine,
263 Farmington Avenue, Farmington, CT 06030-2948, USA

Division of Neonatology, Connecticut Children's Medical Center, Hartford, CT, USA
e-mail: hussain@uchc.edu

N. Maulik and T. Karagiannis (eds.), *Molecular Mechanisms and Physiology of Disease: Implications for Epigenetics and Health*, DOI 10.1007/978-1-4939-0706-9_1,

1.1 Introduction

If genetics and the transmission of the DNA code could be considered the indelible ink that writes on the parchment of inheritance, epigenetics could be thought of as the annotations made by pencil to this document; not surprisingly, erasures and alterations to this penciled code could leave smudges and marks that could alter the reading of this manuscript (Gosden and Feinberg 2007). If genetics is the link of transmission of DNA code from generation to generation, epigenetics is the interpretation of this code in individual cells and tissues of the organism within a particular generation based on its unique environment. Therefore it is not surprising that a single-celled zygote with one set of genetic instructions can differentiate into tissues as varied as bone and brain and that identical twins can differ significantly in their normal phenotypes and susceptibility to diseases.

1.2 Waddington and Historical Background

Waddington in 1952 described the term epigenetics—"the science concerned with the causal analysis of development and the regulative capacities of the early embryo" (Waddington 1952). It was subsequently recognized as a "science of developmental processes in general" by Huxley (1956). Recently, the field of epigenetics has helped provide important clues to determinants of health and disease in animal models and humans. The use of epigenetics is also being made in drug development and design (Marx 2012). Various processes have been found to be responsible for inducing epigenetic effects with the most well known being DNA methylation, histone modifications, and microRNA (miRNA) modulation of transcription.

The discovery of genetic imprinting led to the identification of DNA methylation as the first biological process linked with epigenetic modifications. Genetic imprinting, which distinctly involves the maternal or paternal genome, was first described using gamete transfer experiments with mice in 1984 (McGrath and Solter 1984; Surani et al. 1984); the first imprinted gene *igf2r* was identified soon thereafter in 1991 (Barlow et al. 1991; Bartolomei et al. 1991; DeChiara et al. 1991). Correlation of genetic imprinting with human disease was first reported for Prader–Willi syndrome (PWS) in 1989 (Nicholls et al. 1989); although it was later in 1993 that DNA methylation was identified as the process responsible for genetic imprinting (Li et al. 1993).

Other mechanisms for epigenetic modifications were soon discovered pertaining to histone modifications. The state of nuclear chromatin whether tightly compacted (heterochromatin) or loose (euchromatin) has long been known to be associated with silencing or activation of DNA function and thus important in regulation of cell function (Brown 1966). However, it was in late 1990s that the role of histone proteins in folding of chromatin and altering availability of active regions of genes to transcriptional factor regulation was first suspected (Jones et al. 1998; Nan et al. 1998);

followed soon after by a more definitive review implicating histone modifications in epigenetic regulation (Jenuwein and Allis 2001). Of all the ways histone proteins could be modified (methylation, acetylation), acetylation is perhaps the most important. Histone acetyl transferases (HATs) and histone deacetylases (HDACs) are key enzymes in regulating the balance of lysine acetylation of histones.

In early 2000s, another mechanism of epigenetic regulation was found to be the microRNA-induced modulation of mRNA transcription. microRNAs are evolutionary conserved noncoding nucleotides with 19–24 bases that regulate the translation and degradation of specific mRNA targets based on base pairing to complementary sites mainly in the 3′-untranslated region of the target mRNAs (Ambros 1989; Lagos-Quintana et al. 2001; Lau et al. 2001; Lee and Ambros 2001; He and Hannon 2004). miRNAs are uniquely organized within the genome and can be transcribed either from distinct intergenic loci or as a by-product of a host gene (Chen and Meister 2005; Bartel 2009). With the completion of the ENCODE project involving the "silent" non-gene-linked areas of the human DNA, it is being increasingly recognized that other noncoding RNAs may also play an important role in epigenetic regulation (Maher 2012).

There appears to be a close interaction between various mechanisms of epigenetic regulation. Some of the genes regulated by DNA methylation are for long noncoding RNAs (lncRNA) such as *H19* involved with *IGF2* regulation. DNA methylation can initiate chromatin condensation and histone modifications through methyl-DNA binding (MDB) proteins. Moreover imprinted domains transcribe numerous small noncoding RNAs (sncRNAs defined as <105 base pairs) which include a number of microRNAs (<24 bp). The imprinted domains in humans have the highest density of microRNAs with almost 7 % of microRNAs in humans encoded in this region (Labialle and Cavaille 2011; Girardot et al. 2012). Therefore the strict categorization of epigenetic processes is probably an oversimplification of a complexly interrelated process; but for the purpose of description and classification these processes will be dealt with separately.

1.3 Scope of the Chapter

Mammalian life cycle is characterized by an initial near complete reset of genetic potential (establishing totipotency) at the stages of primordial germ cells (PGCs) and early preimplantation embryo (pre-IE); with later epigenomic reprogramming in post-implantation embryo (post-IE); followed by extended phases of cellular differentiation, unidirectional specific development, and effective lineage-restriction modulated through a complex network of transcription factors (Hackett and Surani 2013). The epigenome is represented by a combination of systems of DNA methylation, histone modifications, and small RNA influences regulating gene expression through mitosis and meiosis of cells in the developing organism. The epigenome plays a significant role in gene–environment interactions and it is likely that there are critical periods in early development where the normal process of establishment of the epigenome puts the organism at higher risk. In this report we plan to use the "critical period" approach in categorizing the epigenetic basis of disease and altered development. Most of the discussion in this chapter is based on DNA methylation changes and effects related to other epigenetic mechanisms such as histone modifications and microRNA will be mentioned briefly as appropriate. A detailed discussion of changes related to histones and microRNA is beyond the scope of this chapter.

1.4 Concept of "Critical Period"

The genetic blueprint of an organism is constantly influenced by its environment for the ultimate expression of the phenotype. The capability of the organism in responding to the environmental influences indicates its "developmental plasticity" (Burggren and Reyna 2011). Developmental plasticity connotes not only a manifestation of genotype–phenotype interaction during development but also defines a predictable series of reaction norms that persist at the individual, population, or species level (Hutchings 2011; Symonds et al. 2009). If the molecular basis of this developmental plasticity is transmittable across generations, it is probably a result of epigenetic modifications of the genome. Critical period of hormonal activity and its epigenetic actions have been well documented for sex differences in the brain (Nugent and McCarthy 2011) and also been implicated in the "Developmental Origins of Health and Disease (DOHaD)" hypothesis (Martin-Gronert and Ozanne 2012). Certain developmental transcription factors that have been shown to be epigenetically programmed by the early environment are PPARα, PDX-1, and HNF4α (Park et al. 2008; Martin-Gronert and Ozanne 2012; Lillycrop et al. 2005; Sandovici et al. 2011). The importance of the timing of insult causing the epigenetic changes is also highlighted in the Dutch Famine Cohort where it was noted that certain genes are affected by insults around the perinatal period (*IL10*, *INISIGF*, *LEP*, *ABCA1*, and *MEG3*) but in other genes such as *GNASAS* epigenetic changes were associated with exposure to famine later in gestation (Lumey et al. 2007).

The effect of a stressor (both in dosage and duration) in modifying genotypic expression varies based on the innate biological processes that are ongoing in that particular organism's development. There are certain periods where complex innate developmental changes provide a window of vulnerability that may be termed "critical periods." During this critical period, the dose and/or duration of the stressor may move the organism to a different trajectory of development. A certain threshold may determine if the change is permanent or temporary. The site of the change, whether epigenetic or non-epigenetic will determine if the change may persist in one or many generations (Burggren and Reyna 2011). Thus there may be a biological basis for the truism attributed to Friedrich Nietzsche: "That which does not kill us makes us stronger"; maybe not quite "stronger" but certainly "more adaptable."

The difference in timing of generation of male and female gametes and the differences that exist in the hormonal milieu in the two sexes during periods of development may be part of the reason that sexual dimorphism is an important consideration in determining the critical period of epigenetic vulnerability (Vige et al. 2008). This is also an important consideration in designing experimental and evaluating epidemiologic studies (Vige et al. 2008).

Another important consideration is the relationship of a particular genetic or epigenetic modification to other interactions within the organism to what is being increasingly referred to as the "interactome" (Venkatesan et al. 2009). Using network-based approaches to understanding systems biology, it is getting increasingly recognized that genes and proteins constitute nodes that link to a network with central and peripheral hubs in which the central hubs are associated with vitally essential elements the abnormalities of which are embryo-lethal and hence do not manifest disease; the peripheral hubs of relatively nonessential elements being compatible with survival are responsible for disease (Barabasi et al. 2011; Barabasi and Oltvai 2004). Since each cell within an individual organism has a unique, time-varying epigenome, an integrated molecular pathological epidemiology (MPE) approach may be needed to fully understand and classify these processes (Ogino et al. 2013). To some extent these above mentioned approaches have been used in this chapter to classify conditions that have an epigenetic basis.

1.5 DNA Methylation Dynamics During Life Cycle

Of the various mechanisms of epigenetic regulation, DNA methylation dynamics during the mammalian life cycle are most well studied (Nafee et al. 2008) (Fig. 1.1). Similar life cycle patterns for histone modifications and noncoding RNA regulation are still under intense investigation. DNA methylation occurs at 5′ of cytosine (5mC) within a CpG dinucleotide sequence and is critical for embryonic development in mammals. DNA methylation plays an important role not only in genomic imprinting but also in gene repression, X chromosome inactivation, and transposon silencing among other cellular processes (Bird 2002). The most common sites within the genome for DNA methylation is the intergenic region and in repetitive sequences (such as satellite repeats, and long and short interspersed nuclear elements—LINEs and SINEs). However, the promoter sequences of genes which are mostly GC-rich are usually unmethylated (Weichenhan and Plass 2013). This may be due to the fact that 5mC in DNA is inherently unstable and tends to deaminate to thymidine thus depleting the bulk of genome of CpG motifs except where they are clustered in regions known as CpG islands (CGIs). The CGIs are usually associated with promoters of genes. Generally most CpG sites within the matured genome are by default methylated irrespective of the genetic context but the CGIs remain unmethylated during development (Meissner et al. 2008; Suzuki and Bird 2008; Hackett and Surani 2013).

The mechanism by which methylated DNA (methylated cytosine—5mC) controls gene expression is by the attraction and binding of MDB proteins and the subsequent chromatin condensation into a transcriptional repressive configuration

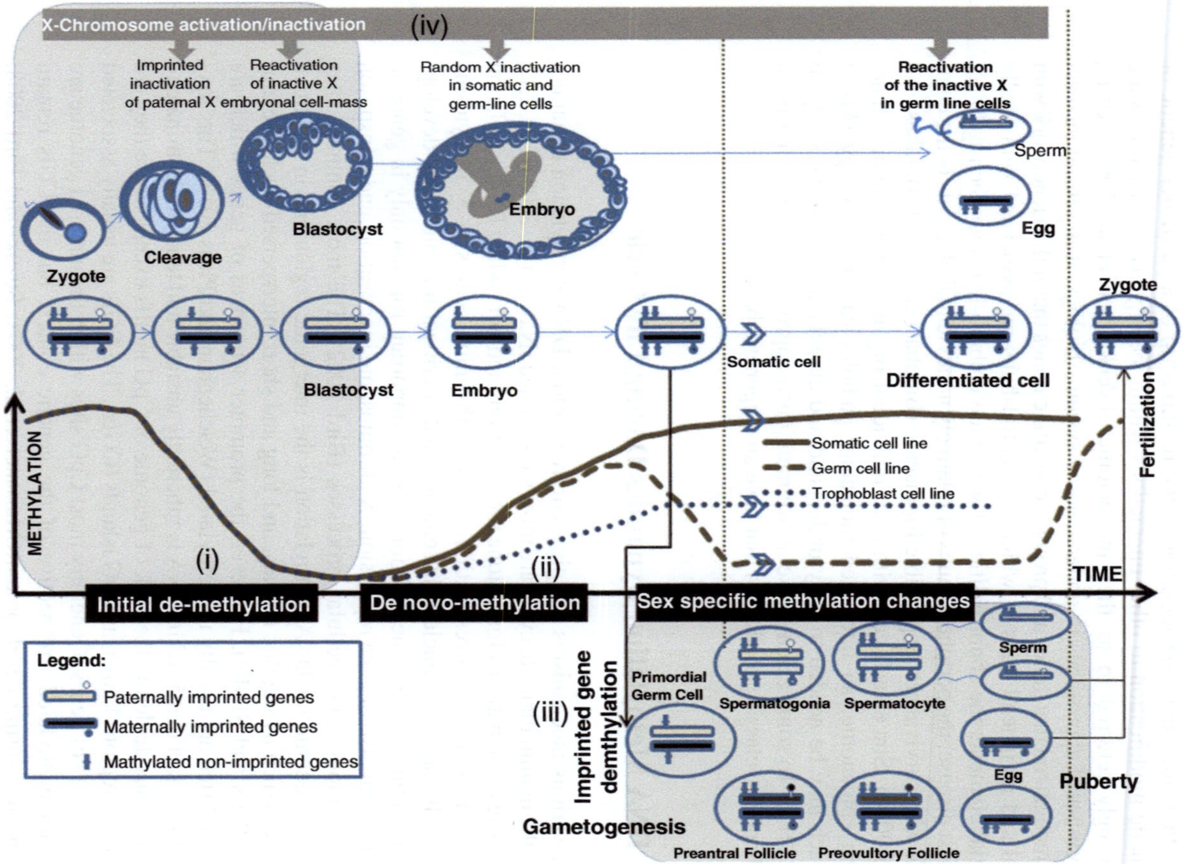

Fig. 1.1 Periods of epigenetic changes human development [adapted from (Nafee et al. 2008)]. The change is methylation status, i.e.,—(i) initial demethylation, (ii) de novo methylation, for somatic, trophoblastic, and (iii) sex-specific methylation changes in germ cell line, are depicted on an arbitrary time line. Also superimposed on the same timeline are timings of (iv) X chromosome activation and inactivation to illustrate how critical periods of vulnerability are concentrated in the early stages of human development [*reprinted with permission from ARS*]

(Bird 1992; Newell-Price et al. 2000). Four mammalian MDB proteins identified so far are MeCP2, Mdb1, Mdb2, and Mdb4 (Lan et al. 2010).

During mammalian development, immediately after fertilization, DNA derived from both gametes undergoes dynamic remodeling to give rise to a globally demethylated state or a metaphoric "clean slate" on which subsequently, a progressively lineage-specific methylation pattern (methylome) is established for the uniquely new individual and then maintained through subsequent mitosis (Hackett and Surani 2013; Wossidlo et al. 2011). The zygote with subsequent stages of cell division and implantation continues with extensive demethylation giving rise to the pluripotent inner-cell mass which has the capability to form embryonic stem (ES) cells in vitro. Embryonic lineage specification then occurs simultaneously with de novo methylation as the organism locks-in more specific cellular identity. At a future point (corresponding to E6.5 in mice and post-conception day 13–15 in humans), the inner cell mass continues to progress with differentiation and cellular specialization of somatic tissue; the cells destined to form PGC undergo another demethylation phase (E12.5 in mice) that lasts until gamete-specific methylome is developed and imprinting of maternal and paternal genes is completed (Hackett and Surani 2013; Horsthemke 2010). Interestingly, in the germline processes remethylation of the genome completes in the male germline before birth (at approximately embryonic day 18.5 in mice) (Davis et al. 2000); but in the female, remethylation of the oocyte does not occur until after birth and puberty when it is initiated every time a crop of follicles is recruited (Reik and Walter 2001). It is also worth noting that during the initial phase of demethylation after fertilization, the imprinted genes of the germline and the gametes do not get demethylated and preserve their methylation imprint. These gametes when they come together at fertilization from different parents to form the zygote start the cycle of demethylation once again. Importantly the erasure and reestablishment of parent of origin-specific imprints is vital for creating totipotency in the zygote (Hajkova et al. 2002).

Throughout cell division, the maintenance or preservation of methylation of DNA is done by the ubiquitously expressed DNA methyltransferase 1 (DNMT1) which methylates the hemimethylated CpG dinucleotides in the nascent DNA after replication (Hermann et al. 2004). Another important factor in maintenance of methylation is the KRAB zinc-finger protein ZFP57 (Li et al. 2008). When needed, for establishment of new methylation patterns, de novo DNA methyltransferases 3A and 3B (DNMT3A and DNMT3B) are utilized and DNMT3L may act as their cofactor (Hata et al. 2002). Demethylation was previously thought to occur passively in the absence of DNMT1 but recently active demethylation systems have been identified. Active demethylation is mediated by Tet-methylcytosine dioxygenases (Tet 1 and Tet 3) that convert 5-methylcytosine by oxidation to 5-hydroxymethylcytosine and then to 5-formylcytosine (5fC). Subsequently, base excision repair pathway and thymine-DNA glycosylase (TDG) may be involved in removal of these oxidation products (He et al. 2011; Tahiliani et al. 2009; Rivera and Ross 2013; Seisenberger et al. 2013).

Given the complexity of DNA methylation processes that occur at various times in the life cycle, it is not surprising that there may be many critical periods of vulnerability where external or inherited factors may alter the methylome and depending on the timing of these alterations, the effect may be seen in the zygote or the inner cell mass or the developing germline or the fully differentiated somatic cell; consequently the ultimate expression of the change may be seen in the whole organism within its lifetime or through the gametes manifest the change in a future generation. A good example of the transgenerational effect is the development of cancer seen in female offspring of women who were exposed to diethylstilbestrol (DES) during the germline development of their fetus in early pregnancy (Li et al. 2003; Walker and Gore 2011).

1.6 Classification of Epigentics Related Disorders in Childhood

Based on our understanding of DNA methylation dynamics during the human life cycle, it is possible to classify disorders of DNA methylation and other epigenetic mechanisms by the critical time period in which the susceptible cells are affected, starting from the PGCs developing in the genital ridge of the embryo to the transmission of these epigenetic marks to the subsequent generation at the point of fertilization (Fig. 1.2). Long-term effects of the epigenetic change would probably depend on the timing of the change and to the potential generation of cells and tissue that particular cell can produce. In the case of PGCs, the effect of epigenetic imprints may last up to the F3 generation of the organism (Skinner et al. 2013); but an epigenetic change in a mature somatic cell will probably be restricted to that cell's specific progeny in the tissue (Teschendorff et al. 2013). The former could result in significant organism-wide changes involving multiple tissue lines and the latter could significantly impact a single tissue as in the development of malignancy. For the purpose of this chapter, mostly DNA methylation-related epigenetic changes are described and changes related to histone modifications and microRNAs will only be mentioned as appropriate.

A comprehensive classification of all disorders associated with epigenetic changes is provided and most of these disorders have an effect on childhood health and disease. Broadly the disorders that result in epigenetic changes may be classified into two main categories:

1. Primary epigenetic disorders—(Table 1.1) where the changes primarily affect the epigenome without any accompanying or predisposing genetic changes. The inheritance of these disorders is variable and the ability for therapeutic manipulation is more robust because the underlying functional proteins involved are not defective but only their regulation is altered.
2. Secondary epigenetic disorders based on primary genetic conditions—(Table 1.2) where epigenetic changes occur because of primary changes in the genome and its gene product. The inheritance of these disorders follows classical Mendelian or chromosomal inheritance patterns. Since a defect in functional protein may be responsible for the changes seen, the ability for therapeutic manipulation is much more limited.

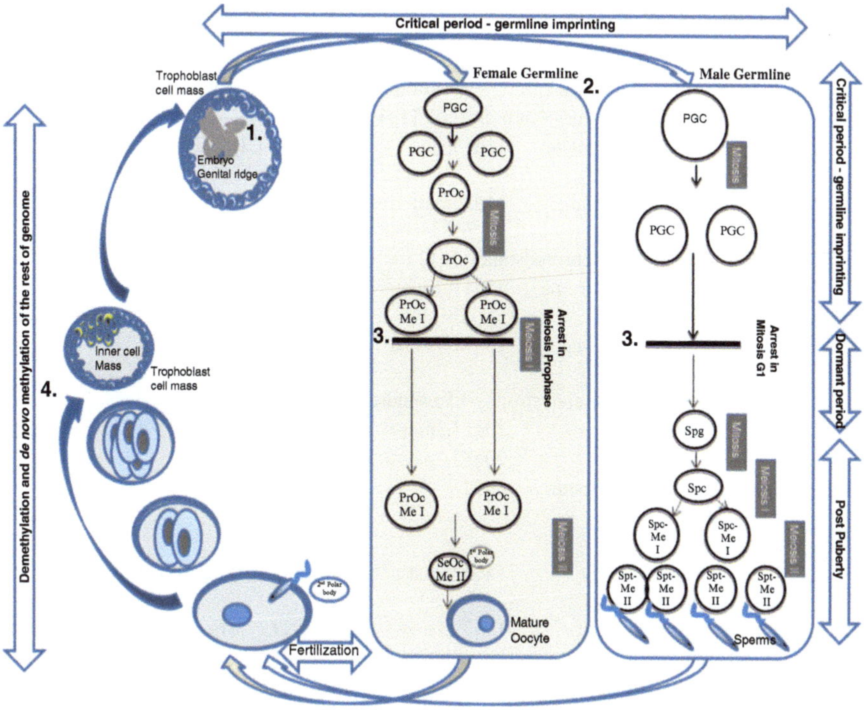

Fig. 1.2 Germline methylation cycle. *1.* Primordial germ cells originate in epiblast cells arising in the posterior primitive streak at E7.5 in mice (corresponding to post-conception days 15–18 in humans). They then migrate, starting to the genital ridge around E8.5 in mice (post-conception day 19 in humans) and complete their migration around E11.5 in mice (post-conception days 32–33 in humans). *2.* It is soon after the migration to the genital ridge that imprinted genes are demethylated and start the process of reestablishment of the imprint with remethylation. This period of DNA demethylation is between E11.5 and E12.5 in mice and corresponds approximately to post-conception days 33–40 in humans. Following that the genomes of the developing gametes are de novo methylated and acquire imprints and this process continues until at least E18.5 in male mice corresponding to post-conceptional day 49–52 in human males. *3.* In the fetal period, there is an arrest of germline development which in males is in G1 phase of mitosis after completion of remethylation but in females the arrest occurs after initiation of meiosis in the prophase of Meiosis I. In females however, the remethylation can occur until the oocyte matures before ovulation; thus going into puberty and beyond. In fact, female secondary oocytes do not complete their meiosis until right at fertilization with the formation of the second polar body. *4.* After fertilization of male and female gametes to form the zygote, the germline imprinted genes do not participate in the genome-wide demethylation and de novo methylation that is critical in establishing of the embryo and trophoblastic cell lines and their further differentiation. With formation of the epiblasts in the embryo the cycle of germline imprinting starts again. Abbreviations used in the figure: *PGC* primordial germ cells, *PrOc* primary oocyte, *PrOc MeI* primary oocyte undergoing Meiosis I, *SeOc MeII* secondary oocyte undergoing Meiosis II, *Spg* spermatogonia, *Spc* spermatocyte, *Spc-MeI* spermatocyte undergoing Meiosis I, *Spt-MeII* spermatids undergoing Meiosis II

Table 1.1 Primary epigenetic disorders

Conditions	Chr	Genes
I.A. Effects at the primordial germ cell stage of development (disorders of imprinting)[a]		
I.A.1. BWS and SRS syndromes (chromosome 11p15)		
I.A.1.1. Beckwith–Weidemann synd. (BWS)	11p15pat	*IGF2*, *H19*, *CDKN1C*, *KVLQT1*, *KCNQ10T1* (*LIT1*)
I.A.1.2. Silver–Russell synd. (SRS)	11p15mat; 7q13	*IGF2*, *H19*, *MEST*, *PEG1*, *CPA4*, *COPG2*, *MESTIT*, *CIT2*/*COPG2IT1*
I.A.2. PWS and Angelman (chromosome 15q11)		
I.A.2.1. Prader–Willi syndrome (PWS)	15q11-13pat	*SNURF-SNRPN*, *NDN*, *MKRN3* *MAGEL2*
I.A.2.2. Angelman syndrome (AS)	15q11-13mat	*UBE3A*
I.A.3. Albright hereditary osteodystrophy-like syndromes		
I.A.3.1. PHP-Ia	20q13.3mat	*GNAS1*
I.A.3.2. PPHP	20q13.3pat	*GNAS1*
I.A.4. Transient neonatal diabetes mellitus type I	6q24	*PLAG1*, *HYMA1*
I.A.5. UPD 14 syndromes		
I.A.5.1. Wang syndrome or UPD(14)pat	14q32.2pat	*DLK1*, *RTL1*, *MEG3* (*GTL2*), *MEG8*
I.A.5.2. Temple syndrome or UPD(14)mat	14q32.2mat	*DLK1*, *RTL1*, *MEG3* (*GTL2*)
I.A.6. Multilocus hypomethylation defects (MHD)		*PLAG1*, *ZAC1*, *ZFP57*, *NLRP2*, *CTCF*, *MBD3*, *SNRPN*, *PEG3*, *NESPAS*, *H19*
I.A.7. Maternal UPD(16)	UPD(16)	
I.A.8. Germline epimutations associated with neoplasia		
I.A.8.1. Lynch syndrome	?	*MLH1*, *MSH2*
I.A.8.2. Familial paragangliomas	11q23	*SDHD*
I.A.8.3. NOEY2 gene-related cancers	?	*NOEY2*
I.B. Effects from conception to completion of embryogenesis including first trimester		
I.B.1. Turner syndrome and selective X^m or X^p effects	Xp	*Xlr*
I.B.2. Epigenetic influences on trophoblast and placenta	UPD(14); 6q24.2	*RTL1*, *PEG11*, *DLK1*, *DIO3*, *NLRP7*, *PLAG1*
I.B.3. Epigenetic effects on somatic cells—early development		
I.B.3.1. Twin growth discordance	Many	*H19*, *IGF2*
I.B.3.2. Prematurity	Many	*NFIX*, *RAPGEEF2*, *MSRB3*
I.B.3.3. Infantile biliary atresia	?	*ITGAL*(*CD11A*), *RASSF1A*, *p16*, *CDH1*, *TFPI2*, *NPTX2*, *APC*
I.B.3.4. Wilms' tumor	5; 11p15	*H16*, *IGF2*, *WT1*, *CTNNB1*, *WTX*, *TP53*
I.B.3.5. Hepatoblastoma	11p15	*H19*
I.B.3.6. Retinoblastoma		*RB1*
I.B.3.7. Others		*H19*, *Igfr2*, *U2af-rs1*
I.C. Effects from mid-late gestation and continuing to postnatal development until adulthood		
I.C.1. Postnatal stress and epigenetics	?	*NR3C1*

(continued)

Table 1.1 (continued)

Conditions	Chr	Genes
I.C.2. Epigenetic basis of adult onset disease	Many	*IL10, GNASAS, INSIGF, LEP, ABCA1, MEG3, PPARα, PDX1, HNF4α*
I.C.3. Malignancy and epigenetics	Many	

By convention human genes are represented in capitalized italics and mouse genes in lower case italics

[a]A number of disorders of imprinting are common with chromosomal rearrangements and uniparental disomy, however, purely epigenetic changes can potentially case the same condition

Abbreviations: *Chr.* chromosomes, X^m maternally derived X chromosome, X^p paternally derived X chromosome, *synd.* syndrome

Table 1.2 Secondary epigenetic disorders due to disorder of genetic DNA

Conditions	Chr	Genes
II.A. Gene abnormalities with DNA methylation effects		
II.A.1. ICF (type 1) syndrome	1, 9, 16 & X	*ZBTB244, SYK, SH3BP5*
II.A.2. Rett syndrome—MeCP2 gene		*MECP2, MKX, CKB, FYN*
II.A.3. Fragile X syndrome	Xq27.3	*FMR1*
II.A.4. X-linked alpha-thalassemia/mental retardation syndrome (ATX-R)	Xq13	*ATRX*
II.A.5. Fascioscapulohumeral dystrophy	4q35	*Multiple genes*
II.A.6. Hereditary sensory and autonomic neuropathy type I (HSAN1)	19p13.2	*Multiple genes*
II.A.7. Autosomal dominant cerebellar ataxia, deafness, and narcolepsy		
II.B. Genetic syndromes causing histone modifications		
II.B.1. Rubinstein–Taybi syndrome	16p13.3	*CREBBP, EP300*
II.B.2. Genitopatellar syndrome (GPS)		*KAT6B*
II.B.3. Say-Barber-Biesecker-Young-Simpson syndrome (SBBYS)		*KAT6B*
II.B.4. Coffin–Lowry syndrome	Xp22.2	*RPS6KA3*
II.B.5. Sotos syndrome	5q35.2	*NSD1*
II.B.6. Weaver syndrome		*EZH2, NSD1*
II.B.7. Brachydactyly-mental retardation syndrome (BDMR)		
II.B.8. Kleefstra syndrome	9q34.3	*EHMT1*
II.B.9. Kabuki syndrome	X	*MLL2, KDM6A*
II.B.10. Siderius X-linked mental retardation syndrome (MRXSSD)	X	*PHF8*
II.B.11. Claes–Jensen X-linked mental retardation syndrome	X	*JARID1C (SMCX)*
II.C. Genetic mutations that affect noncoding RNAs		
II.C.1. Amyotrophic lateral sclerosis		*TARDBP*
II.C.2. DiGeorge syndrome		*DGCR8*
II.C.3. Goiter, multinodular1, with or without Sertoli–Leydig cell tumors	22q11.	*DICER1*
II.D. Chromosome deletion and rearrangements—epigenetic changes:		
II.D.1. 9q subteloric deletion syndrome:	9q	*EHMT1*
II.D.2. 46XY inversion(10)(q11.1;q21.3)	X, 10q	*TRIP8*
II.D.3. Wolf–Hirschhorn syndrome	4p16.3	*WHSC1L1*

By convention human genes represented in capitalized italic and mouse genes in lower case italics

Abbreviations: *Chr* chromosomes

1.7 Primary Epigentic Disorders

The primary disorders of epigenetic regulation may be further classified into three groups based on the cell populations affected at critical periods of epigenetic changes that occur within an individual's lifetime (Table 1.1; Fig. 1.3).

1.7.1 *Germline*

Effects at the primordial germ cell stage of development (*disorders of imprinting*): Most of the disorders in this category may also be caused by chromosomal rearrangements and uniparental disomy (UPD) but purely epigenetic changes can potentially cause a similar condition. Formation of gametes occurs from PGCs and "imprinted genes" are established that do not get erased during the rest of the life in that generation. However, these imprints can be transmitted through subsequent

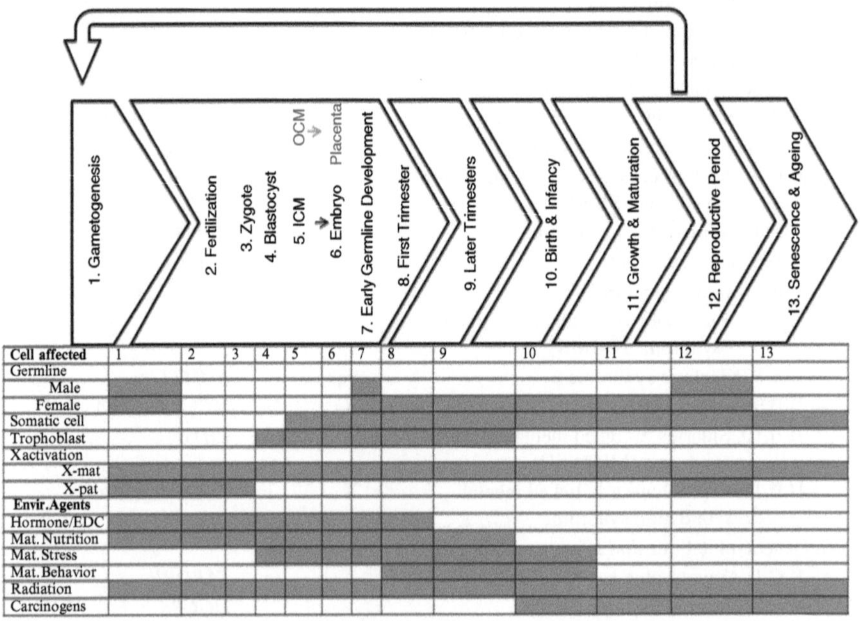

Fig. 1.3 Critical periods of vulnerability: *shaded* areas of the table corresponding to the numbered stages of human development indicate critical periods of vulnerability of different cellular populations affected at different stages of a lifetime. Putative environmental agents implicated in bringing about the epigenetic changes are also listed with similar *shaded* areas indicating presumed time of their maximum effect. Abbreviations: *ICM* inner cell mass, *OCM* outer cell mass, *X-Mat* maternally derived X chromosome, *X-pat* paternally derived X chromosome, *EDC* endocrine disrupting chemicals such as vinclozolin and bisphenol A

generations. The timing of this starts in the later stages of early embryogenesis, corresponding to E11.5–E12.5 in mice (post-conception days 33–40 in humans) and ends by E18.5 in male mice (corresponding to post-conception days 49–52 in humans). In females of both mice and humans on the other hand, this stage may not be completed through puberty and adulthood.

1.7.2 Somatic and Trophoblastic Cells Including X Chromosome Inactivation

Effects from conception to completion of embryogenesis including first trimester: Fertilization of the gametes and formation of the early embryo from conception to most of the first trimester of gestation involves a period of demethylation and de novo methylation of the genome. Most purely epigenetic disorders are part of this group because this is the stage where most of the methylation markers are removed and reestablished (except the imprinted genes which do not change); also the inactivation of the extra X chromosome occurs in females.

1.7.3 Later Somatic Cell

Effects from mid-late gestation and continuing to postnatal development until adulthood: Conditions listed under this category have the most potential for modifications to promote health and prevent or treat disease. The field of study of "Developmental Origins of Health and Disease (DoHaD)" and research related to the origins or cancers and neuropsychiatric disorders are directly influenced by the epigenetic changes that may occur during this stage.

The subsequent part of this chapter will detail the processes involved and the conditions that occur due to abnormalities in the normal process of epigenetic regulation and differentiation.

1.8 Effects at the Primordial Germ Cell Stage of Development (Disorders of Imprinting) (*Potential Multigenerational Effects*)

Human embryo develops sexual identity by about 8 weeks after conception. Determination of gonadal sex is by the differentiation of the bipotential embryonic gonad into male testis or female ovary. Subsequently the action of testicular hormones are involved in the development of male external genitalia and regression of female internal reproductive organs; by default the lack of action of testicular hormones leads to the development of the female reproductive organs (Ludbrook and Harley 2004). PGCs originate in epiblast cells arising in the posterior primitive streak at E7.5 in mice (corresponding to post-conception days 15–18 in humans) (Morgan et al. 2005; O'Rahilly 1979). They then migrate to the genital ridge around E8.5 in mice (post-conception day 19 in humans) and complete their migration around E11.5 in mouse (post-conception days 32–33 in humans). It is soon after the migration to the genital ridge that imprinted genes are demethylated and start the process of reestablishment of the imprint with remethylation. This period of DNA demethylation is between E11.5 and E12.5 in mice and corresponds approximately to post-conception days 33–40 in humans (Morgan et al. 2005; O'Rahilly 1979). Following that the genomes of the developing gametes are de novo methylated and acquire imprints and this process continues until at least E18.5 in male mice corresponding to post-conceptional days 49–52 in human males. In females of both species however, the remethylation can occur until the oocyte matures before ovulation; thus going into puberty and beyond (Morgan et al. 2005; O'Rahilly 1979). Changes occurring in the germline at this stage have the potential to have manifestations into the next two generation of offspring (Manikkam et al. 2012; Skinner et al. 2013).

1.8.1 Mechanism of Imprinting

In contrast to the biallelic expression of most genes, expression of genes subject to genomic imprinting is monoallelic and based on the sex of the transmitting parent. The mechanism of imprinting was discovered from a series of nuclear transplantation experiments in the early 1980s in which it was shown that gene function could differ based on whether a particular set of genes were inherited from the mother or the father (McGrath and Solter 1984; Surani et al. 1984; Cattanach and Kirk 1985). The correlation of these observations to human disease was initially made with the finding of maternal heterodisomy with altered gene imprinting in PWS (Nicholls et al. 1989). Soon other human diseases related to imprinting were identified and currently there is a growing list of conditions that could be classified as imprinting disorders (Eggermann et al. 2013). More and more conditions are being identified because there are at least 150 imprintable genes in the mouse and at least 60 have been identified as human imprintable genes (http://igc.otago.ac.nz/home.html) (Harwell 2013; Horsthemke 2010). Genomic imprinting could be understood as a series of steps where imprints are reset at generation. Imprints are erased in PGCs, established in the germline according to the sex of the embryo and maintained throughout life until the development of PGCs of the next generation. Imprinting marks are inherited from paternal gametes and maintained unchanged during developmental reprogramming and demethylation in early embryogenesis and persist during mitosis in somatic cells through the lifetime. However, during later embryogenesis when PGCs are in the process of migrating to the genital ridge in the development of the new gonad, the imprints are erased and reestablished for the next generation according to the sex of the contributing parent (Eggermann et al. 2013). The process of imprinting is mostly regulated by imprinting centers (ICs) on the same gene on which it has its effect (*cis*-acting). Imprinted genes and ICs seem to occur in clusters and thus far they have been found mostly on chromosomes 6, 7, 11, 14, 15, and 20 in humans. Imprinted genes are typically associated with genes regulating growth, development, and evolutionary survival and are characterized by the presence of differentially methylated regions (DMRs). Imprinted gene DMRs represent parental allele-specific methylation profiles which could be maintained across many generations (Murphy et al. 2012; Park et al. 2012). Epigenetic transitions during gametogenesis are important in their impact on development and disease in infants (Kota and Feil 2010). Most imprint control regions (ICRs) have DNA imprints conferred during oogenesis and could be termed "maternal ICRs"—e.g., *KCNQ1* on chromosome 11, *GNAS* on chromosome 20, and *PWS* on chromosome 15. Conversely, only two could be termed "paternal ICRs"—viz. *IGF2-H19* imprinted domain on chromosome 11 and *DLK1-DIO3* on chromosome 14 (Girardot et al. 2012).

1.8.2 Molecular Changes Involved in Imprinting Disorders

There are four common ways by which imprinting disorders originate and only one of them involves abnormal methylation of specific DNA regions termed epimutation (in contrast to genetic mutations involving DNA sequence). The other three mechanisms involve alteration of genetic material: (a) UPD where both chromosomal homologs are inherited from one parent, (b) microdeletions of specific regions or duplications of a particular DNA segment, and (c) point mutations involving DNA sequences in the imprinted genes. For the purpose of the following discussion on imprinting, no distinction will be made regarding the various mechanisms involved but rather the focus will be on the effect of the change of imprint.

1.8.3 Sex Differences in Imprinting

The process of genetic imprinting though largely similar has important differences in males and females (Fig. 1.2). As mentioned above, the process of imprint erasure from germline cells occurs at the same time in both males and females in early embryogenesis (E11.5–E12.5 in mice corresponding to post-conception days 33–40 in humans). However, the process of gamete formation and thus the process of reestablishment of the methylation imprint differs greatly between male and female embryos. Male germline cells enter into a state of mitotic arrest at this stage (approximately around post-conception days 40–42 in humans) and will reinitiate mitosis and the two phases of meiosis (Meiosis I and Meiosis II) to generate spermatozoa only after puberty. In contrast, female germline cells complete their mitosis and start their Meiosis I stage while still in the developing gonad at the genital ridge, and go into arrest at Prophase of Meiosis I. They stay in this stage until the start of each cycle of oocyte production after puberty when they finally complete Meiosis I and Meiosis II to generate the mature oocyte.

In female germline of the developing oocyte, after erasure at the PGC stage, DNMT3A and DNMT3L (de novo methyltransferases) are required for the reestablishment of maternal imprints (Li and Sasaki 2011) DNMT3B is also needed but is dispensable. The DNMT3A/DNMT3L complex finds its sex-specific targets in one of the three ways: by seeking DNA regions with an 8–10 nucleotide CpG interval, by recognition of unmethylated H3K4 (histone), or by help of KRAB zinc-finger protein ZFP57. In addition, there may also be a role for Nesp transcripts at the *GNAS* locus in oocytes or the presence of lysine H3K4 demethylase KDM1B (Li and Sasaki 2011). In male germline also, DNMT3A and DNMT3L play a central role but the role of ZFP57 and other factors is not prominent. Moreover, the role of an isoform of DNMT1o involved in maintenance of methylation is prominent in the oocyte and conversely the role of the demethylation mechanism related to Tet3 is more prominent in the male pronucleus (Rivera and Ross 2013).

1.8.4 Establishment and Maintenance of Imprinting

Imprint establishment and maintenance play a role in the development of childhood disease (Tomizawa and Sasaki 2012). In preimplantation embryo, DNMT1 is mainly responsible for maintenance of DNA methylation and its isoform DNMT1o is important at the one-cell oocyte level. Other important elements in maintenance of imprinting during the preimplantation embryo wave of demethylation are: ZFP57, PGC7 (Stella), and a methyl-CpG binding protein Mbd3. In the post-implantation period, again DNMT1 is important in somatic lineages. Other factors include differential histone modifications: H3K4me and histone acetylation marks the less-CpG-methylated allele; H3K9me3, H3K20me3, and H2A/H4R3me2 mark the more-CpG-methylated imprinted allele. In the trophoblast lineage of the embryo, however, DNA methylation has less of a role and the role of histone modifications seems to be more prominent. The important histones being H3K9me and H3K27me3; and their effects mediated by G9a and PRC2 (Polycomb repressive complex 2), respectively.

1.8.5 Erasure of Imprinting

Even though imprinted genes evade the wave of demethylation in the early preimplantation development, they have a critical period of demethylation that occurs at the PGC stage of the developing germline in the embryo. Most methylation marks from imprinted genes are erased in PGC of both sexes and then reappear at different stages in the two sexes. In men, the methylation marks on imprinted genes reappear in prospermatogonia, before cells reenter mitosis and this process occurs during early embryogenesis, well before birth. In women this process is delayed until after the initiation of meiosis and the oocytes have started growing in follicles, occurring after birth and attainment of puberty. Therefore it is not surprising that transgenerational inheritance of epigenetic alterations may also be different in men and women (Hitchins et al. 2007; Gosden and Feinberg 2007). Erasure of imprinting in PGC can be active by demethylases, or passive by not methylating during replication. The 10–11 translocation family proteins (Tet1, Tet2, and Tet3) are important in conversion of 5 methylcytosine (5mC) to 5 hydroxymethylcytosine (5hmC) which is an important intermediary in the demethylation process (He et al. 2011).

1.9 Disorders of Imprinting (Table 1.1)

1.9.1 BWS and Silver–Russell Syndromes

The two paradigmatic disorders of imprinting are Beckwith–Weidemann syndrome (BWS) and Silver–Russell syndrome (SRS)—both involving the cluster of genes on chromosome 11p15.5 regulated by two ICRs: the maternally methylated *KCNQ1OT1* ICR (or KvDMR1) located in the 5′ region of *KCNQ1OT1*, and the paternally methylated H19 ICR located between *H19* and *IGF2* (Demars et al. 2011). There is an inverse relationship between *H19* and *IGF2* expressions with *H19* expression having a suppressive effect on *IGF2*. Hypomethylation of *KCNQ1OT1* is the most common finding in BWS or sometimes there is hypermethylation of *H19*. Conversely, in SRS, there is hypomethylation of *H19* ICR (Tomizawa and Sasaki 2012). Paternal UPD is found in BWS and maternal UPD in SRS.

Beckwith–Weidemann Syndrome

BWS, first described in 1964 (Wiedemann 1964), is characterized by macrosomia, macroglossia, omphalocele, and characteristic ear creases and/or pits. Other features include capillary nevus flammeus over the central forehead and eyelids, a large fontanel, prominent eyes with periorbital fullness, accelerated bone age, growth asymmetry, hemihypertrophy, and organomegaly involving the kidneys, liver, pancreas, and spleen. Patients with BWS have an increased intra-abdominal tumor rate (10–20 % of cases) particularly of the kidneys and occasionally the liver. Additional findings include neonatal hypoglycemia, present in more than one-third of cases, cardiovascular defects, and cryptorchidism. Prevalence of this condition is 1:15,000 and about 60 % are due to epigenetic imprinting defects. Genes involved are clustered on a band at chromosome 11p15.5 containing many imprinted genes, both maternal and paternal. This large domain is organized into two separately controlled "imprint control regions (ICRs)"—a telomeric (ICR1) and centromeric (ICR2). Genes included are *IGF2* (paternally expressed), *H19* (maternally expressed),

CDKN1C (maternally expressed), *KVLQT1* (maternally expressed), and *KCNQ1OT1* (*LIT1*) (paternally expressed). The epigenetically influenced modes of expression of each of these genes have been correlated with varying clinical expressions of the syndrome (Bliek et al. 2009b; Butler 2009). The involvement of chromosome 11p15 is mostly epigenetic with hypomethylation of centromeric imprinting center region in about 50 %; although other abnormalities including paternal UPD 11 (15 %), loss of imprinting of *IGF2* (5 %), mutations in *CDKN1C,* or the centromeric imprinting center region (10 %) are also found (Butler 2009).

Silver–Russell Syndrome

SRS was first reported by Silver et al. in 1953 (Silver et al. 1953) and further characterized by Russell in 1954 (Russell 1954). It is characterized by severe prenatal and postnatal growth restriction, a characteristic facial appearance including a small, triangular face, frontal prominence, late closure of the anterior fontanel, immature bone development, somatic growth asymmetry especially of the limbs, and clinodactyly (Russell 1954; Silver et al. 1953). Hypoglycemia is a common feature. Other less constant features are *café au lait* spots, hypospadias, cardiac defects, precocious puberty, and developmental delays. Prevalence of this condition is 1:100,000 to 1:3,000 and about 50 % are due to epigenetic imprinting defects. The most common basis for this syndrome is hypomethylation of telomeric 11p15 imprinting center (40–60 % of cases) (Butler 2009). Maternal duplication of chromosome 11p15 (5 %), maternal UPD 7 (10 %), and 7p duplications or unknown (40 %) are also frequently noted findings. This is thus the first human disorder with imprinting disturbances affecting more than one chromosome (i.e., chromosome 7 and 11) (Butler 2009). Genes associated with this condition are found on chromosome 7p13 band including *MEST* (mesoderm-specific transcript), *PEG1* (paternally expressed gene 1), carboxypeptidase A4 (*CPA4*), coatomer protein complex subunit gamma 2 (*COPG2*), and two imprinted noncoding RNAs (*MESTIT*, *C1T2/COPG2IT1*) (Abu-Amero et al. 2008). On chromosome 15, epimutations reported in SRS are typically due to hypomethylation of the ICR1 domain (Bullman et al. 2008). Detailed review of the various genetic and epigenetic changes and the resultant variation in phenotypes for these conditions are available (Feinberg 2007; Horsthemke and Buiting 2008).

1.9.2 Prader–Willi Syndrome and Angelman Syndrome

PWS and Angelman syndrome (AS) are clinically distinct neurodevelopmental disorders that map to the same imprinted region on human chromosome 15q11-13. This PWS–AS region comprises four protein-coding genes transcribed from the paternal chromosome, i.e., *SNRPN*, *MKRN3*, *MAGEL2,* and *NDN,* and one protein-coding gene (*UBE3A*) transcribed from the maternal chromosome. Angelman syndrome is a neurodevelopmental disorder caused by abnormalities at the ICRs cluster containing *UBE3A* which transcribes the protein E3 ubiquitin-protein ligase that is important for proteosomal degradation of proteins having a role in cell cycle, signal transduction, transcription, and plasticity of synapses.

Prader–Willi Syndrome

PWS was first described by Prader, Labhart, and Willi in 1956 (Prader et al. 1956). It is characterized by infantile hypotonia, early childhood obesity, short stature, small hands and feet, growth hormone deficiency, hypogenitalism/hypogonadism, mental deficiency, behavioral problems (temper tantrums and skin picking), and a characteristic facial appearance with a narrow bifrontal diameter, short upturned nose, triangular mouth, enamel hypoplasia, almond-shaped eyes, and sticky saliva (Butler 2009; Ledbetter et al. 1981; Prader et al. 1956). The feeding difficulties encountered in these patients in infancy later develop into a problem with hyperphagia and early childhood obesity. Prevalence of this condition is 1:25,000 to 1:10,000 and only about 1–5 % are due to epigenetic imprinting defects. Genes involved in the manifestation of PWS are paternally expressed and maternally silenced, located within the chromosome 15q11-q13 region (*SNURF-SNRPN*, *NDN*, *MKRN3,* and *MAGEL2*) and most of them are involved directly or indirectly in brain development and function. The genetic subtypes of maternal disomy 15 (type I) can often be distinguished from those with 15q deletion (type II). Paternal 15q11-q13 deletion is the most common cause of this syndrome in 70 % of cases and maternal UPD is the next common etiology in 25 % of cases. Disorders of imprinting account for only about 5 % of cases. An interesting feature of PWS is that the imprinting defect may represent primary epimutations derived from paternal grandmother. This suggests that incorrect erasure of imprint in the paternal PGCs might have been responsible (Buiting 2010; Buiting et al. 1998; Buiting et al. 2003).

Angelman Syndrome

Angelman syndrome also known as the "happy-puppet syndrome" was first described in 1965 (Angelman 1965). It has a phenotype of seizures, severe mental retardation, ataxia and jerky arm movements, hypopigmentation, inappropriate laughter, lack of speech, microbrachycephaly, maxillary hypoplasia, a large mouth with protruding tongue, prominent nose, and wide-spaced teeth (Williams et al. 1995). It involves the same region of chromosome 15q11-13 but is related to abnormality in expression of one specific imprinted (maternally expressed) gene, i.e., *UBE3A*, an ubiquitin ligase gene involved in early brain development. Prevalence of this condition is 1:20,000 to 1:12,000 and only about 2–3 % are due to epigenetic imprinting defects. The most common cause of AS is de novo maternal deletions involving chromosome 15q11.2-q13 (70 %); with 2 % resulting from paternal UPD and about 2–3 % resulting from epigenetic imprinting defects. There is a large subset of patients (25 % of the remaining) that are the result of mutations in the gene encoding ubiquitin protein ligase E3A (*UBE3A*) (Kishino et al. 1997).

1.9.3 Albright Hereditary Osteodystrophy, Pseudohypoparathyroidism Type Ia and Pseudo-pseudohypoparathyroidism Disorders

Albright hereditary osteodystrophy (AHO) was first described in 1952 (Albright et al. 1952). The typical AHO phenotype is characterized by small stature (final adult height 54–60 in.), moderate obesity, mental deficiency (average IQ of 60), round face with a short nose and short neck, delayed dental eruption and enamel hypoplasia, short metacarpals and metatarsals especially of fourth and fifth digits, short distal phalanx of the thumb, osteoporosis, areas of mineralization in subcutaneous tissues including the basal ganglia, variable hypocalcemia and/or hyperphosphatemia, and seizures. Some cases also show evidence of hypothyroidism, hypogonadism, lens opacity or cataracts, optic atrophy, ocular degeneration, and vertebral anomalies. AHO is usually due to resistance of the action of parathyroid hormone and if so is termed *pseudohypoparathyroidism* (*PHP*). There are two variants of PHP (PHP-Ia and PHP-Ib) depending on the presence of other hormone resistance apart from parathyroid hormone. Nearly all patients with pseudohypoparathyroidism type Ia (PHP-Ia) have mild hypothyroidism, hypogonadism, and abnormal response to growth hormone-releasing hormone, while patients with PHP-Ib (which is not an epigenetic condition) also present with parathyroid hormone-resistance, but lack resistance to other hormones. Another condition *pseudo-pseudohypoparathyroidism* (*PPHP*) has a similar phenotype to PHP-Ia but does not have end-organ resistance to parathyroid hormone. A decrease in activity of a guanine nucleotide-binding signaling (GNAS) protein which is responsible for cAMP coupling and signaling is a common finding in these two conditions.

Epigenetic modifications of *GNAS1* gene and maternal and paternal allele expression differences are known to be responsible for PHP-Ia and PPHP (Lalande 2001; Liu et al. 2000). Specifically, loss of methylation at the maternally methylated ICRs located at the *GNAS exonA/B* in chromosome 20q13.3, that codes for GSα, a stimulatory G-protein subunit, is implicated in PHP-Ia; manifested clinically as end-organ (especially renal) resistance to parathyroid hormone leading to hypocalcemia, hyperphosphotemia, and obesity (Kelsey 2010; Bastepe and Juppner 2005; Bastepe et al. 2001). In contrast, paternally affected changes in the *GNAS* gene are responsible for the clinical findings of PPHP that shares the hypocalcemia and other findings of PHP-Ia except for lack of resistance of the parathyroid hormone receptor activity. PHP-Ia is caused by a mutation resulting in loss of function of the Gs-α isoform of the *GNAS* gene on the maternal allele. This results in expression of the Gs-α protein only from the paternal allele. In contrast, PPHP is caused by mutations resulting in loss of function of the Gs-α isoform of the *GNAS* gene on the paternal allele and resultant expression of the Gs-α isoform only from the maternal allele. Thus PHP-Ia and PPHP can occur in the same family. Of interest is the observation that the autosomal dominant form of PHP-Ib is caused by heterozygous mutations disrupting a long-range imprinting control element of *GNAS* and has not yet been shown to be epigenetically regulated.

The nomenclature of the above mentioned conditions is prone to confusion and therefore to summarize:

- *PHP-Ia*: Has clinical features of AHO, multiple hormone resistance, decreased erythrocyte Gs activity (due to α-subunit), decreased response to parathyroid hormone infusion, and *GNAS1* mutation in the maternal allele (Mantovani and Spada 2006). More than 90 % of times the cause of this condition is related to epigenetic imprinting defects.
- *PPHP*: Has clinical features of AHO without endocrine abnormalities, decreased Gs activity (due to α-subunit), normal response to parathyroid hormone infusion, and a *GNAS1* mutation in the paternal allele (Mantovani and Spada 2006).
- *PHP-Ib*: Do not have features of AHO, no endocrine abnormalities, normal erythrocyte Gs activity, parathyroid infusion response has renal-specific resistance, and *GNAS* locus methylation defect results in specific lack of expression only in renal tissues (Mantovani and Spada 2006).
- *PHP-II*: Do not have features of AHO, have normal erythrocyte Gs activity, and have isolated renal PTH resistance (Mariot et al. 2008).
- Also of note is the condition "McCune–Albright syndrome" which manifests as polyostotic fibrous dysplasia and is a distinct entity from AHO.

1.9.4 Transient Neonatal Diabetes Mellitus Type I

Transient Neonatal Diabetes Mellitus (TNDM) is defined as hyperglycemia presenting within the first 6 months of life. There are three known types (TNDM1, 2, and 3) and TNDM1 is known to be caused by genetic or epigenetic changes at an imprinted locus (genes *PLAG1* and/or *HYMA1*) on chromosome 6q24 (Mackay and Temple 2010). About 95 % of infants with this condition are born with intrauterine growth restriction starting in the third trimester of pregnancy. The clinical onset of TNDM1 in early infancy is followed by spontaneous remission in about half the cases with subsequent relapse in adolescence or early adulthood. Prevalence of this condition is 1:400,000 and about 30 % are due to epigenetic imprinting defects. TNDM1 is unusual among imprinting disorders because it is caused by over-expression of a genetic region. There is over-expression of *PLAG1/HYMA1* due to hypomethylation of the TNDM1 DMR in this condition. About 20 % of patients show a hypomethylation of the *PLAG1/HYMA1* ICRs region which is normally maternally methylated; 50 % show hypomethylation at other regions (some with *ZFP7* mutations) (Mackay et al. 2006; Mackay et al. 2008). The three main genetic etiologies of this condition are: (1) paternal UPD of chromosome 6 (40 %); (2) duplication of the imprinted TNDM1 region on chromosome 6 (32 %); and (3) maternal hypomethylation of the TNDM1 DMR without chromosomal anomaly (28 %). The biology of TNDM1 has interesting correlates to the "thrifty phenotype" hypothesis and may play a role in understanding the origins of adult onset diabetes mellitus (Hales et al. 1991).

1.9.5 UPD-14-Related Syndromes (Wang and Temple Syndromes)

Human chromosome 14q32.2 carries a cluster of imprinted genes including paternally expressed genes (PEGs) such as *DLK1* (delta, Drosophila homolog-like 1), a transmembrane signaling protein which is a growth regulator homologous to proteins in the Notch/delta pathway and *RTL1* (Temple 2007). In this region are also the maternally expressed genes (MEGs) such as *MEG3* (also known as *GTL2*), *RTL1as* (RTL1 antisense), and *MEG8*, together with the intergenic differentially methylated region (IG-DMR) and the *MEG3*-DMR. Consistent with this, paternal and maternal UPD for chromosome 14 (upd(14)pat and upd(14)mat) cause distinct phenotypes. The prevalence of these conditions has not yet been determined.

Wang Syndrome or UPD(14)pat: Paternal Uniparental Disomy 14

Wang syndrome involving chromosome 14q32.2 imprinted region results in a phenotype that is characterized by severe developmental delay along with physical features of facial abnormality, small bell-shaped chest with "coat-hanger" type ribs and abdominal wall defects (omphalocele) (Wang et al. 1991; Butler 2009). There is often a history of placentomegaly and polyhydramnios prior to birth. This could potentially be caused by an epigenetic modification of the same region. Evidence for this comes from a case series involving eight individuals with an UPD(14)pat-like phenotype where in the absence of UPD(14) various deletions and epimutations affecting the imprinted region were identified (Kagami et al. 2008). Excessive *RTL1* gene expression was found to be relevant to UPD(14)pat-like phenotypes (Kagami et al. 2008). In addition to UPD(14)pat, isolated defects in methylation at the *DLK1/GTL2* locus have been noted in some patients (Eggermann 2011).

Temple Syndrome or UPD(14)mat: Maternal Uniparental Disomy 14

Temple syndrome involves the same region of chromosome 14q32.2 with characteristic findings of congenital hypotonia, joint laxity, feeding difficulty, gross motor delay with mild to moderate mental retardation, small hands and feet, early onset of puberty, and truncal obesity. The facial features are characterized by a prominent forehead, a bulbous nasal tip and a short philtrum. There are some clinical features that overlap PWS and therefore screening for UPD(14)mat should be done in all patients with PWS phenotype (Eggermann 2011). About 30 % of cases will show rapid postnatal head growth usually due to hydrocephalus that is arrested spontaneously. Dysmorphic facial features include a prominent forehead, prominent supraorbital ridges, a short philtrum and down-turned corners of the mouth (Temple et al. 1991; Butler 2009). It is usually associated with a history of prenatal and postnatal growth restriction (Butler 2009). In a recent case series of three cases with an UPD(14) mat-like phenotype in the absence of UPD(14) various deletions and epimutations affecting the imprinted region were identified (Kagami et al. 2008). The results, together with evidence from animal data, suggest that the IG-DMR has an important *cis*-acting regulatory function on the maternally inherited chromosome and that decreased *DLK1* and *RTL1* gene expression are relevant to UPD(14)mat-like phenotypes (Kagami et al. 2008). The role of maternally expressed *MEG3* from the same domain is being investigated.

1.9.6 Multilocus Hypomethylation Defects

Recently, a combination of imprinted disorders have been described in the same patient and termed *Multilocus Hypomethylation Defects* (*MHD*) (Mackay et al. 2006). Combination of hypomethylation of the ICR2 of 11p15 (associated with BWS) was found in patients diagnosed with TNDM carrying an epimutation of *PLAG1/ZAC1* (Arima et al. 2005; Mackay et al. 2006; Mackay et al. 2008). This association appears to be most common with aberrant hypomethylation of ICR2 and BWS and less frequently associated with SRS and PHP-Ia (Eggermann et al. 2013). The role of other *trans*-acting factors has been suggested in the origin of MHD. Some of these factors include effects of changes in the *ZFP57*, *NLRP2*, *CTCF*, and *MBD3* genes (Demars and Gicquel 2012). Recent report of a case with hypomethylation of maternally influenced *SNRPN*, *KCNQ1OT1*, *PEG3*, and *NESPAS* loci along with paternally influenced *H19* locus presenting with features of both PWS and BWS is another case in point (Baple et al. 2011). In some reports MHD has been more common in BWS patients born of assisted reproductive technologies (ART) especially one of monozygotic twins but the findings were inconsistent (Lim et al. 2009; Bliek et al. 2009b; Rossignol et al. 2006; Bliek et al. 2009a). A common feature in MHDs is that both maternally and paternally imprinted loci are affected suggesting an origin beyond germline reprogramming. Also, there is a high frequency of mosaicism in these cases suggesting a post-fertilization mechanism. Moreover a more basic problem in regulation of methylation may be involved. These findings highlight the importance of testing for multiple loci when one imprinting disorder is suspected as there may be important ramification of the findings on prognostication and genetic counseling.

1.9.7 Maternal UPD(16)

UPD(16)mat is most commonly seen in the context of a pregnancy with placental insufficiency and intrauterine growth restriction associated with confined placental mosaicism (Eggermann et al. 2004). This condition may be associated with fetal anomalies but no characteristic syndrome has yet been defined (Yong et al. 2002).

1.9.8 Germline Epimutations Associated with Neoplasia

Lynch Syndrome

The finding of germline epimutations in some cases of Lynch syndrome makes the case for inclusion of this condition under this category. Lynch syndrome or hereditary nonpolyposis colorectal cancer (HNPCC) is usually an autosomal dominantly inherited syndrome characterized by susceptibility to early onset colorectal and endometrial cancers, as well as other specific cancers. It is caused by germline mutations in DNA mismatch repair (MMR) genes, mainly affecting *MLH1* and *MSH2* genes. However, besides genetic mutations, constitutional epigenetic silencing of MMR genes *MLH1* and *MSH2* has been recently reported as another possible cause (Crepin et al. 2012; Hitchins et al. 2007; Hitchins et al. 2011). The fact that this condition is sometimes transmissible to the next generation without evidence of changes in genetic sequences suggests that the mechanism is a germline epimutations event (Hitchins et al. 2011; Hitchins et al. 2007).

Familial Paragangliomas

Familial Paragangliomas is characterized by the development of slow-growing tumors of the paraganglionic system especially involving the carotid body in the neck and paraganglia in the abdomen. Tumor development is exclusively seen with paternal transmission of the imprinted gene. Germline heterozygous inactivating mutations are implicated involving genes which encode at least one of the four units of mitochondrial complex II, i.e., succinate dehydrogenase, especially the gene for subunit D (*SDHD*). *SDHD*, located at chromosome 11q23, shows a parent-of-origin effect because the disease is observed almost exclusively when the mutation is transmitted from the father, although some cases of maternal transmission have been reported (Beristain et al. 2013). The most common pattern of change is UPD. A unique feature is the environmental influence of altitude and barometric oxygen tension in the induction of this effect in those males carrying the mutation (Baysal 2004). There is also evidence for tissue-specific imprinted regulation of the *SHDD* gene by a long range epigenetic mechanism which may involve oxygen homeostasis (Baysal 2013).

NOEY2 Gene-Related Cancers

Disorders of imprinting and UPD are implicated in the development of certain ovarian and breast cancers involving the *NOEY2* gene (Yu et al. 1999). The two-hit Knudson's model of tumorigenesis has been used to explain the role of epigenetics in cancer (Sapienza 1991). It is known that only the paternal allele of *NOEY2* is expressed because of maternal imprinting. The second somatic hit silencing the paternal allele in *NOEY2*-linked ovarian cancer could occur through various mechanisms especially "loss of heterozygosity" a process not uncommon in mammalian systems (Makishima and Maciejewski 2011).

1.10 Effects from Conception to Completion of Embrogenesis Including First Trimester: Disorders of Somatic and Trophoblastic Cell Epigenome (Single Generation—Organism Wide Effects, May Affect Only One of Monozygotic Twins)

Epigenetic changes that occur after the formation of the gametes from the germline and their fertilization are important in the development of the early zygote, the silencing of the extra X chromosomes in females, replication of the zygote to form the blastocyst, and then in the differentiation of the inner cell mass that develops into the embryo from the trophoblastic cell mass that develops into the placenta and its membranes. Effects on epigenetic changes at this stage may affect all subsequent cells if the gametes or the zygote are affected before any cell division has occurred. If the effects occur after the first few cell divisions as may happen in case of ART, the effect may have an asymmetry and may involve one of the pair of twins. If the changes occur even later after the blastocyst has differentiated into the inner cell mass and the trophoblast, the effect may be seen only in the fetus or the placenta or show a differential expression in these two cell masses. As mentioned above under imprinted genes, a later effect on the epigenome during the formation and migration of the PGCs may have changes that paradoxically do not affect the F1 generation but affect the subsequent generations. Such germline effects are associated with heritable imprinting disorders and also may be important in the effect of environmental toxins such as DES and Bisphenol that may impact many generations. Therefore the timing of epigenetic change is important and the classification of epigenetic disorders in this chapter is based on the various critical time periods in the life cycle of humans.

1.10.1 The Inactivation of X Chromosome and Related Conditions

Normal X Chromosome Inactivation

Males receive their X chromosome from their mother and hence it is fully expressed with all X-linked genes derived from the mother. But in females, with the presence of two X chromosomes derived one each for maternal and paternal sources, there is need for silencing of one of them with the formation of a dense heterochromatin structure, the Barr body (Lyon 1961). X chromosome inactivation (XCI) is a dosage compensation mechanism that silences most of the genes on one X chromosome in each human female cell and the process of silencing of one of the X chromosomes involves epigenetic modifications including DNA methylation especially at the CGI islands at the promoter regions of the silenced genes (Sharp et al. 2011). Recently interaction of histone modifications and lncRNA (Xist) in the process of XCI has also been defined. *Xist* initiates XCI by spreading in *cis* across the future inactive X chromosome recruiting PRC2 and forming a transcriptionally silent nuclear compartment enriched for repressive chromatin modifications including trimethylation of histone 3 lysine 27 (H3K27me3). This occurs by initial localization of Xist to distal sites across the chromosome by exploiting chromosome conformation, and then spreading to new sites through its ability to modify chromatin structure (Engreitz et al. 2013). The active X is represented as X(a) and the inactive X is represented as X(i). In animals, parent-of-origin-specific methylation differences in XCI are seen but similar findings in humans are unproven (Sharp et al. 2011; Skuse et al. 1997; Sagi et al. 2007; Raefski and O'Neill 2005). However, parent-of-origin expression appears to have a significant effect on neurodevelopmental and behavioral outcomes best illustrated in the case of Turner syndrome.

1.10.2 Turner Syndrome and Effects of Selective Maternal or Paternal X Chromosome Inactivation

It has been postulated that differences in phenotype (physical or behavioral) of Turner syndrome may be due to the existence of imprinted genetic loci from X^m (maternal X) or X^p (paternal X) that is present in the index patient (Sagi et al. 2007). X^p Turner fetuses are more likely to abort and therefore most Turner's born are X^m (60–80 %). Using a mouse model of Turner syndrome, a cluster of three genes show transcriptional repression of paternal alleles; imprinting of these genes Xlr3b, Xlr4b, and Xlr4c appears to be independent of X inactivation (Raefski and O'Neill 2005). Counterparts of these genes in humans have not yet been discovered. Human X^m patients were significantly overweight and this finding was consistent with findings in mice with Xlr gene changes (Ishikawa et al. 1999). As is well known, imprinted genes are related to growth. Moreover, girls lacking the X^p show behavioral socialization problems more frequently than those lacking X^m (Skuse et al. 1997; Lepage et al. 2012). There is speculation that this region of the X chromosome may be responsible for the higher incidence of autism and other similar disorders in males who can have only one X chromosome, i.e., X^m (Lewitus and Kalinka 2013; Keverne 2012). There is a putative imprinted region on the short arm of X that has been implicated (Skuse et al. 1997). In a mouse model of Turner's two X-linked imprintable genes have been identified (Raefski and O'Neill 2005).

1.10.3 Epigenetic Influences on Trophoblastic Tissue and Placenta

Prenatal and perinatal environmental factors influence fetal and placental epigenome in humans (Hogg et al. 2012). The effect of paternal genes on placental growth is well illustrated by the findings in paternal uniparental disomy14 or UPD(14)pat which is associated with placental growth abnormalities (Kagami et al. 2012). Specifically the expression of the *RTL1* also known as the *PEG11* gene is about five times normal in UPD(14)pat. The protein product of this and other related genes *DLK1* and *DIO3* were identified in vascular endothelial cells and pericytes of chorionic villi. These may contribute to the pathological findings of capillary lumens irregularly dilated with thickened endothelium in stem and intermediate villi (not in terminal villi) clinically manifested as placentomegaly and polyhydramnios that are usually seen in this condition (Kagami et al. 2012).

Another piece of evidence comes from the familial recurrent biparental complete hydatidiform mole that occurs due to *NLRP7* mutation. This is a disorder of imprinting in which an empty egg is fertilized with subsequent diploidization of the paternal genome resulting in hyperproliferative vesicular trophoblasts development without any fetus. In some women instead of the *NLRP7* mutation there may be a normal maternal and paternal genome but with epimutation involving loss of methylation at multiple ICRs especially *NLRP7* (El-Maarri et al. 2003; Tomizawa and Sasaki 2012; Kou et al. 2008).

Exposure to in utero infection or inflammation as in chorioamnionitis is associated with changes in *PLAGL1*, an imprinted gene in humans (Liu et al. 2013). *PLAGL1* is located at chromosome 6q24.2 and encodes a zinc-finger transcription factor thought to be involved in growth via *IGF2* signaling. Its role in another epigenetically induced disease TNDM is well established. Placental inflammation's role as risk factor for chronic disease in older children and adults could thus have an epigenetic basis.

There is also evidence of other epigenetic modulations in the placenta that have an effect on outcome of pregnancy and on infant development. The finding of increased placental miRNA-16 has been correlated with adverse infant growth and neurobehavioral scores (Maccani et al. 2013; Maccani et al. 2011). The association of altered placental miRNA in pregnancies complicated by preeclampsia and bisphenol exposure also implicates the role of these epigenetic modulators in fetal development (Pineles et al. 2007; Avissar-Whiting et al. 2010; Maccani et al. 2010).

1.10.4 Epigenetic Influences on Somatic Tissue During Early Development

Twin Growth Discordance

Even among monozygotic twins sharing the same genome, the variation in blood and nutrient supply in the womb results in altered epigenetic profiles. As would be expected, these changes are predominantly seen in growth-related genes. In an extensive study profiling approximately 20,000 CpG sites in 22 monozygotic and 12 dizygotic twin pairs, it was shown that the widest methylation changes were related to genes involved in growth, metabolism, and cardiovascular disease (Gordon et al. 2012). These findings add to the evidence for the "Developmental origins of Health and Disease" hypothesis. Another interesting fact related to the origin of twinning in certain pregnancies relates to the higher incidence of monozygotic twinning in infants born discordant of whom one has BWS (Bliek et al. 2009b). Twinning in BWS seems to affect exclusively female fetuses, is mostly associated with discordance affecting only one of the twins, and is almost exclusively cased by hypomethylation in the affected twin (Mackay and Temple 2010). These observations may be related to the timing of origins of MZ twinning which occurs due to influences between days 3 and 9 of zygotic development. Also of relevance to the occurrence of discordance may be the fact that this period of time coincides with the timing of failure of imprint maintenance in early zygote resulting in different growth compartments with different growth characteristics (Bestor 2003; Mackay and Temple 2010). Thus, twin discordance may not only be caused by epigenetic influences but also affect the later outcome of both the overgrown and the growth-restricted fetuses.

Prematurity

The effect that prematurity and its multisystemic manifestations have on the human epigenome and later onset of disease is being investigated (Stunkel et al. 2012). There is also an implication that prematurity is associated with primary changes in the epigenome during gestation. Altered DNA methylation status of peripheral blood leucocytes with respect to *NFIX*, *RAPGEF2,* and *MSRB3* genes were shown in babies born premature versus term neonates (Lee et al. 2012). The implications of these findings are yet to be determined. It is interesting to note that *NFIX* gene expression is related to skeletal and brain development.

Infantile Biliary Atresia

DNA hypomethylation is noted in human liver tissue with infantile biliary atresia and similar pathology could be replicated in a zebrafish model of biliary atresia by the use of methylation inhibitors in early development (Matthews et al. 2011). A possible mechanism postulated is an initiating event causing inhibition of DNA methylation which leads to *IFN*γ gene activation and altered biliary development and the combination of these two-related processes leading to biliary atresia (Matthews et al. 2011). Abnormal DNA methylation of *ITGAL* (*CD11A*) was seen in peripheral blood lymphocytes obtained from infants with biliary atresia (Dong et al. 2012). The evidence for epigenetic origins is strengthened by the findings of this disorder in discordant monozygotic twins (Fallon et al. 2013; Nakamura and Tanoue 2013). However, no definite human gene modification has been linked so far. Related conditions such as primary sclerosing cholangitis and cholangiocarcinoma may also have similar epigenomic markers. Numerous genes are aberrantly methylated in cholangiocarcinoma including but not limited to *RASSF1A* (27–74 %), *p16* (76–83 %), *CDH1* (22–43 %), *TFPI-2* (40 %), *NPTX2* (40 %), and *APC* (26–46 %) (Timmer et al. 2013).

Wilms' Tumor

Wilms' tumor is also associated with some overgrowth syndromes such as BWS. Wilms' tumor accounts for 8 % of childhood cancers and is the most common childhood kidney malignancy with incidence of 1:10,000 children. Wilms' tumor has been shown to be associated with genetic defects at chromosome 5 loci—*WT1*, *CTNNB1*, *WTX*, *TP53*, and the imprinted 11p15 region especially at the *H19/IGF2* locus. A recent attempt at classification of Wilms' tumor based on its molecular mechanism identified that chromosome 11p15 abnormality accounted for 74 % of all tumors; of these H19 epimutation accounted for 34 % and paternal UPD 11p15 for the other 40 % (Scott et al. 2012). They also found that in sporadic tumors with non-familial origins there were a significant number with bilateral disease (Scott et al. 2012). An explanation given is the possibility that these epimutations occur as early post-zygotic events and can thus be present in both kidneys but absent from other somatic cells. Other lines of evidence also suggest that in some cases of Wilms' tumor aberrant *H19* methylation (epimutation) arises somatically and not in germline unlike PWS and BWS (Moulton et al. 1994; Steenman et al. 1994). The incidence of this tumor in the Japanese is half that of Caucasian children; possibly because of lower incidence of loss of *H19/IGF2* imprinting that is found in the Japanese population (Haruta et al. 2012).

Hepatoblastoma

The *IGF2/H19* imprinted gene domain located at chromosome 11p15 has also been associated with hepatoblastoma seen in BWS (Engel et al. 2000). In tissues from conceptus after the use of ART, higher *H19* DNA methylation levels were found suggesting that the timing of the abnormal methylation event may be post-fertilization (Zechner et al. 2010). The association of poor fetal growth and low-birth weight infants with a 20-fold increase in hepatoblastoma is well described although the reason for this is not yet elucidated (Spector and Birch 2012). The potential role of epigenetic modification of *H19* gene domain in the association of hepatoblastoma with low-birth weight infants is intriguing.

Retinoblastoma

There are many clues to the link that retinoblastoma (*RB*) family of genes and proteins have to the establishment and maintenance of cellular epigenomes. Loss of *RB* family members result in an altered epigenetic landscape, which involves both facultative and constitutive heterochromatin, that leads to genomic instability, loss of differentiation, and may thus promote neoplastic changes (Fiorentino et al. 2013). The genetic basis of retinoblastoma involving the loss of function of the *RB1* tumor suppressor gene has recently been found to have prominent epigenetic basis.

The expression of this disease maybe in fact be based on the concerted action of various epigenetic influences—the most prominent being DNA methylation and the action of microRNAs especially miR-17/20a in tumor inducing roles and miR-34a in a tumor suppressor role (Reis et al. 2012).

Other Genes with Developmental Somatic Epigenetic Changes

Other genes showing somatic epimutations that persist during development are *H19*, *Igf2r*, and *U2af-rs1*. These give rise to abnormal phenotypes (Dean et al. 1998). The rate of somatic epimutations is at least twice as that for genetic mutations and therefore the possibility of its role in disease causation is much higher—however much work needs to be done in this field (Horsthemke 2006).

1.11 Effects from Mid-Late Gestation and Continuing to Postnatal Development Until Adulthood: Disorders of Somatic Cells Epigenome (Single Generation—Tissue-Specific Effects)

Once the newborn infant is separated from the maternal environment of the womb, the changes in the epigenome continue to occur but with different ramifications. These changes are usually tissue and organ specific and have more of a potential for modification and modulation by counter-regulatory environmental factors.

1.11.1 Postnatal Stress and Epigenetics

The effect of stress on epigenetic changes occurs not only within the maternal environment during fetal development but can also occur postnatally in susceptible individuals. Maternal stress may cause minimal or no changes in her own methylation patterns but causes significant changes in methylation patterns of the growing infant (Kinnally et al. 2011; Mulligan et al. 2012). This was demonstrated by finding methylation changes of peripheral blood leucocytes at the *NR3C1* gene from mothers and infants exposed to severe stress in the war-torn Democratic Republic of Congo (Mulligan et al. 2012). *NR3C1* gene encodes the glucocorticoid receptor involved in cell proliferation, differentiation, and thus influences newborn's birth weight. This observation in humans corroborates findings from animal studies showing similar changes not only in peripheral blood cells but also in areas of the brain (Meaney et al. 2007; Wilkinson et al. 2011).

Glucocorticoid receptor changes in the brain have also been associated with child abuse (McGowan et al. 2009). Epigenetic changes in suicide patients who were abused as children were different than those who died without such exposure. Using mouse models of adolescent stress, the role of genetic and epigenetic factors were distinguished. Researchers demonstrated that the specific DNA methylation changes of the tyrosine hydroxylase gene, which regulates dopaminergic transmission and is implicated in psychiatric conditions, were present only in some genetic variants and not in others (Niwa et al. 2013). Therefore the interactions between genetic and epigenetic influences are complex and a critical appraisal of evidence needs to be made before establishing definite cause–effect relationships.

It is important to note that effects of stress may also be transmissible through paternal influences that originate prior to conception (Franklin et al. 2010). Experiments in rats suggest that there may be stress-linked epigenetic marks in the sperm that affect fetal development (Dietz et al. 2011; Nestler 2012). Parental effects of stress exposure on phenotypic variation in offspring are complex and could originate from one or both parents before or after birth (Caldji et al. 2011).

1.11.2 Epigenetic Basis of Adult Onset Disease in Humans

Epigenetic modifications in the womb has been implicated in the "Developmental Origins of Health and Disease (DOHaD) hypothesis" originally proposed by Barker (Barker and Osmond 1987). Evidence for this is growing especially based on the Dutch Famine Birth Cohort Study (www.hongerwinter.nl) (Heijmans et al. 2009). Changes in methylation patterns of specific genes were found to persist decades after the initial prenatal insult. Prenatal exposure to famine especially in the periconception period modified the methylation status of *IL10*, *GNASAS*, *INSIGF*, *LEP*, *ABCA1*, and *MEG3* genes. Some of the modifications for *GNASAS*, *INISIGF*, and *LEP* were different in males and females (Heijmans et al. 2007; Tobi et al. 2009; Heijmans et al. 2009). Abnormal epigenetic programming has also been implicated in obesity-related conditions in mother having growth effects on the child (Ludwig and Currie 2010). The role of proper maternal nutrition in the normal and abnormal epigenetic programming of the fetus is being extensively studied. Evidence of epigenetic alterations in the insulin-receptor-promoter in the hypothalamus of Wistar rats may have implications in understanding human disease (Plagemann et al. 2010). The epigenetic alterations in regulation of transcription factors may explain the widespread effects of a relatively small modification (Martin-Gronert and Ozanne 2012). Some candidate transcription factors that have been implicated in animal studies are *PPARα* (Lillycrop et al. 2005), *PDX-1* (Park et al. 2008), and *HNF4α* (Sandovici et al. 2011).

Another area of active investigation is the role of environmental agents and endocrine disruptors in modulating epigenetic programming in utero. Effect of pollutants such as bisphenol and phthalates has shown to have fetal effects that have footprints of epigenetic alterations persistent into later life. The case for its role in obesity is made in a review by Grun (Grun and Blumberg 2009). Moreover, there is evidence that hypospadias in male infants in relation to endocrine disrupter exposure is associated with altered methylation patterns of the androgen receptor and

DNMT levels in penile foreskin (Vottero et al. 2011). Aberrant hypomethylation has been noted in genes linked to T cell and lymphocyte function-associated antigen-1 in SLE (Lu et al. 2006). Drugs such as procainamide and hydralazine that cause hypomethylation of somatic genes can cause lupus in humans and in animal models (Quddus et al. 1993).

It is well known that maternal obesity is related to obesity in their offspring. It has been found that bariatric surgery to reduce weight of obese women reduces birth weight and the risk of obesity in their offspring born after surgery compared to those born before (Smith et al. 2009; Martin-Gronert and Ozanne 2012). This highlights the importance of understanding the critical period in making changes that can then have a life-long effect on the outcome.

1.11.3 Malignancy and Epigenetics

In 1983, cancer was the first disease described to have altered epigenetic marks. Feinberg et al. demonstrated that colorectal cancers exhibited a global loss of DNA methylation in comparison with their normal counterparts (Feinberg and Vogelstein 1983). This hypomethylation may have caused genomic instability, chromosome rearrangements, and induced aberrant activation of certain genes (Feinberg and Tycko 2004). Subsequent works found that DNA hypermethylation of the promoter region of tumor suppressor genes was also a frequent event in malignancy (Herman 1999). Later, other epigenetic mechanisms displayed abnormal regulation, such as histone modifications (Fraga and Esteller 2005) and microRNA posttranscriptional modulation (Esteller 2008; Friedman et al. 2009; Girardot et al. 2012; Reis et al. 2012). Modification of the cancer epigenome is being used as strategy for treatment of these disorders.

The cancers associated with the *NOEY2* gene were mentioned above under imprinted genes. Apart from a disorder of imprinting, this cancer needs to have a somatic mutation as part of a "two-hit hypothesis" to manifest its effects. The second somatic hit silencing the paternal allele in *NOEY2*-linked ovarian cancer could occur through various mechanisms affecting methylation of the critical regions (Makishima and Maciejewski 2011).

1.12 Secondary Epigenetic Disorders Due to Disorder of Genetic DNA (Mendelian or Chromosomal Transmission Across Generations)

In some instances, the primary problem may be a genetic abnormality that in turn has an effect on the epigenome because of the effect on DNA methyltransferases, methyl-binding domain proteins, HDACs, histone methylases, histone demethylases, and members of the noncoding RNA machinery (Berdasco and Esteller 2013) (Table 1.2). In such instances, the mode of inheritance may be Mendelian or classic chromosomal transmission mechanisms. However, since the end result is an epigenetic alteration, with genetic mechanism still intact for normal protein formation, the potential for modification may have important ramifications for preventive and therapeutic strategies in combating such conditions. The conditions could be further classified by the effect of the genetic abnormality on the resultant epigenetic mechanism, i.e., DNA methylation, histone modification, or microRNA modulation. It is important to realize that some of these conditions may have more than one epigenetic mechanism altered.

1.12.1 Gene Abnormalities with DNA Methylation Effects

ICF (Type 1) Syndrome

In the developing embryo, after the immediate post-zygotic genome-wide demethylation (that occurs in all but the germline cells) there is a wave of de novo methylation catalyzed by DNMT3A and DNMT3B. Absence of DNMT3B expression results in profound developmental defects and is embryo-lethal in mice. In humans homozygous mutations of this gene resulting in decreased activity rather than complete absence of activity leads to a specific phenotype termed the ICF syndrome (acronym for immunodeficiency, centromere instability, and facial anomalies) (Xu et al. 1999; Hansen et al. 1999). ICF syndrome is characterized by immunodeficiency due to decreased B cells and agammaglobulinemia along with centromere instability of chromosomes 1, 16, and 9 and facial anomalies including hypertelorism, flat nasal bridge, epicanthic folds, protrusion of the tongue, low-set ears, and micrognathia (Maraschio et al. 1988; Maraschio et al. 1989). These are caused by DNMT3B deficiency leading to hypomethylation and epigenetic changes in the majority of patients (Type 1 ICF). Although in certain cases, there may be genetic alterations in *ZBTB244* (Type 2 ICF). DNMT3A and DNMT3B are important for the de novo introduction of methyl groups during development (Lana et al. 2012). Hypomethylation of genes related to B-cell receptor-mediated maturation pathway have been identified as the likely cause of decreased number of mature B cells and resulting agammaglobulinemia (Heyn et al. 2012b). An overall decrease in methylation by 41 % was seen in the whole genome with the X chromosome showing a more severe loss of 63 %. This finding leads to the speculation that epigenetic alterations in the X chromosome specifically at loci *SYK* and *SH3BP5* (involved with Burton's Tyrosine Kinase involved with B-cell maturation) leading to altered B-cell function may also play a part in another syndrome with similar phenotype, i.e., congenital agammaglobulinemia (Yamadori et al. 1999; Kurosaki and Hikida 2009). Thus, modification of the hypermethylated genes by generic DNMT inhibitors such as 5-Azacytidine may be a useful strategy in the treatment of patients with Type 1 ICF syndrome (Heyn et al. 2012b).

Rett Syndrome: MeCP2 Gene

Rett syndrome is an X-linked neurodevelopmental disorder characterized by autistic features, epileptic seizures, gait ataxia, and stereotypical hand movements. The usual cause is a de novo mutation of the *MECP2* gene of paternal origin that occurs during spermatogenesis. The effects are due to global changes in neuronal chromatin structure because of global changes in histone methylation patterns (Amir et al. 1999; Skene et al. 2010). In tissue samples from monozygotic twins, differences in DNA methylation between the twins were detected in fibroblasts in the upstream regions of genes involved in brain function and skeletal tissues such as Mohawk Homeobox (*MKX*), brain-type creatine kinase (*CKB*), and FYN tyrosine kinase protooncogene (*FYN*) (Miyake et al. 2013). Recently in animal studies, the role of early glucocorticoid exposure has been shown to modulate the putative genes involved in Rett-like condition in mice and targeting the glucocorticoid system may provide evidence that pharmacological interventions during critical early time windows can persistently improve the behavioral phenotype of Rett mice (De Filippis et al. 2013). The role of *MECP2* in early life environmental modulation of the epigenome is also being increasingly highlighted (De Filippis et al. 2013; Na et al. 2013).

Fragile X Syndrome

Fragile X syndrome (FXS) is the most common heritable form of impaired intellectual ability worldwide. FXS is an X-linked neurodevelopmental disorder of dominant inheritance characterized by cognitive and behavioral difficulties and facial dysmorphism (elongated face, large and protruding ears), which can manifest in mild to severe forms. Affected females exhibit symptoms but usually to a lesser extent, due to the presence of a normal allele (Gallagher and Hallahan 2012). FXS patients have an expansion of a single trinucleotide sequence (CGG) at the promoter of the fragile X mental retardation gene (*FMR1*), mapped at Xq27.3. Healthy individuals carry between 5 and 44 CGG repeats, while affected patients with full mutations carry more than 200 repeats. The expansion of CGG repeats results in the methylation of the affected DNA, which leads in the full mutation alleles to the epigenetic silencing of the *FMR1* and the lack of its product, the fragile X mental retardation protein (FMRP) (Gallagher and Hallahan 2012).

X-Linked Alpha-Thalassemia/Mental Retardation Syndrome (ATR-X Syndrome)

X-linked alpha-thalassemia/mental retardation syndrome (ATR-X syndrome) is one of the syndromes associated with abnormal epigenetic gene regulation. It is a form of X-linked mental retardation that affects males and is characterized by severe mental retardation, a mild form of α-thalassemia, facial and skeletal abnormalities along with autistic behavior. This syndrome is caused by a mutation in the *ATRX* gene on the X chromosome (Xq13), which encodes ATRX protein. The protein has two functionally important domains: an ADD (ATRX-DNMT3-DNMT3L) domain at the N-terminus, and chromatin-remodeling domain in the C-terminal half, where the *ATRX* gene mutations of most ATR-X patients reside. Perturbation in DNA methylation at the heterochromatic loci has been reported in ATR-X patients, and ATRX protein is thought to be involved in the establishment and maintenance of DNA methylation (Gibbons 2006). Based on its various clinical phenotypes, the expressions of many genes, including α-globin genes, seem to be abnormally regulated in ATR-X patients (Wada 2009).

Fascioscapulohumeral Dystrophy

Autosomal dominant facioscapulohumeral muscular dystrophy (FSHD) is characterized by facial and shoulder girdle muscle weakness, with progression of the disease to anterior foreleg, abdominal and pelvic girdle muscles. The typical involvement of facial muscles includes *orbicularis oculi* and *oris* muscles, resulting in impaired palpebral occlusion and transverse smile characteristic of this condition (Tawil et al. 1998; Tawil and Van Der Maarel 2006). FHSD patients can also exhibit extra muscular features such as hearing loss, retinopathy, mental retardation, and epileptic seizures (Tawil and Van Der Maarel 2006). There is good evidence that FSHD is not caused by defects in a single gene; instead, the deregulation of epigenetic mechanisms results in aberrant transcription of multiple disease-related genes (Tawil et al. 1998). Most FSHD patients are linked to molecular rearrangements in the subtelomeric region of chromosome 4 long arm (4q35) which maps to a 3.3-kb tandem-repeated macrosatellite, D4Z4. Healthy individuals carry between 11 and 100 repeats, whereas FSHD patients have a reduced number of copies of D4Z4 that range from 1 to 10 (van Deutekom et al. 1993). The effect of these is not related to altered expression of a protein-coding gene but rather to various forms of epigenetic modulations ranging from DNA methylation to altered chromosomal architecture leading to deregulation of 4q35 region (Sacconi et al. 2012; Neguembor and Gabellini 2010).

Hereditary Sensory and Autonomic Neuropathy Type I

Hereditary sensory and autonomic neuropathy type I (HSAN1) has been found to be related to mutations affecting DNMT1 methyltransferase gene located on chromosome 19p13.2. The resultant effects are premature degradation of mutant proteins, reduced methyltransferase activity, and impaired heterochromatin binding during the G2 cell cycle phase leading to global hypomethylation and site-specific hypermethylation abnormalities (Klein et al. 2011).

Autosomal Dominant Cerebellar Ataxia, Deafness and Narcolepsy

Autosomal dominant cerebellar ataxia, deafness and narcolepsy (ADCA-DN) is characterized by late onset (30–40 years old) cerebellar ataxia, sensory neuronal deafness, narcolepsy-cataplexy, and dementia. DNMT1 abnormalities were found in all patients in a particular study cohort. Abnormalities in this widely expressed DNA methyltransferase result in problems with maintaining methylation patterns in development, and mediating transcriptional repression by HDAC2 binding (Winkelmann et al. 2012).

1.12.2 Genetic Syndromes Causing Histone Modifications

Rubinstein–Taybi Syndrome

Rubinstein–Taybi syndrome (RTS) is a rare human genetic disorder characterized by growth and psychomotor development delay, unusual facial features with down-slanting palpebral fissures, broad nasal bridge, beaked nose and micrognathia, and skeletal abnormalities in extremities, such as radially diverted phalanges and broad and duplicated distal phalanges of thumbs and toes (Rubinstein and Taybi 1963). Many RTS patients have a genetic mutation which has been mapped to chromosome 16p13.3, a genomic region encoding cyclic AMP response element (CREB)-binding protein (*CREBBP*) (Hallam and Bourtchouladze 2006). Its action is probably related at least partly to the role of CREBBP as a histone acetyltransferase and its corresponding ability for epigenetic regulation through histone modification (Wang et al. 2010). Another gene *EP300* with similar function may also be involved (Bartsch et al. 2010; van Belzen et al. 2011).

Genitopatellar Syndrome

Genitopatellar syndrome (GPS) is manifested by patellar aplasia or hypoplasia along with external genital anomalies and severe intellectual disability. Most cases have demonstrated abnormality in the *KAT6B* gene which encodes a member of the *MYST* family of histone acetyltransferases (Campeau et al. 2012; Campeau and Lee 1993). A reduced level of both histone H3 and H4 acetylation has been shown in patient-derived cells suggesting that dysregulation of histone acetylation is the mechanism of disease manifestation (Simpson et al. 2012).

Say-Barber-Biesecker-Young-Simpson Syndrome

Say-Barber-Biesecker-Young-Simpson syndrome (SBBYS) is part of a blepharophimosis-mental retardation syndrome group (Ohdo-like syndromes) (Szakszon et al. 2013; Verloes et al. 2006). It is a multiple congenital malformation syndrome characterized by vertical narrowing and shortening of the palpebral fissures, ptosis, intellectual disability, hypothyroidism, hearing impairment, and dental anomalies. Mutations of the gene encoding the histone acetyltransferase *KAT6B* have been recently identified in individuals affected by SBBYS syndrome (Szakszon et al. 2013).

Coffin–Lowry Syndrome

Coffin–Lowry syndrome is an X-linked disorder characterized by characteristic facial features, severe psychomotor retardation, facial and digital dysmorphism, and progressive skeletal deformations. CLS is caused by mutations in the *RPS6KA3* gene located at Xp22.2, which encodes RSK2, a growth factor-regulated protein kinase (Jacquot et al. 1998a; Jacquot et al. 1998b; Pereira et al. 2010). Histone modification of H3 appears to be a direct or indirect target of RSK2, suggesting that chromatin remodeling might contribute to mitogen-activated protein kinase-regulated gene expression (Huidobro et al. 2013; Sassone-Corsi et al. 1999).

Sotos Syndrome

Sotos syndrome is an overgrowth condition in children, first described as "cerebral gigantism" by Sotos et al. in 1964 (Sotos et al. 1964). It is characterized by large body size and early accelerated growth, advanced bone age, and developmental delay, usually accompanied by learning difficulties. Typical facial features include macrocephaly, prominent jaw, down-slanting palpebral fissures, high hairline with sparse hair growth, malar flushing, a long narrow face, and the head is said to resemble an inverted pear (Tatton-Brown et al. 1993). There are also many other features associated with the syndrome, such as neonatal jaundice, hypotonia, seizures, scoliosis, cardiac defects, and genitourinary anomalies (Huidobro et al. 2013). Haploinsufficiency of the nuclear receptor SET domain-containing protein 1 (*NSD1*) gene located on 5q35.2 has been implicated as the cause of Sotos syndrome (Sohn et al. 2013).

Weaver Syndrome

Weaver syndrome is manifested in young children by the presence of retrognathia and have large, fleshy ears while both children and adults with classic Weaver syndrome are hyperteloric and the eyes are almond shaped (Tatton-Brown and Rahman 2013). Mutations in the *EZH2* gene are most common (Tatton-Brown and Rahman 1993) but there are some *NSD1* defects (more commonly linked with Sotos syndrome) that have been associated as well (Tatton-Brown and Rahman 2013; Rio et al. 2003). Similarly, it shares a number of clinical features with Sotos syndrome but there are certain features that are more commonly present in Weaver's such as—a connective tissue phenotype with soft, loose skin, umbilical hernia and thin, deep-set nails. Other potentially distinguishing features described among individuals with

EZH2 alterations include a deep hoarse voice and camptodactyly of the fingers and/or toes evolving into boutonniere deformities in adulthood (Tatton-Brown and Rahman 2013). *NSD1* and *EZH2* are SET domain-containing histone methyltransferases that play key roles in the regulation of transcription through histone modification and chromatin modeling. *NSD1* which is more commonly associated with Sotos syndrome preferentially methylates lysine residue 36 of histone 3 (H3K36) and is primarily associated with active transcription. In contrast *EZH2* associated with Weaver's syndrome, shows specificity for lysine residue 27 (H3K27) and is associated with transcriptional repression (Tatton-Brown and Rahman 2013).

Brachydactyly-Mental Retardation Syndrome

This condition is characterized by a variety of features, including intellectual disabilities, developmental delays, behavioral abnormalities, sleep disturbance, craniofacial and skeletal abnormalities (including brachydactyly type E), and autism spectrum disorder (Williams et al. 2010). Histone deacetylase 4 (HDAC4) haploinsufficiency is responsible for psychomotor and behavioral abnormalities in combination with the brachydactyly-mental retardation syndrome (BDMR) syndrome-specific facial dysmorphism pattern (Tammachote et al. 2012; Hacihamdioglu et al. 2013; Williams et al. 2010; Villavicencio-Lorini et al. 2013).

Kleefstra Syndrome

Kleefstra syndrome is characterized by developmental delay/intellectual disability (childhood), hypotonia, and distinct facial features. A majority of individuals function in the moderate to severe spectrum of intellectual disability although a few may have mild delay. There is usually severe expressive speech delay with little speech development, but general language development is usually at a higher level, making nonverbal communication possible. Other findings may include heart defects, renal/urologic defects, genital defects in males, severe respiratory infections, epilepsy/febrile seizures, autistic-like features in childhood, and extreme apathy or catatonic-like features after puberty (Kleefstra et al. 1993). The syndrome can be caused either by a microdeletion in chromosomal region 9q34.3 or by a mutation in the euchromatin histone methyltransferase 1 (*EHMT1*) gene which modifies histone function and thus causes epigenetic dysregulation (Willemsen et al. 2012; Kleefstra et al. 2005; Kleefstra et al. 2009).

Kabuki Syndrome

Kabuki syndrome (KS) is characterized by distinctive craniofacial anomalies and multiple malformations including cardiac anomalies, skeletal abnormalities, and mild to moderate intellectual disability (Bokinni 2012; Hannibal et al. 2011). The estimated prevalence is 1 in 32,000. Kabuki syndrome is usually caused by mutations in *MLL2*, a gene that encodes a Trithorax-group histone methyltransferase, a protein important in the epigenetic control of active chromatin states (Hannibal et al. 2011). Recently partial or complete deletions of an X chromosome gene—*KDM6A* have been identified in some cases of Kabuki syndrome. *KDM6A* encodes a histone demethylase that interacts with *MLL2* (Lederer et al. 2012).

Siderius X-Linked Mental Retardation Syndrome

Siderius X-linked Mental Retardation syndrome (MRXSSD) is a condition first described in a case series that have facial clefts associated with X-linked mental retardation (Siderius et al. 1999). The finding of *PHF* gene abnormality with this clinical phenotype suggests an important function of *PHF8* in midline formation and in the development of cognitive abilities (Laumonnier et al. 2005; Abidi et al. 2007). The coded PHF8 protein harbors two functional domains, a PHD finger and a JmjC (Jumonji-like C terminus) domain, implicating it in transcriptional regulation and chromatin remodeling. Histone methylation modulated by *PHF8* plays a critical role in neuronal differentiation (Qiu et al. 2010).

Claes–Jensen X-Linked Mental Retardation Syndrome

Claes–Jensen X-linked mental retardation syndrome is characterized by severe mental retardation, slowly progressive spastic paraplegia, facial hypotonia, and maxillary hypoplasia (Claes et al. 2000). The gene abnormality in this condition maps to the *JARID1C* (Jumonji AT-rich-interactive domain 1C), formerly known as "*SMCX*." The JARID1C protein belongs to the highly conserved ARID protein family. It contains several DNA-binding motifs that link it to transcriptional regulation and chromatin remodeling, processes that are defective in various other forms of mental retardation. Recent studies suggest that *JARID1C* mutations are a relatively common cause of X-linked mental retardation and that this gene might play an important role in human brain function (Jensen et al. 2005).

1.12.3 Genetic Mutations That Affect Noncoding RNAs

Noncoding RNAs are functional RNA molecules that are not translated into protein but contribute to epigenetic regulation via degradation of protein-coding transcripts or by translational repression (Berdasco and Esteller 2013). The best described noncoding RNAs are microRNAs that have been associated with epigenetic regulation (Croce 2009). miRNAs are transcribed as individual units termed primary miRNA which, after processing by the Drosha complex, are exported out of the nucleus by the protein exportin (*XPO5*) and become mature miRNA after undergoing further processing by Dicer and TAR RNA-binding protein 2 (*TARBP2*). They ultimately exert their action via RNA-induced silencing complex (RISC). Genetic changes that affect any of the processes in this sequence of events can impact miRNA level and function.

Amyotrophic Lateral Sclerosis

Amyotrophic lateral sclerosis is an autosomal dominant neurodegenerative disorder affecting the motor neurons in the neural axis resulting in fatal paralysis and respiratory failure (Berdasco and Esteller 2013). Heterozygous mutation in the TAR DNA-binding protein 43 (*TARDBP*), a member of the miRNA machinery in the cell accounts for about 50 % of patients (Ling et al. 2010).

DiGeorge Syndrome

This is characterized by distinct facial features, submucous cleft palate, conotruncal heart defects, thymic aplasia or hypoplasia, neonatal hypocalcemia, poor T cell function, psychiatric and learning disabilities. Hemizygous deletion of chromosome 22q11.2 affecting the *DGCR8* gene (DiGeorge syndrome critical region gene 8) is found (Shiohama et al. 2003). This region encodes an RNA binding-protein which is critical in microRNA processing and release from the nucleus (Han et al. 2006). This has multisystemic effects involving various microRNA types (Berdasco and Esteller 2013).

Goiter, Multinodular1, with or Without Sertoli–Leydig Cell Tumors

This is a form of autosomal dominant multinodular goiter characterized by nodular overgrowth of the thyroid gland and in some females associated with Sertoli–Leydig cell tumors of the ovary. Germline mutations in *DICER1*, a gene that codes for a Raze III endoribonuclease, have been identified in families affected by multinodular goiter and gonadal tumors (Rio Frio et al. 2011). Dysregulation of miRNA because of abnormalities of DICER have been documented and correlated with the phenotype of this condition (Rio Frio et al. 2011; Berdasco and Esteller 2013).

1.12.4 Chromosome Deletion and Rearrangements: Epigenetic Changes

9q Subteloric Deletion Syndrome

The 9q subtelomeric deletion syndrome is characterized in most cases by moderate to severe mental retardation, childhood hypotonia, and facial dysmorphism; some cases that also have congenital heart defects, urogenital defects, epilepsy, and behavioral problems are frequently observed (Kleefstra et al. 2009). The two main etiologies for this condition are submicroscopic 9q34.3 deletion or intragenic *EHMT1* mutations leading to haploinsufficiency of the *EHMT1* gene as mentioned above under Kleefstra syndrome (Greer and Shi 2012; Kleefstra et al. 2009).

46XY Inversion(10)(q11.1;q21.3)

46 XY inversion (10)(q11.1;q21.3) is another example of autism associated with the X chromosome. In this case disruption of the *TRIP8* gene may be implicated (Castermans et al. 2007). This gene exerts its effect through the JMJC domain resulting in histone modifications (Castermans et al. 2007).

Wolf–Hirschhorn Syndrome

Wolf–Hirschhorn syndrome (WHS) is a syndrome caused by a deletion or disruption of the distal region of the short arm of chromosome 4 at the 4p16.3 site. It is characterized by the presence of a peculiar phenotype, delayed growth, delayed psychomotor development, and epilepsy (Blanco-Lago et al. 2013). It may also be a result of chromosome translocation of chromosome (4:14)(p16;q32). In most cases there is abnormality of the histone methyltransferase—WHS candidate 1-like 1 (*WHSC1L1*) gene that results in epigenetic effects through histone modulation (Kang et al. 2013; Sheth et al. 2012).

1.13 Epigenetics, Brain Function and Memory

From the above list of syndromes that have an effect on mental development, autism and other neuronal changes, it is apparent that epigenetics has a major role in human brain functioning in health and disease. Epigenetic mechanisms in the dentate gyrus may be one of the molecular switches of hippocampus-associated memory formation (Reul et al. 2009). Epigenetic modifications are thought to be involved in brain memory and plasticity and therefore may also be important in behavioral disorders (Dulac 2010). This mechanism may involve DNA methylation (Day and Sweatt 2010) or histone modifications (Dulac 2010). Histone methylation by HDAC3 is involved with long-term memory formation (McQuown et al. 2011). Even after birth, there is evidence of behavior and nurture influencing changes in epigenetics (Meaney et al. 2007). Development of pharmacological agents using epigenetic alteration of histones may be by HDAC inhibitors which seem to have an effect on behavior and memory (Abel and Zukin 2008; Vecsey et al. 2007). There is evidence that epigenetic alterations of the brain by the action of histone methyltransferases G9a may be involved in cocaine-induced brain plasticity (Maze et al. 2010).

As a category, many syndromes resulting in mental retardation have epigenetic origins. Some of these have been described elsewhere in the chapter. Mental retardation syndromes have been linked to histone and DNA modifications and to microRNA dysregulation (Iwase and Shi 2011).

1.14 Epigentic Disorders Associated with Assisted Reproductive Technology

An association between assisted reproductive techniques and increased risk for epigenetic defects is increasingly being noted (Katari et al. 2009) but it is not entirely clear if this is due to the technique or the subfertility pattern of the parents (Horsthemke and Buiting 2008). The correlation of ART and higher incidence of disorders of epigenetics can however be explained by the critical period concept where the period of demethylation and de novo methylation corresponds to the period of exposure to ART techniques (Niemitz and Feinberg 2004; Feinberg 2007; Kohda and Ishino 2013). The pattern of methylation changes also provides clues to this association. For example, the hypomethylation of *LIT1* is the most common finding (13/14 cases) in BWS associated with ART compared to non-ART conception where the incidence is only about 33 % (Niemitz and Feinberg 2004). A recent review of 30,959 infants born of IVF in Sweden showed that IVF treatment though not directly associated with autistic disorder was associated with mental retardation; and specific procedures of IVF such as intracytoplasmic sperm injection (ICSI) for paternal infertility were associated with an increased relative risk for both autistic disorder and mental deficit (Sandin et al. 2012). An epigenetic mechanism was implicated based on similar observations in animal studies (Lucifero et al. 2004; Paoloni-Giacobino and Chaillet 2004; De Rycke et al. 2002). ICSI is related to increased methylation problems in offspring as demonstrated in ICSI-generated mice (Xu et al. 2013). Especially relevant may be asymmetries that exist in the patterns of development of male and female zygotes that have a potential to be altered during the ICSI procedure (Kohda and Ishino 2013).

Epigenetic effects in cloned offspring are also much higher than in non-cloned normal conception, and include the following cellular manifestations: (a) X-inactivation errors, (b) improper imprinting of paternal genes, (c) abnormal DNA methylation, (d) abnormal histone acetylation and methylation, (e) failure to activate key pluripotency genes such as *Oct4* (Morgan et al. 2005).

1.14.1 Epigenetic Drift, Cancer, and Aging

Recently the changes associated with time in the landscape of DNA methylation of most of the genome of somatic cells have been characterized and termed "Epigenetic Drift" (Teschendorff et al. 2013). Some of the changes could also be tissue specific and may yield biomarkers that help in monitoring tissue function changes with age and disease. This concept has potential implications in understanding the biology of stem cells, neoplasia, and aging-related changes. Epigenetic drift has been well documented in changes involved with human aging (Fuso et al. 2012; Hannum et al. 2013; Heyn et al. 2012a; Martin 2012; Mendelsohn and Larrick 2013; Teschendorff et al. 2013). This may be related to many mechanisms but the association with telomerase shortening is especially intriguing (Farwell et al. 2000). There is also a large body of evidence suggesting its role in the development of neoplasia; therefore, understanding the mechanisms behind this process may be important in developing strategies for prevention and treatment of cancers (Teschendorff et al. 2013).

1.15 Epigenetics and Multigenerational Disease Processes

From analysis of the Dutch Famine Cohort in the Netherlands, researchers observed transgenerational phenotypic changes in fetal growth and birth weight (Lumey and Stein 1997). Transgenerational effects of epigenetics were initially described in endocrine disruptors in rats causing male infertility into the F3 generation (Anway et al. 2006) and more evidence has been accumulating on this subject (Youngson and Whitelaw 2008). There are mouse models to explain transgenerational maintenance of parent-specific methylation patterns (Park et al. 2012). There is increasing evidence of transgenerational effects of epigenetic changes especially noted in the *MLH1* gene associated with Lynch syndrome which is characterized by multiple MLH1-negative cancers of the colorectum and endometrium and hemiallelic methylation of the *MLH1* in all somatic cells (Goel et al. 2011; Hitchins et al. 2007). In a study involving families of two cases with germline *MLH1* epimutations, it was discovered that the epimutations was transmitted from a mother to her son but that imprint was later erased in his sperms. The affected allele though inherited by three other siblings from these two families studied had reverted to the normal active state (Goel et al. 2011; Hitchins et al. 2007).

Transgenerational effects have been demonstrated on reproductive functions in male and female offspring (up to F3 generation) when F0 gestating rats were exposed to various chemicals during period of embryonic gonadal development (Manikkam et al. 2012; Anway et al. 2006). Specific and consistent DMRs were identified in the sperms of all lineages suggesting the role of epigenetic mechanisms (Manikkam et al. 2012). The process of epigenetic inheritance may be different in men and women and we need to be cognizant of this fact not only in animal experiments but also in analyzing human epidemiological data relating to epigenetics (Hitchins et al. 2011; Hitchins et al. 2007).

1.16 Epigenetics and Evolutionary Relationship of Imprinting to Growth

Genomic imprinting may have evolved about 150 million years ago in a common live-born mammalian ancestor after divergence from egg-laying animals. Subsequently there is evidence of continuing evolution from metatherian (marsupials—where most of the development is extra uterine) to eutherian mammals (where most of the development is intrauterine) (Murphy and Jirtle 2003; Killian et al. 2000). The theory postulates that this is related to the father trying to have the largest possible baby with only one partner (him) even at the cost of mother's life; versus the mother trying to have a smaller infant so that she can have more babies with more partners. The "battle-of-sexes hypothesis" or "conflict theory," where the mother and father are each vying to leave their genes as lasting legacies has been given scientific validity and it has become clear that paternally- and maternally imprinted genes sometimes have opposite effects during fetal development. Paternally imprinted or controlled genes promote fetal growth and maternally imprinted genes suppress it (Haig and Graham 1991; Moore and Haig 1991). Evidence for this "battle-of-sexes hypothesis" mostly comes from animal studies, but it is also suggested by some observations in humans; human triploid fetuses develop a large placenta if the extra genetic material is paternal but placental tissue is sparse if the extra genomic material is maternal (Hall 1990). There are many hypotheses that attempt to explain this phenomenon from an evolutionary perspective but none have gained wide acceptance so far.

1.17 Epigenetics, Biomarkers, and Drug Discovery

Development of biomarkers is an important step in early recognition of disease and in monitoring the progression or regression of a particular condition. There is ongoing progress in identification of biomarkers for epigenetic conditions. The challenge is that most such conditions are time and tissue specific. However, in studies using different human tissues from different ages of life, it was shown that in independent imprinted genes some of the DMR methylation levels were comparable at birth between two tissues—umbilical cord blood and buccal cells. Moreover, within buccal cells, DMR methylation levels were similar at birth and at 1 year of age. Furthermore, comparison to other fetal tissues showed that DMR methylation in one tissue would be comparable to other tissues within the same subject for the markers studied. Therefore, it is possible that buccal cells or other easily obtained tissue from an individual could be used as a biomarker in epidemiological studies (Murphy et al. 2012; Park et al. 2012). Another potential biomarker for fat mass in childhood is the methylation status of the retinoid X receptor in umbilical cord (Godfrey et al. 2011). Certain epigenetic biomarkers have been found that are associated with exposure to certain environmental compounds and persist across generations, thus allowing for assessment of ancestral environmental exposures associated with adult onset disease (Manikkam et al. 2012).

An understanding of critical periods in epigenetic modifications is important in proper targeting of appropriate therapy. A recent example is the use of JMJD2 histone demethylases to epigenetically control herpes virus infection and prevent its reactivation from latency (Liang et al. 2013). Knowing that the JMJD2 family of chromatin-modification enzymes are required for both HSV and hCMV immediate-early (IE) genes essential for viral transcription activators, was key in developing therapeutic strategies for inhibiting these enzymes and has opened up a whole new field of antiviral drug discovery. Similarly, advances in our understanding of the critical points of epigenetic alterations in cancers are opening up possibilities for developing preventative and therapeutic strategies (Yang et al. 2013; Ren et al. 2013b; Ren et al. 2013a; Qiao et al. 2013; Dietel et al. 2013; Andreoli et al. 2013). The understanding of periods of vulnerability of the human brain cells to epigenetic changes is providing opportunities to develop drug therapies for improving brain function and memory (Vecsey et al. 2007; Reul et al. 2009; McQuown et al. 2011; Day and Sweatt 2010).

1.18 Epigenetics: Future Trends

Study of the epigenome is being fast-tracked by a number of other important developments. The human methylome with differences between stem cells and fibroblasts has been cataloged (Lister et al. 2009). Recently the encyclopedia of DNA elements (ENCODE) project has identified and mapped regions of transcription, transcription factor association and correlated them with chromatin structure and histone modifications to help us better understand epigenetic regulation of the genome (Dunham et al. 2012). Altered epigenetic patterns in gene regulatory sequences have thus been revealed by the ENCODE project and the effects are global within the organisms and especially in human cancers (Bernstein et al. 2012). The use of Gene Environment Association Studies (GENEVA) and multi-site genome-wide association study (GWAS) consortia have added significantly to our understanding of epigenetics and its role in human health and disease. The challenge in these studies will be to have a properly defined phenotype (phenotype harmonization) (Bennett et al. 2011). A number of human therapies targeting the epigenome are being tested especially those related to histone modifications effect on improvement of memory, anticancer drugs, and antiviral medications. The therapeutic use of microRNA is being investigated, and modification of DNA methylation in periconception and early embryogenesis are being studied in animal models with great promise. In utero modifications of targeted molecules such as leptins are being investigated to improve birth outcomes (Gluckman and Hanson 2007; Gluckman et al. 2009; Gluckman et al. 2007) Changes in post-weaning diet in mice have been shown to affect genomic imprinting of the *Igf2* locus giving rise to the possibility of nutritional and other therapeutic manipulation of the epigenome for improving health in humans during their entire lifetime (Waterland et al. 2006). Understanding the critical periods of epigenome vulnerability will add greatly to the proper targeting of these potential therapies.

1.19 Conclusions

Our understanding of epigenetic mechanisms in the preservation of health and the development of disease is rapidly expanding. With the advent of new knowledge and discovery of new therapies, it is important to classify epigenetic conditions using a system that can be useful in targeting the developed therapies to critical periods of development where they can have the most impact. The classification of epigenetic conditions presented here helps in determining not only the optimal time in the life cycle but also the mechanism of epigenetic change viz., DNA methylation, histone modification, or microRNA modulation. The fact that in a majority of these disorders the affected protein is not intrinsically flawed but abnormally expressed provides ground for optimism that given the appropriate epigenetic switch, the right amount of that protein might still be restored.

Acknowledgement The support of Tasneem Hussain, M-OTR in the editing and review of this manuscript is greatly appreciated.

References

Abel T, Zukin RS (2008) Epigenetic targets of HDAC inhibition in neurodegenerative and psychiatric disorders. Curr Opin Pharmacol 8:57–64

Abidi F, Miano M, Murray J, Schwartz C (2007) A novel mutation in the PHF8 gene is associated with X-linked mental retardation with cleft lip/cleft palate. Clin Genet 72:19–22

Abu-Amero S, Monk D, Frost J, Preece M, Stanier P, Moore GE (2008) The genetic aetiology of Silver-Russell syndrome. J Med Genet 45:193–199

Albright F, Forbes AP, Henneman PH (1952) Pseudo-pseudohypoparathyroidism. Trans Assoc Am Physicians 65:337–350

Ambros V (1989) A hierarchy of regulatory genes controls a larva-to-adult developmental switch in C. Elegans. Cell 57:49–57

Amir RE, Van Den Veyver IB, Wan M, Tran CQ, Francke U, Zoghbi HY (1999) Rett syndrome is caused by mutations in X-linked MeCP2, encoding methyl-CpG-binding protein 2. Nat Genet 23:185–188

Andreoli F, Barbosa AJ, Parenti MD, Del Rio A (2013) Modulation of epigenetic targets for anticancer therapy: clinicopathological relevance, structural data and drug discovery perspectives. Curr Pharm Des 19:578–613

Angelman H (1965) 'Puppet Children': a report of three cases. Dev Med Child Neurol 7:681–688

Anway MD, Memon MA, Uzumcu M, Skinner MK (2006) Transgenerational effect of the endocrine disruptor vinclozolin on male spermatogenesis. J Androl 27:868–879

Arima T, Kamikihara T, Hayashida T, Kato K, Inoue T, Shirayoshi Y, Oshimura M, Soejima H, Mukai T, Wake N (2005) ZAC, LIT1 (KCNQ1OT1) and P57KIP2 (CDKN1C) are in an imprinted gene network that may play a role in Beckwith-Wiedemann syndrome. Nucleic Acids Res 33:2650–2660

Avissar-Whiting M, Veiga KR, Uhl KM, Maccani MA, Gagne LA, Moen EL, Marsit CJ (2010) Bisphenol A exposure leads to specific microrna alterations in placental cells. Reprod Toxicol 29:401–406

Baple EL, Poole RL, Mansour S, Willoughby C, Temple IK, Docherty LE, Taylor R, Mackay DJ (2011) An atypical case of hypomethylation at multiple imprinted loci. Eur J Hum Genet 19:360–362

Barabasi AL, Oltvai ZN (2004) Network biology: understanding the cell's functional organization. Nat Rev Genet 5:101–113

Barabasi AL, Gulbahce N, Loscalzo J (2011) Network medicine: a network-based approach to human disease. Nat Rev Genet 12:56–68

Barker DJ, Osmond C (1987) Death rates from stroke in England and Wales predicted from past maternal mortality. Br Med J (Clin Res Ed) 295:83–86

Barlow DP, Stoger R, Herrmann BG, Saito K, Schweifer N (1991) The mouse insulin-like growth factor type-2 receptor is imprinted and closely linked to the Tme locus. Nature 349:84–87

Bartel DP (2009) Micrornas: target recognition and regulatory functions. Cell 136:215–233
Bartolomei MS, Zemel S, Tilghman SM (1991) Parental imprinting of the mouse H19 gene. Nature 351:153–155
Bartsch O, Labonte J, Albrecht B, Wieczorek D, Lechno S, Zechner U, Haaf T (2010) Two patients with Ep300 mutations and facial dysmorphism different from the classic Rubinstein-Taybi syndrome. Am J Med Genet A 152a:181–184
Bastepe M, Juppner H (2005) Gnas locus and pseudohypoparathyroidism. Horm Res 63:65–74
Bastepe M, Pincus JE, Sugimoto T, Tojo K, Kanatani M, Azuma Y, Kruse K, Rosenbloom AL, Koshiyama H, Juppner H (2001) Positional dissociation between the genetic mutation responsible for pseudohypoparathyroidism type Ib and the associated methylation defect at exon A/B: evidence for a long-range regulatory element within the imprinted GNAS1 locus. Hum Mol Genet 10:1231–1241
Baysal BE (2004) Genomic imprinting and environment in hereditary paraganglioma. Am J Med Genet C Semin Med Genet 129c:85–90
Baysal BE (2013) Mitochondrial complex II and genomic imprinting in inheritance of paraganglioma tumors. Biochim Biophys Acta 1827:573–577
Bennett SN, Caporaso N, Fitzpatrick AL, Agrawal A, Barnes K, Boyd HA, Cornelis MC, Hansel NN, Heiss G, Heit JA, Kang JH, Kittner SJ, Kraft P, Lowe W, Marazita ML, Monroe KR, Pasquale LR, Ramos EM, Van Dam RM, Udren J, Williams K (2011) Phenotype harmonization and cross-study collaboration in GWAS consortia: the Geneva experience. Genet Epidemiol 35:159–173
Berdasco M, Esteller M (2013) Genetic syndromes caused by mutations in epigenetic genes. Hum Genet 132:359–383
Beristain E, Vicente MA, Guerra I, Gutierrez-Corres FB, Garin I, Perez De Nanclares G (2013) Disomy as the genetic underlying mechanisms of loss of heterozigosity in SDHD-paragangliomas. J Clin Endocrinol Metab 98:E1012–E1016
Bernstein BE, Birney E, Dunham I, Green ED, Gunter C, Snyder M (2012) An integrated encyclopedia of DNA elements in the human genome. Nature 489:57–74
Bestor TH (2003) Imprinting errors and developmental asymmetry. Philos Trans R Soc Lond B Biol Sci 358:1411–1415
Bird A (1992) The essentials of DNA methylation. Cell 70:5–8
Bird A (2002) DNA methylation patterns and epigenetic memory. Genes Dev 16:6–21
Blanco-Lago R, Malaga I, Garcia-Penas JJ, Garcia-Ron A (2013) Wolf-Hirschhorn syndrome. A series of 27 patients: their epidemiological and clinical characteristics. The current situation of the patients and the opinions of their caregivers regarding the diagnostic process. Rev Neurol 57:49–56
Bliek J, Alders M, Maas SM, Oostra RJ, Mackay DM, Van Der Lip K, Callaway JL, Brooks A, Van 't Padje S, Westerveld A, Leschot NJ, Mannens MM (2009a) Lessons from BWS twins: complex maternal and paternal hypomethylation and a common source of haematopoietic stem cells. Eur J Hum Genet 17:1625–1634
Bliek J, Verde G, Callaway J, Maas SM, De Crescenzo A, Sparago A, Cerrato F, Russo S, Ferraiuolo S, Rinaldi MM, Fischetto R, Lalatta F, Giordano L, Ferrari P, Cubellis MV, Larizza L, Temple IK, Mannens MM, Mackay DJ, Riccio A (2009b) Hypomethylation at multiple maternally methylated imprinted regions including PLAGL1 and GNAS loci in Beckwith-Wiedemann syndrome. Eur J Hum Genet 17:611–619
Bokinni Y (2012) Kabuki syndrome revisited. J Hum Genet 57:223–227
Brown SW (1966) Heterochromatin. Science 151:417–425
Buiting K (2010) Prader-Willi syndrome and Angelman syndrome. Am J Med Genet C Semin Med Genet 154c:365–376
Buiting K, Dittrich B, Gross S, Lich C, Farber C, Buchholz T, Smith E, Reis A, Burger J, Nothen MM, Barth-Witte U, Janssen B, Abeliovich D, Lerer I, Van Den Ouweland AM, Halley DJ, Schrander-Stumpel C, Smeets H, Meinecke P, Malcolm S, Gardner A, Lalande M, Nicholls RD, Friend K, Schulze A, Matthijs G, Kokkonen H, Hilbert P, Van Maldergem L, Glover G, Carbonell P, Willems P, Gillessen-Kaesbach G, Horsthemke B (1998) Sporadic

imprinting defects in Prader-Willi syndrome and Angelman syndrome: implications for imprint-switch models, genetic counseling, and prenatal diagnosis. Am J Hum Genet 63:170–180

Buiting K, Gross S, Lich C, Gillessen-Kaesbach G, El-Maarri O, Horsthemke B (2003) Epimutations in Prader-Willi and Angelman syndromes: a molecular study of 136 patients with an imprinting defect. Am J Hum Genet 72:571–577

Bullman H, Lever M, Robinson DO, Mackay DJ, Holder SE, Wakeling EL (2008) Mosaic maternal uniparental disomy of chromosome 11 in a patient with Silver-Russell syndrome. J Med Genet 45:396–399

Burggren WW, Reyna KS (2011) Developmental trajectories, critical windows and phenotypic alteration during cardio-respiratory development. Respir Physiol Neurobiol 178:13–21

Butler MG (2009) Genomic imprinting disorders in humans: a mini-review. J Assist Reprod Genet 26:477–486

Caldji C, Hellstrom IC, Zhang TY, Diorio J, Meaney MJ (2011) Environmental regulation of the neural epigenome. FEBS Lett 585:2049–2058

Campeau PM, Lee BH (1993–2014) KAT6B-related disorders. In: Pagon RA, Adam MP, Bird TD, Dolan CR, Fong CT, Smith RJH, Stephens K (eds) GeneReviews® [Internet]. University of Washington, Seattle, Seattle

Campeau PM, Kim JC, Lu JT, Schwartzentruber JA, Abdul-Rahman OA, Schlaubitz S, Murdock DM, Jiang MM, Lammer EJ, Enns GM, Rhead WJ, Rowland J, Robertson SP, Cormier-Daire V, Bainbridge MN, Yang XJ, Gingras MC, Gibbs RA, Rosenblatt DS, Majewski J, Lee BH (2012) Mutations in KAT6B, encoding a histone acetyltransferase, cause genitopatellar syndrome. Am J Hum Genet 90:282–289

Castermans D, Vermeesch JR, Fryns JP, Steyaert JG, Van De Ven WJ, Creemers JW, Devriendt K (2007) Identification and characterization of the TRIP8 and REEP3 genes on chromosome 10q21.3 as novel candidate genes for autism. Eur J Hum Genet 15:422–431

Cattanach BM, Kirk M (1985) Differential activity of maternally and paternally derived chromosome regions in mice. Nature 315:496–498

Chen PY, Meister G (2005) Microrna-guided posttranscriptional gene regulation. Biol Chem 386:1205–1218

Claes S, Devriendt K, Van Goethem G, Roelen L, Meireleire J, Raeymaekers P, Cassiman JJ, Fryns JP (2000) Novel syndromic form of X-linked complicated spastic paraplegia. Am J Med Genet 94:1–4

Crepin M, Dieu MC, Lejeune S, Escande F, Boidin D, Porchet N, Morin G, Manouvrier S, Mathieu M, Buisine MP (2012) Evidence of constitutional MLH1 epimutation associated to transgenerational inheritance of cancer susceptibility. Hum Mutat 33:180–188

Croce CM (2009) Causes and consequences of microrna dysregulation in cancer. Nat Rev Genet 10:704–714

Davis TL, Yang GJ, Mccarrey JR, Bartolomei MS (2000) The H19 methylation imprint is erased and re-established differentially on the parental alleles during male germ cell development. Hum Mol Genet 9:2885–2894

Day JJ, Sweatt JD (2010) DNA methylation and memory formation. Nat Neurosci 13:1319–1323

De Filippis B, Ricceri L, Fuso A, Laviola G (2013) Neonatal exposure to low dose corticosterone persistently modulates hippocampal mineralocorticoid receptor expression and improves locomotor/exploratory behaviour in a mouse model of Rett syndrome. Neuropharmacology 68:174–183

De Rycke M, Liebaers I, Van Steirteghem A (2002) Epigenetic risks related to assisted reproductive technologies: risk analysis and epigenetic inheritance. Hum Reprod 17:2487–2494

Dean W, Bowden L, Aitchison A, Klose J, Moore T, Meneses JJ, Reik W, Feil R (1998) Altered imprinted gene methylation and expression in completely ES cell-derived mouse fetuses: association with aberrant phenotypes. Development 125:2273–2282

Dechiara TM, Robertson EJ, Efstratiadis A (1991) Parental imprinting of the mouse insulin-like growth factor II gene. Cell 64:849–859

Demars J, Gicquel C (2012) Epigenetic and genetic disturbance of the imprinted 11p15 region in Beckwith-Wiedemann and Silver-Russell syndromes. Clin Genet 81:350–361

Demars J, Le Bouc Y, El-Osta A, Gicquel C (2011) Epigenetic and genetic mechanisms of abnormal 11p15 genomic imprinting in Silver-Russell and Beckwith-Wiedemann syndromes. Curr Med Chem 18:1740–1750

Dietel M, Johrens K, Laffert M, Hummel M, Blaker H, Muller BM, Lehmann A, Denkert C, Heppner FL, Koch A, Sers C, Anagnostopoulos I (2013) Predictive molecular pathology and its role in targeted cancer therapy: a review focussing on clinical relevance. Cancer Gene Ther 20:211–221

Dietz DM, Laplant Q, Watts EL, Hodes GE, Russo SJ, Feng J, Oosting RS, Vialou V, Nestler EJ (2011) Paternal transmission of stress-induced pathologies. Biol Psychiatry 70:408–414

Dong R, Zhao R, Zheng S, Zheng Y, Xiong S, Chu Y (2012) Abnormal DNA methylation of ITGAL (CD11a) in Cd4+ T cells from infants with biliary atresia. Biochem Biophys Res Commun 417:986–990

Dulac C (2010) Brain function and chromatin plasticity. Nature 465:728–735

Dunham I, Kundaje A, Aldred SF, Collins PJ, Davis CA, Doyle F, Epstein CB, Frietze S, Harrow J, Kaul R, Khatun J, Lajoie BR, Landt SG, Lee BK, Pauli F, Rosenbloom KR, Sabo P, Safi A, Sanyal A, Shoresh N, Simon JM, Song L, Trinklein ND, Altshuler RC, Birney E, Brown JB, Cheng C, Djebali S, Dong X, Ernst J, Furey TS, Gerstein M, Giardine B, Greven M, Hardison RC, Harris RS, Herrero J, Hoffman MM, Iyer S, Kelllis M, Kheradpour P, Lassmann T, Li Q, Lin X, Marinov GK, Merkel A, Mortazavi A, Parker SC, Reddy TE, Rozowsky J, Schlesinger F, Thurman RE, Wang J, Ward LD, Whitfield TW, Wilder SP, Wu W, Xi HS, Yip KY, Zhuang J, Bernstein BE, Green ED, Gunter C, Snyder M, Pazin MJ, Lowdon RF, Dillon LA, Adams LB, Kelly CJ, Zhang J, Wexler JR, Good PJ, Feingold EA, Crawford GE, Dekker J, Elinitski L, Farnham PJ, Giddings MC, Gingeras TR, Guigo R, Hubbard TJ, Kellis M, Kent WJ, Lieb JD, Margulies EH, Myers RM, Starnatoyannopoulos JA, Tennebaum SA, Weng Z, White KP, Wold B, Yu Y, Wrobel J, Risk BA, Gunawardena HP, Kuiper HC, Maier CW, Xie L, Chen X, Mikkelsen TS et al (2012) An integrated encyclopedia of DNA elements in the human genome. Nature 489:57–74

Eggermann T (2011) Imprinting disorders in humans. In: Tollefsbol TO (ed) Handbook of epigenetics: the new molecular and medical genetics, 1st edn. Academic, San Diego, CA

Eggermann T, Curtis M, Zerres K, Hughes HE (2004) Maternal uniparental disomy 16 and genetic counseling: new case and survey of published cases. Genet Couns 15:183–190

Eggermann T, Elbracht M, Schroder C, Reutter H, Soellner L, Spengler S, Begemann M (2013) Congenital imprinting disorders: a novel mechanism linking seemingly unrelated disorders. J Pediatr 163(4):1202–1207

El-Maarri O, Seoud M, Coullin P, Herbiniaux U, Oldenburg J, Rouleau G, Slim R (2003) Maternal alleles acquiring paternal methylation patterns in biparental complete hydatidiform moles. Hum Mol Genet 12:1405–1413

Engel JR, Smallwood A, Harper A, Higgins MJ, Oshimura M, Reik W, Schofield PN, Maher ER (2000) Epigenotype-phenotype correlations in Beckwith-Wiedemann syndrome. J Med Genet 37:921–926

Engreitz JM, Pandya-Jones A, Mcdonel P, Shishkin A, Sirokman K, Surka C, Kadri S, Xing J, Goren A, Lander ES, Plath K, Guttman M (2013) The Xist lncRNA exploits three-dimensional genome architecture to spread across the X chromosome. Science 341(6147):1237973

Esteller M (2008) Epigenetics in cancer. N Engl J Med 358:1148–1159

Fallon SC, Chang S, Finegold MJ, Karpen SJ, Brandt ML (2013) Discordant presentation of biliary atresia in premature monozygotic twins. J Pediatr Gastroenterol Nutr 57(4):e22–e23

Farwell DG, Shera KA, Koop JI, Bonnet GA, Matthews CP, Reuther GW, Coltrera MD, Mcdougall JK, Klingelhutz AJ (2000) Genetic and epigenetic changes in human epithelial cells immortalized by telomerase. Am J Pathol 156:1537–1547

Feinberg AP (2007) Phenotypic plasticity and the epigenetics of human disease. Nature 447:433–440

Feinberg AP, Tycko B (2004) The history of cancer epigenetics. Nat Rev Cancer 4:143–153

Feinberg AP, Vogelstein B (1983) Hypomethylation distinguishes genes of some human cancers from their normal counterparts. Nature 301:89–92
Fiorentino FP, Marchesi I, Giordano A (2013) On the role of retinoblastoma family proteins in the establishment and maintenance of the epigenetic landscape. J Cell Physiol 228:276–284
Fraga MF, Esteller M (2005) Towards the human cancer epigenome: a first draft of histone modifications. Cell Cycle 4:1377–1381
Franklin TB, Russig H, Weiss IC, Graff J, Linder N, Michalon A, Vizi S, Mansuy IM (2010) Epigenetic transmission of the impact of early stress across generations. Biol Psychiatry 68:408–415
Friedman JM, Liang G, Liu CC, Wolff EM, Tsai YC, Ye W, Zhou X, Jones PA (2009) The putative tumor suppressor microrna-101 modulates the cancer epigenome by repressing the polycomb group protein Ezh2. Cancer Res 69:2623–2629
Fuso A, Cavallaro RA, Nicolia V, Scarpa S (2012) PSEN1 Promoter demethylation in hyperhomocysteinemic TgCRND8 mice is the culprit, not the consequence. Curr Alzheimer Res 9:527–535
Gallagher A, Hallahan B (2012) Fragile X-associated disorders: a clinical overview. J Neurol 259:401–413
Gibbons R (2006) Alpha thalassaemia-mental retardation, X linked. Orphanet J Rare Dis 1:15
Girardot M, Cavaille J, Feil R (2012) Small regulatory RNAs controlled by genomic imprinting and their contribution to human disease. Epigenetics 7:1341–1348
Gluckman PD, Hanson MA (2007) Developmental plasticity and human disease: research directions. J Intern Med 261:461–471
Gluckman PD, Lillycrop KA, Vickers MH, Pleasants AB, Phillips ES, Beedle AS, Burdge GC, Hanson MA (2007) Metabolic plasticity during mammalian development is directionally dependent on early nutritional status. Proc Natl Acad Sci U S A 104:12796–12800
Gluckman PD, Hanson MA, Bateson P, Beedle AS, Law CM, Bhutta ZA, Anokhin KV, Bougneres P, Chandak GR, Dasgupta P, Smith GD, Ellison PT, Forrester TE, Gilbert SF, Jablonka E, Kaplan H, Prentice AM, Simpson SJ, Uauy R, West-Eberhard MJ (2009) Towards a new developmental synthesis: adaptive developmental plasticity and human disease. Lancet 373:1654–1657
Godfrey KM, Sheppard A, Gluckman PD, Lillycrop KA, Burdge GC, Mclean C, Rodford J, Slater-Jefferies JL, Garratt E, Crozier SR, Emerald BS, Gale CR, Inskip HM, Cooper C, Hanson MA (2011) Epigenetic gene promoter methylation at birth is associated with child's later adiposity. Diabetes 60:1528–1534
Goel A, Nguyen TP, Leung HC, Nagasaka T, Rhees J, Hotchkiss E, Arnold M, Banerji P, Koi M, Kwok CT, Packham D, Lipton L, Boland CR, Ward RL, Hitchins MP (2011) De novo constitutional MLH1 epimutations confer early-onset colorectal cancer in two new sporadic Lynch syndrome cases, with derivation of the epimutation on the paternal allele in one. Int J Cancer 128:869–878
Gordon L, Joo JE, Powell JE, Ollikainen M, Novakovic B, Li X, Andronikos R, Cruickshank MN, Conneely KN, Smith AK, Alisch RS, Morley R, Visscher PM, Craig JM, Saffery R (2012) Neonatal DNA methylation profile in human twins is specified by a complex interplay between intrauterine environmental and genetic factors, subject to tissue-specific influence. Genome Res 22:1395–1406
Gosden RG, Feinberg AP (2007) Genetics and epigenetics—nature's pen-and-pencil set. N Engl J Med 356:731–733
Greer EL, Shi Y (2012) Histone methylation: a dynamic mark in health, disease and inheritance. Nat Rev Genet 13:343–357
Grun F, Blumberg B (2009) Endocrine disrupters as obesogens. Mol Cell Endocrinol 304:19–29
Hacihamdioglu B, Arslan M, Sari E, Kurtcu K, Yesilkaya E (2013) Brachydactyly mental retardation syndrome in differential diagnosis of pseudopseudohypoparathyroidism. J Pediatr Endocrinol Metab 26:793–795
Hackett JA, Surani MA (2013) DNA methylation dynamics during the mammalian life cycle. Philos Trans R Soc Lond B Biol Sci 368:20110328
Haig D, Graham C (1991) Genomic imprinting and the strange case of the insulin-like growth factor II receptor. Cell 64:1045–1046

Hajkova P, Erhardt S, Lane N, Haaf T, El-Maarri O, Reik W, Walter J, Surani MA (2002) Epigenetic reprogramming in mouse primordial germ cells. Mech Dev 117:15–23
Hales CN, Barker DJ, Clark PM, Cox LJ, Fall C, Osmond C, Winter PD (1991) Fetal and infant growth and impaired glucose tolerance at age 64. BMJ 303:1019–1022
Hall JG (1990) Genomic imprinting: review and relevance to human diseases. Am J Hum Genet 46:857–873
Hallam TM, Bourtchouladze R (2006) Rubinstein-Taybi syndrome: molecular findings and therapeutic approaches to improve cognitive dysfunction. Cell Mol Life Sci 63:1725–1735
Han J, Lee Y, Yeom KH, Nam JW, Heo I, Rhee JK, Sohn SY, Cho Y, Zhang BT, Kim VN (2006) Molecular basis for the recognition of primary microRNAs by the Drosha-DGCR8 complex. Cell 125:887–901
Hannibal MC, Buckingham KJ, Ng SB, Ming JE, Beck AE, Mcmillin MJ, Gildersleeve HI, Bigham AW, Tabor HK, Mefford HC, Cook J, Yoshiura K, Matsumoto T, Matsumoto N, Miyake N, Tonoki H, Naritomi K, Kaname T, Nagai T, Ohashi H, Kurosawa K, Hou JW, Ohta T, Liang D, Sudo A, Morris CA, Banka S, Black GC, Clayton-Smith J, Nickerson DA, Zackai EH, Shaikh TH, Donnai D, Niikawa N, Shendure J, Bamshad MJ (2011) Spectrum of Mll2 (Alr) mutations in 110 cases of Kabuki syndrome. Am J Med Genet A 155a:1511–1516
Hannum G, Guinney J, Zhao L, Zhang L, Hughes G, Sadda S, Klotzle B, Bibikova M, Fan JB, Gao Y, Deconde R, Chen M, Rajapakse I, Friend S, Ideker T, Zhang K (2013) Genome-wide methylation profiles reveal quantitative views of human aging rates. Mol Cell 49:359–367
Hansen RS, Wijmenga C, Luo P, Stanek AM, Canfield TK, Weemaes CM, Gartler SM (1999) The DNMT3B DNA methyltransferase gene is mutated in the ICF immunodeficiency syndrome. Proc Natl Acad Sci U S A 96:14412–14417
Haruta M, Arai Y, Watanabe N, Fujiwara Y, Honda S, Ohshima J, Kasai F, Nakadate H, Horie H, Okita H, Hata J, Fukuzawa M, Kaneko Y (2012) Different incidences of epigenetic but not genetic abnormalities between Wilms tumors in Japanese and Caucasian children. Cancer Sci 103:1129–1135
Harwell (2013) Mousebook [Online]. Medical Research Council Harwell. http://Www.Mousebook.Org/Catalog.Php?Catalog=Imprinting. Accessed 10 July 2013
Hata K, Okano M, Lei H, Li E (2002) DnmT3L cooperates with the Dnmt3 family of de novo DNA methyltransferases to establish maternal imprints in mice. Development 129:1983–1993
He L, Hannon GJ (2004) MicroRNAs: small RNAs with a big role in gene regulation. Nat Rev Genet 5:522–531
He YF, Li BZ, Li Z, Liu P, Wang Y, Tang Q, Ding J, Jia Y, Chen Z, Li L, Sun Y, Li X, Dai Q, Song CX, Zhang K, He C, Xu GL (2011) Tet-mediated formation of 5-carboxylcytosine and its excision by TDG in mammalian DNA. Science 333:1303–1307
Heijmans BT, Kremer D, Tobi EW, Boomsma DI, Slagboom PE (2007) Heritable rather than age-related environmental and stochastic factors dominate variation in DNA methylation of the human IGF2/H19 locus. Hum Mol Genet 16:547–554
Heijmans BT, Tobi EW, Lumey LH, Slagboom PE (2009) The epigenome: archive of the prenatal environment. Epigenetics 4:526–531
Herman JG (1999) Hypermethylation of tumor suppressor genes in cancer. Semin Cancer Biol 9:359–367
Hermann A, Gowher H, Jeltsch A (2004) Biochemistry and biology of mammalian DNA methyltransferases. Cell Mol Life Sci 61:2571–2587
Heyn H, Li N, Ferreira HJ, Moran S, Pisano DG, Gomez A, Diez J, Sanchez-Mut JV, Setien F, Carmona FJ, Puca AA, Sayols S, Pujana MA, Serra-Musach J, Iglesias-Platas I, Formiga F, Fernandez AF, Fraga MF, Heath SC, Valencia A, Gut IG, Wang J, Esteller M (2012a) Distinct DNA methylomes of newborns and centenarians. Proc Natl Acad Sci U S A 109:10522–10527
Heyn H, Vidal E, Sayols S, Sanchez-Mut JV, Moran S, Medina I, Sandoval J, Simo-Riudalbas L, Szczesna K, Huertas D, Gatto S, Matarazzo MR, Dopazo J, Esteller M (2012b) Whole-genome bisulfite DNA sequencing of a DNMT3B mutant patient. Epigenetics 7:542–550
Hitchins MP, Wong JJ, Suthers G, Suter CM, Martin DI, Hawkins NJ, Ward RL (2007) Inheritance of a cancer-associated MLH1 germ-line epimutation. N Engl J Med 356:697–705

Hitchins MP, Rapkins RW, Kwok CT, Srivastava S, Wong JJ, Khachigian LM, Polly P, Goldblatt J, Ward RL (2011) Dominantly inherited constitutional epigenetic silencing of MLH1 in a cancer-affected family is linked to a single nucleotide variant within the 5′utr. Cancer Cell 20:200–213

Hogg K, Price EM, Hanna CW, Robinson WP (2012) Prenatal and perinatal environmental influences on the human fetal and placental epigenome. Clin Pharmacol Ther 92:716–726

Horsthemke B (2006) Epimutations in human disease. Curr Top Microbiol Immunol 310:45–59

Horsthemke B (2010) Mechanisms of imprint dysregulation. Am J Med Genet C Semin Med Genet 154c:321–328

Horsthemke B, Buiting K (2008) Genomic imprinting and imprinting defects in humans. Adv Genet 61:225–246

Huidobro C, Fernandez AF, Fraga MF (2013) The role of genetics in the establishment and maintenance of the epigenome. Cell Mol Life Sci 70:1543–1573

Hutchings JA (2011) Old wine in new bottles: reaction norms in salmonid fishes. Heredity (Edinb) 106:421–437

Huxley AF (1956) Epigenetics. Nature 177:807–809

Ishikawa H, Banzai M, Yamauchi T (1999) Developmental retardation of XO mouse embryos at mid-gestation. J Reprod Fertil 115:263–267

Iwase S, Shi Y (2011) Histone and DNA modifications in mental retardation. Prog Drug Res 67:147–173

Jacquot S, Merienne K, De Cesare D, Pannetier S, Mandel JL, Sassone-Corsi P, Hanauer A (1998a) Mutation analysis of the RSK2 gene in Coffin-Lowry patients: extensive allelic heterogeneity and a high rate of de novo mutations. Am J Hum Genet 63:1631–1640

Jacquot S, Merienne K, Pannetier S, Blumenfeld S, Schinzel A, Hanauer A (1998b) Germline mosaicism in Coffin-Lowry syndrome. Eur J Hum Genet 6:578–582

Jensen LR, Amende M, Gurok U, Moser B, Gimmel V, Tzschach A, Janecke AR, Tariverdian G, Chelly J, Fryns JP, Van Esch H, Kleefstra T, Hamel B, Moraine C, Gecz J, Turner G, Reinhardt R, Kalscheuer VM, Ropers HH, Lenzner S (2005) Mutations in the JARID1C gene, which is involved in transcriptional regulation and chromatin remodeling, cause X-linked mental retardation. Am J Hum Genet 76:227–236

Jenuwein T, Allis CD (2001) Translating the histone code. Science 293:1074–1080

Jones PL, Veenstra GJ, Wade PA, Vermaak D, Kass SU, Landsberger N, Strouboulis J, Wolffe AP (1998) Methylated DNA and MeCP2 recruit histone deacetylase to repress transcription. Nat Genet 19:187–191

Kagami M, Sekita Y, Nishimura G, Irie M, Kato F, Okada M, Yamamori S, Kishimoto H, Nakayama M, Tanaka Y, Matsuoka K, Takahashi T, Noguchi M, Masumoto K, Utsunomiya T, Kouzan H, Komatsu Y, Ohashi H, Kurosawa K, Kosaki K, Ferguson-Smith AC, Ishino F, Ogata T (2008) Deletions and epimutations affecting the human 14q32.2 imprinted region in individuals with paternal and maternal Upd(14)-like phenotypes. Nat Genet 40:237–242

Kagami M, Matsuoka K, Nagai T, Yamanaka M, Kurosawa K, Suzumori N, Sekita Y, Miyado M, Matsubara K, Fuke T, Kato F, Fukami M, Ogata T (2012) Paternal uniparental disomy 14 and related disorders: placental gene expression analyses and histological examinations. Epigenetics 7:1142–1150

Kang D, Cho HS, Toyokawa G, Kogure M, Yamane Y, Iwai Y, Hayami S, Tsunoda T, Field HI, Matsuda K, Neal DE, Ponder BA, Maehara Y, Nakamura Y, Hamamoto R (2013) The histone methyltransferase Wolf-Hirschhorn syndrome candidate 1-like 1 (WHSC1L1) is involved in human carcinogenesis. Genes Chromosomes Cancer 52:126–139

Katari S, Turan N, Bibikova M, Erinle O, Chalian R, Foster M, Gaughan JP, Coutifaris C, Sapienza C (2009) DNA methylation and gene expression differences in children conceived in vitro or in vivo. Hum Mol Genet 18:3769–3778

Kelsey G (2010) Imprinting on chromosome 20: tissue-specific imprinting and imprinting mutations in the GNAS locus. Am J Med Genet C Semin Med Genet 154c:377–386

Keverne EB (2012) Significance of epigenetics for understanding brain development, brain evolution and behaviour. Neuroscience 264:207–217

Killian JK, Byrd JC, Jirtle JV, Munday BL, Stoskopf MK, Macdonald RG, Jirtle RL (2000) M6P/IGF2R Imprinting evolution in mammals. Mol Cell 5:707–716

Kinnally EL, Feinberg C, Kim D, Ferguson K, Leibel R, Coplan JD, John Mann J (2011) DNA methylation as a risk factor in the effects of early life stress. Brain Behav Immun 25: 1548–1553

Kishino T, Lalande M, Wagstaff J (1997) UBE3A/E6-AP mutations cause Angelman syndrome. Nat Genet 15:70–73

Kleefstra T, Nillesen WM, Yntema HG (1993–2014) Kleefstra syndrome. In: Pagon RA, Adam MP, Bird TD, Dolan CR, Fong CT, Smith RJH, Stephens K (eds) GeneReviews® [Internet]. University of Washington, Seattle, Seattle

Kleefstra T, Smidt M, Banning MJ, Oudakker AR, Van Esch H, De Brouwer AP, Nillesen W, Sistermans EA, Hamel BC, De Bruijn D, Fryns JP, Yntema HG, Brunner HG, De Vries BB, Van Bokhoven H (2005) Disruption of the gene euchromatin histone methyl transferase1 (Eu-HMTase1) is associated with the 9q34 subtelomeric deletion syndrome. J Med Genet 42:299–306

Kleefstra T, Van Zelst-Stams WA, Nillesen WM, Cormier-Daire V, Houge G, Foulds N, Van Dooren M, Willemsen MH, Pfundt R, Turner A, Wilson M, Mcgaughran J, Rauch A, Zenker M, Adam MP, Innes M, Davies C, Lopez AG, Casalone R, Weber A, Brueton LA, Navarro AD, Bralo MP, Venselaar H, Stegmann SP, Yntema HG, Van Bokhoven H, Brunner HG (2009) Further clinical and molecular delineation of the 9q subtelomeric deletion syndrome supports a major contribution of EHMT1 haploinsufficiency to the core phenotype. J Med Genet 46:598–606

Klein CJ, Botuyan MV, Wu Y, Ward CJ, Nicholson GA, Hammans S, Hojo K, Yamanishi H, Karpf AR, Wallace DC, Simon M, Lander C, Boardman LA, Cunningham JM, Smith GE, Litchy WJ, Boes B, Atkinson EJ, Middha S, Pj BD, Parisi JE, Mer G, Smith DI, Dyck PJ (2011) Mutations in DNMT1 cause hereditary sensory neuropathy with dementia and hearing loss. Nat Genet 43:595–600

Kohda T, Ishino F (2013) Embryo manipulation via assisted reproductive technology and epigenetic asymmetry in mammalian early development. Philos Trans R Soc Lond B Biol Sci 368:20120353

Kota SK, Feil R (2010) Epigenetic transitions in germ cell development and meiosis. Dev Cell 19:675–686

Kou YC, Shao L, Peng HH, Rosetta R, Del Gaudio D, Wagner AF, Al-Hussaini TK, Van Den Veyver IB (2008) A recurrent intragenic genomic duplication, other novel mutations in NLRP7 and imprinting defects in recurrent biparental hydatidiform moles. Mol Hum Reprod 14:33–40

Kurosaki T, Hikida M (2009) Tyrosine kinases and their substrates in B lymphocytes. Immunol Rev 228:132–148

Labialle S, Cavaille J (2011) Do repeated arrays of regulatory small-RNA genes elicit genomic imprinting? Concurrent emergence of large clusters of small non-coding RNAs and genomic imprinting at four evolutionarily distinct eutherian chromosomal loci. Bioessays 33:565–573

Lagos-Quintana M, Rauhut R, Lendeckel W, Tuschl T (2001) Identification of novel genes coding for small expressed RNAs. Science 294:853–858

Lalande M (2001) Imprints of disease at GNAS1. J Clin Invest 107:793–794

Lan J, Hua S, He X, Zhang Y (2010) DNA methyltransferases and methyl-binding proteins of mammals. Acta Biochim Biophys Sin (Shanghai) 42:243–252

Lana E, Megarbane A, Tourriere H, Sarda P, Lefranc G, Claustres M, De Sario A (2012) DNA replication is altered in immunodeficiency centromeric instability facial anomalies (ICF) cells carrying DNMT3B mutations. Eur J Hum Genet 20:1044–1050

Lau NC, Lim LP, Weinstein EG, Bartel DP (2001) An abundant class of tiny RNAs with probable regulatory roles in Caenorhabditis elegans. Science 294:858–862

Laumonnier F, Holbert S, Ronce N, Faravelli F, Lenzner S, Schwartz CE, Lespinasse J, Van Esch H, Lacombe D, Goizet C, Phan-Dinh Tuy F, Van Bokhoven H, Fryns JP, Chelly J, Ropers HH, Moraine C, Hamel BC, Briault S (2005) Mutations in PHF8 are associated with X linked mental retardation and cleft lip/cleft palate. J Med Genet 42:780–786

Ledbetter DH, Riccardi VM, Airhart SD, Strobel RJ, Keenan BS, Crawford JD (1981) Deletions of chromosome 15 as a cause of the Prader-Willi syndrome. N Engl J Med 304:325–329

Lederer D, Grisart B, Digilio MC, Benoit V, Crespin M, Ghariani SC, Maystadt I, Dallapiccola B, Verellen-Dumoulin C (2012) Deletion of KDM6A, a histone demethylase interacting with MLL2, in three patients with Kabuki syndrome. Am J Hum Genet 90:119–124

Lee RC, Ambros V (2001) An extensive class of small RNAs in Caenorhabditis elegans. Science 294:862–864

Lee H, Jaffe AE, Feinberg JI, Tryggvadottir R, Brown S, Montano C, Aryee MJ, Irizarry RA, Herbstman J, Witter FR, Goldman LR, Feinberg AP, Fallin MD (2012) DNA methylation shows genome-wide association of NFIX, RAPGEF2 and MSRB3 with gestational age at birth. Int J Epidemiol 41:188–199

Lepage JF, Hong DS, Hallmayer J, Reiss AL (2012) Genomic imprinting effects on cognitive and social abilities in prepubertal girls with Turner syndrome. J Clin Endocrinol Metab 97:E460–E464

Lewitus E, Kalinka AT (2013) Neocortical development as an evolutionary platform for intragenomic conflict. Front Neuroanat 7:2

Li Y, Sasaki H (2011) Genomic imprinting in mammals: its life cycle, molecular mechanisms and reprogramming. Cell Res 21:466–473

Li E, Beard C, Jaenisch R (1993) Role for DNA methylation in genomic imprinting. Nature 366:362–365

Li S, Hursting SD, Davis BJ, Mclachlan JA, Barrett JC (2003) Environmental exposure, DNA methylation, and gene regulation: lessons from diethylstilbesterol-induced cancers. Ann N Y Acad Sci 983:161–169

Li X, Ito M, Zhou F, Youngson N, Zuo X, Leder P, Ferguson-Smith AC (2008) A maternal-zygotic effect gene, Zfp57, maintains both maternal and paternal imprints. Dev Cell 15:547–557

Liang Y, Vogel JL, Arbuckle JH, Rai G, Jadhav A, Simeonov A, Maloney DJ, Kristie TM (2013) Targeting the JMJD2 histone demethylases to epigenetically control herpesvirus infection and reactivation from latency. Sci Transl Med 5:167ra5

Lillycrop KA, Phillips ES, Jackson AA, Hanson MA, Burdge GC (2005) Dietary protein restriction of pregnant rats induces and folic acid supplementation prevents epigenetic modification of hepatic gene expression in the offspring. J Nutr 135:1382–1386

Lim D, Bowdin SC, Tee L, Kirby GA, Blair E, Fryer A, Lam W, Oley C, Cole T, Brueton LA, Reik W, Macdonald F, Maher ER (2009) Clinical and molecular genetic features of Beckwith-Wiedemann syndrome associated with assisted reproductive technologies. Hum Reprod 24:741–747

Ling SC, Albuquerque CP, Han JS, Lagier-Tourenne C, Tokunaga S, Zhou H, Cleveland DW (2010) ALS-associated mutations in TDP-43 increase its stability and promote TDP-43 complexes with FUS/TLS. Proc Natl Acad Sci U S A 107:13318–13323

Lister R, Pelizzola M, Dowen RH, Hawkins RD, Hon G, Tonti-Filippini J, Nery JR, Lee L, Ye Z, Ngo QM, Edsall L, Antosiewicz-Bourget J, Stewart R, Ruotti V, Millar AH, Thomson JA, Ren B, Ecker JR (2009) Human DNA methylomes at base resolution show widespread epigenomic differences. Nature 462:315–322

Liu J, Litman D, Rosenberg MJ, Yu S, Biesecker LG, Weinstein LS (2000) A GNAS1 imprinting defect in pseudohypoparathyroidism type IB. J Clin Invest 106:1167–1174

Liu Y, Hoyo C, Murphy S, Huang Z, Overcash F, Thompson J, Brown H, Murtha AP (2013) DNA methylation at imprint regulatory regions in preterm birth and infection. Am J Obstet Gynecol 208(395):E1–E7

Lu Q, Qiu X, Hu N, Wen H, Su Y, Richardson BC (2006) Epigenetics, disease, and therapeutic interventions. Ageing Res Rev 5:449–467

Lucifero D, Chaillet JR, Trasler JM (2004) Potential significance of genomic imprinting defects for reproduction and assisted reproductive technology. Hum Reprod Update 10:3–18

Ludbrook LM, Harley VR (2004) Sex determination: a 'Window' of DAX1 activity. Trends Endocrinol Metab 15:116–121

Ludwig DS, Currie J (2010) The association between pregnancy weight gain and birthweight: a within-family comparison. Lancet 376:984–990

Lumey LH, Stein AD (1997) Offspring birth weights after maternal intrauterine undernutrition: a comparison within sibships. Am J Epidemiol 146:810–819

Lumey LH, Stein AD, Kahn HS, Van Der Pal-De Bruin KM, Blauw GJ, Zybert PA, Susser ES (2007) Cohort profile: the Dutch hunger winter families study. Int J Epidemiol 36:1196–1204

Lyon MF (1961) Gene action in the X-chromosome of the mouse (Mus musculus L.). Nature 190:372–373

Maccani MA, Avissar-Whiting M, Banister CE, Mcgonnigal B, Padbury JF, Marsit CJ (2010) Maternal cigarette smoking during pregnancy is associated with downregulation of miR-16, miR-21, and miR-146a in the placenta. Epigenetics 5:583–589

Maccani MA, Padbury JF, Marsit CJ (2011) miR-16 and miR-21 expression in the placenta is associated with fetal growth. PLoS One 6, E21210

Maccani MA, Padbury JF, Lester BM, Knopik VS, Marsit CJ (2013) Placental miRNA expression profiles are associated with measures of infant neurobehavioral outcomes. Pediatr Res 74(3):272–278

Mackay DJ, Temple IK (2010) Transient neonatal diabetes mellitus type 1. Am J Med Genet C Semin Med Genet 154c:335–342

Mackay DJ, Boonen SE, Clayton-Smith J, Goodship J, Hahnemann JM, Kant SG, Njolstad PR, Robin NH, Robinson DO, Siebert R, Shield JP, White HE, Temple IK (2006) A maternal hypomethylation syndrome presenting as transient neonatal diabetes mellitus. Hum Genet 120:262–269

Mackay DJ, Callaway JL, Marks SM, White HE, Acerini CL, Boonen SE, Dayanikli P, Firth HV, Goodship JA, Haemers AP, Hahnemann JM, Kordonouri O, Masoud AF, Oestergaard E, Storr J, Ellard S, Hattersley AT, Robinson DO, Temple IK (2008) Hypomethylation of multiple imprinted loci in individuals with transient neonatal diabetes is associated with mutations in ZFP57. Nat Genet 40:949–951

Maher B (2012) Encode: the human encyclopaedia. Nature 489:46–48

Makishima H, Maciejewski JP (2011) Pathogenesis and consequences of uniparental disomy in cancer. Clin Cancer Res 17:3913–3923

Manikkam M, Guerrero-Bosagna C, Tracey R, Haque MM, Skinner MK (2012) Transgenerational actions of environmental compounds on reproductive disease and identification of epigenetic biomarkers of ancestral exposures. PLoS One 7, E31901

Mantovani G, Spada A (2006) Mutations in the Gs alpha gene causing hormone resistance. Best Pract Res Clin Endocrinol Metab 20:501–513

Maraschio P, Zuffardi O, Dalla Fior T, Tiepolo L (1988) Immunodeficiency, centromeric heterochromatin instability of chromosomes 1, 9, and 16, and facial anomalies: the ICF syndrome. J Med Genet 25:173–180

Maraschio P, Tupler R, Dainotti E, Piantanida M, Cazzola G, Tiepolo L (1989) Differential expression of the ICF (immunodeficiency, centromeric heterochromatin, facial anomalies) mutation in lymphocytes and fibroblasts. J Med Genet 26:452–456

Mariot V, Maupetit-Mehouas S, Sinding C, Kottler ML, Linglart A (2008) A maternal epimutation of GNAS leads to Albright osteodystrophy and parathyroid hormone resistance. J Clin Endocrinol Metab 93:661–665

Martin GM (2012) Stochastic modulations of the pace and patterns of ageing: impacts on quasi-stochastic distributions of multiple geriatric pathologies. Mech Ageing Dev 133:107–111

Martin-Gronert MS, Ozanne SE (2012) Mechanisms underlying the developmental origins of disease. Rev Endocr Metab Disord 13:85–92

Marx V (2012) Epigenetics: reading the second genomic code. Nature 491:143–147

Matthews RP, Eauclaire SF, Mugnier M, Lorent K, Cui S, Ross MM, Zhang Z, Russo P, Pack M (2011) DNA hypomethylation causes bile duct defects in zebrafish and is a distinguishing feature of infantile biliary atresia. Hepatology 53:905–914

Maze I, Covington HE 3rd, Dietz DM, Laplant Q, Renthal W, Russo SJ, Mechanic M, Mouzon E, Neve RL, Haggarty SJ, Ren Y, Sampath SC, Hurd YL, Greengard P, Tarakhovsky A, Schaefer A, Nestler EJ (2010) Essential role of the histone methyltransferase G9a in cocaine-induced plasticity. Science 327:213–216

Mcgowan PO, Sasaki A, D'alessio AC, Dymov S, Labonte B, Szyf M, Turecki G, Meaney MJ (2009) Epigenetic regulation of the glucocorticoid receptor in human brain associates with childhood abuse. Nat Neurosci 12:342–348
Mcgrath J, Solter D (1984) Completion of mouse embryogenesis requires both the maternal and paternal genomes. Cell 37:179–183
Mcquown SC, Barrett RM, Matheos DP, Post RJ, Rogge GA, Alenghat T, Mullican SE, Jones S, Rusche JR, Lazar MA, Wood MA (2011) HDAC3 is a critical negative regulator of long-term memory formation. J Neurosci 31:764–774
Meaney MJ, Szyf M, Seckl JR (2007) Epigenetic mechanisms of perinatal programming of hypothalamic-pituitary-adrenal function and health. Trends Mol Med 13:269–277
Meissner A, Mikkelsen TS, Gu H, Wernig M, Hanna J, Sivachenko A, Zhang X, Bernstein BE, Nusbaum C, Jaffe DB, Gnirke A, Jaenisch R, Lander ES (2008) Genome-scale DNA methylation maps of pluripotent and differentiated cells. Nature 454:766–770
Mendelsohn AR, Larrick JW (2013) Rejuvenation of adult stem cells: is age-associated dysfunction epigenetic? Rejuvenation Res 16:152–157
Miyake K, Yang C, Minakuchi Y, Ohori K, Soutome M, Hirasawa T, Kazuki Y, Adachi N, Suzuki S, Itoh M, Goto YI, Andoh T, Kurosawa H, Oshimura M, Sasaki M, Toyoda A, Kubota T (2013) Comparison of genomic and epigenomic expression in monozygotic twins discordant for Rett syndrome. PLoS One 8, E66729
Moore T, Haig D (1991) Genomic imprinting in mammalian development: a parental tug-of-war. Trends Genet 7:45–49
Morgan HD, Santos F, Green K, Dean W, Reik W (2005) Epigenetic reprogramming in mammals. Hum Mol Genet 14(Spec No 1):R47–R58
Moulton T, Crenshaw T, Hao Y, Moosikasuwan J, Lin N, Dembitzer F, Hensle T, Weiss L, Mcmorrow L, Loew T et al (1994) Epigenetic lesions at the H19 locus in Wilms' tumour patients. Nat Genet 7:440–447
Mulligan CJ, D'errico NC, Stees J, Hughes DA (2012) Methylation changes at Nr3c1 in newborns associate with maternal prenatal stress exposure and newborn birth weight. Epigenetics 7:853–857
Murphy SK, Jirtle RL (2003) Imprinting evolution and the price of silence. Bioessays 25:577–588
Murphy SK, Huang Z, Hoyo C (2012) Differentially methylated regions of imprinted genes in prenatal, perinatal and postnatal human tissues. PLoS One 7, E40924
Na ES, Nelson ED, Kavalali ET, Monteggia LM (2013) The impact of MeCP2 loss- or gain-of-function on synaptic plasticity. Neuropsychopharmacology 38:212–219
Nafee TM, Farrell WE, Carroll WD, Fryer AA, Ismail KM (2008) Epigenetic control of fetal gene expression. BJOG 115:158–168
Nakamura K, Tanoue A (2013) Etiology of biliary atresia as a developmental anomaly: recent advances. J Hepatobiliary Pancreat Sci 20:459–464
Nan X, Ng HH, Johnson CA, Laherty CD, Turner BM, Eisenman RN, Bird A (1998) Transcriptional repression by the methyl-CpG-binding protein MeCP2 involves a histone deacetylase complex. Nature 393:386–389
Neguembor MV, Gabellini D (2010) In junk we trust: repetitive DNA, epigenetics and facioscapulohumeral muscular dystrophy. Epigenomics 2:271–287
Nestler EJ (2012) Epigenetics: stress makes its molecular mark. Nature 490:171–172
Newell-Price J, Clark AJ, King P (2000) Dna methylation and silencing of gene expression. Trends Endocrinol Metab 11:142–148
Nicholls RD, Knoll JH, Butler MG, Karam S, Lalande M (1989) Genetic imprinting suggested by maternal heterodisomy in nondeletion Prader-Willi syndrome. Nature 342:281–285
Niemitz EL, Feinberg AP (2004) Epigenetics and assisted reproductive technology: a call for investigation. Am J Hum Genet 74:599–609
Niwa M, Jaaro-Peled H, Tankou S, Seshadri S, Hikida T, Matsumoto Y, Cascella NG, Kano S, Ozaki N, Nabeshima T, Sawa A (2013) Adolescent stress-induced epigenetic control of dopaminergic neurons via glucocorticoids. Science 339:335–339

Nugent BM, Mccarthy MM (2011) Epigenetic underpinnings of developmental sex differences in the brain. Neuroendocrinology 93:150–158

Ogino S, Lochhead P, Chan AT, Nishihara R, Cho E, Wolpin BM, Meyerhardt JA, Meissner A, Schernhammer ES, Fuchs CS, Giovannucci E (2013) Molecular pathological epidemiology of epigenetics: emerging integrative science to analyze environment, host, and disease. Mod Pathol 26:465–484

O'rahilly R (1979) Early human development and the chief sources of information on staged human embryos. Eur J Obstet Gynecol Reprod Biol 9:273–280

Paoloni-Giacobino A, Chaillet JR (2004) Genomic imprinting and assisted reproduction. Reprod Health 1:6

Park JH, Stoffers DA, Nicholls RD, Simmons RA (2008) Development of type 2 diabetes following intrauterine growth retardation in rats is associated with progressive epigenetic silencing of Pdx1. J Clin Invest 118:2316–2324

Park YJ, Herman H, Gao Y, Lindroth AM, Hu BY, Murphy PJ, Putnam JR, Soloway PD (2012) Sequences sufficient for programming imprinted germline DNA methylation defined. PLoS One 7, E33024

Pereira PM, Schneider A, Pannetier S, Heron D, Hanauer A (2010) Coffin-Lowry syndrome. Eur J Hum Genet 18:627–633

Pineles BL, Romero R, Montenegro D, Tarca AL, Han YM, Kim YM, Draghici S, Espinoza J, Kusanovic JP, Mittal P, Hassan SS, Kim CJ (2007) Distinct subsets of microRNAs are expressed differentially in the human placentas of patients with preeclampsia. Am J Obstet Gynecol 196(261):E1–E6

Plagemann A, Roepke K, Harder T, Brunn M, Harder A, Wittrock-Staar M, Ziska T, Schellong K, Rodekamp E, Melchior K, Dudenhausen JW (2010) Epigenetic malprogramming of the insulin receptor promoter due to developmental overfeeding. J Perinat Med 38:393–400

Prader A, Labhart A, Willi H (1956) Ein syndrom von adipositas, kleinwuchs, kryptorchismus und oligophrenie nach myatonieartigem zustand im neugeborenenalter. Schweiz Med Wochenschr 86:1260–1261

Qiao Z, Ren S, Li W, Wang X, He M, Guo Y, Sun L, He Y, Ge Y, Yu Q (2013) Chidamide, a novel histone deacetylase inhibitor, synergistically enhances gemcitabine cytotoxicity in pancreatic cancer cells. Biochem Biophys Res Commun 434:95–101

Qiu J, Shi G, Jia Y, Li J, Wu M, Dong S, Wong J (2010) The X-linked mental retardation gene PHF8 is a histone demethylase involved in neuronal differentiation. Cell Res 20:908–918

Quddus J, Johnson KJ, Gavalchin J, Amento EP, Chrisp CE, Yung RL, Richardson BC (1993) Treating activated Cd4+ T cells with either of two distinct DNA methyltransferase inhibitors, 5-azacytidine or procainamide, is sufficient to cause a lupus-like disease in syngeneic mice. J Clin Invest 92:38–53

Raefski AS, O'neill MJ (2005) Identification of a cluster of X-linked imprinted genes in mice. Nat Genet 37:620–624

Reik W, Walter J (2001) Genomic imprinting: parental influence on the genome. Nat Rev Genet 2:21–32

Reis AH, Vargas FR, Lemos B (2012) More epigenetic hits than meets the eye: microRNAs and genes associated with the tumorigenesis of retinoblastoma. Front Genet 3:284

Ren J, Chu Y, Zhang Y, Li Y, Cai H, Zhang X, Zhao D, Li Z, Ma H, Li W, Wang H, Wang J, Chen Y, Gao YE, Xiao L, Liu R, Qian J, Liu Y, Shi X, Jiang SW (2013a) Epigenetic interventions increase the radiation sensitivity of cancer cells. Curr Pharm Des. http://www.ncbi.nlm.nih.gov/pubmed/23888958

Ren J, Zhang J, Cai H, Li Y, Zhang Y, Zhang X, Zhao D, Li Z, Ma H, Li W, Wang J, Gao YE, Chen Y, Xiao L, Liu R, Qian J, Liu Y, Li J (2013b) HDAC as a therapeutic target for treatment of endometrial cancers. Curr Pharm Des. http://www.ncbi.nlm.nih.gov/pubmed/23888962

Reul JM, Hesketh SA, Collins A, Mecinas MG (2009) Epigenetic mechanisms in the dentate gyrus act as a molecular switch in hippocampus-associated memory formation. Epigenetics 4:434–439

Rio Frio T, Bahubeshi A, Kanellopoulou C, Hamel N, Niedziela M, Sabbaghian N, Pouchet C, Gilbert L, O'brien PK, Serfas K, Broderick P, Houlston RS, Lesueur F, Bonora E, Muljo S, Schimke RN, Bouron-Dal Soglio D, Arseneau J, Schultz KA, Priest JR, Nguyen VH, Harach HR, Livingston DM, Foulkes WD, Tischkowitz M (2011) DICER1 mutations in familial multinodular goiter with and without ovarian Sertoli-Leydig cell tumors. JAMA 305:68–77

Rio M, Clech L, Amiel J, Faivre L, Lyonnet S, Le Merrer M, Odent S, Lacombe D, Edery P, Brauner R, Raoul O, Gosset P, Prieur M, Vekemans M, Munnich A, Colleaux L, Cormier-Daire V (2003) Spectrum of NSD1 mutations in Sotos and Weaver syndromes. J Med Genet 40:436–440

Rivera RM, Ross JW (2013) Epigenetics in fertilization and preimplantation embryo development. Prog Biophys Mol Biol 113(3):423–432

Rossignol S, Steunou V, Chalas C, Kerjean A, Rigolet M, Viegas-Pequignot E, Jouannet P, Le Bouc Y, Gicquel C (2006) The epigenetic imprinting defect of patients with Beckwith-Wiedemann syndrome born after assisted reproductive technology is not restricted to the 11p15 region. J Med Genet 43:902–907

Rubinstein JH, Taybi H (1963) Broad thumbs and toes and facial abnormalities. A possible mental retardation syndrome. Am J Dis Child 105:588–608

Russell A (1954) A syndrome of intra-uterine dwarfism recognizable at birth with cranio-facial dysostosis, disproportionately short arms, and other anomalies (5 examples). Proc R Soc Med 47:1040–1044

Sacconi S, Camano P, De Greef JC, Lemmers RJ, Salviati L, Boileau P, Lopez De Munain Arregui A, Van Der Maarel SM, Desnuelle C (2012) Patients with a phenotype consistent with facioscapulohumeral muscular dystrophy display genetic and epigenetic heterogeneity. J Med Genet 49:41–46

Sagi L, Zuckerman-Levin N, Gawlik A, Ghizzoni L, Buyukgebiz A, Rakover Y, Bistritzer T, Admoni O, Vottero A, Baruch O, Fares F, Malecka-Tendera E, Hochberg Z (2007) Clinical significance of the parental origin of the X chromosome in Turner syndrome. J Clin Endocrinol Metab 92:846–852

Sandin S, Hultman CM, Kolevzon A, Gross R, Maccabe JH, Reichenberg A (2012) Advancing maternal age is associated with increasing risk for autism: a review and meta-analysis. J Am Acad Child Adolesc Psychiatr 477–486:e471

Sandovici I, Smith NH, Nitert MD, Ackers-Johnson M, Uribe-Lewis S, Ito Y, Jones RH, Marquez VE, Cairns W, Tadayyon M, O'neill LP, Murrell A, Ling C, Constancia M, Ozanne SE (2011) Maternal diet and aging alter the epigenetic control of a promoter-enhancer interaction at the Hnf4a gene in rat pancreatic islets. Proc Natl Acad Sci U S A 108:5449–5454

Sapienza C (1991) Genome imprinting and carcinogenesis. Biochim Biophys Acta 1072:51–61

Sassone-Corsi P, Mizzen CA, Cheung P, Crosio C, Monaco L, Jacquot S, Hanauer A, Allis CD (1999) Requirement of Rsk-2 for epidermal growth factor-activated phosphorylation of histone H3. Science 285:886–891

Scott RH, Murray A, Baskcomb L, Turnbull C, Loveday C, Al-Saadi R, Williams R, Breatnach F, Gerrard M, Hale J, Kohler J, Lapunzina P, Levitt GA, Picton S, Pizer B, Ronghe MD, Traunecker H, Williams D, Kelsey A, Vujanic GM, Sebire NJ, Grundy P, Stiller CA, Pritchard-Jones K, Douglas J, Rahman N (2012) Stratification of Wilms tumor by genetic and epigenetic analysis. Oncotarget 3:327–335

Seisenberger S, Peat JR, Hore TA, Santos F, Dean W, Reik W (2013) Reprogramming DNA methylation in the mammalian life cycle: building and breaking epigenetic barriers. Philos Trans R Soc Lond B Biol Sci 368:20110330

Sharp AJ, Stathaki E, Migliavacca E, Brahmachary M, Montgomery SB, Dupre Y, Antonarakis SE (2011) DNA methylation profiles of human active and inactive X chromosomes. Genome Res 21:1592–1600

Sheth F, Akinde OR, Datar C, Adeteye OV, Sheth J (2012) Genotype-phenotype characterization of Wolf-Hirschhorn syndrome confirmed by fish: case reports. Case Rep Genet 2012:878796

Shiohama A, Sasaki T, Noda S, Minoshima S, Shimizu N (2003) Molecular cloning and expression analysis of a novel gene DGCR8 located in the DiGeorge syndrome chromosomal region. Biochem Biophys Res Commun 304:184–190

Siderius LE, Hamel BC, Van Bokhoven H, De Jager F, Van Den Helm B, Kremer H, Heineman-De Boer JA, Ropers HH, Mariman EC (1999) X-linked mental retardation associated with cleft lip/palate maps to Xp11.3-q21.3. Am J Med Genet 85:216–220

Silver HK, Kiyasu W, George J, Deamer WC (1953) Syndrome of congenital hemihypertrophy, shortness of stature, and elevated urinary gonadotropins. Pediatrics 12:368–376

Simpson MA, Deshpande C, Dafou D, Vissers LE, Woollard WJ, Holder SE, Gillessen-Kaesbach G, Derks R, White SM, Cohen-Snuijf R, Kant SG, Hoefsloot LH, Reardon W, Brunner HG, Bongers EM, Trembath RC (2012) De novo mutations of the gene encoding the histone acetyltransferase KAT6B cause Genitopatellar syndrome. Am J Hum Genet 90:290–294

Skene PJ, Illingworth RS, Webb S, Kerr AR, James KD, Turner DJ, Andrews R, Bird AP (2010) Neuronal MeCP2 is expressed at near histone-octamer levels and globally alters the chromatin state. Mol Cell 37:457–468

Skinner MK, Haque CG, Nilsson E, Bhandari R, Mccarrey JR (2013) Environmentally induced transgenerational epigenetic reprogramming of primordial germ cells and the subsequent germ line. PLoS One 8, E66318

Skuse DH, James RS, Bishop DV, Coppin B, Dalton P, Aamodt-Leeper G, Bacarese-Hamilton M, Creswell C, Mcgurk R, Jacobs PA (1997) Evidence from Turner's syndrome of an imprinted X-linked locus affecting cognitive function. Nature 387:705–708

Smith J, Cianflone K, Biron S, Hould FS, Lebel S, Marceau S, Lescelleur O, Biertho L, Simard S, Kral JG, Marceau P (2009) Effects of maternal surgical weight loss in mothers on intergenerational transmission of obesity. J Clin Endocrinol Metab 94:4275–4283

Sohn YB, Lee CG, Ko JM, Yang JA, Yun JN, Jung EJ, Jin HS, Park SJ, Jeong SY (2013) Clinical and genetic spectrum of 18 unrelated Korean patients with Sotos syndrome: frequent 5q35 microdeletion and identification of four novel NSD1 mutations. J Hum Genet 58:73–77

Sotos JF, Dodge PR, Muirhead D, Crawford JD, Talbot NB (1964) Cerebral gigantism in childhood. A syndrome of excessively rapid growth and acromegalic features and a nonprogressive neurologic disorder. N Engl J Med 271:109–116

Spector LG, Birch J (2012) The epidemiology of hepatoblastoma. Pediatr Blood Cancer 59:776–779

Steenman MJ, Rainier S, Dobry CJ, Grundy P, Horon IL, Feinberg AP (1994) Loss of imprinting of IGF2 is linked to reduced expression and abnormal methylation of H19 in Wilms' tumour. Nat Genet 7:433–439

Stunkel W, Pan H, Chew SB, Tng E, Tan JH, Chen L, Joseph R, Cheong CY, Ong ML, Lee YS, Chong YS, Saw SM, Meaney MJ, Kwek K, Sheppard AM, Gluckman PD, Holbrook JD (2012) Transcriptome changes affecting Hedgehog and cytokine signalling in the umbilical cord: implications for disease risk. PLoS One 7, E39744

Surani MA, Barton SC, Norris ML (1984) Development of reconstituted mouse eggs suggests imprinting of the genome during gametogenesis. Nature 308:548–550

Suzuki MM, Bird A (2008) Dna methylation landscapes: provocative insights from epigenomics. Nat Rev Genet 9:465–476

Symonds ME, Sebert SP, Budge H (2009) The impact of diet during early life and its contribution to later disease: critical checkpoints in development and their long-term consequences for metabolic health. Proc Nutr Soc 68:416–421

Szakszon K, Salpietro C, Kakar N, Knegt AC, Olah E, Dallapiccola B, Borck G (2013) De novo mutations of the gene encoding the histone acetyltransferase KAT6B in two patients with Say-Barber/Biesecker/Young-Simpson syndrome. Am J Med Genet A 161:884–888

Tahiliani M, Koh KP, Shen Y, Pastor WA, Bandukwala H, Brudno Y, Agarwal S, Iyer LM, Liu DR, Aravind L, Rao A (2009) Conversion of 5-methylcytosine to 5-hydroxymethylcytosine in mammalian DNA by MLL partner TET1. Science 324:930–935

Tammachote R, Kingsuwannapong N, Tongkobpetch S, Srichomthong C, Yeetong P, Kingwatanakul P, Monico CG, Suphapeetiporn K, Shotelersuk V (2012) Primary hyperoxaluria type 1 and brachydactyly mental retardation syndrome caused by a novel mutation in AGXT and a terminal deletion of chromosome 2. Am J Med Genet A 158a:2124–2130

Tatton-Brown K, Rahman N (1993–2014) EZH2-related overgrowth. In: Pagon RA, Adam MP, Bird TD, Dolan CR, Fong CT, Smith RJH, Stephens K (eds) GeneReviews® [Internet]. University of Washington, Seattle, Seattle

Tatton-Brown K, Rahman N (2013) The NSD1 and EZH2 overgrowth genes, similarities and differences. Am J Med Genet C Semin Med Genet 163:86–91

Tatton-Brown K, Cole TRP, Rahman N (1993–2014) Sotos syndrome. In: Pagon RA, Adam MP, Bird TD, Dolan CR, Fong CT, Smith RJH, Stephens K (eds) GeneReviews® [Internet]. University of Washington, Seattle, Seattle

Tawil R, Van Der Maarel SM (2006) Facioscapulohumeral muscular dystrophy. Muscle Nerve 34:1–15

Tawil R, Figlewicz DA, Griggs RC, Weiffenbach B (1998) Facioscapulohumeral dystrophy: a distinct regional myopathy with a novel molecular pathogenesis. FSH consortium. Ann Neurol 43:279–282

Temple IK (2007) Imprinting in human disease with special reference to transient neonatal diabetes and Beckwith-Wiedemann syndrome. Endocr Dev 12:113–123

Temple IK, Cockwell A, Hassold T, Pettay D, Jacobs P (1991) Maternal uniparental disomy for chromosome 14. J Med Genet 28:511–514

Teschendorff AE, West J, Beck S (2013) Age-associated epigenetic drift: implications, and a case of epigenetic thrift? Hum Mol Genet 22(R1):R7–R15

Timmer MR, Beuers U, Fockens P, Ponsioen CY, Rauws EA, Wang KK, Krishnadath KK (2013) Genetic and epigenetic abnormalities in primary sclerosing cholangitis-associated cholangiocarcinoma. Inflamm Bowel Dis 19:1789–1797

Tobi EW, Lumey LH, Talens RP, Kremer D, Putter H, Stein AD, Slagboom PE, Heijmans BT (2009) Dna methylation differences after exposure to prenatal famine are common and timing- and sex-specific. Hum Mol Genet 18:4046–4053

Tomizawa S, Sasaki H (2012) Genomic imprinting and its relevance to congenital disease, infertility, molar pregnancy and induced pluripotent stem cell. J Hum Genet 57:84–91

Van Belzen M, Bartsch O, Lacombe D, Peters DJ, Hennekam RC (2011) Rubinstein-Taybi syndrome (Crebbp, Ep300). Eur J Hum Genet 19(Preceeding):118–120

Van Deutekom JC, Wijmenga C, Van Tienhoven EA, Gruter AM, Hewitt JE, Padberg GW, Van Ommen GJ, Hofker MH, Frants RR (1993) Fshd associated DNA rearrangements are due to deletions of integral copies of a 3.2 Kb tandemly repeated unit. Hum Mol Genet 2:2037–2042

Vecsey CG, Hawk JD, Lattal KM, Stein JM, Fabian SA, Attner MA, Cabrera SM, Mcdonough CB, Brindle PK, Abel T, Wood MA (2007) Histone deacetylase inhibitors enhance memory and synaptic plasticity via CREB:CBP-dependent transcriptional activation. J Neurosci 27:6128–6140

Venkatesan K, Rual JF, Vazquez A, Stelzl U, Lemmens I, Hirozane-Kishikawa T, Hao T, Zenkner M, Xin X, Goh KI, Yildirim MA, Simonis N, Heinzmann K, Gebreab F, Sahalie JM, Cevik S, Simon C, De Smet AS, Dann E, Smolyar A, Vinayagam A, Yu H, Szeto D, Borick H, Dricot A, Klitgord N, Murray RR, Lin C, Lalowski M, Timm J, Rau K, Boone C, Braun P, Cusick ME, Roth FP, Hill DE, Tavernier J, Wanker EE, Barabasi AL, Vidal M (2009) An empirical framework for binary interactome mapping. Nat Methods 6:83–90

Verloes A, Bremond-Gignac D, Isidor B, David A, Baumann C, Leroy MA, Stevens R, Gillerot Y, Heron D, Heron B, Benzacken B, Lacombe D, Brunner H, Bitoun P (2006) Blepharophimosis-mental retardation (BMR) syndromes: a proposed clinical classification of the so-called Ohdo syndrome, and delineation of two new BMR syndromes, one X-linked and one autosomal recessive. Am J Med Genet A 140:1285–1296

Vige A, Gallou-Kabani C, Junien C (2008) Sexual dimorphism in non-mendelian inheritance. Pediatr Res 63:340–347

Villavicencio-Lorini P, Klopocki E, Trimborn M, Koll R, Mundlos S, Horn D (2013) Phenotypic variant of brachydactyly-mental retardation syndrome in a family with an inherited interstitial 2q37.3 microdeletion including HDAC4. Eur J Hum Genet 21:743–748

Vottero A, Minari R, Viani I, Tassi F, Bonatti F, Neri TM, Bertolini L, Bernasconi S, Ghizzoni L (2011) Evidence for epigenetic abnormalities of the androgen receptor gene in foreskin from children with hypospadias. J Clin Endocrinol Metab 96(12):E1953–E1962

Wada T (2009) X-linked alpha-thalassemia/mental retardation syndrome. Rinsho Byori 57:382–390

Waddington CH (1952) The epigenetics of birds. University Press, Cambridge

Walker DM, Gore AC (2011) Transgenerational neuroendocrine disruption of reproduction. Nat Rev Endocrinol 7:197–207

Wang JC, Passage MB, Yen PH, Shapiro LJ, Mohandas TK (1991) Uniparental heterodisomy for chromosome 14 in a phenotypically abnormal familial balanced 13/14 robertsonian translocation carrier. Am J Hum Genet 48:1069–1074

Wang J, Weaver IC, Gauthier-Fisher A, Wang H, He L, Yeomans J, Wondisford F, Kaplan DR, Miller FD (2010) CBP histone acetyltransferase activity regulates embryonic neural differentiation in the normal and Rubinstein-Taybi syndrome brain. Dev Cell 18:114–125

Waterland RA, Lin JR, Smith CA, Jirtle RL (2006) Post-weaning diet affects genomic imprinting at the insulin-like growth factor 2 (Igf2) locus. Hum Mol Genet 15:705–716

Weichenhan D, Plass C (2013) The evolving epigenome. Hum Mol Genet 70:R1–R6

Wiedemann HR (1964) Familial malformation complex with umbilical hernia and macroglossia—a “new syndrome”? J Genet Hum 13:223–232

Wilkinson MB, Dias C, Magida J, Mazei-Robison M, Lobo M, Kennedy P, Dietz D, Covington H 3rd, Russo S, Neve R, Ghose S, Tamminga C, Nestler EJ (2011) A novel role of the WNT-dishevelled-GSK3beta signaling cascade in the mouse nucleus accumbens in a social defeat model of depression. J Neurosci 31:9084–9092

Willemsen MH, Vulto-Van Silfhout AT, Nillesen WM, Wissink-Lindhout WM, Van Bokhoven H, Philip N, Berry-Kravis EM, Kini U, Van Ravenswaaij-Arts CM, Delle Chiaie B, Innes AM, Houge G, Kosonen T, Cremer K, Fannemel M, Stray-Pedersen A, Reardon W, Ignatius J, Lachlan K, Mircher C, Helderman Van Den Enden PT, Mastebroek M, Cohn-Hokke PE, Yntema HG, Drunat S, Kleefstra T (2012) Update on Kleefstra syndrome. Mol Syndromol 2:202–212

Williams CA, Angelman H, Clayton-Smith J, Driscoll DJ, Hendrickson JE, Knoll JH, Magenis RE, Schinzel A, Wagstaff J, Whidden EM et al (1995) Angelman syndrome: consensus for diagnostic criteria. Angelman syndrome foundation. Am J Med Genet 56:237–238

Williams SR, Aldred MA, Der Kaloustian VM, Halal F, Gowans G, Mcleod DR, Zondag S, Toriello HV, Magenis RE, Elsea SH (2010) Haploinsufficiency of HDAC4 causes brachydactyly mental retardation syndrome, with brachydactyly type E, developmental delays, and behavioral problems. Am J Hum Genet 87:219–228

Winkelmann J, Lin L, Schormair B, Kornum BR, Faraco J, Plazzi G, Melberg A, Cornelio F, Urban AE, Pizza F, Poli F, Grubert F, Wieland T, Graf E, Hallmayer J, Strom TM, Mignot E (2012) Mutations in DNMT1 cause autosomal dominant cerebellar ataxia, deafness and narcolepsy. Hum Mol Genet 21:2205–2210

Wossidlo M, Nakamura T, Lepikhov K, Marques CJ, Zakhartchenko V, Boiani M, Arand J, Nakano T, Reik W, Walter J (2011) 5-Hydroxymethylcytosine in the mammalian zygote is linked with epigenetic reprogramming. Nat Commun 2:241

Xu GL, Bestor TH, Bourc'his D, Hsieh CL, Tommerup N, Bugge M, Hulten M, Qu X, Russo JJ, Viegas-Pequignot E (1999) Chromosome instability and immunodeficiency syndrome caused by mutations in a DNA methyltransferase gene. Nature 402:187–191

Xu XR, Fu R, Wang LY, Wang N, Zhang F, Le F, Li L, Li LJ, Liu XZ, Zheng YM, Lou HY, Jiang SW, Zhu XM, Huang HF, Jin F (2013) Epigenetic inheritance of paternally expressed imprinted genes in the testes of ICSI mice. Curr Pharm Des. http://www.ncbi.nlm.nih.gov/pubmed/23909805

Yamadori T, Baba Y, Matsushita M, Hashimoto S, Kurosaki M, Kurosaki T, Kishimoto T, Tsukada S (1999) Bruton's tyrosine kinase activity is negatively regulated by Sab, the Btk-SH3 domain-binding protein. Proc Natl Acad Sci U S A 96:6341–6346

Yang S, Xiao X, Jia Y, Liu X, Zhang Y, Devor EJ, Meng X, Thiel KW, Leslie KK (2013) Epigenetic modification restores functional PR expression in endometrial cancer cells. Curr Pharm Des. http://www.ncbi.nlm.nih.gov/pubmed/23888956

Yong PJ, Marion SA, Barrett IJ, Kalousek DK, Robinson WP (2002) Evidence for imprinting on chromosome 16: the effect of uniparental disomy on the outcome of mosaic trisomy 16 pregnancies. Am J Med Genet 112:123–132

Youngson NA, Whitelaw E (2008) Transgenerational epigenetic effects. Annu Rev Genomics Hum Genet 9:233–257

Yu Y, Xu F, Peng H, Fang X, Zhao S, Li Y, Cuevas B, Kuo WL, Gray JW, Siciliano M, Mills GB, Bast RC Jr (1999) NOEY2 (ARHI), an imprinted putative tumor suppressor gene in ovarian and breast carcinomas. Proc Natl Acad Sci U S A 96:214–219

Zechner U, Pliushch G, Schneider E, El Hajj N, Tresch A, Shufaro Y, Seidmann L, Coerdt W, Muller AM, Haaf T (2010) Quantitative methylation analysis of developmentally important genes in human pregnancy losses after art and spontaneous conception. Mol Hum Reprod 16:704–713

Chapter 2
Role of Epigenetics in Neural Differentiation: Implications for Health and Disease

Estela G. Toraño, Agustin F. Fernandez, Rocio G. Urdinguio, and Mario F. Fraga

Abstract Neural differentiation is a complex process that requires highly accurate spatial and temporal regulation by extracellular and intracellular programs. Epigenetic mechanisms, such as DNA methylation, covalent histone posttranscriptional modifications, chromatin organization, and noncoding regulatory RNA, are key regulators of pluripotency maintenance and differentiation. The misregulation of these mechanisms could lead to neurological diseases and cancer.

Keywords DNA methylation • Histone modifications • Epigenetics • Neural differentiation • Neural diseases

2.1 Introduction

The development of the central nervous system (CNS) arises from the external layer of the embryo, the ectoderm (Bohacek et al. 2013). This is a complex and tightly regulated phenomenon which, briefly, consists in the initial formation of the

E.G. Toraño, Ph.D. • A.F. Fernandez, Ph.D. • R.G. Urdinguio, Ph.D.
Cancer Epigenetics Laboratory, Instituto Universitario de Oncología del Principado de Asturias (IUOPA), HUCA, Universidad de Oviedo, Oviedo, Spain
e-mail: estelait@gmail.com; affernandez@hca.es; rgurdinguio@gmail.es

M.F. Fraga, Ph.D. (✉)
Cancer Epigenetics Laboratory, Instituto Universitario de Oncología del Principado de Asturias (IUOPA), HUCA, Universidad de Oviedo, Oviedo, Spain

Department of Immunology and Oncology, National Center for Biotechnology, CNB-CSIC, Cantoblanco, Madrid 28049, Spain
e-mail: mffraga@cnb.csic.es

N. Maulik and T. Karagiannis (eds.), *Molecular Mechanisms and Physiology of Disease: Implications for Epigenetics and Health*, DOI 10.1007/978-1-4939-0706-9_2,

neural tube which then develops to become the brain and spinal cord (Olynik and Rastegar 2012). The CNS is composed by many cell types, including neurons, astrocytes, oligodendrocytes, all of which are formed from the same multipotent precursor cells—neural stem cells (NSCs)—(Dietrich et al. 2006) which have the capacity to self-renew and differentiate in multiple lineages. NSCs are capable of generating specific neurons, of a particular length and at exactly the correct place and time for the requirements of each stage of development; neurons appearing first, and then astrocytes and oligodendrocytes (Olynik and Rastegar 2012). During development, NSCs firstly divide symmetrically to expand, and then start to divide asymmetrically, resulting in a new NSC and a neuron. In later stages of development, NSCs are able to become, in addition to neurons, to astrocytes, and/or to oligodendrocytes while at the same time maintaining their capacity for self-renewal. This differentiation has to be strictly regulated during development, both spatially and temporally, by extracellular cues such as the NOTCH signaling family, TGF-β, FGF, EGF, and FGF2 growth factors, neuregulins (NRG), and by intracellular programs including the expression of homeobox (HB) genes and epigenetic modifications (Mizutani et al. 2007; Namihira et al. 2008). Knowledge of exactly how the molecular determination of NSC differentiation takes place could have major implications for the study of many diseases such as cancer and neurodevelopmental disorders.

The term epigenetics was first introduced by Conrad Waddington in 1942 to explain the variations between genes and their products. However, the word has evolved to incorporate the study of mitotically and/or meiotically stable and heritable changes in gene expression which are not accompanied by changes in the DNA sequence. Epigenetic modifications are crucial for gene expression regulation during the cell cycle, development, differentiation, and in response to environmental or biological variations (Brooks et al. 2010). Epigenetic regulation comprises DNA methylation, covalent histone posttranscriptional modifications (such as methylation, acetylation, ubiquitination, and phosphorylation), chromatin organization, and noncoding regulatory RNA (Bernstein et al. 2007). Epigenetic mechanisms are key regulators of pluripotency maintenance and also of cell fate specification. During their differentiation from embryonic stem cells (ESCs) to NSCs, cells have already acquired epigenetic marks (Meissner et al. 2008). NSC maintenance requires epigenetic mechanisms that allow the inhibition of neuronal and glial cells, whereas differentiation of NSCs requires the elimination of the epigenetic suppression of neural and glial specification genes (Hsieh and Eisch 2010). During the differentiation process, neural genes can become activated due to the increased accessibility of their promoters, whereas pluripotency neural genes are silenced (Hirabayashi and Gotoh 2010) (Fig. 2.1).

In this chapter, we will discuss and summarize the epigenetic changes that occur during differentiation from ESCs to NSCs and then to mature neural and glial cells, and their relation with many neurological disorders.

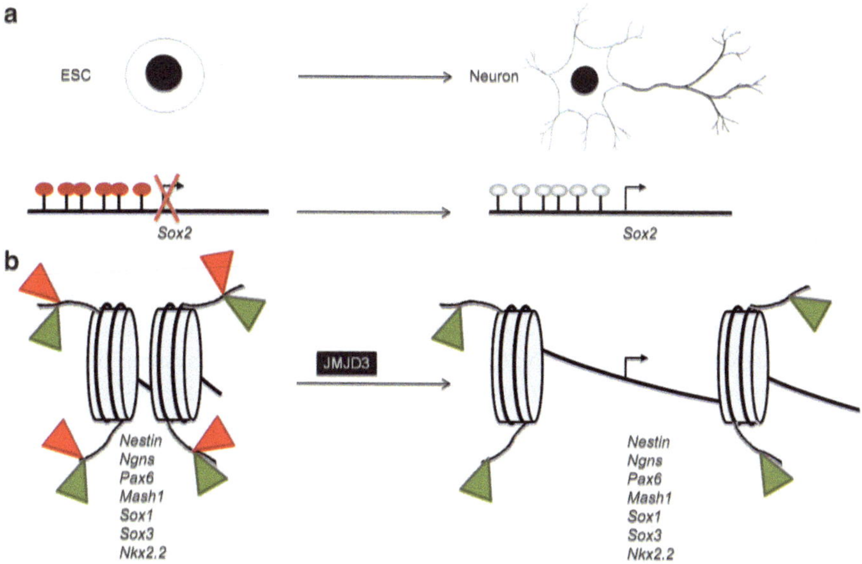

Fig. 2.1 Epigenetic changes during the differentiation process in neural genes. (**a**) During ESC differentiation, neuronal genes such as Sox2 are activated via decreased DNA methylation. (**b**) In ESCs, neural genes carry both the H3K27me3-repressed mark (*red triangle*) and the H3k4me3-activated mark (*green triangle*). During differentiation, JMJD3 acts to remove H3k27me3 on the promoter of many neural genes (*Nestin, Pax6, Ngns, Mash1, Sox1, Sox3, Nkx2.2*), which become expressed

2.2 Epigenetic Mechanisms and Neural Differentiation

2.2.1 DNA Methylation

One of the most extensively studied epigenetic modifications in mammals is DNA methylation, which plays an important role in many biological processes, such as genomic imprinting, X-chromosome inactivation (XCI) during development, regulation of gene expression, and maintenance of epigenetic memory, among others. DNA methylation consists of the covalent addition of a methyl (CH_3) group, from the methyl donor S-adenosylmethionine (SAM) to the 5′ carbon of the pyrimidine ring of the citosine base that precedes guanine (CpG) (Herman and Baylin 2003; Weber et al. 2007). CpGs are mainly associated in clusters called CpG islands and are located at the promoter region of more than 50 % of genes (Bird 1986). In healthy cells, most CpG islands are unmethylated when located at a transcription start site (TSS), and when methylated they are usually associated with silent genes. Indeed, hypermethylation of CpG islands in promoter regions is related to the silencing of many tumor suppressor genes (TSG) in cancer. Conversely, CpG sites located in repetitive and transposon elements, intergenic regions, and gene bodies are usually heavily methylated (Ellis et al. 2009; Kanai 2008). During ESC differentiation, there is an increase in DNA promoter methylation regions (Delcuve et al. 2009) and many pluripotency genes are silenced by DNA methylation (Mohn et al. 2008), while neuronal genes such as *Sox2* are activated via decreased DNA methylation (Sikorska et al. 2008) (Fig. 2.1a).

DNA methylation is carried out by DNA methyltransferases—DNMT1, DNMT3A, and DNMT3B—the first being responsible for the maintenance of the DNA-methylated status following DNA replication and the two latter being de novo methyltransferases (Robertson 2001). High levels of DNMT1 have been found to maintain DNA methylation in NSCs in the embryonic nervous system. In the case of de novo methyltransferases, *DNMT3B* is expressed in embryonic NSCs, whereas DNMT3A is expressed at late developmental stages (Feng et al. 2007). *DNMT3A*-deficient ESCs have increased cell proliferation and premature glial differentiation (Wu et al. 2012). A decrease in DNMT3B causes the failure of neuronal differentiation in vitro, maybe due to the principal expression of this methyltransferase in early embryonic cells and neural progenitors (Bai et al. 2005; Feng et al. 2005; Watanabe et al. 2006).

The association between DNA methylation and gene repression can be mediated by methyl-CpG-binding proteins (MBPs) which recognize methylated DNA, bind to it, and recruit different chromatin remodeling complexes (Defossez and Stancheva 2011). In the brain, MBD1 and MeCP2, two MBPs that contain a methyl-CpG-binding domain (MBD), are strongly expressed and participate in neurodevelopment and plasticity through the regulation of other epigenetic factors (Fan and Hutnick 2005; Jobe et al. 2012).

2.2.2 Histone Tail Posttranslational Modifications

The basic unit of chromatin is the nucleosome. It consists of 147 bp of DNA wrapped twice around two copies of each of the histones H2A, H2B, H3, and H4. The N-terminus histone tails are susceptible to posttranslational modification (PTM) by acetylation, methylation, ubiquitination, phosphorylation, and other processes. PTMs need to be strictly regulated, both spatially and temporally, during development. Depending on the amino acid residues that the histone PTM is attached to, these covalent modifications have profound effects on chromatin organization and, as a consequence, on gene activation and inactivation depending whether transcriptional machinery has greater or lesser accessibility (Delcuve et al. 2009). Together, the different PTMs and the effects they exert are referred to as the "histone code" (Bernstein et al. 2007). Addition of acetyl groups to lysines correspond with the open chromatin state, which is very important in nucleosome formation and chromatin folding. In the case of histone methylation, it can occur at the lysine and arginine residues of histones, and depending on the amino acid residue, the effect can be different. For example, H3K4me3 is associated with gene activation, whereas H3k27me3 and H3k9me3 are inhibitory epigenetic marks. In ESCs, chromatin structure is very open and active and during development, acetylation marks are substantially reduced and there is an overall increase in repressive marks which results in differentiated tissues having a more condensed chromatin structure (Meshorer et al. 2006).

Histone Acetyl Transferases and Histone Deacetylases

Histone acetyl transferases (HATs) are the enzymes responsible for catalyzing the acetylation of lysine residues of histone, and its reversion is carried out by histone deacetylases (HDACs) and both are implicated as regulators of neural-specific gene-expression patterns in the brain (Abel and Zukin 2008). Acetylation is the most widely studied histone modification and plays an important role in gene regulation (MacDonald and Howe 2009) and its regulatory role is evolutionarily conserved: in *Drosophila*, neural differentiation is connected with high acetylation levels whereas low levels are related to glial differentiation (Flici et al. 2011).

HATs comprise three major families: general control non-derepressible 5 (Gcn5)-related *N*-acetyltransferases (GNATs), and p300/CBP and MYST proteins (Lee and Workman 2007). It has been shown that knockdown of these HATs leads to aberrant ESC differentiation although HATs have not been extensively studied in a developmental context, neither in vivo nor in cellular systems. More studies in NSCs and of brain development are necessary to understand the role of histone acetylation in embryonic development (Lilja et al. 2013).

Histone deacetylation is catalyzed by the HDAC enzymes which are critical players in many biological processes, including differentiation (Haberland et al. 2009b). They can regulate stem cell self-renewal and differentiation through the control of a variety of target genes, and also regulate (NSC) differentiation. Also, HDACs can target nonhistone protein targets, such as transcription factors, transcription regulators, signal transduction mediators, DNA repair enzymes, and even each other (Xu et al. 2007). In mammals, there are 18 HDACs, which are classified into four classes depending on sequence identity and domain organization (Dokmanovic et al. 2007). Class I HDACs (1, 2, 3, and 8) are located in the nucleus and are known to have critical functions during early development (Yang and Seto 2008). HDAC1 and HDAC2 act together to maintain neuronal specification. Deletion of either one of these HDACs leads to severe brain abnormalities and postnatal lethality (Montgomery et al. 2009). HDAC1 is enhanced in glial cells in the adult brain, whereas HDAC2 is upregulated in the differentiation of NSCs to different neural lineages (MacDonald and Roskams 2008). Overexpression of *HDAC2* in neurons decreases synaptic plasticity and memory formation (Guan et al. 2009). HDAC1, 2, and 3 inhibit oligodentrocytic differentiation and HDAC2, in addition, inhibits astrocytic differentiation (Montgomery et al. 2009). In *HDAC8* global deletion mice, their development of scull morphogenesis points to it having a unique role in cranial differentiation (Haberland et al. 2009a).

Class II HDACs (4, 5, 6, 7, 9, and 10) have cell type-specific expression and may serve as key regulators of neural development, but their roles are not well defined. For instance, HDAC4, 5, 7, and 9 are upregulated in differentiated NSCs (Ajamian et al. 2003); HDAC5 regulates neuronal differentiation (Schneider et al. 2008) and, along with HDAC4, are involved in neuronal maturation (Majdzadeh et al. 2008). HDAC3 and 5 also participate in NSCs proliferation (Sun et al. 2007). The third class of HDACs, called sirtuins, requires nicotinamide adenine dinucleotide (NAD^+) for their activity, linking them with cell metabolism and redox state (Calvanese and Fraga 2011). Seven sirtuin members, SIRT1-7, have been identified in mammals, with different subcellular locations; SIRT1, 6, and 7 are located in the nucleus, SIRT2 is cytosolic, and SIRT3, 4, and 5 are found in the mitochondria (Verdin et al. 2010). SIRT1 has a role in ESC maintenance, through the epigenetic repression of many developmental genes (Calvanese et al. 2010). It has also been implicated in neuronal differentiation, but its role is not very clear as it has been associated with both activation and inhibition of neural differentiation (Lilja et al. 2013). HDAC11 alone forms HDAC class IV.

HDAC inhibitors (HDACis) are molecules that inhibit HDAC activities, which allow efficient control of gene expression (Kretsovali et al. 2012). They are classified into four different families: the short-chain fatty acids (sodium butyrate, phenylbutyrate, and valproic acid (VPA)), the hyroxamic acids (trichostatin A (TSA) and suberoylanilide hydroxamic acid (SAHA)), the epoxyketones (trapoxin), and the benzamides (Abel and Zukin 2008). Their effects cause transcription alterations which result in no net gain in the number of induced or repressed genes. Sirtuins are not inhibited by compounds such as vorinostat or TSA in contrast to class I and II HDACs (Xu et al. 2007). The administration of HDACis such as VPA can induce differentiation into neural lineage and glial suppression due to the induction of a neurogenic bHLH transcription factor, *NeuroD* (Hsieh et al. 2004). HDACis have potent anticancer activities such as arresting tumor growth, induction of differentiation, and apoptosis (Minucci and Pelicci 2006). For these reasons, they represent a good therapeutic approach for tackling many neurological diseases.

Histone Methyltransferases

Histone methyltransferases (HMTs) catalyze histone methylation of lysine or arginine residues of PTMs. Histone methylation can be associated with either gene silencing or gene activation, depending on the number of methyl groups (1, 2, or 3) and the location of the residue. Examples of repressive marks are H3K9me2, H3K9me3, H3k27me3, and H4K20me3 while H3K4me3 and H3K36me3 are examples of active marks (Mosammaparast and Shi 2010).

Two members of the chromatin remodeling system, Polycomb-group (PcG) and Thithorax-group (TrxG) proteins, are evolutionarily conserved from *Drosophila* to humans and are responsible for the correct expression and regulation of the majority of key developmental genes in ESCs. PcG and TrxG protein complexes have antagonistic functions in that PcG proteins promote heterochromatin formation via H3k27me3 inhibitory epigenetic mark, whereas TrxG proteins have the reverse effect due to the promotion of H3k4me3 associated with gene activation (Bernstein et al. 2006; Ng and Gurdon 2008; Ringrose and Paro 2007; Schuettengruber et al. 2007). PcG proteins play a key role in silencing developmental genes and, as a consequence, in pluripotency maintenance and inhibition of differentiation. They form two polycomb-repressive complexes, PRC1 and -2, of which each contains a different set of core proteins. Both silence an extensive range of key developmental genes in ESCs due to trimethylation at histone 3 lysine 27. In addition, most PcG target genes also carry H3k4me3. This histone signature (H3k27me3 and H3k4me3 together) occurs in regions referred to as bivalent domains, and collectively, means the gene is maintained poised for activation and prepares ESCs for differentiation.

During differentiation, active genes are enriched in H3k4me3 due to TrxG protein complex action, while the demethylase JMJD3 is recruited, which removes the H3K27me3 mark. In contrast, genes that remain silenced retain H3k27me3 and lose H3k4me3 through Rbp2 demethylase, which is recruited by PRC2 complex (Cloos et al. 2008; Pasini et al. 2008; Soshnikova and Duboule 2008). This "bivalent" state is resolved during the differentiation process when genes becomes univalent as a result of neural differentiation, leaving them in an "on" or "off" state of transcription. Among the genes carrying the "bivalent" mark are *Hox* genes (Barber and Rastegar 2010), the master regulators of embryonic development. The genome is composed of 39 *Hox* genes organized in four clusters, *Hoxa, Hoxb, Howc*, and *Hoxd*. They control the exact purpose of each developing tissue in the body and, during neurogenesis, they are responsible for dictating and leading somatogenesis, cellular migration, and axonal direction, and their misregulation leads to disease and cancer (Barber and Rastegar 2010; Oury and Rijli 2007).

In neural differentiation, JMJD3 acts to remove H3k27me3 on the *Nestin* promoter, a neurofilament gene whose activation is a step in the transition from ESC to NSC (Burgold et al. 2008). More examples of neural genes that lose repressive marks during differentiation into neural lineage are paired box gene 6 (*Pax6*), neurogenins (Ngns), *Mash1* (achaete–scute complex homolog 1, or *Ascl1*), SRY-Box 1 (*Sox1*), *Sox3*, and NK2 transcription factor-related locus 2 (*Nkx2.2*) (Hirabayashi and Gotoh 2010; Mikkelsen et al. 2007) (Fig. 2.1b). In addition, during transition from ESC to NSC, a new "bivalent" state is established in functioning genes in terminally differentiated neurons. These neuron-specific genes become poised for expression and lose the H3k27me3 mark in the final differentiation before becoming expressed.

2.2.3 Noncoding RNA

Noncoding RNA (ncRNA) refers to the part of the RNA that is not translated into protein and includes microRNAs (miRNAs), small interfering RNAs (siRNAs), small nucleolar RNAs, and PIWI-interacting RNAs (piRNAs) (Li and Zhao 2008).

They regulate gene expression through the control of chromatin structures, RNA modifications, DNA transcription, and mRNA translation and splicing (Mohamed Ariff et al. 2012). ncRNAs are important executers in epigenetic regulation and, in particular, miRNAs play a role in stem cells maintenance and differentiation through degradation of their target mRNAs (Guo et al. 2010). miRNA activities act in coordination with DNA marks and histone modifications to ensure the correct differentiation of all the cell types in the CNS (Olynik and Rastegar 2012). The most abundant miRNA in both the embryonic and adult CNS is miR-124, whose levels are increased during neuronal differentiation (Makeyev et al. 2007). miR-124 is critical in neurogenesis due to its targeting of *Sox9*, which is essential for multipotent NSC formation and maintenance (Cheng et al. 2009; Scott et al. 2010). Other examples of miRNAs involved in neural differentiation are *miR-9*, which is expressed in neurogenic areas of the brain and controls NSC proliferation and differentiation, and Let-79, which reduces proliferation and induces neural differentiation (Zhao et al. 2009, 2010). The misregulation of these miRNAs is related with cancer and many neurological diseases, such as Alzheimer's and Parkinson's (Junn and Mouradian 2012).

2.3 Epigenetics in Neural Diseases

As a whole, epigenetic mechanisms are thought to be involved in a number of neurological disorders. The inadequate control of proliferation, "poised" state, or the imbalance between HATs and HDACs and the promotion or inhibition of neural differentiation has been associated with many neurological disorders and tumorigenesis of the nervous system. Any malfunction of the epigenetic machinery during neural development could lead to neural diseases and knowledge of how aberrant epigenetic mechanisms take place in such development would provide good opportunities for therapeutic intervention (Fig. 2.2).

2.3.1 Rett Syndrome

Rett syndrome (RTT) is an X-linked dominant neurological disorder that predominantly affects females, with an incidence of 1 in 10,000–15,000 female births. It is characterized by normal development during the first 6–18 months after birth, followed by the appearance of severe problems, including autistic features, epileptic seizures, and poor motor and language skills (Rett 1986). This disorder is principally caused both by mutations and duplications in the *MECP2* gene (Bird 2008; Urdinguio et al. 2009; Van Esch et al. 2005). Although this gene disruption affects all tissues, its deregulation seems to be particularly damaging to brain function (Chen et al. 2001; Guy et al. 2001; Neul et al. 2008). Additionally, XCI is thought to cause a mosaicism of MeCP2 protein expression and differences in penetrance of the

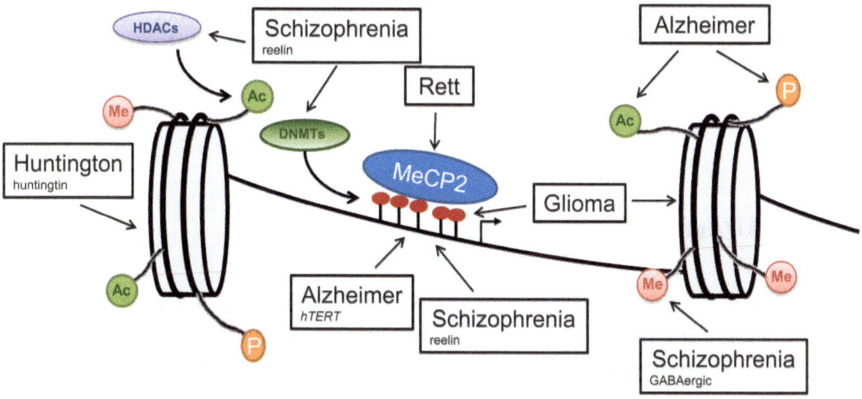

Fig. 2.2 Some epigenetic mechanisms implicated in neural diseases. Rett syndrome is principally caused by a mutation in MeCP2. It has been shown that the DNA methylation of telomerase inverse transcriptase (hTERT) promoter is increased in patients with Alzheimer's disease and an aberrant gene expression related to histone acetylation and phosphorylation has been found. In Huntington disease, huntingtin protein has been found to be modified by many PTMs. In schizophrenia patients, the extracellular matrix protein reelin is reduced; a decrease, which may be regulated by DNA methylation and HDACs, and its expression are increased by DNA methyltransferase. Also, an increase in GABAergic mRNA has been demonstrated and shown to be associated with a rise in H3k4me3. Gliomas present specific epigenetic patterns such as aberrant DNA methylation and changes in histone PTMs and their regulatory genes

symptoms, although a recent study showed that XCI patterns in various peripheral tissues did not differ between RTT discordant monozygotic twins (Miyake et al. 2013).

Currently, many works are focusing on understanding the different facets of MeCP2 function. One of the targets of MeCP2 is the brain-derived neurotrophic factor (*BNDF*) gene, whose protein is synthesized in response to neuronal activity and is essential for appropriate brain function. A recent work using ESCs has studied the maturation process of neurons and found that MeCP2 regulates not only BDNF levels but also the nuclear size and RNA synthesis during the process (Yazdani et al. 2012). Furthermore, RTT mutations in *MECP2* have recently been described which abolish the interaction of MeCP2 with the NCoR/SMRT corepressor, a finding in line with the hypothesis that brain dysfunction in RTT is caused by a loss of the MeCP2 connection between the NCoR/SMRT corepressors and chromatin (Lyst et al. 2013).

Microarray expression analyses have suggested that MeCP2 is able to regulate a wide range of genes in different regions of the brain (Jordan et al. 2007; Tudor et al. 2002; Urdinguio et al. 2009). Although MeCP2 was firstly described as interacting with repressor complexes and inhibiting gene expression, compelling evidence is pointing to MeCP2 being involved in the regulatory action of both activator and repressor functions (Ben-Shachar et al. 2009; Chahrour et al. 2008; Samaco and Neul 2011; Zachariah and Rastegar 2012). This indicates that adequate regulation exerted by MeCP2 is essential for correct brain function. It has been reported that RTT symptoms are, fortunately, reversed by MeCP2 restoration in mouse models of RTT (Bird 2008; Guy et al. 2007), which brings hope for the treatment of this complex disease.

2.3.2 *Alzheimer's Disease*

Alzheimer's disease (AD) is a neurodegenerative disease associated with dementia and shows progressive memory loss and cognitive decline. It is associated with plaques containing amyloid-ß, and neurofibrillary tangles in the brain (Ittner and Gotz 2011). Many genetic risk factors for AD have been identified, although only a few cases of AD can be explained by specific gene mutations. Besides that, the phenotypic discordance between monozygotic twins where one has Alzheimer's might be explained by the existence of epigenetic mechanisms that contribute to the development of this illness (Poulsen et al. 2007). In addition, the fact that reduced neurogenesis is a common feature in AD could be due to failures in the differentiation process, including epigenetic failures. There are some studies that relate epigenetic mechanisms directly with AD: It has been shown that the DNA methylation of telomerase inverse transcriptase (*hTERT*) promoter is increased in patients with AD (Silva et al. 2008) and an aberrant gene expression related to histone acetylation and phosphorylation has also been found (Kilgore et al. 2010).

2.3.3 *Huntington's and Parkinson's Diseases*

Huntington's disease (HD) and Parkinson's disease are neurodegenerative diseases. Despite the fact that the molecular mechanisms implicated in their development appear to be very different, both are late-onset and have been associated with the accumulation of intracellular toxic proteins (Rubinsztein 2006).

HD is a heritable disease characterized by abnormal involuntary movements, cognitive dysfunction, and psychiatric symptoms (Walker 2007). It is caused by an autosomal-dominant mutation in the huntingtin gene (*HTT*), which produces an expansion of a poly-glutamine repeat within the amino terminus of the protein huntingtin (HTT). The mutant form of this protein has been found to interact with HATs, which suggests that epigenetics play a role in HD. Also, this protein associates with HDAC corepressors to repress transcription (Gray 2010, 2011; Steffan et al. 2000). These interactions between HTT protein and the regulation of the histone code can lead to aberrations of gene expression. Accordingly, genome-wide expression profiling patterns in HD patients have shown alterations in mRNA expression (Borovecki et al. 2005). Recently, a study has shown that treatment with HDACis in a mouse model of HD resulted in improved motor function, extended survival, and reduced brain atrophy (Chopra et al. 2012).

Parkinson's disease (PD) is a degenerative disorder of the CNS that affects 1–2 % of the population over the age of 65. The most common symptoms are resting tremor, rigidity, bradykinesia, and postural instability, which result from the loss of neuromelanin containing dopaminergic neurons (Thomas and Beal 2011). Many studies have related PD with specific/particular genetic mutations (Hardy 2010; Nuytemans et al. 2010; Ramirez et al. 2006), but in the last few years there is growing evidence pointing to epigenetic mechanisms contributing to PD development (Coppede 2012). For instance, the expression of the gene frequently altered in PD, alpha-synuclein (*SNCA*), can be regulated by DNA methylation (Matsumoto et al. 2010). Also, treatments with TSA performed in a rat model of PD have shown a neuroprotective action of this epigenetic drug (Monti et al. 2010), and levels of many miRNAs have also been shown to be altered (Gillardon et al. 2008). These studies evidence the relationship between epigenetic mechanism and PD, and more investigations in this area could help to make progress in discovering new targets and designing appropriate therapies.

2.3.4 Schizophrenia

Schizophrenia is a mental disease characterized by a serious disorder of cognition. Common symptoms include delusions, hallucinations, paranoid, bizarre thoughts, social dysfunction, poor motivation, and apathy, among others. Diagnosis is usually in adolescence or later, suggesting that it may will be a neurodevelopmental disorder (Sawa and Snyder 2002). To date, DNA methylation has been examined for only a small number of candidate genes (Roth et al. 2009). For instance, the levels of the extracellular matrix protein reelin are reduced in postmortem brains from patients diagnosed with schizophrenia or bipolar illness with psychosis. This downregulation is thought to be mediated by epigenetic mechanisms given that *Reelin* promoter contains several sites for DNA methylation and HDAC and DNMT inhibitors increase its expression. Furthermore, another gene influencing the GABAergic system, the glutamic acid decarboxylase 1 (GAD1) showed changes in schizophrenia patients related to chromatin remodeling modifications (Abel and Zukin 2008).

2.3.5 Glioma

Glioma is the most common primary brain tumor and causes more than 40 % of all CNS neoplasms. It is well known that aberrant epigenetic mechanisms lead to cancer and glioma progression (Nagarajan and Costello 2009). Gliomas are classified by their state of differentiation and present distinct epigenetic patterns such as aberrant DNA methylation (Martinez et al. 2009), changes in histone PTMs and their regulatory genes (Kreth et al. 2012), and also downregulation and upregulation of miRNAs (Croce 2009).

2.3.6 *Therapeutic Applications*

Targeting histone acetylation could provide benefits for the treatment of many neurological diseases. For example, HDACis might interfere in neurological diseases to provide a protective effect (Chuang et al. 2009). As mentioned earlier, VPA promotes neural differentiation and could have important clinical applications in the treatment of neurological diseases such as epilepsy, bipolar disorders, and serious depression (Blaheta and Cinatl 2002).

2.4 Conclusions

It is true to say that current knowledge of the epigenetic changes that take place during neural development, neural disorders, and cancer development and their clinical potential is quite wide. However, many aspects remain unknown and other aspects need to be fully explored for a truly complete understanding of the development of neural disease and cancer. The use of new technical tools such as high-throughput approaches, and the development of stem cell-based therapies should lead to the identification of new therapeutic targets and result in improvements in quality of life for patients.

References

Abel T, Zukin RS (2008) Epigenetic targets of HDAC inhibition in neurodegenerative and psychiatric disorders. Curr Opin Pharmacol 8:57–64

Ajamian F, Suuronen T, Salminen A, Reeben M (2003) Upregulation of class II histone deacetylases mRNA during neural differentiation of cultured rat hippocampal progenitor cells. Neurosci Lett 346:57–60

Bai S, Ghoshal K, Datta J, Majumder S, Yoon SO, Jacob ST (2005) DNA methyltransferase 3b regulates nerve growth factor-induced differentiation of PC12 cells by recruiting histone deacetylase 2. Mol Cell Biol 25:751–766

Barber BA, Rastegar M (2010) Epigenetic control of Hox genes during neurogenesis, development, and disease. Ann Anat 192:261–274

Ben-Shachar S, Chahrour M, Thaller C, Shaw CA, Zoghbi HY (2009) Mouse models of MeCP2 disorders share gene expression changes in the cerebellum and hypothalamus. Hum Mol Genet 18:2431–2442

Bernstein BE, Mikkelsen TS, Xie X, Kamal M, Huebert DJ, Cuff J, Fry B, Meissner A, Wernig M, Plath K, Jaenisch R, Wagschal A, Feil R, Schreiber SL, Lander ES (2006) A bivalent chromatin structure marks key developmental genes in embryonic stem cells. Cell 125:315–326

Bernstein BE, Meissner A, Lander ES (2007) The mammalian epigenome. Cell 128:669–681

Bird AP (1986) CpG-rich islands and the function of DNA methylation. Nature 321:209–213

Bird A (2008) The methyl-CpG-binding protein MeCP2 and neurological disease. Biochem Soc Trans 36:575–583

Blaheta RA, Cinatl J Jr (2002) Anti-tumor mechanisms of valproate: a novel role for an old drug. Med Res Rev 22:492–511

Bohacek J, Gapp K, Saab BJ, Mansuy IM (2013) Transgenerational epigenetic effects on brain functions. Biol Psychiatry 73:313–320

Borovecki F, Lovrecic L, Zhou J, Jeong H, Then F, Rosas HD, Hersch SM, Hogarth P, Bouzou B, Jensen RV, Krainc D (2005) Genome-wide expression profiling of human blood reveals biomarkers for Huntington's disease. Proc Natl Acad Sci U S A 102:11023–11028

Brooks WH, Le Dantec C, Pers JO, Youinou P, Renaudineau Y (2010) Epigenetics and autoimmunity. J Autoimmun 34:J207–J219

Burgold T, Spreafico F, De Santa F, Totaro MG, Prosperini E, Natoli G, Testa G (2008) The histone H3 lysine 27-specific demethylase Jmjd3 is required for neural commitment. PLoS One 3:e3034

Calvanese V, Fraga MF (2011) SirT1 brings stemness closer to cancer and aging. Aging (Albany NY) 3:162–167

Calvanese V, Lara E, Suarez-Alvarez B, Abu Dawud R, Vazquez-Chantada M, Martinez-Chantar ML, Embade N, Lopez-Nieva P, Horrillo A, Hmadcha A, Soria B, Piazzolla D, Herranz D, Serrano M, Mato JM, Andrews PW, Lopez-Larrea C, Esteller M, Fraga MF (2010) Sirtuin 1 regulation of developmental genes during differentiation of stem cells. Proc Natl Acad Sci U S A 107:13736–13741

Chahrour M, Jung SY, Shaw C, Zhou X, Wong ST, Qin J, Zoghbi HY (2008) MeCP2, a key contributor to neurological disease, activates and represses transcription. Science 320:1224–1229

Chen RZ, Akbarian S, Tudor M, Jaenisch R (2001) Deficiency of methyl-CpG binding protein-2 in CNS neurons results in a Rett-like phenotype in mice. Nat Genet 27:327–331

Cheng LC, Pastrana E, Tavazoie M, Doetsch F (2009) miR-124 regulates adult neurogenesis in the subventricular zone stem cell niche. Nat Neurosci 12:399–408

Chopra V, Quinti L, Kim J, Vollor L, Narayanan KL, Edgerly C, Cipicchio PM, Lauver MA, Choi SH, Silverman RB, Ferrante RJ, Hersch S, Kazantsev AG (2012) The sirtuin 2 inhibitor AK-7 is neuroprotective in Huntington's disease mouse models. Cell Rep 2:1492–1497

Chuang DM, Leng Y, Marinova Z, Kim HJ, Chiu CT (2009) Multiple roles of HDAC inhibition in neurodegenerative conditions. Trends Neurosci 32:591–601

Cloos PA, Christensen J, Agger K, Helin K (2008) Erasing the methyl mark: histone demethylases at the center of cellular differentiation and disease. Genes Dev 22:1115–1140

Coppede F (2012) Genetics and epigenetics of Parkinson's disease. ScientificWorldJournal 2012:489830

Croce CM (2009) Causes and consequences of microRNA dysregulation in cancer. Nat Rev Genet 10:704–714

Defossez PA, Stancheva I (2011) Biological functions of methyl-CpG-binding proteins. Prog Mol Biol Transl Sci 101:377–398

Delcuve GP, Rastegar M, Davie JR (2009) Epigenetic control. J Cell Physiol 219:243–250

Dietrich J, Han R, Yang Y, Mayer-Proschel M, Noble M (2006) CNS progenitor cells and oligodendrocytes are targets of chemotherapeutic agents in vitro and in vivo. J Biol 5:22

Dokmanovic M, Clarke C, Marks PA (2007) Histone deacetylase inhibitors: overview and perspectives. Mol Cancer Res 5:981–989

Ellis L, Atadja PW, Johnstone RW (2009) Epigenetics in cancer: targeting chromatin modifications. Mol Cancer Ther 8:1409–1420

Fan G, Hutnick L (2005) Methyl-CpG binding proteins in the nervous system. Cell Res 15: 255–261

Feng J, Chang H, Li E, Fan G (2005) Dynamic expression of de novo DNA methyltransferases Dnmt3a and Dnmt3b in the central nervous system. J Neurosci Res 79:734–746

Feng J, Fouse S, Fan G (2007) Epigenetic regulation of neural gene expression and neuronal function. Pediatr Res 61:58R–63R

Flici H, Erkosar B, Komonyi O, Karatas OF, Laneve P, Giangrande A (2011) Gcm/Glide-dependent conversion into glia depends on neural stem cell age, but not on division, triggering a chromatin signature that is conserved in vertebrate glia. Development 138:4167–4178

Gillardon F, Mack M, Rist W, Schnack C, Lenter M, Hildebrandt T, Hengerer B (2008) MicroRNA and proteome expression profiling in early-symptomatic alpha-synuclein(A30P)-transgenic mice. Proteomics Clin Appl 2:697–705

Gray SG (2010) Targeting histone deacetylases for the treatment of Huntington's disease. CNS Neurosci Ther 16:348–361

Gray SG (2011) Targeting Huntington's disease through histone deacetylases. Clin Epigenetics 2:257–277

Guan JS, Haggarty SJ, Giacometti E, Dannenberg JH, Joseph N, Gao J, Nieland TJ, Zhou Y, Wang X, Mazitschek R, Bradner JE, Depinho RA, Jaenisch R, Tsai LH (2009) HDAC2 negatively regulates memory formation and synaptic plasticity. Nature 459:55–60

Guo H, Ingolia NT, Weissman JS, Bartel DP (2010) Mammalian microRNAs predominantly act to decrease target mRNA levels. Nature 466:835–840

Guy J, Hendrich B, Holmes M, Martin JE, Bird A (2001) A mouse Mecp2-null mutation causes neurological symptoms that mimic Rett syndrome. Nat Genet 27:322–326

Guy J, Gan J, Selfridge J, Cobb S, Bird A (2007) Reversal of neurological defects in a mouse model of Rett syndrome. Science 315:1143–1147

Haberland M, Mokalled MH, Montgomery RL, Olson EN (2009a) Epigenetic control of skull morphogenesis by histone deacetylase 8. Genes Dev 23:1625–1630

Haberland M, Montgomery RL, Olson EN (2009b) The many roles of histone deacetylases in development and physiology: implications for disease and therapy. Nat Rev Genet 10:32–42

Hardy J (2010) Genetic analysis of pathways to Parkinson disease. Neuron 68:201–206

Herman JG, Baylin SB (2003) Gene silencing in cancer in association with promoter hypermethylation. N Engl J Med 349:2042–2054

Hirabayashi Y, Gotoh Y (2010) Epigenetic control of neural precursor cell fate during development. Nat Rev Neurosci 11:377–388

Hsieh J, Eisch AJ (2010) Epigenetics, hippocampal neurogenesis, and neuropsychiatric disorders: unraveling the genome to understand the mind. Neurobiol Dis 39:73–84

Hsieh J, Nakashima K, Kuwabara T, Mejia E, Gage FH (2004) Histone deacetylase inhibition-mediated neuronal differentiation of multipotent adult neural progenitor cells. Proc Natl Acad Sci U S A 101:16659–16664

Ittner LM, Gotz J (2011) Amyloid-beta and tau–a toxic pas de deux in Alzheimer's disease. Nat Rev Neurosci 12:65–72

Jobe EM, Mcquate AL, Zhao X (2012) Crosstalk among epigenetic pathways regulates neurogenesis. Front Neurosci 6:59

Jordan C, Li HH, Kwan HC, Francke U (2007) Cerebellar gene expression profiles of mouse models for Rett syndrome reveal novel MeCP2 targets. BMC Med Genet 8:36

Junn E, Mouradian MM (2012) MicroRNAs in neurodegenerative diseases and their therapeutic potential. Pharmacol Ther 133:142–150

Kanai Y (2008) Alterations of DNA methylation and clinicopathological diversity of human cancers. Pathol Int 58:544–558

Kilgore M, Miller CA, Fass DM, Hennig KM, Haggarty SJ, Sweatt JD, Rumbaugh G (2010) Inhibitors of class 1 histone deacetylases reverse contextual memory deficits in a mouse model of Alzheimer's disease. Neuropsychopharmacology 35:870–880

Kreth FW, Thon N, Tonn JC (2012) Low-grade gliomas. J Neurosurg 116:468–70; author reply 469–70

Kretsovali A, Hadjimichael C, Charmpilas N (2012) Histone deacetylase inhibitors in cell pluripotency, differentiation, and reprogramming. Stem Cells Int 2012:184154

Lee KK, Workman JL (2007) Histone acetyltransferase complexes: one size doesn't fit all. Nat Rev Mol Cell Biol 8:284–295

Li X, Zhao X (2008) Epigenetic regulation of mammalian stem cells. Stem Cells Dev 17: 1043–1052

Lilja T, Heldring N, Hermanson O (2013) Like a rolling histone: epigenetic regulation of neural stem cells and brain development by factors controlling histone acetylation and methylation. Biochim Biophys Acta 1830:2354–2360

Lyst MJ, Ekiert R, Ebert DH, Merusi C, Nowak J, Selfridge J, Guy J, Kastan NR, Robinson ND, De Lima Alves F, Rappsilber J, Greenberg ME, Bird A (2013) Rett syndrome mutations abolish the interaction of MeCP2 with the NCoR/SMRT co-repressor. Nat Neurosci 16:898–902

Macdonald VE, Howe LJ (2009) Histone acetylation: where to go and how to get there. Epigenetics 4:139–143

Macdonald JL, Roskams AJ (2008) Histone deacetylases 1 and 2 are expressed at distinct stages of neuro-glial development. Dev Dyn 237:2256–2267

Majdzadeh N, Wang L, Morrison BE, Bassel-Duby R, Olson EN, D'Mello SR (2008) HDAC4 inhibits cell-cycle progression and protects neurons from cell death. Dev Neurobiol 68: 1076–1092

Makeyev EV, Zhang J, Carrasco MA, Maniatis T (2007) The MicroRNA miR-124 promotes neuronal differentiation by triggering brain-specific alternative pre-mRNA splicing. Mol Cell 27:435–448

Martinez R, Martin-Subero JI, Rohde V, Kirsch M, Alaminos M, Fernandez AF, Ropero S, Schackert G, Esteller M (2009) A microarray-based DNA methylation study of glioblastoma multiforme. Epigenetics 4:255–264

Matsumoto L, Takuma H, Tamaoka A, Kurisaki H, Date H, Tsuji S, Iwata A (2010) CpG demethylation enhances alpha-synuclein expression and affects the pathogenesis of Parkinson's disease. PLoS One 5:e15522

Meissner A, Mikkelsen TS, Gu H, Wernig M, Hanna J, Sivachenko A, Zhang X, Bernstein BE, Nusbaum C, Jaffe DB, Gnirke A, Jaenisch R, Lander ES (2008) Genome-scale DNA methylation maps of pluripotent and differentiated cells. Nature 454:766–770

Meshorer E, Yellajoshula D, George E, Scambler PJ, Brown DT, Misteli T (2006) Hyperdynamic plasticity of chromatin proteins in pluripotent embryonic stem cells. Dev Cell 10:105–116

Mikkelsen TS, Ku M, Jaffe DB, Issac B, Lieberman E, Giannoukos G, Alvarez P, Brockman W, Kim TK, Koche RP, Lee W, Mendenhall E, O'Donovan A, Presser A, Russ C, Xie X, Meissner A, Wernig M, Jaenisch R, Nusbaum C, Lander ES, Bernstein BE (2007) Genome-wide maps of chromatin state in pluripotent and lineage-committed cells. Nature 448:553–560

Minucci S, Pelicci PG (2006) Histone deacetylase inhibitors and the promise of epigenetic (and more) treatments for cancer. Nat Rev Cancer 6:38–51

Miyake K, Yang C, Minakuchi Y, Ohori K, Soutome M, Hirasawa T, Kazuki Y, Adachi N, Suzuki S, Itoh M, Goto YI, Andoh T, Kurosawa H, Oshimura M, Sasaki M, Toyoda A, Kubota T (2013) Comparison of genomic and epigenomic expression in monozygotic twins discordant for Rett syndrome. PLoS One 8:e66729

Mizutani K, Yoon K, Dang L, Tokunaga A, Gaiano N (2007) Differential Notch signalling distinguishes neural stem cells from intermediate progenitors. Nature 449:351–355

Mohamed Ariff I, Mitra A, Basu A (2012) Epigenetic regulation of self-renewal and fate determination in neural stem cells. J Neurosci Res 90:529–539

Mohn F, Weber M, Rebhan M, Roloff TC, Richter J, Stadler MB, Bibel M, Schubeler D (2008) Lineage-specific polycomb targets and de novo DNA methylation define restriction and potential of neuronal progenitors. Mol Cell 30:755–766

Montgomery RL, Hsieh J, Barbosa AC, Richardson JA, Olson EN (2009) Histone deacetylases 1 and 2 control the progression of neural precursors to neurons during brain development. Proc Natl Acad Sci U S A 106:7876–7881

Monti B, Gatta V, Piretti F, Raffaelli SS, Virgili M, Contestabile A (2010) Valproic acid is neuroprotective in the rotenone rat model of Parkinson's disease: involvement of alpha-synuclein. Neurotox Res 17:130–141

Mosammaparast N, Shi Y (2010) Reversal of histone methylation: biochemical and molecular mechanisms of histone demethylases. Annu Rev Biochem 79:155–179

Nagarajan RP, Costello JF (2009) Epigenetic mechanisms in glioblastoma multiforme. Semin Cancer Biol 19:188–197

Namihira M, Kohyama J, Abematsu M, Nakashima K (2008) Epigenetic mechanisms regulating fate specification of neural stem cells. Philos Trans R Soc Lond B Biol Sci 363:2099–2109

Neul JL, Fang P, Barrish J, Lane J, Caeg EB, Smith EO, Zoghbi H, Percy A, Glaze DG (2008) Specific mutations in methyl-CpG-binding protein 2 confer different severity in Rett syndrome. Neurology 70:1313–1321

Ng RK, Gurdon JB (2008) Epigenetic inheritance of cell differentiation status. Cell Cycle 7:1173–1177

Nuytemans K, Theuns J, Cruts M, Van Broeckhoven C (2010) Genetic etiology of Parkinson disease associated with mutations in the SNCA, PARK2, PINK1, PARK7, and LRRK2 genes: a mutation update. Hum Mutat 31:763–780

Olynik BM, Rastegar M (2012) The genetic and epigenetic journey of embryonic stem cells into mature neural cells. Front Genet 3:81

Oury F, Rijli FM (2007) [Hoxa2: a key gene for the facial somatosensory map]. Med Sci (Paris) 23:247–9

Pasini D, Hansen KH, Christensen J, Agger K, Cloos PA, Helin K (2008) Coordinated regulation of transcriptional repression by the RBP2 H3K4 demethylase and Polycomb-Repressive Complex 2. Genes Dev 22:1345–1355

Poulsen P, Esteller M, Vaag A, Fraga MF (2007) The epigenetic basis of twin discordance in age-related diseases. Pediatr Res 61:38R–42R

Ramirez A, Heimbach A, Grundemann J, Stiller B, Hampshire D, Cid LP, Goebel I, Mubaidin AF, Wriekat AL, Roeper J, Al-Din A, Hillmer AM, Karsak M, Liss B, Woods CG, Behrens MI, Kubisch C (2006) Hereditary parkinsonism with dementia is caused by mutations in ATP13A2, encoding a lysosomal type 5 P-type ATPase. Nat Genet 38:1184–1191

Rett A (1986) Rett syndrome. History and general overview. Am J Med Genet Suppl 1:21–25

Ringrose L, Paro R (2007) Polycomb/Trithorax response elements and epigenetic memory of cell identity. Development 134:223–232

Robertson KD (2001) DNA methylation, methyltransferases, and cancer. Oncogene 20: 3139–3155

Roth TL, Lubin FD, Sodhi M, Kleinman JE (2009) Epigenetic mechanisms in schizophrenia. Biochim Biophys Acta 1790:869–877

Rubinsztein DC (2006) The roles of intracellular protein-degradation pathways in neurodegeneration. Nature 443:780–786

Samaco RC, Neul JL (2011) Complexities of Rett syndrome and MeCP2. J Neurosci 31: 7951–7959

Sawa A, Snyder SH (2002) Schizophrenia: diverse approaches to a complex disease. Science 296:692–695

Schneider JW, Gao Z, Li S, Farooqi M, Tang TS, Bezprozvanny I, Frantz DE, Hsieh J (2008) Small-molecule activation of neuronal cell fate. Nat Chem Biol 4:408–410

Schuettengruber B, Chourrout D, Vervoort M, Leblanc B, Cavalli G (2007) Genome regulation by polycomb and trithorax proteins. Cell 128:735–745

Scott CE, Wynn SL, Sesay A, Cruz C, Cheung M, Gomez Gaviro MV, Booth S, Gao B, Cheah KS, Lovell-Badge R, Briscoe J (2010) SOX9 induces and maintains neural stem cells. Nat Neurosci 13:1181–1189

Sikorska M, Sandhu JK, Deb-Rinker P, Jezierski A, Leblanc J, Charlebois C, Ribecco-Lutkiewicz M, Bani-Yaghoub M, Walker PR (2008) Epigenetic modifications of SOX2 enhancers, SRR1 and SRR2, correlate with in vitro neural differentiation. J Neurosci Res 86:1680–1693

Silva PN, Gigek CO, Leal MF, Bertolucci PH, de Labio RW, Payão SL, Smith Mde A (2008) Promoter methylation analysis of SIRT3, SMARCA5, HTERT and CDH1 genes in aging and Alzheimer's disease. J Alzheimers Dis 13:173–176

Soshnikova N, Duboule D (2008) Epigenetic regulation of Hox gene activation: the waltz of methyls. Bioessays 30:199–202

Steffan JS, Kazantsev A, Spasic-Boskovic O, Greenwald M, Zhu YZ, Gohler H, Wanker EE, Bates GP, Housman DE, Thompson LM (2000) The Huntington's disease protein interacts with p53 and CREB-binding protein and represses transcription. Proc Natl Acad Sci U S A 97: 6763–6768

Sun G, Yu RT, Evans RM, Shi Y (2007) Orphan nuclear receptor TLX recruits histone deacetylases to repress transcription and regulate neural stem cell proliferation. Proc Natl Acad Sci U S A 104:15282–15287

Thomas B, Beal MF (2011) Molecular insights into Parkinson's disease. F1000 Med Rep 3:7
Tudor M, Akbarian S, Chen RZ, Jaenisch R (2002) Transcriptional profiling of a mouse model for Rett syndrome reveals subtle transcriptional changes in the brain. Proc Natl Acad Sci U S A 99:15536–15541
Urdinguio RG, Sanchez-Mut JV, Esteller M (2009) Epigenetic mechanisms in neurological diseases: genes, syndromes, and therapies. Lancet Neurol 8:1056–1072
Van Esch H, Bauters M, Ignatius J, Jansen M, Raynaud M, Hollanders K, Lugtenberg D, Bienvenu T, Jensen LR, Gecz J, Moraine C, Marynen P, Fryns JP, Froyen G (2005) Duplication of the MECP2 region is a frequent cause of severe mental retardation and progressive neurological symptoms in males. Am J Hum Genet 77:442–453
Verdin E, Hirschey MD, Finley LW, HAIGIS MC (2010) Sirtuin regulation of mitochondria: energy production, apoptosis, and signaling. Trends Biochem Sci 35:669–675
Walker FO (2007) Huntington's disease. Lancet 369:218–228
Watanabe D, Uchiyama K, Hanaoka K (2006) Transition of mouse de novo methyltransferases expression from Dnmt3b to Dnmt3a during neural progenitor cell development. Neuroscience 142:727–737
Weber M, Hellmann I, Stadler MB, Ramos L, Paabo S, Rebhan M, Schubeler D (2007) Distribution, silencing potential and evolutionary impact of promoter DNA methylation in the human genome. Nat Genet 39:457–466
Wu Z, Huang K, Yu J, Le T, Namihira M, Liu Y, Zhang J, Xue Z, Cheng L, Fan G (2012) Dnmt3a regulates both proliferation and differentiation of mouse neural stem cells. J Neurosci Res 90:1883–1891
Xu WS, Parmigiani RB, Marks PA (2007) Histone deacetylase inhibitors: molecular mechanisms of action. Oncogene 26:5541–5552
Yang XJ, Seto E (2008) The Rpd3/Hda1 family of lysine deacetylases: from bacteria and yeast to mice and men. Nat Rev Mol Cell Biol 9:206–218
Yazdani M, Deogracias R, Guy J, Poot RA, Bird A, Barde YA (2012) Disease modeling using embryonic stem cells: MeCP2 regulates nuclear size and RNA synthesis in neurons. Stem Cells 30:2128–2139
Zachariah RM, Rastegar M (2012) Linking epigenetics to human disease and Rett syndrome: the emerging novel and challenging concepts in MeCP2 research. Neural Plast 2012:415825
Zhao C, Sun G, Li S, Shi Y (2009) A feedback regulatory loop involving microRNA-9 and nuclear receptor TLX in neural stem cell fate determination. Nat Struct Mol Biol 16:365–371
Zhao C, Sun G, Li S, Lang MF, Yang S, Li W, Shi Y (2010) MicroRNA let-7b regulates neural stem cell proliferation and differentiation by targeting nuclear receptor TLX signaling. Proc Natl Acad Sci U S A 107:1876–1881

Chapter 3
An Overview of Epigenetic Mechanisms in Health and Disease

Claire Westerland and Tom C. Karagiannis

Abstract Epigenetic modifications are emerging as key players in not only the regulation of normal genomic expression patterns, but also their role in disease progression due to their deregulation. An irregular change in DNA methylation, either through aberrant hypermethylation or hypomethylation, may have serious consequences relating to carcinogenesis by potentiating or preventing gene functions. Conversely, the reversal of abnormal hypermethylation has been aided by the advent of DNMT inhibitors to restore tumour suppressor gene function. In addition, the field of histone post-translational modifications (PTMs) is expanding. The best characterised and seemingly most involved in pathogenesis remain acetylation and methylation of amino acid residues within N-terminal tails. As the complexity of the epigenetic language becomes more apparent, it has been found that there is significant crosstalk between modifications, including DNA methylation and histone PTMs. Great promise lies in the development of histone deacetylase inhibiting compounds, which initiate a vast number of effects normalising or eradicating tumour cells.

Keywords Epigenetic modifications • Chromatin remodelling • Acetylation • Methylation • DNA regulation • Histone deacetylases • Histone deacetylase inhibitors

---

C. Westerland
Epigenomic Medicine, Baker IDI Heart and Diabetes Institute, The Alfred Medical Research and Education Precinct, 75 Commercial Road, Melbourne, VIC, Australia

T.C. Karagiannis (✉)
Epigenomic Medicine, Baker IDI Heart and Diabetes Institute, The Alfred Medical Research and Education Precinct, 75 Commercial Road, Melbourne, VIC, Australia

Department of Pathology, The University of Melbourne, Parkville, VIC, Australia
e-mail: tom.karagiannis@bakeridi.edu.au

N. Maulik and T. Karagiannis (eds.), *Molecular Mechanisms and Physiology of Disease: Implications for Epigenetics and Health*, DOI 10.1007/978-1-4939-0706-9_3,

3.1 Introduction

The term epigenetics, coined by Waddington in 1942, is used to describe reversible alterations in gene expression or cellular phenotype caused by modifications other than changes in the underlying DNA sequence and permit inheritance of expression patterns. Epigenetic modifications are laid down in response to environmental or internal stimuli, and help to regulate cellular functions, allowing cell types of different phenotypes to develop from an identical genome. Packaging of DNA is intricate and complex given the importance of folding metres of nucleotide sequences into the nucleus whilst allowing the accessibility of enzymes required for replication and transcription to come into contact with the DNA. In eukaryotes, this is solved by the presence of histones, folded protein chains assembled into a globular nucleosome. The nucleosome contains an octamer of the four pairs of core histones H2A, H2B, H3 and H4 wrapped in 147 base pairs of the right-handed DNA helix, coiled around 1.65 times in a left-handed superhelix (Kouzarides 2007; Luger and Mader 1997; Bishop 2008). Nucleosomes are separated by stretches of linker DNA of varied length, dependent on tissue type and species (Woodcock et al. 2006). Core histones have a globular C-terminal, whilst basic N-terminal tails rich in arginine and lysine project from the core histones, contributing to their affinity for DNA and acting as the target of covalent, post-translational modifications (PTMs) (Brent et al. 2004). The complex involving DNA wrapped around nucleosomes is referred to as chromatin and the tightness of its compaction permits or restricts access to DNA. The primary structure of chromatin is a 10 nm fibre and is further compacted by stacking nucleosomes to form the 30 nm diameter secondary structure (Szerlong and Hansen 2011). Various models have been proposed for the secondary structure, including the zigzag two-start helical ribbon (Bednar et al. 1998), two-start crossed-linker (Williams et al. 1986) and one-start solenoid (Finch and Klug 1976; Notbohm et al. 1979) although there is evidence that more than one form may be present simultaneously (Grigoryev et al. 2011). Epigenetic modifications remodel chromatin and can provide binding sites for other molecules, by histone structural variants, DNA methylation, non-coding RNA and PTMs of histones—which may alter chromatin tightness or be linked to events such as DNA repair (Paull et al. 2000). Post-translational histone modifications include phosphorylation, acetylation, sumoylation, ADP-ribosylation, methylation and ubiquitylation. Due to the heritable nature of epigenetic alterations, changes in DNA methylation, histone methylation and histone acetylation have all been implicated in cancer development, and so will be the focus of this chapter.

3.2 DNA Methylation

DNA methylation is governed by DNA-methyltransferase (DNMT) enzymes utilising an S-adenosylmethionine as a methyl-donor at the fifth carbon within a cytosine base (Goll and Bestor 2005), via a "base flipping" mechanism where the cytosine is moved into an extrahelical position, providing the DNMT access to the base (Huang and MacKerell 2004). For a cytosine to be susceptible to this alteration, it must be present within a CpG dinucleotide which is symmetrical in double-stranded DNA allowing the epigenetic mark to be maintained through DNA replication. DNMT1 is a critical enzyme in this process preferentially binding to hemi-methylated DNA and methylating the other DNA strand (Hermann et al. 2004b). Although renowned for being one of the most stable epigenetic marks, in germ cell development and embryonic differentiation DNA it undergoes active demethylation (Reik et al. 2001; Smith et al. 2012). Cytosine methylation is utilised in the processes of imprinting (Smrzka et al. 1995), X-inactivation (Sharp et al. 2011), tissue-specific cell differentiation (De Smet et al. 1999), silencing of repetitive sequences such as retrotransposons and in tumour development (Smallwood and Kelsey 2011).

CpG dinucleotides can occur within the genome as CpG islands or isolated CpGs and are methylated via different processes. CpG islands (CGI) are 5′ sequences longer than 500 bp, with a CpG content greater than 55 % (Takai and Jones 2002) and are located close to 76 % of promoters within the human genome (Davuluri et al. 2001). Promoter CGIs are associated with nucleosome-free regions and aren't methylated, allowing access for transcriptional enzymes to the DNA (Choi 2010). Most CGI in promoters are resistant to de novo methylation during development, whether expressed or not (Bestor et al. 1992), with less than 20 % methylation (Eckhardt et al. 2006; Weber et al. 2007). CGIs that undergo methylation become transcriptionally silenced (Bird 2002) as methylation blocks the binding of several transcription factors which require contact with the major groove of DNA. CGIs also recruit histone deacetylases (HDACs) or enlist methyl-binding proteins which modulate chromatin resulting in tighter, transcriptionally inactive structures (Bird and Wolffe 1999; Le Guezennec et al. 2006).

Isolated CpGs mainly occur in regions of repeats and transposable sequences (Wilson et al. 2007) which are highly susceptible to cytosine methylation. Eighty percent of these CpGs are found to be methylated in both humans and mice (Antequera 2003) as CpG-poor regions are predisposed to DNA methylation (Bird 2002) due to de novo methylation of nucleosomal DNA by DNMT3b (Hermann et al. 2004a). Hypomethylation of normally methylated sequences, such as those controlling transcription, can allow their expression and subsequently lead to cancer development (Ehrlich 2002). Over 45 % of the genome is composed of retrotransposons, such as LINEs (Smit and Riggs 1996), which have the potential to alter gene sequence and function by inserting themselves (Chen et al. 2012; Muotri et al. 2010). These repetitive sequences are constitutively methylated in normal somatic cells (Steinhoff and Schulz 2004) with more than 90 % of methylated CpGs existing in retrotransposons (Yu et al. 2001), upholding closed chromatin which represses transcription and prevents them from abnormal insertions (Miriam and Mark 2010). Cancer cells have been observed to develop hypomethylated transposons, permitting their transcription and leading to harmful translocation and recombination of chromosomes (Rodriguez et al. 2006; Delgado-Cruzata et al. 2012).

Profiling of cancer methylation patterns show that some TSGs are methylated in multiple cancers, such as $p16^{INK4a}$, responsible for inhibiting cell cycle progression, which is methylated in prostate, bladder, breast, renal and colon cancers (Herman et al. 1995; Gonzalez-Zulueta et al. 1995). Other TSG promoter methylation is seen only in tumours of certain tissues (Esteller et al. 2001), for example CDH1 promoter hypermethylation downregulates E-cadherin in oesophageal cancer, allowing metastasis and invasion (Corn et al. 2001). Similarly, ICAM-1 codes an adhesion molecule and silencing allows colorectal cancer cells to break free and metastasize to the liver (Tachimori et al. 2005). The tumour suppressor gene APC is found to be methylated in hepatocellular carcinoma in 81.7 % of cases with no methylation occurring in the controls (Lee et al. 2003) allowing for a highly accurate marker for tumours. The specific patterns of promoter methylation may allow tumours to be diagnosed as previous studies have shown hypermethylation in these regions in precancerous tumours as an early stage of oncogenesis (Esteller et al. 2000b). In addition, serum levels could potentially be used to calculate the risk of cancers (Esteller et al. 1999; Laird 2003), the stage of tumour progression, metastatic risk or sensitivity to chemotherapeutic agents (Strathdee et al. 1999; Esteller et al. 2000a).

Paradoxically, global hypomethylation is a common feature of cancers also (Choi et al. 2009; Feinberg et al. 2006; Feinberg and Vogelstein 1983), due to activation of oncogenes and instability of both chromosomes and the genome. Lowered global methylation can be associated with a large increase in the risk of developing breast cancer due to genomic instability (Choi et al. 2009), even in families at high risk of developing the disease. Satellite 2 (Sat2) is a specific repetitive sequence normally methylated, but found to be demethylated in many breast tumours (Wu et al. 2012), and these changes, both globally and gene specific, offer the hope of a potential, non-invasive biomarker test (Delgado-Cruzata et al. 2012).

Treatment with nucleoside-analogue DNMT inhibitors such as 5-aza-cytidine (5-AC, azacitidine, Vidaza) and 5-aza-2′-deoxcycitidine (5-AZA-Cde, decitabine, Dacogen) can reverse hypermethylation at promoters and subsequently reactivate transcription of TSGs by causing the removal of nucleosomes. Both of these drugs were first approved by the FDA for the treatment of myelodysplastic syndromes (Kantarjian et al. 2006; Silverman et al. 2002), precursors for leukaemia (Ziemba et al. 2011) and are now in use for the treatment of acute myelogenous leukaemia (AML) (Marks 2012; Francesca et al. 2011). Murine studies indicated that 5-AZA-Cdr is more potent than 5-CR in the treatment of leukaemia, resulting in a 647 % increase in survival, compared to 115 % for the same dose of 5-CR (Momparler et al. 1984). Vulnerability to each drug depends on the nucleoside transporters present, as each drug has different transporter profiles (Damaraju et al. 2012). Further explanation for the differing clinical effectiveness of azacitidine and decitabine lies in that only azacitidine prevents protein synthesis, whereas decitabine is effective at inducing DNA damage, reducing DNMT1 presence and hypomethylating DNA in concentrations 2–10 times lower than azacitidine (Hollenbach et al. 2004).

5-AC is converted intracellularly to its active form, azacytidine triphosphate (Glover and Leyland-Jones 1987; Leone et al. 2002) and is incorporated into RNA and DNA in a 65:35 ratio (Hollenbach et al. 2004). The drug inhibits DNMT1 to cause global hypomethylation, directly induces DNA damage, and its incorporation into RNA interferes with the synthesis of proteins and DNA. Anti-neoplastic effects in myelodysplasic syndromes are due to hypomethylation of promoters reactivating TSGs to allow differentiation, downregulation of hematopoiesis-inducing cytokines and also apoptotic death of rapidly diving cells, including those of myeloid origin (Keating 2011). 5-AC has significant cytotoxicity due to incorporation into nucleic acids, so patients may experience mild side effects associated with its usage, as well as reactions at the injection site (Keating 2011). It is effective in about half (49 %) of patients, improving their quality of life and rates of remission (Silverman et al. 1993), as well as their psychological well-being (Silverman et al. 2002). In a trial comparing subcutaneous 5-AC with conventional treatment such as chemotherapy, 5-AC prolonged survival by a median of 9.4 months (Keating 2011). An esterised nucleoside derivative, 2′, 3′, 5′-triacetyl-5-azacitidine (TAC), has shown promise in trials as a more soluble, stable and bioavailable prodrug form, with lower toxicity and can be taken orally (Ziemba et al. 2011) as both 5-AC and 5-AZA-Cde have short half-lives (Lin et al. 1981; Notari and DeYoung 1975).

5-AZA-Cde's antitumorigenic effects are through a combination of promoting apoptosis and inhibiting the activity of cancer cells, strengthening tight junctions to disrupt invasion (Shin et al. 2012). It significantly depletes the amount of DNMT1 available within eight hours of a single dose (Al-Salihi et al. 2011), to cause global hypomethylation and prevent promoter hypermethylation (Alcazar et al. 2012). At lower concentrations than those required to trigger apoptosis in cancer cells, 5-AZA-Cde promotes differentiation, cell cycle arrest in G_2/M or cycle exit and inhibits proliferation (Qian-Ying et al. 2012; Alcazar et al. 2012). Through promoter demethylation, apoptosis is caused by the downregulation of anti-apoptotic proteins such as Bcl-2, production of reactive oxygen species (ROS), caspase activation (Shin et al. 2012) and also by upregulation of TRAIL (tumour necrosis factor-related apoptosis-inducing ligand), to activate the extrinsic apoptotic pathway (Matias et al. 2013).

3.3 Histone Methylation

In histones, methylation occurs on the amino acid residues; lysine and arginine on the histone tails. Complexity is added by the number of methyl groups that can be attached to these sites—lysine can be mono-, di- or tri-methylated, and arginine can be mono- or di-methylated (Ng et al. 2009). The methyl donor is an S-adenosylmethionine (AdoMet), from which a methyl group is transferred to a lysine's ε-amino (NH_2) group, or arginine's ω- or δ-amino groups. Methylation at both sites can inhibit or activate transcription depending on the context of protein and specific residue within the sequence modified. A variety of domains recognise methylation of lysine or arginine, including the tudor domains, chromodomains, PHD fingers, Ankyrin repeats, WD40 repeats, HEAT domains, PWWP domains and MBT domains, with PHD also able to bind to unmodified lysines (Gardner et al. 2011).

Lysine residues are methylated by histone lysine methyltransferases (KMTs), all but one containing the conserved, catalytic SET domain of 130 amino acids (Zhang and Bruice 2008; Black et al. 2012). SET domain containing enzymes are grouped based on the sequences surrounding it (Xiaodong et al. 2005). KMTs transfer between one and three methyl groups to a lysine residue using cofactor AdoMet as a donor (Guo and Guo 2007), with the cofactor residing on one side of the SET domain and the substrate on the other. A lysine channel brings the side chain into contact with AdoMet, and this channel allows the lysine to be brought into contact with a tyrosine with hydrogen bond formation allowing transfer of the methyl group. Stearic properties of the channel determine how many methyl groups may be added, as a phenylalanine in place of a tyrosine permits larger, already methylated substrate lysines to pass through, allowing further methylation (Bing et al. 2003).

The first of the KMTs to be discovered was the SET domain containing SUV39H1 (KMT1A), which methylates lysine 9 of H3 (Rea et al. 2000), and provided the opportunity to search for other KMTs with homology to the enzymatic (SET) domain. Some evidence for interplay between PTMs lies in the presence within SUV39 family sequences of reader domains for other modifications, with some members harbouring chromodomains or MBDs, for recognition of other methylated lysines or methylated DNA, respectively (Völkel and Angrand 2007). The second class of KMTs, smaller enzymes with a structurally distinct catalytic domain, is so far represented by a single enzyme, KMT4 or Dot1L in humans (Black et al. 2012). The KMTs have high substrate and methylation state specificity, generally binding only one specific lysine within a tail. This specificity is crucial because methylation states provide additional information—for example trimethylation of lysine 27 on histone H3 is associated with X-inactivation, but not dimethylation (Collins et al. 2005).

Methylation is a reversible modification, and the activity of KMTs is opposed by the lysine demethylases (KDMs), which are divided into two families. All KMTs require Adomet as a cofactor and generate formaldehyde and carbon dioxide as by-products of the reaction. The amine oxidases are exemplified by LSD1 (KDM1A), often found in corepressor complexes, which demethylates using an oxidation reaction (Shi et al. 2004). FAD as a necessary cofactor in this class (Black et al. 2012). LSD1's function is dependent on its binding partners, for example activating when complexed with the androgen receptor, or repressing transcription when bound to CoRest (Cloos et al. 2008). The Jumonji domain proteins demethylate via oxidation, and are dependent on α-ketoglutarate, O_2 and Fe(II) (Cloos et al. 2008). A third group of enzymes, the LOX family, demethylates lysines by deamination, an area that must be further investigated (Black et al. 2012).

Arginine residues are methylated at their guanidino group on the terminal nitrogen atom (Bedford 2007) to form monomethylarginine (MMA), symmetric dimethylarginine (SDMA) or asymmetric dimethylarginine (ADMA) (Turner 2005). Each methylation at the guanidino group removes a hydrogen bond donor and changes the arginine's potential-binding partners (Bedford and Clarke 2009b) and argentines' that have been methylated become more hydrophobic (Kleinschmidt et al. 2008). SDMA is a repressive transcriptional mark, whereas ADMA has a role in transcriptional activation, and methyl marks also act in DNA repair, RNA processing and imprinting (Kleinschmidt et al. 2008). With much less specificity than KMTs, protein arginine N-methyltransferases (PRMTs) commonly have multiple arginine targets

(Bedford and Clarke 2009b), the majority of substrates involved with RNA (Bedford and Clarke 2009a). PRMT recruitment to promoters is mediated by coactivators, to modify histones and other protein substrates within chromatin (Kleinschmidt et al. 2008). There are 11 PRMTs found in humans, distributed between three types, containing an AdoMet-binding pocket within a conserved catalytic domain, and the total enzyme ranging from 316 to 956 amino acids in length (Wolf 2009). Type I and II PRMTs first catalyse formation of MMA as an intermediate, and then further methylate it to form ADMA and SDMA, respectively, whereas type III forms a monomethylated residue (Di Lorenzo and Bedford 2011). Tudor domains bind and recognise arginines that are symmetrically dimethylated (Cote and Richard 2005). Type IV enzymes catalyse the methylation of the internal δ nitrogen, but this type of activity has so far only been discovered in yeast (Bedford and Clarke 2009b).

PRMT1 is responsible for 84 % of total cellular protein arginine methyltransferase activity, and 54 % of asymmetric arginine methyltransferase activity (Pal and Sif 2007), expressed in all tissues but found at highest levels in neuronal tissues of embryos where it is involved in neuronal differentiation (Cimato et al. 2002). It regulates cellular location of proteins (Pal and Sif 2007) and despite a wide range of substrates shows a preference for arginines flanked by glycines (Bedford and Clarke 2009b). It is expressed at equal levels across all differentiated cell types, yet relative levels of splice-variant isoforms vary between regular and cancerous breast tissues, providing hope for a new diagnostic tool (Wolf 2009).

PRMT4, also known as CARM1 (coactivator-associated arginine methyltransferase 1), acts in conjunction with PRMT1 and two other coactivator classes, the histone acetyltransferase (HAT) p300/CBP and p160 family, to enhance transcription, although PRMT1 and CARM1 also have repressive roles, and knockdown studies show that their functions partially overlap (Kleinschmidt et al. 2008). It is also a coactivator for transcription factors including p53, and nuclear receptors including the androgen receptors (Shia et al. 2012). CARM1 aids transcription by recruiting other proteins after nuclear receptor binding and recruits RNA splicing factors (Di Lorenzo and Bedford 2011).

PRMT5 has involvement with the Janus tyrosine kinase indicating a role in signal transduction pathways activated by cytokines, and was the first type II enzyme discovered and main type II PRMT in mammals (Di Lorenzo and Bedford 2011). PRMT6 is the smallest of the PRMT family at 316aa and is involved in cellular immunity against HIV-1 by downregulating expression of viral genes. A gene duplication event in PRMT7 leads to possession of two AdoMet-binding cores, but it cannot function without both domains. It has been found to have both type II and III activity, but is dependant on the substrate (Di Lorenzo and Bedford 2011). PRMT8 has 80 % homology to PRMT1, the closest relationship between members of the PRMT family, causing some overlap in substrate specificity. PRMT7 is notable for being the only membrane-bound PRMT (Bedford and Clarke 2009b).

Unmodified arginine residues can be converted into citrulline by a deimination reaction by the peptidylarginine deiminase family (PADIs), whose main substrates include the core histones (Bedford and Clarke 2009a). No enzyme has been found so far that can reverse arginine deimination or methylation, but methylation prevents their deimination, and deimination prevents their methylation—antagonistic activities but not true demethylation (Di Lorenzo and Bedford 2011). Jmjd6 was a purported arginine demethylase, but more recent studies have brought this into question, finding only lysine oxidase activity (Di Lorenzo and Bedford 2011).

3.3.1 Aberrant Histone Methylation in Cancer

Both lysine and arginine methylation are commonly dysregulated in cancers, due to their involvement in transcriptional regulation. PRMT5 overexpression corepresses E-cadherin, allowing epithelial to mesenchymal transition, a key step in metastasis. This role is seen in its abnormally high levels in leukaemia, lymphoma and gastric cancer (Bedford and Clarke 2009a). An isoform of leukaemic fusion protein AML-ETO has some histone methyltransferase activity due to recruitment of PRMT1, which activates cancer-promoting genes via methylation of arginine 3 in H4 in promoters (Shia et al. 2012). CARM1 aids proliferation of breast cancer cells in response to oestrogen, as well as showing elevated levels in castration-resistant lines of prostate cancer (Bedford and Clarke 2009a). LSD1 can function as either a tumour suppressor or an oncogene, dependent on the complex it is involved in, so may be downregulated or overexpressed to carcinogenic effect (DeCarlo and Hadden 2012). Leukaemic fusion proteins resulting from an MLL translocation commonly recruit Dot1L, to encourage proliferation and loss of differentiation in lung tumours (Kim et al. 2011).

3.4 Histone Acetylation

Histone acetylation is a dynamic and reversible process with the level of acetylation reliant on the relative levels of activity of HATs and antagonistic HDACs. This equilibrium regulates cellular activities such as proliferation, differentiation and senescence, and its deregulation is a key factor in pathogenesis.

Specific ε-amino group lysine residues in the N-terminal tails of core histones are acetylated by HATs, neutralising the positive charge of the residue and decreasing the overall electrostatic attraction between nucleosome and DNA, as well as hindering the hydrogen bonding ability of the lysine. Acetylation therefore decreases the level of contact between histones and DNA, as well as altering interaction of histones with other histones in neighbouring nucleosomes, and histone interaction with other chromatin proteins (Lacoste and Coté 2003; Hong et al. 1993). This condenses the chromatin structure, permitting access to DNA by transcription factor complexes and enzymes, and also makes the chromatin more flexible. Strong evidence for the role of lysine acetylation in transcriptional regulation was provided by the early observation that chromatin in actively transcribed genes is normally hyperacetylated at histones, whilst hypoacetylated in silenced regions (Allfrey et al. 1964).

Recent study has expanded the known protein substrates to include the oncoprotein c-MYC, tumour suppressor p53 and HATs themselves in autophosphorylation (Jiang et al. 2012; Sapountzi and Coté 2010) as well as other proteins residing in the nucleus and cytoplasm. Roles for histone acetylation are not limited to transcriptional regulation, as acetylation of lysines 5 and 12 in H4 is necessary to lay down new core histones and assemble nucleosomes (Yang 2004), but it is most commonly associated with transcriptional activation.

The distinction between type A and B HATs lies in their cellular localisation—type A HATs reside in the nucleus, acetylating substrates including core histones and proteins associated with chromatin, whereas type B HATs are found in the cytoplasm where they aid newly synthesised histones' transit into the nucleus via acetylation (Schrump 2009; Roth et al. 2001). The superfamilies of type A HATs are grouped according to homology of domains: MYST, GNAT (Gcn5-related N-acetyltransferase) and p300/CBP (Sterner and Berger 2000). Bromodomains recognise acetylation on histone tails, and these may be part of the structure of complexes containing HDACs, HATs, transcription factors or proteins maintaining chromatin compaction (Wang et al. 2012).

Human members of the MYST family are hMOF, HBO1, MOZ, MORF and TIP60, and all possess greatly homologous catalytic domains (Sapountzi and Coté 2010). TIP60 and HMOF contain chromodomains for recognition of acetylated lysine, whereas MOZ and MORF have protein-binding plant homeodomains (PHD). TIP60 acetylates H4 and transcription factors, and can either act with other enzymes to co-activate (Brady et al. 1999) or co-repress transcription (Nordentoft and Jorgensen 2003) in a gene-dependent manner.

p300 and CBP are functionally redundant tumour suppressors, with high sequence homology (Roth et al. 2001). They acetylate the core histones and transcription factors, targeting genes for active transcription (Martinez-Balbas et al. 1998) showing very similar distributions to promoters—out of 8,707 associated with one or both enzymes there were 222 only associated with p300, and 2,747 with only CBP (Wang et al. 2009).

The GNAT family in humans comprises PCAF and GCN5. Both enzymes require the presence of coenzyme acetyl coA to prevent their inactivation, and will acetylate linker histone H1 only if free from chromatin complexes (Herrera et al. 1997). PCAF (p300/CBP-binding protein-associated factor) derives its name from its interaction with p300/CBP (Sterner and Berger 2000), and autoacetylation localises it to the nucleus. PCAF is found in the PCAF complex, which has a preference for residues within H3 (Turner 2000).

Histone acetylation is a dynamic and reversible process, the level of acetylation reliant on relative levels of activity of HATs and antagonistic HDACs, and the balance vital to regulated development and aberrant activity a key factor in pathogenesis. HDACs remove acetyl groups from lysine residues to restore their positive charge and leaving DNA more tightly wound around core histones, thus less accessible to DNA-binding factors. HDACs are unable to bind to DNA directly, and so must form complexes to allow them to bind and alter DNA, available binding partners playing a key role in expression patterns for a given cell type (Haberland et al. 2009).

Table 3.1 Members of classical HDAC family

Class	Coenzyme	Size (kDal)	Members	Yeast homologue
I	Zn^{+}	22–25	HDACs 1, 2, 3, 8	Rpd3
IIa	Zn^{+}	120–135	HDACs 4, 5, 7, 9	HDAI
IIb	Zn^{+}	120–135	HDACs 6, 10	HDAI
IV	Zn^{+}	120–135	HDAC 11	HDAI

Table 3.2 Classification of the Sirtuins, non-metal-dependent HDAC enzymes

Class	Coenzyme	Size (kDal)	Location	Members	Yeast homologue
I	NAD^{+}	4–50	Nucleus, cytoplasm (1 and 2 only), mitochondria (3 only found here)	SIRT1, SIRT2, SIRT3	Sir2
II	NAD^{+}	4–50	Mitochondria	SIRT4	Sir2
III	NAD^{+}	4–50	Mitochondria	SIRT5	Sir2
IV	NAD^{+}	4–50	Nucleus	SIRT6, SIRT7	Sir2

The 18 characterised human types are divided into classes I, II, III and IV, based on comparison to yeast HDACs (Table 3.1). Classes I, II and IV are the metal-dependent "classical" HDAC family, with a Zn^{2+} cofactor necessary for their active site function (Atadja 2011). The class III enzymes are known as the sirtuins due to similarity to yeast enzyme Sirt2, require cofactor nicotinamide adenine dinucleotide (NAD^{+}) to function, and have little homology to the other HDACs (Atadja 2011) (Table 3.2).

Class I HDACs (HDACs 1, 2, 3 and 8) have a highly conserved N-terminal deacetylase domain and only short N- and C-terminal domains. Their varied roles are seen in their association with both transcriptionally active and inactive chromatin (Wang et al. 2009). Non-histone targets include transcription factors and AMP-activated protein kinase (AMPK), a protein complex involved in cellular metabolic stress pathways (Lin et al. 2012b). All class I HDACs so far have been determined to be part of repressor complexes with the exception of HDAC8, which is found in the cytoplasm of smooth muscle cells, allowing it to acetylate non-nuclear proteins and aid muscle contraction (Witt et al. 2009; Wolfson et al. 2013).

Both HDAC1 and HDAC2 are found in the same three chromatin remodelling complexes, the nucleosome remodelling and deacetylating complex (NuRD), Sin3 and corepressor for element-1 silencing transcription factor (CoREST). The CoRest complex is responsible for the inactivation of neuronal genes in non-neuronal tissues, where their expression would be redundant or even harmful (Ballas et al. 2001). Sin3 is recruited by methylated DNA-binding protein MeCP2 (Jones et al. 1998), whereas MBD2 complexes with NuRD to target it to methylated sequences (Le Guezennec et al. 2006). NuRD also has a subunit, CDH1, with two chromodomains, and includes LSD1 for another dimension of chromatin-modifying capability (Ramirez et al. 2012).

HDAC3 is found in a corepressor complex with NCoR1 (nuclear receptor corepressor) and SMRT (silencing mediator of retinoic and thyroid receptors), corepressors that are necessary for typical embryonic development (Wilson et al. 2006; Oberoi et al. 2011). The binding of both corepressors stimulates deacetylase activity and mediates its targeting to promoter sequences, as isolated HDAC3 is not an active deacetylase (Guenther et al. 2001).

3.4.1 Class II HDACs

The class II HDACs have been found to bind to NCoR/SMRT and recruit the activity of class I HDACs, which may explain initial results that found class II HDAC activity in impure samples—suppression of this binding or purified class II samples have no deacetylase activity (Fischle et al. 2002; Verdin et al. 2003). A shorter splice variant of HDAC 9 that lacks the HDAC domain is equally effective in repression of target genes, suggesting that their inhibition of transcription is through some mechanism other than intrinsic HDAC activity, such as its interaction with HP1 and CTBP, known corepressors of transcription, often using the N-terminal domain (Lahm et al. 2007; Zhang et al. 2001).

The class IIa HDACs have two important domains, the carboxyl-terminal HDAC domain and the amino-terminal protein interaction domain (Fischle et al. 2001). Class IIa HDACs are divergent from the other metal-dependent HDACs by their ability to shuttle between the nucleus and cytoplasm, phosphorylation in the N-terminal domain promotes retention in the cytoplasm in association with 14-3-3 proteins, and nuclear import is in association with MEF2 (McKinsey et al. 2000; Verdin et al. 2003; Yang and Seto 2008). Class IIa expression patterns are specific to tissue types and developmental stages, such as HDAC7 in vascular endothelium during embryogenesis (Chang et al. 2006) and HDAC4 for regulation of chondrocytes at growth plates (Vega et al. 2004). Class IIa have reduced catalytic activity compared to the other HDACs, due to a tyrosine to histidine substitution, so that a transition state in the reaction is no longer effectively stabilised (Bottomley et al. 2008)—reversion to tyrosine in experiments resulted in a greater than 1,000-fold increase in activity (Lahm et al. 2007).

Class IIb consists of HDAC6 and HDAC10, both of which are mainly cytoplasmic. Despite each being complete, HDAC6's catalytic domains are unable to function individually (Verdin et al. 2003). HDAC6 is predominantly localised to the cytoplasm in the absence of a stimulus, where it is the main deacetylase, and its partial translocation to the nucleus occurs in association with cell cycle arrest (Fischle et al. 2001; Yang and Seto 2008). It provides a link between protein acetylation and ubiquitination by forming a complex with other proteins to bind ubiquitinated targets for deacetylation, deacetylation allowing further ubiquitination at these residues (Seigneurin-Berny et al. 2001). When misfolded ubiquitinated proteins accumulate in cells and would otherwise have cytotoxic effects, HDAC6 has a protective role, catalysing aggresome formation by sequestering ubiquitin-marked misfolded proteins and promoting degradation by autophagy (Boyault et al. 2007; Bali et al. 2005). Other cytoplasmic functions include deacetylation of α-tubulin in microtubules during chemotaxis (Hubbert et al. 2002), vital to vascular endothelial cell migration in angiogenesis (Li et al. 2011b). The role of HDAC10, which possesses only one catalytic domain, is yet to be fully elucidated.

3.4.2 *Class IV HDACs*

Whilst the class IV catalytic domain shows similarity to both class I and II, HDAC11, the only member of this class, is mostly like the class I HDACs (Bottomley et al. 2008). In addition to the HDAC domain, it has short N- and C-terminal extensions but is the smallest of the HDAC family, mainly consisting of the catalytic domain. Expression is high only in certain tissues—the testis, brain, kidneys, heart and skeletal muscle, as well as several cancer lines, and predominantly resides in the nucleus of cells (Gao et al. 2002). HDAC6 is highly conserved from yeast to humans, indicating that it may have an important, albeit undiscovered, role (Yang and Seto 2008).

3.4.3 *Class III HDACs: "Sirtuins"*

The Sir2 homologues have functions relating to regulation of metabolism, transcription and apoptosis (Sauve et al. 2006). Within the sirtuin family, enzymes are classed across species according to their deacetylase domains of around 250 amino acids and reliant on NAD^+, which they cleave during deacetylation. They are not sensitive to the classical HDAC inhibitors (HDACi) due to their different active sites and mechanism of catalysis, and are grouped with the ADP-ribosyltransferases (Sauve et al. 2006). In humans class I includes SIRT1, 2 and 3, class II consists of SIRT 4, SIRT 5 is the sole member of class III, and SIRT6 and 7 make up class IV (Roy 2000). A fifth class, class U, is found in gram-positive bacteria (Sauve et al. 2006). Until recently, SIRT5 was believed to only possess weak, limited spectrum deacetylase activity (Du et al. 2011), but recently its roles in post-translational protein modification have been expanded to include much stronger lysine desuccinylation and demalonylation (Peng et al. 2011). Relatively weak deacetylase activity is found in SIRT4-7, suggesting that they too may have other enzymatic functions (Yuan and Marmorstein 2012).

3.4.4 *HDAC Expression/Aberrant Function in Cancer*

Hypoacetylation of tumour suppressor genes due to HDAC overexpression is common to many cancers, and repressive chromatin structures silence certain genes to allow unregulated growth of cells. Class I HDAC upregulation is often correlated with poorer outcomes for patients, knockout studies suggesting that HDACs 1 and 3 are necessary for the abnormal proliferation and longer survival times of cancer cells, whereas loss of HDAC 4 and 7 function had no discernible effects (Keith et al. 2003). Class II isoform expression is often decreased in tumours, high levels of expression being a predictor of better survival (Weichert 2009). HDAC1 and HDAC2 overexpression has the cancer-promoting functions of increasing cell survival, proliferation, and preventing expression of tumour suppressor p21, as well as encouraging loss of differentiation in cancer cells (Weichert et al. 2008a). In colon cancer cells, HDAC3 is upregulated to 2.8 times the level found in normal cells, and the other class I members, HDAC2 and HDAC1, are found in elevated levels compared to normal mucosa cells (Wilson et al. 2006), all three linked to poor prognosis and an advanced stage of disease in multiple cancers (Weichert et al. 2008a, b). In hepatocellular carcinoma, HDAC1 expression levels correlated with loss of differentiation and invasiveness (Xie et al. 2012).

APC loss of function mutation in colon cancer is correlated with HDAC2 expression—elevated levels were found in 82 % of colon tumours, and significant differences observed between tumour and normal colon cells (Zhu et al. 2004). Even mice lacking APC and with no tumours showed higher levels of HDAC2 expression than comparative tissues in wild-type mice (Zhu et al. 2004).

In leukaemias induced by chromosomal translocations, oncogenic transcription factors may recruit HDACs and support their aberrant activity. PML-RARα forms a corepressor complex containing HDAC1 or HDAC2, which targets promoters of select genes in acute promyelocytic leukaemia (APL) to suppress function of tumour suppressor PML (Matsushita et al. 2006). AML-ETO interacts with HDAC1 in AML to cause increased chromatin binding, the pattern of which defines disease subtype and can predict survival outcomes (Tickenbrock et al. 2011).

Sirt1 studies involving cancer report apparently antagonistic roles: under certain circumstances, it acts as a tumour suppressor, but is oncogenic in others. Cancer-promoting effects of overexpression include increased angiogenesis, cell proliferation and survival (Stunkel and Campbell 2011) and anti-apoptotic prevention of Bax translocation to mitochondria (Cea et al. 2011), although it also prevents intestinal polyps from progressing to tumours and inhibits colon cancer cell proliferation (Firestein et al. 2008). This is reflected in the expression patterns of tumours, in which some display overexpression and others down-regulation of Sirt1. Cells lacking SIRT3 exhibit elevated levels of ROS, potentially increasing their chances of developing cancer, as well as higher levels of glycolysis as seen in cancer cells (Giralt and Villarroya 2012).

3.5 Histone Deacetylase Inhibitors in Cancer Therapy

HDACi impair HDAC function by blocking their active sites to impair enzymatic activity. Common structural features include a group to co-ordinate with active site zinc, joined by a hydrocarbon chain to a cap that hydrogen bonds with key residues within the HDAC lysine-binding channel (Monneret 2005). By inhibition of HDAC

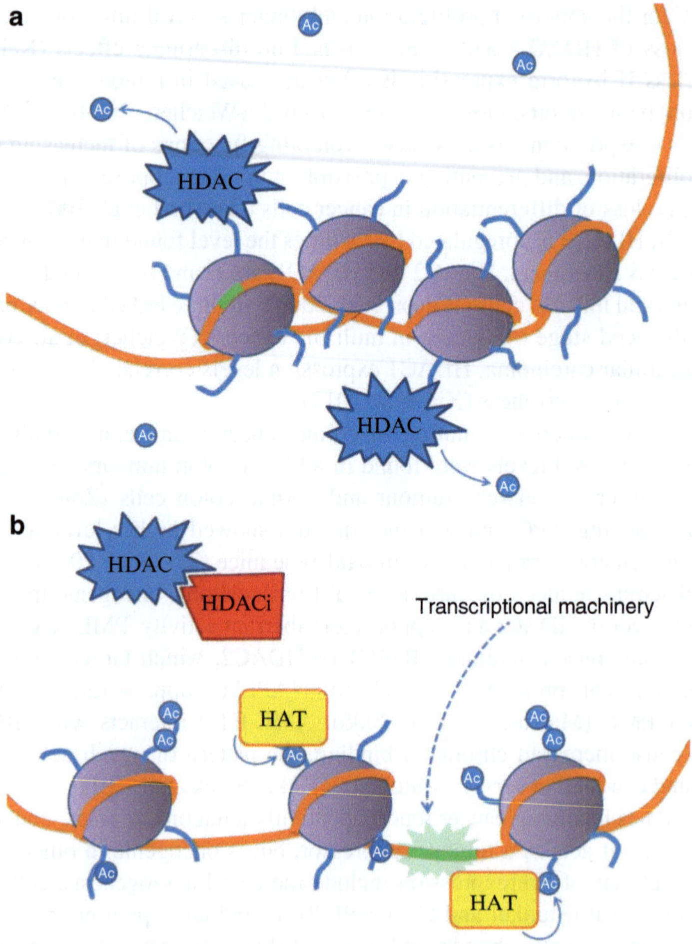

Fig. 3.1 (**a**) Histone deacetylation leads to the formation of tight, repressive chromatin structures, inhibiting access to genes by transcription machinery and preventing transcription. (**b**) Histone deacetylase inhibitors prevent HDAC action. The unopposed activity of HATs leads to histone tail hyperacetylation resulting in open, transcriptionally active chromatin with access to genes (*green*) by transcription complexes

enzymes, HAT activity is unopposed and acetylation at targets increases, opening up chromatin and increasing expression of certain genes such as tumour suppressors (Fig. 3.1). The increased acetylation is not global (Richon et al. 2000) showing selective targeting of genes with altered expression in cancer and leaving normal cells unaffected at doses toxic to cancer cells despite histone hyperacetylation occurring in both subsets (Ungerstedt et al. 2005). Targets for HDACi are the substrates of HATs, including many non-histone proteins, resulting in the diverse range of cellular responses to HDACi treatment including induction of cell differentiation, maturation, cell cycle arrest, senescence and cell death in carcinoma cells (Fig. 3.2).

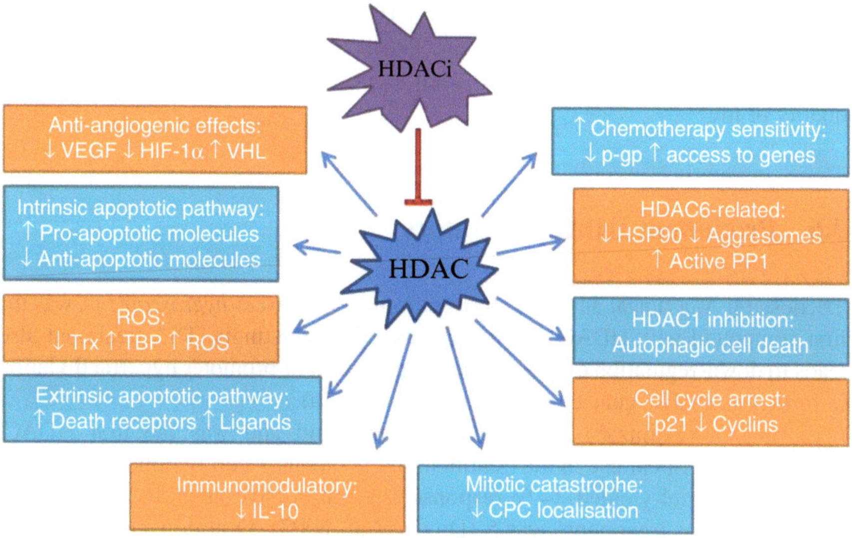

Fig. 3.2 Summary of effects induced by HDACi in malignant cells

These drugs are distributed between six structural classes on the basis of their functional groups—the hydroxamic acids, electrophilic ketones, cyclic peptides, short-chain fatty acids, benzamides and miscellaneous (Newbold et al. 2008). Up to 20 % of genes are shown to have altered expression after HDACi treatment, increased and decreased expression occurring in approximately numbers of genes, influenced by drug concentration, cell line and length of exposure (Licciardi et al. 2012; Xu et al. 2007). The gene expression profile changes resulting from HDACi treatment have some overlap due to shared mechanisms of action (Peart et al. 2005), but some genes are altered in a class-specific manner (Newbold et al. 2008) and resistance profiles accordingly different (Peart et al. 2003). Generalised effects are cell cycle arrest, intrinsic or extrinsic apoptic pathway activation, autophagic cell death, mitotic catastrophe, senescence and ROS generation (Xu et al. 2007).

Short chain fatty acids include butyrate, the first HDACi to be synthesised, but use is restricted by their limited bioavailability, specificity and potency (Sato 2012). The aliphatic chain acts as an analogue of the lysine side chain, blocking the HDAC lysine channel. This class also includes valproic acid (VPA), a class I isoform-selective inhibitor which was previously used as an anti-epileptic. The hydroxamic acid class includes suberoylanilide hydroxamic acid (SAHA), Trichostatin A (TSA), and inhibition is via hydroxamic acid group coordination with zinc in the HDAC active site (Marks 2007). Romidepsin, a naturally derived prodrug, is a member of the cyclic peptide class.

The US Food and Drug Administration (FDA) has approved romidepsin and vorinostat for the treatment of cutaneous T cell lymphoma, assorted conditions in which malignant T cells invade the skin and immunity becomes dysfunctional, resulting in immunosuppresion (Campbell et al. 2010) with romidepsin also approved for peripheral T-cell lymphoma (PTCL), a collection of aggressive disease subtypes caused by invasive, malignant T-cells. Even in cases where treatment is not curative, debilitating symptoms such as pruritis may be eased for improved quality of life.

3.6 Prominent HDAC Inhibitors

Trichostatin A, a fungal antibiotic derived from *Streptomyces hygroscopicus*, was the first compound with HDAC activity discovered. Interestingly, TSA treatment also leads to down-regulation of DNMTs and the reversal of promoter hypermethylation to further normalise gene expression (Li et al. 2011a).

Vorinostat (Zolinza®) is a synthetic compound and was the first HDACi approved for clinical use in the treatment of cutaneous T cell lymphoma (CTCL). The phase II trial yielded 24 % response rate, impressive given that the median number of prior treatments was 5 (Duvic et al. 2007). Oral delivery avoids the risk of line sepsis from skin commensals such as *S. aureus*, a potential cause of fatality as CTCL patients become immunocompromised with disease progression. The drug binds the coenzyme zinc with its hydroxamic acid group, forming hydrogen bonds with residues in the active site of HDAC (Grozinger and Schreiber 2002) and lacks selectivity, inhibiting class I HDACs as well as class IIb HDAC6, and weakly inhibits class IIa (Rangwala et al. 2012). The effect of restoring histone acetylation is highly transient, and lost rapidly when treatment is stopped, allowing recovery from any serious side effects noticed during treatment and normal cells are up to ten times as resistant to its effects as cancer cells, avoiding unnecessary side effects from generalised cytotoxicity (Sarfstein et al. 2010). Resistance arises to vorinostat in the form of overexpression of prosurvival proteins Bcl-2 and Bcl-XL, which combat initiation of the intrinsic apoptotic pathway by preventing damage of the mitochondrial membrane by Bax and Bak (Newbold et al. 2008).

Romidepsin (depsipeptide, FK228, Isotodax®) is a highly potent prodrug, isolated from *Chromobacterium violaceum*. Its disulfide bond must be reduced intracellularly by glutathione to yield the active, less stable form, with an active sulfhydryl group. Glutathione upregulation is normally a determinant of multidrug resistance, but cellular levels can be taken as an indicator of sensitivity to romidepsin (Peart et al. 2003). The prodrug form is hydrophobic and can easily cross cellular membranes, making romidepsin highly effective compared to other HDACi (Monneret 2005). Romidepsin preferentially inhibits class I HDACs (class I selective), but with higher affinity for HDAC1 and 2 over HDACs 4 and 6 (Newbold et al. 2008), and weak inhibition of class II HDACs (Bertino and Otterson 2011). It was approved by the FDA for CTCL in 2009 for intravenous delivery (Sato 2012) following promising results in phase II trials of patients with poor prognostic factors and prior failed treatments (Piekarz et al. 2009), and is the only HDACi approved for PTCL.

In the pivotal phase II trial for approval in the treatment of PTCL, romidepsin yielded responses in 38 % of patients and complete responses in 18 % (Coiffier et al. 2012). Overexpression of Bcl-2 does not confer resistance to romidepsin, although abnormally high levels of Bcl-X_L prevent it from inducing cell death (Newbold et al. 2008). Resistance to HDACi can also arise in cells expressing high levels of P-glycoprotein (P-gp), an efflux protein involved in multidrug resistance, but no resistance has been found to other HDACi by this mechanism (Peart et al. 2003).

3.7 Cell Cycle Arrest

Cyclin-dependent kinase (CDK) inhibitor p21 is normally induced by p53 in response to DNA damage, yet its induction is p53-independent in HDACi treatment and is mediated by increased acetylation at promoter histones (Xu et al. 2007). HDAC1 inhibits p21 expression (Xie et al. 2012); hence, p21 is frequently upregulated with HDACi treatment leading to G_1 cell cycle arrest (Richon et al. 2000). Cyclins, which drive cycle progression, are downregulated contributing to cell cycle arrest (Nebbioso et al. 2005). Arrest in G_1 occurs with low dose treatment and in both G_1 and G_2 at higher doses (Richon et al. 2000) (Fig. 3.3).

3.8 Intrinsic Apoptosis

The intrinsic apoptotic pathway is commonly initiated by HDACi to selectively cause cell death. Mitochondrial membrane permeability allows cytochrome-C release, which stimulates caspase-9 to activate the effector caspases and initiate apoptotic morphology (Xu et al. 2006) (Fig. 3.4). HDACi causes upregulation of Bid, which once cleaved initiates Bax, which causes damage to the mitochondrial membrane, or other pro-apoptotic proteins such as Bim, Bok and Bmf, whilst simultaneously decreasing expression of antagonistic anti-apoptotic molecules such as Bcl-2 and Bcl-X_L, which protect the mitochondrial membrane, and survivin (Xu et al. 2006). Evidence for the necessity of this pathway is seen in cell death mediated by vorinostat where Bim and Bid are necessary for its therapeutic effect in lymphoma (Lindemann et al. 2007).

3.9 Extrinsic Apoptotic Pathways

Ligation of death receptors by their ligands, such as Fas by FasL, DR4/5 by TRAIL and TNFR-1 by TNF, instigates activation of intracellular signalling by caspases 8 and 10 (Xu et al. 2007). Death receptors and ligands are upregulated in the presence

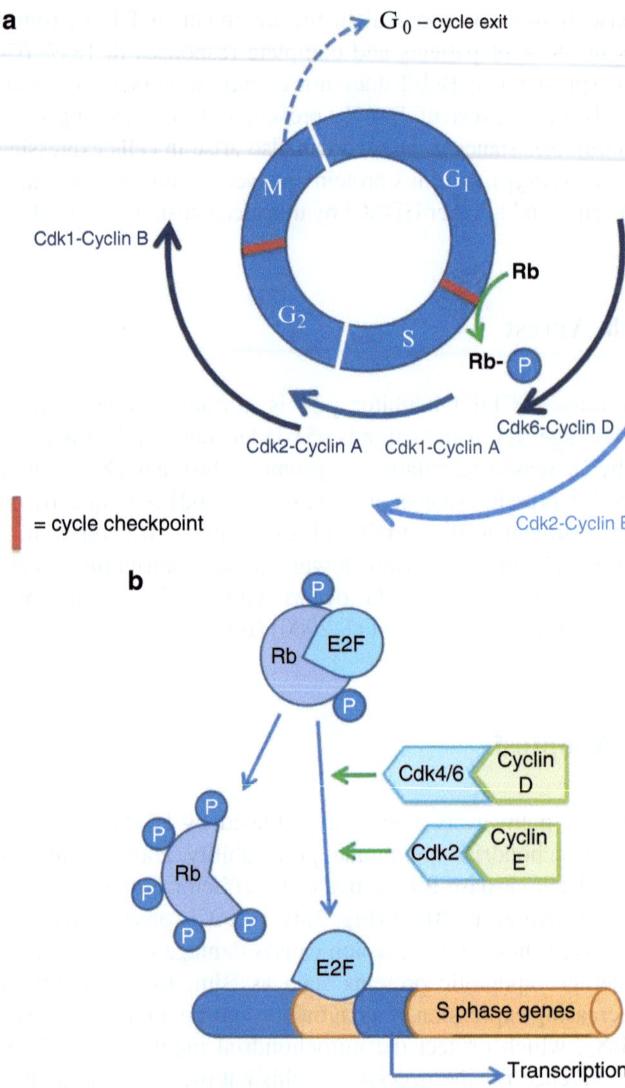

Fig. 3.3 (**a**) The role of varying cyclins and cyclin-dependent kinases in cell cycle progression. (**b**) The function of E2F in transcription of S-phase genes

of HDACi, whilst normal cells are unaffected (Insinga et al. 2005). The ability for TRAIL to induce apoptosis in transformed and tumour cells whilst leaving normal cells unaffected is also of great clinical significance (Pitti et al. 1996). Increased sensitivity to TRAIL-mediated apoptosis is reported with HDACi treatment (Jin et al. 2002). Further potency is added by downregulation of c-FLIP, Bcl-2 and XIAP (inhibitor of apoptosis), molecules that inhibit the death receptor pathway, following HDACi exposure (Rosato and Grant 2005; Peart et al. 2005) (Fig. 3.4).

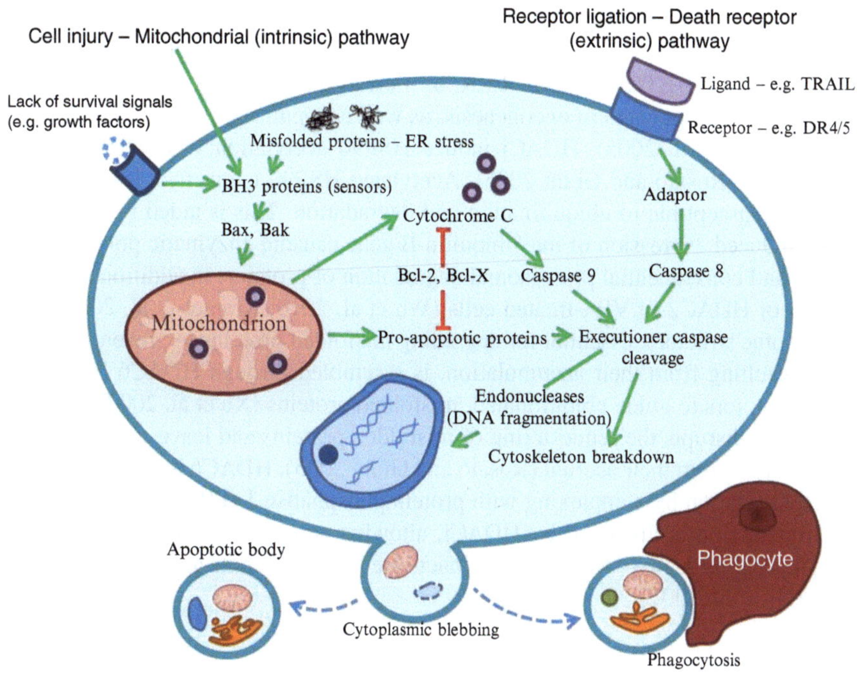

Fig. 3.4 Intrinsic and extrinsic apoptotic pathway convergence

3.10 Reactive Oxygen Species

An increase in ROS occurs in malignant cells treated with HDACi. ROS may be scavenged by the antioxidant thioredoxin (Trx) to prevent DNA, protein and lipid damage. Trx's role in carcinogenesis is multifaceted, also inhibiting signalling and Bim induction by apoptosis signal-regulating kinase (ASK1) (Xu et al. 2007). Trx is present in low levels in many cancers, but HDACi induce greater Trx expression in normal cells. Higher expression of Trx or other free radical scavengers are associated with HDACi resistance in cancer cells (Shao et al. 2001). Vorinostat causes expression of Trx-binding protein (TBP-2) to prevent ROS binding by Trx, which then amplifies the effects of ROS accumulation (Ungerstedt et al. 2005). Caspases are cysteine proteases that trigger cell death in response to intracellular or extracellular stress, such as accumulation of ROS, and are seen to rise in vorinostat-treated cells (Ungerstedt et al. 2005).

3.11 HDAC6 Inhibition-Related Effects

HDAC6 may be a useful target for selective HDACi due to its necessity for tumorigenesis, with cytoskeletal substrates and the role in cellular phenotype transformations induced by oncogenes Ras and ErbB2 (Lee et al. 2008). By physically binding and also by deacetylation, HDAC6 stabilises the chaperone HSP90 to allow it to maintain cellular levels of its client proteins, which include chimeric proteins Bcr-Abl, AML1-ETO and PML-RαR, as well as anti-apoptotic AKT, and c-Ras, which are all implicated in oncogenesis, as well as members of cell signalling cascades (Bali et al. 2005). HDACi induce HSP90 acetylation without altering its expression (Rosato and Grant 2005). Acetylated HSP90 is inactive, leaving client proteins susceptible to ubiquitination and degradation. This is aided by VPA- and TSA-induced expression of the Ubiquitin B gene causing enzymatic polyubiquitylation and consequential proteasomal degradation of proteins, in addition to degradation of HDAC2 in VPA-treated cells (Wu et al. 2010; Kramer et al. 2008). The aggresome structure, essential for degrading misfolded proteins to prevent cytotoxicity resulting from their accumulation, is assembled through HDAC6 binding to dynein motors to guide ubiquitinated, misfolded proteins (Xu et al. 2007). HDAC6 inhibition disrupts the sequestering of misfolded proteins and leaves cells vulnerable to stress from their accrual (Rosato and Grant 2005). HDAC6 influences protein phosphorylation by complexing with protein phosphatise 1 (PP1) to disable it, and this interaction is disrupted by HDACi, allowing active PP1 to dephosphorylates target proteins such as Akt, which is inactivated along with its anti-apoptotic pathway (Xu et al. 2007).

3.12 Anti-angiogenesis

Tumours require extensive, abnormal vasculature to survive the hypoxic conditions encountered as they enlarge, and interruption of their blood supply can halt their growth. After 24 h of treatment with vorinostat, anti-angiogenic protein thrombospondin-1 (TSP-1) increased significantly, whilst expression and secretion of the pro-angiogenic VEGF (vascular epithelial growth factor) was decreased (Duvic et al. 2007). Overexpression of VEGF decreases expression of the anti-angiogenic von Hippel–Lindau (VHL) factor, which ubiquitinates hypoxia inducible factor (HIF-1α) for proteasomal degradation, although HDACi usage restores VHL levels (Kong et al. 2006). Increased transcription of class I HDACs 1, 2 and 3 is induced by hypoxia and activates a pathway leading to transcription of HIF-1α, part of the HIF-1 transcriptional complex that regulates angiogenesis-related genes (Geng et al. 2011) that may also be induced by low oxygen via HDAC7 translocation to the nucleus. HDAC inhibition decreases HIF-1α activation, whilst also acetylating and degrading (Marks 2007). Inhibition of HDACs 4 and 6 disrupts their stabilisation of HIF-1α and leads to ubiquitin-independent proteasomal digestion (Qian et al. 2006), as does disruption of HSP90s chaperone function via HDAC6 inhibition. Decreases in VEGF and HIF-1α expression levels are also seen following treatment with romidepsin (Patrick et al. 2011).

3.13 Mitotic Catastrophe, Autophagy

TSA induces arrest during prometaphase of mitosis by preventing normal chromosome alignment, and interferes with microtubule attachment to kinetochores to cause segregation errors and mitotic catastrophe (Ma et al. 2008; Robbins et al. 2005). In romidepsin treatment, increased acetylation at centromeres interferes with other PTMs characteristic of that region, as well as preventing recruitment of the chromosomal passenger complex (CPC), an essential driver of many events in mitosis and cytokinesis. This generalised disruption of events leads to cell death, cells exhibiting abnormalities such as multiple nuclei and failure of sister chromatids to separate (Zhang et al. 2010).

Autophagy has a two-sided role in cancer: loss of autophagy can result in damaged cells being retained, whereas abnormally occurring autophagy can be cytoprotective for cells under stress, aiding adaptation to hypoxic conditions or chemotherapy, a hallmark of aggressive cancers (Eisenberg-Lerner and Kimchi 2009; Carew et al. 2010). Inhibition of HDAC1 triggers caspase-independent autophagic cell death (Xie et al. 2012).

3.14 Immunomodulation

HDACi have a large toll on macrophage recognition of and phagocytic response to bacteria, as well as decreasing intracellular bacterial killing by inhibiting production of reactive oxygen and nitrogen species by immune cells (Mombelli et al. 2011), resulting in immunosuppresion of already vulnerable patients. HDACi inhibit the activation of T cells and natural killer (NK) cells, but conversely sensitise tumour cells to their cytolytic activities through increasing surface receptors (Schmudde et al. 2008), creating a dilemma that could be resolved by sensitising immune cells to tumour antigens before commencing HDACi treatment as immune cells activated prior to treatment remain active (Schmudde et al. 2010). Romidepsin and vorinostat affect a decrease of 95–99 % in the levels of IL-10, a key cytokine in CTCL, normalising immune function and increasing survival times (Tiffon et al. 2011).

3.15 Side Effects

The lack of specificity of HDACi leads to some side effects including weight loss, fatigue and diarrhoea, as well as haematological toxicities (Witt et al. 2009; Mercurio et al. 2010) Paradoxically, treatment with HDACi was shown to enhance the migration activity of 13/30 cancers studied, and boosted metastasis of tumours in mice (Lin et al. 2012a). It is hoped that specific HDACi may be able to alleviate side effects by narrowing their spectrum of inhibition, and hence minimising the pathways altered.

3.16 Synergistic Effects

Several cytotoxic drugs commonly used in cancer therapy are substrates for the MDR1 (multidrug resistance) gene product p-glycoprotein pump (P-gp), expression of which decreases prognosis and response to chemotherapy by causing efflux of the drug. HDACi strongly decreased MDR1 expression in drug-resistant cells yet increased expression in sensitive cells (El-Khoury et al. 2007). Anti-cancer drugs that target DNA require access to the DNA, much like transcription factors, and are limited by the condensed chromatin state maintained by histone deacetylation; hence, HDACi can aid their access and effectiveness.

3.17 Conclusions

The reversible nature of epigenetic marks makes epigenomic medicine a promising target for many future studies, as deregulation of epigenetic processes is implicated in a diverse range of human conditions, ranging from neurodegenerative conditions to pulmonary disease. The field of epigenetic drugs in cancer treatment is a rapidly evolving and exciting area of research, as is the potential for more specific epigenetic biomarkers to allow non-invasive tracking of tumour progression, metastatic potential and recurrence. Personalised treatment regimes based on epigenetic profiles for tumour susceptibility are a hope for the future, providing better prognoses and avoiding unnecessary cytotoxic side effects of futile therapies. The field of PTMs still requires further study, to reveal all the enzymes involved and classify in more detail interactions between modifications. As the mechanisms of HDAC and their inhibiting compounds are further characterised, the field for more collaborative treatments using the synergistic effects of various classes of epigenomic drugs with chemotherapeutic drugs and other conventional treatments can be explored. Current inhibitors possess limited selectivity, with hope that class- or isoform-specific HDACi will allow more specificity in the pathways induced or disabled, limiting the side effects experienced in patients.

References

Alcazar O, Achberger S, Aldrich W, Hu ZB, Negrotto S, Saunthararajah Y, Triozzi P (2012) Epigenetic regulation by decitabine of melanoma differentiation in vitro and in vivo. Int J Cancer 131:18–29

Allfrey VG, Faulkner R, Mirsky AE (1964) Acetylation and methylation of histones and their possible role in the regulation of RNA synthesis. Proc Natl Acad Sci U S A 51:786–794

Al-Salihi M, Yu M, Burnett DM, Alexander A, Samlowski W, Fitzpatrick FA (2011) The depletion of DNA methyltransferase-1 and the epigenetic effects of 5-aza-2′ deoxycytidine (decitabine) are differentially regulated by cell cycle progression. Epigenetics 6:1021–1028

Antequera F (2003) Structure, function and evolution of CpG island promoters. Cell Mol Life Sci 60:1647–1658

Atadja PW (2011) HDAC inhibitors and cancer therapy. Prog Drug Res 67:175–95

Bali P, Pranpat M, Bradner J, Balasis M, Fiskus W, Guo F, Rocha K, Kumaraswamy S, Boyapalle S, Atadja P, Seto E, Bhalla K (2005) Inhibition of histone deacetylase 6 acetylates and disrupts the chaperone function of heat shock protein 90: a novel basis for antileukemia activity of histone deacetylase inhibitors. J Biol Chem 280:26729–26734

Ballas N, Battaglioli E, Atouf F, Andres ME, Chenoweth J, Anderson ME, Burger C, Moniwa M, Davie JR, Bowers WJ, Federoff HJ, Rose DW, Rosenfeld MG, Brehm P, Mandel G (2001) Regulation of neuronal traits by a novel transcriptional complex. Neuron 31:353–365

Bedford MT (2007) Arginine methylation at a glance. J Cell Sci 120:4243–4246

Bedford MT, Clarke SG (2009) Protein arginine methylation in mammals: who, what, and why. Mol Cell 33:1–13

Bednar J, Horowitz RA, Grigoryev SA, Carruthers LM, Hansen JC, Koster AJ, Woodcock CL (1998) Nucleosomes, linker DNA, and linker histone form a unique structural motif that directs the higher-order folding and compaction of chromatin. Proc Natl Acad Sci U S A 14173

Bertino EM, Otterson GA (2011) Romidepsin: a novel histone deacetylase inhibitor for cancer. Expert Opin Investig Drugs 20:1151–1158

Bestor TH, Gundersen G, Kolsto AB, Prydz H (1992) Cpg islands in mammalian gene promoters are inherently resistant to de novo methylation. Genet Anal Tech Appl 9(2):48–53 [Accessed 12 Dec 2012]

Bing X, Jonathan RW, Steven JG (2003) SET domains and histone methylation. Curr Opin Struct Biol 13:699–705

Bird A (2002) DNA methylation patterns and epigenetic memory. Genes Dev 16:6–21

Bird AP, Wolffe AP (1999) Methylation-induced repression—belts, braces, and chromatin. Cell 99:451–454

Bishop TC (2008) Geometry of the nucleosomal DNA superhelix. Biophys J 95:1007–1017

Black JC, Van Rechem C, Whetstine JR (2012) Histone lysine methylation dynamics: establishment, regulation, and biological impact. Mol Cell 48:491–507

Bottomley MJ, Lo Surdo P, Di Giovine P, Cirillo A, Scarpelli R, Ferrigno F, Jones P, Neddermann P, De Francesco R, Steinkuhler C, Gallinari P, Carfi A (2008) Structural and functional analysis of the human HDAC4 catalytic domain reveals a regulatory structural zinc-binding domain. J Biol Chem 283:26694–26704

Boyault C, Zhang Y, Fritah S, Caron C, Gilquin B, Kwon SH, Garrido C, Yao TP, Vourc'h C, Matthias P, Khochbin S (2007) HDAC6 controls major cell response pathways to cytotoxic accumulation of protein aggregates. Genes Dev 21:2172–2181

Brady ME, Ozanne DM, Gaughan L, Waite I, Cook S, Neal DE, Robson CN (1999) Tip60 is a nuclear hormone receptor coactivator. J Biol Chem 274:17599–17604

Brent B-T, David AW, Robert MF, John TL, Kraus WL, Michelle DW (2004) Specific contributions of histone tails and their acetylation to the mechanical stability of nucleosomes. J Mol Biol 346:135–146

Campbell JJ, Clark RA, Watanabe R, Kupper TS (2010) Sezary syndrome and mycosis fungoides arise from distinct T-cell subsets: a biologic rationale for their distinct clinical behaviors. Blood 116:767–771

Carew JS, Medina EC, Esquivel JA II, Mahalingam D, Swords R, Kelly K, Zhang H, Huang P, Mita AC, Mita MM, Giles FJ, Nawrocki ST (2010) Autophagy inhibition enhances vorinostat-induced apoptosis via ubiquitinated protein accumulation. J Cell Mol Med 14:2448–2459

Cea M, Soncini D, Fruscione F, Raffaghello L, Garuti A, Emionite L, Moran E, Magnone M, Zoppoli G, Reverberi D, Caffa I, Salis A, Cagnetta A, Bergamaschi M, Casciaro S, Pierri I, Damonte G, Ansaldi F, Gobbi M, Pistoia V, Ballestrero A, Patrone F, Bruzzone S, Nencioni A

(2011) Synergistic interactions between HDAC and sirtuin inhibitors in human leukemia cells. PLoS One 6:12p

Chang S, Young BD, Li S, Qi X, Richardson JA, Olson EN (2006) Histone deacetylase 7 maintains vascular integrity by repressing matrix metalloproteinase 10. Cell 126:321–334

Chen L, Dahlstrom JE, Lee SH, Rangasamy D (2012) Naturally occurring endo-siRNA silences LINE-1 retrotransposons in human cells through DNA methylation. Epigenetics 7:758–771

Choi J (2010) Constrasting chromatin organization of CpG islands and exons in the human genome. Genome Biol 11(7):R70

Choi JY, James SR, Link PA, McCann SE, Hong CC, Davis W, Nesline MK, Ambrosone CB, Karpf AR (2009) Association between global DNA hypomethylation in leukocytes and risk of breast cancer. Carcinogenesis 30:1889–1897

Cimato TR, Tang J, Xu Y, Guarnaccia C, Herschman HR, Pongor S, Aletta JM (2002) Nerve growth factor-mediated increases in protein methylation occur predominantly at type I arginine methylation sites and involve protein arginine methyltransferase 1. J Neurosci Res 67:435–442

Cloos PAC, Christensen J, Agger K, Helin K (2008) Erasing the methyl mark: histone demethylases at the center of cellular differentiation and disease. Genes Dev 22:1115–1140

Coiffier B, Pro B, Prince HM, Foss F, Sokol L, Greenwood M, Caballero D, Borchmann P, Morschhauser F, Wilhelm M, Pinter-Brown L, Padmanabhan S, Shustov A, Nichols J, Carroll S, Balser J, Balser B, Horwitz S (2012) Results from a pivotal, open-label, Phase II study of romidepsin in relapsed or refractory peripheral T-cell lymphoma after prior systemic therapy. J Clin Oncol 30:631–636

Collins RE, Tachibana M, Tamaru H, Smith KM, Jia D, Zhang X, Selker EU, Shinkai Y, Cheng X (2005) In vitro and in vivo analyses of a Phe/Tyr switch controlling product specificity of histone lysine methyltransferases. J Biol Chem 280:5563–5570

Corn PG, Heath EI, Heitmiller R, Fogt F, Forastiere AA, Herman JG, Wu T-T (2001) Frequent hypermethylation of the 5′ CpG island of E-cadherin in esophageal adenocarcinoma. Clin Cancer Res 7:2765–2769

Cote J, Richard S (2005) Tudor domains bind symmetrical dimethylated arginines. J Biol Chem 280:28476–28483

Damaraju VL, Mowles D, Yao S, Ng A, Young JD, Cass CE, Tong Z (2012) Role of human nucleoside transporters in the uptake and cytotoxicity of azacitidine and decitabine. Nucleosides Nucleotides Nucleic Acids 31:236–255

Davuluri RV, Grosse I, Zhang MQ (2001) Computational identification of promoters and first exons in the human genome. Nat Genet 29(4):412–7

De Smet C, Lurquin C, Lethe B, Martelange V, Boon T (1999) DNA methylation is the primary silencing mechanism for a set of germ line- and tumor-specific genes with a CpG-rich promoter. Mol Cell Biol 19:7327–7335

Decarlo D, Hadden MK (2012) Oncoepigenomics: making histone lysine methylation count. Eur J Med Chem 56:179–194

Delgado-Cruzata L, Wu HC, Perrin M, Liao YY, Kappil MA, Ferris JS, Flom JD, Yazici H, Santella RM, Terry MB (2012) Global DNA methylation levels in white blood cell DNA from sisters discordant for breast cancer from the New York site of the breast cancer family registry. Epigenetics 7:868–874

Di Lorenzo A, Bedford MT (2011) Histone arginine methylation. FEBS Lett 585:2024–2031

Du JT, Zhou YY, Su XY, Yu JJ, Khan S, Jiang H, Kim J, Woo J, Kim JH, Choi BH, He B, Chen W, Zhang S, Cerione RA, Auwerx J, Hao Q, Lin HN (2011) Sirt5 is a NAD-dependent protein lysine demalonylase and desuccinylase. Science 334:806–809

Duvic M, Talpur R, Ni X, Zhang C, Hazarika P, Kelly C, Chiao JH, Reilly JF, Ricker JL, Richon VM, Frankel SR (2007) Phase 2 trial of oral vorinostat (suberoylanilide hydroxamic acid, SAHA) for refractory cutaneous T-cell lymphoma (CTCL). Blood 109:31–39

Eckhardt F, Lewin J, Cortese R, Rakyan VK, Attwood J, Burger M, Burton J, Cox TV, Davies R, Down TA, Haefliger C, Horton R, Howe K, Jackson DK, Kunde J, Koenig C, Liddle J, Niblett

D, Otto T, Pettett R (2006) DNA methylation profiling of human chromosomes 6, 20 and 22. Nat Genet 38:1378–1385

Ehrlich M (2002) DNA methylation in cancer: too much, but also too little. Oncogene 21:5400

Eisenberg-Lerner A, Kimchi A (2009) The paradox of autophagy and its implication in cancer etiology and therapy. Apoptosis 14:376–391

El-Khoury V, Breuzard G, Fourré N, Dufer J (2007) The histone deacetylase inhibitor trichostatin A downregulates human MDR1 (ABCB1) gene expression by a transcription-dependent mechanism in a drug-resistant small cell lung carcinoma cell line model. Br J Cancer 97:562–573

Esteller M, Sanchez-Cespedes M, Rosell R, Sidransky D, Baylin SB, HERMAN JG (1999) Detection of aberrant promoter hypermethylation of tumor suppressor genes in serum DNA from non-small cell lung cancer patients. Cancer Res 59:67–70

Esteller M, Garcia-Foncillas J, Andion E, Goodman SN, Hidalgo OF, Vanaclocha V, Baylin SB, Herman JG (2000a) Inactivation of the DNA-repair gene MGMT and the clinical response of gliomas to alkylating agents. N Engl J Med 343:1350–1354

Esteller M, Sparks A, Toyota M, Sanchez-Cespedes M, Capella G, Peinado MA, Gonzalez S, Tarafa G, Sidransky D, Meltzer SJ, Baylin SB, Herman JG (2000b) Analysis of adenomatous polyposis coli promoter hypermethylation in human cancer. Cancer Res 60:4366–4371

Esteller M, Corn PG, Baylin SB, Herman JG (2001) A gene hypermethylation profile of human cancer. Cancer Res 72

Feinberg AP, Vogelstein B (1983) Hypomethylation distinguishes genes of some human cancers from their normal counterparts. Nature 301:89–92

Feinberg AP, Ohlsson R, Henikoff S (2006) The epigenetic progenitor origin of human cancer. Nat Rev Genet 7:21–33

Finch JT, Klug A (1976) Solenoidal model for superstructure in chromatin. Proc Natl Acad Sci U S A 1897

Firestein R, Blander G, Michan S, Oberdoerffer P, Ogino S, Campbell J, Bhimavarapu A, Luikenhuis S, DE Cabo R, Fuchs C, Hahn WC, Guarente LP, Sinclair DA (2008) The SIRT1 deacetylase suppresses intestinal tumorigenesis and colon cancer growth. PLoS One 3:1–9

Fischle W, Kiermer V, Dequiedt F, Verdin E (2001) The emerging role of class II histone deacetylases. Biochem Cell Biol 79:337–348

Fischle W, Dequiedt F, Hendzel MJ, Guenther MG, Lazar MA, Voelter W, Verdin E (2002) Enzymatic activity associated with class II HDACs is dependent on a multiprotein complex containing HDAC3 and SMRT/N-CoR. Mol Cell 9:45–57

Francesca B, Elena S, Erico M, Antonella G, Alessandro S, Alberto B, Germano F, Valeria S (2011) Proteomic analysis identifies differentially expressed proteins in AML1/ETO acute myeloid leukemia cells treated with DNMT inhibitors azacitidine and decitabine. Leuk Res 36:607–618

Gao L, Cueto MA, Asselbergs F, Atadja P (2002) Cloning and functional characterization of HDAC11, a novel member of the human histone deacetylase family. J Biol Chem 277: 25748–25755

Gardner KE, Allis CD, Strahl BD (2011) Operating on chromatin, a colorful language where context matters. J Mol Biol 409:36–46

Geng H, Harvey CT, Pittsenbarger J, Liu Q, Beer TM, Xue C, Qian DZ (2011) HDAC4 protein regulates HIF1α protein lysine acetylation and cancer cell response to hypoxia. J Biol Chem 286:38095–38102

Giralt A, Villarroya F (2012) SIRT3, a pivotal actor in mitochondrial functions: metabolism, cell death and aging. Biochem J 444:1–10

Glover AB, Leyland-Jones B (1987) Biochemistry of azacitidine: a review. Cancer Treat Rep 71:959–964

Goll MG, Bestor TH (2005) Eukaryotic cytosine methyltransferases. Annu Rev Biochem 74:481–514

Gonzalez-Zulueta M, Bender CM, Yang AS, Nguyen T, Beart RW, Van Tornout JM, Jones PA (1995) Methylation of the 5′ CpG island of the p16/CDKN2 tumor suppressor gene in normal and transformed human tissues correlates with gene silencing. Cancer Res 55:4531–4535

Grigoryev SA, Arya G, Correll S, Woodcock CL, Schlick T (2011) Evidence for heteromorphic chromatin fibers from analysis of nucleosome interactions. Proc Natl Acad Sci U S A 106:13317–13322

Grozinger CM, Schreiber SL (2002) Deacetylase enzymes: biological functions and the use of small-molecule inhibitors. Chem Biol 9:3–16

Guenther MG, Barak O, Lazar MA (2001) The SMRT and N-CoR corepressors are activating cofactors for histone deacetylase 3. Mol Cell Biol 21:6091–6101

Guo H-B, Guo H (2007) Mechanism of histone methylation catalyzed by protein lysine methyltransferase SET7/9 and origin of product specificity. Proc Natl Acad Sci U S A 104:8797–8802

Haberland M, Montgomery RL, Olson EN (2009) The many roles of histone deacetylases in development and physiology: implications for disease and therapy. Nat Rev Genet 10:32–42

Herman JG, Merlo A, Mao L, Lapidus RG, Issa JP, Davidson NE, Sidransky D, Baylin SB (1995) Inactivation of the CDKN2/p16/MTS1 gene is frequently associated with aberrant DNA methylation in all common human cancers. Cancer Res 55:4525–4530

Hermann A, Gowher H, Jeltsch A (2004a) Biochemistry and biology of mammalian DNA methyltransferases. Cell Mol Life Sci 61:2571–2587

Hermann A, Goyal R, Jeltsch A (2004b) The Dnmt1 DNA-(cytosine-C5)-methyltransferase methylates DNA processively with high preference for hemimethylated target sites. J Biol Chem 279:48350–48359

Herrera JE, Bergel M, Yang X-J, Nakatani Y, Bustin M (1997) The histone acetyltransferase activity of human GCN5 and PCAF is stabilized by coenzymes. J Biol Chem 272:27253–27258

Hollenbach PW, Nguyen AN, Brady H, Williams M, Ning Y, Richard N, Krushel L, Aukerman SL, Heise C, Macbeth KJ (2004) A comparison of azacitidine and decitabine activities in acute myeloid leukemia cell lines. PLoS One 5:1–10

Hong L, Schroth GP, Matthews HR, Yau P, Bradbury EM (1993) Studies of the DNA binding properties of histone H4 amino terminus. Thermal denaturation studies reveal that acetylation markedly reduces the binding constant of the H4 "tail" to DNA. J Biol Chem 268:305–314

Huang N, Mackerell AD Jr (2004) Atomistic view of base flipping in DNA. Philos Trans A Math Phys Eng Sci 362:1439–1460

Hubbert C, Guardiola A, Shao R, Kawaguchi Y, Ito A, Nixon A, Yoshida M, Wang X-F, Yao T-P (2002) HDAC6 is a microtubule-associated deacetylase. Nature 417:455–458

Insinga A, Monestiroli S, Ronzoni S, Gelmetti V, Marchesi F, Viale A, Altucci L, Nervi C, Minucci S, Pelicci PG (2005) Inhibitors of histone deacetylases induce tumor-selective apoptosis through activation of the death receptor pathway. Nat Med 11:71–76

Jiang JF, Lu JY, Lu D, Liang ZJ, Li LC, Ouyang SS, Kong XQ, Jiang HL, Shen BR, Luo C (2012) Investigation of the acetylation mechanism by GCN5 histone acetyltransferase. PLoS One 7:13p

Jin Z, Dicker DT, El-Deiry WS (2002) Enhanced sensitivity of G1 arrested human cancer cells suggests a novel therapeutic strategy using a combination of simvastatin and TRAIL. Cell Cycle 1:82–89

Jones PL, Veenstra GJ, Wade PA, Vermaak D, Kass SU, Landsberger N, Strouboulis J, Wolffe AP (1998) Methylated DNA and MeCP2 recruit histone deacetylase to repress transcription. Nat Genet 19:187–191

Kantarjian H, Issa JPJ, Rosenfeld CS, Bennett JM, Albitar M, Dipersio J, Klimek V, Slack J, De Castro C, Ravandi F, Helmer R, Shen LL, Nimer SD, Leavitt R, Raza A, Saba H (2006) Decitabine improves patient outcomes in myelodysplastic syndromes—resuits of a phase III randomized study. Cancer 106:1794–1803

Keating GM (2011) Azacitidine: a review of its use in the management of myelodysplastic syndromes/acute myeloid leukaemia. Drugs 72:1111–1136

Keith BG, Junling L, Michael JS, Ru-Qi W, Daniel HA, Steven KD (2003) Role of Class I and Class II histone deacetylases in carcinoma cells using siRNA. Biochem Biophys Res Commun 310:529–536

Kim W, Kim R, Park G, Park J-W, Kim J-E (2011) Deficiency of H3K79 histone methyltransferase Dot1-like protein (DOT1L) inhibits cell proliferation. J Biol Chem 287:5588–5599

Kleinschmidt MA, Streubel G, Samans B, Krause M, Bauer U-M (2008) The protein arginine methyltransferases CARM1 and PRMT1 cooperate in gene regulation. Nucleic Acids Res 36:3202–3213

Kong XG, Lin Z, Liang DM, Fath D, Sang NL, Caro J (2006) Histone deacetylase inhibitors induce VHL and ubiquitin-independent proteasomal degradation of hypoxia-inducible factor 1 alpha. Mol Cell Biol 26:2019–2028

Kouzarides T (2007) Chromatin modifications and their function. Cell 128:693–705

Kramer OH, Muller S, Reichardt S, Heinzel T (2008) Mechanism for ubiquitylation of the leukemia fusion proteins AML1-ETO and PML-RAR alpha. FASEB J 22

Lacoste N, Coté J (2003) Le code épigénétique des histones. Med Sci 19:955–959

Lahm A, Paolini C, Pallaoro M, Nardi MC, Jones P, Neddermann P, Sambucini S, Bottomley MJ, Lo Surdo P, Carfí A, Koch U, De Francesco R, Steinkühler C, Gallinari P (2007) Unraveling the hidden catalytic activity of vertebrate class IIa histone deacetylases. Proc Natl Acad Sci U S A 104:17335–17340

Laird PW (2003) The power and the promise of DNA methylation markers. Nat Rev Cancer 3:253–266

Le Guezennec X, Vermeulen M, Brinkman A, Hoeijmakers W, Cohen A, Lasonder E, Stunnenberg H (2006) MBD2/NuRD and MBD3/NuRD, two distinct complexes with different biochemical and functional properties. Mol Cell Biol 26:843–851

Lee S, Lee HJ, Kim JH, Lee HS, Jang JJ, Kang GH (2003) Aberrant CpG island hypermethylation along multistep hepatocarcinogenesis. Am J Pathol 163:1371–1378

Lee Y-S, Lim K-H, Guo X, Kawaguchi Y, Gao Y, Barrientos T, Ordentlich P, Wang X-F, Counter CM, Yao T-P (2008) The cytoplasmic deacetylase HDAC6 is required for efficient oncogenic tumorigenesis. Cancer Res 68:7561–7569

Leone G, Teofili L, Voso MT, Lubbert M (2002) DNA methylation and demethylating drugs in myelodysplastic syndromes and secondary leukemias. Haematologica 87:1324–1341

Li C-T, Hsiao Y-M, Wu T-C, Lin Y-W, Yeh K-T, Ko J-L (2011a) Vorinostat, SAHA, represses telomerase activity via epigenetic regulation of telomerase reverse transcriptase in non-small cell lung cancer cells. J Cell Biochem 112:3044–3053

Li DW, Xie SB, Ren Y, Huo LH, Gao JM, Cui DD, Liu M, ZHOU J (2011b) Microtubule-associated deacetylase HDAC6 promotes angiogenesis by regulating cell migration in an EB1-dependent manner. Protein Cell 2:150–160

Licciardi PV, Kwa FAA, Ververis K, Di Costanzo N, Balcerczyk A, Tang ML, El-Osta A, Karagiannis TC (2012) Influence of natural and synthetic histone deacetylase inhibitors on chromatin. Antioxid Redox Signal 17:340–354

Lin KT, Momparler RL, Rivard GE (1981) High-performance liquid chromatographic analysis of chemical stability of 5-aza-2′-deoxycytidine. J Pharm Sci 70:1228–1232

Lin K, Wang Y, Chen C, Ho C, Su W, Jou Y (2012a) HDAC inhibitors augmented cell migration and metastasis through induction of PKCs leading to identification of low toxicity modalities for combination cancer therapy. Clin Cancer Res 18:4691–4701

Lin Y-Y, Kiihl S, Suhail Y, Liu S-Y, Chou Y-H, Kuang Z, Lu J-Y, Khor CN, Lin C-L, Bader JS, Irizarry R, Boeke JD (2012b) Functional dissection of lysine deacetylases reveals that HDAC1 and p300 regulate AMPK. Nature 482:251–255

Lindemann RK, Newbold A, Whitecross KF, Cluse LA, Frew AJ, Ellis L, Williams S, Wiegmans AP, Dear AE, Scott CL, Pellegrini M, Wei A, Richon VM, Marks PA, Lowe SW, Smyth MJ, Johnstone RW (2007) Analysis of the apoptotic and therapeutic activities of histone deacetylase inhibitors by using a mouse model of B cell lymphoma. Proc Natl Acad Sci U S A 104:8071

Luger K, Mader AW (1997) Crystal structure of the nucleosome core particle at 2.8 A resolution. Nature 389:251

Ma Y, Cai S, Lu Q, Lu X, Jiang Q, Zhou J, Zhang C (2008) Inhibition of protein deacetylation by trichostatin A impairs microtubule-kinetochore attachment. Cell Mol Life Sci 65:3100–3109

Marks PA (2007) Discovery and development of SAHA as an anticancer agent. Oncogene 26:1351–1356

Marks PW (2012) Decitabine for acute myeloid leukemia. Expert Rev Anticancer Ther 12:299–305

Martinez-Balbas MA, Bannister AJ, Martin K, Haus-Seuffert P, Meisterernst M, Kouzarides T (1998) The acetyltransferase activity of CBP stimulates transcription. EMBO J 17:2886–2893

Matsushita H, Scaglioni PP, Bhaumik M, Rego EM, Cai LF, Majid SM, Miyachi H, Kakizuka A, Miller WH, Pandolfi PP (2006) In vivo analysis of the role of aberrant histone deacetylase recruitment and RAR alpha blockade in the pathogenesis of acute promyelocytic leukemia. J Exp Med 203:821–828

Mckinsey TA, Zhang CL, Lu J, Olson EN (2000) Signal-dependent nuclear export of a histone deacetylase regulates muscle differentiation. Nature 408:106–111

Mercurio C, Minucci S, Pelicci PG (2010) Histone deacetylases and epigenetic therapies of hematological malignancies. Pharmacol Res 62:18–34

Miriam KK, Mark AB (2010) Review: a mobile threat to genome stability: the impact of non-LTR retrotransposons upon the human genome. Semin Cancer Biol 20:211–221

Mombelli M, Lugrin J, Rubino I, Chanson AL, Giddey M, Calandra T, Roger T (2011) Histone deacetylase inhibitors impair antibacterial defenses of macrophages. J Infect Dis 204:1367–1374

Momparler RL, Momparler LF, Samson J (1984) Comparison of the antileukemic activity of 5-AZA-2′-deoxycytidine, 1-beta-D-arabinofuranosylcytosine and 5-azacytidine against L1210 leukemia. Leuk Res 8:1043–1049

Monneret C (2005) Histone deacetylase inhibitors. Eur J Med Chem 40:1–13

Muotri AR, Marchetto MC, Coufal NG, Oefner R, Yeo G, Nakashima K, Gage FH (2010) L1 retrotransposition in neurons is modulated by MeCP2. Nature 468:443–446

Nebbioso A, Clarke N, Voltz E, Germain E, Ambrosino C, Bontempo P, Alvarez R, Schiavone EM, Ferrara F, Bresciani F, Weisz A, De Lera AR, Gronemeyer H, Altucci L (2005) Tumor-selective action of HDAC inhibitors involves TRAIL induction in acute myeloid leukemia cells. Nat Med 11:77–84

Newbold A, Lindemann RK, Cluse LA, Whitecross KF, Dear AE, Johnstone RW (2008) Characterisation of the novel apoptotic and therapeutic activities of the histone deacetylase inhibitor romidepsin. Mol Cancer Ther 7:1066–1079

Ng SS, Yue WW, Oppermann U, Klose RJ (2009) Dynamic protein methylation in chromatin biology. Cell Mol Life Sci 66:407–422

Nordentoft I, Jorgensen P (2003) The acetyltransferase 60 kDa trans-acting regulatory protein of HIV type 1-interacting protein (Tip60) interacts with the translocation E26 transforming-specific leukaemia gene (TEL) and functions as a transcriptional co-repressor. Biochem J 374:165–173

Notari RE, Deyoung JL (1975) Kinetics and mechanisms of degradation of the antileukemic agent 5-azacytidine in aqueous solutions. J Pharm Sci 64:1148–1157

Notbohm H, Hollandt H, Meissner J, Harbers E (1979) Low angle X-ray scattering studies of chromatin in different solvents; analysis by comparison with computer-simulated scattering curves. Int J Biol Macromol 1:180–184

Oberoi J, Fairall L, Watson PJ, Yang JC, Czimmerer Z, Kampmann T, Goult BT, Greenwood JA, Gooch JT, Kallenberger BC, Nagy L, Neuhaus D, Schwabe JW (2011) Structural basis for the assembly of the SMRT/NCoR core transcriptional repression machinery. Nat Struct Mol Biol 18:177–184

Pal S, Sif S (2007) Interplay between chromatin remodelers and protein arginine methyltransferases. J Cell Physiol 213:306–315

Patrick I, Eleftherios PS, Michael S, Daniel F, André F (2011) Endometriosis: inhibition of transcription, expression, and secretion of the vascular epithelial growth factor in human epithelial endometriotic cells by romidepsin. Fertil Steril 95:1579–1583

Paull TT, Rogakou EP, Yamazaki V, Kirchgessner CU, Gellert M, Bonner WM (2000) A critical role for histone H2AX in recruitment of repair factors to nuclear foci after DNA damage. Curr Biol 10:886–895

Peart MJ, Tainton KM, Ruefli AA, Dear AE, Sedelies KA, O'reilly LA, Waterhouse NJ, Trapani JA, Johnstone RW (2003) Novel mechanisms of apoptosis induced by histone deacetylase inhibitors. Cancer Res 63:4460–4471

Peart MJ, Smyth GK, Van Laar RK, Bowtell DD, Richon VM, Marks PA, Holloway AJ, Johnstone RW (2005) Identification and functional significance of genes regulated by structurally different histone deacetylase inhibitors. Proc Natl Acad Sci U S A 102:3697–3702

Peng C, Lu Z, Xie Z, Cheng Z, Chen Y, Tan M, Luo H, Zhang Y, He W, Yang K, Zwaans BM, Tishkoff D, Ho L, Lombard D, He TC, Dai J, Verdin E, Ye Y, Zhao Y (2011) The first identification of lysine malonylation substrates and its regulatory enzyme. Mol Cell Proteomics 10(12):M111

Piekarz RL, Frye R, Turner M, Wright JJ, Allen SL, Kirschbaum MH, Zain J, Prince HM, Leonard JP, Geskin LJ, Reeder C, Joske D, Figg WD, Gardner ER, Steinberg SM, Jaffe ES, Stetler-Stevenson M, Lade S, Fojo AT, Bates SE (2009) Phase II multi-institutional trial of the histone deacetylase inhibitor romidepsin as monotherapy for patients with cutaneous T-cell lymphoma. J Clin Oncol 27:5410–5417

Pitti RM, Marsters SA, Ruppert S, Donahue CJ, Moore A, Ashkenazi A (1996) Induction of apoptosis by Apo-2 ligand, a new member of the tumor necrosis factor cytokine family. J Biol Chem 271:12687–12690

Qian DZ, Kachhap SK, Collis SJ, Verheul HMW, Carducci MA, Atadja P, Pili R (2006) Class II histone deacetylases are associated with VHL-independent regulation of hypoxia-inducible factor 1α. Cancer Res 66:8814–8821

Qian-Ying L, Da-Wei C, Li-Ping X, Rong-Qing Z, Hong-Zhong W (2012) Decitabine, independent of apoptosis, exerts its cytotoxic effects on cell growth in melanoma cells. Environ Toxicol Pharmacol 32:423–429

Ramirez J, Dege C, Kutateladze TG, Hagman J (2012) MBD2 and multiple domains of CHD4 are required for transcriptional repression by Mi-2/NuRD complexes. Mol Cell Biol 32:5078–5088

Rangwala S, Duvic M, Chunlei Z (2012) Trends in the treatment of cutaneous T-cell lymphoma—critical evaluation and perspectives on vorinostat. Blood Lymph Cancer Targets Ther 2:17–27

Rea S, Eisenhaber F, O'Carroll D, Strahl BD, Sun ZW, Schmid M, Opravil S, Mechtler K, Ponting CP, Allis CD, Jenuwein T (2000) Regulation of chromatin structure by site-specific histone H3 methyltransferases. Nature 406:593

Reik W, Dean W, Walter JR (2001) Epigenetic reprogramming in mammalian development. Science 1089

Richon VM, Sandhoff TW, Rifkind RA, Marks PA (2000) Histone deacetylase inhibitor selectively induces p21WAF1 expression and gene-associated histone acetylation. Proc Natl Acad Sci U S A 97:10014–10019

Robbins AR, Jablonski SA, Yen TJ, Yoda K, Robey R, Bates SE, Sackett DL (2005) Inhibitors of histone deacetylases alter kinetochore assembly by disrupting pericentromeric heterochromatin. Cell Cycle 4:717–726

Rodriguez J, Frigola J, Vendrell E, Risques RA, Fraga MF, Morales C, Moreno V, Esteller M, Capellà G, Ribas M, Peinado MA (2006) Chromosomal instability correlates with genome-wide DNA demethylation in human primary colorectal cancers. Cancer Res 66:8462–8468

Rosato RR, Grant S (2005) Histone deacetylase inhibitors: insights into mechanisms of lethality. Expert Opin Ther Targets 9:809–824

Roth SY, Denu JM, Allis CD (2001) Histone acetyltransferases. Annu Rev Biochem 70:81

Roy AF (2000) Phylogenetic classification of prokaryotic and eukaryotic Sir2-like proteins. Biochem Biophys Res Commun 273:793–798

Sapountzi V, Coté J (2010) MYST-family histone acetyltransferases: beyond chromatin. Cell Mol Life Sci 68:1147–1156

Sarfstein R, Bruchim I, Fishman A, Werner H (2010) The mechanism of action of the histone deacetylase inhibitor vorinostat involves interaction with the insulin-like growth factor signaling pathway. PLoS One 6:1–12

Sato A (2012) Vorinostat approved in Japan for treatment of cutaneous T-cell lymphomas: status and prospects. Onco Targets Ther 5:67–75

Sauve AA, Wolberger C, Schramm VL, Boeke JD (2006) The biochemistry of sirtuins. Annu Rev Biochem 75:435–465

Schmudde M, Braun A, Pende D, Sonnemann J, Klier U, Beck JF, Moretta L, Broker BM (2008) Histone deacetylase inhibitors sensitize tumour cells for cytotoxic effects of natural killer cells. Cancer Lett 272:110–121

Schmudde M, Friebe E, Sonnemann J, Beck JF, Broker BM (2010) Histone deacetylase inhibitors prevent activation of tumour-reactive NK cells and T cells but do not interfere with their cytolytic effector functions. Cancer Lett 295:173–181

Schrump DS (2009) Cytotoxicity mediated by histone deacetylase inhibitors in cancer cells: mechanisms and potential clinical implications. Clin Cancer Res 15:3947–3957

Seigneurin-Berny D, Verdel A, Curtet S, Lemercier C, Garin J, Rousseaux S, Khochbin S (2001) Identification of components of the murine histone deacetylase 6 complex: link between acetylation and ubiquitination signaling pathways. Mol Cell Biol 21:8035–8044

Shao L-E, Diccianni MB, Tanaka T, Gribi R, Yu AL, Pullen JD, Camitta BM, Yu J (2001) Thioredoxin expression in primary T-cell acute lymphoblastic leukemia and its therapeutic implication. Cancer Res 61:7333–7338

Sharp AJ, Stathaki E, Migliavacca E, Brahmachary M, Montgomery SB, Dupre Y, Antonarakis SE (2011) DNA methylation profiles of human active and inactive X chromosomes. Genome Res 21:1592–1600

Shi Y, Lan F, Matson C, Mulligan P, Whetstine JR, Cole PA, Casero RA, Shi Y (2004) Histone demethylation mediated by the nuclear amine oxidase homolog LSD1. Cell 119:941–953

Shia WJ, Okumura AJ, Yan M, Sarkeshik A, Lo MC, Matsuura S, Komeno Y, Zhao X, Nimer SD, Yates JR III, Zhang DE (2012) PRMT1 interacts with AML1-ETO to promote its transcriptional activation and progenitor cell proliferative potential. Blood 119:4953–4962

Shin DY, Kim GY, Kim CG, Kim WJ, Kang HS, Choi YH (2012) Anti-invasive effects of decitabine, a DNA methyltransferase inhibitor, through tightening of tight junctions and inhibition of matrix metalloproteinase activities in AGS human gastric carcinoma cells. Oncol Rep 28:1043–1050

Silverman LR, Holland JF, Weinberg RS, Alter BP, Davis RB, Ellison RR, Demakos EP, Cornell CJ Jr, Carey RW, Schiffer C (1993) Effects of treatment with 5-azacytidine on the in vivo and in vitro hematopoiesis in patients with myelodysplastic syndromes. Leukemia 7(Suppl 1):21–9

Silverman LR, Demakos EP, Peterson BL, Kornblith AB, Holland JC, Odchimar-Reissig R, Stone RM, Nelson D, Powell BL, DeCastro CM, Ellerton J, Larson RA, Schiffer CA, Holland JF (2002) Randomized controlled trial of azacitidine in patients with the myelodysplastic syndrome: a study of the cancer and leukemia group B. J Clin Oncol 20:2429–2440

Smallwood SA, Kelsey G (2011) De novo DNA methylation: a germ cell perspective. Trends Genet 28:33–42

Smit AFA, Riggs AD (1996) Tiggers and other DNA transposon fossils in the human genome. Proc Natl Acad Sci U S A 93(4):1443–8 [Online, Accessed 12 Dec 2012]

Smith ZD, Chan MM, Mikkelsen TS, Gu H, Gnirke A, Regev A, Meissner A (2012) A unique regulatory phase of DNA methylation in the early mammalian embryo. Nature 484:339–344

Smrzka OW, Faé I, Stöger R, Kurzbauer R, Fischer GF, Henn T, Weith A, Barlow DP (1995) Conservation of a maternal-specific methylation signal at the human Igf2r locus. Hum Mol Genet 4:1945–1952

Matias S, Fabio S, Arantxa G, Gianmaria F, Mauro R, Oronza AB, Isabella P, Piergiuseppe P, Luciano Di C, Saverio M (2013) The DNA demethylating agent decitabine activates the TRAIL pathway and induces apoptosis in acute myeloid leukemia. Biochim Biophys Acta 1832:114–120

Steinhoff C, Schulz WA (2004) Transcriptional regulation of the human LINE-1 retrotransposon L1.2B. Mol Genet Genomics 270:394–402

Sterner D, Berger S (2000) Acetylation of histones and transcription-related factors. Microbiol Mol Biol Rev 64:435–461

Strathdee G, Mackean MJ, Illand M, Brown R (1999) A role for methylation of the hMLH1 promoter in loss of hMLH1 expression and drug resistance in ovarian cancer. Oncogene 18:2335–2341

Stunkel W, Campbell RM (2011) Sirtuin 1 (SIRT1): the misunderstood HDAC. J Biomol Screen 16:1153–1169
Szerlong HJ, Hansen JC (2011) Nucleosome distribution and linker DNA: connecting nuclear function to dynamic chromatin structure. Biochem Cell Biol 89:24–34
Tachimori A, Yamada N, Sakate Y, Yashiro M, Maeda K, Ohira M, Nishino H, Hirakawa K (2005) Up regulation of ICAM-1 gene expression inhibits tumour growth and liver metastasis in colorectal carcinoma. Eur J Cancer 41:1802–1810
Takai D, Jones PA (2002) Comprehensive analysis of CpG islands in human chromosomes 21 and 22. Proc Natl Acad Sci U S A 3740
Tickenbrock L, Klein HU, Trento C, Hascher A, Göllner S, Bäumer N, Kuss R, Agrawal S, Bug G, Serve H, Thiede C, Ehninger G, Stadt UZ, McClelland M, Wang Y, Becker A, Koschmieder S, Berdel WE, Dugas M, Müller-Tidow C (2011) Increased HDAC1 deposition at hematopoietic promoters in AML and its association with patient survival. Leuk Res 35:620–625
Tiffon C, Adams J, van der Fits L, Wen S, Townsend P, Ganesan A, Hodges E, Vermeer M, Packham G (2011) The histone deacetylase inhibitors vorinostat and romidepsin downmodulate IL-10 expression in cutaneous T-cell lymphoma cells. Br J Pharmacol 162:1590–1602
Turner BM (2000) Histone acetylation and an epigenetic code. Bioessays 22(9):836–45 [Online, Accessed 11 Dec 2012]
Turner BM (2005) Reading signals on the nucleosome with a new nomenclature for modified histones. Nat Struct Mol Biol 12:110–112
Ungerstedt JS, Sowa Y, Xu WS, Shao Y, Dokmanovic M, Perez G, Ngo L, Holmgren A, Jiang X, Marks PA (2005) Role of thioredoxin in the response of normal and transformed cells to histone deacetylase inhibitors. Proc Natl Acad Sci U S A 102:673–678
Vega RB, Matsuda K, Oh J, Barbosa AC, Yang X, Meadows E, McAnally J, Pomajzl C, Shelton JM, Richardson JA, Karsenty G, Olson EN (2004) Histone deacetylase 4 controls chondrocyte hypertrophy during skeletogenesis. Cell 119:555–566
Verdin E, Dequiedt F, Kasler HG (2003) Class II histone deacetylases: versatile regulators. Trends Genet 19:286–293
Völkel P, Angrand P-O (2007) The control of histone lysine methylation in epigenetic regulation. Biochimie 89:1–20
Wang Z, Zang C, Cui K, Schones DE, Barski A, Peng W, Zhao K (2009) Genome-wide mapping of HATs and HDACs reveals distinct functions in active and inactive genes. Cell 138:1019–1031
Wang RR, Li Q, Helfer CM, Jiao J, You JX (2012) Bromodomain protein Brd4 associated with acetylated chromatin is important for maintenance of higher-order chromatin structure. J Biol Chem 287:10738–10752
Weber M, Hellmann I, Stadler MB, Ramos L, Paabo S, Rebhan M, Schubeler D (2007) Distribution, silencing potential and evolutionary impact of promoter DNA methylation in the human genome. Nat Genet 39:457–466
Weichert W (2009) HDAC expression and clinical prognosis in human malignancies. Cancer Lett 280:168–176
Weichert W, Röske A, Gekeler V, Beckers T, Stephan C, Jung K, Fritzsche FR, Niesporek S, Denkert C, Dietel M, Kristiansen G (2008a) Histone deacetylases 1, 2 and 3 are highly expressed in prostate cancer and HDAC2 expression is associated with shorter PSA relapse time after radical prostatectomy. Br J Cancer 98:604–610
Weichert W, Röske A, Niesporek S, Noske A, Buckendahl AC, Dietel M, Gekeler V, Boehm M, Beckers T, Denkert C (2008b) Class I histone deacetylase expression has independent prognostic impact in human colorectal cancer: specific role of class I histone deacetylases in vitro and in vivo. Clin Cancer Res 14:1669–1677
Williams SP, Athey BD, Muglia LJ, Schappe RS, Gough AH, Langmore JP (1986) Chromatin fibers are left-handed double helices with diameter and mass per unit length that depend on linker length. Biophys J 49:233–248
Wilson AJ, Byun DS, Popova N, Murray LB, L'Italien K, Sowa Y, Arango D, Velcich A, Augenlicht LH, Mariadason JM (2006) Histone deacetylase 3 (HDAC3) and other class I HDACs regulate

colon cell maturation and p21 expression and are deregulated in human colon cancer. J Biol Chem 281:13548–13558

Wilson AS, Power BE, Molloy PL (2007) DNA hypomethylation and human diseases. Biochim Biophys Acta 1775:138–162

Witt O, Deubzer HE, Milde T, Oehme I (2009) HDAC family: what are the cancer relevant targets? Cancer Lett 277

Wolf SS (2009) The protein arginine methyltransferase family: an update about function, new perspectives and the physiological role in humans. Cell Mol Life Sci 66:2109–2121

Wolfson NA, Pitcairn CA, Fierke CA (2013) HDAC8 Substrates: histones and beyond. Biopolymers 99:112–126

Woodcock C, Skoultchi A, Fan Y (2006) Role of linker histone in chromatin structure and function: H1 stoichiometry and nucleosome repeat length. Chromosome Res 14:17–25

Wu P, Tian Y, Chen G, Wang B, Gui L, Xi L, Ma X, Fang Y, Zhu T, Wang D, Meng L, Xu G, Wang S, Ma D, Zhou J (2010) Ubiquitin B: an essential mediator of trichostatin A-induced tumor-selective killing in human cancer cells. Cell Death Differ 17:109–118

Wu HC, Delgado-Cruzata L, Flom JD, Perrin M, Liao Y, Ferris JS, Santella RM, Terry MB (2012) Repetitive element DNA methylation levels in white blood cell DNA from sisters discordant for breast cancer from the New York site of the Breast Cancer Family Registry. Carcinogenesis 33:1946–1952

Xiaodong C, Collins RE, Xing Z (2005) Structural and sequence motifs of protein (histone) methylation enzymes. Annu Rev Biophys Biomol Struct 34:267–294

Xie HJ, Noh JH, Kim JK, Jung KH, Eun JW, Bae HJ, Kim MG, Chang YG, Lee JY, Park H, Nam SW (2012) HDAC1 inactivation induces mitotic defect and caspase-independent autophagic cell death in liver cancer. PLoS One 7:e34265–e34265

Xu W, Ngo L, Perez G, Dokmanovic M, Marks PA (2006) Intrinsic apoptotic and thioredoxin pathways in human prostate cancer cell response to histone deacetylase inhibitor. Proc Natl Acad Sci U S A 103:15540–15545

Xu WS, Parmigiani RB, Marks PA (2007) Histone deacetylase inhibitors: molecular mechanisms of action. Oncogene 26:5541–52

Yang XJ (2004) The diverse superfamily of lysine acetyltransferases and their roles in leukemia and other diseases. Nucleic Acids Res 32:959–976

Yang X-J, Seto E (2008) The Rpd3/Hda1 family of lysine deacetylases: from bacteria and yeast to mice and men. Nat Rev Mol Cell Biol 9:206–218

Yu F, Zingler N, Schumann G, Stratling WH (2001) Methyl-CpG-binding protein 2 represses LINE-1 expression and retrotransposition but not Alu transcription. Nucleic Acids Res 29:4493–4501

Yuan H, Marmorstein R (2012) Structural basis for sirtuin activity and inhibition. J Biol Chem 287:42428–42435

Zhang X, Bruice TC (2008) Enzymatic mechanism and product specificity of SET-domain protein lysine methyltransferases. Proc Natl Acad Sci U S A 105:5728–5732

Zhang CL, Mckinsey TA, Lu JR, Olson EN (2001) Association of COOH-terminal-binding protein (CtBP) and MEF2-interacting transcription repressor (MITR) contributes to transcriptional repression of the MEF2 transcription factor. J Biol Chem 276:35–39

Zhang X, Zhang Z, Chen G, Zhao M, Wang D, Zhang X, Du Z, Xu Y, Yu X (2010) FK228 induces mitotic catastrophe in A549 cells by mistargeting chromosomal passenger complex localization through changing centromeric H3K9 hypoacetylation. Acta Biochim Biophys Sin (Shanghai) 42:677–687

Zhu P, Martin E, Mengwasser J, Schlag P, Janssen KP, Göttlicher M (2004) Induction of HDAC2 expression upon loss of APC in colorectal tumorigenesis. Cancer Cell 5:455–463

Ziemba A, Hayes E, Freeman BB III, Tao Y, Pizzorno G (2011) Development of an oral form of azacytidine: 2′3′25′ triacetyl-5-azacytidine. Chemother Res Pract 2011:965826

Chapter 4
Epigenetics: Role of Histone Proteases in Cellular Functions and Diseases

Papita Mandal, Naveen Verma, Gajendra K. Azad, Vikash Singh, Upendarrao Golla, and Raghuvir S. Tomar

Abstract In the past few decades, exciting advances have been made to understand the epigenetic regulation of chromatin structure and function. There has been tremendous progress in the identification and study of protein complexes of catalytic nature, which reversibly modify chromatin (DNA as well as histone proteins) during various nuclear processes that involves DNA. Histone proteins undergo post-translational modifications (PTMs) like acetylation, phosphorylation, ubiquitination, methylation, and proteolytic clipping. However, the proteolytic clipping of histone tails is not as well understood as other covalent modifications. In some cases, the proteolytic processing, particularly of histone H3 and H1, has been considered as a physiologically regulated event. For example, in *Tetrahymena*, six amino acids are removed from the NH_2-terminus of histone H3 in transcriptionally silent micronuclei. Similarly, during viral infection of foot-and-mouth disease virus, H3 has been reported to be cleaved between Leu20 and Ala21 from the NH_2-terminus. Lately, in parallel to the emergence of the "histone code" hypothesis, there has been substantial excitement in the field of site-specific proteolytic processing of some of the core histones. A chromatin-bound proteolytic activity with unique specificity for histone H2A has long been identified and characterized in quite detail. Recently, human Cathepsin L and an unidentified protease in yeast and another in chicken liver have been shown to cleave H3 from NH_2-terminus. Such processing of histones has the potential to regulate chromatin dynamics to an extent that makes it physiologically relevant and crucial. This comprehensive review will shed light on advancements made so far on proteolytic processing of histones and future directions of study. Here we discuss the biochemical properties and biological functions of histone proteolysis in transcription, viral diseases, stem cell differentiation, and sporulation.

P. Mandal • N. Verma, M.S. • G.K. Azad, M.S. • V. Singh, M.S. • U. Golla, M.S.
R.S. Tomar, Ph.D. (✉)
Laboratory of Chromatin Biology, Indian Institute of Science Education and Research, ITI (Gas Rahat) Building, Bhopal 462 023, India
e-mail: papita@iiserb.ac.in; naveen.7.2.1990@gmail.com; gajendra@iiserb.ac.in; vikash@iiserb.ac.in; upendarrao@iiserb.ac.in; rst@iiserb.ac.in

N. Maulik and T. Karagiannis (eds.), *Molecular Mechanisms and Physiology of Disease: Implications for Epigenetics and Health*, DOI 10.1007/978-1-4939-0706-9_4,

Keywords Protease • Histones • Post-translation modifications • Epigenetics • Diseases

4.1 Introduction

In eukaryotic cells, genes are complexed with core histones and other chromosomal proteins to form chromatin (Luger et al. 1997; Wolffe and Guschin 2000; Workman and Kingston 1998). The basic repeating unit of chromatin, the nucleosome (Wolffe and Kurumizaka 1998; Wolffe and Guschin 2000), includes two copies of each of the four core histones (H2A, H2B, H3, and H4) wrapped around by 146 base pairs of DNA (Neely and Workman 2002; Luger et al. 2012; Andrews and Luger 2011; Richmond et al. 1984). With the aid of linker histone H1, the nucleosomes are further packaged into 30 nm fibers with six nucleosomes per turn in a spiral or solenoidal arrangement to form chromatin (Osipova et al. 1990; Hayes et al. 1990; Pachov et al. 2011; Woodcock et al. 2006; Woodcock and Dimitrov 2001). The precise organization of chromatin is very critical for many cellular processes, including transcription (Workman and Kingston 1998; Cosgrove 2012; Adams and Workman 1993; Ajiro and Allis 2002; Steger et al. 1998; Jansen et al. 2012; Saunders et al. 1990; Workman and Abmayr 2004), replication (Vignali et al. 2000; Travers et al. 2012), repair (Lukas et al. 2011; Soria et al. 2012; Widlak et al. 2006; Vaissiere and Herceg 2010), recombination, chromosomal segregation (Berchowitz et al. 2009), etc.

As an important component of the nucleosome, each core histone is composed of a structured three helix domain called histone fold (Arents and Moudrianakis 1995; Kukimoto et al. 2004; Simon et al. 2011; Bando et al. 1997) and two unstructured tails (Luger et al. 2000; Luger and Richmond 1998; Ouzounis and Kyrpides 1996; Simon et al. 2011; Depken and Schiessel 2009; Zlatanova and van Holde 1992). The histone tails regulate gene expression by affecting the dynamics of chromatin structure (Wolffe and Guschin 2000; Widom 1998; Chodaparambil et al. 2006). The dynamic changes in chromatin structure are directly influenced by PTMs (Imhof and Becker 2001) such as acetylation (Carrozza et al. 2003; Eberharter and Becker 2002), phosphorylation (Cerutti and Casas-Mollano 2009), poly(ADP-ribosylation) (Burzio et al. 1979; Poirier and Savard 1980), ubiquitination (Ma et al. 2011; Chandrasekharan et al. 2009, 2010; Shilatifard 2006), and methylation (Li et al. 2009; Lu et al. 2008) of specific amino acids within the tails of histones as shown in Fig. 4.1. The "histone code" hypothesis predicts that a preexisting modification affects subsequent modifications on histone tails which serve to recruit different proteins or protein complexes to regulate diverse chromatin functions such as gene expression, DNA replication, and chromosomal segregation (Chakravarthy et al. 2005). The removal of these tails (and consequently the crucial residues) thus can drastically affect the modifications present on the histones.

The highly basic histone tails are predicted to be less structured than the central histone-fold regions and are believed to interact with the negatively charged DNA backbone or with other chromatin-associated proteins (Bharath et al. 2002;

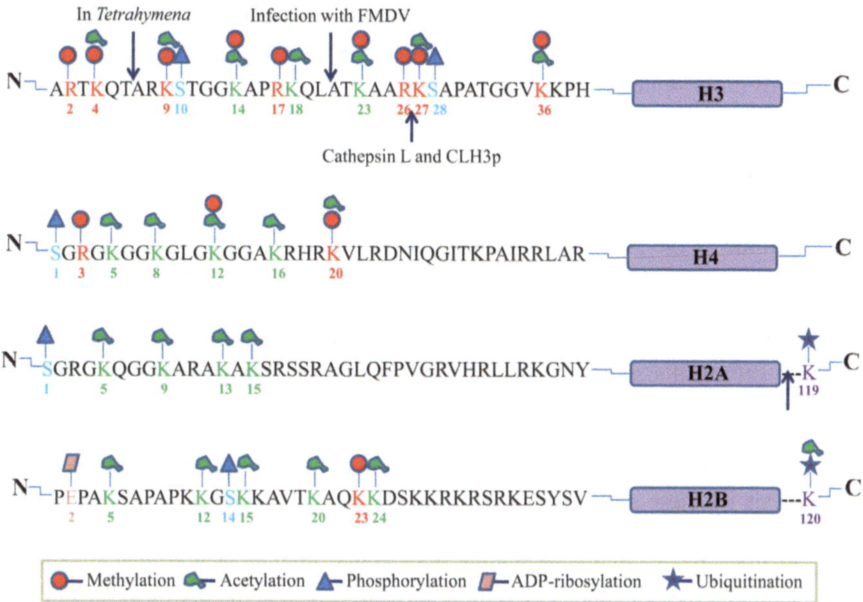

Fig. 4.1 Post-translational modifications and proteolytic cleavage of histone tails. Histones undergo various reversible or irriversible covalent post-translational modifications (PTMs) like acetylation, methylation, phosphorylation, ubiquitination, sumoylation, and proteolytic processing during development and disease pathogenesis. The sequence-specific or PTM-specific cleavage sites on histone tails by various proteases are marked along with the different PTMs

Vogler et al. 2010). The tails of histones are also thought to be involved in nucleosome–nucleosome interaction for establishing transcriptionally repressive chromatin, referred to as heterochromatin. The covalent PTMs of histone tails may alter their interaction with DNA, with chromatin-associated proteins, and with other regulatory proteins that may be required for various cellular processes.

In general, cellular proteins are continuously synthesized and degraded (Pedersen et al. 1986; Ward and Richardson 1991). Balance between synthesis and degradation of proteins is one of the major regulatory factors behind several biological processes. The homeostasis of proteins in cells is determined by the equilibrium between the rates of synthesis and degradation. Proteases are key molecules for regulation of turnover of proteins and different physiological processes such as apoptosis, blood coagulation, cell cycle control, and removal of histone proteins during fertilization (Magdalena and Adam 2012; Morin et al. 2012).

The NH_2-termini of histones are exposed and labile to proteolysis and portions of certain histone tails undergo cleavage at precise stages in the cell cycle, development, differentiation, aging, sporulation, and viral infection. The enzymatic degradation of histones has been described in different tissues and organisms by different research groups (Bartley and Chalkley 1970; Falk et al. 1990; Elia and Moudrianakis 1988; Eickbush et al. 1976, 1988; Chong et al. 1974; Grigera and Tisminetzky 1984;

Davie et al. 1986; Sakai et al. 1987; Tesar and Marquardt 1990). Several investigators have reported the presence of proteolytic activity capable of degrading histones as well as nonhistone proteins in chromatin from a variety of sources such as calf thymus (Lipinska and Klyszejko-Stefanowicz 1980), rabbit thymus, *Xenopus* liver, rat liver (Lipinska et al. 1980; Garrels et al. 1972; Ramponi et al. 1978), rat ventral prostate, rat macrophages, cycad pollen (Brandt and Von Holt 1975), chicken liver (Mandal et al. 2012), fish, mouse (Duncan et al. 2008), and yeast (Santos-Rosa et al. 2009). Some of the proteases and their specific cleavage sites for proteolytic processing are represented in Fig. 4.1.

Also, over the past 30 years, a number of investigators reported the presence of proteolytic activity in isolated nuclei and chromatin. EUO gene of an obligate intracellular pathogen *Chlamydia trachomatis* has been shown to encode a histone H1-specific protease (HSP) (Kaul et al. 1997). The above reports clearly indicate that the proteolytic processing of histones is a universally conserved process. But, the exact mechanism of histone removal/degradation as well as the necessity of proteolysis by these molecules is not clear. However, two possibilities exist: first, the histones may be degraded directly from the chromatin and then replaced by fresh histones from a pool and second, the histones may be in equilibrium between chromatin and the pool and degradation may take place in the pool itself under certain physiological conditions.

4.2 Classification of Histone Proteases

4.2.1 Classification Based on pH

The histone proteases can be classified according to the pH range found optimum for the in vitro activity of the proteases.

Acidic proteases: Degrade histones at acidic pH
Neutral proteases: Degrade histones at neutral pH
Alkaline proteases: Degrade histones at alkaline pH

4.2.2 Classification Based on Specificity

Histone-Specific Proteases

Histone specificity refers to the property of the histone-clipping proteases by virtue of which they cleave only a particular histone and not any other. The HSP activities which have been discovered and characterized till now are described below.

A histone H2A-specific protease activity was discovered by Eickbush et al. in 1976 (Eickbush et al. 1976) in the chromatin of calf thymus. During the analysis of prepared total histones, one extra protein band was seen which migrated between histone H2A and H4 during SDS–polyacrylamide gel electrophoresis. The appearance of this extra band was correlated with the disappearance of the histone H2A band during the lengthy extraction procedure of histones, suggesting the existence of a histone H2A protease. The chromatin of calf liver and brain were also tested for the presence of H2A cleavage activity under similar conditions but no such activity could be demonstrated with brain chromatin. Later, by partial amino acid sequencing of the extra band, the cleavage site was mapped between Val114 and Leu115 at the C-terminus of histone H2A. Since its discovery, extensive work has been carried out to characterize the histone H2A-specific protease activity in detail.

The protease responsible for proteolysis of histone H2A has been partially purified from calf thymus chromatin. The partial purification procedure involves the extraction of chromatin with a salt concentration at which histone H2A and H2B are released from the chromatin suggesting that H2A-specific protease is bound tightly with chromatin. The isolated H2A protease has been used as a unique probe for the analysis of histone octamer structure. In vitro histone octamer reconstitution and subsequent gel filtration studies have shown that C-terminally cleaved H2A (cH2A) can form dimers with H2B but this dimer has substantially lowered affinity for H3:H4 tetramer. Therefore, it is suggested that the removal of 15C-terminal residues from H2A would result in a decrease in octamer stability. It has also been studied that H2A present in histone octamer which is complexed with DNA is not accessible to H2A-protease. On the other hand, the H2A protease cleaves histone H2A present in H2A:H2B dimers bound to DNA and the proteolysis of H2A is inhibited if the enzyme is treated with H3:H4 tetramer-DNA complex prior to the addition of H2A:H2B dimer-DNA complex. Further, it was demonstrated that under physiological conditions the H2A-specific protease binds primarily to the highly basic NH_2-terminal tail domain of H3:H4 tetramers and this binding lowers the effective concentration of the enzyme available for cleavage of H2A. The H2A-specific protease cleaves octameric H2A when tails of the H3:H4 tetramers are modified by acetylation suggesting that the activity of this H2A-specific protease is regulated by PTMs of the tail domains of histone H3:H4 tetramers (Elia and Moudrianakis 1988).

A chromatin-bound HSP has also been purified (Dyson and Walker 1984) from calf thymus chromatin and it is speculated that this protease may play a role in the regulation of gene expression or turnover of the protein. The EUO gene of *Chlamydia trachomatis* has been shown to encode an HSP. A neutral protease B (Tsurugi and Ogata 1982) specific to histone H1 has been partially purified (Tsurugi and Ogata 1986) from rat liver chromatin by chromatography through sepharose 6B and DEAE-Sephadex columns. The neutral protease B was found to degrade all the core histones. However, the histone H1 was specifically digested only when it was complexed with an equal amount of DNA. A nucleotide and pyrophosphate-dependent histone H1-specific proteolytic activity has been demonstrated in permeabilized human lymphocytes (Surowy and Berger 1983). Specific proteases for histone H4 and H2B, if they exist, are yet to be discovered.

In mammalian kidney cells (BHK cells) infected with foot-and-mouth disease virus (FMDV) (Grigera and Tisminetzky 1984), histone H3 is selectively degraded, and the product (named as Pi) migrates just above H4 in SDS–polyacrylamide gels. In these cells, the degradation of H3 was blocked by cycloheximide treatment immediately after the infection. But when the protein synthesis inhibitor was added 2 h post-infection, H3 cleavage was observed which suggests that the H3-specific protease was translated immediately after infection. The conversion of H3 to Pi has also been observed in IB-RS2 (a swine-derived cell line) cells when infected with FMDV but not when infected with Herpes Simplex Virus (HSV) suggesting that the conversion of H3 to Pi is caused by the FMDV virus and not by the host.

Further, it was demonstrated that the conversion of H3 to Pi is catalyzed by the FMDV 3C protease and the cleavage site has been mapped between Leu20 and Ala21 from the N-terminal end (Falk et al. 1990). The processed H3 (Pi), where N-terminal is truncated, lacks most of the lysine residues responsible for acetylation, and may thus be unable to maintain the conformation required for chromatin to be transcriptionally active; and this cleavage would therefore shut off the host cell transcription (Tesar and Marquardt 1990).

Brandt and Von Holt (1975) had also observed the proteolytic degradation of histones H3 and H4 during the extraction of histones from chromatin of cycad pollen by 0.25 N HCl. They mapped the cleavage site in histone H3 between Lys23 and Ala24, and discussed that selective cleavage of certain histones could also provide a mechanism for the regulation of gene expression.

Histone Nonspecific Proteases

Histone subtype nonspecific proteases having protease activity similar to trypsin have also been reported by several investigators, which degrade most of the histones nonspecifically (Harvima et al. 1988). However the susceptibility of all the five histones to nonspecific proteolysis is critically dependent upon whether histones are free or in association with DNA. In the nucleohistone complex, histone H1 is the first protein susceptible for proteolysis, whereas the core histones remain almost resistant. But once they are dissociated from DNA, the H1 becomes resistant to proteolysis and core histones are rapidly degraded suggesting that the specificity of the proteases is dependent on the accessibility and conformation of the histones. Such a histone subtype nonspecific proteolytic activity which is tightly bound to chromatin was observed in calf thymus (Bartley and Chalkley 1970). Another neutral protease of the same nature with a molecular weight of 200 KDa, purified from rat liver chromatin, also degrades histones and nonhistone proteins (Chong et al. 1974). A possible role for this endogenous chromatin-bound protease has been suggested in gene activation and in the removal of histones from DNA during spermatogenesis. An entirely different protease activity associated with avian erythroid chromatin was detected which was capable of degrading histones of chromatin at pH 3, and was less active at pH 9 but had no activity at neutral pH (Harlow and Wells 1975). In calf thymus, a calcium-activated neutral protease (CANP) was

Table 4.1 List of proteases identified and their source organism

S. No	Type of histone protease	Origin	References
1	H3 protease	Tetrahymena	Allis et al. (1980)
2	H3 protease	Viral	Falk et al. (1990), Tesar and Marquardt (1990)
3	H3 protease	Yeast	Santos-Rosa et al. (2009)
4	H3 protease	Chicken	Mandal et al. (2012)
5	H2A protease	Calf thymus	Lipinska and Klyszejko-Stefanowicz (1980)
6	Cathepsin L H3 protease	Mouse	Duncan et al. (2008)
7	H1-like protease	Tetrahymena	Allis et al. (1984)

reported (Sakai et al. 1987) which cleaves histone H2A, H2B, and H3 at certain sequences. Further, a protease activity specific to histones from sperm was found in the egg extract of sea urchin (Suzuki et al. 1990). This protease was purified and characterized with the monomeric molecular weight of 28 kDa and named as SPKK protease because of its unique specificity to the SPKK motif which is present in sperm histones H2B and H1. These histones from sea urchin sperm are rich in the SPKK sequence which is not present in usual histones. It has been discussed that this SPKK protease may play a role in the unpacking of sperm chromatin and transcriptional activation of male origin genes during fertilization (Suzuki et al. 1990). A list of histone proteases identified in different organisms is shown in Table 4.1.

4.3 Significance of Proteolytic Processing of Histone Tails

Although substantial information is available now on histone proteases, their roles in vivo are yet to be ascertained. Different investigators have speculated different roles of histone proteases on the basis of certain experimental analyses. The NH_2-terminal residues of histone tails are subjected to a wide array of PTMs including acetylation, methylation, phosphorylation, citrullination, ubiquitination, ADP-ribosylation, sumoylation, and proteolytic processing (Gardner et al. 2011). The combinatorial action of these modifications regulates critical DNA processes including replication, repair, and transcription. These modifications have been correlated with a variety of human diseases including arthritis, cancer, heart disease, diabetes, and neurodegenerative disorders (Portela and Esteller 2010). Almost all of these PTMs are quite well established, except for proteolytic processing which is poorly understood. Majority of the investigators have discussed the possible role of histone proteases either in the turnover of histones, the derepression of genes, or in the removal of histones during spermatogenesis. Recently, Jenuwein and Allis (2001) have hypothesized in their "histone code" hypothesis that histone proteases might play a role in erasing methylation marks from the tails of histones by proteolysis. Some of the speculations and evidences about the role of histone proteases are described here.

Cleavage of histone H3 has been observed in BHK-21 cells upon treatment with FMDV. Later it was shown that FMDV encodes a 3C protease which induces the cleavage of twenty amino acids from nuclear histone H3. Since the acetylation of NH_2-terminal tails of histone H3 and H4 is known to occur for the activation of transcription, the cleavage of the N-terminal domain of histone H3 during infection could be an efficient way for FMDV to switch off host cell transcription (Falk et al. 1990).

Chromatin undergoes structural and chemical changes during development and aging (Moindrot et al. 2012; O'Sullivan and Karlseder 2012). Recently, it has been shown that Cathepsin L-dependent proteolytic cleavage of NH_2-terminus of histone H3 is required for stem cell differentiation (Duncan et al. 2008) and a yet unidentified H3 protease is responsible for the expression of genes required for sporulation in yeast (Santos-Rosa et al. 2009). Progesterone hormone-induced proteolytic processing of histone H3 in Japanese quail has also been observed (Mahendra and Kanungo 2000). Also, it has been shown that tissues having high cell turnover like thymus and intestinal mucosa exhibit a greater rate of proteolysis of nucleohistone compared to other tissues (Bartley and Chalkley 1970). Several other investigators have also drawn similar conclusions.

In another experiment, the mass of the erythrocyte nuclei was increased, chromatin dispersed, and RNA synthesis resumed when chick erythrocytes (which contain dormant nuclei with condensed chromatin) were fused with HeLa cells suggesting the presence of some factors in the cytoplasm of HeLa cells that reactivate inactive erythrocyte nuclei. The reactivation of fused erythrocyte nuclei was found to be suppressed in the presence of protease inhibitors which suggests indirectly that cellular proteases might play a role in reactivation of erythrocyte nuclei (Johnson and Harris 1969; Van der Veer and Bootsma 1982). The role of histone proteolysis has also been postulated in the inactivation of Tetrahymena micronuclei during growth and development (Van der Veer and Bootsma 1982).

The biological implication of proteolytic processing would be that the selective cleavage of certain histones might provide a mechanism for the regulation of genes at the transcriptional level. For example, proteolytic cleavage of histones in highly packed chromatin may result in unfolding of a particular region. Perhaps the cell retains the enzyme in a sequestered form until it is required. For example, it is known that H2A protease is able to either open the octamer or shift the equilibrium from an octamer toward an H3-H4 tetramer and cH2A-H2B dimer in solution. Thus, the enzyme may be able to completely destabilize the nucleosome in vivo (Eickbush et al. 1988; Elia and Moudrianakis 1988; Watson and Moudrianakis 1982).

During vertebrate spermatogenesis and other specialized developmental situations, nearly complete removal of histones from the genome is known to occur. Therefore, a role of histone proteases has also been proposed during spermatogenesis when nucleohistone is transformed into the nucleoprotamine complex.

It has been observed that a scaffold-associated protease is able to degrade the histone H1 selectively in the presence of DNA containing single-strand breaks induced by gamma-radiation or DNaseI treatment. This experiment suggests that HSP activated by gamma-irradiated DNA may be involved in the regulation of the access of repair enzymes to the damaged portions of DNA within chromatin (Gaziev and Kutsyi 1988, 1992).

4.4 Possible Regulatory Mechanism of Proteolytic Processing of Histones

The physiological significance of proteolytic processing of histone tails is not clear. However, following are the possibilities that could regulate or be regulated by proteolytic clipping of histones: PTMs of certain amino acids, viral infections, aging, DNA damage, differentiation, gene expression, and apoptosis. Specific PTMs of an enzyme or histones can alter the specificity of proteolytic enzymes. The susceptibility of histones to proteolysis in chromatin could be regulated by the addition or removal of acetyl or methyl groups on lysine residues near the site of cleavage. For example, in mammalian systems, Cathepsin L has been shown to cleave histone H3 during stem cell differentiation (Duncan et al. 2008). In yeast, a similar activity has been identified which is required for the expression of sporulation-specific genes (Santos-Rosa et al. 2009). Both these H3 cleaving activities are regulated by PTMs of histone H3.

The modification of certain lysine residues may either make the lysine residues unrecognizable to proteases thereby preventing a cleavage at that site or may lead to a change in the interaction of histones with DNA or proteins. For example, upon viral infection by FMDV (Falk et al. 1990; Tesar and Marquardt 1990), cellular transcription was found to be inhibited which has been strongly correlated with the NH_2-terminal proteolytic processing of histone H3. The perturbation of cross-talk between different PTMs by proteolytic processing may be correlated with the maintenance of different transcriptional states and be tightly regulated with the help of different histone PTMs and modifiers as shown in Fig. 4.2. The transcriptionally active chromatin differs from inactive chromatin by acetylated lysine residues at the NH_2-termini of core histones. Therefore, the NH_2-terminally truncated H3 lacks most of the lysine residues for acetylation and may thus be unable to maintain the conformation required for chromatin to be transcriptionally active. Similarly, the NH_2-terminal proteolytic processing of histone H3 during aging may also be responsible for the decline in gene expression (Mandal et al. 2012).

4.5 Conclusion

Since the histones are such an indispensable part of chromatin, any major changes in their sequence or structure may lead to dramatic effects on the cellular metabolism. Although not much is known about the different histone proteases and their

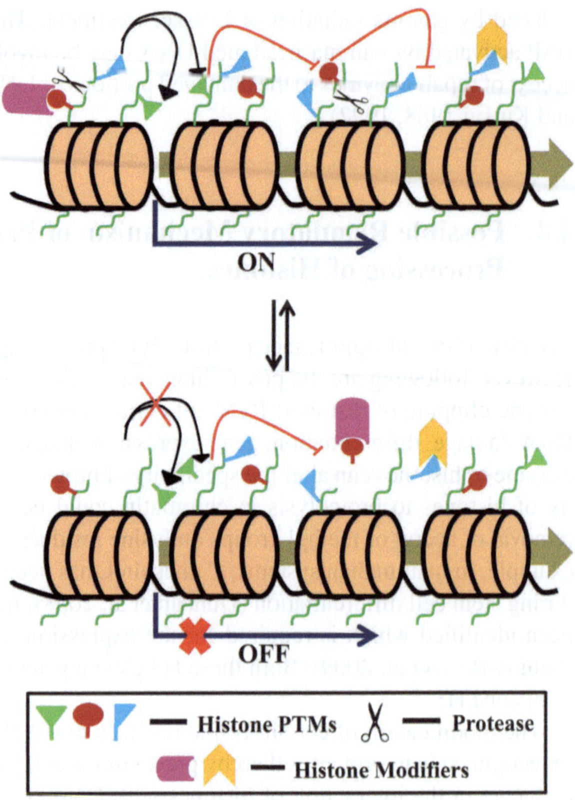

Fig. 4.2 Regulation of gene expression by proteases. Proteolytic processing of histone tails by proteases results in the loss of various post-translational modifications and the cross-talk between these modifications which regulate gene expression and epigenetic fate. Proteases define actual or potential transcription states directly or indirectly with the help of various chromatin modifiers and resultant PTMs during development and disease pathogenesis. "Compatible" modifications (which facilitate other modifications to occur and/or can coexist) are represented by *black arrows* and "incompatible" modifications (which negatively affect other modification and/or cannot coexist) are shown in *red*

role in the cellular processes, one can be sure that the phenomenon has a significant relevance and importance albeit in a temporally or spatially regulated manner in an organism. Extensive research needs to be carried out in this field in order to reveal more yet unidentified histone proteases and their functions.

References

Adams CC, Workman JL (1993) Nucleosome displacement in transcription. Cell 72:305–308

Ajiro K, Allis CD (2002) Histone code hypothesis. Tanpakushitsu Kakusan Koso 47:753–760

Allis CD, Allen RL, Wiggins JC, Chicoine LG, Richman R (1984) Proteolytic processing of h1-like histones in chromatin: a physiologically and developmentally regulated event in Tetrahymena micronuclei. J Cell Biol 99:1669–1677

Allis CD, Bowen JK, Abraham GN, Glover CV, Gorovsky MA (1980) Proteolytic processing of histone H3 in chromatin: a physiologically regulated event in Tetrahymena micronuclei. Cell 20:55–64

Andrews AJ, Luger K (2011) Nucleosome structure(s) and stability: variations on a theme. Annu Rev Biophys 40:99–117

Arents G, Moudrianakis EN (1995) The histone fold: a ubiquitous architectural motif utilized in DNA compaction and protein dimerization. Proc Natl Acad Sci U S A 92:11170–11174

Bando M, Ijuin S, Hasegawa S, Horikoshi M (1997) The involvement of the histone fold motifs in the mutual interaction between human TAF(II)80 and TAF(II)22. J Biochem 121:591–597
Bartley J, Chalkley R (1970) Further studies of a thymus nucleohistone-associated protease. J Biol Chem 245:4286–4292
Berchowitz LE, Hanlon SE, Lieb JD, Copenhaver GP (2009) A positive but complex association between meiotic double-strand break hotspots and open chromatin in Saccharomyces cerevisiae. Genome Res 19:2245–2257
Bharath MM, Chandra NR, Rao MR (2002) Prediction of an HMG-box fold in the C-terminal domain of histone H1: insights into its role in DNA condensation. Proteins 49:71–81
Brandt WF, Von Holt C (1975) Isolation and characterization of the histones from cycad pollen. FEBS Lett 51:84–87
Burzio LO, Riquelme PT, Koide SS (1979) ADP ribosylation of rat liver nucleosomal core histones. J Biol Chem 254:3029–3037
Carrozza MJ, Utley RT, Workman JL, Cote J (2003) The diverse functions of histone acetyltransferase complexes. Trends Genet 19:321–329
Cerutti H, Casas-Mollano JA (2009) Histone H3 phosphorylation: universal code or lineage specific dialects? Epigenetics 4:71–75
Chakravarthy S, Park YJ, Chodaparambil J, Edayathumangalam RS, Luger K (2005) Structure and dynamic properties of nucleosome core particles. FEBS Lett 579:895–898
Chandrasekharan MB, Huang F, Sun ZW (2009) Ubiquitination of histone H2B regulates chromatin dynamics by enhancing nucleosome stability. Proc Natl Acad Sci U S A 106:16686–16691
Chandrasekharan MB, Huang F, Sun ZW (2010) Histone H2B ubiquitination and beyond: Regulation of nucleosome stability, chromatin dynamics and the trans-histone H3 methylation. Epigenetics 5:460–468
Chodaparambil JV, Edayathumangalam RS, Bao Y, Park YJ, Luger K (2006) Nucleosome structure and function. Ernst Schering Res Found Workshop 29–46
Chong MT, Garrard WT, Bonner J (1974) Purification and properties of a neutral protease from rat liver chromatin. Biochemistry 13:5128–5134
Cosgrove MS (2012) Writers and readers: deconvoluting the harmonic complexity of the histone code. Nat Struct Mol Biol 19:739–740
Davie JR, Numerow L, Delcuve GP (1986) The nonhistone chromosomal protein, H2A-specific protease, is selectively associated with nucleosomes containing histone H1. J Biol Chem 261:10410–10416
Depken M, Schiessel H (2009) Nucleosome shape dictates chromatin fiber structure. Biophys J 96:777–784
Duncan EM, Muratore-Schroeder TL, Cook RG, Garcia BA, Shabanowitz J, Hunt DF, Allis CD (2008) Cathepsin L proteolytically processes histone H3 during mouse embryonic stem cell differentiation. Cell 135:284–294
Dyson M, Walker JM (1984) Chromatin associated protease from calf thymus. Int J Pept Protein Res 24:201–207
Eberharter A, Becker PB (2002) Histone acetylation: a switch between repressive and permissive chromatin. Second in review series on chromatin dynamics. EMBO Rep 3:224–229
Eickbush TH, Godfrey JE, Elia MC, Moudrianakis EN (1988) H2a-specific proteolysis as a unique probe in the analysis of the histone octamer. J Biol Chem 263:18972–18978
Eickbush TH, Watson DK, Moudrianakis EN (1976) A chromatin-bound proteolytic activity with unique specificity for histone H2A. Cell 9:785–792
Elia MC, Moudrianakis EN (1988) Regulation of H2a-specific proteolysis by the histone H3:H4 tetramer. J Biol Chem 263:9958–9964
Falk MM, Grigera PR, Bergmann IE, Zibert A, Multhaup G, Beck E (1990) Foot-and-mouth disease virus protease 3C induces specific proteolytic cleavage of host cell histone H3. J Virol 64:748–756
Gardner KE, Allis CD, Strahl BD (2011) Operating on chromatin, a colorful language where context matters. J Mol Biol 409:36–46

Garrels JI, Elgin SC, Bonner J (1972) A histone protease of rat liver chromatin. Biochem Biophys Res Commun 46:545–551

Gaziev AI, Kutsyi MP (1988) Histone H1-specific proteinase is associated with the nuclear matrix and is activated by DNA-containing breaks or denatured sites. Dokl Akad Nauk SSSR 299:240–242

Gaziev AI, Kutsyi MP (1992) Gamma-irradiated DNA activates histone H1-specific proteinase of rat liver nuclei. Int J Radiat Biol 61:169–174

Grigera PR, Tisminetzky SG (1984) Histone H3 modification in BHK cells infected with foot-and-mouth disease virus. Virology 136:10–19

Harlow R, Wells JR (1975) Histone proteases of avian erythroid cells. J Cell Sci 18:217–225

Harvima RJ, Yabe K, Fraki JE, Fukuyama K, Epstein WL (1988) Hydrolysis of histones by proteinases. Biochem J 250:859–864

Hayes JJ, Tullius TD, Wolffe AP (1990) The structure of DNA in a nucleosome. Proc Natl Acad Sci U S A 87:7405–7409

Imhof A, Becker PB (2001) Modifications of the histone N-terminal domains. Evidence for an "epigenetic code"? Mol Biotechnol 17:1–13

Jansen A, Van Der Zande E, Meert W, Fink GR, Verstrepen KJ (2012) Distal chromatin structure influences local nucleosome positions and gene expression. Nucleic Acids Res 40:3870–3885

Jenuwein T, Allis CD (2001) Translating the histone code. Science 293:1074–1080

Johnson RT, Harris H (1969) DNA synthesis and mitosis in fused cells. II. HeLa-chick erythrocyte heterokaryons. J Cell Sci 5:625–643

Kaul R, Hoang A, Yau P, Bradbury EM, Wenman WM (1997) The chlamydial EUO gene encodes a histone H1-specific protease. J Bacteriol 179:5928–5934

Kukimoto I, Elderkin S, Grimaldi M, Oelgeschlager T, Varga-Weisz PD (2004) The histone-fold protein complex CHRAC-15/17 enhances nucleosome sliding and assembly mediated by ACF. Mol Cell 13:265–277

Li B, Jackson J, Simon MD, Fleharty B, Gogol M, Seidel C, Workman JL, Shilatifard A (2009) Histone H3 lysine 36 dimethylation (H3K36me2) is sufficient to recruit the Rpd3s histone deacetylase complex and to repress spurious transcription. J Biol Chem 284:7970–7976

Lipinska A, Klyszejko-Stefanowicz L (1980) The activity of chromatin-bound protease extracted selectively with histone H2B from calf thymus and rat liver. Int J Biochem 11:299–303

Lipinska A, Krawczyk Z, Krajewska W, Klyszejko-Stefanowicz L, Chorazy M (1980) Activity of chromatin-bound protease in histone fractions from rat liver and Morris hepatoma. Neoplasma 27:409–413

Lu X, Simon MD, Chodaparambil JV, Hansen JC, Shokat KM, Luger K (2008) The effect of H3K79 dimethylation and H4K20 trimethylation on nucleosome and chromatin structure. Nat Struct Mol Biol 15:1122–1124

Luger K, Dechassa ML, Tremethick DJ (2012) New insights into nucleosome and chromatin structure: an ordered state or a disordered affair? Nat Rev Mol Cell Biol 13:436–447

Luger K, Mader A, Sargent DF, Richmond TJ (2000) The Atomic Structure of the Nucleosome Core Particle. J Biomol Struct Dyn 17:185–188

Luger K, Mader AW, Richmond RK, Sargent DF, Richmond TJ (1997) Crystal structure of the nucleosome core particle at 2.8 A resolution. Nature 389:251–260

Luger K, Richmond TJ (1998) The histone tails of the nucleosome. Curr Opin Genet Dev 8: 140–146

Lukas J, Lukas C, Bartek J (2011) More than just a focus: The chromatin response to DNA damage and its role in genome integrity maintenance. Nat Cell Biol 13:1161–1169

Ma MK, Heath C, Hair A, West AG (2011) Histone crosstalk directed by H2B ubiquitination is required for chromatin boundary integrity. PLoS Genet 7:e1002175

Magdalena W, Adam L (2012) Future of protease activity assays. Curr Pharm Des 19(6):1062–7

Mahendra G, Kanungo MS (2000) Age-related and steroid induced changes in the histones of the quail liver. Arch Gerontol Geriatr 30:109–114

Mandal P, Azad GK, Tomar RS (2012) Identification of a novel histone H3 specific protease activity in nuclei of chicken liver. Biochem Biophys Res Commun 421:261–267

Moindrot B, Bouvet P, Mongelard F (2012) Chromatin structure and organization: the relation with gene expression during development and disease. Subcell Biochem 61:373–396

Morin V, Sanchez-Rubio A, Aze A, Iribarren C, Fayet C, Desdevises Y, Garcia-Huidobro J, Imschenetzky M, Puchi M, Geneviere AM (2012) The protease degrading sperm histones post-fertilization in sea urchin eggs is a nuclear cathepsin L that is further required for embryo development. PLoS One 7:e46850

Neely KE, Workman JL (2002) The complexity of chromatin remodeling and its links to cancer. Biochim Biophys Acta 1603:19–29

O'Sullivan RJ, Karlseder J (2012) The great unravelling: chromatin as a modulator of the aging process. Trends Biochem Sci 37:466–476

Osipova TN, Karpova EV, Vorob'ev VI (1990) Chromatin higher-order structure: two-start double superhelix formed by zig-zag shaped nucleosome chain with folded linker DNA. J Biomol Struct Dyn 8:11–22

Ouzounis CA, Kyrpides NC (1996) The core histone fold: limits to functional versatility. J Mol Evol 43:541–542

Pachov GV, Gabdoulline RR, Wade RC (2011) On the structure and dynamics of the complex of the nucleosome and the linker histone. Nucleic Acids Res 39:5255–5263

Pedersen P, Seeman T, Hasselgren PO (1986) Protein synthesis and degradation in liver tissue following induction of septic peritonitis in rats. Acta Chir Scand 152:29–34

Poirier GG, Savard P (1980) ADP-ribosylation of pancreatic histone H1 and of other histones. Can J Biochem 58:509–515

Portela A, Esteller M (2010) Epigenetic modifications and human disease. Nat Biotechnol 28: 1057–1068

Ramponi G, Nassi P, Liguri G, Cappugi G, Grisolia S (1978) Purification and properties of a histone-specific protease from rat liver chromatin: effect on acylated histones. FEBS Lett 90: 228–232

Richmond TJ, Finch JT, Rushton B, Rhodes D, Klug A (1984) Structure of the nucleosome core particle at 7 A resolution. Nature 311:532–537

Sakai K, Akanuma H, Imahori K, Kawashima S (1987) A unique specificity of a calcium activated neutral protease indicated in histone hydrolysis. J Biochem 101:911–918

Santos-Rosa H, Kirmizis A, Nelson C, Bartke T, Saksouk N, Cote J, Kouzarides T (2009) Histone H3 tail clipping regulates gene expression. Nat Struct Mol Biol 16:17–22

Saunders MJ, Yeh E, Grunstein M, Bloom K (1990) Nucleosome depletion alters the chromatin structure of Saccharomyces cerevisiae centromeres. Mol Cell Biol 10:5721–5727

Shilatifard A (2006) Chromatin modifications by methylation and ubiquitination: implications in the regulation of gene expression. Annu Rev Biochem 75:243–269

Simon M, North JA, Shimko JC, Forties RA, Ferdinand MB, Manohar M, Zhang M, Fishel R, Ottesen JJ, Poirier MG (2011) Histone fold modifications control nucleosome unwrapping and disassembly. Proc Natl Acad Sci U S A 108:12711–12716

Soria G, Polo SE, Almouzni G (2012) Prime, repair, restore: the active role of chromatin in the DNA damage response. Mol Cell 46:722–734

Steger DJ, Utley RT, Grant PA, John S, Eberharter A, Cote J, Owen-Hughes T, Ikeda K, Workman JL (1998) Regulation of transcription by multisubunit complexes that alter nucleosome structure. Cold Spring Harb Symp Quant Biol 63:483–491

Surowy CS, Berger NA (1983) Nucleotide-stimulated proteolysis of histone H1. Proc Natl Acad Sci U S A 80:5510–5514

Suzuki M, Sugiura M, Ebashi S (1990) Sea urchin protease specific to the SPKK motif in histone. J Biochem 108:347–355

Tesar M, Marquardt O (1990) Foot-and-mouth disease virus protease 3C inhibits cellular transcription and mediates cleavage of histone H3. Virology 174:364–374

Travers AA, Vaillant C, Arneodo A, Muskhelishvili G (2012) DNA structure, nucleosome placement and chromatin remodelling: a perspective. Biochem Soc Trans 40:335–340

Tsurugi K, Ogata K (1982) Studies on the serine proteases associated with rat liver chromatin. J Biochem 92:1369–1381

Tsurugi K, Ogata K (1986) Effects of DNA and urea on the specificity for H1 histone of the neutral protease B partially-purified from rat liver chromatin. J Biochem 99:237–241

Vaissiere T, Herceg Z (2010) Histone code in the cross-talk during DNA damage signaling. Cell Res 20:113–115

Van der Veer E, Bootsma D (1982) Repair DNA synthesis in heterokaryons during reactivation of chick erythrocytes fused with human diploid fibroblasts or HeLa cells. Exp Cell Res 138:469–474

Vignali M, Hassan AH, Neely KE, Workman JL (2000) ATP-dependent chromatin-remodeling complexes. Mol Cell Biol 20:1899–1910

Vogler C, Huber C, Waldmann T, Ettig R, Braun L, Izzo A, Daujat S, Chassignet I, Lopez-Contreras AJ, Fernandez-Capetillo O, Dundr M, Rippe K, Längst G, Schneider R (2010) Histone H2A C-terminus regulates chromatin dynamics, remodeling, and histone H1 binding. PLoS Genet 6:e1001234

Ward W, Richardson A (1991) Effect of age on liver protein synthesis and degradation. Hepatology 14:935–948

Watson DK, Moudrianakis EN (1982) Histone-dependent reconstitution and nucleosomal localization of a nonhistone chromosomal protein: the H2A-specific protease. Biochemistry 21:248–256

Widlak P, Pietrowska M, Lanuszewska J (2006) The role of chromatin proteins in DNA damage recognition and repair. Histochem Cell Biol 125:119–126

Widom J (1998) Structure, dynamics, and function of chromatin in vitro. Annu Rev Biophys Biomol Struct 27:285–327

Wolffe AP, Guschin D (2000) Review: chromatin structural features and targets that regulate transcription. J Struct Biol 129:102–122

Wolffe AP, Kurumizaka H (1998) The nucleosome: a powerful regulator of transcription. Prog Nucleic Acid Res Mol Biol 61:379–422

Woodcock CL, Dimitrov S (2001) Higher-order structure of chromatin and chromosomes. Curr Opin Genet Dev 11:130–135

Woodcock CL, Skoultchi AI, Fan Y (2006) Role of linker histone in chromatin structure and function: H1 stoichiometry and nucleosome repeat length. Chromosome Res 14:17–25

Workman JL, Abmayr SM (2004) Histone H3 variants and modifications on transcribed genes. Proc Natl Acad Sci U S A 101:1429–1430

Workman JL, Kingston RE (1998) Alteration of nucleosome structure as a mechanism of transcriptional regulation. Annu Rev Biochem 67:545–579

Zlatanova JS, van Holde KE (1992) Chromatin loops and transcriptional regulation. Crit Rev Eukaryot Gene Expr 2:211–224

Chapter 5
Anti-inflammatory Effects of Probiotics and Their Metabolites: Possible Role for Epigenetic Effects

Nurşen Türker, Zheng Quan Toh, Tom C. Karagiannis, and Paul V. Licciardi

Abstract It may well be the case that an important part of future medicine is directed not only in the development of novel therapies which restore gut microbiota but also in the administration of preventative strategies through dietary intervention. Dietary changes contribute to 57 % of the total structural variation in gut microbiota, whereas genetic differences attribute no more than 12 %. As such a diet that is rich in complex carbohydrates not only promotes a healthy gut microbiota but also boosts the production of immunomodulatory short chain fatty acids (SCFA) which have HDAC inhibitory properties. Histone deacetylase inhibitors (HDACi) have been observed across a wide range of naturally derived compounds which include sulforaphane from broccoli, diallyl disulfide from garlic, and curcumin from turmeric. Identification of potent anti-inflammatory effects of dietary compounds is worth investigating, particularly due to wide accessibility by the general public. Although our current knowledge of restoring gut microbiota through administration of probiotics and their metabolites and naturally derived HDACi is not complete. Future research is required to understand the mechanistic actions and pharmacokinetics involved in dietary HDACi and probiotics to aid in future developments of therapeutic modalities for the treatment of chronic inflammatory diseases.

N. Türker • Z.Q. Toh
Allergy and Immune Disorders Laboratory, Murdoch Children's Research Institute, Royal Children's Hospital, Melbourne, VIC, Australia

T. Karagiannis (✉)
Epigenomic Medicine, Baker IDI Heart and Diabetes Institute, The Alfred Medical Research and Education Precinct, 75 Commercial Road, Melbourne, VIC, Australia

Department of Pathology, The University of Melbourne, Parkville, VIC, Australia
e-mail: tom.karagiannis@bakeridi.edu.au

P.V. Licciardi
Allergy and Immune Disorders Laboratory, Murdoch Children's Research Institute, Royal Children's Hospital, Melbourne, VIC, Australia

Department of Paediatrics, The University of Melbourne, Parkville, VIC, Australia

N. Maulik and T. Karagiannis (eds.), *Molecular Mechanisms and Physiology of Disease: Implications for Epigenetics and Health*, DOI 10.1007/978-1-4939-0706-9_5,

In this chapter we discuss the inflammasomes related in health and disease and outline current and novel therapies for the treatment of chronic inflammatory diseases including allergies, inflammatory bowel diseases, metabolic syndromes, and autism spectrum disorders.

Keywords Probiotics • Metabolites • Short chain fatty acids • Histone deacetylase inhibitors • Inflammation

5.1 Chronic Inflammatory Diseases

The discovery of the pro-inflammatory transcription factor NF-κB in 1988 has since led to the central hypothesis linking divergent conditions such as arthritis, diabetes, asthma, dementia, and cancer through an underlying common denominator of chronic inflammation (Baltimore 2009). Chronic inflammatory diseases are a major cause of morbidity and mortality as they account for 60 % of deaths worldwide (World Health Organisation 2005). The insidious nature of chronic inflammation enables extensive damage to take place over several years before any major clinical symptoms become apparent. This places inflammation as a major focus of medical research and demands a greater understanding on the mechanisms associated with its deviation from an otherwise "sterile" state (Hussain and Harris 2007). Several risk factors contribute towards the pathogenesis of chronic inflammatory disease, with "Western lifestyle" as the predominant view. This modern way of life is characterized by a reduced exposure to particular microbes, constant encounter with environmental pollutants, and an unhealthy diet consisting of low dietary fiber and an increased intake of refined sugar, animal proteins, and unsaturated fats. Strong evidence suggests that dysbiosis of gut microbiota has been linked with chronic inflammatory diseases (Donato et al. 2010). Part of the concern with "Western lifestyle" lies with the fact that the industrialization process, and in particular, changes to our diet, has taken place much faster than our ancient genomes could adapt (Cordain et al. 2005).

5.2 Inflammatory Process

Inflammation constitutes an important part of host defense and is the first step taken in response to infection or tissue injury. Resolution of this process is pivotal in the return towards homeostasis in which cellular debris is cleared and tissue repair is resolved. The immune system is split into two different sub-divisions that are intertwined, yet unique in their ability to defend against foreign antigens. The first line of defense is the innate immune system which is a nonspecific rapid response to the invading pathogen. Initiation of innate immunity is reliant upon the identification of pathogens through germ line-encoded proteins called pattern recognition receptors

(PRRs). PRRs of innate immune cells sense pathogen-associated molecular patterns (PAMPs) which are conserved regions present on many microbes but absent on host cells (Janeway 1989). PAMPs can include cell wall components such as lipopolysaccharide (LPS) and lipoteichoic acid (LTA) as well as bacterial DNA (CpG). There are four major classes of PRRs that have been identified and these are categorized in accordance to their ligand specificity, function, and localization (Doughty 2011). PRRs include the Toll-like receptors (TLRs), nucleotide binding and oligomerization domain like receptors (NLRs), retinoic-acid inducible gene-1 helicase-like receptors (RLRs), and C-type lectin receptors (CLRs). TLRs mainly recognize PAMPs and are located on the plasma membrane as well as internal membranes (endosome, lysosome, endoplasmic reticulum) (Stutz et al. 2009). CLRs are membrane bound on macrophages; however, they mediate their effects through binding directly with the pathogen followed by internalization (Taylor et al. 2005). NLRs and RLRs are cytosolic receptors that detect PAMPs from within the cytoplasm; however, they are dissimilar in that NLRs are also able to detect endogenous host-derived signals referred to as damage-associated molecular patterns (DAMPs), whereas RLRs only detect cytoplasmic viral RNA (Onoguchi et al. 2011). The interplay between the diverse PRR families results in a holistic brigade in the defense against invading pathogens.

Upon interacting with the specific ligand, one of the modes of action of PRR engagement includes stimulation of receptor-mediated signaling pathways such as NF-κB, through adaptor proteins such as MyD88, Mal, Trif, and TRAM (Creagh and O'Neill 2006). The ubiquitous NFκB is imperative to the modulation of genes that are relevant in inflammatory diseases, apoptosis, and cancer (Sun and Roland 2002). NF-κB is able to regulate the transcription of pro-inflammatory mediators, therefore can target a wide class of genes including that of adhesion molecules, cytokines (TNF-α, IL-1β, IL-6, IL-1β, IL-18), and chemokines (IL-8, MIP-1β, migration inhibitory factor (MIF), monocyte chemotactic protein 1 (MCP-1)). Increased transcription of these genes perpetuates the inflammatory response and promotes cell survival (Nishikori 2005). Potent inducers of NF-κB include pro-inflammatory cytokines such as TNF-α, IL-1β as well as LPS, reactive oxygen species (ROS), and others (Tran et al. 1997). The wide range of stimuli that initiates this signal transduction pathway denotes the broad impacts of NF-κB; hence it is for this reason there is tight regulation of the multiple steps involved in its transcriptional activity (Baldwin 1996).

Cytokines are pleiotropic molecules that elicit their effects during the inflammatory response. In sterile inflammation they clear the injurious stimuli and maintain homeostasis; however, when dysregulated they contribute to the pathogenesis behind a plethora of conditions. The main inflammatory-associated cytokines include TNF-α, IL-1β, and IL-6 amongst many others and they are produced by a variety of cells, the most important source being that of monocytes and macrophages at the site of inflammation (Gabay 2006). Of these, TNF-α and IL-1β are extremely potent inflammatory molecules that mediate inflammation induced by bacterial LPS (Feghali and Wright 1997). These two cytokines can further exert secondary effects by stimulating IL-6 synthesis which depending upon the circumstances can have either inflammatory or anti-inflammatory effects. One suggested mechanism for this is through the inhibition of TNF-α and activation of the anti-inflammatory cytokine IL-10 (Starkie et al. 2003).

Additionally, there is a small subset of chemotatic cytokines called chemokines that function to recruit leukocytes in infection, inflammation, and tissue injury (Moser 2003). Despite its commonly used nomenclature following that of cytokines, the first chemokine to be discovered was IL-8 (Holmes et al. 1991). IL-8 is the most comprehensively studied chemokine and serves a wide range of pro-inflammatory effects that include stimulation of neutrophil degranulation and increased expression of cell adhesion molecules (Oppenheim et al. 1991). MCP-1 is yet another chemokine that is produced by macrophages and endothelial cells. This molecule orchestrates the acute and chronic inflammatory response at the site of injury by recruiting monocytes, memory T cells, and dendritic cells. As such, this chemokine has been implicated in the pathogenesis of diseases characterized by monocytic infiltrates associated with psoriasis, rheumatoid arthritis, and atherosclerosis (Xia and Sui 2009). Upon recruiting immune cells another important component is that of maintaining the cells at the site of inflammation. Of particular interest to this role is that of the chemokine, MIF which is said to prolong the life span and activity of monocytes/macrophages (Baugh 2002). Even with all this insight into cytokines and chemokines, a lot of research still needs to be executed in order to illuminate further on the various roles and regulatory mechanisms of these molecules.

In contrast to other cytokines that are transcriptionally induced and secreted during the inflammatory process, the expression of IL-1β and IL-18 requires processing from precursor molecules (pro-IL-1β and pro-IL-18). This processing entails the activation of caspase-1 through a multiprotein complex referred to as the "inflammasome." Caspase-1 is an intracellular cysteine protease that exists as an inactive zymogen in macrophages and dendritic cells, whereas it is present in its active form in circulating human monocytes (Netea et al. 2009). Upon proteolytic activation caspase-1 cleaves the precursors pro-IL-1β and pro-IL-18 into their biologically active cytokines. Inflammasomes are categorized into two broad families: the NLR family which includes nucleotide binding and oligomerization domain (NOD or NACHT) and leucine rich repeat containing (LRR) and the PYHIN family which includes a pyrin domain (PYD) and hemopoietic expression, interferon-inducibility, nuclear localization domain (HIN) (Proell et al. 2008).

In particular it is the NLR family that has gained much attention since mutations in NLR genes have been linked with several auto-inflammatory diseases (Hoffman and Brydges 2011; Inohara et al. 2005). Of the 22 members currently present within the NLR family, all the members share common features of a centrally located NOD which mediates self-oligomerization and activation of NLRs, C terminal LRRs which mediate autorepression and ligand sensing, and an N terminal which mediates protein–protein interactions for initiating downstream signaling (Fig. 5.1) (Lee 2007). It is the variation in the N terminal that additionally subcategorizes the NLR members into three different groups. The largest group being the PYD-containing group, therefore taking the name NLRP (also known as NALP); another group is the CARD-containing group, which takes the name NLRC; and

Fig. 5.1 Schematic representation of NLR structure

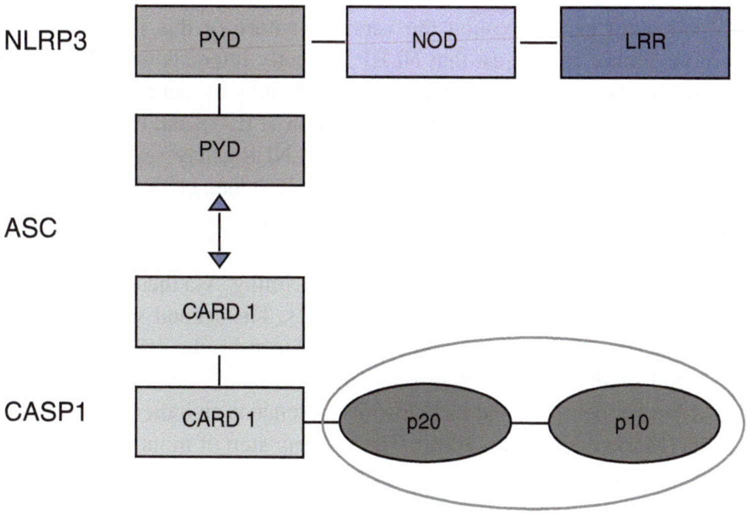

Fig. 5.2 Critical components during assembly of the NLRP complex

finally there is an baculoviral apoptosis inhibitory protein-containing group that is referred to as Naip (Chen et al. 2009).

The largest group of NLRs, the NLRPs [NLR (NOD-like receptor)-related proteins] consist of various proteins that have been extensively studied in the field of inflammasomes. A common feature of the NLRPs is their ability to recruit the adaptor protein, apoptosis-associated peck-like protein containing CARD (ASC) through homophilic PYD–PYD interactions. It is the CARD domain of the ASC which interacts with, and leads to the subsequent activation of caspase-1. Nevertheless, slight variations exist in accordance to the specific NLRP protein in question; for instance, in the case of NLRP1, recruitment of caspase-1 occurs directly via CARD homotypic interaction and in NLRP3 indirectly through CARD–CARD interaction (Martinon Fabio 2004). This can be attributed to the differing constituents of the inflammasome complex between the various NLRP proteins. The NLRP1 inflammasome can be assembled with the lack of the adaptor protein ASC; however, the presence of this molecule is known to enhance inflammasome activity. Oligomerization of the inflammasome facilitates recruitment of the precursor enzyme, pro-caspase 1 which is converted to the active form caspase-1 resulting in two subunits, a 20-kDa (p20) and a 10-kDa (p10) (Fig. 5.2). Ultimately the establishment of the inflammasome is crucial in the processing and maturation of the cytokines IL-1β and IL-18 by caspase-1 (Mariathasan et al. 2004).

Depending on the NLR in the complex, inflammasomes are equipped with the ability to respond to a wide range of signals. NLRP1 can sense lethal toxin from anthrax, NLRC4 responds to flagellin, and the most highly studied inflammasome complex NLRP3 (also known as NALP3) responds to a variety of pathogenic, endogenous, and environmental signals (Rathinam et al. 2012). The broad range of NLRP3 activators makes explicit the versatile nature of this inflammasome. Yet there is no evidence to suggest that NLRP3 directly interacts with the various ligands. As such it has been speculated that NLRP3 may in fact require two separate signals which control the points for the production of IL-1β and IL-18. This two part pathway differentially regulated by TLRs and NLRs may serve as a safeguard against unnecessary production of IL-1β and IL-18 (Chen et al. 2009).

The first signal requires TLR activation of NF-κB which is stimulated by PAMPs such as LPS, LTA, or bacterial nucleic acids. NF-κB expression leads to the production of pro-IL-1β and pro-IL-18. It is said that "priming" via these microbial stimuli is necessary for subsequent activation of NLRP3. The second signal directly activates the NLRP3 inflammasome and can be induced by the effect of extracellular ATP, bacterial toxins, or particulate substances (Franchi et al. 2012). ATP was the first DAMP to be identified and its cytosolic presence is an indicator of stressed and dying cells (Mariathasan et al. 2006). The priming step of monocytes with PAMPs releases ATP which serves as the signal for the assembly of the NLRP3 inflammasome (Stutz et al. 2009). Extracellular ATP results in the stimulation of the ATP-gated $P2X_7$-like receptor; this in turn causes the opening of the hemichannel protein pannexin-1. Activation of this $P2X_7$ receptor induces a complete collapse of ionic gradients that switches cytosol potassium (K^+) from high to low concentration (Kahlenberg and Dubyak 2004). The drop in K^+ is also mimicked by the action of pore-forming toxins such as listeriolysin O or hemolysins that punch holes on the cell membrane. Potassium efflux is known as one of three models through which NLRP3 inflammasome formation can take place. This is in line with studies that demonstrate the presence of high extracellular K^+ concentration inhibiting the activation of the NLRP3 inflammasome (Pétrilli et al. 2007; Franchi et al. 2007).

Another model for NLRP3 activation is based on the phagocytosis of crystalline and particulate substances. Uric acid was one of the first crystals shown to engage with NLRP3 and it is the aetiological agent responsible for gout (Martinon et al. 2006); then airborne pollutants such as asbestos and silica were discovered, the inhalation of which can potentiate towards lung diseases (Dostert et al. 2008). A similar mechanism is also applied for the adjuvant properties of alum, a particulate substance associated with Alzheimer's disease (Halle et al. 2008). Phagocytic cells contain many lysosomes which endocytose and break down into small materials and cellular debris. However, with the case of the crystals and particulate substances, their large size makes it difficult for the lysosomes to catabolise. For this reason they undergo "frustrated" phagocytosis and remain at the surface of the lysosome causing phagolysosomal destabilization. This leads to the release of lysosomal proteases, predominantly cathepsins B leading to NLRP3 inflammasome activation.

The generation of ROS triggered by ATP and particulate matter posits a further model for NLRP3 inflammasome activation. This model is supported by the fact that many NLRP3 agonists are capable of inducing ROS (Dostert et al. 2008). A study with the redox-sensitive protein thioredoxin has shown to bind and activate NLRP3 after the production of ROS by NLRP3 activators (Zhou et al. 2010). However this model is subject to controversy, since another report has found that ROS inhibition does not affect NLRP3 activation per se, but rather negatively interferes with the priming step of NLRP3 inflammasome (Bauernfeind et al. 2011).

Despite several theories for NLRP3 inflammasome activation has surfaced, none of the aforementioned provides a unifying mechanism of action. It may be the case that the dynamic nature of this inflammasome requires an integration of various cellular signals in order to regulate the inflammatory response. In addition to the maturation of the potent pro-inflammatory cytokines IL-1β and IL-18, inflammasomes can also initiate a crucial form of cell death referred to as "pyroptosis". This process forms an important part of tissue repair, as pyroptosis kills the infected cell and eliminates the replicative niche required to perpetuate further damage to the host (Jones and Weiss 2011). The complexity of inflammatory regulation is understandable from the point of view that there are severe risks for the host both from a lack of sufficient response as well as excessive response to a pathogen.

It has only been a decade since the discovery of inflammasomes and yet several studies have already produced evidence to suggest that the production of IL-1β is not exclusive to the canonical caspase-1 inflammasome pathway. Unlike IL-18, studies have demonstrated that IL-1β is able to be produced in an inflammasome independent pathway relying only on processing via caspase-1 (Denes et al. 2012) but also in its absence via caspase-8 or an unknown mechanism which does not support the current inflammasome paradigm (Gringhuis 2012; Mayer-Barber et al. 2010). This provides impetus to further investigate the possibility of other pathways that may be involved in the secretion of IL-1β and IL-18.

5.3 Inflammasomes in Health and Disease

Inflammasomes are widely recognized as being involved in the protective role of infection clearance; however, they are also responsible for the pathogenesis of a range of disorders. The dysregulation of IL-1β and IL-18 production is the driving force behind intestinal inflammation, metabolic diseases, and auto-inflammatory syndromes (Davis et al. 2011). In auto-inflammatory disorders the monocyte-macrophage is the dysfunctional cell, which causes the aberrant inflammatory response. This is different from autoimmune diseases in which T cells are responsible for the underlying pathogenesis. As such delineating the mechanisms that lead to the production of bioactive IL-1β and IL-18 have been the focus of intense research in the past few years.

Indeed NLRs have been implicated in microbial sensing, downstream signaling, and the initiation of the antimicrobial response (Elinav et al. 2011a). Growing support for this role has been established through in vivo infection models

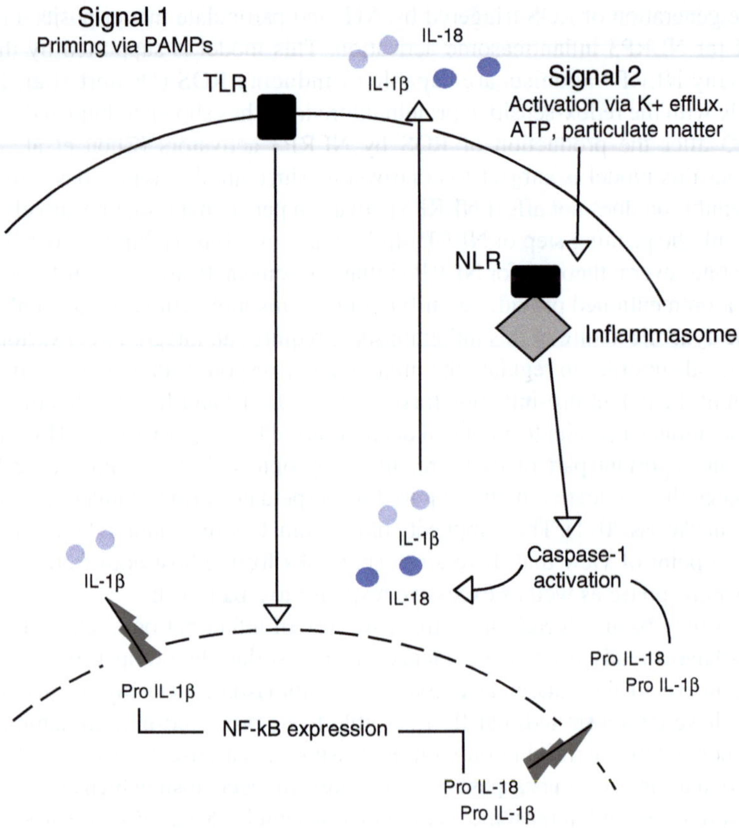

Fig. 5.3 Activation of the inflammasome pathway

demonstrating the involvement of inflammasomes as part of host defense (van de Veerdonk et al. 2011). Pneumolysin (PLY), a key virulence factor from *Streptococcus pneumoniae*, as well as a novel NLRP3 activator, has shown upon interaction an enhanced secretion of IL-1β and the subsequent development of protective immunity against *S. pneumoniae* respiratory infections (McNeela et al. 2010). Identification of the primary mechanisms associated with PLY pathogenesis has opened up an avenue for the design of pneumococcal vaccines. Other studies have shown that caspase-1 deficient mice challenged with the cytosolic bacterium *Listeria monocytogenes* are unable to clear the infectious burden relative to that of wild-type mice. Yet interestingly the caspase-1 deficient mice show no impairment in the adaptive response, highlighting the critical role of inflammasomes as part of innate immunity (Tsuji et al. 2004). In addition to knockout models of the various inflammasome components, another study has focused on TLR-2 deficient macrophages and their response to *Francisella novicada* infection compared to wild-type mice (Jones and Weiss 2011). The presence of TLR-2 enables rapid inflammasome activation aiding in infection clearance through a more profound response by producing higher levels of IL-18 and pyroptotic cell death (Fig. 5.3).

Several other studies have also confirmed the protective role of IL-18 in intestinal homeostasis. This was investigated in animal models via treatment with dextran sodium sulfate (DSS), a polysaccharide that is toxic to the colonic epithelium and capable of inducing colitis. Knockout mice that were deficient in various components of the NLRP3 inflammasome (i.e., caspase-1, NLRP3) revealed weight loss, intestinal bleeding, and greater susceptibility to colitis relative to DSS-treated, wild-type mice (Dupaul-Chicoine et al. 2010; Zaki et al. 2010). It was later determined that serum derived from DSS-treated wild-type mice had detectable IL-18 that was crucial in mediating intestinal homeostasis. When caspase-1 deficient (−/−) mice treated with DSS were given exogenous IL-18, increased weight gain and improved survival rates were observed (Zaki et al. 2010). Therefore, IL-18 production by intestinal epithelial cells is central to the protective role of NLRP3 against intestinal inflammation (Reuter and Pizarro 2004). Other inflammasomes such as NLRP6 have also been identified as a key regulator of the colonic microbial ecology and important in the maintenance of colonic homeostasis and prevention of auto-inflammation through modulating IL-18 (Elinav et al. 2011b). Furthermore a recent study has also demonstrated that inflammasomes carry out an important protective role in response to severe injury (Osuka et al. 2012).

Perhaps the most intriguing finding is that of the involvement of NLRP3 in metabolic disorders such as impaired insulin sensitivity, obesity, diabetes, and atherosclerosis in which chronic inflammation is known as the etiological component responsible for pathogenesis (Strowig et al. 2012). These conditions pose as a burgeoning concern in affluent countries with the WHO estimating that 1.4 billion adults are overweight and 300 million are clinically obese (World Health Organisation 2002). Increased intake of omega-6 fatty acids through Western diet potentiates the risk for chronic inflammatory disease as omega-6 fatty acids are paralleled with an increase in the production of pro-inflammatory prostaglandins E2 and leukotriene B4 (James et al. 2000). In fat tissue, lipids such as palmitate and ceramide participate as ligands responsible for NLRP3 activation, causing the production of IL-1β which results in a reduction in fat oxidation and subsequent insulin resistance. This prompts the pancreas to produce more insulin and the hormone Islet Amyloid Polypeptide (IAPP). Unfortunately this hormone has a natural tendency to form amyloid, which is detected by the macrophages within the pancreas and further activates NLRP3. Ultimately excessive production of IL-1β causes the death of β cells and insulin resistance; a cycle which in an attempt to be protective, accumulates greater damage (Wen et al. 2012). Another mechanism through which NLRP3 is activated in the obese phenotype is through an increased amount of saturated fatty acids and activating ER-stress pathways through triggering calcium signals and free radicals (Vandanmagsar et al. 2011). Hence identification of the involvement of inflammasomes in metabolic disorders will facilitate the design of NLRP-specific drugs as is the case with "glyburide" which inhibits IL-1β secretion by interfering with caspase-1 activity. This serves to prevent IL-1β secretion by antagonizing the $P2X_7$ receptor and interfering with caspase-1 activity (Lamkanfi et al. 2009).

Gain of function mutations in the gene encoding NLRP3 causes spontaneous activation of the inflammasome responsible for several auto-inflammatory disorders which include familial cold auto-inflammatory syndrome (FCAS), Muckle–Wells syndrome (MWS), and neonatal-onset multisystem inflammatory disease (NOMID). These are a rare group of disorders which are characterized by recurrent episodes of fever and rash with no known aetiology. Upon treatment of patients with IL-1β blocking agents there is a rapid cessation of symptoms (Ting et al. 2006).

Taken together these studies allow an enhanced appreciation of the hosts effort to integrate different inflammasome components and spatiotemporally distant PRRs as part of innate immunity. The extensive role of inflammasomes in the pathogenesis of various inflammatory diseases establishes a compelling avenue for future therapeutic intervention.

5.4 Current Therapies for Chronic Inflammatory Disease

In the past decade the most significant breakthrough in the treatment of inflammatory diseases has come about through anti-cytokine strategies. Improved knowledge on the molecular pathology of chronic inflammatory conditions has established that blocking of pro-inflammatory cytokines such as IL-1 or TNF-α can be used as therapy for a variety of conditions. While this may serve to treat excessive inflammation from taking place, it also weakens the immune system resulting in an increased susceptibility towards opportunistic infections (van der Meer 2005). Additionally, such therapies are not suitable for use by all, as one- and two-thirds of patients respond poorly or can relapse once the treatment ceases to have any beneficial effect (Gaestel et al. 2009). Other commonly used inflammatory treatments include nonsteroidal anti-inflammatory drugs (NSAIDs) which work by blocking the COX enzymes that produce prostaglandins. These are widely used for pain relief, fever, or rheumatoid arthritis. Nevertheless prolonged use of NSAIDs has various toxicities such as renal dysfunction, liver abnormalities, and gastrointestinal complications which include dyspepsia and stomach ulcers (Bush et al. 1991). In fact, alongside *Helicobacter pylori* infection, the vast majority of peptic ulcers are believed to have originated from the chronic consumption of NSAID (Sung et al. 2000). NSAID inhibits mucosal synthesis which increases the risk of colonization by pathogenic bacteria, such as *H. pylori*. This interaction ultimately aggravates the toxic effect of NSAIDs; particularly it is the colonization of the mucosa which can stimulate the release of pro-inflammatory cytokine and chemokines evoking a more rigorous inflammatory response (Lazzaroni and Bianchi Porro 2001). NSAIDs work by blocking both COX-1 (protective) and COX-2 (inflammatory disease) enzymes, eliminating the protective function of COX-1 within the stomach. As such, this prompted the development of specific COX-2 inhibitors such as Vioxx and Celebrex; however, the dilemma now faced is that prolonged use increases the likelihood of heart attacks, thrombosis, and stroke (Day 2004). On the basis of the short comings present with conventional anti-inflammatory medicine, there has been a growing interest into the use of alternative therapies to counteract these issues.

5.5 Novel Therapies for Chronic Inflammatory Disease

The microbiome of the gastrointestinal tract represents the largest source of microbial stimulation and has over time evolved a key role in the development of proper immune function. Central to the protective role of beneficial bacteria within the gut include efficient digestion of food (Hooper et al. 2002), maintaining epithelial barrier integrity (Artis 2008), effective fat metabolism (Bäckhed et al. 2004), preventing uptake of foreign antigens (Sanz and De Palma 2009), and modulating immune response (Puren et al. 1999). Of particular importance is the crosstalk which takes place between a functional epithelial barrier and the underlying immune cells (Hill and Artis 2010). A breach in the epithelial barrier results in the exposure of resident inflammatory cells (e.g., macrophages, mast cells, dendritic cells) to both pathogenic and nonpathogenic bacteria, which if unresolved can initiate the slow progression from chronic inflammation to cancer (Karin et al. 2006). The gastrointestinal mucosa depicts a battlefield in which tolerance to normal intestinal bacteria is constantly challenged with foreign antigens introduced through the diet. Hence the gut is constantly in a state of mild homeostatic inflammation and deviations from this tightly regulated equilibrium may serve as an initiating step towards pathogenesis (Mowat and Bain 2011).

A diverse population of bacteria contribute to this internal armada, with a particular dominance of the Firmicutes and Bacteroidetes phyla (Eckburg et al. 2005). Controversy exists in relation to the ratio of Firmicutes and Bacteroidetes phyla within a dysbiotic gut (Mariat et al. 2009); however, diet is suggested to be a major factor in this outcome. Diet has been implicated as one of the key Western lifestyle factors which contribute to the increased incidence of chronic inflammatory diseases in developed countries (Musso et al. 2010). Several studies have assessed the contribution of dietary factors on gut microbiota and the subsequent biological effect on the host. An epidemiological study comparing the influence of long-term dietary habits between a rural African cohort of children with that of urban-raised Europeans demonstrated a stark contrast in gut composition (De Filippo et al. 2010). The African cohort contained lower levels of Firmicutes and greater amounts of the beneficial Bacteroidetes bacteria mainly *Prevottela* and *Xylanibacter*, which were completely absent in the European cohort. Bacteriocides bacteria are strong producers of short chain fatty acids (SCFA) which are immunomodulatory molecules produced following consumption of fiber through fermentation with commensal bacteria. It is for these reasons that communities in which the consumption of high fiber diets is common, such as those in developing countries, have a lower incidence of inflammatory disease. Another experiment showed a similar effect in gnotobiotic mice (germ-free) that were switched from a low fat, plant polysaccharide-rich diet to a high-fat/high-sugar "Western" diet (Turnbaugh et al. 2009b). Hence this makes explicit that dietary choices impact on the gut microbiota and the important role of dysbiosis in the progression towards disease. Evidence for this can be observed across numerous inflammatory-associated conditions such as allergies (Kirjavainen et al. 2002), inflammatory bowel diseases (Tamboli et al. 2004), metabolic syndromes (Tremaroli and Backhed 2012), and autism spectrum disorders (Parracho et al. 2005). Knowledge into the underlying pathogenesis which connects these conditions has paved the way for novel therapeutic interventions aimed at restoring gut microbiota. Thereupon attention has focused on the use of probiotic supplements as it offers the most physiologic and least toxic approach in beneficially altering the gut microbiota in the treatment of inflammatory diseases (Gussler and Graham 2011).

Probiotics are defined as "live microorganisms, which when administered in adequate amounts confer a health benefit on the host" (FAO/WHO 2002). Lactic acid bacteria are the first microbes identified for use in probiotic supplements (Ljungh 2006). Yet the beneficial effects of lactic acid bacteria date back to the times of Professor Elie Metchnikoff, who in 1908 observed the health benefits of consuming fermented dairy products in Bulgarian peasants. This diet replenished the commensal bacteria within the gut and therefore conferred beneficial effects such as better digestion and an enhanced immune system (Anukam and Reid 2007). The resurgence in the importance of diet and gut microbiota has subsequently led to intensified research into different species and strain-specific characteristics of the various probiotics which have become available (Luyer et al. 2005). Probiotics have been associated with important clinical benefits in conditions characterized with gut dysbiosis such as allergies (Michail 2009), inflammatory bowel disease (Zocco et al. 2006), autism spectrum disorders (Parracho et al. 2005), and the more recently identified metabolic diseases (Mallappa et al. 2012) and cardiovascular disease (Kumar et al. 2012). A great deal of evidence pertaining to human and animal studies has demonstrated that the immunomodulatory properties of probiotics function through an enhancement of epithelial barrier, competitive adherence to mucosal lining and immunomodulation (Ohland and MacNaughton 2010). The mechanisms which underlie these various protective responses remain to be elucidated; however, studies demonstrate an important role for innate immunity (Pagnini et al. 2010).

One of the most vital functions of the commensal probiotics lies in their ability to convert dietary fiber (complex carbohydrates) into immunomodulatory SCFA. Production of SCFA is directly linked with the amount of carbohydrate that is consumed and the composition of gut microbiota. It is for this reason that an increased consumption of dietary fiber such as fruits and vegetables is encouraged, as it promotes fermentation of SCFA which drives the immune system away from aberrant inflammation. Numerous studies demonstrate that a deficiency in SCFA is linked with inflammatory diseases (Madsen 2011). Metagenomic screenings demonstrate that the genes responsible for carbohydrate fermentation in the gut are one of the most highly conserved functions of the gut microbiome (Tasse et al. 2010). Dietary fiber escapes digestion in the upper gastrointestinal tract and undergoes bacterial fermentation in the colon: a process which leads to the formation of immunomodulatory SCFAs. The three main SCFAs (acetate, propionate, and butyrate) exert different effects on colonocytes which positively influence gut morphology and function (Scheppach 1994). These probiotic metabolites are major energy substrates for colonocytes (butyrate) or can be used as substrates for lipid synthesis in the liver (acetate, propionate) (Tremaroli and Backhed 2012). Butyrate has particularly gained attention due to its ability to modulate the physiologic roles of proliferation, differentiation, and gene repression (Säemann et al. 2000). Under normal conditions it can stimulate the growth of colonic mucosa, and in malignant cells it has an antiproliferative effect (Bailón et al. 2010). Given the important role that SCFAs, like butyrate, have within the gut; numerous studies have taken to utilizing probiotic metabolites as a therapeutic modality in the treatment of inflammatory diseases.

Despite a great deal of knowledge substantiating the influence of dietary factors in health and disease, the underlying mechanisms responsible remain largely confounding. Nonetheless the role of "epigenetics" is fast becoming the avenue for delineating such mechanisms, as it combines the impact of environmental factors (e.g., nutrition, lifestyle) on the genome and the subsequent influence on the health status of the organism. Epigenetics involves heritable changes which alter gene expression without hindering the DNA sequence. It relies upon posttranslational modifications which alter the accessibility of transcription factors to genes and to areas of gene promoters; thus regulating the outcome of gene expression. Such modifications include (1) histone acetylation, in which a relaxed chromatin structure enables activation of transcription factors (Andoh et al. 1999a) and (2) DNA methylation of CpG islands in promoter region of genes, this is associated with gene silencing (Razin and Kantor 2005). Taken together, epigenetics can be involved in both normal development and disease progression, as will be explained further in the context of gut microbiota.

5.6 Allergic Disease

Developed countries exhibit epidemic proportions of allergic diseases with 40 % of the population in Australia, New Zealand, and the United States demonstrating cases of asthma while the incidence of eczema has doubled and even tripled in some industrialized countries (Graham-Rowe 2011). In another Australian study a survey of more than 250 childcare centers showed that 85–90 % have at least one child with a food allergy (Hadley 2006). Mounting evidence suggests that infants that suffer from allergies have an altered gut microbiota composition. Studies have shown, in comparison to healthy controls children with allergies show an increase in coliforms and *Staphylococcus aureus* in the gut and a deficiency in the beneficial lactobacilli and bifidobacteria biota (Björkstén et al. 1999). As such, numerous studies have demonstrated the role probiotics play in alleviating allergic inflammation. The clinical safety of probiotics has been demonstrated across a variety of strains, since probiotic bacteria are generally selected from the commensal flora of healthy gut microbiota (Borriello et al. 2003).

Promising results have been observed with *Lactobacillus rhamnosus* GG (LGG) supplementation in atopic children with an increase of the anti-inflammatory cytokine IL-10 (Pessi et al. 2000) and enhanced IFN-γ production in children with cow-milk allergies (Pohjavuori et al. 2004). Bifidobacterial supplementation of infants with atopic eczema demonstrates reduction in the *Escherichia coli* as well as increased number of bacteroides numbers during weaning (Kirjavainen et al. 2002). Furthermore early pattern of SCFA levels demonstrates the propensity of infants to develop allergies later in life (Böttcher et al. 2000). For instance when the levels of i-butyric and i-valeric acids are decreased at 1 year of age, this is linked with higher reports of food allergies in 3 years time (Sandin et al. 2009). As such, microbial stimulation of the gut microflora during those formative years is of utmost importance for the establishment of a balanced immune system.

5.7 Inflammatory Bowel Diseases

A recent systemic review carried out on the incidence of inflammatory bowel diseases (IBD) has demonstrated an emergence of global disease (Molodecky et al. 2012). IBD which include Crohn's disease and ulcerative colitis have been of particular interest as studies suggest these conditions are the predecessors of colorectal carcinomas (Ray 2012). The gut microbiota of IBD patients are characterized with a decrease in lactobacilli and bifidobacteria and an increase in *E. coli* and clostridia, as such this dysbiosis presents an avenue for probiotics to intervene (Manichanh et al. 2012).

An accumulating body of evidence suggests that probiotics are able to induce anti-inflammatory effects via inhibition of the inflammatory transcription factor NF-κB (Donato et al. 2010). Administration of the commensal probiotics *Bifidobacterium breve* (*B. breve*) and *LGG* diminishes the expression of inflammatory bowel disease causing factors IL-17, IL-23, and CD40 expression (Ghadimi et al. 2012). Further investigations showed that *B. breve* and LGG modulate their anti-inflammatory effects via epigenetic changes. Histone acetylation mediates LPS-induced NF-κB activation which is responsible for other genes involved in perpetuating the inflammatory response. This data provides evidence to show that probiotics are able to repress conserved inflammatory pathways via restricting access to key transcription factor NF-κB, while concurrently increasing DNA methylation.

As previously alluded to, an important aspect of innate immunity is the role that NLRP3 inflammasomes play in intestinal homeostasis. One study established that the *Lactobacillus* strains *Lactobacillus delbrueckii* subspecies *bulgaricus*, and *Lactobacillus gasseri* are able to induce NLRP3 activation in porcine gastrointestinal lymphoid-associated tissue (GALT) (Tohno 2011). The key point here is that these two strains are able to activate TLR-2 and TLR-9, making them suitable candidates for the priming step necessary for NLRP3 activation. This demonstrates that the role of probiotics may not necessarily involve a repression of the inflammatory process. In this instance, NLRP3 activation is said to restore physiologic inflammation and hence can induce protective effects by improving the hosts defense mechanisms. While similar studies with human NLRP3 have not yet surfaced, the impacts of the porcine model should not be underscored since it has a much greater concordance with the human protein than that of commonly used mouse models. In another experiment the probiotic *Bifidobacterium animalis* coupled with a high fiber diet demonstrated that the reduction in TNF-α was correlated with TLR-2 signaling (Trevisi et al. 2008). This provides key evidence to suggest the role of probiotic interventions may actuate their protective role through TLR-signaling pathway.

The SCFA are known to play an important role in modulating the inflammatory process with the gut. Research into the molecular mechanisms through which SCFA achieve their immunomodulatory effects have identified two key mechanisms: (1) G-protein-coupled receptors (GPR43 and 41) located on innate immune cells (Sina et al. 2009) and (2) histone deacetylase inhibitors (HDACi) (Vinolo et al. 2011b).

In a GPR43 knockout mouse model, the chemotactic effect of SCFA was abolished in comparison to that of wild-type mice (Vinolo et al. 2011a). Consistent with this outcome, studies in mice that lack the GPR43 receptor results in an inability to mount an appropriate inflammatory response in models of colitis, arthritis, and asthma (Sina et al. 2009; Maslowski et al. 2009). Therefore the GPR43 receptor is suggested to be pivotal in the chemotactic signal of neutrophils following SCFA binding. As such this indicates that butyrate may have important effects in the regulation of mucosal immunity through modulation of the cytokine and/or chemokine response (Säemann et al. 2000).

Research into the molecular mechanisms underlying the anti-inflammatory effects of butyrate has focused on modulation of pro-inflammatory cytokines and inhibition of the transcription factor NF-κB (Andoh et al. 2003). In particular, IL-8 has been used as marker of intestinal inflammation. Treatment with butyrate in a TNF-α-induced IL-8 mouse model found that there was a marked decrease not only in IL-8 levels but also in the transcriptional activity of NF-κB (Andoh et al. 1999b). While a number of studies have confirmed these findings, (Fusunyan et al. 1998; Park et al. 2007) other studies have been inconclusive (Diakos et al. 2002; Bailón et al. 2010). Differences in these studies in terms of the methodologies used, culture conditions, and various cell types examined make conclusions about the effectiveness of butyrate difficult.

High fiber diets significantly increase IL-18 levels in the blood compared to low fiber diet. IL-18 is produced via the inflammasome pathway. High fiber is associated with SCFA production and GPR43/41 as well as an increase in IL-18 levels in the blood. Since NLRP3 and NLRP6 have been particularly implicated in gut homeostasis, a link can be formed between level of GPCR and inflammasome activation (Korecka and Arulampalam 2012).

It is proposed that one mechanism for the immunomodulatory effects of SCFAs is through inhibition of histone deacetylation. Of the three SCFA produced by commensal bacteria, butyrate is the most potent HDACi, coming before propionate and the least potent being acetate. Histone hyperacetylation results in a more relaxed chromatin structure making it easy for transcription factors such as NF-κB to be targeted by various anti-inflammatory mediators (Andoh et al. 1999a). However, the epigenetic effects of butyrate are not limited to increasing global histone acetylation; it can also have effects on DNA methylation (Spurling et al. 2008).

In addition to butyrate, other probiotic metabolites produced by commensal probiotics have been investigated for their role during inflammatory bowel diseases. *Bacteriodes fragilis* is a commensal bacteria which colonize the lower gastrointestinal tract and has been observed to play a protective role in an in vivo experimental colitis model induced by *Helicobacter hepaticus*. The beneficial effect of this bacterium is attributed to the production of a single microbial molecule referred to as "polysaccharide A" (PSA) (Mazmanian et al. 2008). PSA exhibits anti-inflammatory effect through inhibiting cells of the IL-17 lineage produced by intestinal immune cells and also through induction of IL-10 (Troy 2010). Hence alongside SCFA, probiotics metabolites like PSA further demonstrate the symbiotic interaction between commensal bacteria and host immune system.

5.8 Metabolic Syndromes

Metabolic syndromes which include obesity, diabetes, and cardiovascular disorders (CVDs) are now increasingly being accepted as a consequence of aberrant gut microbiota modulating host metabolism. Obese individuals contain lower abundance of Bacteroidetes and higher abundance of Firmicutes (Ley et al. 2006). Metagenomic analysis of obese mice supports that a shift in microbial composition can potentiate the obese phenotype. Relative to lean wild type, obese mice were identified to have an increased expression of genes responsible for more efficient extraction of energy from food and therefore stimulating lipogenesis (Turnbaugh et al. 2009a). Yet the impact of host genetics can be delineated as a separate factor in the obesity epidemic, as studies demonstrate that gnotobiotic mice are able to develop adiposity through transplantation of microbiota from obese mice (Turnbaugh et al. 2008). The risks associated with obesity are tremendous and are linked with increased likelihood of conditions such as diabetes and CVDs. Hence the ideal strategy would be to consume a low-fat diet and lose weight; however, due to difficulty in compliance by obese patients (Freedman et al. 2001) strategies implementing the use of probiotics to restore gut microbiota have captured the likes of probiotic treatments.

Intake of probiotic yogurt containing *L. acidophilus* and *Bifidobacterium lactis* in a double-blind placebo-controlled trial (DBPC) has shown promising results as it improves blood glucose and antioxidant status in T2DM patients (Ejtahed et al. 2012). Another important DBPC trial has shown that the probiotic *Lactobacillus gasseri* SBT2055 lowers the effect of abdominal adiposity, body weight suggesting favorable potential for probiotic intervention in metabolic disorders (Kadooka et al. 2010). Another study has identified the protective roles of all three SCFA butyrate, propionate, and acetate against diet-induced obesity and insulin resistance (Lin et al. 2012). Despite this exciting new insight, the host signaling for the immunomodulatory effects of SCFA still remain open ended with one study suggesting the role of GPR41 receptor as necessary for efficient energy harvest from the diet (Samuel et al. 2008). Dietary supplementation of butyrate in diet-induced obese mice prevents the development of insulin resistance and obesity (Cefalu et al. 2009). At the molecular level, the HDACi properties of butyrate may be responsible for the increased mRNA expression of peroxisome proliferator-activated receptor-γ coactivator (PGC-1α), which participates in the regulation of carbohydrate and lipid metabolism (Liang and Ward 2006).

5.9 Autism Spectrum Disorders

Autism spectrum disorders are a complex set of neurological conditions affecting children across three systems which include behavior, imagination, and communication (Wing 1997). Autistic children contain greater amounts of *Clostridium* species in their microbiota and subsequently suffer from gastrointestinal inflammation which includes common dilemmas such as constipation, abdominal discomfort, and diarrhea (Parracho et al. 2005). Hence the shift to correct this dysbiotic gut to alleviate some of these symptoms through probiotic supplements. Numerous studies have found positive influences on gut health and overall behavior of autistic children with use of probiotic treatments (Kidd 2003; Blades 2000). In a DBPC trial treatment using *Lactobacillus plantarum*, treatment significantly increased numbers of the beneficial lactobacilli/enterococci in the faecal microbiota of autistic children (Parracho et al. 2010). Interestingly autistic children actually have higher levels of SCFA present with their faecal microbiota than that of healthy controls. One paper suggests that in particular the SCFA propionate is responsible for the gut disturbances seen in autism (MacFabe 2012), hence making it essential to narrow down strain-specific probiotics, while another study proposes the need for further investigation to decipher the underlying mechanism for increased SCFA in autistic children (Wang et al. 2012).

Human neurodevelopment can be adversely impacted by exposure to toxic substances from the environment such as the heavy metals, mercury and copper. The physiologic roles of eliminating these toxins are reliant upon the expression of the Zn-dependant metal binding protein metallothionein. Western diets are stripped of essential minerals such as zinc and selenium, and this interferes with the production of metallothionein and hinders the metabolic process required to eliminate heavy metals from the body. The buildup of heavy metals in the body causes oxidative stress which inhibits methionine synthase; this enzyme subsequently induces epigenetic alterations which involve a decrease in DNA and histone methylation. In the case of autism, diet also have impacts on neurodevelopment. Pregnant mothers, who took vitamin supplements including vitamin B6, B12, and magnesium, reduced the risk of autism in children. Folate and other dietary methyl donors alter epigenetic regulation of gene expression in their children, thereby reducing the risk of autism (Schmidt et al. 2011). Since dietary factors induce aberrant epigenetic changes leading to gut inflammation in autistic children, and the potential of probiotics in alleviating symptoms, it can be inferred that perhaps this is taking place through epigenetic changes.

5.10 Conclusion

Given the importance of our diet in the contribution of structural variation in gut microbiota, an important factor to consider in the development of novel therapies which restore gut microbiota now includes the administration of preventative medicine through dietary interventions. A diet which heavily includes complex carbohydrates promotes healthy gut microbiota and boosts the production of immunomodulatory

SCFA which have HDAC inhibitory properties. HDACi derived naturally from the diet include sulforaphane from broccoli, diallyldisulfide from garlic, and curcumin from turmeric.

Nevertheless, further research into the epigenetic effects of probiotics and other dietary substances is required. Different microbial communities in individuals may be one factor that causes variation in the outcome of probiotic or probiotic metabolite treatment. The challenge for future research is to use this information to optimize probiotic/dietary therapy to improve human health and prevent microbiota-associated diseases, such as allergies, IBD, metabolic syndrome, and autism. Although our current knowledge of restoring gut microbiota through administration of probiotics and their metabolites is not complete, the idea is tantalizing and opens forth a new era of therapeutic modalities for the treatment of chronic inflammatory diseases.

References

Andoh A, Fujiyama Y, Hata K, Araki Y, Takaya H, Shimida M, Bamba T (1999a) Counter-regulatory effect of sodium butyrate on tumour necrosis factor-alpha (Tnf-A)-induced complement C3 and factor B biosynthesis in human intestinal epithelial cells. Clin Exp Immunol 118:23–29

Andoh A, Bamba T, Sasaki M (1999b) Physiological and anti-inflammatory roles of dietary fiber and butyrate in intestinal functions. J Parenter Enteral Nutr 23:S70–S73

Andoh A, Tsujikawa T, Fujiyama Y (2003) Role of dietary fiber and short-chain fatty acids in the colon. Curr Pharm Des 9:347–358

Anukam KC, Reid G (2007) Probiotics: 100 years (1907–2007) after Elie Metchnikoff's observation. In: Méndez-Vilas A (ed) Communicating current research and educational topics and trends in applied microbiology. Spain, Formatex.Org, pp 466–474

Artis D (2008) Epithelial-cell recognition of commensal bacteria and maintenance of immune homeostasis in the gut. Nat Rev Immunol 8:411–420

Bäckhed F, Ding H, Wang T, Hooper LV, Koh GY, Nagy A, Semenkovich CF, Gordon JI (2004) The gut microbiota as an environmental factor that regulates fat storage. Proc Natl Acad Sci U S A 101:15718–15723

Bailón E, Cueto-Sola M, Utrilla P, Rodríguez-Cabezas ME, Garrido-Mesa N, Zarzuelo A, Xaus J, Gálvez J, Comalada M (2010) Butyrate in vitro immune-modulatory effects might be mediated through a proliferation-related induction of apoptosis. Immunobiology 215:863–873

Baldwin A (1996) The Nf-KappaB and I-kappaB proteins: new discoveries and insights. Annu Rev Immunol 14:649–683

Baltimore D (2009) Discovering Nf-kB. Cold Spring Harbor Perspectives Biol 1. Article: a000026

Bauernfeind F, Bartok E, Rieger A, Franchi L, Núñez G, Hornung V (2011) Cutting edge: reactive oxygen species inhibitors block priming, but not activation, of the Nlrp3 inflammasome. J Immunol 187:613–617

Baugh JA (2002) Macrophage migration inhibitory factor. Crit Care Med 30:S27

Björkstén B, Naaber P, Sepp E, Mikelsaar M (1999) The intestinal microflora in allergic estonian and swedish 2-year-old children. Clin Exp Allergy 29:342–346

Blades M (2000) Autism: an interesting dietary case history. Nutr Food Sci 30:137

Borriello SP, Hammes WP, Holzapfel W, Marteau P, Schrezenmeir J, Vaara M, Valtonen V (2003) Safety of probiotics that contain lactobacilli or bifidobacteria. Clin Infect Dis 36:775–780

Böttcher MF, Nordin EK, Sandin A, Midtvedt T, Björkstén B (2000) Microflora-associated characteristics in faeces from allergic and nonallergic infants. Clin Exp Allergy 30:1590–1596

Bush T, Shlotzhauer TL, Imai K (1991) Nonsteroidal anti-inflammatory drugs proposed guidelines for monitoring toxicity. West J Med 155:39–42
Cefalu WT, Gao Z, Lefevre M, Martin RJ, Ward RE, Ye J, Yin J, Zhang J (2009) Butyrate improves insulin sensitivity and increases energy expenditure in mice. Diabetes 58(7):1509–1517
Chen G, Shaw MH, Kim Y-G, Nuñez G (2009) Nod-like receptors: role in innate immunity and inflammatory disease. Annu Rev Pathol 4:365–398
Cordain L, Eaton SB, Sebastian A, Mann N, Lindeberg S, Watkins BA, O'keefe JH, Brand-Miller J (2005) Origins and evolution of the western diet: health implications for the 21st century. Am J Clin Nutr 81:341–354
Creagh EM, O'Neill LA (2006) Tlrs, Nlrs And Rlrs: a trinity of pathogen sensors that co-operate in innate immunity. Trends Immunol 27:352–357
Davis BK, Wen H, Ting JPY (2011) The inflammasome Nlrs in immunity, inflammation, and associated diseases. Annu Rev Immunol 29:707–735
Day RO (2004) The vascular effects of cox-2 selective inhibitors. Aust Prescr 27:142–145
De Filippo C, Cavalieri D, Di Paola M, Ramazzotti M, Poullet JB, Massart S, Collini S, Pieraccini G, Lionetti P (2010) Impact of diet in shaping gut microbiota revealed by a comparative study in children from Europe and rural africa. Proc Natl Acad Sci U S A 107:14691–14696
Denes A, Lopez-Castejon G, Brough D (2012) Caspase-1: is Il-1 just the tip of the Iceberg [Quest]. Cell Death Dis 3:E338
Diakos C, Prieschl EE, Säemann M, Novotny V, Böhmig G, Csonga R, Baumruker T, Zlabinger GJ (2002) Novel mode of interference with nuclear factor of activated T-cells regulation in T-cells by the bacterial metaboliten-butyrate. J Biol Chem 277:24243–24251
Donato KA, Gareau MG, Wang YJJ, Sherman PM (2010) Lactobacillus rhamnosus Gg attenuates interferon-Γ and tumour necrosis factor-a-induced barrier dysfunction and pro-inflammatory signalling. Microbiology 156:3288–3297
Dostert C, Pétrilli V, Bruggen RV, Steele C, Mossman BT, Tschopp J (2008) Innate immune activation through Nalp3 inflammasome sensing of asbestos and silica. Science 320:674–677
Doughty L (2011) Pathogen associated molecular patterns, pattern recognition receptors and pediatric sepsis. Open Inflamm J 4:31–48
Dupaul-Chicoine J, Yeretssian G, Doiron K, Bergstrom KSB, Mcintire CR, Leblanc PM, Meunier C, Turbide C, Gros P, Beauchemin N, Vallance BA, Saleh M (2010) Control of intestinal homeostasis, colitis, and colitis-associated colorectal cancer by the inflammatory caspases. Immunity 32:367–378
Eckburg PB, Bik EM, Bernstein CN, Purdom E, Dethlefsen L, Sargent M, Gill SR, Nelson KE, Relman DA (2005) Diversity of the human intestinal microbial flora. Science 308:1635–1638
Ejtahed HS, Mohtadi-Nia J, Homayouni-Rad A, Niafar M, Asghari-Jafarabadi M, Mofid V (2012) Probiotic yogurt improves antioxidant status in type 2 diabetic patients. Nutrition 28:539–543
Elinav E, Strowig T, Henao-Mejia J, Flavell RA (2011a) Regulation of the antimicrobial response by Nlr proteins. Immunity 34:665–679
Elinav E, Strowig T, Kau AL, Henao-Mejia J, Thaiss CA, Booth CJ, Peaper DR, Bertin J, Eisenbarth SC, Gordon JI, Flavell RA (2011b) Nlrp6 inflammasome regulates colonic microbial ecology and risk for colitis. Cell 145:745–757
FAO/WHO (2002) Guidelines for the evaluation of probiotics in food. FAO/WHO, London
Feghali CA, Wright TM (1997) Cytokines in acute and chronic inflammation. Front Biosci 2:12–26
Franchi L, Kanneganti T-D, Dubyak GR, Núñez G (2007) Differential requirement of P2x7 receptor and intracellular K+for caspase-1 activation induced by intracellular and extracellular bacteria. J Biol Chem 282:18810–18818
Franchi L, Munoz-Planillo R, Nunez G (2012) Sensing and reacting to microbes through the inflammasomes. Nat Immunol 13:325–332
Freedman MR, Janet K, Kennedy E (2001) Popular diets: a scientific review. Obes Res 9:1S–5S
Fusunyan RD, Quinn JJ, Ohno Y, Macdermott RP, Sanderson IR (1998) Butyrate enhances interleukin (Il)-8 secretion by intestinal epithelial cells in response to Il-1[Beta] and lipopolysaccharide. Pediatr Res 43:84–90

Gabay C (2006) Interleukin-6 and chronic inflammation. Arthritis Res Ther 8(Suppl 2):S3–S3
Gaestel M, Kotlyarov A, Kracht M (2009) Targeting innate immunity protein kinase signalling in inflammation. Nat Rev Drug Discov 8:480–499
Ghadimi D, Helwig U, Schrezenmeir J, Heller KJ, De Vrese M (2012) Epigenetic imprinting by commensal probiotics inhibits the Il-23/Il-17 axis in an in vitro model of the intestinal mucosal immune system. J Leukoc Biol 92:895–911
Graham-Rowe D (2011) Lifestyle: when allergies go west. Nature 479:S2–S4
Gringhuis SI (2012) Dectin-1 is an extracellular pathogen sensor for the induction and processing of Il-1 beta via a noncanonical caspase-8 inflammasome. Nat Immunol 13:246
Gussler J, Graham A (2011) Inflammation and nutrition in chronic disease report of the 111th Abbott nutrition research conference. Abbott Nutrition
Hadley C (2006) Food allergies on the rise? EMBO Rep 7:1080–1083
Halle A, Hornung V, Petzold GC, Stewart CR, Monks BG, Reinheckel T, Fitzgerald KA, Latz E, Moore KJ, Golenbock DT (2008) The Nalp3 inflammasome is involved in the innate immune response to amyloid-beta. Nat Immunol 9:857–865
Hill DA, Artis D (2010) Intestinal bacteria and the regulation of immune cell homeostasis. Annu Rev Immunol 28:623–667
Hoffman HM, Brydges SD (2011) Genetic and molecular basis of inflammasome-mediated disease. J Biol Chem 286:10889–10896
Holmes W, Lee J, Kuang W, Rice G, Wood W (1991) Structure and functional expression of a human interleukin-8 receptor. Science 253:1278–1280
Hooper LV, Midtvedt T, Gordon JI (2002) How host-microbial interactions shape the nutrient environment of the mammalian intestine. Annu Rev Nutr 22:283–307
Hussain SP, Harris CC (2007) Inflammation and cancer: an ancient link with novel potentials. Int J Cancer 121:2373–2380
Inohara N, Chamaillard M, Mcdonald C, Nez GN (2005) Nod-Lrr proteins: role in host-microbial interactions and inflammatory disease. Annu Rev Biochem 74:355–383
James MJ, Gibson RA, Cleland LG (2000) Dietary polyunsaturated fatty acids and inflammatory mediator production. Am J Clin Nutr 71:343–348
Janeway CA Jr (1989) Approaching the asymptote? Evolution and revolution in immunology. Cold Spring Harb Symp Quant Biol 54:1–13
Jones CL, Weiss DS (2011) Tlr2 signaling contributes to rapid inflammasome activation during *F. novicida* infection. PLoS One 6, E20609
Kadooka Y, Sato M, Imaizumi K, Ogawa A, Ikuyama K, Akai Y, Okano M, Kagoshima M, Tsuchida T (2010) Regulation of abdominal adiposity by probiotics (Lactobacillus gasseri Sbt 2055) in adults with obese tendencies in a randomized controlled trial. Eur J Clin Nutr 64:636–643
Kahlenberg JM, Dubyak GR (2004) Mechanisms of caspase-1 activation by P2x7 receptor-mediated K+release. Am J Physiol 286:C1100–C1108
Karin M, Lawrence T, Nizet V (2006) Innate immunity gone awry: linking microbial infections to chronic inflammation and cancer. Cell 124:823–835
Kidd PM (2003) An approach to the nutritional management of autism. Altern Ther Health Med 9:22
Kirjavainen PV, Arvola T, Salminen SJ, Isolauri E (2002) Aberrant composition of gut microbiota of allergic infants: a target of bifidobacterial therapy at weaning? Gut 51:51–55
Korecka A, Arulampalam V (2012) The gut microbiome: scourge, sentinel or spectator? J Oral Microbiol 4:1–14
Kumar M, Nagpal R, Kumar R, Hemalatha R, Verma V, Kumar A, Chakraborty C, Singh B, Marotta F, Jain S, Yadav H (2012) Cholesterol-lowering probiotics as potential biotherapeutics for metabolic diseases. Exp Diabetes Res 2012:902917–902917
Lamkanfi M, Mueller JL, Vitari AC, Misaghi S, Fedorova A, Deshayes K, Lee WP, Hoffman HM, Dixit VM (2009) Glyburide inhibits the cryopyrin/Nalp3 inflammasome. J Cell Biol 187:61–70

Lazzaroni M, Bianchi Porro G (2001) Helicobacter pylori and nsaid gastropathy. Aliment Pharmacol Ther 15:22–27

Lee Ms KY (2007) Signaling pathways downstream of pattern-recognition receptors and their cross talk. Annu Rev Biochem 76:447–480

Ley RE, Turnbaugh PJ, Klein S, Gordon JI (2006) Microbial ecology: human gut microbes associated with obesity. Nature 444:1022–1023

Liang H, Ward WF (2006) Pgc-1α: a key regulator of energy metabolism. Adv Physiol Educ 30:145–151

Lin HV, Frassetto A, Kowalik EJ Jr, Nawrocki AR, Lu MM, Kosinski JR, Hubert JA, Szeto D, Yao X, Forrest G, Marsh DJ (2012) Butyrate and propionate protect against diet-induced obesity and regulate gut hormones via free fatty acid receptor 3-independent mechanisms. PLoS One 7, E35240

Ljungh A (2006) Lactic acid bacteria as probiotics. Curr Issues Intest Microbiol 7:73

Luyer MD, Buurman WA, Hadfoune MH, Speelmans G, Knol J, Jacobs JA, Dejong CHC, Vriesema AJM, Greve JWM (2005) Strain-specific effects of probiotics on gut barrier integrity following hemorrhagic shock. Infect Immun 73:3686–3692

Macfabe DF (2012) Short-chain fatty acid fermentation products of the gut microbiome: implications in autism spectrum disorders. Microb Ecol Health Dis 23: doi: 10.3402/mehd.v23i0.19260

Madsen K (2011) Western diet as a trigger for inflammatory bowel disease. Inflammation and nutrition in chronic disease report on 111th abbott nutrition research conference, pp. 1–11

Mallappa RH, Rokana N, Duary RK, Panwar H, Batish VK, Grover S (2012) Management of metabolic syndrome through probiotic and prebiotic interventions. Indian J Endocr Metab 16:20–27

Manichanh C, Borruel N, Casellas F, Guarner F (2012) The gut microbiota in Ibd. Nat Rev Gastroenterol Hepatol 9:599–608

Mariat D, Firmesse O, Levenez F, Guimaraes V, Sokol H, Dore J, Corthier G, Furet J-P (2009) The firmicutes/bacteroidetes ratio of the human microbiota changes with age. BMC Microbiol 9:123

Mariathasan S, Newton K, Monack DM, Vucic D, French DM, Lee WP, Roose-Girma M, Erickson S, Dixit VM (2004) Differential activation of the inflammasome by caspase-1 adaptors asc and Ipaf. Nature 430:213–218

Mariathasan S, Weiss DS, Newton K, Mcbride J, O'rourke K, Roose-Girma M, Lee WP, Weinrauch Y, Monack DM, Dixit VM (2006) Cryopyrin activates the inflammasome in response to toxins and atp. Nature 440:228–232

Martinon F, Petrilli V, Mayor A, Tardivel A, Tschopp J (2006) Gout-associated uric acid crystals activate the Nalp3 inflammasome. Nature 440:237–241

Martinon Fabio JT (2004) Inflammatory caspases: linking an intracellular innate immune system to autoinflammatory diseases. Cell 117:561–574

Maslowski KM, Vieira AT, Ng A, Kranich J, Sierro F, Di Y, Schilter HC, Rolph MS, Mackay F, Artis D, Xavier RJ, Teixeira MM, Mackay CR (2009) Regulation of inflammatory responses by gut microbiota and chemoattractant receptor Gpr43. Nature 461:1282–1286

Mayer-Barber KD, Barber DL, Shenderov K, White SD, Wilson MS, Cheever A, Kugler D, Hieny S, Caspar P, Núñez G, Schlueter D, Flavell RA, Sutterwala FS, Sher A (2010) Cutting edge: caspase-1 independent Il-1β production is critical for host resistance to mycobacterium tuberculosis and does not require Tlr signaling in vivo. J Immunol 184:3326–3330

Mazmanian SK, Round JL, Kasper DL (2008) A microbial symbiosis factor prevents intestinal inflammatory disease. Nature 453:620–625

Mcneela EA, Burke Á, Neill DR, Baxter C, Fernandes VE, Ferreira D, Smeaton S, El-Rachkidy R, Mcloughlin RM, Mori A, Moran B, Fitzgerald KA, Tschopp J, Pétrilli V, Andrew PW, Kadioglu A, Lavelle EC (2010) Pneumolysin activates the Nlrp3 inflammasome and promotes proinflammatory cytokines independently of Tlr4. PLoS Pathog 6, E1001191

Michail S (2009) The role of probiotics in allergic diseases. Allergy Asthma Clin Immunol 5:5

Molodecky NA, Soon IS, Rabi DM, Ghali WA, Ferris M, Chernoff G, Benchimol EI, Panaccione R, Ghosh S, Barkema HW, Kaplan GG (2012) Increasing incidence and prevalence of the inflammatory bowel diseases with time, based on systematic review. Gastroenterology 142:46–54

Moser B (2003) Chemokines: role in immune cell traffic. Eur Cytokine Netw 4:204–210
Mowat AM, Bain CC (2011) Mucosal macrophages in intestinal homeostasis and inflammation. J Innate Immun 3:550–564
Musso G, Gambino R, Cassader M (2010) Obesity, diabetes, and gut microbiota: the hygiene hypothesis expanded? Diabetes Care 33:2277–2284
Netea MG, Nold-Petry CA, Nold MF, Joosten LAB, Opitz B, Van Der Meer JHM, Van De Veerdonk FL, Ferwerda G, Heinhuis B, Devesa I, Funk CJ, Mason RJ, Kullberg BJ, Rubartelli A, Van Der Meer JWM, Dinarello CA (2009) Differential requirement for the activation of the inflammasome for processing and release of Il-1β in monocytes and macrophages. Blood 113:2324–2335
Nishikori M (2005) Classical and alternative Nf-κB activation pathways and their roles in lymphoid malignancies. J Clin Exp Hematopathol 45:15–24
Ohland CL, Macnaughton WK (2010) Probiotic bacteria and intestinal epithelial barrier function. Am J Physiol 298:G807–G819
Onoguchi K, Yoneyama M, Fujit T (2011) Retinoic acid-inducible gene-I-like receptors. J Interferon Cytokine Res 31:27–31
Oppenheim JJ, Zachariae COC, Mukaida N, Matsushima K (1991) Properties of the novel proinflammatory supergene "Intercrine" cytokine family. Annu Rev Immunol 9:617–648
Osuka A, Hanschen M, Stoecklein V, Lederer JA (2012) A protective role for inflammasome activation following injury. Shock 37:47–55. doi:10.1097/Shk.0b013e318234f7ff
Pagnini C, Saeed R, Bamias G, Arseneau KO, Pizarro TT, Cominelli F (2010) Probiotics promote gut health through stimulation of epithelial innate immunity. Proc Natl Acad Sci 107:454–459
Park J-S, Lee E-J, Lee J-C, Kim W-K, Kim H-S (2007) Anti-inflammatory effects of short chain fatty acids in Ifn-Γ-stimulated raw 264.7 murine macrophage cells: involvement of Nf-kB and Erk signaling pathways. Int Immunopharmacol 7:70–77
Parracho HM, Bingham MO, Gibson GR, Mccartney AL (2005) Differences between the gut microflora of children with autistic spectrum disorders and that of healthy children. J Med Microbiol 54:987–991
Parracho HMRT, Gibson GR, Knott F, Bosscher D, Kleerebezem M, Mccartney AL (2010) A double-blind, placebo-controlled, crossover-designed probiotic feeding study in children diagnosed with autistic spectrum disorders. Int J Prob Preb 5:69–74
Pétrilli V, Papin S, Dostert C, Mayor A, Martinon F, Tschopp J (2007) Activation of the Nalp3 inflammasome is triggered by low intracellular potassium concentration. Cell Death Differ 14:1583–1589
Pessi T, Sütas Y, Hurme M, Isolauri E (2000) Interleukin-10 generation in atopic children following oral Lactobacillus rhamnosus Gg. Clin Exp Allergy 30:1804–1808
Pohjavuori E, Viljanen M, Korpela R, Kuitunen M, Tiittanen M, Vaarala O, Savilahti E (2004) Lactobacillus Gg effect in increasing Ifn-Γ production in infants with cow's milk allergy. J Allergy Clin Immunol 114:131–136
Proell M, Riedl SJ, Fritz JH, Rojas AM, Schwarzenbacher R (2008) The nod-like receptor (Nlr) family: a tale of similarities and differences. PLoS One 3:E2119
Puren AJ, Fantuzzi G, Dinarello CA (1999) Gene expression, synthesis, and secretion of interleukin 18 and interleukin 1β are differentially regulated in human blood mononuclear cells and mouse spleen cells. Proc Natl Acad Sci 96:2256–2261
Rathinam VAK, Vanaja SK, Fitzgerald KA (2012) Regulation of inflammasome signaling. Nat Immunol 13:333–332
Ray K (2012) Gut microbiota: colorectal cancer—driven by inflammation and gut bacteria? Nat Rev Gastroenterol Hepatol 9:558–558
Razin A, Kantor B (2005) Dna methylation in epigenetic control of gene expression. Prog Mol Subcell Biol 38:151–167
Reuter BK, Pizarro TT (2004) Commentary: the role of the Il-18 system and other members of the Il-1r/Tlr superfamily in innate mucosal immunity and the pathogenesis of inflammatory bowel disease: friend or foe? Eur J Immunol 34:2347–2355

Säemann MD, Böhmig GA, Österreicher CH, Burtscher H, Parolini O, Diakos C, Stöckl J, Hörl WH, Zlabinger GJ (2000) Anti-inflammatory effects of sodium butyrate on human monocytes: potent inhibition of Il-12 and up-regulation of Il-10 production. FASEB J 14:2380–2382
Samuel BS, Shaito A, Motoike T, Rey FE, Backhed F, Manchester JK, Hammer RE, Williams SC, Crowley J, Yanagisawa M, Gordon JI (2008) Effects of the gut microbiota on host adiposity are modulated by the short-chain fatty-acid binding G protein-coupled receptor, Gpr41. Proc Natl Acad Sci 105:16767–16772
Sandin A, Bråbäck L, Norin E, Björkstén B (2009) Faecal short chain fatty acid pattern and allergy in early childhood. Acta Paediatr 98:823–827
Sanz Y, De Palma G (2009) Gut microbiota and probiotics in modulation of epithelium and gut-associated lymphoid tissue function. Int Rev Immunol 28:397–413
Scheppach W (1994) Effects of short chain fatty acids on gut morphology and function. Gut 35:S35–S38
Schmidt RJ, Hansen RL, Hartiala J, Allayee H, Schmidt LC, Tancredi DJ, Tassone F, Hertz-Picciotto I (2011) Prenatal vitamins, one-carbon metabolism gene variants, and risk for autism. Epidemiology 22:476–485. doi:10.1097/Ede.0b013e31821d0e30
Sina C, Gavrilova O, Förster M, Till A, Derer S, Hildebrand F, Raabe B, Chalaris A, Scheller J, Rehmann A, Franke A, Ott S, Häsler R, Nikolaus S, Fölsch UR, Rose-John S, Jiang H-P, Li J, Schreiber S, Rosenstiel P (2009) G protein-coupled receptor 43 is essential for neutrophil recruitment during intestinal inflammation. J Immunol 183:7514–7522
Spurling CC, Suhl JA, Boucher N, Nelson CE, Rosenberg DW, Giardina C (2008) The short chain fatty acid butyrate induces promoter demethylation and reactivation of Rarβ2 in colon cancer cells. Nutr Cancer 60:692–702
Starkie R, Ostrowski SR, Jauffred S, Febbraio M, Pedersen BK (2003) Exercise and Il-6 infusion inhibit endotoxin-induced Tnf-A production in humans. FASEB J 17:884–886
Strowig T, Henao-Mejia J, Elinav E, Flavell R (2012) Inflammasomes in health and disease. Nature 481:278–286
Stutz A, Golenbock DT, Latz E (2009) Inflammasomes: too big to miss. J Clin Invest 119:3502–3511
Sun Z, Roland A (2002) Nf-[Kappa]B activation and inhibition: a review. Shock 18:99–106
Sung J, Russell R, Yeomans N, Chan F, Chen S, Fock K, Goh K, Kullavanijaya P, Kimura K, Lau C, Louw J, Sollano J, Triadiafalopulos G, Xiao S, Brooks P (2000) Non-steroidal anti-inflammatory drug toxicity in the upper gastrointestinal tract. J Gastroenterol Hepatol 15:G58–G68
Tamboli CP, Neut C, Desreumaux P, Colombel JF (2004) Dysbiosis in inflammatory bowel disease. Gut 53:1–4
Tasse L, Bercovici J, Pizzut-Serin S, Robe P, Tap J, Klopp C, Cantarel BL, Coutinho PM, Henrissat B, Leclerc M, Dore J, Monsan P, Remaud-Simeon M, Potocki-Veronese G (2010) Functional metagenomics to mine the human gut microbiome for dietary fiber catabolic enzymes. Genome Res 20:1605–1612
Taylor PR, Martinez-Pomares L, Stacey M, Lin H-H, Brown GD, Gordon S (2005) Macrophage receptors and immune recognition. Annu Rev Immunol 23:901–944
Ting JPY, Kastner DL, Hoffman HM (2006) Caterpillers, pyrin and hereditary immunological disorders. Nat Rev Immunol 6:183–195
Tohno M (2011) Immunobiotic Lactobacillus strains augment Nlrp3 expression in newborn and adult porcine gut-associated lymphoid tissues. Vet Immunol Immunopathol 144:410
Tran K, Merika M, Thanos D (1997) Distinct functional properties of Ikappab beta and Ikappab alpha. Mol Cell Biol 17:5386–5399
Tremaroli V, Backhed F (2012) Functional interactions between the gut microbiota and host metabolism. Nature 489:242–249
Trevisi P, De Filippi S, Minieri L, Mazzoni M, Modesto M, Biavati B, Bosi P (2008) Effect of fructo-oligosaccharides and different doses of bifidobacterium animalis in a weaning diet on bacterial translocation and toll-like receptor gene expression in pigs. Nutrition 24:1023–1029

Troy EB (2010) Beneficial effects of bacteroides fragilis polysaccharides on the immune system. Front Biosci 15:25
Tsuji NM, Tsutsui H, Seki E, Kuida K, Okamura H, Nakanishi K, Flavell RA (2004) Roles of caspase-1 in listeria infection in mice. Int Immunol 16:335–343
Turnbaugh PJ, Bäckhed F, Fulton L, Gordon JI (2008) Diet-induced obesity is linked to marked but reversible alterations in the mouse distal gut microbiome. Cell Host Microbe 3:213–223
Turnbaugh PJ, Hamady M, Yatsunenko T, Cantarel BL, Duncan A, Ley RE, Sogin ML, Jones WJ, Roe BA, Affourtit JP, Egholm M, Henrissat B, Heath AC, Knight R, Gordon JI (2009a) A core gut microbiome in obese and lean twins. Nature 457:480–484
Turnbaugh PJ, Ridaura VK, Faith JJ, Rey FE, Knight R, Gordon JI (2009b) The effect of diet on the human gut microbiome: a metagenomic analysis in humanized gnotobiotic mice. Sci Transl Med 1:6ra14
Van De Veerdonk FL, Netea MG, Dinarello CA, Joosten LAB (2011) Inflammasome activation and Il-1b and Il-18 processing during infection. Trends Immunol 32:110–116
Van Der Meer JWM (2005) Side effects of anticytokine strategies. Neth J Med 63:78
Vandanmagsar B, Youm Y-H, Ravussin A, Galgani JE, Stadler K, Mynatt RL, Ravussin E, Stephens JM, Dixit VD (2011) The Nlrp3 inflammasome instigates obesity-induced inflammation and insulin resistance. Nat Med 17:179–188
Vinolo MAR, Ferguson GJ, Kulkarni S, Damoulakis G, Anderson K, Bohlooly-Y M, Stephens L, Hawkins PT, Curi R (2011a) Scfas induce mouse neutrophil chemotaxis through the Gpr43 receptor. PLoS One 6, E21205
Vinolo MAR, Rodrigues HG, Nachbar RT, Curi R (2011b) Regulation of inflammation by short chain fatty acids. Nutrients 3:858–876
Wang L, Christophersen C, Sorich M, Gerber J, Angley M, Conlon M (2012) Elevated fecal short chain fatty acid and ammonia concentrations in children with autism spectrum disorder. Dig Dis Sci 57:2096–2102
Wen H, Ting JPY, O'Neill LAJ (2012) A role for the Nlrp3 inflammasome in metabolic diseases—did Warburg miss inflammation [Quest]. Nat Immunol 13:352–357
Wing L (1997) The autistic spectrum. Lancet 350:1761–1766
World Health Organisation (2002) The world health report: reducing risks, promoting healthy life. Who, Geneva
World Health Organisation (2005) Preventing chronic diseases. a vital investment: who global report. Who, Geneva
Xia M, Sui Z (2009) Recent developments In Ccr2 antagonists. Expert Opin Ther Pat 19:295–303
Zaki MH, Boyd KL, Vogel P, Kastan MB, Lamkanfi M, Kanneganti T-D (2010) The Nlrp3 inflammasome protects against loss of epithelial integrity and mortality during experimental colitis. Immunity 32:379–391
Zhou R, Tardivel A, Thorens B, Choi I, Tschopp J (2010) Thioredoxin-interacting protein links oxidative stress to inflammasome activation. Nat Immunol 11:136–140
Zocco MA, Dal Verme LZ, Cremonini F, Piscaglia AC, Nista EC, Candelli M, Novi M, Rigante D, Cazzato IA, Ojetti V, Armuzzi A, Gasbarrini G, Gasbarrini A (2006) Efficacy of Lactobacillus Gg in maintaining remission of ulcerative colitis. Aliment Pharmacol Ther 23:1567–1574

Chapter 6
Epigenetics of Autoimmune Diseases

Fabio Coppedè and Lucia Migliore

Abstract This chapter provides several examples of epigenetic deregulation in autoimmune diseases, a heterogeneous group of human conditions characterized by a deregulated immune response against the body own organs and tissues. Early studies based on the candidate gene approach have been flanked by genome-wide screenings in the last few years, revealing global changes in DNA methylation or histone tail modifications, as well as deregulated methylation and/or expression of hundreds of genes and microRNAs in cells from patients affected by those disorders. This chapter will focus on epigenetic deregulations observed in systemic lupus erythematosus, rheumatoid arthritis, Sjögren's syndrome, psoriasis, multiple sclerosis, systemic sclerosis, and autoimmune thyroid diseases, even though epigenetic modifications are increasingly being observed in many other autoimmune diseases. By contrast, only a few environmental factors have been shown or suspected to induce the observed epigenetic changes. Epigenetic drugs and RNA silencing experiments have often reversed autoimmune disease-like phenotypes in rodents or cell cultures, leading researchers to debate on their potential use in the treatment of these human conditions.

Keywords Epigenetics • Autoimmune diseases • Systemic lupus erythematosus • Rheumatoid arthritis • Sjögren's syndrome • Psoriasis • Multiple sclerosis • Systemic sclerosis

---

F. Coppedè (✉) • L. Migliore
Department of Translational Research and New Technologies in Medicine and Surgery, Division of Medical Genetics, University of Pisa, Via Roma 55, Pisa, Italy
e-mail: fabio.coppede@med.unipi.it

N. Maulik and T. Karagiannis (eds.), *Molecular Mechanisms and Physiology of Disease: Implications for Epigenetics and Health*, DOI 10.1007/978-1-4939-0706-9_6,

6.1 Introduction

Autoimmune diseases include over 100 conditions characterized by an inappropriate immune response against the body own tissues and falling into two general types: systemic autoimmune diseases that damage many organs and systems such as systemic lupus erythematosus (SLE), rheumatoid arthritis (RA), Sjögren's syndrome (SjS), and systemic sclerosis or scleroderma (SSc) among others, and diseases characterized by an autoimmune response against a specific organ or tissue such as psoriasis, Hashimoto's thyroiditis (HT), Graves' disease (GD), and many others (Selmi 2012). It is now clear that while many of these pathologies can be influenced by heritable factors, also environmental factors such as drugs, ultraviolet light, infectious agents, and diet play a role, and recent genome-wide association studies revealed that genomics alone cannot fully explain the individual susceptibility to those disorders (Selmi 2012). Increasing evidence points to additional mechanisms linking the individual susceptibility with environmental factors, and epigenetics is increasingly recognized as one of the most promising missing links (De Santis and Selmi 2012). Indeed, since epigenetic mechanisms are sensitive to external stimuli, several environmental effects on immune responses could be mediated by epigenetic changes (Costenbader et al. 2012).

The term epigenetics comprises heritable and reversible modifications that alter gene expression without resulting from direct changes in the primary DNA sequence. Several epigenetic mechanisms are known, including DNA methylation, covalent modifications of histone tails, and nucleosome positioning, all interacting to determine chromatin folding and the relative accessibility of a given genetic locus to activating and suppressing transcription factors (Martín-Subero 2011). Noncoding RNAs affecting gene expression are also largely recognized as epigenetic mechanisms (Esteller 2011). All those mechanisms are exhaustively described in Chap. 1 of this book.

Most of the studies performed so far have focused on epigenetic modifications in SLE and rheumatoid arthritis (Quintero-Ronderos and Montoya-Ortiz 2012), but there is increasing evidence of epigenetic changes in other autoimmune disorders such as SjS, multiple sclerosis, systemic sclerosis, psoriasis, autoimmune thyroid diseases (AITDs), and others (Meda et al. 2011; Zhang et al. 2011; Quintero-Ronderos and Montoya-Ortiz 2012). This chapter will provide a summary of the most relevant evidence of epigenetic deregulation in autoimmune disorders.

6.2 Systemic Lupus Erythematosus

SLE is a systemic autoimmune disorder characterized by the production of autoantibodies directed against nuclear self-antigens. The disease can target many organs, including the skin and joints, but also the heart, the kidneys, the nervous system, and others. It afflicts both sexes but occurs more frequently in women. Genetic association studies, genome-wide association studies, and discordance in disease inheritance in twins, have revealed that genetics alone does not completely account for disease heritability (Deapen et al. 1992; Cunninghame Graham 2009). Early studies of SLE epigenetics showed that CD4+ T cells treated with the DNA methylation inhibitor 5-azacytidine respond to the presentation of self antigens (Richardson 1986), and their injection in mice caused a lupus-like syndrome (Quddus et al. 1993). Those studies suggested that impairments of DNA methylation might be involved in autoimmunity.

6.2.1 DNA Methylation in SLE

DNA methylation represents one of the most studied epigenetic mechanisms for gene regulation and consists of the addition of a methyl group to the 5′ position of the cytosine pyrimidine ring (5-methylcytosine) mediated by DNA methyltransferase enzymes (DNMTs) using *S*-adenosylmethione (SAM) as the methyl donor compound (Jones 2012). There are multiple families of DNMTs in mammals. Among them DNMT1 is primarily involved in the maintenance of DNA methylation patterns during development and cell division, whereas DNMT3a and DNMT3b are the de novo methyltransferases and establish DNA methylation patterns during early development (Jones and Liang 2009). Methylated DNA can be specifically recognized by a set of proteins called methyl-CpG binding proteins (MBPs), including MECP2 (methyl-CpG binding protein 2) and MBD proteins (methyl-CpG binding domain proteins), that contain a transcription repression domain to interact with other proteins and enhance DNA methylation-mediated transcriptional repression (Fournier et al. 2012).

Following the observation that inhibitors of DNA methylation were able to induce autoimmune reactions and lupus-like symptoms in animals (Richardson 1986, Quddus et al. 1993), several investigators analyzed DNA methylation levels in SLE patients (Table 6.1). One of the most replicated findings is a global DNA hypomethylation observed in the CD4+ T cells of these patients (Balada et al. 2007a). DNA hypomethylation has been often linked to a reduction of DNMT1 mRNA levels in those cells (Zhu et al. 2011; Qin et al. 2013), but data on DNMT1, DNMT3a, and DNMT3b expression levels are still conflicting (Balada et al. 2008; Liu et al. 2011; Zhu et al. 2011). Also an increased expression of MBD2 and an inverse correlation between MBD2 expression and DNA methylation levels was often observed in CD4+ T cells of SLE patients (Balada et al. 2007b; Qin et al. 2013).

Apart from studies on global DNA methylation levels, the analysis of gene-specific promoter methylation by means of candidate gene approaches led researchers to identify several genes that are hypomethylated and therefore over-expressed in CD4+ T cells of SLE patients, some examples are *ITGAL*, *CD40LG*, *PRF1*, *TNFSF7*(*CD70*), and *KIR* family genes (reviewed in Hughes and Sawalha 2011). Many of those genes encode proteins involved in immune function and inflammation, and it is believed that their overexpression leads to lupus T cell autoreactivity and subsequent induction of autoreactive B cell immunoglobulin production

Table 6.1 Some examples of epigenetic deregulation in systemic lupus erythematosus

DNA methylation		
Human CD4+ T cells	Global hypomethylation	Balada et al. (2007a)
	Reduced DNMT1 and increased MBD2 levels	Zhu et al. (2011), Qin et al. (2013)
	Demethylated genes: *ITGAL, CD40LG, PRF1, TNFSF7(CD70), KIR2DL4*	Hughes and Sawalha (2011)
	LINE-1 hypomethylation	Nakkuntod et al. (2011)
	Whole-genome approaches: 232 hypomethylated genes and 104 hypermethylated genes	Jeffries et al. (2011)
Histone tail modifications		
Animal models: MRL/lpr mice	Global hypoacetylation of histone H3 and H4	Garcia et al. (2005)
	Overexpression of histone deacetylase SIRT1	Hu et al. (2009)
	Treatment with HDACi reversed the lupus-like phenotype	Reilly et al. (2011)
Human CD4+ T cells	Global hypoacetylation of histone H3 and H4	Hu et al. (2008)
	Overexpression of histone deacetylase SIRT1	
	Global H3K9 hypomethylation	
MicroRNA expression		
Human CD4+ T cells	Upregulation of miR-21, miR-148a, miR-126, miR-224	Amarilyo and La Cava (2012), Lu et al. (2013)
	Downregulation of miR-142-3p, miR142-5p, miR-145, miR-146a	Chan et al. (2012), Ding et al. (2012), Lu et al. (2013)

(Hughes and Sawalha 2011). Recently, Balada and coworkers determined the expression levels of *ITGAL*, *PRF1*, *KIR2DL4*, *TNFSF7*(*CD70*), and *CD40LG* genes in CD4+ T cells of patients with SLE and performed correlations with the global DNA methylation status and the levels of DNMT and MBD proteins. SLE patients had significantly elevated transcript levels of *ITGAL*, *PRF1*, and *TNFSF7*(*CD70*), and those levels correlated with global DNA hypomethylation as well as with the expression of most of the DNA methylation-related genes, reinforcing the hypothesis of an epigenetic deregulated network in SLE (Balada et al. 2012).

The analysis of methylation of repetitive elements in SLE patients revealed hypomethylation of LINE-1 but not Alu in CD4+ T lymphocytes, CD8+ T lymphocytes, and B lymphocytes (Nakkuntod et al. 2011).

The candidate gene approach analysis has been paralleled in recent years by whole-genome methylation studies that are revealing hundreds of genes differentially methylated and expressed in SLE patients with respect to controls. One of such approaches performed on monozygotic twins discordant for SLE featured widespread changes in the DNA methylation status of a significant number of genes associated with immune function that occurred in parallel with a global decrease in the 5-methylcytosine content in the affected twins (Javierre et al. 2010). A subsequent genome-wide DNA methylation analysis in CD4+ T cells from SLE patients revealed 236 hypomethylated and 105 hypermethylated CpG sites (representing 232 and 104 genes, respectively). Hypomethylated genes in lupus T cells included, among others, many involved in autoimmunity, while genes involved in folate biosynthesis, required for SAM production and DNA methylation, resulted hypermethylated (Jeffries et al. 2011). Another whole-genome methylation analysis of SLE revealed that hypomethylation of interleukin *IL10* and interleukin receptor *IL1R2* promoters is associated with disease activity (Lin et al. 2012).

6.2.2 Histone Tail Modifications in SLE

The chromatin state represents another important modulator of gene expression profiles. Histone tail acetylation on lysine residues is mediated by histone acetyltransferases (HATs) and represents one of the most studied modifications associated with chromatin relaxation and transcriptional activation (Berger 2007). Another frequently studied modification of histone tails is methylation on either lysine or arginine residues, mediated by protein methyltransferases. Methylation of histone tails can be associated with either condensation or relaxation of the chromatin structure, since several sites for methylation are present on each tail thus allowing many combinations (Martin and Zhang 2005). Most of our current knowledge on histone tail modifications in SLE derives from studies based upon in vitro cell cultures and in vivo studies in murine models of lupus demonstrating that histone deacetylase inhibitors (HDACi) reversed the expression of multiple genes involved in autoimmunity and SLE pathogenesis (reviewed in Reilly et al. 2011). For example a global hypoacetylation of histone H3 and H4 was observed in a mouse model of lupus (MRL/lpr mice) compared to control mice, and the administration of the HDACi trichostatin A reversed histone hypoacetylation with improvement of disease phenotype (Garcia et al. 2005). Others observed an aberrant expression pattern of HATs and histone deacetylases (the enzymes responsible for histone deacetylation) in CD4+ T cells of MRL/lpr mice, among which the overexpression of histone deacetylase SIRT1 was implicated in lupus pathogenesis (Hu et al. 2009). Suppression of SIRT1 expression by means of RNA silencing in the animals resulted in an increase of global histone H3 and H4 acetylation levels and mitigated the disease-related phenotype (Hu et al. 2009).

Studies in humans revealed global histone H3 and H4 hypoacetylation and increased SIRT1 mRNA levels in active lupus CD4+ T cells of SLE patients compared with controls, as well as global histone H3K9 hypomethylation in both active and inactive lupus CD4+ T cells (Hu et al. 2008). There is also evidence that aberrant histone modifications within the *TNFSF7(CD70)* promoter may contribute to the development of lupus by increasing CD70 expression in CD4+ T cells (Zhou et al. 2011). More than 100 different auto-antibodies to nuclear antigens were found in patients with SLE, some of which recognize insoluble nuclear antigens (chromatin, DNA, histones, and RNA), leading researchers to formulate the hypothesis of a potential relationship between auto-antibodies production in SLE with changes in epigenetic patterns, such as DNA methylation and histone tail modifications (Thabet et al. 2012).

6.2.3 RNA-Mediated Epigenetic Mechanisms in SLE

Among noncoding RNAs, microRNA (miRNAs) are a group of small noncoding RNAs of about 22 nucleotides in length that bind to the 3′ untranslated region (3′-UTR) of target mRNAs and mediate their posttranscriptional regulation leading to either degradation or translational inhibition, depending on the degree of sequence complementarities. MiRNAs target about 60 % of all genes (Sato et al. 2011), and a complex network of interactions exists among miRNAs and other epigenetic mechanisms, such as DNA methylation and histone modification processes, to organize the whole gene expression profile.

In 2007 Dai and coworkers published the first report of a difference in miRNA expression between SLE patients and healthy controls (Dai et al. 2007). Since then studies profiling miRNA expression in blood cells, body fluids, and target tissues from SLE patients revealed unique miRNA signatures when compared with healthy individuals or those with other autoimmune diseases (Amarilyo and La Cava 2012; Shen et al. 2012). Among over-expressed miRNAs in lupus CD4+ T cells, miR-21, miR-148a, and miR-126 lead to DNMT1 downregulation by directly targeting its transcript (miR-148a and miR-126) or transcripts of genes that operate in the Ras–MAPK pathway upstream of DNMT1 (miR-21), and resulting in DNMT1 inhibition, DNA hypomethylation, and altered expression of genes involved in both pro-inflammatory and anti-inflammatory processes (Amarilyo and La Cava 2012). Other miRNAs were found to be up-regulated or down-regulated in cells of SLE patients; some examples include miR-142-3p and miR-142-5p that are down-regulated in SLE CD4+ T cells, causing T cell over-activation and B cell hyperstimulation (Ding et al. 2012). By contrast miR-146a, a negative regulator in immune and inflammatory responses, is down-regulated in SLE patients (Chan et al. 2012). A recent study profiled the expression of 270 human miRNAs in T cells from five SLE patients and five healthy controls, and identified under-expressed miR-145 and over-expressed miR-224 as well as down-regulated expression of their target genes linked to accelerated T cell activation-induced cell death (Lu et al. 2013).

6.2.4 Environmental Factors and Their Potential Epigenetic Properties in SLE

The demethylating agent 5-azacytidine was the first drug to be identified to cause lupus-like symptoms in rodents by altering DNA methylation levels (Richardson 1986; Quddus et al. 1993). Since then procainamide and hydralazine also have been suspected of causing SLE by inhibiting DNA methylation and inducing T cells autoreactivity. Procainamide is a competitive inhibitor of DNMT1 enzymatic activity and hydralazine inhibits T and B cell ERK pathway (Cornacchia et al. 1988). Other chemicals, such as air pollutants, have been suspected to act on DNA methylation and other epigenetic mechanisms in SLE (De Santis and Selmi 2012). Among physical agents, it has been suggested that UV light might induce the overexpression of autoimmunity-related genes through aberrant T cell DNA demethylation (Li et al. 2010). Additional factors suspected to epigenetically contribute to the incidence of autoimmune diseases are increasing age and infectious agents (De Santis and Selmi 2012).

6.3 Rheumatoid Arthritis

RA is a systemic autoimmune disease primarily characterized by chronic inflammation of the joints and ultimately leading to joint destruction. Both genetic and environmental factors are involved in disease pathogenesis, but increasing evidence (Klein et al. 2012) supports a role for epigenetic modifications (Table 6.2). RA synovial fibroblasts (RASFs) play a major role in the initiation and perpetuation of the disease; they are the most common cell type at the site of invasion and active contributors in joint damage due to their ability to secrete cytokines, chemokines, and joint-damaging enzymes. Moreover, RASFs show tumoral behavior including invasiveness and resistance to apoptosis. Epigenetic mechanisms have been largely investigated as contributors of RASFs aggressiveness and those cells are the best-characterized ones for epigenetic alterations in RA (Klein et al. 2012; Nakano et al. 2013).

6.3.1 DNA Methylation in RA

In 1991, Corvetta and coworkers observed reduced global DNA methylation in peripheral blood, synovial mononuclear cells, and synovial tissues from RA patients (Corvetta et al. 1991). Others observed hypomethylation and overexpression of LINE 1 retrotransposable elements in RASFs, affecting the expression of other genes likely contributing to cell activation (Neidhart et al. 2000; Kuchen et al. 2004). A subsequent study confirmed those previous findings and revealed that proliferating RASFs were deficient in DNMT1, and that the demethylating agent 5-azacytidine reproduced the activated phenotype of RASFs in normal synovial fibroblasts, with upregulation of over 100 genes including growth factors and receptors, extracellular matrix proteins, adhesion molecules, and matrix-degrading enzymes (Karouzakis et al. 2009). The search for specific genes regulated by DNA methylation in RASFs revealed other demethylated and over-expressed ones,

Table 6.2 Some examples of epigenetic deregulation in rheumatoid arthritis

DNA methylation		
Human synovial fibroblasts (RASFs)	Global hypomethylation	Corvetta et al. (1991), Karouzakis et al. 2009
	LINE-1 hypomethylation	Neidhart et al. 2000, Kuchen et al. 2004
	Reduced DNMT1 levels	Karouzakis et al. (2009)
	Demethylated genes: *CXCL12*	Karouzakis et al. (2011)
	Hypermethylated genes: *DR3*	Takami et al. (2006)
	Whole-genome approaches: 207 hypomethylated or hypermethylated genes	Nakano et al. (2013)
Human CD4+ T cells	Demethylated genes: *CD40LG*	Liao et al. (2012)
Human peripheral blood mononuclear cells	Demethylated genes: *IL-6, IL-10*	Nile et al. (2008), Fu et al. (2011)
Histone tail modifications		
Human synovial fibroblasts (RASFs)	Overexpression of histone deacetylase HDAC1	Horiuchi et al. (2009)
Human peripheral blood mononuclear cells	Overexpression of histone deacetylase HDAC1	Gillespie et al. (2012)
Animal models	Treatment with HDACi reversed the RA-like phenotype	Cantley et al. (2012) De Santis and Selmi (2012)
MicroRNA expression		
RA synovial cells	Upregulation of miR-146a, miR-155, miR-223	Amarilyo and La Cava (2012), Lu et al. (2013)

such as for example *CXCL12* that contributes to the expression of matrix metalloproteinases (Karouzakis et al. 2011). Other genes were found to be hypermethylated in those cells, including the promoter of the death receptor 3 (*DR3*) gene, whose downregulation in RASFs was linked to resistance to apoptosis (Takami et al. 2006). Impairments of DNA methylation were also observed in peripheral blood mononuclear cells of RA patients, some examples are demethylation of CpG sites in interleukin-6 (*IL-6*) and interleukin-10 (*IL-10*) gene promoters (Nile et al. 2008; Fu et al. 2011). The incidence of both RA and SLE is higher in females than in males, and the *CD40LG* gene on the X chromosome was found to be demethylated and overexpressed in CD4+ T cells from female RA and SLE patients (Lian et al. 2012; Liao et al. 2012).

A recent genome-wide approach in RASFs revealed 207 hypermethylated or hypomethylated genes, with hypomethylation increased in multiple pathways related to cell migration (Nakano et al. 2013), and recent studies in RA animal models showed an increased expression of MeCP2 in synovium and fibroblast-like synoviocytes, suggesting that MeCP2 could participate in RA pathogenesis through silencing of certain genes (Miao et al. 2013).

6.3.2 *Histone Tail Modifications in RA*

Several investigators observed increased overexpression and activity of histone deacetylases, and particularly of HDAC1, in RASFs and peripheral blood mononuclear cells of RA patients, suggesting a role for histone tail modifications in disease pathogenesis (Horiuchi et al. 2009; Gillespie et al. 2012). Moreover, HDACi, such as for example trichostatin A, were potent inhibitors of tumor necrosis factor and IL-6 production in those cells (Gillespie et al. 2012; Grabiec et al. 2012). There is also evidence from studies in vitro and in animal models that HDACi have the potential to suppress bone destruction in chronic inflammatory diseases such as RA (Cantley et al. 2012). These are only some of many examples showing anti-inflammatory properties of HDACi in RA models, whose beneficial effects are exerted through reduced production of cytokines, chemokines, and related receptors (De Santis and Selmi 2012).

6.3.3 *RNA-Mediated Epigenetic Mechanisms in RA*

MiRNAs have been largely investigated in the pathogenesis of RA, and some of them, such as for example miR-146a, miR-155, and miR-223, are of particular interest in disease pathogenesis (Ammari et al. 2013). Mir-146a is a negative regulator in immune and inflammatory responses up-regulated in several tissues of RA patients, including RASFs and peripheral blood mononuclear cells, and associated with tumor necrosis factor alpha production and disease activity (Xu et al. 2012). MiR-155 is up-regulated in synovial membrane and synovial fluid macrophages from RA patients (Kurowska-Stolarska et al. 2011), and has a powerful regulatory potential in a wide variety of immune cells through targeting specific mRNAs (Leng et al. 2011). MiR-223 is intensely expressed in RA synovium, and its overexpression suppresses osteoclastogenesis in vitro (Shibuya et al. 2013). Those miRNAs are currently investigated as potential therapeutic targets in RA, and recent integrated analyses of DNA methylation and miRNA expression profiling in RASFs are revealing novel markers of DNA methylation and sets of miRNAs that are controlled by DNA methylation, as well as genes that are regulated by DNA methylation and are targeted by miRNAs with a potential use as clinical markers (de la Rica et al. 2013).

6.3.4 *Environmental Factors and Their Potential Epigenetic Properties in RA*

Among environmental factors, cigarette smoke condensate was shown to up-regulate gene and protein expression of pro-inflammatory cytokines in human fibroblast-like synoviocytes (Shizu et al. 2008), and tobacco smoke is recognized among environmental RA risk factors (Karlson and Deane 2012). However, an epigenetic effect of tobacco smoke in RA is at present only speculative (De Santis and Selmi 2012).

6.4 Other Autoimmune Diseases

Epigenetic studies in autoimmune diseases other than SLE and RA are increasing in recent years (Tables 6.3, 6.4, 6.5, 6.6, and 6.7). Within this paragraph we describe some of the most recent examples.

6.4.1 Epigenetics of SjS

SjS is a systemic autoimmune disease characterized by chronic inflammation leading to reduced secretion of the exocrine salivary and lacrimal glands. Epigenetic studies in SjS are still in their infancy (Table 6.3); however, hypomethylation and overexpression of *TNFSF7(CD70)* were observed in CD4+ T cells of SjS patients (Yin et al. 2010), and hypermethylation of *BP230*, coding for a protein involved in the anchorage of salivary gland cells, was observed in labial salivary glands in SjS (González et al. 2011). The methylation profile of the gene coding for the interferon regulatory factor 5 (*IRF5*) was investigated in CD4+ T cells, B lymphocytes, and monocytes from patients with SjS, but the observed methylation levels were similar to those observed in cells from controls (Gestermann et al. 2012).

Abnormal distribution of aquaporin 5 (AQP5) in salivary gland acini is likely to contribute to the deficiency of fluid secretion in SjS, and the tumor necrosis factor alpha plays an important role in the destruction of acinar structures in exocrine glands,

Table 6.3 Some examples of epigenetic deregulation in Sjögren's syndrome

DNA methylation		
Human CD4+ T cells	Demethylated genes: *TNFSF7(CD70)*	Yin et al. (2010)
	Hypermethylated genes: *BP230*	González et al. (2011)
Histone tail modifications		
Human salivary gland cells	Deacetylation of histone H4 in the promoter of *AQP5* gene and inhibition of aquaporin 5 expression	Yamamura et al. (2012)
MicroRNA expression		
Human salivary gland cells	Deregulation of miR-547 and miR-768-3p expression: the expression of miR-768-3p increases, whereas the expression of miR-574 decreases with increasing focus scores	Alevizos et al. (2011)
Human peripheral blood mononuclear cells	Upregulation of miR-146a and miR-146b	Pauley et al. (2011), Zilahi et al. (2012)

Table 6.4 Some examples of epigenetic deregulation in psoriasis

DNA methylation		
Human skin samples	Demethylated genes: *SHP-1*	Ruchusatsawat et al. (2006)
	Whole-genome approaches: differential methylation of 1,108 sites between normal and psoriatic tissues	Roberson et al. (2012)
Human CD4+ T cells	Whole-genome approaches: hypomethylation of 26 regions of the genome, most of them pericentromeric	Han et al. (2012)
	Hypermethylation of 121 genes on the X chromosome	Han et al. (2012)
Human ematopoietic cells	Hypomethylation of genes coding for p16, p21, and p53	Zhang et al. (2007, 2009)
Histone tail modifications		
Human skin samples	Overexpression of histone deacetylase HDAC1	Tovar-Castillo et al. (2007)
Human peripheral blood mononuclear cells	Global histone H4 hypoacetylation	Zhang et al. (2011)
MicroRNA expression		
Human skin samples	Deregulation of 98 canonical and 15 noncanonical miRNAs, including upregulation of miR-203, miR-21, and miR-31	Joyce et al. (2011), Xia et al. (2013)

Table 6.5 Some examples of epigenetic deregulation in multiple sclerosis

DNA methylation		
Human white matter samples	Demethylated genes: *PAD2*	Mastronardi et al. (2007)
Human peripheral blood mononuclear cells	Demethylated genes: *PAD2*	Calabrese et al. (2012)
Histone tail modifications		
Human white matter samples	Increased histone H3 acetylation	Pedre et al. (2011)
MicroRNA expression		
Human blood cells	Upregulation of miR-21, miR-146a, miR-146b, and miR-326	Fenoglio et al. (2012)
Human brain regions	Upregulation of miR-155, miR-326, and miR-34a	Fenoglio et al. (2012)
Human B lymphocytes	Dowregulation of 49 miRNAs	Sievers et al. (2012)

and inhibits *AQP5* gene expression in human salivary gland acinar cells by suppression of acetylation of histone H4 in the promoter region (Yamamura et al. 2012).

Also some miRNAs were found to be deregulated in SjS salivary glands (miR-547 and miR-768-3p) and/or in peripheral mononuclear cells (miR-146a and miR-146b) (Alevizos et al. 2011; Pauley et al. 2011; Zilahi et al. 2012). Moreover, a recent study has shown that the SjS antigen B is a pre-miRNA-binding protein that regulates miRNA processing in vitro (Liang et al. 2013).

Table 6.6 Some examples of epigenetic deregulation in systemic sclerosis

DNA methylation		
Human scleroderma fibroblasts	Hypermethylation of *FLI1* gene	Wang et al. (2006)
Human CD4+ T cells	Global hypomethylation	Lei et al. (2009)
	Reduced DNMT1, MBD3, and MBD4 mRNA levels	Lei et al. (2009)
	Demethylated genes: *CD40LG, TNFSF7(CD70)*	Jiang et al. (2012) Lian et al. (2012)
Human peripheral blood mononuclear cells	Different methylation profiles of genes on the X chromosome in monozygotic twins discordant for the disease	Selmi et al. (2012)
Histone tail modifications		
Human skin tissues	Overexpression of histone acetyltransferase p300, and histone H4 acetylation of the *COL1A2* locus	Ghosh et al. (2013)
Cultured fibroblasts	Involvement of histone H3 methylation on lysine 27 in regulation of fibroblast activation	Krämer et al. (2012)
MicroRNA expression		
Human skin tissues	Upregulation of miR-23b, and let-7	Li et al. (2012)
	Dowregulation of miR-125b, miR-133a, miR-206, and miR-140-5p	
Human skin tissues and fibroblasts	Upregulation of miR-21	Maurer et al. (2010), Zhu et al. 2012
	Dowregulation of miR-145 and miR-29	
Human dermal fibroblasts	Dowregulation of miR-150	Honda et al. (2013)

Table 6.7 Some examples of epigenetic deregulation in autoimmune thyroid diseases

DNA methylation		
Blood DNA	*MTHFR* C677T polymorphism associated with reduced risk of Graves' disease in women	Mao et al. (2010)
	DNMT1 32204GG genotype associated with DNA hypomethylation and response to treatment in Graves' disease	Arakawa et al. (2012)
	MTRR A66G polymorphism associated with the severity of Hashimoto's thyroiditis	Arakawa et al. (2012)
MicroRNA expression		
Human peripheral blood mononuclear cells	Downregulation of miR-154*, miR-376b, and miR-431* in early stages Graves' disease	Liu et al. (2012)
Human thyroid tissues	Downregulation of miR-146a1 in Graves' disease	Bernecker et al. (2012)
	Downregulation of miR-155_2 and upregulation of miR-200a1 in Hashimoto's thyroiditis	Bernecker et al. (2012)

6.4.2 Epigenetics of Psoriasis

Psoriasis is an organ-specific autoimmune disease triggered by an active immune system causing cells to build up rapidly on the surface of the skin, resulting in thick, white, silvery, or red patches that are sometimes painful. The pathology of psoriasis is complex, involving both genetic and environmental components (Zhang et al. 2012), and increasing evidence supports a role for epigenetic modifications (Table 6.4). Early studies on DNA methylation revealed *SHP-1* promoter methylation in normal epithelial tissues and demethylation in psoriasis. SHP-1 is a tyrosine phosphatase and has been proposed as a candidate tumor suppressor gene in lymphoma, leukemia, and other cancers, as it functions as an antagonist to the growth-promoting and oncogenic potentials of tyrosine kinases (Ruchusatsawat et al. 2006). A reduced proliferative activity has been detected in the hematopoietic cells from patients with psoriasis and linked to hypomethylation of the genes coding for p16, p21, and p53 (Zhang et al. 2007, 2009). More recent genome-wide approaches are revealing hundreds of novel methylation markers of the disease, thereby strengthening the contribution of epigenetics in psoriasis. The methylation levels at 27,578 CpG sites in skin samples from individuals with psoriasis and unaffected individuals revealed different methylation of 1,108 sites (Roberson et al. 2012). Similarly, differences in DNA methylation were found in CD4+ T cells of monozygotic twins discordant for psoriasis (Gervin et al. 2012). A genome-wide DNA methylation profiling of naïve CD4+ T cells showed distinct hypomethylation in 26 regions of the genome ranging in size from 10 to 70 kb (most of them pericentromeric) in patients with psoriasis with respect to healthy controls. Conversely, the promoter regions of 121 genes, and particularly of immune-related genes, on the X chromosome were hypermethylated in psoriasis patient T cells compared to those from healthy controls (Han et al. 2012).

Concerning histone tail modifications, the HDAC-1 mRNA resulted over-expressed in psoriatic skin samples compared with skin specimens from healthy subjects (Tovar-Castillo et al. 2007). Moreover, global histone H4 hypoacetylation was observed in peripheral blood mononuclear cells from psoriasis patients, and there was a negative correlation between the degree of histone H4 acetylation and disease activity (Zhang et al. 2011).

A comprehensive analysis of the normal and psoriatic skin miRNAome with next-generation sequencing revealed 80 known and 18 novel miRNAs that were differentially expressed in psoriatic skin. Of particular significance was the 2.7-fold upregulation of a novel miRNA derived from the antisense strand of the miR-203 locus, which plays a role in epithelial differentiation. Other differentially expressed miRNAs included hematopoietic-specific miRNAs such as miR-142-3p and miR-223/223*, and angiogenic miRNAs such as miR-21, miR-378, miR-100, and miR-31, which was the most highly up-regulated miRNA in psoriatic skin (Joyce et al. 2011). Subsequent functional studies of those miRNAs revealed that miR-21 suppresses apoptosis in activated T cells, and thus, overexpression of miR-21 may contribute to T cell-derived psoriatic skin inflammation (Meisgen et al. 2012), while miR-31 modulates inflammatory cytokine and chemokine production in keratinocytes via targeting serine/threonine kinase 40 (Xu et al. 2013). Moreover, the analysis of more than 670 million qualified reads from 67 small RNA libraries, revealed 21 novel, noncanonical miRNAs (3 small nuclear RNA-derived, 2 tRNA-derived miRNAs, and 16 miRtrons) and 39 novel endo-siRNAs that were expressed in skin, and 15 of them were significantly differentially expressed in psoriatic versus normal skin (Xia et al. 2013).

6.4.3 Epigenetics of Multiple Sclerosis

Multiple sclerosis is an autoimmune demyelinating disease and a common cause of neurodegeneration and disability in young adults. Disease discordance in monozygotic twins indicates environmental importance in its pathogenesis, but a genome-wide DNA methylation study in CD4+ lymphocytes of monozygotic twins discordant for MS failed to find significant differences, thereby dampening research expectations (Baranzini et al. 2010). However, the promoter of the peptidyl arginine-deiminase 2 (*PAD2*) gene was hypomethylated in the white matter from MS patients, resulting in increased synthesis of PAD2 protein that is responsible for the increased amount of citrullinated myelin basic protein, which in turn results in loss of myelin stability in MS brain (Mastronardi et al. 2007). Similar results were observed in peripheral blood mononuclear cells of MS patients, where *PAD2* overexpression was associated with promoter demethylation (Calabrese et al. 2012).

If data on DNA methylation alterations are scarce in MS, an increased histone H3 acetylation associated with increased levels of transcriptional inhibitors of oligodendrocyte differentiation was observed in the white matter of patients with chronic MS (Pedre et al. 2011), and a number of miRNAs have been found to be dysregulated in blood cells from MS patients, in brain lesions, as well as in biological fluids such as serum and plasma (Table 6.5). Some examples are miR-326 that was found to be up-regulated in MS blood and promoted T-helper CD4+ cells differentiation, miR-21, miR-146a and miR-146b up-regulated in peripheral blood mononuclear cells of MS patients as compared with controls, and miR-155, miR-326, and miR-34a that were found to be up-regulated in active MS brain lesions and targeted CD47, a regulatory membrane protein (reviewed in Fenoglio et al. 2012). These are only some of several examples of miRNAs deregulation in MS tissues, and recent large-scale studies are revealing dozens of novel markers, such as for example an expression profiling of 1,059 miRNAs in B lymphocytes that revealed 49 miRNAs down-regulated in untreated MS patients compared with healthy controls (Sievers et al. 2012). A recent integration of miRNAs databases revealed that the miRNAs associated with MS according to different studies are able or predicted to target about 1,500 different genes many of which play a role in T cell activation and signaling, or have transcription factor activity (Angerstein et al. 2012).

Among environmental factors considerable evidence has linked past Epstein–Barr virus (EBV) infection to an increased risk of MS, and, since a complete silencing of the EBV genome in memory B cells is under epigenetic control via DNA methylation and histone tail modifications, some authors have suggested that an epigenetic dysregulation of the EBV latency might contribute to the development of MS and other autoimmune diseases (Niller et al. 2011).

6.4.4 Epigenetics of Systemic Sclerosis

SSc is a systemic autoimmune disease characterized by deposition of collagen in the skin and, less commonly, in other tissues with progressive vasculopathy. Early studies on DNA methylation (Table 6.6) revealed association between enhanced type I collagen expression and epigenetic repression (hypermethylation) of the *FLI1* gene in scleroderma fibroblasts (Wang et al. 2006). Subsequent studies revealed that CD4+ T cell DNA from patients with SSc was significantly hypomethylated relative to controls, and DNMT1, MBD3, and MBD4 mRNAs were significantly decreased in the SSc group (Lei et al. 2009). Demethylation of *TNFSF7(CD70)* was observed to contribute to CD70 overexpression in CD4+ T cells from patients with SSc (Jiang et al. 2012). Moreover, SSc occurs more frequently in females than males, suggesting that epigenetic modifications of genes on the X chromosome might be involved. Particularly, demethylation of *CD40LG* regulatory elements on the inactive X chromosome contributed to CD40L overexpression in CD4+ T cells from female patients with SSc, but no significant difference was observed in the expression of CD40L between male patients with SSc and male control subjects (Lian et al. 2012). A recent methylation profile of all X chromosome genes in peripheral blood mononuclear cells from monozygotic twins discordant for SSc revealed sites with an elevated probability to be consistently hypermethylated ($n=18$) or hypomethylated ($n=25$) in affected twins. Identified genes include transcription factors and surface antigens, and pathway analysis suggests their involvement in cell proliferation, apoptosis, inflammation, and oxidative stress (Selmi et al. 2012).

Increasing evidence suggests the involvement of histone tail modifications in fibrosis (Table 6.6), the hallmark of SSc, characterized by a persistent fibroblast activation triggered by transforming growth factor-β (TGF-β). Indeed, it was observed that the expression of the HAT p300 is markedly elevated in SSc skin biopsies and is induced by TGF-β in explanted normal skin fibroblasts. Moreover, TGF-β enhanced both p300 recruitment and histone H4 acetylation at the *COL1A2* (collagen, type I, α2) locus, suggesting that p300-mediated histone acetylation could represent a fundamental epigenetic mechanism in fibrogenesis (Ghosh et al. 2013). Similarly, inhibition of trimethylation of histone H3 on lysine 27 (H3K27me3), induced by treatment with 3-deazaneplanocin A, stimulated the release of collagen in cultured fibroblasts in a time and dose-dependent manner and was sufficient to induce fibrosis, suggesting that trimethylation of histone H3 on lysine 27 acts as a negative regulator of fibroblast activation (Krämer et al. 2012).

An increasing number of miRNAs was found to be deregulated in SSc samples (Table 6.6). For example, a miRNA array analysis in skin tissues from SSc patients and healthy controls revealed 24 miRNAs that were differentially expressed in patients with SSc and six miRNAs that may be correlated with the pathogenesis of SSc. Particularly, miR-23b and let-7 were up-regulated, while miR-125b, miR-133a, miR-206, and miR-140-5p were down-regulated (Li et al. 2012). Others observed that in comparison with the normal skin tissues, miRNAs were aberrantly expressed in limited cutaneous scleroderma and diffuse cutaneous scleroderma skin tissues, and identified six miRNAs whose expressions were correlated with SSc fibrosis: miR-21, miR-31, miR-146, miR-503, miR-145, and miR-29b. Particularly, the study confirmed that miR-21 was increased whereas miR-145 and miR-29b were decreased both in the skin tissues and in fibroblasts. As predicted target genes, *SMAD7*, *SAMD3*, and *COL1A1* were regulated by these tree miRNAs (Zhu et al. 2012). Previous results had shown that miR-29a was strongly down-regulated in SSc fibroblasts and skin sections as compared with the healthy controls, and that this miRNA acts as a key regulator of collagen expression in SSc (Maurer et al. 2010). Overall, miRNA-29 is a recently discovered class of miRNAs which is related to fibrotic disease and a potential therapeutic target for systemic sclerosis (Peng et al. 2012). More recently, it was found that miR-150 downregulation contributes to the constitutive type I collagen overexpression in SSc dermal fibroblasts via the induction of integrin β3 (Honda et al. 2013).

6.4.5 Epigenetics of AITDs

AITDs comprise Graves' disease and Hashimoto's thyroiditis, both organ-specific autoimmune diseases characterized by female preponderance, and in which the autoimmune attack of the thyroid takes place by infiltration of lymphocytes of the glandule. A possible role of skewed X chromosome inactivation, mediated by epigenetic mechanisms, has been suggested in the etiology of AITD to partially explain the female preponderance (Brix et al. 2005; Chabchoub et al. 2009).

A few studies have been performed to clarify the association between factors regulating DNA methylation and the prognosis of AITDs (Table 6.7). Particularly, those studies focused on polymorphisms in genes encoding DNMTs, methylenetetrahydrofolate reductase (MTHFR), and methionine synthase reductase (MTRR), which are all enzymes essential for DNA methylation reactions. The *MTHFR* C677T polymorphism was associated with reduced GD risk in women (Mao et al. 2010), while the *DNMT1* 32204GG genotype was correlated with DNA hypomethylation and with the intractability of GD, and the *MTRR* 66AA genotype with the severity of HD (Arakawa et al. 2012). Albeit in their infancy, those studies suggest that those genes might account for AITD susceptibility, severity, and response to treatment, partially mediated by changes in DNA methylation (Mao et al. 2010; Arakawa et al. 2012).

Also the few available studies on miRNA profiling in AITD tissues (Table 6.7) suggest deregulated networks in those disorders. Liu and coworkers showed that the expression of miR-154*, miR-376b, and miR-431* was suppressed in peripheral blood mononuclear cells from initial GD patients, and that their expression levels were recovered in GD patients in remission (Liu et al. 2012). Another group showed that miR-146a1 was significantly decreased in the thyroid tissue of GD patients, in comparison with the control group (Bernecker et al. 2012). Similarly, miR-155_2 was significantly decreased and miR-200a1 was significantly increased in the thyroid of HT patients, with respect to the control tissues (Bernecker et al. 2012). Albeit preliminary, those studies suggest a potential role of miRNA deregulations in AITDs that warrants further research.

6.5 Concluding Remarks

In the present chapter we described some examples of epigenetic deregulation in human autoimmune diseases. This field of research has gained tremendous attention in the last 2–3 years and it is now emerging that epigenetic modifications play a role, or are supposed to do it, in several autoimmune disorders, including but not limited to those detailed in this chapter. Indeed, evidence of an epigenetic contribution is increasing also in inflammatory bowel diseases (Jenke and Zilbauer 2012), type 1 diabetes (Dang et al. 2013), immune thrombocytopenic purpura (Khorshied and El-Ghamrawy 2012), and many other inflammatory and/or autoimmune diseases. Despite this, only few environmental factors have been suggested to epigenetically contribute to those disorders, some examples are drugs, air pollutants, ultraviolet light, cigarette smoke, and microbial infections, but for most of them the epigenetic link is still only speculative (De Santis and Selmi 2012). Several authors have suggested that epigenetic deregulations of genes on the X chromosome might account for gender differences, i.e., female predominance, in the incidence of many autoimmune diseases (Lian et al. 2012; Liao et al. 2012), and age-related epigenetic changes might also be of interest (De Santis and Selmi 2012).

Early epigenetic studies in autoimmune diseases, based on the candidate gene approach, have been paralleled and/or replaced in recent years by whole-genome approaches, that are revealing dozens, or even hundreds of genes or miRNAs that are deregulated in the affected tissues as well as in peripheral tissues of the patients (Tables 6.1–6.7). This is leading to a better understanding of the networks involved in disease pathogenesis, thereby opening the way for potential diagnostic and prognostic tools, as well as for epigenetic interventions based on miRNA silencing or chromatin remodeling agents, such as HDAi (Garchow et al. 2011; Reilly et al. 2011).

References

Alevizos I, Alexander S, Turner RJ, Illei GG (2011) MicroRNA expression profiles as biomarkers of minor salivary gland inflammation and dysfunction in Sjögren's syndrome. Arthritis Rheum 63:535–544

Amarilyo G, La Cava A (2012) miRNA in systemic lupus erythematosus. Clin Immunol 144:26–31

Ammari M, Jorgensen C, Apparailly F (2013) Impact of microRNAs on the understanding and treatment of rheumatoid arthritis. Curr Opin Rheumatol 25:225–233

Angerstein C, Hecker M, Paap BK, Koczan D, Thamilarasan M, Thiesen HJ, Zettl UK (2012) Integration of microRNA databases to study microRNAs associated with multiple sclerosis. Mol Neurobiol 45:520–535

Arakawa Y, Watanabe M, Inoue N, Sarumaru M, Hidaka Y, Iwatani Y (2012) Association of polymorphisms in DNMT1, DNMT3A, DNMT3B, MTHFR and MTRR genes with global DNA methylation levels and prognosis of autoimmune thyroid disease. Clin Exp Immunol 170:194–201

Balada E, Ordi-Ros J, Vilardell-Tarrés M (2007a) DNA methylation and systemic lupus erythematosus. Ann N Y Acad Sci 1108:127–136

Balada E, Ordi-Ros J, Serrano-Acedo S, Martinez-Lostao L, Vilardell-Tarrés M (2007b) Transcript overexpression of the MBD2 and MBD4 genes in CD4+ T cells from systemic lupus erythematosus patients. J Leukoc Biol 81:1609–1616

Balada E, Ordi-Ros J, Serrano-Acedo S, Martinez-Lostao L, Rosa-Leyva M, Vilardell-Tarrés M (2008) Transcript levels of DNA methyltransferases DNMT1, DNMT3A and DNMT3B in CD4+ T cells from patients with systemic lupus erythematosus. Immunology 124:339–347

Balada E, Castro-Marrero J, Felip L, Ordi-Ros J, Vilardell-Tarrés M (2012) Associations between the expression of epigenetically regulated genes and the expression of DNMTs and MBDs in systemic lupus erythematosus. PLoS One 7:e45897

Baranzini SE, Mudge J, Van Velkinburgh JC, Khankhanian P, Khrebtukova I, Miller NA, Zhang L, Farmer AD, Bell CJ, Kim RW, May GD, Woodward JE, Caillier SJ, Mcelroy JP, Gomez R, Pando MJ, Clendenen LE, Ganusova EE, Schilkey FD, Ramaraj T, Khan OA, Huntley JJ, Luo S, Kwok PY, Wu TD, Schroth GP, Oksenberg JR, Hauser SL, Kingsmore SF (2010) Genome, epigenome and RNA sequences of monozygotic twins discordant for multiple sclerosis. Nature 464:1351–1356

Berger SL (2007) The complex language of chromatin regulation during transcription. Nature 447:407–412

Bernecker C, Lenz L, Ostapczuk MS, Schinner S, Willenberg H, Ehlers M, Vordenbäumen S, Feldkamp J, Schott M (2012) MicroRNAs miR-146a1, miR-155_2, and miR-200a1 are regulated in autoimmune thyroid diseases. Thyroid 22:1294–1295

Brix TH, Knudsen GP, Kristiansen M, Kyvik KO, Orstavik KH, Hegedüs L (2005) High frequency of skewed X-chromosome inactivation in females with autoimmune thyroid disease: a possible explanation for the female predisposition to thyroid autoimmunity. J Clin Endocrinol Metab 90:5949–5953

Calabrese R, Zampieri M, Mechelli R, Annibali V, Guastafierro T, Ciccarone F, Coarelli G, Umeton R, Salvetti M, Caiafa P (2012) Methylation-dependent PAD2 upregulation in multiple sclerosis peripheral blood. Mult Scler 18:299–304

Cantley MD, Bartold PM, Fairlie DP, Rainsford KD, Haynes DR (2012) Histone deacetylase inhibitors as suppressors of bone destruction in inflammatory diseases. J Pharm Pharmacol 64:763–774

Chabchoub G, Uz E, Maalej A, Mustafa CA, Rebai A, Mnif M, Bahloul Z, Farid NR, Ozcelik T, Ayadi H (2009) Analysis of skewed X-chromosome inactivation in females with rheumatoid arthritis and autoimmune thyroid diseases. Arthritis Res Ther 11:R106

Chan EK, Ceribelli A, Satoh M (2012) MicroRNA-146a in autoimmunity and innate immune responses. Ann Rheum Dis 72(Suppl 2):S90–S95

Cornacchia E, Golbus J, Maybaum J, Strahler J, Hanash S, Richardson B (1988) Hydralazine and procainamide inhibit T cell DNA methylation and induce autoreactivity. J Immunol 140:2197–2200

Corvetta A, Della Bitta R, Luchetti MM, Pomponio G (1991) 5-Methylcytosine content of DNA in blood, synovial mononuclear cells and synovial tissue from patients affected by autoimmune rheumatic diseases. J Chromatogr 566:481–491

Costenbader KH, Gay S, Alarcón-Riquelme ME, Iaccarino L, Doria A (2012) Genes, epigenetic regulation and environmental factors: which is the most relevant in developing autoimmune diseases? Autoimmun Rev 11:604–609

Cunninghame Graham DS (2009) Genome-wide association studies in systemic lupus erythematosus: a perspective. Arthritis Res Ther 11:119

Dai Y, Huang YS, Tang M, Lv TY, Hu CX, Tan YH, Xu ZM, Yin YB (2007) Microarray analysis of microRNA expression in peripheral blood cells of systemic lupus erythematosus patients. Lupus 16:939–946

Dang MN, Buzzetti R, Pozzilli P (2013) Epigenetics in autoimmune diseases with focus on type 1 diabetes. Diabetes Metab Res Rev 29:8–18

De La Rica L, Urquiza JM, Gómez-Cabrero D, Islam AB, López-Bigas N, Tegnér J, Toes RE, Ballestar E (2013) Identification of novel markers in rheumatoid arthritis through integrated analysis of DNA methylation and microRNA expression. J Autoimmun 41:6–16

De Santis M, Selmi C (2012) The therapeutic potential of epigenetics in autoimmune diseases. Clin Rev Allergy Immunol 42:92–101

Deapen D, Escalante A, Weinrib L, Horwitz D, Bachman B, Roy-Burman P, Walker A, Mack TM (1992) A revised estimate of twin concordance in systemic lupus erythematosus. Arthritis Rheum 35:311–318

Ding S, Liang Y, Zhao M, Liang G, Long H, Zhao S, Wang Y, Yin H, Zhang P, Zhang Q, Lu Q (2012) Decreased microRNA-142-3p/5p expression causes CD4+ T cell activation and B cell hyperstimulation in systemic lupus erythematosus. Arthritis Rheum 64:2953–2963

Esteller M (2011) Non-coding RNAs in human disease. Nat Rev Genet 12:861–874

Fenoglio C, Ridolfi E, Galimberti D, Scarpini E (2012) MicroRNAs as active players in the pathogenesis of multiple sclerosis. Int J Mol Sci 13:13227–13239

Fournier A, Sasai N, Nakao M, Defossez PA (2012) Role of methyl-binding proteins in chromatin organization and epigenome maintenance. Brief Funct Genomics 11:251–264

Fu LH, Ma CL, Cong B, Li SJ, Chen HY, Zhang JG (2011) Hypomethylation of proximal CpG motif of interleukin-10 promoter regulates its expression in human rheumatoid arthritis. Acta Pharmacol Sin 32:1373–1380

Garchow BG, Bartulos Encinas O, Leung YT, Tsao PY, Eisenberg RA, Caricchio R, Obad S, Petri A, Kauppinen S, Kiriakidou M (2011) Silencing of microRNA-21 in vivo ameliorates autoimmune splenomegaly in lupus mice. EMBO Mol Med 3:605–615

Garcia BA, Busby SA, Shabanowitz J, Hunt DF, Mishra N (2005) Resetting the epigenetic histone code in the MRL-lpr/lpr mouse model of lupus by histone deacetylase inhibition. J Proteome Res 6:2032–2042

Gervin K, Vigeland MD, Mattingsdal M, Hammerø M, Nygård H, Olsen AO, Brandt I, Harris JR, Undlien DE, Lyle R (2012) DNA methylation and gene expression changes in monozygotic twins discordant for psoriasis: identification of epigenetically dysregulated genes. PLoS Genet 8:e1002454

Gestermann N, Koutero M, Belkhir R, Tost J, Mariette X, Miceli-Richard C (2012) Methylation profile of the promoter region of IRF5 in primary Sjögren's syndrome. Eur Cytokine Netw 23:166–172

Ghosh AK, Bhattacharyya S, Lafyatis R, Farina G, Yu J, Thimmapaya B, Wei J, Varga J (2013) p300 is elevated in systemic sclerosis and its expression is positively regulated by TGF-β: epigenetic feed-forward amplification of fibrosis. J Invest Dermatol 133:1302–1310

Gillespie J, Savic S, Wong C, Hempshall A, Inman M, Emery P, Grigg R, McDermott MF (2012) Histone deacetylases are dysregulated in rheumatoid arthritis and a novel histone deacetylase 3-selective inhibitor reduces interleukin-6 production by peripheral blood mononuclear cells from rheumatoid arthritis patients. Arthritis Rheum 64:418–422

González S, Aguilera S, Alliende C, Urzúa U, Quest AF, Herrera L, Molina C, Hermoso M, Ewert P, Brito M, Romo R, Leyton C, Pérez P, González MJ (2011) Alterations in type I hemidesmosome components suggestive of epigenetic control in the salivary glands of patients with Sjögren's syndrome. Arthritis Rheum 63:1106–1115

Grabiec AM, Korchynskyi O, Tak PP, Reedquist KA (2012) Histone deacetylase inhibitors suppress rheumatoid arthritis fibroblast-like synoviocyte and macrophage IL-6 production by accelerating mRNA decay. Ann Rheum Dis 71:424–431

Han J, Park SG, Bae JB, Choi J, Lyu JM, Park SH, Kim HS, Kim YJ, Kim S, Kim TY (2012) The characteristics of genome-wide DNA methylation in naïve CD4+ T cells of patients with psoriasisor atopic dermatitis. Biochem Biophys Res Commun 422:157–163

Honda N, Jinnin M, Kira-Etoh T, Makino K, Kajihara I, Makino T, Fukushima S, Inoue Y, Okamoto Y, Hasegawa M, Fujimoto M, Ihn H (2013) miR-150 down-regulation contributes

to the constitutive type I collagen overexpression in scleroderma dermal fibroblasts via the induction of integrin β3. Am J Pathol 182:206–216
Horiuchi M, Morinobu A, Chin T, Sakai Y, Kurosaka M, Kumagai S (2009) Expression and function of histone deacetylases in rheumatoid arthritis synovial fibroblasts. J Rheumatol 36:1580–1589
Hu N, Qiu X, Luo Y, Yuan J, Li Y, Lei W, Zhang G, Zhou Y, Su Y, Lu Q (2008) Abnormal histone modification patterns in lupus CD4+ T cells. J Rheumatol 35:804–810
Hu N, Long H, Zhao M, Yin H, Lu Q (2009) Aberrant expression pattern of histone acetylation modifiers and mitigation of lupus by SIRT1-siRNA in MRL/lpr mice. Scand J Rheumatol 38:464–471
Hughes T, Sawalha AH (2011) The role of epigenetic variation in the pathogenesis of systemic lupus erythematosus. Arthritis Res Ther 13:245
Javierre BM, Fernandez AF, Richter J, Al-Shahrour F, Martin-Subero JI, Rodriguez-Ubreva J, Berdasco M, Fraga MF, O'hanlon TP, Rider LG, Jacinto FV, Lopez-Longo FJ, Dopazo J, Forn M, Peinado MA, Carreño L, Sawalha AH, Harley JB, Siebert R, Esteller M, Miller FW, Ballestar E (2010) Changes in the pattern of DNA methylation associate with twin discordance in systemic lupus erythematosus. Genome Res 20:170–179
Jeffries MA, Dozmorov M, Tang Y, Merrill JT, Wren JD, Sawalha AH (2011) Genome-wide DNA methylation patterns in CD4+ T cells from patients with systemic lupus erythematosus. Epigenetics 6:593–601
Jenke AC, Zilbauer M (2012) Epigenetics in inflammatory bowel disease. Curr Opin Gastroenterol 28:577–584
Jiang H, Xiao R, Lian X, Kanekura T, Luo Y, Yin Y, Zhang G, Yang Y, Wang Y, Zhao M, Lu Q (2012) Demethylation of TNFSF7 contributes to CD70 overexpression in CD4+ T cells from patients with systemic sclerosis. Clin Immunol 143:39–44
Jones PA, Liang G (2009) Rethinking how DNA methylation patterns are maintained. Nat Rev Genet 10:805–811
Jones PA (2012) Functions of DNA methylation: islands, start sites, gene bodies and beyond. Nat Rev Genet 13:484–492
Joyce CE, Zhou X, Xia J, Ryan C, Thrash B, Menter A, Zhang W, Bowcock AM (2011) Deep sequencing of small RNAs from human skin reveals major alterations in the psoriasis miRNAome. Hum Mol Genet 20:4025–4040
Karlson EW, Deane K (2012) Environmental and gene-environment interactions and risk of rheumatoid arthritis. Rheum Dis Clin North Am 38:405–426
Karouzakis E, Gay RE, Michel BA, Gay S, Neidhart M (2009) DNA hypomethylation in rheumatoid arthritis synovial fibroblasts. Arthritis Rheum 60:3613–3622
Karouzakis E, Rengel Y, Jüngel A, Kolling C, Gay RE, Michel BA, Tak PP, Gay S, Neidhart M, Ospelt C (2011) DNA methylation regulates the expression of CXCL12 in rheumatoid arthritis synovial fibroblasts. Genes Immun 12:643–652
Khorshied MM, El-Ghamrawy MK (2012) DNA methyltransferase 3B (DNMT3B–579G>T) promotor polymorphism and the susceptibility to pediatric immune thrombocytopenic purpura in Egypt. Gene 511:34–37
Klein K, Ospelt C, Gay S (2012) Epigenetic contributions in the development of rheumatoid arthritis. Arthritis Res Ther 14:227
Krämer M, Dees C, Huang J, Schlottmann I, Palumbo-Zerr K, Zerr P, Gelse K, Beyer C, Distler A, Marquez VE, Distler O, Schett G, Distler JH (2012) Inhibition of H3K27 histone trimethylation activates fibroblasts and induces fibrosis. Ann Rheum Dis 72:614–620
Kuchen S, Seemayer CA, Rethage J, Von Knoch R, Kuenzler P, Beat AM, Gay RE, Gay S, Neidhart M (2004) The L1 retroelement-related p40 protein induces p38delta MAP kinase. Autoimmunity 37:57–65
Kurowska-Stolarska M, Alivernini S, Ballantine LE, Asquith DL, Millar NL, Gilchrist DS, Reilly J, Ierna M, Fraser AR, Stolarski B, McSharry C, Hueber AJ, Baxter D, Hunter J, Gay S, Liew FY, Mcinnes IB (2011) MicroRNA-155 as a proinflammatory regulator in clinical and experimental arthritis. Proc Natl Acad Sci U S A 108:11193–11198

Lei W, Luo Y, Lei W, Luo Y, Yan K, Zhao S, Li Y, Qiu X, Zhou Y, Long H, Zhao M, Liang Y, Su Y, Lu Q (2009) Abnormal DNA methylation in CD4+ T cells from patients with systemic lupus erythematosus, systemic sclerosis, and dermatomyositis. Scand J Rheumatol 38:369–374

Leng RX, Pan HF, Qin WZ, Chen GM, Ye DQ (2011) Role of microRNA-155 in autoimmunity. Cytokine Growth Factor Rev 22:141–147

Li H, Yang R, Fan X, Gu T, Zhao Z, Chang D, Wang W (2012) MicroRNA array analysis of microRNAs related to systemic scleroderma. Rheumatol Int 32:307–313

Li Y, Zhao M, Yin H, Gao F, Wu X, Luo Y, Zhao S, Zhang X, Su Y, Hu N, Long H, Richardson B, Lu Q (2010) Overexpression of the growth arrest and DNA damage-induced 45alpha gene contributes to autoimmunity by promoting DNA demethylation in lupus T cells. Arthritis Rheum 62:1438–1447

Lian X, Xiao R, Hu X, Kanekura T, Jiang H, Li Y, Wang Y, Yang Y, Zhao M, Lu Q (2012) DNA demethylation of CD40l in CD4+ T cells from women with systemic sclerosis: a possible explanation for female susceptibility. Arthritis Rheum 64:2338–2345

Liang C, Xiong K, Szulwach KE, Zhang Y, Wang Z, Peng J, Fu M, Jin P, Suzuki HI, Liu Q (2013) Sjogren syndrome antigen B (SSB)/La promotes global microRNA expression by binding microRNA precursors through stem-loop recognition. J Biol Chem 288:723–736

Liao J, Liang G, Xie S, Zhao H, Zuo X, Li F, Chen J, Zhao M, Chan TM, Lu Q (2012) CD40L demethylation in CD4(+) T cells from women with rheumatoid arthritis. Clin Immunol 145:13–18

Lin SY, Hsieh SC, Lin YC, Lee CN, Tsai MH, Lai LC, Chuang EY, Chen PC, Hung CC, Chen LY, Hsieh WS, Niu DM, Su YN, Ho HN (2012) A whole genome methylation analysis of systemic lupus erythematosus: hypomethylation of the IL10 and IL1R2 promoters is associated with disease activity. Genes Immun 13:214–220

Liu CC, Ou TT, Wu CC, Li RN, Lin YC, Lin CH, Tsai WC, Liu HW, Yen JH (2011) Global DNA methylation, DNMT1, and MBD2 in patients with systemic lupus erythematosus. Lupus 20:131–136

Liu R, Ma X, Xu L, Wang D, Jiang X, Zhu W, Cui B, Ning G, Lin D, Wang S (2012) Differential microRNA expression in peripheral blood mononuclear cells from Graves' disease patients. J Clin Endocrinol Metab 97:E968–E972

Lu MC, Lai NS, Chen HC, Yu HC, Huang KY, Tung CH, Huang HB, Yu CL (2013) Decreased microRNA(miR)-145 and increased miR-224 expression in T cells from patients with systemic lupuserythematosus involved in lupusimmunopathogenesis. Clin Exp Immunol 171:91–99

Mao R, Fan Y, Zuo L, Geng D, Meng F, Zhu J, Li Q, Qiao H, Jin Y, Bai J, Fu S (2010) Association study between methylenetetrahydrofolate reductase gene polymorphisms and Graves' disease. Cell Biochem Funct 28:585–590

Martin C, Zhang Y (2005) The diverse functions of histone lysine methylation. Nat Rev Mol Cell Biol 6:838–849

Martín-Subero JI (2011) How epigenomics brings phenotype into being. Pediatr Endocrinol Rev 9:506–510

Mastronardi FG, Noor A, Wood DD, Paton T, Moscarello MA (2007) Peptidyl argininedeiminase 2 CpG island in multiple sclerosis white matter is hypomethylated. J Neurosci Res 85:2006–2016

Maurer B, Stanczyk J, Jüngel A, Akhmetshina A, Trenkmann M, Brock M, Kowal-Bielecka O, Gay RE, Michel BA, Distler JH, Gay S, Distler O (2010) MicroRNA-29, a key regulator of collagen expression in systemic sclerosis. Arthritis Rheum 62:1733–1743

Meda F, Folci M, Baccarelli A, Selmi C (2011) The epigenetics of autoimmunity. Cell Mol Immunol 8:226–236

Meisgen F, Xu N, Wei T, Janson PC, Obad S, Broom O, Nagy N, Kauppinen S, Keméy L, Ståhle M, Pivarcsi A, Sonkoly E (2012) MiR-21 is up-regulated in psoriasis and suppresses T cell apoptosis. Exp Dermatol 21:312–314

Miao CG, Yang YY, He X, Li J (2013) New advances of DNA methylation and histone modifications in rheumatoid arthritis, with special emphasis on MeCP2. Cell Signal 25:1118–1125

Nakano K, Boyle DL, Firestein GS (2013) Regulation of DNA methylation in rheumatoid arthritis synoviocytes. J Immunol 190:1297–1303

Nakkuntod J, Avihingsanon Y, Mutirangura A, Hirankarn N (2011) Hypomethylation of LINE-1 but not Alu in lymphocyte subsets of systemic lupus erythematosus patients. Clin Chim Acta 412:1457–1461

Neidhart M, Rethage J, Kuchen S, Künzler P, Crowl RM, Billingham ME, Gay RE, Gay S (2000) Retrotransposable L1 elements expressed in rheumatoid arthritis synovial tissue: association with genomic DNA hypomethylation and influence on gene expression. Arthritis Rheum 43:2634–2647

Nile CJ, Read RC, Akil M, Duff GW, Wilson AG (2008) Methylationstatus of a single CpG site in the IL6 promoter is related to IL6 messenger RNA levels and rheumatoid arthritis. Arthritis Rheum 58:2686–2693

Niller HH, Wolf H, Ay E, Minarovits J (2011) Epigenetic dysregulation of epstein-barr virus latency and development of autoimmune disease. Adv Exp Med Biol 711:82–102

Pauley KM, Stewart CM, Gauna AE, Dupre LC, Kuklani R, Chan AL, Pauley BA, Reeves WH, Chan EK, Cha S (2011) Altered miR-146a expression in Sjögren's syndrome and its functional role in innate immunity. Eur J Immunol 41:2029–2039

Pedre X, Mastronardi F, Bruck W, López-Rodas G, Kuhlmann T, Casaccia P (2011) Changed histone acetylation patterns in normal-appearing white matter and early multiple sclerosis lesions. J Neurosci 31:3435–3445

Peng WJ, Tao JH, Mei B, Chen B, Li BZ, Yang GJ, Zhang Q, Yao H, Wang BX, He Q, Wang J (2012) MicroRNA-29: a potential therapeutic target for systemic sclerosis. Expert Opin Ther Targets 16:875–879

Qin HH, Zhu XH, Liang J, Yang YS, Wang SS, Shi WM, Xu JH (2013) Associations between aberrant DNA methylation and transcript levels of DNMT1 and MBD2 in CD4+ T cells from patients with systemic lupus erythematosus. Australas J Dermatol 54:90–95

Quddus J, Johnson KJ, Gavalchin J, Amento EP, Chrisp CR, Young RL, Richardson BC (1993) Treating activated CD4+ T cells with either of two distinct DNA methyltransferase inhibitors, 5-azacytidine or procainamide, is sufficient to cause a lupus-like disease in syngeneic mice. J Clin Invest 92:38–53

Quintero-Ronderos P, Montoya-Ortiz G (2012) Epigenetics and autoimmune diseases. Autoimmune Dis 2012:593720

Reilly CM, Regna N, Mishra N (2011) HDAC inhibition in lupus models. Mol Med 17:417–425

Richardson B (1986) Effect of an inhibitor of DNA methylation on T cells. II. 5-azacytidine induces selfreactivity in antigen-specific T4+ cells. Hum Immunol 17:456–470

Roberson ED, Liu Y, Ryan C, Joyce CE, Duan S, Cao L, Martin A, Liao W, Menter A, Bowcock AM (2012) A subset of methylated CpG sites differentiate psoriatic from normal skin. J Invest Dermatol 132:583–592

Ruchusatsawat K, Wongpiyabovorn J, Shuangshoti S, Hirankarn N, Mutirangura A (2006) SHP-1 promoter 2 methylation in normal epithelial tissues and demethylation in psoriasis. J Mol Med 84:175–182

Sato F, Tsuchiya S, Meltzer SJ, Shimizu K (2011) MicroRNAs and epigenetics. FEBS J 278: 1598–1609

Selmi C (2012) Autoimmunity in 2011. Clin Rev Allergy Immunol 43:194–206

Selmi C, Feghali-Bostwick CA, Lleo A, Lombardi SA, De Santis M, Cavaciocchi F, Zammataro L, Mitchell MM, Lasalle JM, Medsger TJR, Gershwin ME (2012) X chromosome gene methylation in peripheral lymphocytes from monozygotic twins discordant for scleroderma. Clin Exp Immunol 169:253–262

Shen N, Liang D, Tang Y, De Vries N, Tak PP (2012) MicroRNAs-novel regulators of systemic lupus erythematosus pathogenesis. Nat Rev Rheumatol 8:701–709

Shibuya H, Nakasa T, Adachi N, Nagata Y, Ishikawa M, Deie M, Suzuki O, Ochi M (2013) Overexpression of microRNA-223 in rheumatoid arthritis synovium controls osteoclast differentiation. Mod Rheumatol 23:674–685

Shizu M, Itoh Y, Sunahara R, Chujo S, Hayashi H, Ide Y, Takii T, Koshiko M, Chung SW, Hayakawa K, Miyazawa K, Hirose K, Onozaki K (2008) Cigarette smoke condensate upregulates the gene and protein expression of proinflammatory cytokines in human fibroblast-like synoviocyte line. J Interferon Cytokine Res 28:509–521

Sievers C, Meira M, Hoffmann F, Fontoura P, Kappos L, Lindberg RL (2012) Altered microRNA expression in B lymphocytes in multiple sclerosis: towards a better understanding of treatment effects. Clin Immunol 144:70–79

Takami N, Osawa K, Miura Y, Komai K, Taniguchi M, Shiraishi M, Sato K, Iguchi T, Shiozawa K, Hashiramoto A, Shiozawa S (2006) Hypermethylated promoter region of DR3, the death receptor 3 gene, in rheumatoid arthritis synovial cells. Arthritis Rheum 54:779–787

Thabet Y, Cañas F, Ghedira I, Youinou P, Mageed RA, Renaudineau Y (2012) Altered patterns of epigenetic changes in systemic lupus erythematosus and auto-antibody production: is there a link? J Autoimmun 39:154–160

Tovar-Castillo LE, Cancino-Díaz JC, García-Vázquez F, Cancino-Gómez FG, León-Dorantes G, Blancas-González F, Jiménez-Zamudio L, García-Latorre E, Cancino-Díaz ME (2007) Under-expression of VHL and over-expression of HDAC-1, HIF-1alpha, LL-37, and IAP-2 in affected skin biopsies of patients with psoriasis. Int J Dermatol 46:239–246

Wang Y, Fan PS, Kahaleh B (2006) Association between enhanced type I collagen expression and epigenetic repression of the FLI1 gene in scleroderma fibroblasts. Arthritis Rheum 54:2271–2279

Xia J, Joyce CE, Bowcock AM, Zhang W (2013) Noncanonical microRNAs and endogenous siRNAs in normal and psoriatic human skin. Hum Mol Genet 22:737–748

Xu N, Meisgen F, Butler LM, Han G, Wang XJ, Söderberg-Nauclér C, Ståhle M, Pivarcsi A, Sonkoly E (2013) MicroRNA-31 is overexpressed in psoriasis and modulates inflammatory cytokine and chemokine production in keratinocytes via targeting serine/threonine kinase 40. J Immunol 190:678–688

Xu WD, Lu MM, Pan HF, Ye DQ (2012) Association of microRNA-146a with autoimmune diseases. Inflammation 35:1525–1529

Yamamura Y, Motegi K, Kani K, Takano H, Momota Y, Aota K, Yamanoi T, Azuma M (2012) TNF-α inhibits aquaporin 5 expression in human salivary gland acinar cells via suppression of histone H4 acetylation. J Cell Mol Med 16:1766–1775

Yin H, Zhao M, Wu X, Gao F, Luo Y, Ma L, Liu S, Zhang G, Chen J, Li F, Zuo X, Lu Q (2010) Hypomethylation and overexpression of CD70 (TNFSF7) in CD4+ T cells of patients with primary Sjögren's syndrome. J Dermatol Sci 59:198–203

Zhang K, Zhang R, Li X, Yin G, Niu X, Hou R (2007) The mRNA expression and promoter methylation status of the p16 gene in colony-forming cells with high proliferative potential in patients with psoriasis. Clin Exp Dermatol 32:702–708

Zhang K, Zhang R, Li X, Yin G, Niu X (2009) Promoter methylationstatus of p15 and p21 genes in HPP-CFCs of bone marrow of patients with psoriasis. Eur J Dermatol 19:141–146

Zhang P, Su Y, Zhao M, Huang W, Lu Q (2011) Abnormal histonemodifications in PBMCs from patients with psoriasis vulgaris. Eur J Dermatol 21:527–552

Zhang P, Su Y, Lu Q (2012) Epigenetics and psoriasis. J Eur Acad Dermatol Venereol 26:399–403

Zhou Y, Qiu X, Luo Y, Yuan J, Li Y, Zhong Q, Zhao M, Lu Q (2011) Histone modifications and methyl-CpG-binding domain protein levels at the TNFSF7 (CD70) promoter in SLE CD4+ T cells. Lupus 20:1365–1371

Zhu H, Li Y, Qu S, Luo H, Zhou Y, Wang Y, Zhao H, You Y, Xiao X, Zuo X (2012) MicroRNA expression abnormalities in limited cutaneous scleroderma and diffuse cutaneous scleroderma. J Clin Immunol 32:514–522

Zhu X, Liang J, Li F, Yang Y, Xiang L, Xu J (2011) Analysis of associations between the patterns of global DNA hypomethylation and expression of DNA methyltransferase in patients with systemic lupus erythematosus. Int J Dermatol 50:697–704

Zilahi E, Tarr T, Papp G, Griger Z, Sipka S, Zeher M (2012) Increased microRNA-146a/b, TRAF6 gene and decreased IRAK1 gene expressions in the peripheral mononuclear cells of patients with Sjögren's syndrome. Immunol Lett 141:165–168

Chapter 7
The Effect of Nutrition and Exercise on Epigenetics and the Development of Cardiovascular Disease

Thomas F. Whayne Jr.

Abstract Epigenetics is defined as the study of heritable alterations in gene expression or cellular phenotype. The term defines the difference from just a genetic approach. A more precise definition is that epigenetics is all the meiotically and mitotically inherited changes in gene expression that are not encoded in the deoxyribonucleic acid (DNA) sequence itself. Major epigenetic mechanisms are modifications of histone proteins in chromatin and DNA methylation (which does not alter the DNA sequence). There is increasing evidence for the involvement of epigenetics in human disease such as inflammatory disease and cancer. Other chronic diseases are also susceptible to epigenetic modification such as metabolic diseases including obesity, metabolic syndrome, and diabetes mellitus. There is much evidence for the modification of epigenetics by nutrition and exercise. Through these modifications, there is infinite potential for benefit for the fetus, the newborn, and the individual as well as population effects. Association with cardiovascular (CV) disease including coronary heart disease (CHD) and peripheral arterial disease is evident through epigenetic relationships and modification by major CV risk factors such as tobacco abuse. Aging itself may be altered by epigenetic modification. Knowledge of the subject and its relevance is in a very preliminary stage.

Keywords Epigenetics • Epigenome • Chromatin • Flavonoids • Histone • Methylation

Abbreviations

ASC	Apoptosis-associated speck-like protein containing a CARD
CARD	Caspase recruitment domain

T.F. Whayne Jr., M.D., Ph.D., F.A.C.C. (✉)
Division of Cardiovascular Medicine, Gill Heart Institute, University of Kentucky, 326 Wethington Building, 900 South Limestone Street, Lexington, KY 40536-0200, USA
e-mail: twhayn0@uky.edu

N. Maulik and T. Karagiannis (eds.), *Molecular Mechanisms and Physiology of Disease: Implications for Epigenetics and Health*, DOI 10.1007/978-1-4939-0706-9_7,

CV	Cardiovascular
DM	Diabetes mellitus
DNA	Deoxyribonucleic acid
HDAC	Histone deacetylase
PAR	Poly(ADP-ribosylation)
RNA	Ribonucleic acid
ROS	Reactive oxygen species
SAM	S-adenosylmethionine

7.1 Introduction

Epigenetics is not a subject of great familiarity to the practicing clinician and yet it has tremendous relevance to human health. This chapter defines the subject in practical terms. The possible relevance of epigenetics for the development of cardiovascular (CV) disease is discussed. As part of this, the resultant effects of nutrition and exercise on CV disease is considered through their potential favorable modification of epigenetics. The implications for improved health separate from specific genetic inheritance are potentially incredible, including benefits for the fetus, newborn, and the same individual.

7.2 Definition of Epigenetics

Epigenetics is defined as the study of heritable alterations in gene expression or cellular phenotype. The term epigenetics was originally coined by Conrad Waddington in the 1940s to define the difference from just a genetic approach. Originally, epigenetics referred to how genes and their products brought the phenotype into being (Jablonka and Lamb 2002). A precise definition of the principles of epigenetics is all the meiotically and mitotically inherited changes in gene expression that are not encoded in the deoxyribonucleic acid (DNA) sequence itself (Wilson 2008) (Fig. 7.1). Epigenetic modifications of chromatin and DNA are important negative and positive factors for controlling the expressed genome via gene transcription. Two major epigenetic mechanisms are the posttranslational modification of histone proteins in chromatin and DNA methylation. These are regulated by different, but nevertheless coupled, pathways. These phenomena are reversible. The epigenetic state is a central regulator of cellular development and cellular activation. Related to this and supported by increasing levels of evidence, there appears to be a key role played by epigenetics in human disease such as in inflammatory disease and in cancers. Various cancers have an association with altered epigenetic profiles. This can lead to an altered expression of different genes involved in cell growth and/or differentiation (Wilson 2008). The increase in autoimmune and neoplastic diseases observed in advanced age can be associated with an

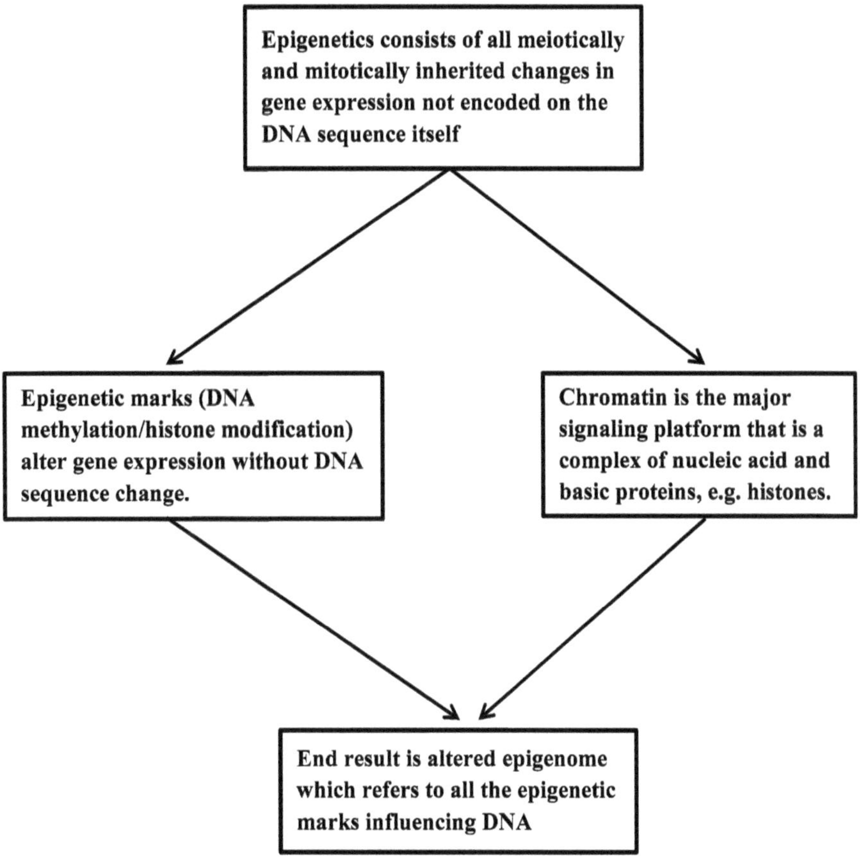

Fig. 7.1 Principles of epigenetics

altered epigenetic state as a possible explanation. Now, the primary concern is with the mechanisms by which cells are committed to a particular form or function and how that function, or structural state, is transmitted in cell lineages. It is important for its practical significance for medicine, agriculture, and conservation of the species and also for its implications for heredity and evolution. Specifically, these changes are specified as caused by mechanisms not involving modification of the underlying DNA sequence (Scheen and Junien 2012). In contrast to the genetic background inherited from parents, epigenetics considers changes that constitute the memory of previous events, even in utero. Later exposure to a hostile environment could therefore result in various pathologies, such as different complex chronic diseases. An example of this occurrence can involve metabolic diseases such as obesity, metabolic syndrome, and type 2 diabetes mellitus (DM). The original definition of epigenetics involves inheritable but reversible phenomena that affect gene expression without altering base pairs. The eukaryotic (eukaryote refers to an organism whose cells have complex structures enclosed within membranes) genome is the same throughout all somatic cells in an organism but specific structures and functions discern one type of cell from another. The explanation of these differences is due to unique gene expression patterns in cells as determined during cell differentiation (Tammen et al. 2013). These cell-specific gene expression patterns can be altered by the environment that an organism is exposed to, resulting in environmentally mediated changes in expression patterns, explained by complex modifications to DNA, histone proteins, and degree of DNA packaging. These are known as epigenetic marks. The specific molecular mechanisms that effect epigenetic regulation include DNA methylation, chromatin modifications, and histone modifications (Cheung and Lau 2005).

Specific definitions relevant to epigenetics and the epigenome are essential. The genome can be defined as one haploid set of chromosomes with the genes contained therein, i.e., the genetic material of an organism. Chromatin is a complex of nucleic acid and basic proteins, such as histones, in eukaryotic cells, dispersed in the nucleus and condensed into chromosomes in mitosis and meiosis. Chromatin is the physiologically relevant substrate for all genetic processes inside the nuclei of eukaryotic cells. Dynamic alterations in local and global organization of this chromatin appear to be key regulators of the function of the genome (Fischle et al. 2003). Multiple signals from inside and outside cells appear to converge on this major signaling platform. Numerous posttranslational modifications of histones, the main protein components of chromatin, have been analyzed and defined. These so-called marks appear to be key mediators of the functional activity of the genome and act in response to signaling pathways that come from "upstream." In contrast to the genome, the term epigenome refers to all of the epigenetic marks influencing the DNA in a single cell. There are multiple situations of cross talk between various components of this complicated regulatory system that are being more and more defined regarding these epigenetic circuits. DNA that is packaged in chromatin is the physiologically relevant base for all DNA-dependent functions inside the nuclei of eukaryotic cells (Nemeth and Langst 2004). Highly compacted structures of DNA and histones tend to repress DNA-dependent processes. Therefore, the concept of dynamic chromatin with two apparently contradictory functions: that of tight compaction and that of free accessibility to DNA. Chromatin is a highly regulated nucleoprotein complex with genetic material structured throughout (Campos and Reinberg 2009). The chromatin can be mobilized to effect cellular processes which include transcription, cell division, differentiation, and DNA repair. In eukaryotes, the chromatin core is composed of nucleosomes, known as repetitive histone octamer units usually enfolded by 147 base pairs of DNA. The DNA is arranged and indexed through these nucleosomal structures to adjust local chromatin compaction and accessibility. Histones are nucleosomal proteins consisting of a histone-fold domain with DNA-wrapping properties. Histones can also be viewed as DNA-binding proteins. An epigenetic trait is a stably inherited phenotype resulting from changes in a chromosome without alterations in the DNA sequence. Histone variants contribute to intrinsic and extrinsic properties of the nucleosome particle by establishing specialized chromatin structures. The existence of well-positioned nucleosomes in certain regions of the genome raises the possibility that histones may be epigenetic components. With epigenetics considered the study of heritable changes in gene expression not mediated at the level of the DNA sequence, molecular mechanisms that mediate epigenetic regulation include DNA methylation and modifications of the chromatin and its contained histones (Cheung and Lau 2005). Key histone-modifying enzymes and the biological functions of many histone posttranslational modifications have begun to be elucidated.

7.3 Altered Function and Inheritance via the Epigenome

The year 2009 marked the 150th anniversary of the publication of Charles Darwin's "On the Origin of Species" (Koonin and Wolf 2009). Darwin attributed importance to random, undirected change that provided material for natural selection. On the other hand, 2009 marked the 200th anniversary of "Philosophie Zoologique," published by Jean-Bapteste Lamarck. Lamarck believed that evolution is modified by nonrandom beneficial changes in phenotype, especially those associated with the use of organs. Lamarck believed these changes were inheritable. Much of his theory was considered controversial and untenable but more simplistic considerations of altered evolution may actually have had their basic beginning with Lamarckian inheritance which theorized that an organism can pass on to its offspring, characteristics that it acquired during its lifetime (Handel and Ramagopalan 2010; Jablonka and Lamb 1989). The work of Lamarck actually has been labeled by some as soft inheritance with no relevance to modern evolutionary theory (Handel and Ramagopalan 2010). However now, some of his early concepts are having a quiet resurgence as the increasing complexity of epigenetic theories of inheritance are considered with their provision for potential mechanisms for environmental influences to be passed from parents to offspring. As compared to Darwin, the basic ideas of Lamarck appear to have more of a potential basic relationship to modern epigenetics.

Epigenetics involves changes to marks on the genome that are copied from one generation of cells to the next (Mathers 2008). The functional history of a gene in one generation may alter its expression in a subsequent generation (Jablonka and Lamb 1989). The epigenome consists of the sum total of the epigenetic marks influencing the DNA in a single cell (Greaves et al. 2012). Gene expression can be altered without involving changes in the primary DNA sequence (Mathers 2008). Such marks involve DNA methylation and posttranslational modifications of histone tails that protrude from nucleosome cores, and these modifications include acetylation, methylation, phosphorylation, and ubiquitination. A hypothesis is that altered epigenetic marking is how environmental effects, including nutrition, diet, and exercise, are received and recorded by the genome. Some of these epigenetic marks can be remembered through multiple cell generations, resulting in altered gene expression and cell function without alteration of the primary DNA. This allows variation of the phenotype in the presence of a fixed genotype. A perfect example of this involves monozygotic twins who may show increasing epigenetic differences with age and differing lifestyles (Mathers 2008). There is increasing evidence of the sensitivity of the epigenome to environmental influence in the individual and then transmission across future generations of the species (Franklin and Mansuy 2010; Suter et al. 2011).

Variations in epigenetic markings may also explain individual variation in risk of disease in response to intervention such as with nutrition. Alterations in the epigenome may be as significant to the development of human disease as are traditional mutations (Wilson 2008). Alleles with different epigenetic marks in the same nucleus can even interact to modify the epigenetics of one or both alleles (Greaves et al. 2012). This is especially evident when two divergent epigenomes are associated in a hybrid in which the methylation patterns of one allele may be changed to resemble those of the other allele using processes such as trans-chromosomal methylation and trans-chromosomal demethylation. Such interaction may be common in many biological systems. Multiple lifestyle factors such as diet, obesity, physical activity, tobacco abuse, excess alcohol consumption, environmental pollutants, psychological stress, and night shift work have been considered with possible relevance to modification of epigenetic patterns with studies especially centering on DNA methylation (Alegria-Torres et al. 2011). There is also increasing evidence in animal studies that adds support to environmental modification of epigenetics as a cause of increased susceptibility to disease (Jirtle and Skinner 2007). A key feature that highlights epigenetic modification from genetic change as related to disease is the potential reversibility of epigenetic modification, reinforcing the idea that our genes are not our absolute destiny (Stein 2012) and that our favorable epigenetic change can benefit our disease risk.

An interesting issue about epigenetics has been raised regarding human in vitro reproduction. The possible relationship of epigenetic reprogramming to in vitro embryo culture, immature sperm cells, and nuclear transfer has been considered as a concern (De Rycke et al. 2002). Various investigators have suggested an increased incidence of epigenetic abnormalities in children conceived by in vitro methodology but others have refuted such allegations. Thus far, irrefutable epigenetic alterations as a result of in vitro fertilization have not yet been demonstrated (Dupont et al. 2009) but certainly, the issue must be kept open considering potential effects on the epigenome.

7.4 Nutritional Modification of Epigenetics

There is growing evidence that epigenetic mechanisms can mediate nutrition effects and be responsible for development of common complex or chronic diseases (McKay and Mathers 2011). This occurs through the changes caused by epigenetics to marks on the genome that are then copied from one cell generation to the next, altering gene expression while not changing DNA sequence. Nutritional factors can have a profound effect on gene expression via epigenetic mechanisms (Fig. 7.2). Wheatley et al., tested the hypothesis of modulation of adipose gene expression by

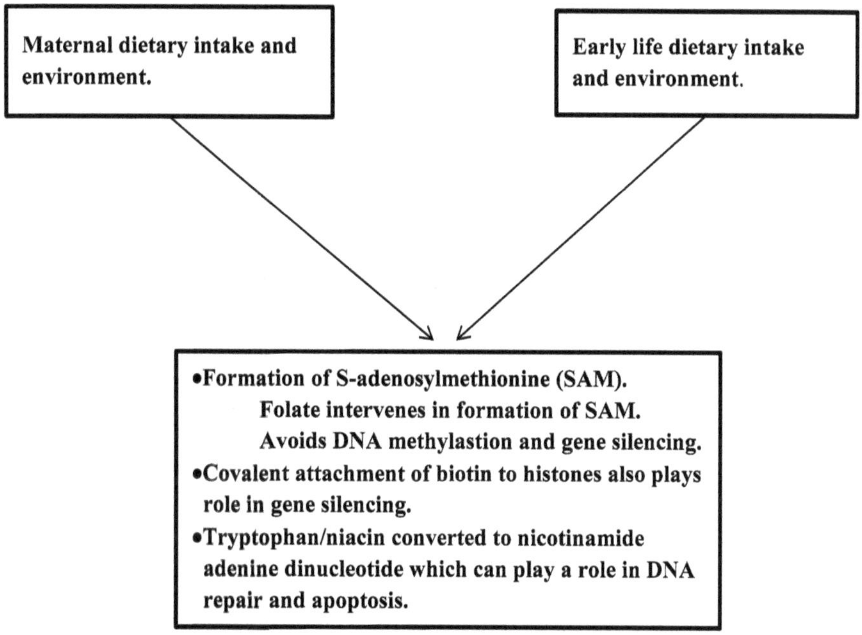

Fig. 7.2 Nutritional modification of epigenetics

obesity reversal through calorie restriction in 48 female mice (Wheatley et al. 2011). Initially, a diet-induced obesity regimen was administered for 8 weeks. Calorie restriction was then compared to a treadmill exercise regimen over 8 weeks. Both methods reduced adiposity by 35–40 % and serum leptin levels by 80 %. However, only calorie restriction increased adiponectin and insulin sensitivity. Gene expression microarray analysis of visceral white adipose tissue revealed 209 genes responsive to both calorie restriction and exercise. On the other hand, calorie restriction alone altered the expression of an additional 496 genes whereas only 20 genes were uniquely affected by exercise. Of genes responsive to calorie restriction, 17 are related to carbohydrate metabolism and glucose transport. This included glucose transporter 4, of which, calorie restriction significantly increased histone 4 acetylation, which was consistent with differential alteration of adipose transcription with calorie restriction.

Extensive epidemiologic and experimental data show that early less-than-optimal nutrition can have health consequences several decades later. On the other hand, good nutrition with, for example, adequate flavonoids and folates in the human diet with alteration of DNA methylation may modify future risk of human colon cancer and CV disease (Duthie 2011). The hypothesis that epigenetic mechanisms may link nutritional imbalances with increased disease risk has been gaining progressive acceptance through the years (Jimenez-Chillaron et al. 2012). A good example is type 2 DM, previously thought to occur mainly in older adults, but both its occurrence plus obesity has increased recently in children. It has been shown in small animal studies that significant dietary changes at different life cycle stages can result in major effects on fat mass or pancreatic function, over a relatively short term (Symonds 2009). However, changes in the human population are much more gradual. Nevertheless, epigenetic mechanisms that regulate pancreatic insulin secretion can be altered by extreme dietary changes in humans in early life, although it is difficult to establish the relative contribution of diet and changes in body mass to DM. It appears to be the complex interactions among food components with various factors including histone modifications, DNA methylation, noncoding ribonucleic acid (RNA) expression, and chromatin remodeling that lead to a dynamic regulation of gene expression resulting in control of the cellular phenotype (Milagro et al. 2013). As would be expected, the perinatal period is when the greatest modifications of phenotype can occur through contributions to developmental programming. Nevertheless, there is evidence that there is also a nutritional influence on epigenetic regulation during adulthood. Just as in type 2 DM, epigenetic pattern changes are associated with hypertension, atherosclerosis, various metabolic disorders, obesity, and weight-loss outcomes. Non-nutritional risk factors usually associated with obesity are also involved in these epigenetic modifications, especially hyperglycemia, inflammation, hypoxia, and oxidative stress. A major focus now is the study of possible innovative therapies based on nutritional or pharmacological agents that can modify epigenetic marks. In the undeveloped world, the most advanced medical therapies may not be available for intrauterine growth problems but much less expensive optimal nutrition during pregnancy may enhance fetal growth and development as well as alleviate the burden of maternal morbidity and mortality in low- and middle-income countries (Wu et al. 2012).

Epigenetics is central to genome structure and function. There is variation in epigenetic status involving individuals with increasing awareness in the importance of this in health and disease (Haggarty 2012). These mechanisms include DNA methylation, histone modification, and regulation by noncoding RNAs. There is much evidence for nutritional alteration of the epigenome and substrate for epigenetic reaction, which include acetyl and methyl groups, both central to nutritional metabolism. Nutritional epigenetics can aid in the clarification of how nutrition can influence health by means of direct effects on the genome. DNA methylation is the most extensively studied means of epigenetic gene modification (Anderson et al. 2012). Apparently, DNA methylation is labile in response to nutrition as well as various environmental effects. As gene expression is altered with resultant diverse phenotypes, there is a potential for a change or increase in disease risk. The major methyl donor for DNA methylation is S-adenosylmethionine (SAM), which results from the so-called one-carbon metabolism. This metabolism can be catalyzed by several enzymes in the presence of certain dietary micronutrients such as folate, choline, betaine (trimethylglycine), and various B vitamins. Therefore, the interest in nutrition and DNA methylation, especially backed by animal evidence; human epidemiological evidence is much less comprehensive. How nutrients can play essential roles in epigenetic events can be described as follows (Oommen et al. 2005). Folate intervenes in the formation of SAM, which then serves as a methyl donor for methylation of cytosines in DNA; such methylation is associated with gene silencing.

Covalent attachment of biotin to histones also plays a role in gene silencing and in any cellular response to DNA damage. Tryptophan and niacin are converted to nicotinamide adenine dinucleotide, which is a substrate for poly(ADP-ribosylation) (PAR) of histones and other DNA-binding proteins. This PAR plays a role in DNA repair and apoptosis. Epigenetic alteration appears to be a mechanism to mediate the effect of early-life environmental exposures and gene–environment interactions on the development of disease later as an adult (Hong and Wang 2012). Maternal dietary intake of fat, folate, protein, and total energy intake can alter epigenetic regulation of specific genes in offspring, resulting in altered tissue function (Burdge et al. 2012).

Personalized nutrition involves the adjustment of food and diet tailored to individual needs and preferences (Rubio-Aliaga et al. 2012). Nutritional guidance that is evidence based and that promotes health requires assessment of bioavailability, bioactivity, and bioefficacy of nutrients. Essential to this are nutritional biomarkers such as serum retinol, zinc, ferritin, and folate. Nutrigenomics may be a promising approach to identify new nutritional biomarkers. This can be viewed as a part of "foodomics" or nutriceuticals, which can be thought of as a comprehensive approach to use food science directed at the improvement of human nutrition and human health (Capozzi and Bordoni 2013). An example of a widespread but still personalized nutritional health problem is vitamin D deficiency, one that varies by regions of the world and by the individual (Whayne 2011). There appears little downside to increasing vitamin D intake and recent observations suggest that the influences of vitamin D on epigenetics may be important in utero for reducing chronic disease later in life, in addition to the multiple possible health benefits (Hossein-Nezhad and Holick 2012), including cardiovascular (Whayne 2011), associated with vitamin D.

7.5 Exercise Modification of Epigenetics

Physical exercise can be viewed as a modulator of epigenetics. Although there is evidence of a positive influence of physical exercise on epigenetic mechanisms and an improvement in health, much clarification of the links between exercise and epigenetics remains to be made (Sanchis-Gomar et al. 2012). There is even concern about a negative influence of excessive and persistent physical exercise on health. Individual physical adaptation to environmental conditions could alter epigenetics and affect gene expression. However, the majority opinion is that long-term repetitive strenuous exercise has a positive effect on health, reduction in aging, and a decreased incidence of cancer, all via epigenetic mechanisms (Fig. 7.3).

Various possible mechanisms are associated with a benefit of exercise on epigenetics. The apoptosis-associated speck-like protein containing a caspase recruitment domain (CARD) is abbreviated as ASC (Hasegawa et al. 2009). ASC was originally identified as a protein that forms large aggregates or the so-called specks in some human leukemia cells treated with chemotherapy and as the protein product of a gene that is silenced in cancer cells by DNA methylation. As a result, ASC has been

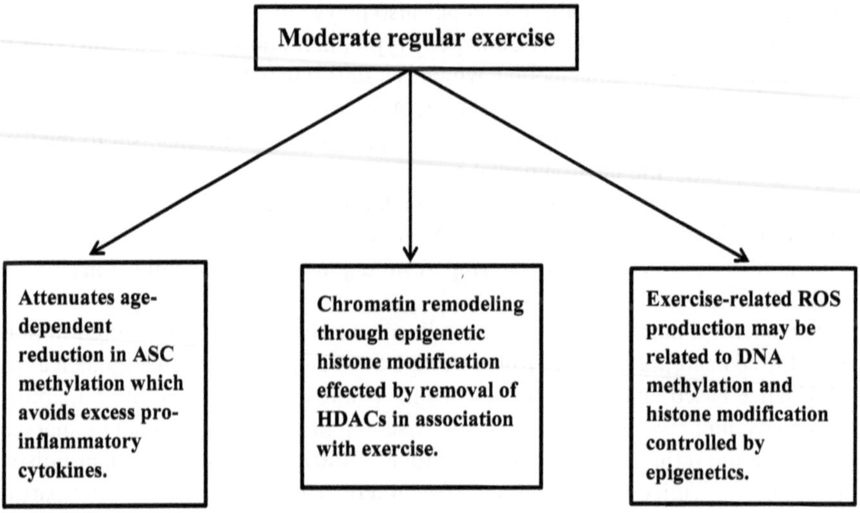

Fig. 7.3 Exercise modification of epigenetics

implicated in apoptosis and tumor suppression, potentially beneficial in cancer chemotherapy since epigenetic silencing of tumor suppressor and pro-apoptotic genes is one of the mechanisms by which resistance to cancer chemotherapy develops (Gordian et al. 2009). However, this is not the case when normal metabolism is present. Chronic moderate exercise has been reported to decrease potentially harmful pro-inflammatory cytokines (Nakajima et al. 2010). There is data that such moderate regular exercise attenuates an age-dependent reduction in ASC methylation, with implied suppression of excess pro-inflammatory cytokines through decreased ASC expression mediated by methylation. Therefore, by maintenance of a higher level of ASC methylation in older age groups, exercise can beneficially suppress excess harmful, pro-inflammatory cytokines by means of decreased ASC expression.

Skeletal muscle adaptations to exercise play an essential role in mediating the potential health benefits of exercise. This occurs in part through changes in skeletal muscle gene expression. The precise mechanisms by which skeletal muscle gene expression occurs in response to exercise are not known. Nevertheless, chromatin remodeling through epigenetic histone modification has appeared as a critical regulatory mechanism that controls gene expression in general. Class IIa histone deacetylases (HDACs) are enzymes that suppress histone acetylation, and these enzymes have been associated with exercise adaptation. In a study involving 60 min of cycling, McGee et al., found that global histone 3 acetylation was increased at lysine 36, a site associated with transcriptional elongation (McGee et al. 2009). The authors also found that $HDAC_4$ and $HDAC_5$ were exported from the nucleus during exercise, such that their transcriptional repressive function was removed. In addition, there was activation of two different protein kinases in response to exercise and these two kinases induce phosphorylation-dependent class IIa HDAC nuclear export. Thus, they considered that their data delineated a signaling pathway that might be associated with skeletal muscle adaptation to exercise. Other accumulating data suggest that exercise of moderate intensity has health promoting effects that are systemic and complex, undoubtedly involving regulation of redox homeostasis and signaling (Radak et al. 2013). Physical exercise causes an elevated generation of reactive oxygen species (ROS). ROS are important modulators of muscle contraction, antioxidant protection, and repair of damage from oxidation, all of which at moderate levels generate physiological responses. Several factors involved in mitochondrial biogenesis are modulated by exercise-associated redox changes. Certain endogenous thiol antioxidants, such as glutathione and thioredoxin, are modulated with exercise-related high oxygen consumption and ROS generation and control cellular function through redox-sensitive signaling and protein interactions. ROS may also play a role in exercise-induced angiogenesis. Also, exercise-related ROS production may be related to DNA methylation and histone modification, thereby creating heritable conditions controlled by epigenetics.

7.6 Epigenetics and Cardiovascular Disease

7.6.1 Coronary Heart Disease/Peripheral Arterial Disease

Individuals who were small at birth with associated poor growth rates in infancy have an increased risk of adult CV disease, osteoporosis, and type 2 DM (Godfrey et al. 2011). This increased risk especially applies if restricted early growth was followed by increased childhood weight gain. These associations involve epigenetic processes that modify the phenotype of an offspring and reflect developmental responses of the fetus and/or infant based on environment (Fig. 7.4). There are also influences from maternal characteristics that include diet, body composition, stress levels, and exercise levels. Study of exercise-induced heart and vascular bed adaptations highlights the different mechanical and metabolic stimuli that may cause short- and long-term adaptations of these CV tissues. Exercise results in increased free radicals while at the same time it improves antioxidative capacity. There is an associated shift in the cellular oxidative stress balance and, in addition, various signal cascades that mediate physiological and pathophysiological cardiac and peripheral vascular adaptations (Bloch et al. 2012). In doing this, exercise alters the molecular composition of the extracellular matrix. This, in turn, plays a role in various signaling cascades. Subsequently, epigenetic modulation may occur in CV tissues, even if just indirectly linked to exercise. Accumulating evidence supports epigenetic modulation being affected by exercise; physical activity may modify the functional genome in cardiac and peripheral vascular beds, comparable to other well-described phenomena such as diet and inflammation. It has been possible to identify multiple epigenetic signatures of CV disease-related environmental

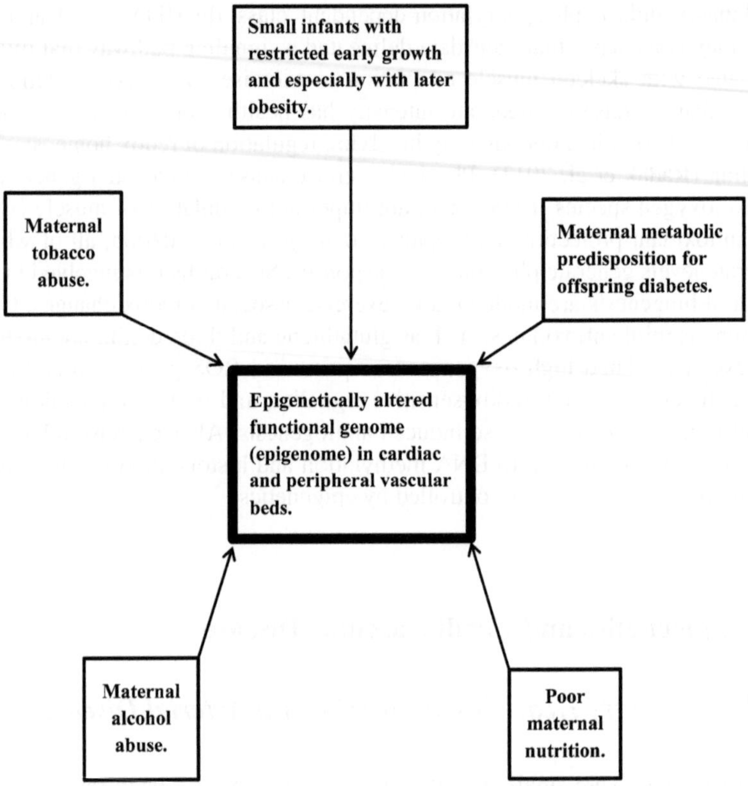

Fig. 7.4 Relationship of infant epigenetics to cardiovascular disease

exposure. Some of these signatures of epigenetic dysregulation can be detected in peripheral blood samples, some within even a few hours of exposure (Baccarelli and Ghosh 2012). Such a situation bodes well for future preventive and therapeutic strategies regarding CV disease.

7.6.2 Diabetes Mellitus

It is very likely that type 2 DM involves epigenetic mechanisms and not just simple genetic markers. Various energy-sensing signaling pathways in type 2 DM have been shown to play significant roles in inflammation, glucolipotoxicity, dysfunction of mitochondria, and oxidative stress, all of which have an association with insulin resistance and type 2 DM (Goh and Sum 2010). These signaling pathways may also regulate gene expression, playing a role in the epigenetic machinery while providing an explanation for how metabolism switches on or off, depending on the availability of food. In addition, there is evidence regarding adaptation to exercise that also links type 2 DM to decreased physical activity (Goh and Sum 2010). Although insulin resistance is common during obesity and aging in both animals and humans, it is actually the inability to compensate for insulin resistance that causes the progression to type 2 DM (Gilbert and Liu 2012). This failure to compensate is due to insulin secretory dysfunction as well as a significant apoptosis of normally functioning β-cells. Our knowledge of the molecular mechanisms that cause this β-cell dysfunction is limited but recent findings suggest that epigenetic changes in response to environmental stimuli may play an important role. Nutrition, oxidative stress, and inflammation may all have their influence and DNA methylation and histone modifications may play a role (Gilbert and Liu 2012). Specific knowledge could lead to possible therapeutic targets for prevention and treatment of DM.

Epidemiological studies suggest that the perinatal environment can cause a predisposition to the development of obesity and type 2 DM in infants. During gestation, malnutrition, obesity, type 1 or type 2 DM, psychological stress, and pharmacological stress in the mother can all promote obesity and DM in the infant (Levin 2008). On the other hand, maternal exercise can ameliorate the occurrence of obesity and DM in offspring. Appropriate perinatal health measures may serve to favorably alter the worldwide epidemic of obesity and DM. Skeletal muscle is a major metabolic organ. As such, it has an important role to play in glucose metabolism including insulin sensitivity and the function of insulin. Telomeres are the ends of eukaryote chromosomes thought to play an important role in cell deterioration including deterioration with age (Heidinger et al. 2012). In individuals of the same age, telomere length is thought to be directly related to potential longevity. A reflection of exposure of muscle myocytes to harmful environmental factors may be reflected by muscle telomere length in association with which chromosomal end telomere shortening appears to make chromosomes more susceptible to damage. In terms of assessing this, there is supportive evidence that leucocyte telomere length may reflect muscle myocyte telomere length when assessing the epidemiology of type 2 DM (Ahmad et al. 2012).

7.6.3 Aging

How aging occurs is not well understood but mechanisms under consideration include inflammation, oxidative damage, dysfunction of mitochondria, change in various neuronal circuits, and altered apoptosis (Kaliman et al. 2011). Epigenetics appears to play an important role in the possible mechanisms of aging. Nutrition has a long-term importance that can affect a subsequent generation. Chronic disease development can be accelerated and, of course, affect life span. Beneficial calorie restriction plays a major nutritional role in contributing to longevity by favorably modulating epigenetic mechanisms such as DNA methylation and histone modification (Ribaric 2012). The overall effect of various adaptive changes may be to delay age-related change and result in prolonged survival. Another mechanism in addition to appropriate nutrition and calorie intake appears to be a combination with exercise; there is also much interest in the polyphenol resveratrol as a supplement. Lifestyle interventions including exercise could open new directions in a beneficial change in aging and the diseases frequently linked to that process (Kaliman et al. 2011). There is a free-radical theory that increased oxidation caused by exercise can be harmful yet in general, exercise appears beneficial to the aging process. As part of this, an epigenetic oxidative redox shift aging theory has been proposed in which a sedentary lifestyle causes a redox shift with oxidation and impaired function of mitochondria. Breaking such a possible vicious cycle with an appropriate beneficial reductive redox shift may favorably modify and delay aging (Brewer 2010).

7.6.4 Tobacco Abuse

The problem of a diagnosis of tobacco use disorder and a possible effect on epigenetics as well as direct cell damage should be considered. Maternal tobacco abuse has been associated with altered methylation of placental cytochrome P450, family 1, subfamily A, polypeptide A (CYP1A1) gene restriction, and therefore fetal growth restriction. Alteration in placental gene expression and DNA methylation over the epigenome has also been associated with tobacco abuse in the mother. Thus, such perinatal exposure can cause significant changes in gene expression involving various pathways (Suter et al. 2011). Additional epigenetic modifications from a subsequent or new lifetime exposure to tobacco smoke are easily envisioned including secondhand smoke.

7.6.5 Metabolic Syndrome and Obesity

Obesity as part of the metabolic syndrome is associated with skeletal muscle defects contributing to insulin resistance and decreased fatty acid oxidation in muscle. There is associated dysfunctional metabolism with an epigenetic origin and possibly a genetic one, especially with severe obesity (Houmard et al. 2011). It has been shown that gastric bypass surgery with associated weight loss can favorably modify metabolism with a resultant increase in insulin sensitivity. Obesogens are substances that alter the regulation of energy balance to favor weight gain and obesity (Grun and Blumberg 2009). Such obesogen exposure can alter the epigenome of multipotent stromal stem cells in the fetus, causing a bias toward adipocyte production (Janesick and Blumberg 2011). Normal homeostatic mechanisms that play a role in weight control are altered by these obesogens such that even with a normal diet and exercise, there is a predisposition to weight gain. Despite this, fat cannot be accumulated unless there is a greater caloric intake than caloric expenditure (Grun and Blumberg 2009). Prenatal or perinatal exposure to obesogenic endocrine disrupting chemicals has been shown to cause a predisposition to store more fat starting at the onset of life. In mice, a high-fat diet exposure during pregnancy has been associated with epigenetic alteration of the expression of adipocytokine genes (Masuyama and Hiramatsu 2012). The offspring of such mice were shown to have higher blood pressure, worse glucose tolerance, higher triglycerides, higher leptin levels, and significantly lower adiponectin expression in white adipose tissue. This was an association with lower acetylation and higher methylation levels of histone H3 at lysine 9 of the adipose-tissue promoter of adiponectin. Excess exposure to estrogen in the uterus or in early life can increase the likelihood of subsequent obesity and metabolic syndrome. Such exposure can result in epigenetic changes with a resultant predisposition to excess weight (Janesick and Blumberg 2012). On the other hand, in the adult, estrogen treatment tends to decrease obesity (Grun and Blumberg 2009). Therefore, estrogen exposure during sensitive windows of development may cause a lifetime problem in maintaining an optimal weight, exacerbated by continued poor nutrition and suboptimal exercise. Common obesogen exposures involve antidiabetic thiazolidinediones, such as rosiglitazone and pioglitazone, despite their insulin-sensitizing effect (Grun and Blumberg 2009). Organotins are a class of organic pollutants widely used in industry and an example is tributyltin, which can cause potential harm as a man-made obesogen contributing to obesity and unfavorably altering human health (Grun 2010).

7.6.6 Statin Myopathy

Some patients with a statin myopathy appear to have their susceptibility due to a preexisting subclinical inherited muscular disorder or a genetic alteration in statin uptake proteins involving encoding by solute carrier, organic anion transporter family, member 1B1 (SLCO1B1), or the enzyme system involving cytochrome P450. It has also been considered that genes affecting the perception of pain and vascular receptor polymorphism might also be able to contribute to statin myopathy (Ghatak et al. 2010). It is easy to then also postulate a relationship of statin myopathy to epigenetics.

7.6.7 Flavonoids

Flavonoids are polyphenolic phytochemicals that have been shown to provide many beneficial effects for human health (Gilbert and Liu 2010). Specific flavonoids, which are the largest class of polyphenols, include three of the four main subgroups consisting of isoflavones, flavonols, and catechins (Valls et al. 2009), and have been given much attention because of their ability to modify the activity of chromatin-modifying enzymes that thereby result in epigenetic modifications (Gilbert and Liu 2010). An example is epigallocatechin-3-gallate, which has been shown to inhibit the activity of histone acetyltransferase and DNA methyltransferase.

7.7 Pharmacoepigenomics

Pharmacological therapy targeted at the epigenome is a major new therapeutic concept. Epigenomics is the field that focuses on nongenomic modifications that can influence gene expression. Pharmacoepigenomics involves epigenetic modification by medications and has the significant promise of being able to offer more and more personalized medicines (Manolopoulos et al. 2011). Up to this point in time, pharmacoepigenomics has mainly been concerned with cancer pharmacotherapy but there is accumulating data on individual responses to oral antidiabetic treatment that may lead to beneficial therapies. In cancer, epigenetic deregulation has been shown to occur early in carcinogenesis and this deregulation is potentially reversible (Ong et al. 2011). This reversibility may result in intervention strategies that target the epigenome for both cancer treatment and cancer prevention. Epigenetic therapy in cancer is directed at reversing the abnormalities that follow disruption of the balanced epigenetic signaling, using both natural compounds and synthetic molecules that are active on specific epigenetic targets (Mai and Altucci 2009). Such so-called epi-drugs consist of inhibitors of DNA methyltransferases, HDACs, histone acetyltransferases, histone methyltransferases, and histone demethylases. The silencing of tumor suppressors during malignant transformation offers the rationale for the utilization of chromatin-remodeling agents in cancer chemotherapy (Dario et al. 2008). Both the DNA methyltransferase inhibitors and the HDAC inhibitors have shown promise in cancer chemotherapy (Peedicayil 2006). There will undoubtedly be an expansion of such therapy. Some emerging biomarkers may have value in monitoring drug effects and in defining molecular signatures of response, toxicity, and effective dose (Kalebic 2003). Hopefully this promise of pharmacoepigenomics in cancer and early-on in diabetes mellitus will ultimately extend to CV disease.

7.8 Conclusions

Environmental modification of the epigenome including the effects of nutrition and exercise plays a major role in the maintenance of health and the development of disease, including CV disease. The major risk factors associated and the results of their presence can be favorably altered by factors such as good nutrition and exercise. Understanding this and how it affects the fetus, the newborn, and even the individual subject and patient later in life, may lead to major health benefits.

References

Ahmad S, Heraclides A, Sun Q, Elgzyri T, Ronn T, Ling C, Isomaa B, Eriksson KF, Groop L, Franks PW, Hansson O (2012) Telomere length in blood and skeletal muscle in relation to measures of glycaemia and insulinaemia. Diabet Med 29:e377–e381

Alegria-Torres JA, Baccarelli A, Bollati V (2011) Epigenetics and lifestyle. Epigenomics 3:267–277

Anderson OS, Sant KE, Dolinoy DC (2012) Nutrition and epigenetics: an interplay of dietary methyl donors, one-carbon metabolism and DNA methylation. J Nutr Biochem 23:853–859

Baccarelli A, Ghosh S (2012) Environmental exposures, epigenetics and cardiovascular disease. Curr Opin Clin Nutr Metab Care 15:323–329

Bloch W, Suhr F, Zimmer P (2012) Molecular mechanisms of exercise-induced cardiovascular adaptations. Influence of epigenetics, mechanotransduction and free radicals. Herz 37:508–515

Brewer GJ (2010) Epigenetic oxidative redox shift (EORS) theory of aging unifies the free radical and insulin signaling theories. Exp Gerontol 45:173–179

Burdge GC, Hoile SP, Lillycrop KA (2012) Epigenetics: are there implications for personalised nutrition? Curr Opin Clin Nutr Metab Care 15:442–447

Campos EI, Reinberg D (2009) Histones: annotating chromatin. Annu Rev Genet 43:559–599

Capozzi F, Bordoni A (2013) Foodomics: a new comprehensive approach to food and nutrition. Genes Nutr 8(1):1–4

Cheung P, Lau P (2005) Epigenetic regulation by histone methylation and histone variants. Mol Endocrinol 19:563–573

Dario LS, Rosa MA, Mariela E, Roberto G, Caterina C (2008) Chromatin remodeling agents for cancer therapy. Rev Recent Clin Trials 3:192–203

De Rycke M, Liebaers I, Van Steirteghem A (2002) Epigenetic risks related to assisted reproductive technologies: risk analysis and epigenetic inheritance. Hum Reprod 17:2487–2494

Dupont C, Armant DR, Brenner CA (2009) Epigenetics: definition, mechanisms and clinical perspective. Semin Reprod Med 27:351–357

Duthie SJ (2011) Epigenetic modifications and human pathologies: cancer and CVD. Proc Nutr Soc 70:47–56

Fischle W, Wang Y, Allis CD (2003) Histone and chromatin cross-talk. Curr Opin Cell Biol 15:172–183

Franklin TB, Mansuy IM (2010) Epigenetic inheritance in mammals: evidence for the impact of adverse environmental effects. Neurobiol Dis 39:61–65

Ghatak A, Faheem O, Thompson PD (2010) The genetics of statin-induced myopathy. Atherosclerosis 210:337–343

Gilbert ER, Liu D (2010) Flavonoids influence epigenetic-modifying enzyme activity: structure—function relationships and the therapeutic potential for cancer. Curr Med Chem 17:1756–1768

Gilbert ER, Liu D (2012) Epigenetics: the missing link to understanding beta-cell dysfunction in the pathogenesis of type 2 diabetes. Epigenetics 7:841–852

Godfrey KM, Inskip HM, Hanson MA (2011) The long-term effects of prenatal development on growth and metabolism. Semin Reprod Med 29:257–265

Goh KP, Sum CF (2010) Connecting the dots: molecular and epigenetic mechanisms in type 2 diabetes. Curr Diabetes Rev 6:255–265

Gordian E, Ramachandran K, Singal R (2009) Methylation mediated silencing of TMS1 in breast cancer and its potential contribution to docetaxel cytotoxicity. Anticancer Res 29:3207–3210

Greaves I, Groszmann M, Dennis ES, Peacock WJ (2012) Trans-chromosomal methylation. Epigenetics 7:800–805

Grun F (2010) Obesogens. Curr Opin Endocrinol Diabetes Obes 17:453–459

Grun F, Blumberg B (2009) Minireview: the case for obesogens. Mol Endocrinol 23:1127–1134

Haggarty P (2012) Nutrition and the epigenome. Prog Mol Biol Transl Sci 108:427–446

Handel AE, Ramagopalan SV (2010) Is Lamarckian evolution relevant to medicine? BMC Med Genet 11:73

Hasegawa M, Imamura R, Motani K, Nishiuchi T, Matsumoto N, Kinoshita T, Suda T (2009) Mechanism and repertoire of ASC-mediated gene expression. J Immunol 182:7655–7662

Heidinger BJ, Blount JD, Boner W, Griffiths K, Metcalfe NB, Monaghan P (2012) Telomere length in early life predicts lifespan. Proc Natl Acad Sci U S A 109:1743–1748

Hong X, Wang X (2012) Early life precursors, epigenetics, and the development of food allergy. Semin Immunopathol 34:655–669

Hossein-Nezhad A, Holick MF (2012) Optimize dietary intake of vitamin D: an epigenetic perspective. Curr Opin Clin Nutr Metab Care 15:567–579
Houmard JA, Pories WJ, Dohm GL (2011) Is there a metabolic program in the skeletal muscle of obese individuals? J Obes 2011:250496
Jablonka E, Lamb MJ (1989) The inheritance of acquired epigenetic variations. J Theor Biol 139:69–83
Jablonka E, Lamb MJ (2002) The changing concept of epigenetics. Ann N Y Acad Sci 981:82–96
Janesick A, Blumberg B (2011) Endocrine disrupting chemicals and the developmental programming of adipogenesis and obesity. Birth Defects Res C Embryo Today 93:34–50
Janesick A, Blumberg B (2012) Obesogens, stem cells and the developmental programming of obesity. Int J Androl 35:437–448
Jimenez-Chillaron JC, Diaz R, Martinez D, Pentinat T, Ramon-Krauel M, Ribo S, Plosch T (2012) The role of nutrition on epigenetic modifications and their implications on health. Biochimie 94:2242–2263
Jirtle RL, Skinner MK (2007) Environmental epigenomics and disease susceptibility. Nat Rev Genet 8:253–262
Kalebic T (2003) Epigenetic transitions: towards therapeutic targets. Expert Opin Ther Targets 7:693–699
Kaliman P, Parrizas M, Lalanza JF, Camins A, Escorihuela RM, Pallas M (2011) Neurophysiological and epigenetic effects of physical exercise on the aging process. Ageing Res Rev 10:475–486
Koonin EV, Wolf YI (2009) Is evolution Darwinian or/and Lamarckian? Biol Direct 4:42
Levin BE (2008) Epigenetic influences on food intake and physical activity level: review of animal studies. Obesity (Silver Spring) 16(Suppl 3):S51–S54
Mai A, Altucci L (2009) Epi-drugs to fight cancer: from chemistry to cancer treatment, the road ahead. Int J Biochem Cell Biol 41:199–213
Manolopoulos VG, Ragia G, Tavridou A (2011) Pharmacogenomics of oral antidiabetic medications: current data and pharmacoepigenomic perspective. Pharmacogenomics 12:1161–1191
Masuyama H, Hiramatsu Y (2012) Effects of a high-fat diet exposure in utero on the metabolic syndrome-like phenomenon in mouse offspring through epigenetic changes in adipocytokine gene expression. Endocrinology 153:2823–2830
Mathers JC (2008) Session 2: personalised nutrition. Epigenomics: a basis for understanding individual differences? Proc Nutr Soc 67:390–394
Mcgee SL, Fairlie E, Garnham AP, Hargreaves M (2009) Exercise-induced histone modifications in human skeletal muscle. J Physiol 587:5951–5958
McKay JA, Mathers JC (2011) Diet induced epigenetic changes and their implications for health. Acta Physiol (Oxf) 202:103–118
Milagro FI, Mansego ML, De Miguel C, Martinez JA (2013) Dietary factors, epigenetic modifications and obesity outcomes: Progresses and perspectives. Mol Aspects Med 34(4):782–812
Nakajima K, Takeoka M, Mori M, Hashimoto S, Sakurai A, Nose H, Higuchi K, Itano N, Shiohara M, Oh T, Taniguchi S (2010) Exercise effects on methylation of ASC gene. Int J Sports Med 31:671–675
Nemeth A, Langst G (2004) Chromatin higher order structure: opening up chromatin for transcription. Brief Funct Genomic Proteomic 2:334–343
Ong TP, Moreno FS, Ross SA (2011) Targeting the epigenome with bioactive food components for cancer prevention. J Nutrigenet Nutrigenomics 4:275–292
Oommen AM, Griffin JB, Sarath G, Zempleni J (2005) Roles for nutrients in epigenetic events. J Nutr Biochem 16:74–77
Peedicayil J (2006) Epigenetic therapy—a new development in pharmacology. Indian J Med Res 123:17–24
Radak Z, Zhao Z, Koltai E, Ohno H, Atalay M (2013) Oxygen consumption and usage during physical exercise: the balance between oxidative stress and ROS-dependent adaptive signaling. Antioxid Redox Signal 18(10):1208–1246
Ribaric S (2012) Diet and aging. Oxid Med Cell Longev 2012:741468

Rubio-Aliaga I, Kochhar S, Silva-Zolezzi I (2012) Biomarkers of nutrient bioactivity and efficacy: a route toward personalized nutrition. J Clin Gastroenterol 46:545–554

Sanchis-Gomar F, Garcia-Gimenez JL, Perez-Quilis C, Gomez-Cabrera MC, Pallardo FV, Lippi G (2012) Physical exercise as an epigenetic modulator. Eustress, the "positive stress" as an effector of gene expression. J Strength Cond Res 26(12):3469–3472

Scheen AJ, Junien C (2012) Epigenetics, interface between environment and genes: role in complex diseases. Rev Med Liege 67:250–257

Stein RA (2012) Epigenetics and environmental exposures. J Epidemiol Community Health 66:8–13

Suter M, Ma J, Harris A, Patterson L, Brown KA, Shope C, Showalter L, Abramovici A, Aagaard-Tillery KM (2011) Maternal tobacco use modestly alters correlated epigenome-wide placental DNA methylation and gene expression. Epigenetics 6:1284–1294

Symonds ME (2009) Conference on "Multidisciplinary approaches to nutritional problems". Symposium on "Diabetes and health". Nutrition and its contribution to obesity and diabetes: a life-course approach to disease prevention? Proc Nutr Soc 68:71–77

Tammen SA, Friso S, Choi SW (2013) Epigenetics: the link between nature and nurture. Mol Aspects Med 34(4):753–764

Valls J, Millan S, Marti MP, Borras E, Arola L (2009) Advanced separation methods of food anthocyanins, isoflavones and flavanols. J Chromatogr A 1216:7143–7172

Whayne TF Jr (2011) Vitamin d: popular cardiovascular supplement but benefit must be evaluated. Int J Angiol 20:63–72

Wheatley KE, Nogueira LM, Perkins SN, Hursting SD (2011) Differential effects of calorie restriction and exercise on the adipose transcriptome in diet-induced obese mice. J Obes 2011:265417

Wilson AG (2008) Epigenetic regulation of gene expression in the inflammatory response and relevance to common diseases. J Periodontol 79:1514–1519

Wu G, Imhoff-Kunsch B, Girard AW (2012) Biological mechanisms for nutritional regulation of maternal health and fetal development. Paediatr Perinat Epidemiol 26(Suppl 1):4–26

Chapter 8
Epigenetic Events Associated with Obesity and Diabetes

Ernesto Burgio and Lucia Migliore

Abstract Obesity is becoming a major public health concern. During the last years, genetic and epigenetic factors have been supposed to contribute to increase (or decrease) the susceptibility to gain weight and to develop obesity-related comorbidities. Metabolic syndrome, defined by a combination of disturbed glucose and insulin metabolism, central obesity, dyslipidemia, and hypertension, is considered to be a risk factor for type 2 diabetes and cardiovascular disease. The role of genetic factors involved in the etiology of human obesity is beyond question. Moreover there is evidence that the current epidemic of obesity and diabetes is environmentally driven. Studies during the past decade have indicated that normal metabolic regulation during adulthood not only requires a good matching of energy intake with energy expenditure, but also is influenced by fetal and postnatal environments. Epidemiological studies and experimental models show that maternal nutritional constraint during pregnancy alters the metabolic phenotype of the offspring and that this can be passed to subsequent generations. Recent researches in a number of laboratories all over the world suggest the continuous increase in the environment and food chains of "obesogens," above all of endocrine disruptors, i.e., chemicals that interfere with many homeostatic mechanisms, altering the regulation of energy balance, promoting fat accumulation, adipogenesis, and weight gain. Finally epigenetic marks could be useful to personalize nutrition, to early detect those individuals with more risk to develop metabolic disorders or to better respond to a treatment.

E. Burgio (✉)
European Cancer and Environment Research Institute (ECERI), Bruxelles, Belgium

ISDE International Society of Doctors for Environment Scientific Office,
Via della Fioraia 17/19-52100, Arezzo, Italy
e-mail: erburg@libero.it

L. Migliore
Department of Translational Research and New Technologies in Medicine and Surgery,
Division of Medical Genetics, University of Pisa, Via S. Maria, 55-56126, Pisa, Italy

N. Maulik and T. Karagiannis (eds.), *Molecular Mechanisms and Physiology of Disease: Implications for Epigenetics and Health*, DOI 10.1007/978-1-4939-0706-9_8,

Keywords Epigenetics • Obesity • Diabetes • Metabolic syndrome • Obesogens

8.1 Introduction

It would not make sense to introduce the major theme of the genetic and epigenetic origins of obesity without trying to make a point about an issue which is nowadays considered the most serious and common consequence of improper nutrition and, more generally, of unhealthy lifestyles that characterize our time: the "obesity pandemic," that could become the most severe public health emergency of the twenty-first century.

Thousands of articles in the scientific press, popular magazines, and newspapers have been written about the increased prevalence of obesity in the past few years. Over the past decade, approximately 22,300 articles that have the word obesity in the title have been reviewed by PubMed; by inserting in the search the articles with the word obesity in the abstract, the total number is nearly tripled (Freudenberg 2011).

Current epidemiological data show a dramatic increase in obesity in industrialized countries over the past 20 years: its prevalence has trebled in both men and women (currently 25–30 %), with the largest increases seen in adolescents and young adults. Alarmingly, similar trends are also apparent in children, and the problem has also begun to affect the developing countries (Bartolomucci et al. 2012).

In fact, the increase was really dramatic, which justifies the use of the term "pandemic": recent data from the 1999–2000 National Health and Nutrition Examination Survey (NHANES) show that almost 65 % of the adult population in the United States is overweight, having a body mass index (BMI) greater than 25 kg/m^2, compared to 56 % seen in NHANES III, conducted between 1988 and 1994. Above all, rates of overweight have almost tripled since the first NHANES (1971–1974). The prevalence of obesity, defined as BMI greater than 30 kg/m^2, has increased dramatically from 23 to 31 % over the same time period.

Children are not immune to the epidemic: as the full extent of the increase in weight in the population has been recognized, the significant involvement of children and adolescents has become evident (Rubenstein 2005), with the prevalence of overweight in pediatric age up by 36 % during the same period (St-Onge et al. 2003).

What is especially important to note is that up to 30 years ago obesity was rarely considered by pediatricians, and only in the context of rare genetic disorders, such as Prader–Willi syndrome. Worryingly, obese children are at increased risk for becoming obese adults (Whitaker et al. 1997), and more frequently subject to many obesity-related health conditions once restricted to adults (Dietz 1998). More than 60 % of children who are overweight before puberty will be overweight in early adulthood, and the rising rates of childhood obesity have rendered type 2 diabetes, once an "adult-onset" disease, almost unheard of in kids, relatively common between adolescents and even in children, especially in certain ethnic groups (Hu 2011). This is of particular concern, because a reduction in the average age at which noncommunicable diseases become apparent could greatly increase the burden on health services that will have to provide treatment during much of the adult life of these patients.

In fact obesity and overweight are known to have adverse health effects, and to impact the risk and prognosis for a number of serious medical conditions such as type 2 diabetes, hyperinsulinemia, insulin resistance, coronary heart disease, high blood pressure, stroke, gout, liver disease, asthma and pulmonary problems, gall bladder disease, kidney disease, reproductive problems, osteoarthritis, and some forms of cancer (Mokdad et al. 2003; Irigaray et al. 2007).

Most countries are experiencing dramatic increases both in obesity and in diabetes. As an example, the prevalence of overweight individuals in China doubled in women and almost tripled in men from 1989 to 1997 (Bell et al. 2001), with an explosive increase in diabetes prevalence within a relatively short time: in 1980, less than 1 % of Chinese adults had the disease; by 2008, the prevalence reached nearly 10 % (Yang et al. 2010).

The real question we should be asking ourselves is: what are the causes of such a huge epidemiological change that has been rightly defined in terms of "quasi-infectious" pandemic? Some authors acknowledged that the current epidemic of obesity should be considered in the context of globalization as a communicable rather than noncommunicable process, a "socially contagious feature of globalization" (Bornstein et al. 2008), while others spoke in terms of evolutionary change: in fact this is the first time that an entire species faces a dramatic change of its phenotype (Bartolomucci et al. 2012).

The conventional wisdom holds that obesity is the result of a positive energy balance, i.e., of an imbalance between energy intake and expenditure (too many calories in and too few calories burned) (Stubbs and Lee 2004): in this perspective, the current pandemic should be basically ascribed to the recent (from an evolutionary perspective) adoption of a sedentary lifestyle, coupled with the high availability of foods with high caloric content (Lopez and Knudson 2012). These features accompanied the past decades with unprecedented transitions in our lifestyle: in less than 50 years, we have become an "obese species." For the first time in human history, the number of people who are obese and overweight surpassed the number of them who are underweight (Speakman and O'Rahilly 2012).

Even if this basic concept retains its validity, it is increasingly evident that obesity is not simply a product of overeating and lack of exercise, but the result of a prolonged disturbance in the homeostatic regulation of energy metabolism that favors triglyceride storage and adipocyte hypertrophy. Moreover, this accumulation of fat and/or mobilization of lipids from adipose depots are controlled by hormonal regulation of appetite and satiety, glucose levels, basal metabolic rate, metabolic set points, number, size, and metabolic activity of adipocytes (Grün and Blumberg 2009a).

It is in this context that research has offered, in recent years, the most significant results, demonstrating that this altered metabolic regulation is due, at least in part, to a deregulation of the hormonal circuits, which ultimately leads to an increase in food intake, triglyceride storage, adipocyte hyperplasia, and hypertrophy; and that a whole series of "obesogens"—i.e., of molecules (especially endocrine disruptors) scattered in food chains—could interfere with this very complex and fine tuned circuits, altering regulation of energy balance and favoring weight gain and obesity (Grün and Blumberg 2009b).

8.2 Obesity and its Related Pathology: The Metabolic Disorder

As a matter of fact, many known comorbidities observed in obese adults are now observed more frequently in youth: insulin resistance (IR, Sinha et al. 2002) and type 2 diabetes (T2D, Libman et al. 2003) representing a dramatic example. Overweight and obesity in youth are also associated with various risk factors for cardiovascular disease and have been shown to be associated with the early development of atherosclerotic lesions (Goran et al. 2003).

There is strong evidence to suggest an association between obesity and poor mental health in teenagers, the overweight children having a fivefold increased risk for low health-related quality of life, a risk similar to that observed in children affected by cancer (Schwimmer et al. 2003). The economic burden of childhood obesity has also increased threefold in the past 20 years, reaching $127 million per year (Wang and Dietz 2002).

Impaired glucose tolerance and type 2 diabetes are now largely recognized as the metabolic key features of the clinical disorders clustered in the metabolic syndrome (MS), including central obesity, arterial hypertension, prothrombotic and proinflammatory states, ovarian polycystosis, hyperuricemia, and nonalcoholic fatty liver disease.

8.3 Pathophysiology of Adipose Tissue and Obesity

New observations, theories, and empirical data have forced endocrinologists, scientists, immunologists, and the general public to change their long-held beliefs about physiology and pathophysiology of the adipose tissue. Nevertheless there is no doubt that we are only at the beginning of a long scientific path towards a better understanding of the links among obesity and diabetes, cardiovascular dysfunction, cancer, and a myriad of other diseases.

Some key points to consider are as follows:

1. Adipose tissue plays a critical role in energy homeostasis, not only in storing triglycerides, but also secreting adipokines that control immunity and neuroendocrine function (Ahima 2006)
2. Adipokines released by fat cells are beneficial in health unless the body accumulates too much fat (especially visceral fat), resulting in abnormal signaling that could actually increase the progression of the disease (Fain 2010)
3. Obese adipose tissue shows features characteristic of active local inflammation (Das 2001) which is involved in the progression to insulin resistance (Hirosumi et al. 2002) and diabetes (Antuna-Puente et al. 2008)
4. Fat distribution plays an important role in health (Lafontan and Berlan 2003)

High caloric intake and/or decreased energy expenditure results in a state of positive energy balance. This normally drives an increase in adipose tissue mass by two distinct mechanisms. First, excess energy is stored as additional triglycerides in existing adipocytes resulting in enlargement of these cells, which is called hypertrophy. A low generation rate of new adipocytes associates with adipose hypertrophy, which is linked to low insulin sensitivity and high circulating insulin levels. Secondly, if the number of fat cells is not sufficient to store increasing amounts of triglycerides, new adipocytes are generated by adipogenesis of mesenchymal precursor cells in a process described as hyperplasia (Arner et al. 2010). As long as adipose tissue is able to increase the capacity to store excessive energy by hypertrophy and/or hyperplasia, metabolic abnormalities are rare. However, if adipose tissue is unable to recruit new fat cells, the size of the existing adipocytes may abnormally increase, which is known to be associated with whole-body insulin resistance (Klöting et al. 2010).

8.3.1 The Differentiation of Adipose Tissue

The development of obesity is characterized not only by increased storage of lipids in existing fat cells but also by the generation of new adipocytes from progenitor cells, a process called adipogenesis. Firstly during determination, multipotent mesenchymal stem cells commit to preadipocytes. These cells exhibit similar morphology compared with stem cells; however, they are committed to the adipogenic lineage and are not longer able to transform into osteoblasts, myocytes, or chondrocytes. Secondly, during differentiation, preadipocytes become mature fat cells.

Adipocyte differentiation from mesenchymal stem cells is tightly regulated at a molecular level by several transcription factors. WNT signaling molecules are important key regulatory factors that play a unique role in the determination of multipotent mesenchymal stem cells into preadipocytes. Furthermore, increasing evidence suggests that in later stages of adipogenesis, WNT inhibitors are necessary to maintain WNT signaling in an inactive state. These effects at the cellular and molecular level are crucial in the pathogenesis of obesity and type 2 diabetes, as illustrated both by animal models and by several genetic studies in humans (Laudes 2011).

Adipose tissue is not just a passive lipid repository. Adipose depots also function as active endocrine organs that participate in the body's feedback system that fine-tunes the regulation of appetite and the metabolic integration between organs and inflammatory responses.

The etiology of obesity varies, reflecting many possible points of perturbation in the regulatory pathways that maintain fat homeostasis. A multitude of factors influence whether an individual will develop obesity. Genetic, nutritional, and environmental factors are known to impact hunger and satiety, basal metabolic rate, carbohydrate and lipid flux, and the regulation of adipocyte proliferation and differentiation and developmental programming of metabolic set points (Grün and Blumberg 2009b).

8.4 Genetic Factors in Obesity

Although monogenic obesity syndromes are rare, genetic variation is assumed to play an important role in determining the interindividual differences in susceptibility or resistance to the current "obesogenic" environment, which is characterized by easy access to high-calorie, high-fat food, and reduced energy expenditure (Swinburn et al. 1999). The genetic contribution to obesity and diabetes (Vimaleswaran and Loos 2010) has been established through family (Permutt et al. 2005), twin (Stunkard et al. 1986a), and adoption studies. Twin studies have shown that genetic factors explain 40–80 % of the variance in BMI and in risk of obesity (Herskind et al. 1996), while lower heritabilities have been reported for family (20–50 %) (Rice et al. 1999) and adoption (20–60 %) (Stunkard et al. 1986b) studies. The higher concordance of type 2 diabetes in monozygotic twins (50–70 %) compared with dizygotic twins (20–37 %) provides evidence of a genetic contribution to this condition (Poulsen et al. 1999). Further evidence of a genetic component comes from studies on family history of type 2 diabetes. While the lifetime risk of developing type 2 diabetes is 7 % in the general population, this risk is four to sixfolds higher (30–40 %) in individuals who had one parent with type 2 diabetes and tenfold (70 %) if both parents had diabetes (Köbberling and Tillil 1982).

Anyway obesity is most definitely a multifactorial or complex disease as it is caused by a complex interaction between genetic, behavioral, and environmental factors. As mentioned earlier, the conventional theory holds that obesity is the result of a positive energy balance, due to overeating and high caloric fatty diets combined with a sedentary lifestyle on a background of genetic predisposition for the disease. However, although much attention has been paid on these factors, including the need to incorporate healthy foods in our diets and more exercise into our lifestyle, these factors cannot alone explain the alarming rise in obesity (Newbold et al. 2009).

Until the 1990s, fat cells or "adipocytes" were considered to be just storage depots for excess metabolic fuel. However, following the discovery of an adipocyte-derived hormone termed "leptin" (Zhang et al. 1994) that communicates energy reserve information from adipocytes to other organs of the body including the central nervous system, a new appreciation emerged that these "fat storage cells" actually function as an endocrine organ (Collins 2005).

The identification of the hypothalamic leptin–melanocortin signaling pathway as a critical regulator in energy homeostasis and food intake has been essential for genetic research. Novel loci or DNA sequences from this pathway potentially involved in the pathogenesis of obesity have been recently discovered, by mutation analysis, candidate gene and genome-wide association studies (GWASs), as well as copy number analysis. Their role in monogenic and complex forms of obesity is gradually clarifying.

Anyway, despite the heritability estimates ranging between 40 and 70 % and despite intense efforts to identify genetic variants that predispose to obesity and type 2 diabetes, using a candidate gene approach and genome-wide linkage studies, progress has been, until recently, slow and success limited (Vimaleswaran and Loos 2010) and the list of common obesity susceptibility variants by the currently published GWASs only explains a small proportion of the individual variation in risk.

Monogenic forms of obesity refer to a highly penetrant form of the disease resulting from mutations in, or deletions of single genes (Mendelian conditions). To date, there are eight well-established monogenic obesity genes: leptin (*LEP*), leptin receptor (*LEPR*), proopiomelanocortin (*POMC*), prohormone convertase 1 (*PCSK1*), melanocortin 4 receptor (*MC4R*), single-minded homologue 1 (*SIM1*), brain-derived neurotrophic factor (*BDNF*), and neurotrophic tyrosine kinase receptor type 2 (*NTRK2*). Mutations in these eight genes are known to cause early onset obesity and hyperphagia and may account for up to 10 % of severely obese children (D'Angelo and Koiffmann 2012).

Several of the likely causal genes in predisposition to obesity are highly expressed or known to act in the central nervous system (CNS) and thus are thought to be involved in obesity susceptibility via CNS-mediated effects (Choquet and Meyre 2011).

A recent work has shown expression of fourteen likely causal obesity risk genes (*FTO*, *MC4R*, *BDNF*, *TMEM18*, *KCTD15*, *NEGR1*, *NRXN3*, *ETV5*, *MTCH2*, *SEC16B*, *TFAP2B*, *GNPDA2*, *FAIM2*, and *LYPLAL1*) in the hypothalamus of both obese and lean rats, which either support or bring new evidence for a potential central effect of these genes on energy homeostasis (Schmid et al. 2012).

Another observation reinforcing the role of genes involved in the central regulation of food intake in obesity predisposition is that so far, three obesity susceptibility loci are located near genes (*MC4R*, *SH2B1*, and *BDNF*) that have already been shown to carry deleterious mutations disrupting hypothalamic functions and leading to monogenic forms of early-onset obesity with hyperphagia as a common feature.

It was recently reported that in a GWASs study on copy number variants (CNVs), individuals with extreme phenotypes were found to carry a number of large and rare CNVs (specifically deletions on chromosome 16p11.2); besides developmental delay, this was also associated to obesity (Bachmann-Gagescu et al. 2010).

Mounting evidence supports a role for haplo insufficiency of SH2B1 in the obesity phenotype of patients with the 220 kb deletion: it encodes an adaptor protein involved in the LEP and insulin signaling (D'Angelo and Koiffmann 2012).

Several studies have been performed to link polymorphisms of susceptibility genes to obesity-related traits that could take into account interindividual differences. To date, more than 150 genetic loci are associated with the development of monogenic, syndromic, or multifactorial forms of type 2 diabetes or obesity (Drong et al. 2012). Even if we may induce from these outcomes that the genetic variability concerning the leptin–melanocortin pathway to be of paramount importance, there is still a lot of heritability that currently cannot be explained.

8.5 Environmental Factors in Obesity and Obesity-Associated Disorders

There is growing agreement among experts that the environment, rather than biology, is driving the epidemic of obesity and diabetes (Hill and Peters 1998). There is no sign that the rapid increase in obesity seen over the past two decades is abating: that is why there is an urgent need to push back against the environmental forces that are producing gradual weight gain in the population (Hill et al. 2003). Some of the putative contributors to the obesity epidemic that have been recently reviewed (McAllister et al. 2009) include infections, sleep debt, reduction in variability of ambient temperatures, obesogen chemicals (chiefly endocrine disruptors), increasing maternal age, greater fecundity among people with higher adiposity, assortative mating, pharmaceutical iatrogenesis, intrauterine and intergenerational effects, and epigenetic mechanisms.

Along with the increasing worldwide incidence of obesity-associated disorders, research has recently unraveled important pathways reciprocally connecting metabolism with the immune system. We already discussed that the development of obesity is a complex process involving genetic susceptibility and environmental factors, which both remain only partially understood. In such a context, gut microbiota is being increasingly recognized as an important factor connecting genes, environment, and immune system. Genomic and environmental factors at the basis of mutual host–microbiota interactions have been intensely investigated with metagenomic and metabolomic approaches in the last 5 years (Musso et al. 2010).

We live in a microbial world (Whitman et al. 1998). Coevolution, coadaptation, and codependency are all features of our relationships with our indigenous microbiota (Dethlefsen et al. 2007). The human gut is a lush microbial ecosystem containing about 100 trillion microorganisms and up to 500–1,000 different species, whose collective genome, the "metagenome," contains 100-fold more genes than the entire human genome (Xu and Gordon 2003). The symbiosis of our extended genome plays a role in host homeostasis and energy extraction from diet: Gordon and colleagues proposed that the microbiota from obese subjects specifically increases the energy harvested from the diet, providing an extra energy to the host (Ley et al. 2006).

While the human genome is inherited, the human microbiome is acquired from the environment anew every generation. Infants obtain initial microbes from the mother during vaginal birth (Dominguez-Bello et al. 2010) and the microbiome establishes during first year of life, bacterial abundances increasing ~ 6 orders of magnitude within the first weeks of life, becoming more adult-like within the first year (Palmer et al. 2007). Microbiota composition is unique to each body site (each body site (e.g., gut, skin, oral, nasal, urogenital) is home to a unique community), with few differences (over time and by gender), but continues to change over a lifetime, especially from mid-age to elderly (Claesson et al. 2011).

Metagenomic studies demonstrated that certain mixes of gut microbiota might protect or predispose the host to obesity (Tsai and Coyle 2009).

The gut microbiota contributes to host metabolism by several mechanisms including increased energy harvest from the diet, modulation of lipid metabolism, altered endocrine function, and increased inflammatory tone. There is evidence that gut microbiota has a role in the regulation of energy homeostasis and fat storage. Interactions among microorganisms in the gut appear to have an important role in host energy homeostasis, with hydrogen-oxidizing methanogens enhancing the metabolism of fermentative bacteria (DiBaise et al. 2008). Differences in caloric extraction of ingested food substances may be due to the composition of the gut microbiota, suggesting that the metabolic activities of the gut microbiota facilitate the extraction of calories from ingested dietary substances and help to store these calories in host adipose tissue for later use. There are many studies describing an altered microbiological colonization in the gut of obese subjects: data indicate that obese subjects have more bacteria belonging to Firmicutes and relatively less bacteria belonging to Bacteroidetes phyla (Bäckhed et al. 2004; Ley et al. 2006). Additionally, germ-free mice are resistant to diet-induced obesity caused by consumption of a high-fat/high-sugar "Western" diet (Bäckhed et al. 2007). The gut microbiota could thus be considered to be an environmental factor that modulates obesity and other metabolic diseases (Greiner and Bäckhed 2011).

On the other hand, variations in gut microbiota are likely to affect human toxicodynamics and increase individual exposure to obesogenic and diabetogenic chemicals (Ghanim et al. 2009): in fact the toxicology and pharmacology literature suggests that interindividual variations in gut microbiota may affect chemical metabolism via direct activation of chemicals, depletion of metabolites needed for biotransformation, alteration of host biotransformation enzyme activities, changes in enterohepatic circulation, altered bioavailability of environmental chemicals and/or antioxidants from food, and alterations in gut motility and barrier function. On this basis we can state that gut microbiota composition likely affects obesity and diabetes, as does exposure to environmental chemicals.

Indeed, accumulating data suggest an important role for toxicology in the etiology of obesity (Newbold et al. 2008). Despite the potential importance of endocrine disruptors in the pathogenesis of metabolic diseases, the contribution of synthetic chemical exposure to the obesity and diabetes epidemics remains largely unrecognized and underappreciated, even though emerging data support a biologically plausible causative link between continuously increasing production of synthetic organic chemicals and the two "pandemics" rates (Neel and Sargis 2011).

Above all, the Obesogen Theory (Grün and Blumberg 2006) has proposed that in our time a caloric surplus is not the only way to gain weight: in fact some chemicals, termed obesogens, could shift lipid homeostasis, acting epigenetically on endocrine pathways, increasing size and number of adipocytes, decreasing the tendency to oxidize fatty acids, etc. (Grün and Blumberg 2009a) and contributing to the epidemic of obesity and related metabolic disorder (Decherf and Demeneix 2011).

During the last years, the concept of obesogens was extended to include substances that may modify metabolic balance at the central, hypothalamic level. Two prime candidates are tributyltin (TBT) and tetrabromobisphenol A (TBBPA), widespread pollutants able to interfere with hypothalamic gene regulations (Janesick and Blumberg 2011) (see also paragraph 9).

In fact the most important and worrying issue concerning not only the obesogens theory, but more in general the main problem of the origins of the current pandemic of obesity and type 2 diabetes, is the continuously growing prenatal and neonatal exposure to a lot of molecules that can interfere with the epigenetic programming of tissues and organs, increasing the propensity towards obesity in adult life (and even, as we will see in the next paragraph) in the next generations (Kirchner et al. 2010).

8.6 The Developmental Origin of Health and Diseases: Fetal Programming and Adult Obesity

The rapid increase in the incidence of chronic noncommunicable diseases over the past two decades cannot be explained solely by genetic and adult lifestyle factors. There is now considerable evidence that the fetal and early postnatal environment strongly influences the risk of developing such diseases in later life. Human studies have shown that low birth weight is associated with an increased risk of CVD, type 2 diabetes, obesity, and hypertension, although recent studies have shown that overnutrition in early life can also increase susceptibility to future metabolic disease. The mechanism by which the maternal nutritional environment induces such changes is beginning to be understood and involves the altered epigenetic regulation of specific genes. The association between poor intrauterine growth and increased risk of disease in later life may reflect a mismatch between the future environment "predicted" by the embryo/fetus, based on signals from the mother during gestation, and the actual environment experienced in later life (Lillycrop 2011).

Early nutrition affects adult metabolism in humans and other mammals, potentially via persistent alterations in DNA methylation. A dozen years ago Waterland and Jirtle showed that dietary methyl supplementation of a female of yellow agouti (A(vy)) mice (which harbors a transposable element in the agouti gene), with extra folic acid, vitamin B (12), choline, and betaine alters the phenotype of the offspring via increased CpG methylation at the A(vy) locus and that the epigenetic metastability which confers this liability is due to the A(vy) transposable element (Waterland and Jirtle 2003). These findings had a major impact, suggesting on the one hand that dietary supplementation, long presumed to be purely beneficial, could have unintended deleterious influences on the establishment of gene regulation in humans, secondly, that even small changes in nutrition during pregnancy may have dramatic, lifelong consequences for the fetus, interfering with the epigenetic programming of organs and tissues. Some years later, it was established that in rats, the altered methylation of specific gene promoters and the consequently altered metabolic phenotype in the liver induced in the F1 generation by maternal protein restriction during pregnancy is transmitted to the F2 generation, showing a mechanism for the transmission of induced phenotypes between generations (Burdge et al. 2007). In human subjects, Heijmans and coworkers reported hypomethylation of the imprinted insulin-like growth factor-2 gene in genomic DNA isolated from whole blood of individuals who were exposed to famine in utero during the Dutch Hunger Winter, compared to unexposed same-sex siblings (Heijmans et al. 2008).

The same group also found that the insulin-like growth factor promoter was hypomethylated in individuals whose mothers were periconceptually exposed to famine, while IL-10, LEP, ATP-binding cassette A1, and the guanine nucleotide-binding protein were hypermethylated (Tobi et al. 2009).

These well-known studies established that a nutritional challenge in early life can result in a change in DNA methylation which is detectable even 60 years later, suggesting that also in humans, as in the animal studies, early-life environmental changes can induce long-term alterations in the epigenetic regulation of genes.

On the other hand, alterations in pre-/perinatal nutrition could predispose towards obesity and other associated diseases such as T2D, particularly in an environment with high availability of energy-dense food. Data from a variety of animal models have supported a link between the perinatal nutritional environment and the programming of energy balance "set points." Epidemiological studies in humans suggest that during pregnancy and lactation maternal malnutrition, obesity, type 1 and type 2 diabetes, and psychological, immunological, and pharmacological stressors may increase the incidence of obesity and type 2 diabetes in the offspring (Levin 2009). Normal birth diet can decrease the negative effects of some of these prenatal factors, but maternal diet high in fat, diabetes, and newborn augmented access to food all promote the development of obesity and metabolic syndrome in the offspring (Levin 2006).

Although the mechanisms underlying this metabolic imprinting require further elucidation, the evidence accumulated to date indicates that perinatal hormones represent key signals that program CNS (hypothalamic development and function) and exert lasting effects on body weight regulation and glucose homeostasis. Plagemann and coworkers hypothesized that early overfeeding may alter DNA methylation patterns of hypothalamic promoter regions of genes critically involved in the lifelong regulation of food intake and body weight (Plagemann et al. 2009).

Peripheral hormones represent important signals that regulate adiposity as well as CNS circuits that control food intake. The best characterized hormonal adiposity signals are insulin and LEP. Besides playing an important role in the regulation of energy balance and neuroendocrine functions in mature animals, LEP acts early in life as a developmental signal that promotes the formation of metabolic pathways. Insulin also appears to exert important influences on the development of hypothalamic circuits that regulate energy homeostasis (Bouret 2009). Epidemiological, clinical, as well as experimental data indicate that insulin, when occurring in elevated concentrations during perinatal life, may program by itself the development of obesity and diabetes (Plagemann 2008).

Little is known about the mechanisms underlying inheritance of disease risk relating to these high nutrition pathways. There might be effects on adipocyte differentiation (Spalding et al. 2008) or a prenatal influence on hepatic fat deposition in later adulthood via effects on mitochondrial function (Bruce et al. 2009). A high-fat diet might adversely affect the redox state (Anderson et al. 2009), thereby causing coronary endothelial dysfunction (Galili et al. 2007).

However although such effects could be produced in a mother consuming a high-fat diet, it is not known whether oxidative stress affects offspring and placental function. Epigenetic mechanisms might be involved, as the offspring of dams receiving a high-fat diet show changes in the microRNA expression patterns, in particular those associated with insulin-like growth factors and methyl transferases (Godfrey et al. 2011, Zhang et al. 2009). An association between the methylation status of specific genes (in particular hypermethylation of *RXRα*, an essential gene for adipogenesis) in human umbilical cord tissue and the subsequent development of childhood adiposity was recently found in two longitudinal cohorts (Godfrey et al. 2011).

As already mentioned, the environmental obesogen hypothesis holds that prenatal or early life exposure to endocrine disrupting chemicals could predispose to increased fat mass and obesity. Obesogens could alter the epigenome of mesenchymal stem cells, biasing them towards the adipocyte lineage at the expense of osteoblasts. Hence, humans exposed to obesogens during early life might have an altered stem cell compartment, which is preprogrammed towards an adipogenic fate. This results in a higher steady state number of adipocytes and a lifelong endeavor to maintain a healthy weight, which can be exacerbated by social influences that promote poor diet and inadequate exercise (Janesick and Blumberg 2011).

Finally, it is useful to remind that the "Developmental Origins of Health and Disease hypothesis" and inheritance-oriented investigations concerning gene–nutrient interactions on energy homoeostasis and metabolic functions have suggested that inflammation could be not only a comorbidity of obesity but also a cause (Martínez et al. 2012).

8.7 Epigenetic Modifications in Obesity

The most extensively studied epigenetic modifications in humans are DNA methylation, covalent histone tail modifications, nucleosome remodeling, and RNA-mediated targeting.

Feinberg et al. have identified four variably methylated regions (VMRs) that show covariation with BMI and are located in or near genes implicated in regulating body weight or diabetes (Feinberg et al. 2010).

Obesity was found associated to changes in promoter DNA methylation patterns of some genes, such as serotonin transporter gene, that has a critical role in regulating food intake, body weight, and energy balance (Zhao et al. 2013). In a recent review, Milagro and coworkers reviewed studies showing altered methylation and/or histone acetylation levels in genes involved in appetite regulation (such as *LEP*, neuropeptide Y, *POMC*, *MC4R*), glucose homeostasis (insulin, insulin receptor), adipogenesis (adiponectin), feeding and fasting regulation (*FTO*: fat mass and obesity associated), glucose homeostasis (*IGF2* Insulin-like growth factor 2), body weight homeostasis (*CEBPA*: CCAAT/enhancer-binding protein C/EBP, alpha), lipid storage (*FASN*: fatty acid synthase) but also in more general metabolic processes (vitamin metabolic process, inflammatory responses, oxidative stress) (Milagro et al. 2013).

Also type 2 diabetes, hypertension, atherosclerosis, and other metabolic disorders have been repeatedly associated to changes in epigenetic patterns. Many studies dealing with comparisons between individuals with obesity-associated disorders and control subjects have been performed. An excess of differentially methylated sites in genomic regions of subjects with or without type 2 diabetes mellitus, including a CpG site in the first intron of the *FTO* gene, has been described (Toperoff et al. 2009). A similar approach has revealed that 187 genes were differentially methylated between diabetes patients with end stage renal disease and diabetic patients without nephropathy (Sapienza et al. 2011).

Much progress in identifying epigenetic changes induced by (or inducing) obesity has already been made, with candidate and genome-wide approaches. Studies dealing with whole-genome methylation appears irrelevant and usually lacks of prognostic value in metabolic diseases because a low sensibility. On the contrary studies using more focused approaches (gene-specific methylation) have found that *LEP* and *TNF-alpha* methylation levels could be used as epigenetic biomarkers concerning the response to diet (Cordero et al. 2011). In this sense, although it would be of greater interest studying the baseline epigenetic differences in obesity-related tissues, such as adipose tissue or liver, peripheral blood mononuclear cells (or otherwise white cells) are preferentially used because of simplicity, speed, accuracy, and ease of access (Widschwendter et al. 2008).

Another interesting point concerning the search for epigenetic biomarkers in obesity is the possibility of identifying those individuals that have more susceptibility to develop metabolic impairments with respect to those that are less prone (Milagro et al. 2013).

8.7.1 Endocrine Disruptors and Obesity

We have already mentioned the Obesogen Theory: among the substances, which undoubtedly play a key role in this context, are the so-called endocrine disruptors. In fact a huge amount of data coming from hundreds of laboratories all around the world have shown that a great variety of environmental endocrine disrupting chemicals can influence adipogenesis and obesity, interfering with the adipose tissue biology; modifying hormone synthesis, transport, and metabolism; deregulating central hypothalamic–pituitary–adrenal axis and derailing many homeostatic mechanisms essential to weight control (Grün and Blumberg 2007).

Many endocrine disrupting compounds alter fat cell differentiation or function, initiating or exacerbating deregulation of homeostatic controls (Prior and Armitage 2009). Their impact on adipose tissue may occur through direct modulation of lipogenesis, lipolysis, and adipogenesis, or indirectly by affecting food consumption and LEP secretion, targeting the CNS or lipid homeostasis in liver (Diamanti-Kandarakis et al. 2009).

These "obesogens" may perturb various endocrine axes, generally targeting nuclear receptors (NR), including sex steroid receptors, the retinoic acid receptors (RXR), the gamma peroxisome proliferator receptor (PPARγ), or the glucocorticoid receptor (GR), all affecting directly or indirectly adipocyte physiology and more generally the regulation of energy homeostasis (Somm et al. 2009).

Examples of endocrine-disrupting chemicals comprise heavy metals, solvents, pesticides, PCB, organic phosphates, phthalates, organotins, diethylstilbestrol (DES), and bisphenol A (BPA). They include either mimetic substances of lipophilic hormones, such as bisphenol A, TBT, or inhibitors of endogenous hormone metabolism (e.g., TBT action on aromatase activity). Consequently, the sites of action are varied and the interactions very complex, especially for compounds like organotins (chemical compounds based on tin with hydrocarbon substituents) that have multiple molecular targets. In many cases, dose–response curves are not monotonic, but exhibit changing phenotypes across different dose ranges (as seen with phytoestrogens and DES). This is of great concern, since it is becoming increasingly clear that when no monotonic dose–response curves occur, the effects of low doses cannot be predicted by the effects observed at high doses: in fact many EDCs act as metabolic toxicants at high doses, while at lower levels, more similar to environmental exposures, the effects may be quite different and even paradoxical (Vandenberg et al. 2012).

Additional complexity is also introduced by timing of exposure, gender, and genetic predisposition. Developmental exposure represents a limited window of heightened sensitivity where long-term effects distant from the initial insult can be established, and possibly only in a limited subset of the population. This delay in response and the experimental difficulty in establishing cause and effect for environmental factors may provide a partial answer to the underappreciated role that chemical obesogens might play.

The increased obesity risk due to prenatal maternal smoking certainly provides a proof-of-concept that long-term deregulation of metabolic homeostasis is relevant at a population level (Power and Jefferis 2002).

Epigenetic changes from obesogenic exposures are currently poorly understood. This will become an area for future intensive research efforts given their potential for long lasting, transgenerational effects (Grün and Blumberg 2009b). As for the possible epigenetic and transgenerational effects of EDCs, the most studied molecule is bisphenol A, a synthetic chemical and weak estrogen agonist found in food and beverage containers, baby bottles, and dental materials. For what concerns bisphenol A, the most famous experiment was made, a decade ago, on the Agouti mouse: BPA-induced hypomethylation and increased expression of the Agouti gene in prenatally exposed mice (early developmental stages generally represent the period of greatest sensitivity to these chemicals) led to the birth of mice characterized by yellow rather than brown fur, as well as by tendency to develop obesity, diabetes, and tumors (Dolinoy 2008). Furthermore, rodent mothers with the agouti phenotype were more likely to have offspring with that phenotype in the second generation. This important study constitutes a proof of the fact that prenatal exposure to synthetic estrogen agonists such as BPA can affect the epigenome and thereby lead to endocrinological consequences (Fleisch et al. 2012).

Recent observations demonstrated that widely diffuse environmental compounds such a mixture of plastic-derived compounds, BPA and phthalates, and a hydrocarbon mixture involving jet fuel (JP-8) can promote epigenetic transgenerational inheritance of adult onset disease, including obesity. Gestating F0 generation female rats were transiently exposed during the fetal gonadal development period a hydrocarbon mixture involving jet fuel (JP-8). The direct exposure F1 generation had an increased incidence of kidney abnormalities in both females and males, prostate and pubertal abnormalities in males, and primordial follicle loss and polycystic ovarian disease in females. The first transgenerational generation is the F3 generation, and the jet fuel lineage had an increased incidence of primordial follicle loss and polycystic ovarian disease in females, and obesity in both females and males. Moreover analysis of the jet fuel lineage F3 generation sperm epigenome identified 33 differential DNA methylation regions, termed epimutations (Tracey et al. 2013).

Similarly gestating F0 generation female rats were exposed to the plastic mixture during embryonic days 8–14 of gonadal sex determination, and the incidence of adult onset disease was evaluated in F1 and F3 generation rats. Significant increases in the incidence of total disease/abnormalities in F1 and F3 generation male and female animals from plastics lineages were found. In particular, in the F3 generation animals, pubertal abnormalities, testis disease, obesity, and ovarian disease (primary ovarian insufficiency and polycystic ovaries) were increased (Manikkam et al. 2013).

Much remains to be discovered about the possible molecular mechanisms characterizing environmental obesogens and their overall significance for the epidemic of obesity and T2D. However, given the data already available on obesogens, the chemical effects we experience daily and the multiple targets with which they might interfere, it seems quite likely that obesogen exposure can play an important role in obesity epidemic.

8.7.2 Dietary Factors and Obesity

As we have repeatedly stressed throughout this chapter, many epidemiological studies in humans and experimental studies in animals suggest that maternal under nutrition, obesity and diabetes during gestation and lactation, maternal and fetal stress, embryo–fetal exposure to endocrine disruptors, and other obesogens can all induce early-life nutritional programming and produce obesity in offspring.

There are several nutritional events in pregnancy and lactation, such as energetic deprivation, protein restriction and excess fat, which may determine a cluster of disorders affecting energy efficiency in the offspring as well as different metabolic pathways, which are mediated by epigenetics encompassing the chromatin information encrypted by DNA methylation patterns, histone covalent modifications, and noncoding RNA or microRNAs. Epigenetic mechanisms may be boosted or impaired by dietary and environmental factors in the mother, intergenerationally or transiently transmitted, and could be involved in the obesity and inflammation susceptibility in the offspring (Martínez et al. 2012).

Although prenatal period is the time of highest phenotypic plasticity, contributing largely to developmental programming, also during infancy and adulthood there is evidence of nutritional influence on epigenetic regulation. In the last years, a number of studies have connected different dietary patterns, nutrients, and food components with epigenetic processes that regulate gene expression and may contribute to an increased susceptibility to obesity and other metabolic diseases. The large differences in DNA methylation observed between human preadipocytes and mature adipocytes suggest that epigenetics plays an important role in the process of adipocyte differentiation. Obesity and its related complications have been repeatedly associated with epigenetic alterations: methylation changes in blood leukocyte DNA have been observed in obese adolescents, whereas in type 2 diabetes, a great number of genes present pathologic methylation patterns in muscles, which was suggested to contribute to the onset of insulin resistance (Milagro et al. 2013).

Accumulating literature shows that diet in particular can influence the biochemical pathways of methylation processes by modulating the availability of methyl donors, including folate, choline, and methionine, as well as through their effects on methyltransferase activity. However the epigenetic mechanisms that modulate adipocyte differentiation, hepatic steatosis, insulin signaling, appetite regulation and insulin secretion, as well as important bioenergetic pathways, such as lipolysis, fat oxidation, and glucose uptake, and oxidation are still only partially known.

Finally, it is important to remind that obesity shares with most chronic diseases the presence of an inflammatory component, which is reflected in increased circulating levels of pro-inflammatory proteins and occurs not only in adults but also in adolescents and children, which accounts for the development of metabolic disease and other associated health alterations. The possible role of nutrition in this condition of systemic subacute chronic inflammation has still to be elucidated, even if many studies have already shown that a nutrient excess and adipocyte expansion trigger endoplasmic reticulum stress; that abnormally elevated blood lipid levels, including NEFA, can induce inflammation through various mechanisms (such as modulation of adipokine production or activation of Toll-like receptors); and that hypoxia occurring in hypertrophied adipose tissue stimulates the expression of inflammatory genes and activates immune cells (de Heredia et al. 2012).

8.8 Epigenetic Therapy in Obesity

There is compelling and growing evidence about the applications of epigenetic drugs as a novel therapy of diabetes, as several epigenetic mechanisms have been reported to control adipogenic differentiation and influence energy metabolism. Moreover there are evidences that many nutritional factors could act by modulating DNA methylation or histone modifications and some of them might be used in obesity therapy due, at least in part, to their epigenetic mechanisms. This outcome is more evident in relation to the methyl donors (folate, methionine, choline, and vitamin B12), especially when maternal diet is supplemented. Methyl donors are of critical importance during fetal development, when they can interfere with DNA methylation and influence neural precursor cell proliferation and brain development. Even in the adult population, there are differences in the health outcomes due to methyl donor deficiency (fatty liver, insulin resistance) in function of common genetic variants (Milagro et al. 2013).

Inhibitors of two classes of epigenetic enzymes, the DNA methylation inhibitors (DNMTs) and histone deacetylase inhibitors (HDACs), have already demonstrated utility as molecularly targeted chemotherapeutic agents for specific cancers, and are approved drugs for these indications (Copeland et al. 2010). Thus an important research field concerns the applications of drugs targeting epigenetic enzymes, such as HDACs, or DNMTs as a novel therapy for obesity and related pathologies. Although in vitro and animal studies show encouraging results in different aspects (i.e., HDAC inhibitors enhance β-cell differentiation and survival and insulin signaling), there is a lack of clinical studies aimed at determining diabetes-specific epigenetic profiles and a precise assessment of the epigenome of endocrine pancreatic cells, for example of insulin-secreting β cells, in the nondiabetic state and at early and late diabetic states (Bramswig and Kaestner 2012). Furthermore we should not forget that most histone-modifying enzymes and DNA methyltransferases lack of specificity; they have a broad target range, being expressed in many different tissues. This increases the likelihood of "off-target" effects, and will present a significant challenge for drug development. Moreover, a careful evaluation of the potential secondary effects is mandatory. The use of these drugs in obesity must reinforce even more the study of the potential side effects that epigenetic drugs could induce in the different organs and tissues.

Other food components such as polyphenols and organosulfur compounds, several fatty acids (particularly PUFA), minerals, and vitamins that have been positively experimented in cancer prevention and treatment (owing to their antioxidative and anti-inflammatory properties) could act as epigenetic therapeutic agents in obesity and related disorders, acting at single or multiple sites in the adipocyte life cycle associated with apoptosis, adipogenesis, and lipolysis (Milagro et al. 2013).

8.9 Concluding Remarks

A better understanding of how perinatal hormones (particularly insulin and LEP) exert their neurotrophic effects may open new avenues for understanding pre- and perinatally acquired predisposition to obesity and diabetes. Furthermore, a more detailed determination of whether hypothalamic misprogramming can be reversed, and the definition of the precise limits of the critical period for plasticity may provide new preventive and/or therapeutic opportunities.

Consequently, there are no convincing estimates of the extent to which individual differences in the risk of and obesity reflect epigenetic variation. To address this, the field is now preparing to carry out the kinds of large-scale, global studies of epigenetic marks that will provide a more comprehensive and systematic view of the contribution of epigenetics to disease pathogenesis. By analogy with the genetic equivalent, these have been termed epigenome-wide association studies. Some of these studies are targeting disease relevant tissues (such as subcutaneous fat or specific blood-cell constituents), whereas others are, for pragmatic reasons, focused on whole blood (Drong et al. 2012).

To conclude, according to the scenario depicted in Fig. 8.1, we can affirm that prevention is likely to be the only strategy for stemming the tide of the obesity epidemic (Levin 2006). Many researchers have stressed that prevention is not only possible but is the most realistic and cost-effective approach for dealing born with

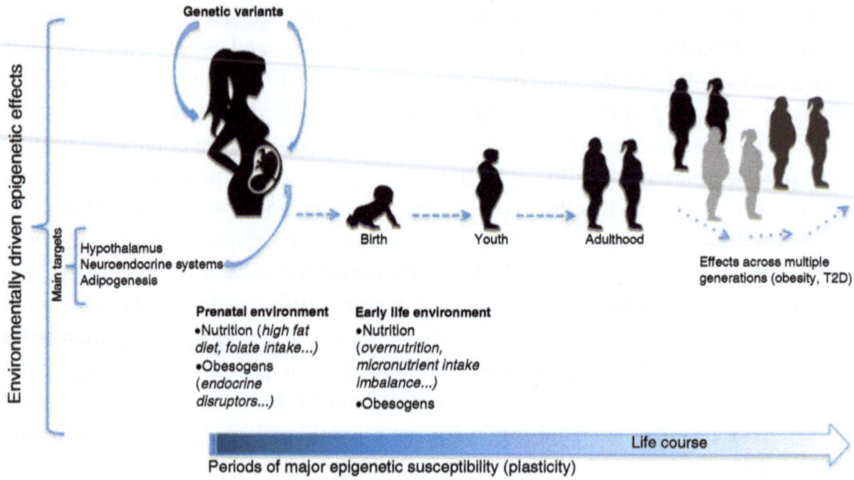

Fig. 8.1 Importance of environmentally driven epigenetic effects during life course and consequences across generations

childhood obesity and with adult obesity (Lobstein et al. 2004). At primary care level, practitioners can have a role providing information on healthy eating and physical activity to all members of the family as well as supporting effective parenting skills generally tabulated into three categories (behavioral counseling, screening, and prophylaxis) (Dietz and Gortmaker 2001). It appears to be very significant that some papers suggest that overweight prevention should begin before and during pregnancy (maternal diet, maternal pre-pregnancy obesity, maternal smoking before and during pregnancy), and in the perinatal period, with the identification and avoidance of factors that can produce permanent, adverse alterations in neural pathways controlling energy homeostasis (Salsberry and Reagan 2005). As a recent review convincingly states: "If the disease component of obesity lies not in adipose tissue itself, but in the interaction between adipose tissue biology and our modern industrialized environment, efforts to combat obesity would be much more effective if they prioritized 'external' environmental change rather than attempting to manipulate 'internal' biology through pharmaceutical or behavioral means" (Wells 2012).

References

Ahima RS (2006) Adipose tissue as an endocrine organ. Obesity (Silver Spring) 14:242S–249S

Anderson EJ, Lustig ME, Boyle KE, Woodlief TL, Kane DA, Lin CT, Price JW 3rd, Kang L, Rabinovitch PS, Szeto HH, Houmard JA, Cortright RN, Wasserman DH, Neufer PD (2009) Mitochondrial H2O2 emission and cellular redox state link excess fat intake to insulin resistance in both rodents and humans. J Clin Invest 119:573–581

Antuna-Puente B, Feve B, Fellahi S, Bastard JP (2008) Adipokines: the missing link between insulin resistance and obesity. Diabetes Metab 34:2–11

Arner E, Westermark PO, Spalding KL, Britton T, Rydén M, Frisén J, Bernard S, Arner P (2010) Adipocyte turnover: relevance to human adipose tissue morphology. Diabetes 59:105–109

Bachmann-Gagescu R, Mefford HC, Cowan C, Glew GM, Hing AV, Wallace S, Bader PI, Hamati A, Reitnauer PJ, Smith R, Stockton DW, Muhle H, Helbig I, Eichler EE, Ballif BC, Rosenfeld J, Tsuchiya KD (2010) Recurrent 200-kb deletions of 16p11.2 that include the *SH2B1* gene are associated with developmental delay and obesity. Genet Med 12:641–647

Bäckhed F, Manchester JK, Semenkovich CF, Gordon JI (2007) Mechanisms underlying the resistance to diet-induced obesity in germ-free mice. Proc Natl Acad Sci U S A 104:979–984

Bäckhed F, Ding H, Wang T, Hooper LV, Koh GY, Nagy A, Semenkovich CF, Gordon JI (2004) The gut microbiota as an environmental factor that regulates fat storage. Proc Natl Acad Sci U S A 101:15718–15723

Bartolomucci A, Parmigiani S, Rodgers RJ, Vidal-Puig A, Allan SE, Siegel V (2012) The Obese Species: a special issue on obesity and metabolic disorders. Foreword. Dis Model Mech 5:563–564

Bell AC, Ge K, Popkin BM (2001) Weight gain and its predictors in Chinese adults. Int J Obes Relat Metab Disord 25:1079–1086

Bornstein SR, Ehrhart-Bornstein M, Wong ML, Licinio J (2008) Is the worldwide epidemic of obesity a communicable feature of globalization? Exp Clin Endocrinol Diabetes 116:S30–S32

Bouret SG (2009) Early life origins of obesity: role of hypothalamic programming. J Pediatr Gastroenterol Nutr 48:S31–S38

Bramswig NC, Kaestner KH (2012) Epigenetics and diabetes treatment: an unrealized promise? Trends Endocrinol Metab 23:286–291

Bruce KD, Cagampang FR, Argenton M, Zhang J, Ethirajan PL, Burdge GC, Bateman AC, Clough GF, Poston L, Hanson MA, Mcconnell JM, Byrne CD (2009) Maternal high-fat feeding primes steatohepatitis in adult mice offspring, involving mitochondrial dysfunction and altered lipogenesis gene expression. Hepatology 50:1796–1808

Burdge GC, Slater-Jefferies J, Torrens C, Phillips ES, Hanson MA, Lillycrop KA (2007) Dietary protein restriction of pregnant rats in the F0 generation induces altered methylation of hepatic gene promoters in the adult male offspring in the F1 and F2 generations. Br J Nutr 97:435–439

Choquet H, Meyre D (2011) Molecular basis of obesity: current status and future prospects. Curr Genomics 12:154–168

Claesson MJ, Cusack S, O'sullivan O, Greene-Diniz R, De Weerd H, Flannery E, Marchesi JR, Falush D, Dinan T, Fitzgerald G, Stanton C, Van Sinderen D, O'Connor M, Harnedy N, O'connor K, Henry C, O'Mahony D, Fitzgerald AP, Shanahan F, Twomey C, Hill C, Ross RP, O'Toole PW (2011) Composition, variability, and temporal stability of the intestinal microbiota of the elderly. Proc Natl Acad Sci U S A 108:4586–4591

Collins S (2005) Overview of clinical perspectives and mechanisms of obesity. Birth Defects Res A Clin Mol Teratol 73:470–471

Copeland RA, Olhava EJ, Scott MP (2010) Targeting epigenetic enzymes for drug discovery. Curr Opin Chem Biol 14:505–510

Cordero P, Campion J, Milagro FI, Goyenechea E, Steemburgo T, Javierre BM, Martinez JA (2011) Leptin and TNF-alpha promoter methylation levels measured by MSP could predict the response to a low-calorie diet. J Physiol Biochem 67:463–470

D'Angelo CS, Koiffmann CP (2012) Copy number variants in obesity-related syndromes: review and perspectives on novel molecular approaches. J Obes 2012:845480

Das UN (2001) Is obesity an inflammatory condition? Nutrition 17:953–966

De Heredia FP, Gómez-Martínez S, Marcos A (2012) Obesity, inflammation and the immune system. Proc Nutr Soc 71:332–338

Decherf S, Demeneix BA (2011) The obesogen hypothesis: a shift of focus from the periphery to the hypothalamus. J Toxicol Environ Health B Crit Rev 14:423–448

Dethlefsen L, Mcfall-Ngai M, Relman DA (2007) An ecological and evolutionary perspective on human-microbe mutualism and disease. Nature 449:811–818

Diamanti-Kandarakis E, Bourguignon JP, Giudice LC, Hauser R, Prins GS, Soto AM, Zoeller RT, Gore AC (2009) Endocrine-disrupting chemicals: an Endocrine Society scientific statement. Endocr Rev 30:293–342
Dibaise JK, Zhang H, Crowell MD, Krajmalnik-Brown R, Decker GA, Rittmann BE (2008) Gut microbiota and its possible relationship with obesity. Mayo Clin Proc 83:460–469
Dietz WH (1998) Health consequences of obesity in youth: childhood predictors of adult disease. Pediatrics 101:518–525
Dietz WH, Gortmaker SL (2001) Preventing obesity in children and adolescents. Annu Rev Public Health 22:337–353
Dolinoy DC (2008) The agouti mouse model: an epigenetic biosensor for nutritional and environmental alterations on the fetal epigenome. Nutr Rev 66:S7–S11
Dominguez-Bello MG, Costello EK, Contreras M, Magris M, Hidalgo G, Fierer N, Knight R (2010) Delivery mode shapes the acquisition and structure of the initial microbiota across multiple body habitats in newborns. Proc Natl Acad Sci U S A 107:11971–11975
Drong AW, Lindgren CM, McCarthy MI (2012) The genetic and epigenetic basis of type 2 diabetes and obesity. Clin Pharmacol Ther 92:707–715
Fain JN (2010) Release of inflammatory mediators by human adipose tissue is enhanced in obesity and primarily by the nonfat cells: a review. Mediators Inflamm 2010:513948
Feinberg AP, Irizarry RA, Fradin D, Aryee MJ, Murakami P, Aspelund T, Eiriksdottir G, Harris TB, Launer L, Gudnason V, Fallin MD (2010) Personalized epigenomic signatures that are stable over time and covary with body mass index. Sci Transl Med 2:49ra67
Fleisch AF, Wright RO, Baccarelli AA (2012) Environmental epigenetics: a role in endocrine disease? J Mol Endocrinol 49:R61–R67
Freudenberg N (2011) The social science of obesity. Lancet 37:760
Galili O, Versari D, Sattler KJ, Olson ML, Mannheim D, McConnell JP, Chade AR, Lerman LO, Lerman A (2007) Early experimental obesity is associated with coronary endothelial dysfunction and oxidative stress. Am J Physiol Heart Circ Physiol 292:H904–H911
Ghanim H, Abuaysheh S, Sia CL, Korzeniewski K, Chaudhuri A, Fernandez-Real JM, Dandona P (2009) Increase in plasma endotoxin concentrations and the expression of toll-like receptors and suppressor of cytokine signaling-3 in mononuclear cells after a high-fat, high-carbohydrate meal: implications for insulin resistance. Diabetes Care 32:2281–2287
Godfrey KM, Sheppard A, Gluckman PD, Lillycrop K, Burdge GC, Mclean C, Rodford J, Slater-Jefferies JL, Garratt E, Crozier SR, Emerald BS, Gale CR, Inskip HM, Cooper C, Hanson MA (2011) Epigenetic gene promoter methylation at birth is associated with child's later adiposity. Diabetes 60:1528–1534
Goran MI, Ball GDC, Cruz ML (2003) Obesity and risk of type 2 diabetes and cardiovascular disease in children and adolescents. J Clin Endocrinol Metab 88:1417–1427
Greiner T, Bäckhed F (2011) Effects of the gut microbiota on obesity and glucose homeostasis. Trends Endocrinol Metab 22:117–123
Grün F, Blumberg B (2006) Environmental obesogens: organotins and endocrine disruption via nuclear receptor signaling. Endocrinology 147:S50–S55
Grün F, Blumberg B (2007) Perturbed nuclear receptor signaling by environmental obesogens as emerging factors in the obesity crisis. Rev Endocr Metab Disord 8:161–171
Grün F, Blumberg B (2009a) Minireview: the case for obesogens. Mol Endocrinol 23:1127–1134
Grün F, Blumberg B (2009b) Endocrine disrupters as obesogens. Mol Cell Endocrinol 304:19–29
Heijmans BT, Tobi EW, Stein AD, Putter H, Blauw GJ, Susser ES, Slagboom PE, Lumey LH (2008) Persistent epigenetic differences associated with prenatal exposure to famine in humans. Proc Natl Acad Sci U S A 105:17046–17049
Herskind AM, Mcgue M, Iachine IA, Holm N, Sørensen TI, Harvald B, Vaupel JW (1996) Untangling genetic influences on smoking, body mass index and longevity: a multivariate study of 2464 Danish twins followed for 28 years. Hum Genet 98:467–475
Hill JO, Peters JC (1998) Environmental contributions to the obesity epidemic. Science 280:1371–1374
Hill JO, Wyatt HR, Reed GW, Peters JC (2003) Obesity and the environment: where do we go from here? Science 299:853–855

Hirosumi J, Tuncman G, Chang L, Görgün CZ, Uysal KT, Maeda K, Karin M, Hotamisligil GS (2002) A central role for JNK in obesity and insulin resistance. Nature 420:333–336
Hu FB (2011) Globalization of diabetes: the role of diet, lifestyle, and genes. Diabetes Care 34:1249–1257
Irigaray P, Newby JA, Lacomme S, Belpomme D (2007) Overweight/obesity and cancer genesis: more than a biological link. Biomed Pharmacother 61:665–678
Janesick A, Blumberg B (2011) Endocrine disrupting chemicals and the developmental programming of adipogenesis and obesity. Birth Defects Res C Embryo Today 93:34–50
Kirchner S, Kieu T, Chow C, Casey S, Blumberg B (2010) Prenatal exposure to the environmental obesogen tributyltin predisposes multipotent stem cells to become adipocytes. Mol Endocrinol 24:526–539
Klöting N, Fasshauer M, Dietrich A, Kovacs P, Schön MR, Kern M, Stumvoll M, Blüher M (2010) Insulin-sensitive obesity. Am J Physiol Endocrinol Metab 299:E506–E515
Köbberling J, Tillil H (1982) Empirical risk figures for first degree relatives of non-insulin dependent diabetics. In: Köbberling J, Tattersall R (eds) The genetics of diabetes mellitus. Academic, London
Lafontan M, Berlan M (2003) Do regional differences in adipocyte biology provide new pathophysiological insights? Trends Pharmacol Sci 24:276–283
Laudes M (2011) Role of WNT signalling in the determination of human mesenchymal stem cells into preadipocytes. J Mol Endocrinol 46:R65–R72
Levin BE (2006) Metabolic imprinting: critical impact of the perinatal environment on the regulation of energy homeostasis. Philos Trans R Soc Lond B Biol Sci 361:1107–1121
Levin BE (2009) Synergy of nature and nurture in the development of childhood obesity. Int J Obes (Lond) 33:S53–S56
Ley RE, Turnbaugh PJ, Klein S, Gordon JI (2006) Microbial ecology: human gut microbes associated with obesity. Nature 444:1022–1023
Libman IM, Pietropaolo M, Arslanian SA, Laporte RE, Becker DL (2003) Changing prevalence of overweight children and adolescents at onset of insulin-treated diabetes. Diabetes Care 26:2871–2875
Lillycrop KA (2011) Effect of maternal diet on the epigenome: implications for human metabolic disease. Proc Nutr Soc 70:64–72
Lobstein T, Bauer L, Uauy R (2004) Obesity in children and young people: a crisis in public health. Obes Rev 5:4–104
Lopez KN, Knudson JD (2012) Obesity: from the agricultural revolution to the contemporary pediatric epidemic. Congenit Heart Dis 7:189–199
Manikkam M, Tracey R, Guerrero-Bosagna C, Skinner MK (2013) Plastics derived endocrine disruptors (BPA, DEHP and DBP) induce epigenetic transgenerational inheritance of obesity, reproductive disease and sperm epimutations. PLoS One 8(1):e55387
Martínez JA, Cordero P, Campión J, Milagro FI (2012) Interplay of early-life nutritional programming on obesity, inflammation and epigenetic outcomes. Proc Nutr Soc 71(2):276–283
Mcallister EJ, Dhurandhar NV, Keith SW, Aronne LJ, Barger J, Baskin M, Benca RM, Biggio J, Boggiano MM, Eisenmann JC, Elobeid M, Fontaine KR, Gluckman P, Hanlon EC, Katzmarzyk P, Pietrobelli A, Redden DT, Ruden DM, Wang C, Waterland RA, Wright SM, Allison DB (2009) Ten putative contributors to the obesity epidemic. Crit Rev Food Sci Nutr 49:868–913
Milagro FI, Mansego ML, De Miguel C, Martínez JA (2013) Dietary factors, epigenetic modifications and obesity outcomes: progresses and perspectives. Mol Aspects Med 34(4):782–812
Mokdad AH, Ford ES, Bowman BA, Dietz WH, Vinicor F, Bales VS, Marks JS (2003) Prevalence of obesity, diabetes, and obesity-related health risk factors, 2001. JAMA 289:76–79
Musso G, Gambino R, Cassader M (2010) Obesity, diabetes, and gut microbiota: the hygiene hypothesis expanded? Diabetes Care 33:2277–2284
Neel BA, Sargis RM (2011) The paradox of progress: environmental disruption of metabolism and the diabetes epidemic. Diabetes 60:1838–1848
Newbold RR, Padilla-Banks E, Jefferson WN (2009) Environmental estrogens and obesity. Mol Cell Endocrinol 304:84–89

Newbold RR, Padilla-Banks E, Jefferson WN, Heindel JJ (2008) Effects of endocrine disruptors on obesity. Int J Androl 31:201–208
Palmer C, Bik EM, Digiulio DB, Relman DA, Brown PO (2007) Development of the human infant intestinal microbiota. PLoS Biol 5:e177
Permutt MA, Wasson J, Cox N (2005) Genetic epidemiology of diabetes. J Clin Invest 115:1431–1439
Plagemann A (2008) A matter of insulin: developmental programming of body weight regulation. J Matern Fetal Neonatal Med 21:143–148
Plagemann A, Harder T, Brunn M, Harder A, Roepke K, Wittrock-Staar M, Ziska T, Schellong K, Rodekamp E, Melchior K, Dudenhausen JW (2009) Hypothalamic proopiomelanocortin promoter methylation becomes altered by early overfeeding: an epigenetic model of obesity and the metabolic syndrome. J Physiol 587:4963–4976
Poulsen P, Kyvik KO, Vaag A, Beck-Nielsen H (1999) Heritability of type II (non-insulin-dependent) diabetes mellitus and abnormal glucose tolerance—a population-based twin study. Diabetologia 42:139–145
Power C, Jefferis BJ (2002) Fetal environment and subsequent obesity: a study of maternal smoking. Int J Epidemiol 31:413–419
Prior LJ, Armitage JA (2009) Neonatal overfeeding leads to developmental programming of adult obesity: you are what you ate. J Physiol 587:2419
Rice T, Pérusse L, Bouchard C, Rao DC (1999) Familial aggregation of body mass index and subcutaneous fat measures in the longitudinal Québec family study. Genet Epidemiol 16:316–334
Rubenstein AH (2005) Obesity: a modern epidemic. Trans Am Clin Climatol Assoc 116:103–113
Salsberry PJ, Reagan PB (2005) Dynamics of early childhood overweight. Pediatrics 116:1329–1338
Sapienza C, Lee J, Powell J, Erinle O, Yafai F, Reichert J, Siraj ES, Madaio M (2011) DNA methylation profiling identifies epigenetic differences between diabetes patients with ESRD and diabetes patients without nephropathy. Epigenetics 6:20–28
Schmid PM, Heid I, Buechler C, Steege A, Resch M, Birner C, Endemann DH, Riegger GA, Luchner A (2012) Expression of fourteen novel obesity-related genes in Zucker diabetic fatty rats. Cardiovasc Diabetol 11:48
Schwimmer JB, Burwinkle TM, Varni JW (2003) Health-related quality of life severely obese children and adolescents. JAMA 289:1813–1819
Sinha R, Fisch G, Teague B, Tamborlane WV, Banyas B, Allen K, Savoye M, Rieger V, Taksali S, Barbetta G, Sherwin RS, Caprio S (2002) Prevalence of impaired glucose tolerance among children and adolescents with marked obesity. N Engl J Med 346:802–810
Somm E, Schwitzgebel VM, Toulotte A, Cederroth CR, Combescure C, Nef S, Aubert ML, Hüppi PS (2009) Perinatal exposure to bisphenol a alters early adipogenesis in the rat. Environ Health Perspect 117:1549–1555
Spalding KL, Arner E, Westermark PO, Bernard S, Buchholz BA, Bergmann O, Blomqvist L, Hoffstedt J, Näslund E, Britton T, Concha H, Hassan M, Rydén M, Frisén J, Arner P (2008) Dynamics of fat cell turnover in humans. Nature 453:783–787
Speakman JR, O'Rahilly S (2012) Fat: an evolving issue. Dis Model Mech 5:569–573
St-Onge MP, Keller KL, Heymsfield SB (2003) Changes in childhood food consumption patterns: a cause for concern in light of increasing body weights. Am J Clin Nutr 78:1068–1073
Stubbs CO, Lee AJ (2004) The obesity epidemic: both energy intake and physical activity contribute. Med J Aust 181:489–491
Stunkard AJ, Foch TT, Hrubec Z (1986a) A twin study of human obesity. JAMA 256:51–54
Stunkard AJ, Sørensen TI, Hanis C, Teasdale TW, Chakraborty R, Schull WJ, Schulsinger F (1986b) An adoption study of human obesity. N Engl J Med 314:193–198
Swinburn B, Egger G, Raza F (1999) Dissecting obesogenic environments: the development and application of a framework foridentifying and prioritizing environmental interventions for obesity. Prev Med 29:563–570
Tobi EW, Lumey LH, Talens RP, Kremer D, Putter H, Stein AD, Slagboom PE, Heijmans BT (2009) DNA methylation differences after exposure to prenatal famine are common and timing- and sex-specific. Hum Mol Genet 18:4046–4053

Toperoff G, Aran D, Kark JD, Rosenberg M, Dubnikov T, Nissan B, Wainstein J, Friedlander Y, Levy-Lahad E, Glaser B, Hellman A (2009) Genome-wide survey reveals predisposing diabetes type 2-related DNA methylation variations in human peripheral blood. Hum Mol Genet 21:371–383

Tracey R, Manikkam M, Guerrero-Bosagna C, Skinner MK (2013) Hydrocarbons (jet fuel JP-8) induce epigenetic transgenerational inheritance of obesity, reproductive disease and sperm epimutations. Reprod Toxicol 36:104–116

Tsai F, Coyle WJ (2009) The microbiome and obesity: is obesity linked to our gut flora? Curr Gastroenterol Rep 11:307–313

Vandenberg LN, Colborn T, Hayes TB, Heindel JJ, Jacobs DR Jr, Lee DH, Shioda T, Soto AM, Vom Saal FS, Welshons WV, Zoeller RT, Myers JP (2012) Hormones and endocrine-disrupting chemicals: low-dose effects and nonmonotonic dose responses. Endocr Rev 33:378–455

Vimaleswaran KS, Loos RJ (2010) Progress in the genetics of common obesity and type 2 diabetes. Expert Rev Mol Med 12:e7

Wang G, Dietz WH (2002) Economic burden of obesity in youths aged 6 to 17 years: 1979–1999. Pediatrics 109:E81-1

Waterland RA, Jirtle RL (2003) Transposable elements: targets for early nutritional effects on epigenetic gene regulation. Mol Cell Biol 23:5293–5300

Wells JC (2012) The evolution of human adiposity and obesity: where did it all go wrong? Dis Model Mech 5:595–607

Whitaker RC, Wright JA, Pepe MS, Seidel KD, Dietz WH (1997) Predicting obesity in young adulthood from childhood and parental obesity. N Engl J Med 337:869–873

Whitman WB, Coleman DC, Wiebe WJ (1998) Prokaryotes: the unseen majority. Proc Natl Acad Sci U S A 95:6578–6583

Widschwendter M, Apostolidou S, Raum E, Rothenbacher D, Fiegl H, Menon U, Stegmaier C, Jacobs IJ, Brenner H (2008) Epigenotyping in peripheral blood cell DNA and breast cancer risk: a proof of principle study. PLoS One 3(7):e2656

Xu J, Gordon JI (2003) Inaugural article: honor thy symbionts. Proc Natl Acad Sci U S A 100:10452–10459

Yang W, Lu J, Weng J, Jia W, Ji L, Xiao J, Shan Z, Liu J, Tian H, Ji Q, Zhu D, Ge J, Lin L, Chen L, Guo X, Zhao Z, Li Q, Zhou Z, Shan G, He J (2010) China National Diabetes and Metabolic Disorders Study Group. Prevalence of diabetes among men and women in China. N Engl J Med 362:1090–1110

Zhang J, Zhang F, Didelot X, Bruce KD, Cagampang FR, Vatish M, Hanson M, Lehnert H, Ceriello A, Byrne CD (2009) Maternal high fat diet during pregnancy and lactation alters hepatic expression of insulin like growth factor-2 and key microRNAs in the adult offspring. BMC Genomics 10:478

Zhang Y, Proenca R, Maffei M, Barone M, Leopold L, Friedman JM (1994) Positional cloning of the mouse obese gene and its human homologue. Nature 372:425–432

Zhao J, Goldberg J, Vaccarino V (2013) Promoter methylation of serotonin transporter gene is associated with obesity measures: a monozygotic twin study. Int J Obes (Lond) 37(1):140–145

Chapter 9
Molecular Mechanisms in the Development and Progression of Asthma: The Role of Epigenetic Regulation and the Airway Epithelium

Stephanie Tortorella, Simon G. Royce, and Tom C. Karagiannis

Abstract Asthma is increasingly recognised as a heterogeneous disease, with multiple phenotypes that differ in severity, pathology, therapeutic response and long-term outcome. A combination of genetic, epigenetic and environmental factors are thought to contribute to the molecular diversity of the disease, with sensitisation (T-cell differentiation) dependent on the local microenvironment and the nature of the invading pathogen. A strong body of evidence exists associating numerous environmental and genetic components in asthma development, with multiple asthma genes involved independently (through the inheritance of polymorphisms) or through the interaction with the environment to increase risk. However, the inability to reproduce inheritance patterns and the dramatic increase in incidence over the last decade provides strong evidence that changes in the environment have activated a pre-existing susceptibility, including the alteration in epigenetic regulation, to play an important role in disease. The role of epigenetic regulation and modulation in the

---

Simon G. Royce and Tom C. Karagiannis contributed equally with all other contributors.

S. Tortorella (✉)
Epigenomic Medicine, Baker IDI Heart and Diabetes Institute, The Alfred Medical Research and Education Precinct, 75 Commercial Road, Melbourne, VIC, Australia
e-mail: stephanie.tortorella@bakeridi.edu.au

S.G. Royce
Epigenomic Medicine, Baker IDI Heart and Diabetes Institute, The Alfred Medical Research and Education Precinct, 75 Commercial Road, Melbourne, VIC, Australia

Department of Pathology, The University of Melbourne, Parkville, VIC, Australia

Allergy and Immune Disorders, Murdoch Children's Research Institute, Parkville, VIC, Australia
e-mail: simon.royce@mcri.edu.au

T.C. Karagiannis
Epigenomic Medicine, Baker IDI Heart and Diabetes Institute, The Alfred Medical Research and Education Precinct, 75 Commercial Road, Melbourne, VIC, Australia

Department of Pathology, The University of Melbourne, Parkville, VIC, Australia
e-mail: tom.karagiannis@bakeridi.edu.au

N. Maulik and T. Karagiannis (eds.), *Molecular Mechanisms and Physiology of Disease: Implications for Epigenetics and Health*, DOI 10.1007/978-1-4939-0706-9_9,

development of asthma and allergy has been widely speculated. Interestingly, factors known to be involved in disease susceptibility including genetic predisposition and exposure to environmental stimuli (in utero and post-natal) have been explored as factors involved in the mechanisms associated with the epigenome. Thus, it is proposed that modification of the epigenome in the regulation of important pathways, including those involved in asthma-associated gene expression and T-cell differentiation play a direct role in disease. In addition, current research focuses on the central role of the airway epithelium in asthma development and progression. Inherently defective in disease, the mechanisms associated with epithelial dysfunction, including the increased susceptibility to injury and the inability to activate normal repair processes are yet to be completely elucidated. Trefoil factor 2 (TFF2), previously shown to be upregulated in asthma and involved in airway epithelial restitution fails to protect the epithelium from pathogen-induced injury. By focusing on the role of epigenetic mechanisms, the epithelium and TFF2 in asthma pathogenesis, this chapter highlights their potential as targets in future therapeutic research.

Keywords Asthma • Chromatin modifications • Trefoil factor • Airway hyperresponsiveness • Epigenetics

9.1 Introduction

The prevalence, complexity and severity of allergic disorders, including asthma continues to increase on a global scale (Pawankar et al. 2012). According to the World Health Organisation, 300 million people worldwide are afflicted with asthma, a disease that remains a heavy burden on health-care systems as these numbers steadily rise (Weinberg 2011). The dramatic increase in incidence over the last 20 years provides strong evidence that changes in the environment (in utero and post-natal pathogen exposure) have activated a pre-existing susceptibility, including the alteration in epigenetic regulation, to play an important role in disease development (Yang and Schwartz; Yuyama et al. 2002). In addition, the airway epithelium has proven to be central in its progression and pathogenesis, providing a link between the development of the disease and the clinically significant events characteristic of human asthma (Hackett and Knight 2007). By focusing on trefoil factor 2 (TFF2), a protein shown to be upregulated in asthma and involved in airway epithelial restitution, from an endogenous mechanistic aspect highlights the requirement to elucidate the role of the epithelium in asthma, and provides information into targeting the epithelium in future therapeutic research.

9.2 Asthma, Airway Remodelling and Disease Phenotypes

To date, asthma is defined as an inflammatory respiratory disease, characterised by sudden, chronic symptoms of wheezing, sputum production, variable reversible airflow limitation and airway hyperresponsiveness (AHR) (Bousquet et al. 2000).

As a highly heterogeneous disease of the respiratory tract, current management strategies are inadequate in their ability to prevent the development of the disease. Anti-inflammatory corticosteroids are the most effective and therefore, most widely prescribed therapeutic agent in response to active inflammation during exacerbations (Murata and Ling). However, a subset of patients with chronic, severe disease may be classified as corticosteroid resistant, as clinical symptoms persist despite high dosage levels (Barnes and Adcock 2009).

Molecular heterogeneity between human asthma patients, including differences observed in the immune response following allergen exposure, is thought to contribute to corticosteroid-resistance and the inability to establish the exact mechanisms involved in disease development and progression (Woodruff et al. 2009). Allergic asthma is characterised by the production of Th_2 cells and its associated cytokines in response to allergen exposure in atopic individuals, contributing to 75–80 % of all asthmatic patients (Holgate 2008) (Fig. 9.1). Interestingly however, it has been shown that over 40 % of the Western population is classified as atopic with only approximately 7 % of those expressing atopy in the form of asthma (Beasley et al. 1989). Recent developments also suggest that a significant proportion of severe, allergic asthma cases are driven by alternative inflammatory pathways (Prescott 2006), including that driven by Th_1 and Th_{17} cells (Fig. 9.1). Although, atopic asthma is managed well with corticosteroids generally, it has been shown that cases mediated by Th_1- and Th_{17}-immune pathways are largely corticosteroid resistant, with persistence of clinically relevant symptoms including AHR (Robins et al.; Yang et al. 2009; McKinley et al. 2008; Cui et al. 2005). These pathways appear to derive from a common naive precursor cell, whose differentiation pathway is determined by cytokine, environmental, genetic and epigenetic signals during primary antigenic stimulation and sensitisation (Bluestone et al. 1995; Abbas et al. 1996; O'Garra et al. 1998). Collectively, the inherent immune and phenotypic heterogeneity observed in human asthma highlights the importance of understanding the mechanisms involved in the development of disease, with rationale centred on establishing a common factor in disease pathogenesis for future, target-based therapeutic research. In this chapter, two pathways will be explored in detail—the role of epigenetic mechanisms in the development of asthma, and the involvement of the airway epithelium in disease progression.

A number of pathologically significant events are also thought to contribute to the limitations in current asthma treatment, with the inability of corticosteroids to reverse and/or prevent airway remodelling observed in both childhood and adult cases of chronic, severe disease (Holgate et al.). The structural alterations associated include epithelial goblet cell metaplasia, collagen deposition and thickening of the subepithelial lamina reticularis with increased matrix deposition, smooth muscle hyperplasia and hypertrophy, and angiogenesis (Vignola et al. 2000). The exact mechanisms that contribute to the process of airway remodelling in asthma are yet to be elucidated, with current knowledge indicating an interrelationship with inflammation and AHR. Until recently, these structural alterations have been considered to be a secondary phenomenon, developing late in disease progression as a direct consequence of persistent inflammation (Fedorov et al. 2005). Although this is observed

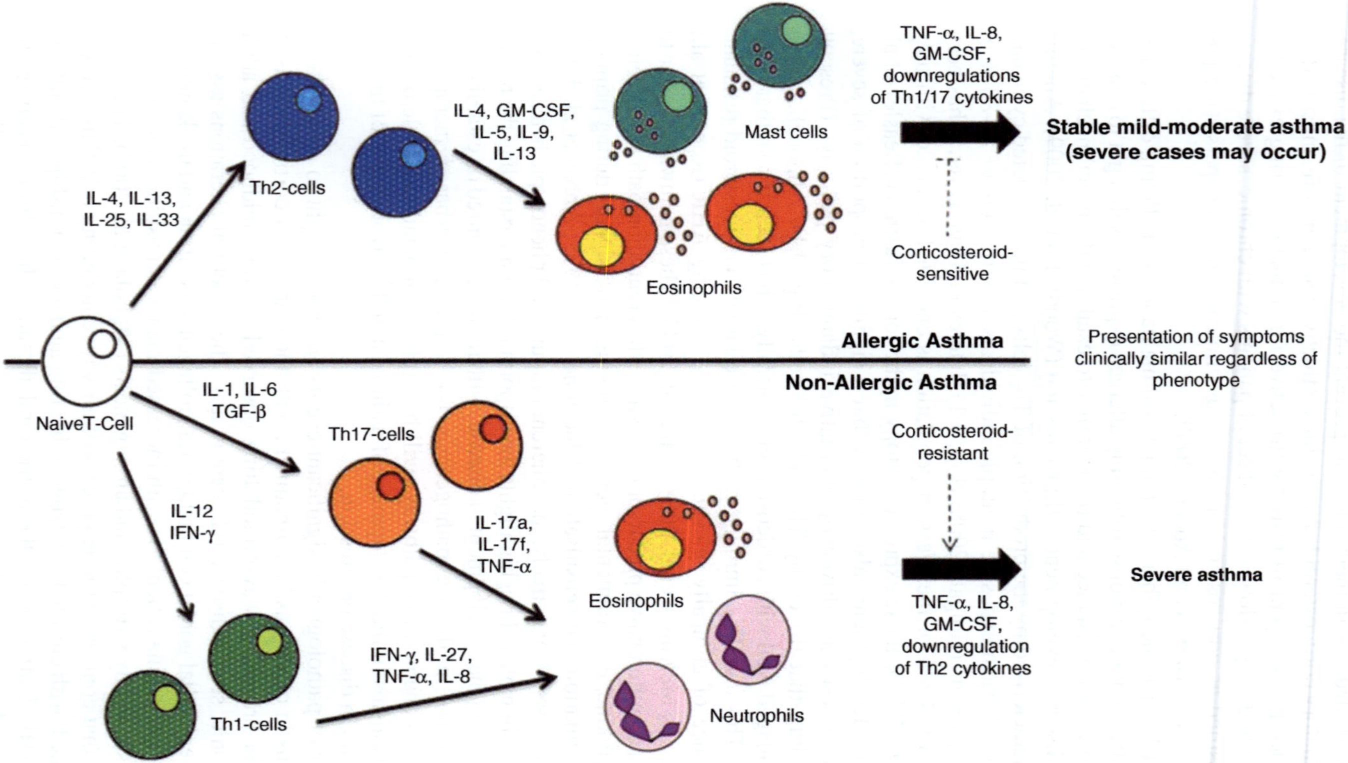

Fig. 9.1 The phenotypic subtypes of asthma. During early sensitisation, caused by a combination of epigenetic, genetic and environmental factors, T-cells undergo differentiation dependent on the local microenvironment and the nature of the invading pathogen. Exposure to allergens including house dust mite and pollens direct naive T-cells to differentiate along a Th_2 pathway, with individuals developing atopy. This type of asthma which involves the recruitment of mast cells and eosinophils is most common, and also the most likely to be controlled with current corticosteroid treatment. Conversely, individuals' exposure to pollutants and irritants usually in utero and post-natal development cause naive T-cells to differentiate into Th_1 of Th_{17} cells. Recruitment of neutrophils in this immune pathway causes a non-allergic asthma phenotype, with most cases persisting into adulthood, and more severe with corticosteroid-resistance

in a subset of patients with late-onset disease, airway remodelling has also been shown to be a consistent feature of childhood asthma independent of inflammation, with no requirement for concurrent eosinophil infiltration of the airway tissue (Payne et al. 2003; Prescott 2006). In order to explain this observation, it is proposed that the alterations in airway structure are, at least in part, regulated through the ability of epithelial cells to communicate with the underlying mesenchyme to maintain and propagate remodelling and inflammatory responses throughout the airway wall (Davies 2009; Hackett 2012).

9.3 Epigenetic Mechanisms and the Development of Asthma

There is a strong body of evidence associating numerous environmental and genetic components in asthma development. Changes in the primary DNA sequence of genes involved in both immune and inflammatory pathways are either independently associated with asthma, or interact with the environment to affect disease risk. However, due to the irreproducibility of inheritance patterns and the dramatic increase in asthma incidence in a relatively short period of time, the ability of these factors to affect and be affected by epigenetic mechanisms have brought about an alternative explanation for disease development (Fig. 9.2).

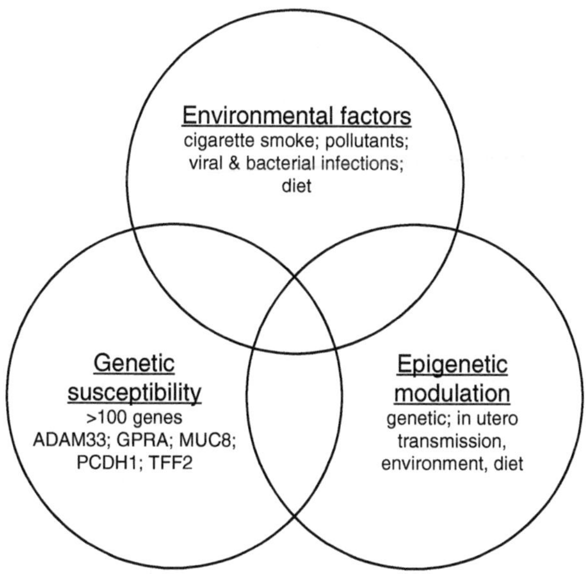

Fig. 9.2 Venn diagram illustrating the relationship between environmental exposures, epigenetic modulation and genetic susceptibility in the development of asthma. These factors are interrelated, with many of the environmental factors displayed also affecting the epigenome. In addition, gene expression including those genes known to be involved in asthma susceptibility, including ADAM33 is subject to epigenetic modulation. A combination of these factors is thought to cause disease development, with the exact contribution of each factor yet to be elucidated

9.3.1 Genetic Basis of Asthma: Heritability and Susceptibility

A complex heritable disease, asthma has a strong familial connection (36–79 % heritability) with a non-Mendelian pattern of inheritance and polymorphisms in more than 100 genes (Ober and Yao 2011; Vercelli 2008). The genes involved in asthma susceptibility may be categorised into four main groups: (1) genes associated with immunoregulation, (2) genes associated with Th-cell differentiation, (3) genes associated with epithelial structure and function and (4) genes associated with lung function, remodelling and disease severity (Vercelli 2008). In addition to these genes, positional cloning studies have found other gene polymorphisms involved in asthma development and susceptibility including ADAM33 and TFF2 (Zhang et al. 2012; Allen et al. 2003). Interestingly, asthma shows a parent-of-origin transmission of inheritance, with an affected mother significantly more likely to transmit the disease than an affected father (Demenais et al. 2001). Similar to the mode of inheritance observed for multiple epigenetic mechanisms, a number of known genes associated with asthma are transmitted in a parent-of-origin fashion, including the *FCERIB* locus (Sandford et al. 2000) and the *Spink5*gene (Liu et al. 2009). Yet the inability to replicate these associations between differing studies, the diverse nature of genes subject to asthma- and allergy-related polymorphisms and the finding that genetic susceptibility may explain the cause of disease in only a subset of patients, implicate other mechanisms, such as epigenetics, in the development and progression of disease.

In order to appreciate the potential importance of epigenetic modulation in the development of asthma, numerous studies have established that DNA methylation is involved in the differential expression of disease-associated genes *ADAM33*, *ALOX12* and *PTGDR* (Yang et al. 2008; Morales et al. 2012; Isidoro-García et al. 2011). Polymorphisms of *ADAM33*, found by means of positional cloning, are strongly associated with asthma and AHR (Van Eerdewegh et al. 2002). Through the analysis of bronchial biopsy samples from human subjects, it was found that the *ADAM33* gene contains a regulatory CpG island within its promoter (Yang et al. 2008). Bisulfite sequencing demonstrated that this region is subject to epigenetic (methylation) modulation, with the region hypomethylated in *ADAM33*-expressing fibroblasts and hypermethylated in epithelial cell which fail to express the gene. Recent advances in epigenetic research in asthma have also linked methylation modulation of *ALOX12* gene expression in children with or without persistent wheeze (Morales et al. 2012). Furthermore, interplay between genetic polymorphisms and methylation levels were observed, with genetic variation influencing *ALOX12* methylation. In another study, expression of the asthma susceptibility gene *PTGDR* is shown to be regulated by epigenetic mechanisms, with allergic asthma patients exhibiting a distinct methylation pattern (demethylation of promoter) as compared to controls (Isidoro-García et al. 2011). Assessment of the important asthma susceptibility locus 17q21 also presents the interrelationship between genetic polymorphisms and epigenetic regulation (Berlivet et al. 2012). Using lymphoblastoid cell lines, this study showed that the locus harbours three genes—*ZPBP2*, *GSDMB* and *ORMDL3* which exhibit allele-specific differences in gene expression levels determined by distinct polymorphism and DNA methylation profiles in vitro. Although the exact role of *ADAM33*, *ALOX12*, *PTGDR* and locus 17q21 in asthma development is yet to be completely elucidated, the link between their expression regulated by DNA methylation, and their involvement in disease susceptibility provides rationale to further research the epigenetic mechanisms involved in asthma pathogenesis.

9.3.2 Gene–Environment Interactions in Asthma Development and the Role of Epigenetic Regulation

Exposure to various environmental pollutants and pathogens in utero and in postnatal (early childhood) development is thought to be a central factor in asthma pathogenesis. Besides studies involving the complex interaction between genetic susceptibility and the environment, many of the factors associated with asthma development including exposure to cigarette smoke, air pollutants and microbial allergens have been shown to be involved in epigenetic modulation.

As the most important risk factor for asthma development in children, and its known ability to alter epigenetic mechanisms, extensive research can be found involving the exposure to cigarette smoke and its role in disease pathogenesis (Lodrup Carlsen and Carlsen 2001). Direct inhalation of cigarette smoke components, including 7,12-dimethylbenz{a}anthracene (DMBA) was found to alter DNA methylation patterns in mouse lung (Phillips and Goodman 2009). Several studies have also shown that in utero cigarette smoke exposure affects DNA methylation in multiple tissues, including the placenta which resulted in CpG methylation and expression of numerous genes involved in the oxidative stress pathway in smokers, compared to levels measured in placenta from non-smokers (Suter et al. 2011). In addition to this, an association between in utero exposure to cigarette smoke and the development of asthma has been extensively researched (Xepapadaki et al. 2009; Lee et al. 2012). Genes involved in airway inflammation, such as *AXL* and *PTPRO* were found to be methylated in response to in utero cigarette smoke exposure in buccal DNA of the offspring (Breton et al. 2009). Furthermore, studies have demonstrated histone modification associations between environmental exposure and asthma risk. A significant reduction of HDAC2 expression in the airways of rats and different strains of mice following cigarette smoke exposure produced similar results to HDAC2 expression levels observed relatively in asthmatic smokers (Yang et al. 2006; Cosio et al. 2004). Adenuga et al., found that exposure to cigarette smoke to mice caused a decrease in HDAC1 and HDAC2 activity in the lung through the ability of cigarette components to induce phosphorylation and serine/threonine proteasomal degradation (Adenuga et al. 2009). Collectively, the link between epigenetic modulation with cigarette smoke exposure and asthma risk is displayed, with many of the epigenetic mechanisms observed in smokers, also observed in asthma patients.

Air pollution may also impact asthma development via pollutant-induced and pollutant-regulated epigenetic modulation. Research demonstrates that in utero exposure to airborne polycyclic aromatic hydrocarbons (PAH) correlates to asthma in offspring, caused by DNA methylation of CpG regions in the *ACSL3* promoter (Perera et al. 2009). In addition, 3-year PAH exposure was associated with higher methylation of *FoxP3*, reduced Treg function and increased asthma severity in children with asthma (Nadeau et al. 2010). The association between in utero exposure to dichlorodiphenyldichloroethylene (DDE) and asthma in children (Sunyer et al. 2005) has been established, with levels of DDE in cord blood within the Menorca cohort inversely correlated with DNA methylation in one of the *ALOX12* (previously described) CpG regions (Morales et al. 2012). Furthermore, diesel exhaust particulate (DEP) matter has been reported to induce airway inflammation and exacerbate asthma in vivo (Cao et al. 2007). This same study also indicates that DEP exposure induces COX-2 gene expression through the acetylation of histone H4, the degradation of HDAC1 and the recruitment of histone acetyltransferase (HAT) p300 in association with the *COX2* promoter. Children exposed to particulate air pollution display reduced methylation in a CpG site of the *NOS2* gene promoter region (Salam et al. 2012). Involved in nitric oxide synthesis, promoter methylation in *NOS2* effects exhaled nitric oxide (FeNO) with higher methylation in *ARG1* and *ARG2* also associated with lower FeNO, and a strong correlation produced in children with asthma (Breton et al. 2011). Recent studies have also demonstrated that lower methylation of the *NOS2* and *IL-6* gene promoter in nasal epithelium is associated with higher FeNO in children with asthma (Baccarelli et al. 2012).

Early life bacterial and viral infections have thought to contribute to asthma development, with subsequent infections resulting in disease exacerbations following initial sensitization and exposure. In a murine *Aspergillus fumigatus*-sensitised murine model, exposure to *A. fumigatus* in conjunction with DEP resulted in methylation of several CpG sites at the *Ifng* promoter and demethylation at one IL-4 promoter with methylation modulation correlating significantly with changes in IgE expression and production (Liu et al. 2008). Stimulation of human CD4+ T-cells derived from asthma patients with the allergen components of house dust mite led to demethylation in several CpG sites in the IL-4 promoter and increased expression levels of IL-4 in supernatant when compared to those control non-atopic subjects (Kwon et al. 2008). Stimulation of multiple T-cell and monocytic cell lines with lipopolysaccharide enhanced IL-8 release through toll-like receptor-4, which induces histone H3 and H4 acetylation at the *IL-8* promoter (Tsaprouni et al. 2007).

In addition to the research involved in establishing the role of epigenetic modulation in enhancing asthma susceptibility, environmental factors including the exposure to specific bacterial and viral infections, may protect from disease development. Exposure to farm life in early development (in utero and post-natal) significantly reduces the risk of asthma and allergy (Schaub et al. 2009; Ege et al. 2011). An abundance of bacteria and their components, including endotoxin is found in such an environment, with the risk of asthma development decreasing with exposure to high concentrations of endotoxin, and the microbial diversity of the environment (Ege et al. 2011). Furthermore, cord blood from neonates subjected to an in utero and post-natal rural environment possessed increased numbers and activity of Treg cells, *FoxP3* demethylation and increased *FoxP3* gene expression when compared to levels observed in non-rural neonates (Schaub et al. 2009). Another study found that when pregnant mice were exposed to farm-derived gram-negative bacterium *Acinetobacter lwoffii* F78, and subsequently challenged offspring with ovalbumin (OVA) to induce an allergic airways disease phenotype, supernatant from splenic mononuclear cells had significantly increased levels of IFN-γ and reduction of IL-4, IL-5 and IL-13 expression (Brand et al. 2011). In utero exposure to protective bacteria was also demonstrated to be mediated through the modifications of histone H4 at the *Ifng* promoter in mice, with reduced acetylation of H4 in those offspring unexposed and prevention of this reduced acetylation in those exposed. Inhibition of acetylation in H4 was found to increase airway inflammation and AHR in OVA-challenged mice through its ability to regulate IFN-γ production.

9.3.3 In Utero

Exposures including those previously mentioned have been known to modulate the epigenome and enhance susceptibility to asthma development in offspring. Further investigation of these exposures, including that of the maternal diet has been shown to be risk factors in disease pathogenesis. Hollingsworth et al. displayed in a murine model the importance of dietary methyl donors in disease, with exposure to these donors associated with increased allergic airway inflammation (Hollingsworth et al. 2008). The same study aimed to explain such a phenomenon, with airway inflammation thought to be mediated by the increased methylation of *RUNX3* gene. Given the finding that *RUNX3*-deficient mice possess elevated levels of serum IgE and increased AHR (Fainaru et al. 2005), inhibition of *RUNX3* gene expression through DNA methylation may be associated with asthma development. Further research is required to fully elucidate the role of this gene in disease.

9.3.4 Epigenetic Regulation and Modulation of the Immune System in Allergy and Asthma

As an immune-mediated disease, asthma is characterised by the ability of T-cells to differentiate mainly towards a Th_2-phenotype (observed in allergic asthma), although other subtypes including Th_1 and Th_{17} may be involved in non-allergic cases (Lloyd and Hessel 2010). As discussed previously, these cases are thought to be more severe, non-resoluting and resistant to corticosteroid treatment. Important pathways in the immune response, such as T-cell differentiation and Treg function have been shown to be regulated by epigenetic mechanisms (Jones and Chen 2006; White et al. 2006). Multiple studies have shown that differentiation of naive T-cells into mature T cells is accompanied by changes in both methylation status and chromatin structure in cytokine genes dependent on the subtype (Table 9.1)

Table 9.1 Epigenetic regulation of T-cell differentiation in allergy and asthma

Pathway	Epigenetic finding(s)	Reference
Th_1	Progressive demethylation of *Ifng* promoter	White et al. (2006)
	Methylation of intergenic region between IL-4 and IL-13	Lee et al. (2002)
	Acetylation of H3 and H4 in *Ifng* locus	Fields et al. (2002)
	Binding of Tbet to *Ifng* promoter induces promoter activity (acetylation) and dislocates HDAC complexes (enables HATs to bind)	Chang et al. (2008)
Th_2	Appearance of DNase I hypersensitive sites and demethylation around sites within IL-4 and IL-13 promoters	Santangelo et al. (2002)
	Increased methylation of *Ifng* promoter	Jones and Chen (2006)
	Demethylation of RAD50-hypersensitive site 7 within Th_2 cytokine LCR	Kim et al. (2007)
	Acetylation of H3 and H4 in IL-13/IL-4 region	Avni et al. (2002)
	GATA3 enhances acetylation in IL-4 locus	Fields et al. (2002)
	Chromatin remodelling of Th_2 cytokine LCR important in airway inflammation and AHR	Koh et al. (2010)
	Binding of HDAC1 to IL-4 gene locus; HDAC1 important in airway inflammation	Grausenburger et al. (2010)
	STAT6 (important in IL-4 signalling transcription) susceptible to DNA methylation	Kim et al. (2010)
	Histone methylase SUV39H1 participates in trimethylation of H3K9 modifies binding sites for HP1α (promotes transcriptional silencing); controls Th_2 lineage stability	Allan et al. (2012)
Th_{17}	Acetylation of histone H3 at IL-17a/IL-17f gene promoters and multiple conserved non-coding sequences in locus	Akimzhanov et al. (2007)
	H3K4 methylation, but no repressive H3K27 modifications at IL-17a/IL-17f gene promoters	Wei et al. (2009)
	Chromatin structure of IL17a-IL17f, *Ifng* and *Rorc* loci not stable in Th_{17} cells; TGFβ-induced increase in IL17a/IL17f expression reflected in enhanced H3K4 methylation across gene locus	Mukasa et al. (2010)
Treg	Demethylation of *FoxP3* gene promoter in Tregs	Janson et al. (2008)
	First intronic CpG region in *FoxP3* gene has decreased methylation following TGF-β signalling; after TCR signalling increased binding to cyclic-AMP response element-binding protein/activating transcription factors leads to increased *FoxP3* expression	Kim et al. (2007)
	FoxP3 methylation profile distinguishes natural Tregs (found in thymus) from those induced by TGF-β (found in periphery); upstream enhancer in *FoxP3* gene demethylated in natural Tregs (with acetylation of H3) and methylated in TGF-β-induced Tregs	Lal et al. (2009)
	IL-6-induced STAT3-dependent methylation of upstream *FoxP3* enhancer, suppressing development of Tregs	Doganci et al. (2005)
	Antigen-specific memory Th_2 cells may redifferentiation into functional $FoxP3^+$ Treg cells (regulated by TGF-β in the presence of retinoic acid and rapamycin)	Kim et al. (2010)

(Lovinsky-Desir et al. 2012). Maturation of naive T-cells into Th_2 cells involves the modification of epigenomic structure of *IL-4* and *IL-13* genes, while differentiation into Th_1 cells is accompanied by epigenomic changes in the structure of the IFN-γ locus (Agarwal et al. 1998).

Specifically, research suggests that Th_1-cell differentiation is accompanied by the progressive demethylation of CpG sites in the *Ifng* promoter (White et al. 2006), in addition to methylation of a highly conserved DNase I-hypersensitive region at the 3′ end of the IL-4 locus (Lee et al. 2002). Reports also indicate that in Th_1 cells, hyperacetylation of histones H3 and H4 occur at the *Ifng* locus via a Stat4 and T-bet-dependent mechanism (Chang et al. 2008), with no such pattern observed at the IL-4 locus (Fields et al. 2002). Chang et al. also found that inhibition of histone deacetylase (HDAC) through the T-bet-dependent removal of Sin3A-HDAC complexes, in naive T cells stimulates acquisition of H4 acetylation, *Ifng* transcription and ultimately, Th_1 differentiation (Chang et al. 2008).

Conversely, Th_2-cell differentiation is accompanied by the appearance of DNase I hypersensitive sites with demethylation around these sites within the *IL-4* and *IL-13* promoters (Santangelo et al. 2002), and the increased methylation of the *Ifng* promoter (Jones and Chen 2006). To detail the pathway involved in Th_2 differentiation, Kim et al. 2010) reports that STAT6, an important gene in signalling *IL-4* transcription, possesses enhanced susceptibility to DNA methylation (Kim and Lee 2011). Another study indicates that the RAD50-hypersensitive site 7 within the Th_2 cytokine locus control region (LCR) is subject to demethylation in a STAT6-dependent manner and only in cells stimulated under conditions similar to that during Th_2 differentiation (Kim et al. 2007). Chromatin modifications are also thought to contribute to the epigenetic mechanisms regulating differentiation of naive T-cells into Th_2-cells. Acetylation of histones H3 and H4 in the *IL-4* and *IL-13* region (Avni et al. 2002) through *GATA3* regulation is found to be important in transcription of Th_2-associated genes (Fields et al. 2002). Furthermore, chromatin remodelling of the Th_2 LCR gene is important in the regulation and coordination of Th_2 cytokine production, airway inflammation and AHR (Koh et al.). The importance of individual HDACs in T-cell differentiation has also been established, with a loss of HDAC1 displaying an increase in airway inflammation, mucus hypersecretion and AHR (Grausenburger et al. 2010). It was also demonstrated in this study that HDAC1 is recruited to the *IL-4* gene locus, indicating a role of HDAC1 in regulating IL-4 production. Recent advances have indicated that there are functionally relevant epigenetic pathways involved in T-cell subset differentiation. Allan et al. explored the role of the SUV39H1-H3K9me3-HP1α silencing pathway in the control of Th_2-lineage stability (Allan et al. 2012). This pathway involves the histone methylase SUV29H1, which regulates the methylation of histone H3 on the HP1α locus, promotes transcriptional silencing of Th_1 loci, and ensures naive T-cells follow a pathway towards Th_2 differentiation.

A role for Th_{17} cells and their associated cytokines have been implicated in the immunopathology of asthma (Doe et al. 2010). Similar to that observed in Th_1- and Th_2-differentiation, Th_{17}-cell differentiation is found to be regulated, at least in part, by epigenetic mechanisms. TGF-β and IL-6/IL-21 have been shown to induce Th_{17}

cell differentiation (Mangan et al. 2006), with gene transcription regulated by *RORγt* and *Stat5* (Ivanov et al. 2006; Laurence et al. 2007). Modulations in chromatin structure of Th_{17}-associated genes have been established in multiple studies. In one report, methylation of histone H3K4 independent of any repressive H3K27 modifications are found at the *IL17a/IL17f* gene promoters (Wei et al. 2009) in Th_{17} differentiation. Another study observed an increase in histone H3 acetylation at the *IL17a/IL17f* gene promoters as a result of the Th_{17} lineage pathway (Akimzhanov et al. 2007). Both papers showed significant H3K27 methylation within the *Ifng* promoter, with an absence of H3 acetylation and H3K4 methylation also observed in Th_{17}-cells (Wei et al. 2009; Akimzhanov et al. 2007). Mukasa et al., established that chromatin structure of the gene loci involved in Th_{17} cell differentiation, including *IL17a/IL17f*, *Ifng* and *Rorc*, are not stable, but change in response to environmental stimuli (Mukasa et al.). These authors also report that TGF-β (an important cytokine implicated in asthma pathogenesis) induces an increase in *IL17a/IL17f* gene expression through the enhanced methylation of histone H3K4 across the locus. The ability of the local cellular environment, in conjugation with chromatin remodelling, to regulate Th_{17} cell differentiation may indicate an important relationship between disease states and physiological conditions, although this requires extensively more research.

Regulatory T-cells (Tregs) are central in the maintenance of self-tolerance, with the transcription factor FoxP3 crucial for the regulation and expression of active Tregs (Ansel et al. 2003). Through the regulation of FoxP3 by epigenetic mechanisms, the investigation of Treg cell differentiation has been explored by multiple studies. In short, complete demethylation of the conserved *FoxP3* promoter region provides stability to FoxP3 expression and commitment to the Treg phenotype in humans (Janson et al. 2008). Regulation of the methylation profile in *FoxP3* gene is dependent on STAT3, a downstream target for IL-6 during allergic airway inflammation in a murine model following OVA sensitisation (Doganci et al. 2005). Interestingly, the FoxP3 methylation profile characterises and distinguishes natural Tregs (found in the thymus) from those induced by TGF-β (found in the periphery) (Lal et al. 2009). Specifically, this paper reports that the upstream enhancer in the *FoxP3* gene is demethylated in natural Tregs (with acetylation of histone H3) and methylated in TGF-β-induced Tregs. TGF-β, a major cytokine implicated in the pathogenesis of asthma including the progression to airway remodelling, has been shown to have the ability to modulate the methylation profile of *FoxP3* by decreasing methylation in the first intronic CpG region of the gene (Kim and Leonard 2007). In addition, following T-cell receptor signalling, an increased binding to cyclic-AMP response element-binding protein/activating transcription factor was observed, and in this study led to increased expression of *FoxP3* gene. In the context of allergic asthma, authors used purified T-cells from the spleen and lymph nodes of mice induced to display an allergic airways disease phenotype (through the administration of OVA) (Kim et al. 2010). Results indicated that memory Th_2 cells had the ability to redifferentiate into functional $FoxP3^+$ Treg cells by TGF-β when stimulated in the presence of retinoic acid and rapamycin.

The role of epigenetics in the development of asthma and allergy has been widely speculated, with research aimed to establish its significance. Interestingly, factors known to be involved in asthma aetiology including, genetic susceptibility and exposure to environmental stimuli (in utero and post-natal), have been explored as factors involved in modulating the epigenome. Due to this, it is proposed that epigenetic modulation and regulation of important pathways, including those involved in gene expression and T-cell differentiation play a direct role in disease development. Figure 9.2 summarises the interrelationship between epigenetic mechanisms, environmental stimuli and genetic susceptibility, with all three factors understood to possess a role, at least in part, to asthma pathogenesis. However, the exact role of each factor and the precise relationship between such factors are yet to be completely established and require further investigation.

9.4 Mechanisms Involved in the Development and Progression of Asthma: The Airway Epithelium

Under normal circumstances, the airway epithelium containing ciliated columnar, mucus-secreting goblet and surfactant-secreting, progenitor Clara cells forms a highly regulated and impermeable barrier to environmental insults during respiration (Xiao et al. 2011). This barrier serves to maintain tissue homeostasis, and when compromised by environmental pathogens activate an immunological response that aims to protect the underlying lung tissue (Davies 2009). The involvement of the epithelium in asthma pathogenesis is thought to be of primary significance, and may not simply be a consequence of chronic inflammation (Hackett and Knight 2007). During disease exacerbations, an inherently impaired epithelial barrier renders the airway susceptible to infection, which in turn stimulates sensitisation and immune activation to physiologically innocuous pathogens (Xiao et al. 2011) (Fig. 9.3). The mechanisms, which cause the epithelium to be inherently defective both in its increased susceptibility to injury and inability to activate normal repair processes, are inadequately defined. It may be plausible however, that epigenetic modulation in association with exposure to environmental stimuli may render the epithelium inherently defective during in utero and post-natal development as previously described (Hammad and Lambrecht 2008). The direct and intimate involvement of the epithelium during asthma development and disease progression is thus, an attractive target for current and future research.

Evidence to suggest that the airway epithelium is inherently defective in human asthma patients involves global gene expression profiling of the epithelium from human disease biopsy samples in comparison to that of healthy control samples (Kicic et al. 2010). The primary disruption of epithelial tight junctions allows pathogens to infiltrate the underlying airway wall, resulting in a complex interaction with immune and inflammatory cells (Holgate 2007). Despite the passage of airway epithelial cells of human asthma patients several times, and the separation of these cells from any inflammatory cells or mediators for a prolonged period, it has been

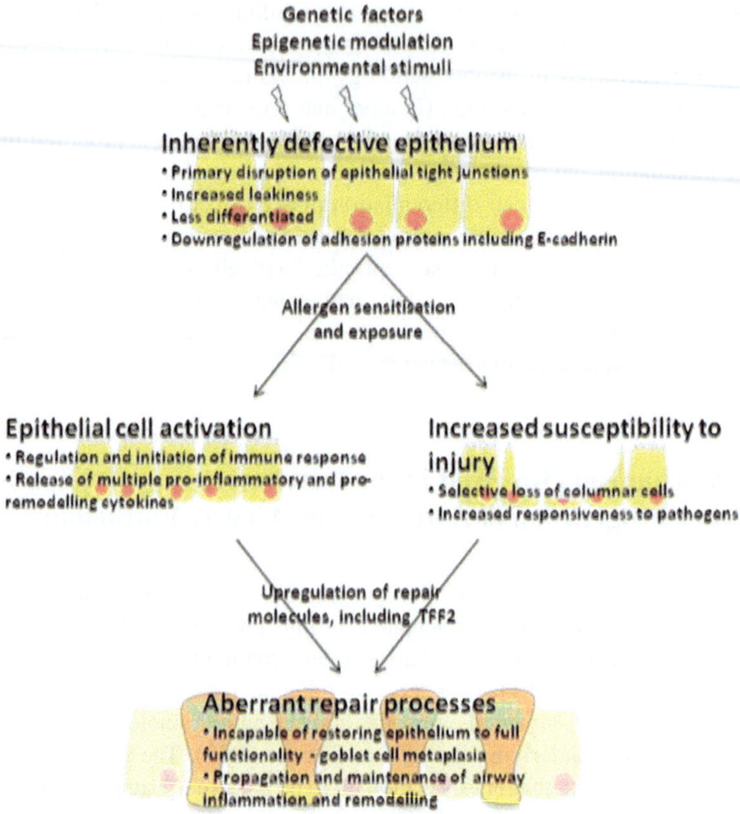

Fig. 9.3 The central role of the epithelium in asthma development and progression. An inherently defective epithelium thought to be caused by a combination of genetic, epigenetic and environmental factors in asthma is subject to allergen and/or pollutant sensitisation in utero and/or in post-natal development. Sensitisation and subsequent exposure to these allergens induces epithelial cell activation, with these cells possessing increased susceptibility to injury. Through the activation of epithelial cells, and the direct role of injury to the immune response following pathogen invasion, these cells in addition to other inflammatory cells including fibroblasts, macrophages, mast cells and eosinophils release multiple pro-inflammatory and pro-remodelling growth factors and cytokines in order to eliminate the incriminating pathogen and initiate repair responses in an attempt to restore the epithelium to full functionality. An upregulation of repair molecules including trefoil factor 2, a peptide involved in epithelial restitution leads to an aberrant repair process. Goblet cell metaplasia is one such consequence of this aberrant response. Collectively, these events characteristic of the inherently defective epithelium in asthma are important for the propagation and maintenance of inflammation and remodelling throughout the airway wall, with a negative feedback loop perpetuating disease

demonstrated that these cells are unable to form effectively functioning tight junctions (Wan et al. 2000). Furthermore results from the same study measured a marked reduction in transepithelial resistance, indicating an increased leakiness. Additional studies suggest that cells appear to be less differentiated (Kicic et al. 2010) and adhesion proteins including E-cadherin, essential for epithelial structure and cell-to-cell communication, are downregulated within the diseased epithelium as compared to control samples (Trautmann et al. 2005). An increased susceptibility of epithelial cells to oxidant-induced damage and apoptosis (Bucchieri et al. 2002), and the abnormal expression of several pro-inflammatory transcription factors (Sampath et al. 1999) have also been reported in asthma patients compared to healthy controls. These findings support the hypothesis that fundamental alterations in the epithelial barrier function contribute to the onset and progression of asthma (Kicic et al. 2010).

Activation of epithelial cells following environmental stimuli exposure is a key event in the recognition of pathogens that coordinate the subsequent immune response. The involvement of dendritic cells in conjunction with the inherently dysfunctional epithelium is thought to be central in allergen sensitisation and presentation of allergen to circulating T-cells (Lambrecht and Hammad 2009). Isolation of dendritic cells from the airways of mice exposed to allergen and transferred to naive mice induced the production of Th_2 cells which were specific to house dust mite, which was involved in the initial sensitisation process (Hammad et al.). On the other hand, depletion of dendritic cells from the airway of both naive or allergen-sensitised mice resulted in the inability to initiate the production of Th_2 cells or the development of airway inflammation when these mice were exposed to the allergen in question (van Rijt et al. 2005). Upregulation of important cytokines, including thymic stromal lymphopoietin (TSLP), granulocyte-macrophage colony-stimulating factor (GM-CSF), IL-25 and IL-33, are important in the initiation of the immune response and allergen sensitisation in asthma. Mice exposed to OVA with the overexpression of GM-CSF induce spontaneous sensitisation and a Th_2-related response (Stampfli et al. 1998). Furthermore, a deficiency in GM-CSF causes a failure to sensitise mice to house dust mite allergen (Cates et al. 2004), and attenuates the ability of DEP to increase allergic sensitisation (Ohta et al. 1999). Interestingly, when human epithelial cells collected from asthma patients were cultured, results indicated that these cells continuously overproduce GM-CSF (Ritz et al. 2002) independent of allergen exposure, which suggests that the production of GM-CSF may be epigenetically regulated in disease (Ritz et al. 2002). Yet once initiated and activated, the epithelial response to environmental stimuli are maintained through the downstream effects of the initial immune response, including the production of important pro-inflammatory and pro-remodelling cytokines, providing a negative feedback loop that can perpetuate damage to the epithelium, and ultimately affect disease progression and severity.

Due to the innate dysregulation and intrinsic defection of the airway epithelium both biochemically and functionary in asthma, it has been shown that these cells have an increased susceptibility to injury. Ultrastructural changes, characterised by the sloughing of columnar cells, and goblet cell metaplasia with increased mucus secretion and airway plugging occur in the epithelium in association with damage and provide evidence of dysfunctional activation in the presence of a predetermined stimuli (Hamilton et al. 2001). Selective loss of columnar cells is observed in sufferers of asthma and, when compared with that of non-asthmatic subjects, may be increased by 15–45 % (Zhou et al. 2011). An increased susceptibility to injury is thought to be caused by the predetermined capacity of epithelial cells to activate pathways involved in the immune response. Regulation and promotion of epithelial cell activation has been shown to be caused by a combination of early life exposure to allergen and downstream signalling cascades (Lambrecht and Hammad 2012). Viral infection of the airway epithelium with respiratory syncytial virus (RSV) or cigarette smoke exposure induces the upregulation of toll-like receptor 4 (TLR4) expression and promotes its localisation to the cell membrane, which in turn increases the responsiveness of epithelial cells to endotoxin (Pace et al. 2008; Monick et al. 2003). In addition, the role of the transcription factor nuclear factor-κB (NF-κB) in the expression of multiple inflammatory cytokines and the activation of the epithelium has been established, with mice deficient in the NF-κB subunits p50 or p65 displaying reduced responses to endotoxin and exposure to other allergens (Yang et al. 1998). Newer studies also exhibited the ability of constitutively activated NF-κB within epithelial cells to activate dendritic cells, cause injury to the epithelium itself and promote sensitisation to OVA in a murine model (Ather et al. 2011). Conversely, inhibition of epithelial expression of NF-κB in mice reduced the recruitment of Th_2 cells in the lung and attenuated airway remodelling (Broide et al. 2005). This increased susceptibility to damage, coupled with defective epithelium activation, results in the induction of an ineffective and impaired repair phenotype (Holgate 1998).

Under normal circumstances, a damaged epithelium is able to activate a self-repair process with efficiency (Erjeflt and Persson 1997). Injury to the airway epithelium initiates a repair pathway involving the migration, proliferation and differentiation of neighbouring undamaged cells (Holgate 2008). The ultimate outcome of such a process involves resolution and return of the tissue to its normal structure and function (Davies 2009). However, it has become increasingly apparent that this normal repair process is compromised and rendered defective in the epithelium of asthma patients. Most interesting is the finding that the expression of a large number of genes involved in the epithelial repair process was reduced (Zhang et al. 2012). Additionally, airway epithelial cells have been shown to proliferate at a faster rate than control cells, with asthma samples displaying a dramatic impairment in wound healing ability (Stevens et al. 2008). In detailed studies, epithelial wounds in airway tissue from asthma patients were unable to close after 10 days, with a subset of samples found to possess a maximum wound closure of less than 70 % 30 days post-wound induction. During normal repair processes, progenitor Clara cells have been shown to proliferate and undergo phenotypic differentiation to re-establish the integrity and functional arrangement of the epithelium (Crosby and Waters 2010). However, with chronic Clara cell depletion as observed in human samples and induced allergic airway disease (AAD) in mice, the ineffectiveness of the repair process during injury is highlighted (Perl et al. 2011). Downregulation of the primary extracellular protein fibronectin observed in diseased epithelial cells demonstrates the significant impairment of repair processes in wound healing (Kicic et al. 2010) as this protein profoundly influences the survival, proliferation and differentiation of epithelial cells in physiological conditions (Zahm et al. 1991). Knocking down of fibronectin expression in healthy airway epithelial cells in vitro resulted in a significant impairment in wound healing, with addition of fibronectin to the same cells reversing the impairment, and restoring the capacity of these cells to self-repair. Furthermore, a normal response to epithelial injury includes the upregulation of receptors that drive proliferation and repair, including members of the epithelial growth factor receptor (EGFR) family (Le Cras et al. 2011). Despite the finding that expression of EGFR is markedly increased, especially in areas of columnar cell loss (Amishima et al. 1998), this upregulation does not correlate with the proliferative response of repairing cells (Puddicombe et al. 2000). Accordingly, this data demonstrates that the response to injury and the repair process that ensues is dysregulated in the epithelium of patients diagnosed with asthma.

Following environmental stimuli-induced injury, the epithelium provides the microenvironment for persistent and chronic inflammation and the irreversible structural alterations of airway remodelling (Xiao et al. 2011). The interactions and signalling observed between epithelial, mesenchymal, neural and ECM cells are necessary to initiate numerous functions in the lung, allowing an exchange of information between these elements in response to various stimuli (Evans et al. 1999). The activation of aberrant repair pathways and the production of numerous growth factors and cytokines are thought to be central to the severity, pathology, therapeutic response and long-term outcome of disease (Holgate 1998). The mechanisms, which cause the epithelium to be inherently defective both in its increased susceptibility to injury and inability to activate normal repair processes, are inadequately defined. A combination of environmental, genetic and epigenetic factors is thought to contribute, yet their exact involvement is yet to be elucidated. The difficulty in establishing the complex interaction between these factors in the development of disease has led research to focus on targeting asthma progression and the ability to attenuate and/or reverse epithelial dysfunction.

9.5 Trefoil Factor 2

TFF2 is one of the three known mammalian trefoil peptides (White 2001). Small and protease resistant, these proteins are well established as protective molecules in the gut (Franic et al. 2005). Specifically, TFF2 is produced by intestinal epithelium and promotes repair by mediating epithelial restitution and inhibiting apoptosis (Hoffmann 2005). Studies utilising methods of gene disruption to generate mice deficient in TFF2 indicate that this protein promotes gastric mucosal healing through the inhibition of acid secretion and stimulation of mucosal proliferation (White 2001). Interestingly, TFF2 is upregulated in diverse pathologic conditions of the gastrointestinal tract, including at the sites of gastric and duodenal ulceration, and Crohn's disease (Wright 1993). Yet despite this upregulation, the dissolution of epithelial integrity is a consistent feature of these gastrointestinal conditions. In an attempt to explain the cause of such a phenomenon, research suggests that while short-term upregulation of TFF2 appears to be beneficial in epithelial repair processes, chronic upregulation may in fact contribute to pathologically significant events, including the progression to gastric cancer formation (White 2001). With some reports demonstrating an upregulation of TFF2 expression in the gastric tissue of chronically infected H. Pylori mice (Nomura et al. 2005) and in patients with H. Pylori-associated chronic gastritis and gastric cancer (Hu et al. 2003; Leung et al. 2002). Nevertheless, the mechanisms involved in the upregulation of TFF2 in numerous chronic pathological conditions and tumour formation have yet to be established in the gastrointestinal tract setting, with its functional capacity questioned as this observed upregulation is unable to prevent epithelial injury and mediate a normal repair response that re-establishes normal barrier structure and function. Further studies are required to understand this phenomenon in the gastrointestinal tract, which would further benefit research in understanding the role of TFF2 in the airway.

The role of TFF2 in the lung is less established, with its expression only more recently documented in the epithelial cells of human and animal lung (Kuperman et al. 2005; Nikolaidis et al. 2006). Studies suggest that TFF2 in the lung possesses a similar protective function to that expressed in the gut. TFF2 is found to have the capacity to promote cell migration of human epithelial cells and activation of rapid repair mechanisms in response to injurious stimuli in the lung (Oertel et al. 2001). Furthermore, TFF2 has been observed to enhance airway epithelial cell survival during restitution processes by inhibiting apoptosis and promoting angiogenesis (Hoffmann 2005). It is therefore plausible to hypothesise that TFF2 plays a key role in the initiation and/or progression of airway remodelling in asthma (Royce et al. 2011). However, the exact role of TFF2 in pathogenesis and its contribution to such structural alterations in this disease remains to be elucidated.

Previously, studies have demonstrated the rapid induction of TFF2 by mucus positive airway epithelial cells with expression not contributing to the regulation of the inflammatory response in a mouse model of acute airway inflammation (Nikolaidis et al. 2006). However recent reports suggest that TFF2 has an important role in the regulation of IL-33 at mucosal surfaces and the development of the type 2 allergic immune response in the lung following allergen exposure (Wills Karp et al. 2012). Specifically, TFF2 promotes IL-33 release from lung epithelia and alveolar macrophages initiated be an injurious event, which in turn is thought to be required for the production of major inflammatory cytokines IL-4 and IL-13, and AHR. Thus in light of these recent findings, TFF2 may have the capacity to mediate both inflammatory events and epithelial restitution processes in the lung in response to allergic insult.

Although expression of TFF2 is increased in asthma, it does not lead to an appropriate repair response and restitution of epithelium to normal functionality (Holgate 2000). Upregulation of TFF2 in asthma fails to prevent the progression of airway inflammation and epithelial remodelling changes suggesting that this expression change may be insufficient to prevent chronic epithelial injury and/or promote normal repair processes. Furthermore, the upregulation of pro-inflammatory and pro-remodelling mediators may override the potential benefits associated with increased TFF2 expression. Alternatively, there may be a failure to downregulate TFF2 expression after restitution, with the consequence that epithelial cells are inappropriately 'held' in a repair phenotype (Holgate 2000). Similar to research concerning the gastrointestinal tract, further studies are required to establish the exact mechanisms that upregulate TFF2 in asthma and the role of this upregulation in disease.

9.6 Novel Therapeutic Strategies in Asthma

As support for the role of epigenetic modulation in allergy and asthma continues to strengthen, the capacity to alter these changes has proven to be an attractive target for future therapeutic research. The clinical utility of HDAC inhibitors (HDACi) in oncology are relatively well characterised, with the mechanisms associated with their anticancer effects involving cell death and apoptosis, differentiation, decreased migration and invasion, and cell cycle arrest (Bolden et al. 2006). The role of HDACs in asthma has also been widely reported, with HDAC expression and activity of class I enzymes HDAC1 and HDAC2, decreased in bronchial biopsies from patients with asthma compared with normal subjects (Ito et al. 2002). In addition, decreased HDAC activity was implicated in alveolar macrophages and peripheral blood mononuclear cells in patients with asthma than in control subjects (Cosio et al. 2004). Conversely both studies report an increase in HAT activity in patients with asthma (Ito et al. 2002; Cosio et al. 2004). Interestingly, the molecular mechanisms and actions of corticosteroids involves downregulating the expression of multiple inflammatory genes by reversing the increased HAT activity observed in asthma, through the recruitment of HDAC2 (Barnes 2006). Furthermore, corticosteroid insensitivity (resistance) has been correlated with a reduction in HDAC activity in patients with asthma (Hew et al. 2006). Collectively, these findings suggest that HDACi may have a therapeutic benefit in reducing the severity of disease exacerbations through their ability to activate HDACs in order to attenuate the expression of numerous inflammatory genes.

Preliminary animal studies involving a number of general HDACi have indicated a potential for their use in asthma. In one report, mice sensitised and subsequently challenged with OVA to exhibit an AAD phenotype were administered with a representative HDACi, Trichostatin A (TSA) (Choi et al. 2005). Results indicate that TSA attenuates AHR and inflammation following methacholine administration. However, a recent study involving mice sensitised and challenged with *A. fumigates* disputes the finding that TSA has anti-inflammatory properties (Banerjee et al. 2012). In contrast, the ability of TSA to inhibit drug-induced bronchoconstriction was validated in this study in both the murine model and in human lung samples. Inhibition of drug-induced constriction in in vitro studies involving the administration of suberoylanilide hydroxamic acid (SAHA) to isolated guinea pig tracheal rings has also been published (Assem et al. 2008). The administration of the broad-spectrum HDACi, valproic acid was administered to AAD mice (induced through OVA sensitisation and exposure) in order to evaluate its capacity to attenuate the important characteristics of asthma (Royce et al.). Although valproic acid treatment was found not to affect inflammatory cell or infiltrate counts, administration resulted in reduced epithelial thickness and subepithelial collagen deposition when compared to vehicle-treated mice. In confirmation of previous findings, valproic acid also attenuated methacholine-induced AHR. Taken together, these results show the ability of HDACi to reverse drug-induced bronchoconstriction, with further research required to establish the ability of this class of drugs in reducing airway inflammation and remodelling changes.

The epithelium has been shown to be central in asthma pathogenesis, through its ability to activate the immune response, and propagate and maintain airway inflammation and remodelling. Through its ability to connect the molecular mechanisms involved in the development and progression of asthma pathogenesis, the epithelium is an attractive target for therapeutic research. Involved in epithelial restitution in the airway, TFF2 has been shown to be upregulated in asthma with endogenous expression levels seemingly inadequate to protect and/or reverse epithelial damage caused by invading pathogens. Interestingly, TFF2 has been shown to be important in the regulation of airway remodelling in two murine models of AAD (Royce et al.). In this study, TFF2-deficient AAD mice (AAD induced by either OVA or *A. fumigates*) exhibited increased goblet cell metaplasia, increased subepithelial fibrosis deposition (in both models) and increased epithelial thickness (in *A. fumigates* model). Thus, it is shown that TFF2 has the capacity to attenuate important epithelial and subepithelial remodelling events, characteristic of chronic, severe disease. A pilot study examining the effect of exogenous TFF2 treatment in a chronic OVA AAD model in mice support the contention that endogenous TFF2 fails to prevent epithelial injury and promote reparation (Royce et al. 2012). Despite no significant difference in the inflammatory response reported between TFF2-treated and untreated chronic AAD mice, important remodelling changes including goblet cell metaplasia, and lamina reticularis thickness was significantly attenuated with such treatment. Furthermore, bronchial epithelial apoptosis and AHR was significantly reduced following exogenous TFF2 treatment. Similar to those findings involving the administration of HDACi to animal models of AAD, further investigation is required to establish the capability of exogenous TFF2 in diminishing the inflammatory response, decreasing AHR and attenuating the structural alterations that constitute airway remodelling.

9.7 Conclusion

As prevalence, complexity and severity of asthma continues to increase on a global scale, despite advances in medicine and the understanding of disease, the importance of elucidating its development and pathogenesis is emphasised. The molecular heterogeneity of the disease is thought to be contributed to a combination of genetic, epigenetic and environmental factors. The role of genetic susceptibility and inheritance, and the exposure to various environmental pathogens in utero and in post-natal development have been widely studied in the acquisition of an asthma phenotype.

Epigenetic regulation, modulated through interactions with genetic and environmental factors, has been explored in an attempt to explain the dramatic increase in incidence over the last decade. Modulation of the epigenome in the regulation of important asthma-associated pathways, including gene expression and T-cell differentiation is thought to play a direct role in conferring asthma susceptibility, development and progression. In addition, an inherently defective epithelium is thought to be of primary significance. It is hypothesised that the epithelium has the capacity to propagate and maintain remodelling and inflammatory changes, through its altered communication with the underlying mesenchyme. TFF2, known to be involved in epithelial restitution has previously been shown to be upregulated in asthma. Despite this, increased expression fails to lead to complete restoration of the epithelium to normal functionality. By understanding the limited knowledge available, we aim to establish the mechanisms involved in the pathogenesis of asthma, and more specifically determine the roles of epigenetic modulation, the epithelium and TFF2 in disease. This chapter discussed the inadequacy of current research and provided the rationale to focus on targeting these mechanisms for future therapies.

Acknowledgements The support of the Australian Institute of Nuclear Science and Engineering is acknowledged. TCK was the recipient of AINSE awards. TCK is a Future Fellow and Epigenomic Medicine Laboratory is supported by the Australian Research Council. Supported in part by the Victorian Government's Operational Infrastructure Support Program.

References

Abbas AK, Murphy KM, Sher A (1996) Functional diversity of helper T lymphocytes. Nature 383(6603):787–793

Adenuga D, Yao H, March TH, Seagrave J, Rahman I (2009) Histone deacetylase 2 is phosphorylated, ubiquitinated, and degraded by cigarette smoke. Am J Respir Cell Mol Biol 40: 464–473

Agarwal S, Rao A (1998) Modulation of chromatin structure regulates cytokine gene expression during T cell differentiation. Immunity 9(6):765–775

Akimzhanov AM, Yang XO, Dong C (2007) Chromatin remodeling of interleukin-17 (IL-17)-IL-17F cytokine gene locus during inflammatory helper T cell differentiation. J Biol Chem 282: 5969–5972

Allan RS, Zueva E, Cammas F, Schreiber HA, Masson V, Belz GT, Roche D, Maison C, Quivy J-P, Almouzni G, Amigorena S (2012) An epigenetic silencing pathway controlling T helper 2 cell lineage commitment. Nature 487:249–253

Allen M, Heinzmann A, Noguchi E, Abecasis G, Broxholme J, Ponting C, Bhattacharyya S, Tinsley J, Zhang Y, Holt R, Jones EY, Lench N, Carey A, Jones H, Dickens N, Dimon C, Nicholls R, Baker C, Xue L, Townsend E, Kabesch M, Weiland S, Carr D, Von Mutius E, Adcock I, Barnes P, Lathrop GM, Edwards M, Moffatt M, Cookson WOCM (2003) Positional cloning of a novel gene influencing asthma from chromosome 2q14. Nat Genet 35:258–263

Amishima M et al (1998) Expression of epidermal growth factor and epidermal growth factor receptor immunoreactivity in the asthmatic human airway. Am J Respir Crit Care Med 157(6):1907–1912

Ansel KM, Lee DU, Rao A (2003) An epigenetic view of helper T cell differentiation. Nat Immunol 4:616–623

Assem E-SK et al (2008) Effects of a selection of histone deacetylase inhibitors on mast cell activation and airway and colonic smooth muscle contraction. Int Immunopharmacol 8(13):1793–1801

Ather JL et al (2011) Airway epithelial NF-kappaB activation promotes allergic sensitization to an innocuous inhaled antigen. Am J Respir Cell Mol Biol 44(5):631–638

Avni O, Lee D, Macian F, Szabo S, Glimcher L, Rao A (2002) T(H) cell differentiation is accompanied by dynamic changes in histone acetylation of cytokine genes. Nat Immunol 3:643–651

Baccarelli A, Rusconi F, Bollati V, Catelan D, Accetta G, Hou L, Barbone F, Bertazzi PA, Biggeri A (2012) Nasal cell DNA methylation, inflammation, lung function and wheezing in children with asthma. Epigenomics 4:91–100

Banerjee A et al (2012) Trichostatin A abrogates airway constriction, but not inflammation, in murine and human asthma models. Am J Respir Cell Mol Biol 46(2):132–138

Barnes PJ (2006) How corticosteroids control inflammation: quintiles prize lecture 2005. Br J Pharmacol 148(3):245–254

Barnes PJ, Adcock IM (2009) Glucocorticoid resistance in inflammatory diseases. Lancet 373(9678):1905–1917

Beasley R et al (1989) Cellular events in the bronchi in mild asthma and after bronchial provocation. Am Rev Respir Dis 139(3):806–817

Berlivet S, Moussette S, Ouimet M, Verlaan D, Koka V, Al Tuwaijri A, Kwan T, Sinnett D, Pastinen T, Naumova A (2012) Interaction between genetic and epigenetic variation defines gene expression patterns at the asthma-associated locus 17q12-q21 in lymphoblastoid cell lines. Hum Genet 131:1161–1171

Bluestone JA et al (1995) TCR gamma delta cells: a specialized T-cell subset in the immune system. Annu Rev Cell Dev Biol 11:307–353

Bolden JE, Peart MJ, Johnstone RW (2006) Anticancer activities of histone deacetylase inhibitors. Nat Rev Drug Discov 5(9):769–784

Bousquet J et al (2000) Asthma: from bronchoconstriction to airways inflammation and remodeling. Am J Respir Crit Care Med 161(5):1720–1745

Brand S, Teich R, Dicke T, Harb H, Yildirim AÃ, Tost JR, Schneider-Stock R, Waterland RA, Bauer U-M, Von Mutius E, Garn H, Pfefferle PI, Renz H (2011) Epigenetic regulation in murine offspring as a novel mechanism for transmaternal asthma protection induced by microbes. J Allergy Clin Immunol 128:618–625, e7

Breton CV, Byun H-M, Wenten M, Pan F, Yang A, Gilliland FD (2009) Prenatal tobacco smoke exposure affects global and gene-specific DNA methylation. Am J Respir Crit Care Med 180:462–467

Breton CV, Byun H-M, Wang X, Salam MT, Siegmund K, Gilliland FD (2011) DNA methylation in the arginase-nitric oxide synthase pathway is associated with exhaled nitric oxide in children with asthma. Am J Respir Crit Care Med 184:191–197

Broide DH et al (2005) Allergen-induced peribronchial fibrosis and mucus production mediated by IκB kinase β-dependent genes in airway epithelium. Proc Natl Acad Sci USA 102(49):17723–17728

Bucchieri F et al (2002) Asthmatic bronchial epithelium is more susceptible to oxidant-induced apoptosis. Am J Respir Cell Mol Biol 27(2):179–185

Cao D, Bromberg PA, Samet JM (2007) COX-2 expression induced by diesel particles involves chromatin modification and degradation of HDAC1. Am J Respir Cell Mol Biol 37:232–239

Cates EC et al (2004) Intranasal exposure of mice to house dust mite elicits allergic airway inflammation via a GM-CSF-mediated mechanism. J Immunol 173(10):6384–6392

Chang S, Collins PL, Aune TM (2008) T-bet dependent removal of Sin3A-histone deacetylase complexes at the Ifng locus drives Th1 differentiation. J Immunol 181:8372–8381

Choi JH et al (2005) Trichostatin A attenuates airway inflammation in mouse asthma model. Clin Exp Allergy 35(1):89–96

Cosio BG, Mann B, Ito K, Jazrawi E, Barnes PJ, Chung KF, Adcock IM (2004) Histone acetylase and deacetylase activity in alveolar macrophages and blood mononocytes in asthma. Am J Respir Crit Care Med 170:141–147

Crosby L, Waters C (2010) Epithelial repair mechanisms in the lung. Am J Physiol Lung Cell Mol Physiol 298(6):L715–L731

Cui J et al (2005) TH1-mediated airway hyperresponsiveness independent of neutrophilic inflammation. J Allergy Clin Immunol 115(2):309–315

Davies D (2009) The role of the epithelium in airway remodeling in asthma. Proc Am Thorac Soc 6(8):678–682

Demenais F, Chaudru V, Martinez M (2001) Detection of parent-of-origin effects for atopy by model-free and model-based linkage analyses. Genet Epidemiol 21(Suppl 1):S186–S191

Doe C, Bafadhel M, Siddiqui S, Desai D, Mistry V, Rugman P, Mccormick M, Woods J, May R, Sleeman MA, Anderson IK, Brightling CE (2010) Expression of the T helper 17-associated cytokines IL-17A and IL-17F in asthma and COPD. Chest 138:1140–1147

Doganci A, Eigenbrod T, Krug N, De Sanctis GT, Hausding M, Erpenbeck VJ, Haddad E-B, Schmitt E, Bopp T, Kallen K-J, Herz U, Schmitt S, Luft C, Hecht O, Hohlfeld JM, Ito H, Nishimoto N, Yoshizaki K, Kishimoto T, Rose-John S, Renz H, Neurath MF, Galle PR, Finotto S (2005) The IL-6R alpha chain controls lung CD4+CD25+ Treg development and function during allergic airway inflammation in vivo. J Clin Invest 115:313–325

Ege MJ, Strachan DP, Cookson WO, Moffatt MF, Gut I, Lathrop M, Kabesch M, Genuneit J, Büchele G, Sozanska B, Boznanski A, Cullinan P, Horak E, Bieli C, Braun-Fahrländer C, Heederik D, Von Mutius E (2011) Gene-environment interaction for childhood asthma and exposure to farming in Central Europe. J Allergy Clin Immunol 127:138–144, e4

Erjeflt JS, Persson CG (1997) Airway epithelial repair: breathtakingly quick and multipotentially pathogenic. Thorax 52(11):1010–1012

Evans MJ et al (1999) The attenuated fibroblast sheath of the respiratory tract epithelial-mesenchymal trophic unit. Am J Respir Cell Mol Biol 21(6):655–657

Fainaru O, Shseyov D, Hantisteanu S, Groner Y (2005) Accelerated chemokine receptor 7-mediated dendritic cell migration in Runx3 knockout mice and the spontaneous development of asthma-like disease. Proc Natl Acad Sci U S A 102:10598–10603

Fedorov IA et al (2005) Epithelial stress and structural remodelling in childhood asthma. Thorax 60(5):389–394

Fields PE, Kim ST, Flavell RA (2002) Cutting edge: changes in histone acetylation at the IL-4 and IFN-gamma loci accompany Th1/Th2 differentiation. J Immunol 169:647–650

Franic TV et al (2005) Reciprocal changes in trefoil 1 and 2 expression in stomachs of mice with gastric unit hypertrophy and inflammation. J Pathol 207(1):43–52

Grausenburger R et al (2010) Conditional deletion of histone deacetylase 1 in T cells leads to enhanced airway inflammation and increased Th2 cytokine production. J Immunol 185(6):3489–3497

Hackett T-L (2012) Epithelial-mesenchymal transition in the pathophysiology of airway remodelling in asthma. Curr Opin Allergy Clin Immunol 12(1):53–59

Hackett T-L, Knight DA (2007) The role of epithelial injury and repair in the origins of asthma. Curr Opin Allergy Clin Immunol 7(1):63–68

Hamilton LM et al (2001) The bronchial epithelium in asthma–much more than a passive barrier. Monaldi Arch Chest Dis 56(1):48–54

Hammad H, Lambrecht BN (2008) Dendritic cells and epithelial cells: linking innate and adaptive immunity in asthma. Nat Rev Immunol 8(3):193–204

Hew M et al (2006) Relative corticosteroid insensitivity of peripheral blood mononuclear cells in severe asthma. Am J Respir Crit Care Med 174(2):134

Hoffmann W (2005) Trefoil factors TFF (trefoil factor family) peptide-triggered signals promoting mucosal restitution. Cell Mol Life Sci 62(24):2932–2938

Holgate ST (1998) The inflammation-repair cycle in asthma: the pivotal role of the airway epithelium. Clin Exp Allergy 28(suppl 5):97–103

Holgate ST (2000) The bronchial epithelial origins of asthma. Chem Immunol 78:62–71

Holgate S (2007) Epithelium dysfunction in asthma. J Allergy Clin Immunol 120(6):1233–1244

Holgate ST (2008) Pathogenesis of asthma. Clin Exp Allergy 38(6):872–897

Hollingsworth JW, Maruoka S, Boon K, Garantziotis S, Li Z, Tomfohr J, Bailey N, Potts EN, Whitehead G, Brass DM, Schwartz DA (2008) In utero supplementation with methyl donors enhances allergic airway disease in mice. J Clin Invest 118:3462–3469

Hu G-Y et al (2003) Expression of TFF2 and Helicobacter pylori infection in carcinogenesis of gastric mucosa. World J Gastroenterol 9(5):910–914

Isidoro-García M, Sanz C, García-Solaesa V, Pascual M, Pescador DB, Lorente F, Dávila I (2011) PTGDR gene in asthma: a functional, genetic, and epigenetic study. Allergy 66:1553–1562

Ito K et al (2002) Expression and activity of histone deacetylases in human asthmatic airways. Am J Respir Crit Care Med 166(3):392–396

Ivanov II, Mckenzie BS, Zhou L, Tadokoro CE, Lepelley A, Lafaille JJ, Cua DJ, Littman DR (2006) The orphan nuclear receptor RORgammat directs the differentiation program of proinflammatory IL-17+ T helper cells. Cell 126:1121–1133

Janson PCJ, Winerdal ME, Marits P, Thörn M, Ohlsson R, Winqvist O (2008) FOXP3 promoter demethylation reveals the committed Treg population in humans. PLoS One 3:e1612

Jones B, Chen J (2006) Inhibition of IFN-gamma transcription by site-specific methylation during T helper cell development. EMBO J 25(11):2443–2452

Kicic A et al (2010) Decreased fibronectin production significantly contributes to dysregulated repair of asthmatic epithelium. Am J Respir Crit Care Med 181(9):889–898

Kim B-S et al (2010) Conversion of Th2 memory cells into Foxp3+ regulatory T cells suppressing Th2-mediated allergic asthma. Proc Natl Acad Sci 107(19):8742–8747

Kim S-H, Lee C-E (2011) Counter-regulation mechanism of IL-4 and IFN-α signal transduction through cytosolic retention of the pY-STAT6:pY-STAT2:p48 complex. Eur J Immunol 41:461–472

Kim H-P, Leonard WJ (2007) CREB/ATF-dependent T cell receptor-induced FoxP3 gene expression: a role for DNA methylation. J Exp Med 204:1543–1551

Kim ST, Fields PE, Flavell RA (2007) Demethylation of a specific hypersensitive site in the Th2 locus control region. Proc Natl Acad Sci U S A 104:17052–17057

Koh BH, Hwang SS, Kim JY, Lee W, Kang M-J, Lee CG, Park J-W, Flavell RA, Lee GR (2010) Th2 LCR is essential for regulation of Th2 cytokine genes and for pathogenesis of allergic asthma. Proc Natl Acad Sci 107:10614–10619

Kuperman D et al (2005) Dissecting asthma using focused transgenic modeling and functional genomics. J Allergy Clin Immunol 116(2):305–311

Kwon N-H, Kim J-S, Lee J-Y, Oh M-J, Choi D-C (2008) DNA methylation and the expression of IL-4 and IFN-gamma promoter genes in patients with bronchial asthma. J Clin Immunol 28: 139–146

Lal G, Zhang N, Van Der Touw W, Ding Y, Ju W, Bottinger EP, Reid SP, Levy DE, Bromberg JS (2009) Epigenetic regulation of Foxp3 expression in regulatory T cells by DNA methylation. J Immunol 182:259–273

Lambrecht BN, Hammad H (2009) Biology of lung dendritic cells at the origin of asthma. Immunity 31(3):412–424

Lambrecht BN, Hammad H (2012) The airway epithelium in asthma. Nat Med 18(5):684–692

Lane N et al (2010) Regulation in chronic obstructive pulmonary disease: the role of regulatory T-cells and Th17 cells. Clin Sci 119:75–86

Laurence A, Tato CM, Davidson TS, Kanno Y, Chen Z, Yao Z, Blank RB, Meylan FO, Siegel R, Hennighausen L, Shevach EM, O'Shea JJ (2007) Interleukin-2 signaling via STAT5 constrains T helper 17 cell generation. Immunity 26:371–381

Le Cras T et al (2011) Epithelial EGF receptor signaling mediates airway hyperreactivity and remodeling in a mouse model of chronic asthma. Am J Physiol Lung Cell Mol Physiol 300(3):L414–L421

Lee DU, Agarwal S, Rao A (2002) Th2 lineage commitment and efficient IL-4 production involves extended demethylation of the IL-4 gene. Immunity 16:649–660

Lee SL, Lam TH, Leung TH, Wong WH, Schooling M, Leung GM, Lau YL (2012) Foetal exposure to maternal passive smoking is associated with childhood asthma, allergic rhinitis, and eczema. Scientific World Journal 2012:542983

Leung W et al (2002) Expression of trefoil peptides (TFF1, TFF2, and TFF3) in gastric carcinomas, intestinal metaplasia, and non-neoplastic gastric tissues. J Pathol 197(5):582–588

Liu J, Ballaney M, Al-Alem U, Quan C, Jin X, Perera F, Chen L-C, Miller RL (2008) Combined inhaled diesel exhaust particles and allergen exposure alter methylation of T helper genes and IgE production in vivo. Toxicol Sci 102:76–81

Liu Q, Xia Y, Zhang W, Li J, Wang P, Li H, Wei C, Gong Y (2009) A functional polymorphism in the SPINK5 gene is associated with asthma in a Chinese Han population. BMC Med Genet 10:59

Lloyd CM, Hessel EM (2010) Functions of T cells in asthma: more than just TH2 cells. Nat Rev Immunol 10(12):838–848

Lodrup Carlsen KC, Carlsen K-H (2001) Effects of maternal and early tobacco exposure on the development of asthma and airway hyperreactivity. Curr Opin Allergy Clin Immunol 1: 139–143

Lovinsky-Desir S, Miller RL (2012) Epigenetics, asthma, and allergic diseases: a review of the latest advancements. Curr Allergy Asthma Rep 12(3):211–220

Mangan PR, Harrington LE, O'Quinn DB, Helms WS, Bullard DC, Elson CO, Hatton RD, Wahl SM, Schoeb TR, Weaver CT (2006) Transforming growth factor-[beta] induces development of the TH17 lineage. Nature 441:231–234

McKinley L et al (2008) TH17 cells mediate steroid-resistant airway inflammation and airway hyperresponsiveness in mice. J Immunol 181(6):4089–4097

Monick MM et al (2003) Respiratory syncytial virus up-regulates TLR4 and sensitizes airway epithelial cells to endotoxin. J Biol Chem 278(52):53035–53044

Morales E, Bustamante M, Vilahur N, Escaramis G, Montfort M, De Cid R, Garcia-Esteban R, Torrent M, Estivill X, Grimalt JO, Sunyer J (2012) DNA hypomethylation at ALOX12 is associated with persistent wheezing in childhood. Am J Respir Crit Care Med 185:937–943

Mukasa R et al (2010) Epigenetic instability of cytokine and transcription factor gene loci underlies plasticity of the T helper 17 cell lineage. Immunity 32(5):616–627

Murata A, Ling PM (2012) Asthma diagnosis and management. Emerg Med Clin North Am 30(2):203–222

Nadeau K, Mcdonald-Hyman C, Noth EM, Pratt B, Hammond SK, Balmes J, Tager I (2010) Ambient air pollution impairs regulatory T-cell function in asthma. J Allergy Clin Immunol 126:845–852, e10

Nikolaidis N et al (2006) Allergen induced TFF2 is expressed by mucus-producing airway epithelial cells but is not a major regulator of inflammatory responses in the murine lung. Exp Lung Res 32(10):483–497

Nomura S et al (2005) Alterations in gastric mucosal lineages induced by acute oxyntic atrophy in wild-type and gastrin-deficient mice. Am J Physiol Gastrointest Liver Physiol 288(2):G362–G375

Ober C, Yao T-C (2011) The genetics of asthma and allergic disease: a 21st century perspective. Immunol Rev 242:10–30

Oertel M et al (2001) Trefoil factor family-peptides promote migration of human bronchial epithelial cells: synergistic effect with epidermal growth factor. Am J Respir Cell Mol Biol 25(4):418–424

O'Garra A (1998) Cytokines induce the development of functionally heterogeneous T helper cell subsets. Immunity 8(3):275–283

Ohta K et al (1999) Diesel exhaust particulate induces airway hyperresponsiveness in a murine model: essential role of GM-CSF. J Allergy Clin Immunol 104(5):1024–1030

Pace E et al (2008) Cigarette smoke increases Toll-like receptor 4 and modifies lipopolysaccharide-mediated responses in airway epithelial cells. Immunology 124(3):401–411

Pawankar R et al (2012) Allergic diseases and asthma: a major global health concern. Curr Opin Allergy Clin Immunol 12(1):39–41

Payne DNR et al (2003) Early thickening of the reticular basement membrane in children with difficult asthma. Am J Respir Crit Care Med 167(1):78–82

Perera F, Tang W-Y, Herbstman J, Tang D, Levin L, Miller R, Ho S-M (2009) Relation of DNA methylation of 5′-CpG island of ACSL3 to transplacental exposure to airborne polycyclic aromatic hydrocarbons and childhood asthma. PLoS One 4:e4488

Perl A-K, Riethmacher D, Whitsett J (2011) Conditional depletion of airway progenitor cells induces peribronchiolar fibrosis. Am J Respir Crit Care Med 183(4):511–521

Phillips JM, Goodman JI (2009) Inhalation of cigarette smoke induces regions of altered DNA methylation (RAMs) in SENCAR mouse lung. Toxicology 260:7–15

Prescott SL (2006) The development of respiratory inflammation in children. Paediatr Respir Rev 7(2):89–96

Puddicombe SM et al (2000) Involvement of the epidermal growth factor receptor in epithelial repair in asthma. FASEB J 14(10):1362–1374

Ritz SA et al (2002) On the generation of allergic airway diseases: from GM-CSF to Kyoto. Trends Immunol 23(8):396–402

Royce S et al (2009) Effect of extracellular matrix composition on airway epithelial cell and fibroblast structure: implications for airway remodeling in asthma. Ann Allergy Asthma Immunol 102(3):238–246

Royce S et al (2011) Trefoil factor 2 regulates airway remodeling in animal models of asthma. J Asthma 48(7):653–659

Royce SG et al (2013) Trefoil factor 2 reverses airway remodeling changes in allergic airways disease. Am J Respir Cell Mol Biol 48(1):135–144

Salam MT, Byun H-M, Lurmann F, Breton CV, Wang X, Eckel SP, Gilliland FD (2012) Genetic and epigenetic variations in inducible nitric oxide synthase promoter, particulate pollution, and exhaled nitric oxide levels in children. J Allergy Clin Immunol 129:232–239, e7

Sampath D et al (1999) Constitutive activation of an epithelial signal transducer and activator of transcription (STAT) pathway in asthma. J Clin Invest 103(9):1353–1361

Sandford AJ, Chagani T, Zhu S, Weir TD, Bai TR, Spinelli JJ, Fitzgerald JM, Behbehani NA, Tan WC, Paré PD (2000) Polymorphisms in the IL4, IL4RA, and FCERIB genes and asthma severity. J Allergy Clin Immunol 106:135–140

Santangelo S, Cousins DJ, Winkelmann NEE, Staynov DZ (2002) DNA methylation changes at human Th2 cytokine genes coincide with DNase I hypersensitive site formation during CD4+ T cell differentiation. J Immunol 169:1893–1903

Schaub B, Liu J, Höppler S, Schleich I, Huehn J, Olek S, Wieczorek G, Illi S, Von Mutius E (2009) Maternal farm exposure modulates neonatal immune mechanisms through regulatory T cells. J Allergy Clin Immunol 123:774–782, e5

Stämpfli MR et al (1998) GM-CSF transgene expression in the airway allows aerosolized ovalbumin to induce allergic sensitization in mice. J Clin Invest 102(9):1704

Stevens PT et al (2008) Dysregulated repair in asthmatic paediatric airway epithelial cells: the role of plasminogen activator inhibitor-1. Clin Exp Allergy 38(12):1901–1910

Sunyer J, Torrent M, Muñoz-Ortiz L, Ribas-Fitó NR et al (2005) Prenatal dichlorodiphenyldichloroethylene (DDE) and asthma in children. Environ Health Perspect 113:1787–1790

Suter M, Ma J, Harris AS, Patterson L, Brown KA, Shope C, Showalter L, Abramovici A, Aagaard-Tillery KM (2011) Maternal tobacco use modestly alters correlated epigenome-wide placental DNA methylation and gene expression. Epigenetics 6:1284–1294

Trautmann A et al (2005) Apoptosis and loss of adhesion of bronchial epithelial cells in asthma. Int Arch Allergy Immunol 138(2):142–150

Tsaprouni LG, Ito K, Adcock IM, Punchard N (2007) Suppression of lipopolysaccharide- and tumour necrosis factor-α-induced interleukin (IL)-8 expression by glucocorticoids involves changes in IL-8 promoter acetylation. Clin Exp Immunol 150:151–157

Van Eerdewegh P, Little RD, Dupuis J, Del Mastro RG, Falls K, Simon J, Torrey D, Pandit S, Mckenny J, Braunschweiger K, Walsh A, Liu Z, Hayward B, Folz C, Manning SP, Bawa A, Saracino L, Thackston M, Benchekroun Y, Capparell N, Wang M, Adair R, Feng Y, Dubois J, Fitzgerald MG, Huang H, Gibson R, Allen KM, Pedan A, Danzig MR, Umland SP, Egan RW, Cuss FM, Rorke S, Clough JB, Holloway JW, Holgate ST, Keith TP (2002) Association of the ADAM33 gene with asthma and bronchial hyperresponsiveness. Nature 418:426–430

van Rijt LS et al (2005) In vivo depletion of lung CD11c+ dendritic cells during allergen challenge abrogates the characteristic features of asthma. J Exp Med 201(6):981–991
Vercelli D (2008) Discovering susceptibility genes for asthma and allergy. Nat Rev Immunol 8:169–182
Vignola AM, Kips J, Bousquet J (2000) Tissue remodeling as a feature of persistent asthma. J Allergy Clin Immunol 105(6):1041–1053
Wan H et al (2000) Quantitative structural and biochemical analyses of tight junction dynamics following exposure of epithelial cells to house dust mite allergen Der p 1. Clin Exp Allergy 30(5):685–698
Wei G, Wei L, Zhu J, Zang C, Hu-Li J, Yao Z, Cui K, Kanno Y, Roh T-Y, Watford WT, Schones DE, Peng W, Sun H-W, Paul WE, O'shea JJ, Zhao K (2009) Global mapping of H3K4me3 and H3K27me3 reveals specificity and plasticity in lineage fate determination of differentiating CD4+ T cells. Immunity 30:155–167
Weinberg EG (2011) The WAO white book on allergy 2011-2012: review article. Curr Allergy Clin Immunol 24(3):156–157
White SR (2001) Trefoil peptides in airway epithelium: an important addition to the plethora of peptides. Am J Respir Cell Mol Biol 25(4):401–404
White GP, Hollams EM, Yerkovich ST, Bosco A, Holt BJ, Bassami MR, Kusel M, Sly PD, Holt PG (2006) CpG methylation patterns in the IFNγ promoter in naive T cells: variations during Th1 and Th2 differentiation and between atopics and non-atopics. Pediatr Allergy Immunol 17:557–564
Wills Karp M et al (2012) Trefoil factor 2 rapidly induces interleukin 33 to promote type 2 immunity during allergic asthma and hookworm infection. J Exp Med 209(3):607–622
Woodruff PG et al (2009) T-helper type 2–driven inflammation defines major subphenotypes of asthma. Am J Respir Crit Care Med 180(5):388
Wright NA (1993) Trefoil peptides and the gut. Gut 34(5):577–579
Xepapadaki P, Manios Y, Liarigkovinos T, Grammatikaki E, Douladiris N, Kortsalioudaki C, Papadopoulos NG (2009) Association of passive exposure of pregnant women to environmental tobacco smoke with asthma symptoms in children. Pediatr Allergy Immunol 20:423–429
Xiao C et al (2011) Defective epithelial barrier function in asthma. J Allergy Clin Immunol 128(3):549–556, e1
Yang IV, Schwartz DA (2011) Epigenetic control of gene expression in the lung. Am J Respir Crit Care Med 183(10):1295
Yang M, Kumar RK, Foster PS (2009) Pathogenesis of steroid-resistant airway hyperresponsiveness: interaction between IFN-gamma and TLR4/MyD88 pathways. J Immunol 182(8):5107–5115
Yang L et al (1998) Essential role of nuclear factor kB in the induction of eosinophilia in allergic airway inflammation. J Exp Med 188(9):1739–1750
Yang S-R, Chida AS, Bauter MR, Shafiq N, Seweryniak K, Maggirwar SB, Kilty I, Rahman I (2006) Cigarette smoke induces proinflammatory cytokine release by activation of NF-kappaB and posttranslational modifications of histone deacetylase in macrophages. Am J Physiol Lung Cell Mol Physiol 291:L46–L57
Yang Y, Haitchi HM, Cakebread J, Sammut D, Harvey A, Powell RM, Holloway JW, Howarth P, Holgate ST, Davies DE (2008) Epigenetic mechanisms silence a disintegrin and metalloprotease 33 expression in bronchial epithelial cells. J Allergy Clin Immunol 121:1393–1399, e14
Yuyama N et al (2002) Analysis of novel disease-related genes in bronchial asthma. Cytokine 19(6):287–296
Zahm JM, Chevillard M, Puchelle E (1991) Wound repair of human surface respiratory epithelium. Am J Respir Cell Mol Biol 5(3):242–248
Zhang Y, Moffatt M, Cookson WOC (2012) Genetic and genomic approaches to asthma: new insights for the origins. Curr Opin Pulm Med 18:6–13
Zhou C et al (2011) Epithelial apoptosis and loss in airways of children with asthma. J Asthma 48(4):358–365

Chapter 10
The Significance of Nanoparticles in Medicine and Their Potential Application in Asthma

Stephanie Tortorella and Tom C. Karagiannis

Abstract In an attempt to diagnose and treat highly complex and often heterogeneous diseases, research aims to utilise the modifiable properties of nano-sized particles. Properties such as size, shape, charge, hydrophobicity, and surface chemistry may be altered in order to facilitate and promote targeted cellular uptake. Following the first FDA-approved nanotherapeutic in 1990, more than 40 have been marketed worldwide with multiple nano-based medicines currently in development. Despite promising results, translation from pre-clinical experimentation to a clinical setting has proven to be difficult. In theory, nanoparticles are designed to possess characteristics which address many of the challenges associated with current clinical practices, such as low toxicity, stability, biocompatibility, favourable distribution within target tissue, and beneficial pharmacokinetic profiles. However, the complexity in the identification of the ideal properties which result in such characteristics is inherent of any therapeutic research, especially one as novel and relatively progressive. The development of nanoparticles for localised and systemic delivery to the lung in the treatment of respiratory disease also shows great potential. Due to the highly efficient clearance mechanisms in the lung, the ability for therapeutics to successfully deposit in the respiratory tract is a major challenge. Yet a correlation between exposure to environmentally and occupationally derived ultrafine (nano-sized) particles and respiratory disease has been established. By confirming that ultrafine particles have the capacity to deposit in parts of the lower respiratory tract to elicit a response albeit toxic, such epidemiological studies provide rationale for the

S. Tortorella
Epigenomic Medicine, Baker IDI Heart and Diabetes Institute, The Alfred Medical Research and Education Precinct, 75 Commercial Road, Melbourne, VIC, Australia

T.C. Karagiannis (✉)
Epigenomic Medicine, Baker IDI Heart and Diabetes Institute, The Alfred Medical Research and Education Precinct, 75 Commercial Road, Melbourne, VIC, Australia

Department of Pathology, The University of Melbourne, Parkville, VIC, Australia
e-mail: tom.karagiannis@bakeridi.edu.au

N. Maulik and T. Karagiannis (eds.), *Molecular Mechanisms and Physiology of Disease: Implications for Epigenetics and Health*, DOI 10.1007/978-1-4939-0706-9_10,

development of nano-based pulmonary therapeutics. Although there has been little effort in designing nanoparticle systems for the treatment of lung disease including asthma, current research involves the development of nanocarriers for clinically relevant asthma drugs and antigen (for immunotherapy). With this, continued advancements in the understanding of human disease including asthma, coupled with knowledge regarding interactions between nanoparticle and cell/tissue systems, are required and provide the platform for nano-based therapeutic and diagnostic research.

Keywords Nanoparticles • Drug delivery • Nanomedicine • Tissue regeneration • Diagnostics • Biomedical imaging • Asthma • Respiratory diseases

10.1 Introduction

The application of nanoparticles in medicine involves diagnosis, treatment, disease prevention, reparation following injury, pain relief and the preservation and improvement of human health, using knowledge at the molecular level (Koutsopoulos 2012). Nanoparticles, sized between 1 and 200 nm are defined as small molecular units of compound which behave with distinct biological activity (Jiang et al. 2008). Personalisation of medicine and the individualisation of patient care have become the prevailing theory in therapeutic and diagnostic research, as scientists aim to treat highly complex and heterogeneous diseases, including asthma. In an attempt to apply such a theory to current medical practice, nanoparticles are thought to have the potential to address many of the associated challenges. It requires the development of new materials by engineering molecules at the nanoscale which will interact with the diseased cells, organelles, and/or tissues (Koutsopoulos 2012). By focusing on the use of nanoparticles in the treatment of obstructive respiratory diseases including asthma, this chapter highlights the potential of nanomedicine, and provides the rationale to continue research in this highly evolving and exciting field.

10.2 Nanoparticles in Medicine

The use of nanotechnology in medicine, coupled with the continued advancement in the understanding of human disease, has become the proposed basis in a number of therapeutic and diagnostic applications. The most important of these include the use of nanoparticles in (1) drug delivery, (2) tissue regeneration and (3) diagnostic/imaging applications.

The properties of nanoparticles in medicine promote and facilitate rapid cellular uptake. Furthermore, the thorough knowledge of nanoparticle chemistry is essential for engineering nanoparticles for useful therapeutic and imaging applications. Such nanoparticles should possess properties such as low toxicity, stability, biocompatibility, favourable distribution within tissue and beneficial pharmacokinetic

Table 10.1 Summary of nanoparticle properties and their ideal nature for cellular uptake

Property	Description	Reference
Size	Optimal size of 40–50 nm nanoparticles in cellular uptake	Jiang et al. (2008), Thorek and Tsourkas (2008)
Shape	Determined by the target receptor/cell/molecule/DNA sequence; dictates ability in specific cellular uptake and alter the time within cell	Chithrani and Chan (2007), Qiu et al. (2010), Gratton et al. (2008)
Charge	Neutral nanoparticles minimise non-specific cellular interactions	Verma and Stellacci (2010), Thorek and Tsourkas (2008), Wang et al. (2010)
	Evidence of cellular uptake of negatively charged nanoparticles despite unfavourable interaction with negatively charged cell membrane	
	Positively charged nanoparticles have greatest efficiency in cellular uptake (primarily researched for carriers in drug- and gene- delivery)	
Affinity to water	Hydrophilic and water-soluble nanoparticles are ideal	Verma and Stellacci (2010)
Particle surface	Ligands bound to nanoparticle surface in order to bind to target dependent on the nature, structure and function of the target	Kim et al. (2009), Perrault et al. (2009)

profiles (Kim et al. 2013b). Through the understanding of how the physiochemical properties of nanomaterials, such as size, shape, charge, affinity to water, protein adsorption and particle surface (Stark 2011) affect their interaction with biological systems, their development to become clinically significant nanomedicines may be achieved. Interestingly, the greater surface area per mass compared with larger particles of the same material or chemistry renders nanoparticles more biologically active (Oberdorster et al. 2005). Table 10.1 provides a summary of the most important properties that may be manipulated in the design process of nanoparticle engineering.

The physical and chemical structure of nanoparticles is important in determining its propensity to enter and/or bind to target cells, with the capacity to interact with their biological machinery and elicit a response. Although it was demonstrated that all nanoparticles within the 2–100 nm size range facilitated a change in the signalling processes essential for cellular function, 40- and 50 nm nanoparticles are found to have the greatest effect (Jiang et al. 2008). Indeed, this size was established to be the optimal diameter for spherical nanoparticles, including those with a core of gold, silica and carbon (Chithrani and Chan 2007; Lu et al. 2009; Jin et al. 2009). Each particle acts as a scaffold, with its shape and orientation determining its interactions with the target and influencing uptake into cells (Albanese et al. 2012). The synthesis and ordering of particles larger than 100 nm with differing structural shape, based on total cellular uptake determined that rods have the highest capacity for uptake, followed by sphere, cylinders and cubes (Gratton et al. 2008). Studies using gold nanoparticles of 100 nm and smaller (ideal size for medical applications), however presented results where spherical particles possess a significant advantage over rods (Qiu et al. 2010; Chithrani and Chan 2007). More specifically, it was found that spherical gold nanoparticles of similar size entered cells with 500 % more efficiency, compared to their rod-shaped counterparts due to the greater time required for internalisation (Jiang et al. 2008). In addition to particle size and shape, its charge relative to the microenvironment, is important in determining its fate. While neutral nanoparticles are ideal in the prevention of unwanted and/or toxic cellular–nanoparticle interactions, most charged nanoparticles are responsible for active cellular interactions (Verma and Stellacci 2010). When compared to neutrally- or negatively charged nanoparticles, positively charged nanoparticles exhibit a faster uptake rate (Thorek and Tsourkas 2008). Due to the slight negative charge of the cell membrane, it is suggested that cellular uptake of positively charged nanoparticles is driven by electrostatic attractions (Wang et al. 2010), with these nanoparticles favouring adhesion onto the cell surface. Conversely, clearance of such nanoparticles in an in vivo setting highlights the difficulty in producing a nanoparticle with ideal properties. For instance, neutral nanoparticles persist in the body, while positively charged nanoparticles are cleared most quickly from the blood (Cedervall et al. 2007). This rapid clearance may be avoided with the addition of poly(ethylene) glycol (PEG) to the surface of nanoparticles, which gives them the ability to evade opsonisation regardless of surface charge (Perrault et al. 2009).

The interaction between nanoparticles and their surrounding microenvironment is determined by the particle surface and the nature of the cellular target, in effect influencing cellular uptake, gene expression and toxicity (Albanese et al. 2012). The surface chemistry and its functionalisation provide an effective way to control the interface between the nanoparticle and the biological system (Kim et al. 2013b) through the capacity to determine and control receptor binding (Lynch et al. 2007). A wide variety of synthetic and natural ligands have been attached to the surface of nanoparticles in order to dictate the cellular response. A study examined gold nanoparticles, with surface modifications based on the addition of oligonucleotides and their ability to be internalised by murine endothelial cells (Rosi et al. 2006). Although negatively charged (previously demonstrated to be poorly internalised) these nanoparticles were shown to be readily taken up by the cells with the capacity to regulate gene expression. Further analysis using a fluorescence-based assay method revealed absorption of serum proteins on the nanoparticle surface through electrostatic and hydrophobic interaction with the cell membrane (Giljohann et al. 2007).

Due to the exploitable nature of each property in the development and engineering of nanoparticles, uptake within the cell and the location of deposition may be determined. Furthermore, nanoparticle design establishes their behaviour and the direct or indirect interaction with the biological target. These interactions may be benign, beneficial or lead to dysfunction in proteins, genes and cells (Kim et al. 2013b). Nanoparticles may enter cells via two major endocytic mechanisms: phagocytosis and pinocytosis (Conner and Schmid 2003). The way in which cellular uptake occurs is strongly dependent on nanoparticle surface chemistry, as well as specific cell surface receptors (Kim et al. 2013b). The most important cellular uptake mechanism for application in nanomedicine is micropinocytosis which occurs in all cell types, but varies according to the accompanying mediator (Kumari et al. 2010). Two distinct pathways exist; the first involves micropinocytosis mediated by

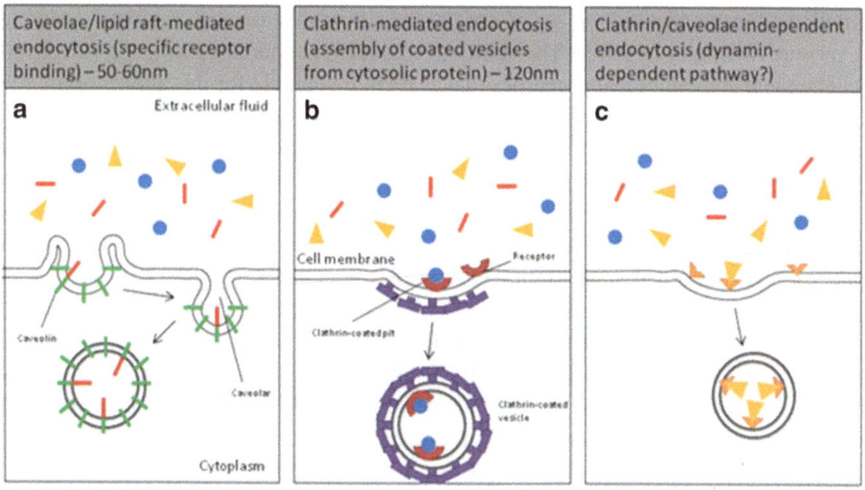

Fig. 10.1 Important cellular uptake mechanisms in nanomedicine. Cellular uptake of nanoparticles via three major endocytic (micropinocytosis) mechanisms differing in accompanying mediator: (**a**) caveolae/lipid raft-dependent, (**b**) clathrin-mediated and (**c**) clathrin/caveolar independent (unknown mediator). A dynamin-dependent pathway has been proposed as a possible independent transport mechanism. The uptake mechanism used is largely dependent on the interaction between cell type and/or state, and the physical and/or chemical nature of the nanoparticle itself

clathrin, caveolae/lipid raft and/or independently of clathrin/caveolae (Fig. 10.1). A dynamin-dependent endocytosis pathway has been proposed as a possible independent transport mechanism (Mayor and Pagano 2007). Interestingly, a recent study produced results that indicated a distinct difference between the way in which an array of cationic gold nanoparticles (differing in surface chemistry) entered cancer and normal cells (Kim et al. 2013b). For example the gold nanoparticle labelled NP2 by the authors, entered the cancer cell line HeLa via a caveolae-dependent pathway, in comparison to the normal cells (MCF10A cell line) which utilised both a scavenger receptor-dependent and dynamin-dependent pathway for its uptake. Through this discovery, it has been shown that there is a capacity to engineer nanoparticles based on the target cell uptake mechanisms in order to increase selectivity for medical applications.

10.2.1 Nanoparticles in Drug- and Gene-Delivery

The transportation of therapeutic compounds to a target site has proven to be difficult with conventional methods. Systemic drug administration is characterised by limited efficacy, poor pharmacokinetics and lack of selectivity (Nevozhay et al. 2007). It is proposed that these limitations may be overcome by controlling drug delivery, with nanoparticle drug carriers shown to have the greatest potential.

It allows for the specific release of drug at the site of disease with minimal systemic exposure (May and Li 2013). The ability to engineer nanoparticles with specific physicochemical and biological properties makes them favourable for this application.

Utilising knowledge of disease pathogenesis at the molecular scale, nanoparticle drug carriers may target specific cell types while preventing unnecessary treatment of healthy tissue, and thus in turn, minimise toxicity (Paulo et al. 2011). Nanoparticles may have the capacity to protect a drug, protein or gene from enzymatic degradation and pH hydrolytic effect (Akagi et al. 2011), allowing its contents to be delivered in a complete functional state to integrate or elicit a cellular response (Slowing et al. 2007). For instance, composition of γ-PGA nanoparticles encapsulating ovalbumin (OVA) has been shown to affect cellular uptake and intracellular degradation (Akagi et al. 2011). Soluble OVA taken up by RAW264 cells was degraded significantly faster than the OVA within the nanocarrier. In turn, this capability enhances drug concentration in target tissues, and therefore lowers the dose of drug required (Wilczewska et al. 2012). An example involves the administration of curcumin as a potential anticancer drug (Strimpakos and Sharma 2008). A dose–response study of curcumin reports that its detection in plasma was found to be in only two of twenty-four healthy subjects following dosage of 10–12 g oral curcumin (Lao et al. 2006). Yet it has been established that an intake of more than 8 g per day was not acceptable to patients (Cheng et al. 2001). A study conducted using nano-based drug delivery of curcumin (Theracurmin) demonstrates a ninefold increase in bioavailability in a rat model (Shaikh et al. 2009). Administration of Theracurmin (150 mg) to healthy human subjects observed plasma curcumin levels comparable to the level after intake of 8 g of conventional curcumin without increasing toxicity (Kanai et al. 2012).

It is also proposed that nanocarriers may aid in overcoming drug resistance by concealing the therapeutic compound from cellular efflux or intracellular destruction (Wang et al. 2010). Edelfosine (ET), a prototype of a promising class of anticancer drugs known as synthetic alkyl-lysophospholipids, has been shown to selectively target tumour cells through its ability to induce a rapid apoptotic response (Mollinedo et al. 1997; Selivanov et al. 2010). Despite this, in vitro studies using the cell line K562 established from a patient with chronic myeloid leukaemia have shown resistance to the drug. In support, other authors demonstrate a slower internalisation rate of ET in resistant cells (K562) than in those cells sensitive to ET (HL-60) (Tsutsumi et al. 1998). A recent study using lipid nanoparticles as a drug carrier for ET produced positive results, which indicate that using a lipid nanocarrier allowed for the internalisation of the drug in both sensitive and resistant leukaemia cells (Lasa-Saracíbar et al. 2013). Moreover, the lipid nanoparticles preserve the apoptotic effect of ET through caspase activation and other unknown pathways in both ET-sensitive and -resistant cells.

Immunomodulatory and stem cell therapies may also be enhanced by nanoparticle drug carriers of compounds that would otherwise be limited due to cellular and biological constraints (Das et al. 2008; Maia et al. 2010). Tolerance acquisition to rapamycin, used as an immunosuppressive drug following allograft transplantation (Tullius et al. 2001), has emerged due to long-term administration (Heesom et al. 1998). Its mechanism of action involves the inhibition of T cell proliferation in response to growth factors, and inhibits B cell activation (Abraham and Wiederrecht 1996). The drug is also known to affect dendritic cell function by inhibiting antigen uptake (Monti et al. 2003) and inducing apoptosis in these cells (Woltman et al. 2003). The encapsulation of rapamycin in poly(D,L-lactic-co-glycolic acid) (PLGA)-nanoparticles has been reported to have the capacity to downregulate ICAM-1, an important adhesion molecule for T cell polarisation (Das et al. 2008). In addition, the delivery of rapamycin by a nanocarrier allows for the maintenance of an immunosuppressive state when compared to free rapamycin. Retinoic acid, important in the neuronal cell differentiation from progenitor cells (Kim et al. 2009), has been shown to have poor water solubility, and a complex transportation system to deliver it to the cell nucleus (Guan et al. 2001). Maia et al., engineered polyethylenimine (PEI) and dextran sulphate nanoparticles to act as a nanocarrier for retinoic acid and determined their capacity to deliver the molecule to neural stem cells (Maia et al. 2010). They report effective loading of retinoic acid into their formulation, with rapid uptake by immature murine neuronal cells leading to their differentiation into neurons. It was also shown that the concentration of retinoic acid required to elicit a neurological response was significantly lower when the molecule was internalised in nanoparticles when compared to its free form. Although the benefits of nanocarriers in drug- and gene-delivery seem to be infinite, the difficulty in the translation of in vitro and in vivo studies to a clinical setting should not be underestimated. However due to the developments already made in this field, continued research into the delivery of therapeutic compounds using nanoparticles is vital for possible future applications.

10.2.2 Tissue Regeneration Using Nanoscale Subcomponents

Through the incorporation of genes, proteins and/or cells within nano-sized particle systems, it is proposed that human tissue may be regenerated following disease-induced injury (Koutsopoulos 2012). The ultimate goal in this area centres on the engineering of a nano-based scaffold that not only provides structural support for cell integration but also has the capacity to regulate cell proliferation, differentiation and migration to form functional tissue (McMahon et al. 2012).

In order to achieve this, it is accepted that the scaffold must mimic the physiological and mechanical properties of the native microenvironment (Koutsopoulos 2012). The best illustration of this involves the regeneration of the skin's dermal layers following injury. The dermis is a collagenous layer, composed of extracellular matrix (ECM) fibres which are sized at the nanoscale (30–130 nm) (Nelson and Tien 2006). ECM also functions as a modulator of cell proliferation, migration, differentiation and apoptosis, with the capacity to provide structural support and facilitate nutrient transport and diffusion (Vracko 1974). Through the production of a nanoparticle scaffold that emulates physiological conditions and tissue properties (such as ECM architecture), the engineered system may promote cell proliferation, migration and attachment with clinically beneficial applications, assisting the processes of healing and regeneration (Kim et al. 2013a). These systems for the regeneration of the dermis have been investigated at a preliminary stage with positive results. The fabrication of mesh fibre scaffolds of poly(caprolactone) (PCL) through a method of electrospinning have been engineered and tested for its ability to mimic natural ECM and to examine cellular behaviour and the interaction between cells and this nanofibre matrix (Venugopal and Ramakrishna 2005). It was found that this scaffold aided fibroblast proliferation in comparison with control tissue culture plates after 72 h. In summary, this investigation established that a PCL-coated matrix may be suitable as a biodegradable scaffold for fibroblast proliferation and migration, with the ability to support the attachment and growth of human dermal fibroblasts in vitro. A similar experiment was conducted using dextran/PLGA electrospun nanofibre scaffolds, with mouse dermal fibroblasts demonstrating cellular attachment, migration and survival within the scaffold itself (Pan et al. 2006). Naturally derived nanofibre scaffolds have also been tested with comparable results, including those engineered from silk fibroin nanofibres (Min et al. 2004) and chitin (Noh et al. 2006) with human keratinocytes and fibroblasts.

10.2.3 Diagnostic and Bio-Imaging Applications of Nanoparticles

The advancement in the use of imaging techniques to identify, monitor and treat disease has allowed for the relatively non-invasive, quantitative and real-time visualisation of tissue, its morphology and function. In order to reveal physiological and pathological changes at the cellular level, imaging is required to be highly sensitive and specific.

Through the utilisation of the general properties associated with nanomedicine, nanoparticles have been engineered as probes and contrasting agents in the imaging process. Nanoparticle design and synthesis, especially its core composition, is based upon the principles of the imaging technique and its method of detection. For example, nanoparticles used in magnetic resonance imaging (MRI) require paramagnetic or superparamagnetic materials to induce contrast, utilising this knowledge through the labelling of such particles with gadolinium ions (Mulder et al. 2006) or iron oxide cores (McCarthy et al. 2007). Conversely, for computed tomography (CT) scanning which requires the attenuation of X-rays, nanoparticle contrasting agents are based on elements such as gold (Kim et al. 2007), bismuth (Rabin et al. 2006) or iodine (Hyafil et al. 2007). It is also rationalised that due to the size of nanoparticles, their capacity for wider distribution with less hindrance by biological barriers may aid in the identification of disease which remains undetected when more crude imaging methods are used. The ability to modify the physiochemical properties of nanoparticles, through a number of functional substitutions, has allowed for their use as probes to enhance tissue contrast and investigate specific biological changes in vivo (Minchin and Martin 2009). To demonstrate, the imaging of fine vessels in mice by MRI using gadolinium-conjugated polyamidoamine nanoparticles showed that the total number of gadolinium substitutions is important to determine optimal contrast (Sato et al. 2001). Additionally, nanoparticles engineered to cross the blood brain barrier with functionalised carbohydrate, and an iron oxide core (van Kasteren et al. 2009), have the capacity to image important components and processes of the brain rendered defective in many neurological diseases.

Exploitation of knowledge in disease pathology permits engineering of nanoparticles with a functional surface to target specifically. For instance, nanoparticles deliberately modified with surface folate have been shown to accumulate in cancer cells that over-express a folate transporter (Rossin et al. 2005). Rossin et al. demonstrated this folate targeting with ^{64}Cu-labelled nanoparticles for positron emission topography (PET) imaging of tumour xenografts in mice. Similarly, folate-conjugated paramagnetic iron oxide nanoparticles for MRI have been developed with comparable results (Choi et al. 2004). More recently, nanoparticles conjugated with folic acid and 4-ethylnyl-*N*-ethyl-1, 8 were tested for the targeting of the folate receptor on human cancer cell lines injected in mice (Hou et al. 2011). Results verified the use of folate-targeting nanoparticles in tumour targeting and imaging, with specificity and accumulation in tumour site indicated. Another example involves the addition of tripeptide arginine-glycine-aspartic acid (RGD) molecules on the surface of nanoparticles in order to target sites of angiogenesis, utilising the interaction between RGD and over-expressed (in cancer) integrins and cell surface proteins including fibronectin. Through the production and application of RGD-labelled perfluorocarbon nanoparticles to tumour tissue in mice, the amount of angiogenesis prior to and following treatment may be quantified (Schmieder et al. 2008). RGD-targeted CdSe/ZnS quantum dots for MRI have also been developed (Mulder et al. 2009). Detection of early stage tumour development has also been reported with the use of superparamagnetic iron oxide, PDA nanoparticles conjugated with RGD peptide (Lin et al. 2012). Target specificity and high affinity allow these nanoparticles to differentiate the expression level of integrin receptor on several cell lines and tumours by in vitro and in vivo MRI when compared to normal cellular material.

In addition to the ability to engineer nanoparticles based on cellular targeting, research developing multimodal imaging nanoparticles for the detection by more than one imaging technique may be useful in diagnostic and treatment settings (Minchin and Martin 2009). Research conducted to generate a combined imaging technique has tested the use of both quantum dots with paramagnetic materials for fluorescent imaging and MRI in order to image-activated endothelial cells by nanoparticle-tagging integrin-binding proteins (Mulder et al. 2005). Through the interchangeable substitution of tagging ligand, it is proposed that this imaging technique may be useful in a range of pathological conditions, including inflammation, apoptosis, atherosclerosis and angiogenesis. Furthermore, the production of iron oxide nanoparticles for MRI that are tagged with the near-infrared fluorophore Cy5.5 for florescence detection are found to have a high affinity to metalloproteinase-2 when conjugated with the venom of the scorpion *Leiurus quinquestriatus* (Veiseh et al. 2005). Since metalloproteinase-2 has been shown to be over-expressed in several human brain tumour subtypes, differentiation from surrounding normal brain tissue may be observed. Recent advancements indicate that multimodal imaging may also be used in understanding disease progression and behaviour. Visible and near-infrared light emitting quantum dots in conjugation with superparamagnetic iron oxide nanoparticles were developed for fluorescence imaging and MRI techniques to monitor the migration of dendritic cells in mice (Mackay et al. 2011). By using a combinational imaging technology, with one possessing high sensitivity and the other with high resolution, many of the limitations involved with individual approaches are overcome and additional information regarding disease state may be obtained (Jennings and Long 2009).

10.3 Clinical Use of Nanotechnology-Based Advancements: Current and Near Future Impact

Since the first marketed nanotherapeutic was approved in 1990, more than 40 have been approved worldwide (Schutz et al. 2013). The most successful application to date involves the ability to selectively target malignant cells with cytotoxic drugs, and subsequently protect healthy tissue from harm (Bourzac 2012). Over 20 % of therapeutic nanoparticle systems in clinical use have been developed for the treatment of cancer (Schutz et al. 2013). The earliest and perhaps most documented example of this is the nanomedicine, Doxil. Doxorubicin, an anthracycline antibiotic is used in the treatment of a wide variety of cancers including, breast, ovarian, sarcomas, lymphomas and acute leukaemias (Speth et al. 1988). Despite the capacity to treat such a range of malignancies, the administration of doxorubicin is dose-limiting with the drug found to accumulate in the heart causing cardiotoxicity (Carvalho et al. 2014). Doxil, a doxorubicin-carrying nanomedicine, utilises previous knowledge in order to limit such detrimental effects on cardiac tissue. It has been established that 100-nm particles are unable to diffuse through healthy blood vessels due to their size, and yet have the capacity to escape the highly permeable tumour vasculature (Gabizon and Martin 1997). Approved in 1995, Doxil utilised such knowledge with doxorubicin particles loaded into lipid bubbles in order to create a drug-delivering nanoparticle system (Barenholz 2012). At the site of the tumour, the accumulated drug is released from its carrier and attacks the surrounding cells. Incidence of congestive heart failure is reduced to one-third in patients given Doxil in comparison to conventional doxorubicin (O'Brien et al. 2004). Nanoparticles are also currently used in diagnostic applications with iron oxide nanoparticles used for MRI and gold nanoparticles for bio-imaging (Schutz et al. 2013).

The process of RNA interference (siRNA), where small amounts of RNA are administered to silence crucial disease-related genes, has proven to be one of the most promising applications for nanomedicine. The processing of siRNA at the cytoplasm or cellular organelles elicits a rapid response as compared to DNA, which requires uptake into the cell nucleus. Recently, synthetic siRNA-based drugs for the silencing of target genes, have been applied in various diseases in vivo including human papillomavirus (Niu et al. 2006), ovarian (Halder et al. 2006) and bone cancer (Takeshita et al. 2005). Engineering nanoparticles in order to carry siRNA requires properties which facilitate cellular uptake and endosomal release as naked siRNA is quickly degraded by enzymes and is unable to cross cellular membranes due to its high negative charge (Paulo et al. 2011). Although delivery of siRNA has been shown to be more efficient using a viral vector as compared to their non-viral counterparts, pre-clinical studies have shown that this type of delivery system has the capacity to induce inflammation (Thomas et al. 2007). Due to this adverse effect, optimisation of delivery in the form of non-viral nanoparticles has become the focus for current research. In vitro and in vivo studies have already exhibited promising results, with Yano et al., successfully delivering the siRNA, B717 (specific for the oncogene bcl-2), to the tumour site in a mouse model of liver metastases using a novel cationic liposome, LIC-101 (Yano et al. 2004). The nanoparticle complex also exhibited anti-tumour activity, with the inhibition of bcl-2 protein and subsequent tumour growth. Phase I clinical trials of a PEG-grafted monolamellar liposome (SNALP; TKM-080301) have also displayed significant anti-tumour effect in a range of solid tumours with liver metastases through the delivery of siRNA specific for the PLK1 gene (http://clinicaltrials.gov).

Multiple nanomedicines are currently in developmental stages, with some exhibiting much promise in clinical trials. Unlike Doxil, which uses a simple lipid-based drug carrier for doxorubicin, the current approach utilises the capacity of polymers to be engineered in order to develop nanomedicines based on disease pathogenesis. Following from Doxil, Celsion have developed a temperature-sensitive liposome nanocarrier for doxorubicin (Thermodox), It is currently in phase III clinical trials for hepatocellular carcinoma and phase II trials for breast cancer and colorectal liver metastases (May and Li 2013). Rather than the passive release of doxorubicin in Doxil, Thermodox has shown temperature-dependent triggered drug release in the range of 39–42 °C (Li et al. 2013). Designed to release the drug at these hyperthermic temperatures causing high localised drug concentration at the tumour site, Thermodox remains stable at physiological body temperatures (May and Li 2013). Multiple pre-clinical studies with xenograft models (colon HCT116, squamous cell FaDu, prostate PC-3, ovarian SKOV-3) demonstrate an improvement in efficacy as compared to free doxorubicin and Doxil (Yarmolenko et al. 2010). Despite these promising results, there is always an opportunity for improvement in nanocarrier formulation.

10.4 Challenges Associated with the Application of Nanoparticle Theory to Medicine

Despite the success of Doxil and other nano-based therapies/diagnostic tools, coupled with the recent advancements in nanomedicine, the ability to improve drug delivery has proven to be difficult. In the case of Doxil, which is passively excluded from healthy tissue due to its size, the drug fails to penetrate the tumour but instead clusters at its perimeter (Bourzac 2012). To date most research has been conducted in vitro and in vivo, with little known about the efficacy of nano-based therapeutics in human systems.

The translation of research into a clinical setting requires much improvement, with positive results in pre-clinical testing not always indicative of results in human trials. According to recent studies, only 20–30 % of Phase I drugs successfully pass through to Phase II/III and from this low percentage, 40–50 % of those drugs in Phase III actually obtain approval (DiMasi et al. 2010). This low success rate may be contributed to a number of factors including failure in toxicology studies revealed during human trials, and an inability to demonstrate a significant increase in efficacy when compared to the current protocols (Schutz et al. 2013). Indeed, nanomaterial itself may be toxic and have the potential to affect the functional capacity of the liver, kidney, lungs, heart, vascular system and/or immune system (Schutz et al. 2013). For example, the Phase II/III clinical trial of a polymer-based doxorubicin nanocarrier (Livatag) was suspended following the presentation of severe pulmonary adverse affects in a subset of patients (Heidel and Davis 2011). It has since been granted to resume the clinical trial, with the investigation ongoing. Another major challenge is to engineer nanocarriers with significantly improved levels of efficacy as compared to the current drug delivery method. When compared to efficacy data for Doxil (Gabizon et al. 1994), researchers who developed a micelle nanocarrier for doxorubicin (NK911) to treat solid tumours showed a 100-fold decrease in plasma concentration, and a 400-fold increase in plasma clearance (Matsumura et al. 2004). This paper suggests that NK911 is less stable in plasma than Doxil, and is not delivered to tumours more efficiently than the current treatment although promising results were observed at a dose of 50 mg m^{-2}. Despite these findings in Phase I clinical trials, a phase II clinical trial of NK911 for the treatment of metastatic pancreatic cancer. The drug, however, did not progress from phase II trials because it could not be demonstrated to have a significant therapeutic benefit as compared to Doxil.

Furthermore, advancements in understanding the interactions between nanoparticles, the drug or molecule in question, and different cell types are essential in order to engineer nanoparticles with specificity to its cellular and more specifically, intracellular target. Figure 10.2 provides a summarised representation of the challenges in the design and engineering of nanoparticle systems prior to the clinical testing phase. As previously identified, the formulation of nanoparticles with optimal conditions and properties, including size, charge, shape and functional moieties is trial and error at this stage, with knowledge of disease at the molecular level essential. Yet, the exact mechanisms which contribute to such disease and the important pathways in its pathogenesis are not always clearly defined in order to form the basis for targeting. Furthermore, research is required into understanding the nature of the intended molecule and its interaction with the nanocarrier. Absorption and/or linkage of the molecule in relation to the nanoparticle must be studied to determine dose (concentration of drug), ideal transportation in order for release and the subsequent elicitation of the desired response. For the capability to target a specific cell type, as well as targeting those cells in a particular state, the evaluation of how nanoparticles enter the cell and its intended target organelle (i.e. cytoplasm, nucleus) is important. With that, new ways in which to monitor such interactions may be useful to overcome the challenges associated with preliminary design, and the inability to translate novel nanoparticle systems to clinical relevancy.

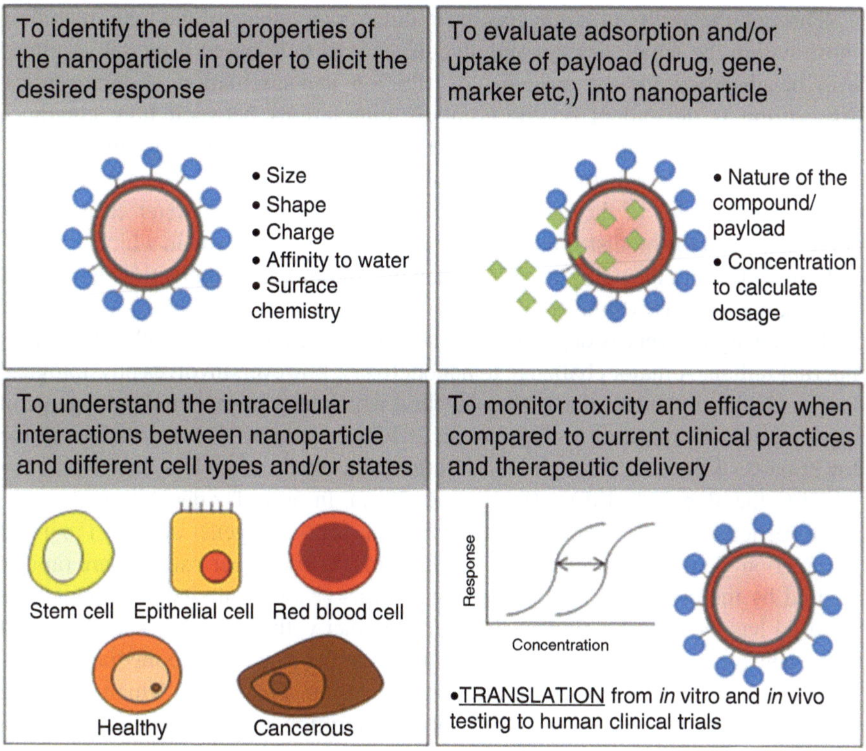

Fig. 10.2 Challenges associated with experimental application and clinical translation of nanoparticle theory in medicine (adapted from Paulo et al. 2011). Pre-experimental design is a necessary step in order to identify optimal properties for a nanoparticle. Acquiring knowledge of the molecular and cellular mechanisms of disease, coupled with the understanding of nanoparticle theory is essential to elicit the desired response. Furthermore good nanoparticle design, including the capacity to uptake and/or adsorb the pay load with efficiency, will aid in subsequent pre-clinical and clinical challenges. Understanding the intracellular interactions between different cells and/or cellular states will also benefit nanoparticle design, specificity and potentially, efficacy. In order to translate pre-clinical data (in vitro and in vivo studies) to human trials, monitoring efficacy and toxicity as compared to current practices and therapeutic delivery is a major challenge, with promising results in cell culture and animals rarely translating to human settings

10.5 Progression in Pulmonary Delivery of Nanoparticles

The lung provides a large surface area for increased biological efficacy, avoidance of hepatic metabolism and better local and systemic action (Kaur et al. 2012). The development of nanoparticles for localised non-invasive delivery to the lung in the treatment of respiratory disease shows great potential in theoretical and preliminary studies. It allows for the relative uniform distribution of drug throughout the pulmonary tissue, improvement in pharmacokinetics, sustained drug release, improvement in patient compliance, decrease in the risk of unwanted systemic effects and has the capacity to induce cellular uptake (Mansour et al. 2009; Sung et al. 2007).

Through the ability to target specific cellular components of the lung, including those within the small airways and alveoli, nanoparticle-based drug delivery may also be absorbed for the systemic distribution and circulation of therapeutics. Absorption is dependent on the complex interactions between lung clearance mechanisms and the molecular characteristics of airborne particles. Interestingly, nanoparticles possess distinct behavioural properties that are significantly different to larger molecules of the same chemical composition (Oberdorster et al. 2005). Size, shape and the biodistribution within the lung dictate the way in which nanoparticles interact with the pulmonary system, while the surface composition determines the ensuing response (Sung et al. 2007).

Pulmonary delivery is dependent on the nanoparticle itself, as well as the lung's internal milieu. A major challenge to nanodelivery, however, involves physiological barriers including mucosal and alveolar fluid which may be rendered pathogenic in some respiratory disease (Mahajan et al. 2010). Particles of 1–5 μm deposit in the lower parts of the tract through inertial impaction, while those less than 1 μm may travel to the alveoli to diffuse (Sung et al. 2007). In order for deposition to occur, these particles must be designed (in the case of drug- and gene- delivery) to avoid elimination via numerous clearance mechanisms. While larger particles are rapidly cleared by the cilia of the upper respiratory tract (Kaur et al. 2012), nanoparticle material has the capacity to deposit and/or interact with the lung tissue. The most important for nano-sized molecules include alveolar macrophage-mediated clearance, endocytosis and absorption into the circulation. The way in which a particle enters respiratory cells, such as diffusion, absorption and/or deposition, is also dependent on the molecule itself. Factors such as solubility, size, ionisation state and surface area also contribute to its bioavailability. Smaller-sized nanoparticles are less impeded by the physiological barriers of the lung when compared to larger nanoparticles. It has been shown that those particles with a diameter of 120 nm have the capacity to move through the sputum layer of patients with cystic fibrosis, whereas the mobility of larger ones (diameters 270–560 nm) are significantly reduced (Rytting et al. 2008). In addition, drugs with a lower pH are largely uncharged which allow them to cross the alveolar mucosal membranes freely and produce faster absorption compared to charged drugs with a higher pH (Labiris and Dolovich 2003). Lipophilic drugs have also been shown to be rapidly uptaken by cells of the pulmonary system due to their capacity to interact with the mucus layer (Sung et al. 2007). Through the continued research in understanding such mechanisms in the lung, and the advancement of nanoparticle systems that utilise this knowledge, the basis for a method in the administration of therapeutics to disease specific cells and/or tissues may be established.

Different strategies have been developed in order to facilitate and enhance drug absorption via the mucosal pulmonary route for localised and/or systemic effect (Kaur et al. 2012). The use of a nanocarrier for the delivery of a drug to the lung has been previously limited due to an inefficiency in the deposition of nanoparticles within the alveoli, with the majority of the inhaled dose eliminated through exhalation (Sung et al. 2007). Improvement on inhalation devices, including nebulisers, metered dose inhalers and dry powder inhalers may aid in delivery to the lower respiratory tract (Willis et al. 2012). However, the numerous challenges associated with the direct delivery of nanoparticles to the lung tissue have limited the success of their application in clinical drug administration to date. Systemic routes, including oral and intravenous, continue to be the most practised method for the administration of lung-targeted drugs. In the case of nanomedicine, intravenous administration is most relevant, with delivery directed to the general circulation with the intent of transporting drug to the lung (Wei and Zhao 2014). Although this method is passive as compared to the direct pulmonary method via inhalation and may cause adverse systemic effects, many of the challenges observed through the efficient clearance methods of the lung and difficulties in drug administration to the lower respiratory tract are eliminated. Both intravenous and pulmonary delivery methods are thus important in the development of lung-targeting nanoparticle systems.

10.6 Exposure to Environmental- and Occupational-Nanoparticles, and Its Role in Airway Disease

Although the amount of nano-sized particles measured in ambient air varies from relatively low (0.5–2 μg/m^3) (Hughes et al. 1980) to significantly high (54–137.5 μg/m^3) (Rao et al. 2012), exposure to any such pollutants is thought to be of toxicological significance. In addition to those ultrafine particles formed during manufacturing and combustion, many nano-sized particles are used to create products such as cosmetics, sunscreens, electronics, tyres and fuel cells (Oberdorster et al. 2005). Continuous exposure to various environmental pollutants and pathogens is thought to be a central factor in the development and progression of multiple respiratory diseases, including asthma (Kauffmann and Demenais 2012). Both naturally derived and industrially produced nano-sized particles have the capacity to enter the airway, and efficiently deposit through diffusional mechanisms in all regions of the respiratory tract (Oberdorster et al. 2005). Systemic exposure may also be achieved through the pulmonary route. Indeed, the same properties (described previously) that make nanoparticle theory so attractive for development in nanomedicine may also prove to be toxic when various ultrafine particles interact with respiratory cells.

In order to advance the understanding of nanoparticle toxicity for both its contribution to human disease and its effect on nanoparticle engineering, results on the direct involvement of these particles in ambient air have been widely reported. Studies including epidemiological studies, controlled human clinical studies and in vivo/in vitro research have produced results indicating a significant association between exposure to environmental and occupational nanoparticle, and development of disease. Exposure to such particles has been associated with pulmonary inflammation, immunomodulation and adverse systemic effects including blood hypercoagulability, contributing to cardiovascular disease (Oberdorster 2001; Mossman et al. 2007). Several epidemiologic studies have found correlations with exposure to ambient nano-sized particles and adverse respiratory disease (von Klot et al. 2002; Pekkanen et al. 1997; Penttinen et al. 2006). The evaluation of deposition rates and toxicology of carbon-based ultrafine particles has also been conducted in controlled clinical settings (Anderson et al. 1990; Frampton et al. 2004; Pietropaoli et al. 2004; Chalupa et al. 2004). Although clear associations can be made between ambient air nano-sized particle exposure and respiratory disease in human studies, the pathogenic mechanisms involved in such an association may only be established in vitro and in vivo. By understanding this relationship at a molecular level, research into both disease pathogenesis and nanomedical advancements can be made utilising such knowledge for therapeutic research.

Titanium dioxide (TiO_2) nanoparticles, widely used in the manufacturing of a number of products, have been shown to induce the production of reactive oxygen species, cytotoxicity and genotoxicity in vitro (Bhattacharya et al. 2009). In addition, in vivo studies in mice have resulted in the findings that instillation of nano-TiO_2 can induce severe emphysema-like lung injury in mice, while intraperitoneal injections lead to interstitial pneumonia, thrombosis of pulmonary vasculature and systemic lesion formation (Chen et al. 2006, 2009). In conjunction, these nanoparticles impair reparation of the airway epithelium through its ability to inhibit adhesion of Clara cells to fibronectin (Zarogiannis et al. 2013). Exposure to TiO_2 or gold (Au) nanoparticles and their ability to modulate a diisocyanate-induced asthma in a murine model have also been investigated (Hussain et al. 2011). Results demonstrate that exposure to these nanoparticles exacerbates lung inflammation with a significant increase in inflammatory cell counts (including neutrophils and macrophages) and pro-inflammatory cytokines (including MMP-9).

Another example involves exposure to elemental carbon nano-sized particles, which have been shown to have a potential adjuvant effect on allergic airway inflammation (Inoue et al. 2008). Although shown to relatively inert, carbon nanoparticles represent a major component of ambient air pollution and diesel exhaust (Hildemann et al. 1994). Alessandrini et al. have produced multiple papers in an attempt to establish the exact mechanisms involved in the association between ultrafine carbon particles and airway inflammation. Mice exposed to carbon nanoparticles were subsequently sensitised to OVA, an allergen well established to induce airway inflammation. Inhalation of these particles caused a significant increase in inflammatory cell infiltrate, and pro-inflammatory cytokines IL-4, IL-5 and IL-13 when compared to those OVA-sensitised mice not exposed to the nano-sized particles (Alessandrini et al. 2006). This same study also reported an increase in mucus production, peribronchiolar and perivascular inflammation and enhanced airway hyperresponsiveness. Following these preliminary studies, exposure to carbon nanoparticles in OVA-sensitised and -challenged mice displayed increased allergen-induced lung lipid peroxidation and NF-KB activation in comparison to non-exposed OVA mice, demonstrating a critical role of carbon ultrafine particles for the induction of oxidative stress (Alessandrini et al. 2009). In confirmation of these studies, others report similar findings (Takano et al. 1998; Inoue et al. 2007).

The particle concentration in ambient air is dominated by ultrafine particles that have been shown to contribute to the morbidity of respiratory disease as discussed. Due to their ability to deposit in the lower respiratory tract to promote airway inflammation, allergic sensitisation and airway hyperresponsiveness (Chalupa et al. 2004), research involving the elucidation of the interaction between particle and lung tissue is important with immediate clinical and regulatory consequences. It also provides a theoretical basis for the engineering of nanomedicines as nano-sized matter have the capacity to enter pulmonary circulation and induce a response. While this response is pathogenic, engineering nanoparticles with therapeutic benefits based on previously determined properties, and drug delivery capabilities may in fact utilise the knowledge found in these toxicity studies. Interestingly, a recent publication has found that inert 50 nm polystyrene nanoparticles possess the capacity to induce resistance in mice to OVA-induced sensitisation (Hardy et al. 2012). This study demonstrates that their pulmonary administration in mice inhibits the generation of allergic airway inflammation and the induction of Th2-based immunity. While the majority of studies report pathogenic and very toxic consequences for nanoparticle exposure, this study is the first of its kind to demonstrate that not all nanoparticles necessarily promote and/or exacerbate airway inflammation. It is with these findings, coupled with the capacity to design nanoparticles with specific properties that provides rationale for the future development of a nano-based approach in the treatment of asthma and other pulmonary lung disease.

10.7 Nanodelivery and the Therapeutic Potential of Nanoparticles in Asthma

Although the theoretical basis for nanoparticle-based therapy for respiratory disease, including asthma is rationalised through previous studies, the ability to translate into clinical or even pre-clinical success has proven to be difficult. The engineering of nanoparticles for therapeutic benefit in lung disease requires further knowledge involved in the interactions between the nanoparticles themselves, and respiratory, immune and inflammatory cells which characterise disease. Over the last decade, there has been little effort in designing nanoparticle systems for the treatment of chronic obstructive lung disease as research focuses on the development of targeted nanoparticles for lung cancer (Vij 2012). Currently, applications in nanoparticle theory for the potential treatment in asthma involve nano-based drug delivery and the deliberate modulation of immunity (important in allergic asthma).

Nanocarriers may be employed in order to control the duration of local or systemic drug/gene release and activity in the lung, affecting their bioavailability and distribution (Kaur et al. 2012). Their ability to carry therapeutic compounds to the lower respiratory tract for deposition and site-specific targeting make nano-based drug delivery an attractive path for therapeutic research. For the elucidation of potential drug carriers for the successful delivery of asthma therapeutics to the site of inflammation and remodelling, current research uses clinically relevant drugs. For example, corticosteroids are used for the treatment of active disease by attenuating the inflammatory response following antigen exposure and severe exacerbations (Shefrin and Goldman 2009). Dexamethasone acetate (DEX)-loaded solid lipid nanoparticles have been administered to mice by an intravenous route in order to target the lung (Xiang et al. 2007). Preliminary results in vitro showed that these nanoparticles exhibit a rapid release mechanism, followed by a gradual release of DEX. In mice, when compared to free DEX, DEX-loaded nanoparticles allowed for reduced uptake by liver and spleen macrophages, and a significantly higher uptake by the macrophages in the lung. By directly targeting the lung, a nanocarrier has the potential to limit corticosteroid dosage and unwanted systemic effects. Another paper, using betamethasone disodium phosphate (BP)-loaded polymeric nanoparticles to evaluate the effects of its intravenous administration in OVA-sensitised mice (Matsuo et al. 2009), found sustained anti-inflammatory activity. By showing a significant decrease in the number of inflammatory cellular infiltrate, as well as in the amount of important mediators IL-13 and IL-4, this paper produced results indicating a strong, rapid, long-lasting therapeutic benefit. Furthermore, a reduction in the number of circulating neutrophils may also indicate a potential for these nanoparticles to treat non-allergic asthma, which is largely thought to be corticosteroid-resistant (Mann and Chung 2006). Liposome-encapsulated steroids have also been developed, with Konduri et al., demonstrating the improved efficacy in liposomal budesonide for the treatment of OVA-induced allergic airway disease in mice (Konduri et al. 2003). In addition to the administration of corticosteroids for the treatment of severe active asthma, acute exacerbations may be prevented using short-acting bronchodilators. Bhavna et al., aimed to improve the efficacy of the β_2-agonist salbutamol administration through the engineering of nano-sized particles (Bhavna et al. 2009). In vitro and in vivo studies in healthy human subjects showed a 2.3-fold increase in total lung deposition as compared to the currently used micronized salbutamol. Collectively, these studies establish the requirement for further research into nanoparticle systems for drug delivery in asthma, and the potential for such to improve efficacy, limit adverse systemic effects and reduce dosage levels.

A recent study involving the production of biodegradable hydroxybenzyl alcohol (HBA)-incorporated polyoxalate (HPOX) nanoparticles as a novel treatment in airway inflammatory disease was tested in cell culture and in vivo (Yoo et al. 2013). This report suggests that these nanoparticles have the capacity to inhibit expression of iNOS, COX-2 and IL-1β (important markers of inflammation and oxidative stress) in LPS-stimulated macrophages as compared to those macrophages treated with free HBA. In addition, OVA-sensitised mice treated with HPOX nanoparticles (intratracheal administration) resulted in significant reductions in OVA-induced airway inflammation in comparison to untreated sensitised mice. HPOX nanoparticles are thought to exert their anti-inflammatory effects by inhibiting leukocyte migration, the production of pro-inflammatory mediators such as IL-4, COX-2 and iNOS, and reducing the number of circulating eosinophils. Although showing the capacity to be of therapeutic benefit independent of any conventional antiasthmatic drug, the next step in HPOX research is to establish its capacity as a nanocarrier.

Due to the limitations in the administration of siRNA for therapeutic benefit, including the inefficiency in cellular targeting, rapid degradation by nucleases and elimination (Dykxhoorn and Lieberman 2006), current research involving siRNA-based therapeutics focus on the possibility of local administration and/or delivery using nanoparticle systems. Success in the local administration of siRNA has already been achieved in the eye, targeting age-related macular degeneration and diabetic retinopathy (Watts and Corey 2010). Several obstacles encountered in systemic delivery of siRNA may be overcome by local administration, including delivery to the lung via inhalation (Durcan et al. 2008). Pulmonary delivery provides direct access to the airway epithelium which is affected in lung diseases such as asthma (Merkel and Kissel 2011). Although local administration may prove to be more efficacious than systemic delivery of siRNA for pulmonary delivery, the ability to translate this into a disease setting may prove difficult. Mucus hypersecretion and airway inflammation, hallmark characteristics of asthma, act as barriers even when siRNA-based drugs are locally delivered. To overcome this, nanoparticle systems designed to evade such disease mechanisms could potentially increase delivery of siRNA to affected areas. In vivo testing in mice have shown that the delivery of siRNA using a nanocarrier is up to ten times more effective in treating respiratory disease as compared to the current method of delivery, which involves the direct administration of free siRNA to the lung (Akinc et al. 2008). The most widely researched nanocarriers for siRNA delivery in the lung and in respiratory disease are chitosan-based, due to its capacity to adhere to mucus. An example of chitosan nanoparticles used in vivo involves the intranasal delivery of siRNA targeted towards the NS1 gene in modulating respiratory syncytial virus (Zhang et al. 2005). Zhang et al., report a significant decrease in viral titers in the lung and decreased inflammation and airway reactivity compared to the untreated control.

Atopy is the most established attributable factor to the development of asthma in children (Arbes et al. 2007). Exposure to a number of airborne pathogens has the capacity in these patients to activate an allergic inflammatory response, mediated by a Th2 cell-dominated immune pathway (Krug et al. 1996). Allergen-specific immunotherapy (SIT) involves the administration of allergen at increasing concentrations in order to obtain a state of hyposensitisation by limiting exacerbations following natural allergen exposure (De Souza Reboucas et al. 2012). Through the modulation of the immune system by altering T cell differentiation to the desired profile (Th1, Th2, Th17 and/or Treg), desensitisation may occur. In the case of allergy, SIT aims to increase the Th1/Th0 ratio, and thereby decrease Th2/Th0 (Alvaro et al. 2013). The role of immunotherapy in the management and treatment of atopic asthma remains controversial. While some human trials produce positive results, concluding that SIT can cause a reduction in asthma symptoms, and an improvement in airway hyperresponsiveness (Abramson et al. 2010), others are critical (Bousquet et al. 2011) as current approaches show limitations in long-term efficacy (or tolerance) and safety (risk of serious allergic reaction and anaphylaxis) (De Souza Reboucas et al. 2012). Through the knowledge collected involving the use of nanoparticles in drug- and gene-delivery, and the ability to manipulate their design for targeting purposes, nanoparticle-based allergen-delivery systems have received much interest as potential adjuvants for SIT (Broos et al. 2010). Some may be designed to activate an immune response by displaying receptors and ligands physiologically expressed on dendritic cells in order to activate antigen presenting cells and inducing T cell activation and differentiation (Moon et al. 2012). Others,

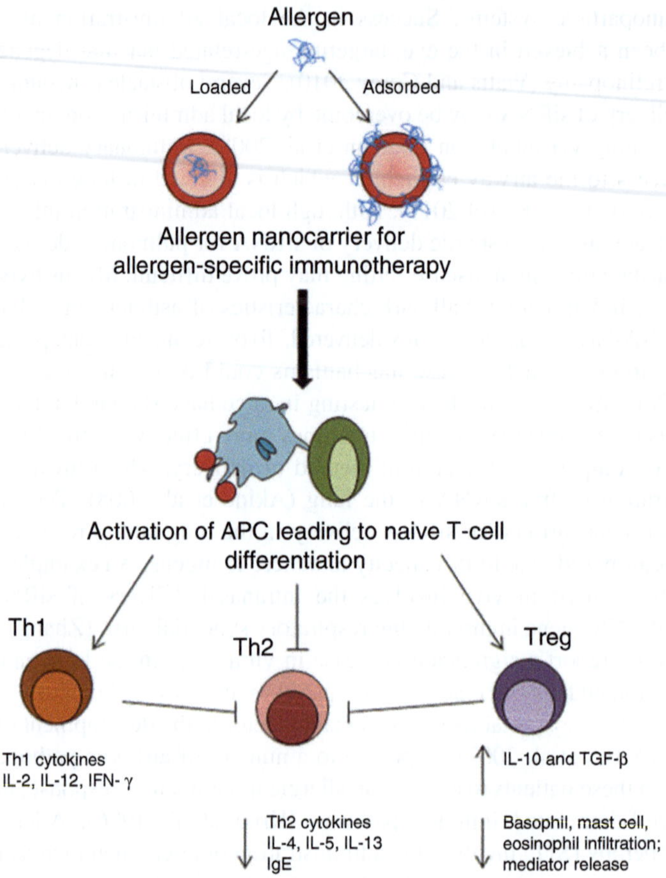

Fig. 10.3 Allergen-specific immunotherapy utilising nanoparticle theory for allergen-delivery (adapted from De Souza Reboucas et al. 2012). Immunomodulation in allergic disease, including asthma following successful delivery of allergen via nanoparticle-based systems to sensitisation tissue and/or cells, is indicated by *arrows*; *truncated arrows* indicate inhibitory effects. As a largely Th2-mediated disease, asthma immunotherapy modulates naive T cell differentiation to inhibit Th2-cell production and its associated cytokines (IgE included). In order to produce a state of desensitisation and/or tolerance, these T cells are induced to follow a Th1 and/or Treg differentiation pathway

perhaps the most important for SIT involve the delivery of antigen to their target cell and/or tissue. Figure 10.3 illustrates the immunomodulation of allergic disease using a theoretical nanoparticle-based immunotherapy. Preliminary studies have used a range of nanoparticle systems to test their capacity to deliver antigen to the target immune cells for allergen desensitisation. For example, PLGA nanoparticles loaded with Bet v1 (major allergen of birch pollen) have been shown to reduce the Th2 predominant response, by increasing IgG2a levels, IFN-γ abd IL-10 levels (Scholl et al. 2004, 2006). In addition, poly(gamma-glutamic acid) nanoparticles were evaluated for their ability to encapsulate OVA and elicit an immune response in HL-60 cells (Akagi et al. 2005). It was also found in subsequent studies that these nanoparticles were able to activate human monocyte-derived dendritic cells and stimulate inflammatory cytokine production, as well as upregulate immunomodulatory mediators involved in efficient T cell differentiation (Broos et al. 2010). The use of poly(vinylpyrrolidone) nanoparticles was also tested with the successful encapsulation of the antigen *Aspergillus fumigatus* (Madan et al. 1997). This study found a sustained increase in IgG levels for approximately 12 weeks in comparison with the observation of sustained IgG levels for 7 days in mice immunised with free antigen. With these results, many of the limitations of current immunotherapy trials including adverse systemic reactions, and the inability to induce long-term desensitisation have been shown, albeit in preliminary in vitro and in vivo, to be overcome with nanoparticle-based antigen delivery.

10.8 Conclusion

The application of nanotechnology in medicine, including its use in drug delivery, tissue regeneration and diagnostic/imaging techniques, is dependent on a number of factors in order to become clinically relevant. The ability to manipulate nanoparticle properties such as size, shape, charge, affinity to water and perhaps most importantly surface functionality, in order to target cellular or disease components with specificity (Albanese et al. 2012), has become a prevailing theory for the treatment of highly complex disease. Nanocarriers for drug- and gene-delivery are engineered in order to facilitate cellular uptake with the capacity to enhance compound concentration to the disease site by limiting systemic exposure and evading physiological elimination mechanisms (Kim et al. 2013b). By achieving this, nanoparticle delivery systems are able to improve efficacy, limit the potential for adverse effects (and rejection, in the case of tissue regeneration) and increase specificity to disease components. Despite the approval of more than 40 nanotherapeutics worldwide (Schutz et al. 2013), and the numerous nano-based medicines currently in clinical trials, the ability to improve delivery has proven to be difficult. Challenges in both the design phase and the subsequent translation into human systems have limited the success in producing nanomedicines for clinical use.

The development of nanoparticles for localised delivery to the lung in the treatment of respiratory disease shows great potential, although systemic administration continues to be the most practised method (Kaur et al. 2012). This is due, at least in part to the highly complex interactions between pulmonary cells and nanoparticles. As the interface between the outside environment and the body's internal milieu, the respiratory system has numerous mechanisms for the elimination of airborne pathogens. For pulmonary delivery, nanoparticles must be designed to evade these mechanisms in order to deposit in the lung efficiently. A correlation between exposure to environmental- and occupational-nanoparticles including titanium oxide and carbon ultrafine particles, and respiratory disease has been widely reported, and provides rationale for further investigation of nanoparticles in a therapeutic capacity (Kauffmann and Demenais 2012). Over the last decade, there has been little effort in designing nanoparticles for the treatment of chronic obstructive lung disease, including asthma. Promising results using nano-based delivery of clinically relevant asthma drugs, including corticosteroids and bronchodilators (Xiang et al. 2007; Bhavna et al. 2009), show the potential for nanoparticles to distribute drugs to the lower respiratory tract, improve efficacy and provide extended therapeutic benefit. Additionally, delivery of antigen for immune desensitisation in allergen immunotherapy studies has also been of interest (De Souza Reboucas et al. 2012). Theoretically, nanoparticles may have the capacity to deliver antigen to the lung more effectively (and with less potential for serious systemic reactions and anaphylaxis) in order to modulate the immune system in allergic asthma.

Utilising knowledge of complex disease and the ever evolving understanding of nanomedicine, novel approaches for the treatment of asthma may be designed, and perhaps in the future, become a reality.

Acknowledgements The support of the Australian Institute of Nuclear Science and Engineering is acknowledged. TCK was the recipient of AINSE awards. TCK is a Future Fellow and Epigenomic Medicine Laboratory is supported by the Australian Research Council. Supported in part by the Victorian Government's Operational Infrastructure Support Program.

References

Abraham RT, Wiederrecht GJ (1996) Immunopharmacology of rapamycin. Annu Rev Immunol 14:483–510

Abramson M, Puy R, Weiner J (2010) Injection allergen immunotherapy for asthma. Cochrane Database Syst Rev (8):CD001186

Akagi T, Kaneko T, Kida T, Akashi M (2005) Preparation and characterization of biodegradable nanoparticles based on poly(gamma-glutamic acid) with l-phenylalanine as a protein carrier. J Control Release 108:226–236

Akagi T, Shima F, Akashi M (2011) Intracellular degradation and distribution of protein-encapsulated amphiphilic poly(amino acid) nanoparticles. Biomaterials 32:4959–4967

Akinc A, Zumbuehl A, Goldberg M, Leshchiner E, Busini V, Hossain N, Bacallado S, Nguyen D, Fuller J, Alvarez R, Borodovsky A, Borland T, Constien R, de Fougerolles A, Dorkin JR, Narayanannair Jayaprakash K, Jayaraman M, John M, Koteliansky V, Manoharan M, Nechev L, Qin J, Racie T, Raitcheva D, Rajeev K, Sah DWY, Soutschek JR, Toudjarska I, Vornlocher H-P, Zimmermann T, Langer R, Anderson D (2008) A combinatorial library of lipid-like materials for delivery of RNAi therapeutics. Nat Biotechnol 26:561–569

Albanese A, Tang PS, Chan WCW (2012) The effect of nanoparticle size, shape, and surface chemistry on biological systems. Annu Rev Biomed Eng 14:1–16

Alessandrini F, Schulz H, Takenaka S, Lentner B, Karg E, Behrendt H, Jakob T (2006) Effects of ultrafine carbon particle inhalation on allergic inflammation of the lung. J Allergy Clin Immunol 117:824–830

Alessandrini F, Beck-Speier I, Krappmann D, Weichenmeier I, Takenaka S, Karg E, Kloo B, Schulz H, Jakob T, Mempel M, Behrendt H (2009) Role of oxidative stress in ultrafine particle-induced exacerbation of allergic lung inflammation. Am J Respir Crit Care Med 179:984–991

Alvaro M, Sancha J, Larramona H, Lucas JM, Mesa M, Tabar AI, Martinez-Cañavate A (2013) Allergen-specific immunotherapy: update on immunological mechanisms. Allergol Immunopathol 41(4):265–272

Anderson PJ, Wilson JD, Hiller FC (1990) Respiratory tract deposition of ultrafine particles in subjects with obstructive or restrictive lung disease. Chest 97:1115–1120

Arbes S, Gergen P, Vaughn B, Zeldin D (2007) Asthma cases attributable to atopy: results from the Third National Health and Nutrition Examination Survey. J Allergy Clin Immunol 120:1139–1145

Barenholz Y (2012) Doxil®—the first FDA-approved nano-drug: lessons learned. J Control Release 160:117–134

Bhattacharya K, Davoren M, Boertz J, Schins R, Hoffmann E, Dopp E (2009) Titanium dioxide nanoparticles induce oxidative stress and DNA-adduct formation but not DNA-breakage in human lung cells. Part Fibre Toxicol 6:17

Bhavna, Ahmad FJ, Mittal G, Jain GK, Malhotra G, Khar RK, Bhatnagar A (2009) Nano-salbutamol dry powder inhalation: a new approach for treating broncho-constrictive conditions. Eur J Pharm Biopharm 71:282–291

Bourzac K (2012) Nanotechnology: carrying drugs. Nature 491:S58–S60

Bousquet P, Calderón MS, Demoly P, Larenas DSE, Passalacqua G, Bachert C, Brozek J, Canonica GW, Casale T, Fonseca J, Dahl R, Durham S, Merk H, Worm M, Wahn U, Zuberbier T, Schünemann H, Bousquet J (2011) The Consolidated Standards of Reporting Trials (CONSORT) Statement applied to allergen-specific immunotherapy with inhalant allergens: a Global Allergy and Asthma European Network (GA(2)LEN) article. J Allergy Clin Immunol 127:49–56, 56.e1

Broos S, Lundberg K, Akagi T, Kadowaki K, Akashi M, Greiff L, Borrebaeck CAK, Lindstedt M (2010) Immunomodulatory nanoparticles as adjuvants and allergen-delivery system to human dendritic cells: implications for specific immunotherapy. Vaccine 28:5075–5085

Carvalho FS, Burgeiro A, Garcia R, Moreno AJ, Carvalho RA, Oliveira PJ (2014) Doxorubicin-induced cardiotoxicity: from bioenergetic failure and cell death to cardiomyopathy. Med Res Rev 34(1):106–135

Cedervall T, Lynch I, Lindman S, Berggård T, Thulin E, Nilsson H, Dawson KA, Linse S (2007) Understanding the nanoparticle-protein corona using methods to quantify exchange rates and affinities of proteins for nanoparticles. Proc Natl Acad Sci 104:2050–2055

Chalupa D, Morrow P, Oberdörster G, Utell M, Frampton M (2004) Ultrafine particle deposition in subjects with asthma. Environ Health Perspect 112:879–882

Chen H-W, Su S-F, Chien C-T, Lin W-H, Yu S-L, Chou C-C, Chen JJW, Yang P-C (2006) Titanium dioxide nanoparticles induce emphysema-like lung injury in mice. FASEB J 20:2393–2395

Chen J, Dong X, Zhao J, Tang G (2009) In vivo acute toxicity of titanium dioxide nanoparticles to mice after intraperitioneal injection. J Appl Toxicol 29:330–337

Cheng AL, Hsu CH, Lin JK, Hsu MM, Ho YF, Shen TS, Ko JY, Lin BR, Ming Shiang W, Yu HS, Jee SH, Chen GS, Chen TM, Chen CA, Lai MK, Pu YS, Pan MH, Wang YJ, Tsai CC, Hsieh CY (2001) Phase I clinical trial of curcumin, a chemopreventive agent, in patients with high-risk or pre-malignant lesions. Anticancer Res 21:2895–2900

Chithrani BD, Chan WCW (2007) Elucidating the mechanism of cellular uptake and removal of protein-coated gold nanoparticles of different sizes and shapes. Nano Lett 7:1542–1550

Choi H, Choi SR, Zhou R, Kung HF, Chen IW (2004) Iron oxide nanoparticles as magnetic resonance contrast agent for tumor imaging via folate receptor-targeted delivery1. Acad Radiol 11:996–1004

Conner SD, Schmid SL (2003) Regulated portals of entry into the cell. Nature 422:37–44

Das S, Haddadi A, Veniamin S, Samuel J (2008) Delivery of rapamycin-loaded nanoparticle down regulates ICAM-1 expression and maintains an immunosuppressive profile in human CD34+ progenitor-derived dendritic cells. J Biomed Mater Res A 85A:983–992

de Souza Reboucas J, Esparza I, Ferrer M, Sanz M, Irache J, Gamazo C (2012) Nanoparticle adjuvants and delivery systems for allergen immunotherapy. J Biomed Biotechnol 2012:474605

Dimasi JA, Feldman L, Seckler A, Wilson A (2010) Trends in risks associated with new drug development: success rates for investigational drugs. Clin Pharmacol Ther 87:272–277

Durcan N, Murphy C, Cryan S-A (2008) Inhalable siRNA: potential as a therapeutic agent in the lungs. Mol Pharm 5:559–566

Dykxhoorn DM, Lieberman J (2006) Running interference: prospects and obstacles to using small interfering RNAs as small molecule drugs. Annu Rev Biomed Eng 8:377–402

Frampton M, Utell M, Zareba W, Oberdörster GN, Cox C, Huang L-S, Morrow P, Lee FE, Chalupa D, Frasier L, Speers D, Stewart J (2004) Effects of exposure to ultrafine carbon particles in healthy subjects and subjects with asthma. Res Rep Health Eff Inst 126:1–47

Gabizon A, Martin F (1997) Polyethylene glycol-coated (pegylated) liposomal doxorubicin. Drugs 54:15–21

Gabizon A, Isacson R, Libson E, Kaufman B, Uziely B, Catane R, Ben Dor CG, Rabello E, Cass Y, Peretz T (1994) Clinical studies of liposome-encapsulated doxorubicin. Acta Oncol 33:779–786

Giljohann DA, Seferos DS, Patel PC, Millstone JE, Rosi NL, Mirkin CA (2007) Oligonucleotide loading determines cellular uptake of DNA-modified gold nanoparticles. Nano Lett 7:3818–3821

Gratton SEA, Ropp PA, Pohlhaus PD, Luft JC, Madden VJ, Napier ME, Desimone JM (2008) The effect of particle design on cellular internalization pathways. Proc Natl Acad Sci 105:11613–11618

Guan K, Chang H, Rolletschek A, Wobus AM (2001) Embryonic stem cell-derived neurogenesis. Retinoic acid induction and lineage selection of neuronal cells. Cell Tissue Res 305:171–176

Halder J, Kamat A, Landen C, Han L, Lutgendorf S, Lin Y, Merritt W, Jennings N, Chavez Reyes A, Coleman R, Gershenson D, Schmandt R, Cole S, Lopez Berestein G, Sood A (2006) Focal adhesion kinase targeting using in vivo short interfering RNA delivery in neutral liposomes for ovarian carcinoma therapy. Clin Cancer Res 12:4916–4924

Hardy C, Lemasurier J, Belz G, Scalzo Inguanti K, Yao J, Xiang S, Kanellakis P, Bobik A, Strickland D, Rolland J, O'Hehir R, Plebanski M (2012) Inert 50-nm polystyrene nanoparticles that modify pulmonary dendritic cell function and inhibit allergic airway inflammation. J Immunol 188:1431–1441

Heesom KJ, Avison MB, Diggle TA, Denton RM (1998) Insulin-stimulated kinase from rat fat cells that phosphorylates initiation factor 4E-binding protein 1 on the rapamycin-insensitive site (serine-111). Biochem J 336:39–48

Heidel J, Davis M (2011) Clinical developments in nanotechnology for cancer therapy. Pharm Res 28:187–199

Hildemann LM, Klinedinst DB, Klouda GA, Currie LA, Cass GR (1994) Sources of urban contemporary carbon aerosol. Environ Sci Technol 28:1565–1576

Hou J, Zhang Q, Li X, Tang Y, Cao M-R, Bai F, Shi Q, Yang C-H, Kong D-L, Bai G (2011) Synthesis of novel folate conjugated fluorescent nanoparticles for tumor imaging. J Biomed Mater Res A 99A:684–689

Hughes TJ, Pellizzari E, Little L, Sparacino C, Kolber A (1980) Ambient air pollutants: collection, chemical characterization and mutagenicity testing. Mutat Res 76:51–83

Hussain S, Vanoirbeek JAJ, Luyts K, de Vooght V, Verbeken E, Thomassen LCJ, Martens JA, Dinsdale D, Boland S, Marano F, Nemery B, Hoet PHM (2011) Lung exposure to nanoparticles modulates an asthmatic response in a mouse model. Eur Respir J 37:299–309

Hyafil F, Cornily J-C, Feig JE, Gordon R, Vucic E, Amirbekian V, Fisher EA, Fuster V, Feldman LJ, Fayad ZA (2007) Noninvasive detection of macrophages using a nanoparticle contrast agent for computed tomography. Nat Med 13:636–641

Inoue K, Takano H, Yanagisawa R, Sakurai M, Abe S, Yoshino S, Yamaki K, Yoshikawa T (2007) Effects of nanoparticles on lung physiology in the presence or absence of antigen. Int J Immunopathol Pharmacol 20:737–744

Inoue K-I, Koike E, Yanagisawa R, Takano H (2008) Impact of diesel exhaust particles on Th2 response in the lung in asthmatic mice. J Clin Biochem Nutr 43:199–200

Jennings LE, Long NJ (2009) 'Two is better than one'-probes for dual-modality molecular imaging. Chem Commun (Camb) 28(24):3511–3524

Jiang W, Kim BYS, Rutka J, Chan WCW (2008) Nanoparticle-mediated cellular response is size-dependent. Nat Nanotechnol 3:145–150

Jin H, Heller DA, Sharma R, Strano MS (2009) Size-dependent cellular uptake and expulsion of single-walled carbon nanotubes: single particle tracking and a generic uptake model for nanoparticles. ACS Nano 3:149–158

Kanai M, Imaizumi A, Otsuka Y, Sasaki H, Hashiguchi M, Tsujiko K, Matsumoto S, Ishiguro H, Chiba T (2012) Dose-escalation and pharmacokinetic study of nanoparticle curcumin, a potential anticancer agent with improved bioavailability, in healthy human volunteers. Cancer Chemother Pharmacol 69:65–70

Kauffmann F, Demenais F (2012) Gene-environment interactions in asthma and allergic diseases: challenges and perspectives. J Allergy Clin Immunol 130:1229–1240

Kaur G, Narang RK, Rath G, Goyal AK (2012) Advances in pulmonary delivery of nanoparticles. Artif Cells Blood Substit Immobil Biotechnol 40:75–96

Kim D, Park S, Lee JH, Jeong YY, Jon S (2007) Antibiofouling polymer-coated gold nanoparticles as a contrast agent for in vivo X-ray computed tomography imaging. J Am Chem Soc 129:7661–7665

Kim M-J, Ahn K, Park S-H, Kang H-J, Jang B, Oh S-J, Jeong Y-J, Heo J-I, Suh J-G, Lim S, Ko Y-J, Huh S-O, Kim S, Park J-B, Kim J, Jo S, Lee J-Y (2009) SIRT1 regulates tyrosine hydroxylase expression and differentiation of neuroblastoma cells via FOXO3a. FEBS Lett 583:1183–1188

Kim HN, Jiao A, Hwang NS, Kim MS, Kang DH, Kim D-H, Suh K-Y (2013a) Nanotopography-guided tissue engineering and regenerative medicine. Adv Drug Deliv Rev 65(4):536–558

Kim ST, Saha K, Kim C, Rotello VM (2013b) The role of surface functionality in determining nanoparticle cytotoxicity. Acc Chem Res 46(3):681–691

Konduri K, Nandedkar S, Dãnes N, Suzara V, Artwohl J, Bunte R, Gangadharam PRJ (2003) Efficacy of liposomal budesonide in experimental asthma. J Allergy Clin Immunol 111:321–327

Koutsopoulos S (2012) Molecular fabrications of smart nanobiomaterials and applications in personalized medicine. Adv Drug Deliv Rev 64:1459–1476

Krug N, Madden J, Redington AE, Lackie P, Djukanovic R, Schauer U, Holgate ST, Frew AJ, Howarth PH (1996) T-cell cytokine profile evaluated at the single cell level in BAL and blood in allergic asthma. Am J Respir Cell Mol Biol 14:319–326

Kumari S, Mg S, Mayor S (2010) Endocytosis unplugged: multiple ways to enter the cell. Cell Res 20:256–275

Labiris NR, Dolovich MB (2003) Pulmonary drug delivery. Part I: physiological factors affecting therapeutic effectiveness of aerosolized medications. Br J Clin Pharmacol 56:588–599

Lao C, Ruffin M, Normolle D, Heath D, Murray S, Bailey J, Boggs M, Crowell J, Rock C, Brenner D (2006) Dose escalation of a curcuminoid formulation. BMC Complement Altern Med 6:10

Lasa-Saracíbar B, Estella-Hermoso de Mendoza A, Mollinedo F, Odero MD, Blanco-Príeto MJ (2013) Edelfosine lipid nanosystems overcome drug resistance in leukemic cell lines. Cancer Lett 334(2):302–310

Li L, ten Hagen TLM, Hossann M, Süss R, van Rhoon G, Eggermont AMM, Haemmerich D, Koning G (2013) Mild hyperthermia triggered doxorubicin release from optimized stealth thermosensitive liposomes improves intratumoral drug delivery and efficacy. J Control Release 168:142–150

Lin R-Y, Dayananda K, Chen T-J, Chen C-Y, Liu G-C, Lin K-L, Wang Y-M (2012) Targeted RGD nanoparticles for highly sensitive in vivo integrin receptor imaging. Contrast Media Mol Imaging 7:7–18

Lu F, Wu S-H, Hung Y, Mou C-Y (2009) Size effect on cell uptake in well-suspended, uniform mesoporous silica nanoparticles. Small 5:1408–1413

Lynch I, Cedervall T, Lundqvist M, Cabaleiro-Lago C, Linse S, Dawson KA (2007) The nanoparticle-protein complex as a biological entity; a complex fluids and surface science challenge for the 21st century. Adv Colloid Interface Sci 134–135:167–174

Mackay PS, Kremers G-J, Kobukai S, Cobb JG, Kuley A, Rosenthal SJ, Koktysh DS, Gore JC, Pham W (2011) Multimodal imaging of dendritic cells using a novel hybrid magneto-optical nanoprobe. Nanomedicine 7:489–496

Madan T, Munshi N, de TK, Maitra A, Usha Sarma P, Aggarwal SS (1997) Biodegradable nanoparticles as a sustained release system for the antigens/allergens of Aspergillus fumigatus: preparation and characterisation. Int J Pharm 159:135–147
Mahajan S, Roy I, Xu G, Yong K-T, Ding H, Aalinkeel R, Reynolds J, Sykes D, Nair B, Lin E, Prasad P, Schwartz S (2010) Enhancing the delivery of anti retroviral drug "Saquinavir" across the blood brain barrier using nanoparticles. Curr HIV Res 8:396–404
Maia J, Santos T, Aday S, Agasse F, Cortes L, Malva JO, Bernardino L, Ferreira L (2010) Controlling the neuronal differentiation of stem cells by the intracellular delivery of retinoic acid-loaded nanoparticles. ACS Nano 5:97–106
Mann B, Chung K (2006) Blood neutrophil activation markers in severe asthma: lack of inhibition by prednisolone therapy. Respir Res 7:59
Mansour H, Rhee Y-S, Wu X (2009) Nanomedicine in pulmonary delivery. Int J Nanomedicine 4:299–319
Matsumura Y, Hamaguchi T, Ura T, Muro K, Yamada Y, Shimada Y, Shirao K, Okusaka T, Ueno H, Ikeda M, Watanabe N (2004) Phase I clinical trial and pharmacokinetic evaluation of NK911, a micelle-encapsulated doxorubicin. Br J Cancer 91:1775–1781
Matsuo Y, Ishihara T, Ishizaki J, Miyamoto K-I, Higaki M, Yamashita N (2009) Effect of betamethasone phosphate loaded polymeric nanoparticles on a murine asthma model. Cell Immunol 260:33–38
May JP, Li S-D (2013) Hyperthermia-induced drug targeting. Expert Opin Drug Deliv 10:511–527
Mayor S, Pagano RE (2007) Pathways of clathrin-independent endocytosis. Nat Rev Mol Cell Biol 8:603–612
Mccarthy JR, Kelly KA, Sun EY, Weissleder R (2007) Targeted delivery of multifunctional magnetic nanoparticles. Nanomedicine 2:153–167
McMahon RE, Wang L, Skoracki R, Mathur AB (2012) Development of nanomaterials for bone repair and regeneration. J Biomed Mater Res B Appl Biomater 101B:387–397
Merkel OM, Kissel T (2011) Nonviral pulmonary delivery of siRNA. Acc Chem Res 45:961–970
Min B-M, Lee G, Kim SH, Nam YS, Lee TS, Park WH (2004) Electrospinning of silk fibroin nanofibers and its effect on the adhesion and spreading of normal human keratinocytes and fibroblasts in vitro. Biomaterials 25:1289–1297
Minchin R, Martin D (2009) Nanoparticles for molecular imaging—an overview. Endocrinology 151:474–481
Mollinedo F, Fernández-Luna JL, Gajate C, Martín-Martín B, Benito A, Martínez-Dalmau R, Modolell M (1997) Selective induction of apoptosis in cancer cells by the ether lipid ET-18-OCH3 (Edelfosine): molecular structure requirements, cellular uptake, and protection by Bcl-2 and Bcl-XL. Cancer Res 57:1320–1328
Monti P, Mercalli A, Leone B, Valerio D, Allavena P, Piemonti L (2003) Rapamycin impairs antigen uptake of human dendritic cells. Transplantation 75:137–145
Moon J, Huang B, Irvine D (2012) Engineering nano- and microparticles to tune immunity. Adv Mater 24:3724–3746
Mossman B, Borm P, Castranova V, Costa D, Donaldson K, Kleeberger S (2007) Mechanisms of action of inhaled fibers, particles and nanoparticles in lung and cardiovascular diseases. Part Fibre Toxicol 4:4
Mulder WJM, Koole R, Brandwijk RJ, Storm G, Chin PTK, Strijkers GJ, de Mello DC, Nicolay K, Griffioen AW (2005) Quantum dots with a paramagnetic coating as a bimodal molecular imaging probe. Nano Lett 6:1–6
Mulder WJM, Strijkers GJ, van Tilborg GAF, Griffioen AW, Nicolay K (2006) Lipid-based nanoparticles for contrast-enhanced MRI and molecular imaging. NMR Biomed 19:142–164
Mulder WM, Castermans K, Beijnum J, Oude Egbrink MA, Chin PK, Fayad Z, Löwik CW, Kaijzel E, Que I, Storm G, Strijkers G, Griffioen A, Nicolay K (2009) Molecular imaging of tumor angiogenesis using alphavbeta3-integrin targeted multimodal quantum dots. Angiogenesis 12:17–24
Nelson CM, Tien J (2006) Microstructured extracellular matrices in tissue engineering and development. Curr Opin Biotechnol 17:518–523

Nevozhay D, Kańska U, Budzyńska R, Boratyński J (2007) [Current status of research on conjugates and related drug delivery systems in the treatment of cancer and other diseases]. Postepy Hig Med Dosw (Online) 61:350–360

Niu XY, Peng ZL, Duan WQ, Wang H, Wang P (2006) Inhibition of HPV 16 E6 oncogene expression by RNA interference in vitro and in vivo. Int J Gynecol Cancer 16:743–751

Noh HK, Lee SW, Kim J-M, Oh J-E, Kim K-H, Chung C-P, Choi S-C, Park WH, Min B-M (2006) Electrospinning of chitin nanofibers: degradation behavior and cellular response to normal human keratinocytes and fibroblasts. Biomaterials 27:3934–3944

O'Brien MER, Wigler N, Inbar M, Rosso R, Grischke E, Santoro A, Catane R, Kieback DG, Tomczak P, Ackland SP, Orlandi F, Mellars L, Alland L, Tendler C (2004) Reduced cardiotoxicity and comparable efficacy in a phase III trial of pegylated liposomal doxorubicin HCl (CAELYX™/Doxil®) versus conventional doxorubicin for first-line treatment of metastatic breast cancer. Ann Oncol 15:440–449

Oberdorster G (2001) Pulmonary effects of inhaled ultrafine particles. Int Arch Occup Environ Health 74:1–8

Oberdorster G, Oberdorster E, Oberdorster J (2005) Nanotoxicology: an emerging discipline evolving from studies of ultrafine particles. Environ Health Perspect 113:823–839

Pan H, Jiang H, Chen W (2006) Interaction of dermal fibroblasts with electrospun composite polymer scaffolds prepared from dextran and poly lactide-co-glycolide. Biomaterials 27:3209–3220

Paulo CSO, Pires Das Neves R, Ferreira L (2011) Nanoparticles for intracellular-targeted drug delivery. Nanotechnology 22:494002

Pekkanen J, Timonen KL, Ruuskanen J, Reponen A, Mirme A (1997) Effects of ultrafine and fine particles in urban air on peak expiratory flow among children with asthmatic symptoms. Environ Res 74:24–33

Penttinen P, Vallius M, Tiittanen P, Ruuskanen J, Pekkanen J (2006) Source-specific fine particles in urban air and respiratory function among adult asthmatics. Inhal Toxicol 18:191–198

Perrault SD, Walkey C, Jennings T, Fischer HC, Chan WCW (2009) Mediating tumor targeting efficiency of nanoparticles through design. Nano Lett 9:1909–1915

Pietropaoli A, Frampton M, Hyde R, Morrow P, Oberdörster G, Cox C, Speers D, Frasier L, Chalupa D, Huang L-S, Utell M (2004) Pulmonary function, diffusing capacity, and inflammation in healthy and asthmatic subjects exposed to ultrafine particles. Inhal Toxicol 16(Suppl 1):59–72

Qiu Y, Liu Y, Wang L, Xu L, Bai R, Ji Y, Wu X, Zhao Y, Li Y, Chen C (2010) Surface chemistry and aspect ratio mediated cellular uptake of Au nanorods. Biomaterials 31:7606–7619

Rabin O, Manuel Perez J, Grimm J, Wojtkiewicz G, Weissleder R (2006) An X-ray computed tomography imaging agent based on long-circulating bismuth sulphide nanoparticles. Nat Mater 5:118–122

Rao BP, Srivastava A, Yasmin F, Ray S, Gupta N, Chauhan C, Rao CVC, Wate SR (2012) Particle size distribution of ambient aerosols in an industrial area. Bull Environ Contam Toxicol 88:717–721

Rosi NL, Giljohann DA, Thaxton CS, Lytton-Jean AKR, Han MS, Mirkin CA (2006) Oligonucleotide-modified gold nanoparticles for intracellular gene regulation. Science 312:1027–1030

Rossin R, Pan D, Qi K, Turner JL, Sun X, Wooley KL, Welch MJ (2005) 64Cu-labeled folate-conjugated shell cross-linked nanoparticles for tumor imaging and radiotherapy: synthesis, radiolabeling, and biologic evaluation. J Nucl Med 46:1210–1218

Rytting E, Nguyen J, Wang X, Kissel T (2008) Biodegradable polymeric nanocarriers for pulmonary drug delivery. Expert Opin Drug Deliv 5:629–639

Sato N, Kobayashi H, Hiraga A, Saga T, Togashi K, Konishi J, Brechbiel MW (2001) Pharmacokinetics and enhancement patterns of macromolecular MR contrast agents with various sizes of polyamidoamine dendrimer cores. Magn Reson Med 46:1169–1173

Schmieder AH, Caruthers SD, Zhang H, Williams TA, Robertson JD, Wickline SA, Lanza GM (2008) Three-dimensional MR mapping of angiogenesis with alpha5beta1(alpha nu beta3)-targeted theranostic nanoparticles in the MDA-MB-435 xenograft mouse model. FASEB J 22:4179–4189

Scholl I, Weissenböck A, Förster-Waldl E, Untersmayr E, Walter F, Willheim M, Boltz Nitulescu G, Scheiner O, Gabor F, Jensen Jarolim E (2004) Allergen-loaded biodegradable poly(D, L-lactic-co-glycolic) acid nanoparticles down-regulate an ongoing Th2 response in the BALB/c mouse model. Clin Exp Allergy 34:315–321

Scholl I, Kopp T, Bohle B, Jensen Jarolim E (2006) Biodegradable PLGA particles for improved systemic and mucosal treatment of Type I allergy. Immunol Allergy Clin North Am 26:349–364, ix

Schutz C, Juillerat Jeanneret L, Mueller H, Lynch I, Riediker M (2013) Therapeutic nanoparticles in clinics and under clinical evaluation. Nanomedicine 8:449–467

Selivanov V, Vizán P, Mollinedo F, Fan TWM, Lee PWN, Cascante M (2010) Edelfosine-induced metabolic changes in cancer cells that precede the overproduction of reactive oxygen species and apoptosis. BMC Syst Biol 4:135

Shaikh J, Ankola DD, Beniwal V, Singh D, Kumar MNVR (2009) Nanoparticle encapsulation improves oral bioavailability of curcumin by at least 9-fold when compared to curcumin administered with piperine as absorption enhancer. Eur J Pharm Sci 37:223–230

Shefrin AE, Goldman RD (2009) Use of dexamethasone and prednisone in acute asthma exacerbations in pediatric patients. Can Fam Physician 55:704–706

Slowing II, Trewyn BG, Lin VSY (2007) Mesoporous silica nanoparticles for intracellular delivery of membrane-impermeable proteins. J Am Chem Soc 129:8845–8849

Speth PAJ, Hoesel QGCM, Haanen C (1988) Clinical pharmacokinetics of doxorubicin. Clin Pharmacokinet 15:15–31

Stark WJ (2011) Nanoparticles in biological systems. Angew Chem Int Ed 50:1242–1258

Strimpakos A, Sharma R (2008) Curcumin: preventive and therapeutic properties in laboratory studies and clinical trials. Antioxid Redox Signal 10:511–545

Sung J, Pulliam B, Edwards D (2007) Nanoparticles for drug delivery to the lungs. Trends Biotechnol 25:563–570

Takano H, Ichinose T, Miyabara Y, Shibuya T, Lim HB, Yoshikawa T, Sagai M (1998) Inhalation of diesel exhaust enhances allergen-related eosinophil recruitment and airway hyperresponsiveness in mice. Toxicol Appl Pharmacol 150:328–337

Takeshita F, Minakuchi Y, Nagahara S, Honma K, Sasaki H, Hirai K, Teratani T, Namatame N, Yamamoto Y, Hanai K, Kato T, Sano A, Ochiya T (2005) Efficient delivery of small interfering RNA to bone-metastatic tumors by using atelocollagen in vivo. Proc Natl Acad Sci U S A 102:12177–12182

Thomas M, Lu J, Chen J, Klibanov A (2007) Non-viral siRNA delivery to the lung. Adv Drug Deliv Rev 59:124–133

Thorek DLJ, Tsourkas A (2008) Size, charge and concentration dependent uptake of iron oxide particles by non-phagocytic cells. Biomaterials 29:3583–3590

Tsutsumi T, Tokumura A, Kitazawa S (1998) Undifferentiated HL-60 cells internalize an antitumor alkyl ether phospholipid more rapidly than resistant K562 cells. Biochim Biophys Acta 1390:73–84

Tullius SG, Reutzel-Selke A, Nieminen-Kelhä M, Jonas S, Pratschke J, Bechstein WO, Neuhaus P, Volk HD (2001) Tolerance induction by the graft itself. Transplant Proc 33:2317–2318

van Kasteren SI, Campbell SJ, Serres SB, Anthony DC, Sibson NR, Davis BG (2009) Glyconanoparticles allow pre-symptomatic in vivo imaging of brain disease. Proc Natl Acad Sci 106:18–23

Veiseh O, Sun C, Gunn J, Kohler N, Gabikian P, Lee D, Bhattarai N, Ellenbogen R, Sze R, Hallahan A, Olson J, Zhang M (2005) Optical and MRI multifunctional nanoprobe for targeting gliomas. Nano Lett 5:1003–1008

Venugopal J, Ramakrishna S (2005) Biocompatible nanofiber matrices for the engineering of a dermal substitute for skin regeneration. Tissue Eng 11:847–854

Verma A, Stellacci F (2010) Effect of surface properties on nanoparticle-cell interactions. Small 6:12–21

Vij N (2012) Synthesis and evaluation of airway targeted PLGA nanoparticles for drug delivery in obstructive lung diseases. Methods Mol Biol 906:303–310 von Klot S, Wölke G, Tuch T, Heinrich J, Dockery DW, Schwartz J, Kreyling WG, Wichmann HE, Peters A (2002) Increased asthma medication use in association with ambient fine and ultrafine particles. Eur Respir J 20:691–702

Vracko R (1974) Basal lamina scaffold-anatomy and significance for maintenance of orderly tissue structure. Am J Pathol 77:314–346

Wang J, Tian S, Petros RA, Napier ME, Desimone JM (2010) The complex role of multivalency in nanoparticles targeting the transferrin receptor for cancer therapies. J Am Chem Soc 132:11306–11313

Watts JK, Corey DR (2010) Clinical status of duplex RNA. Bioorg Med Chem Lett 20:3203–3207

Wei Y, Zhao L (2014) Passive lung-targeted drug delivery systems via intravenous administration. Pharm Dev Technol 19(2):129–136

Wilczewska A, Niemirowicz K, Markiewicz K, Car H (2012) Nanoparticles as drug delivery systems. Pharmacol Rep 64:1020–1037

Willis L, Hayes D, Mansour H (2012) Therapeutic liposomal dry powder inhalation aerosols for targeted lung delivery. Lung 190:251–262

Woltman A, van der Kooij SW, Coffer P, Offringa R, Daha M, van Kooten C (2003) Rapamycin specifically interferes with GM-CSF signaling in human dendritic cells, leading to apoptosis via increased p27KIP1 expression. Blood 101:1439–1445

Xiang Q-Y, Wang M-T, Chen F, Gong T, Jian Y-L, Zhang Z-R, Huang Y (2007) Lung-targeting delivery of dexamethasone acetate loaded solid lipid nanoparticles. Arch Pharm Res 30:519–525

Yano J, Hirabayashi K, Nakagawa S-I, Yamaguchi T, Nogawa M, Kashimori I, Naito H, Kitagawa H, Ishiyama K, Ohgi T, Irimura T (2004) Antitumor activity of small interfering RNA/cationic liposome complex in mouse models of cancer. Clin Cancer Res 10:7721–7726

Yarmolenko P, Zhao Y, Landon C, Spasojevic I, Yuan F, Needham D, Viglianti B, Dewhirst M (2010) Comparative effects of thermosensitive doxorubicin-containing liposomes and hyperthermia in human and murine tumours. Int J Hyperthermia 26:485–498

Yoo D, Guk K, Kim H, Khang G, Wu D, Lee D (2013) Antioxidant polymeric nanoparticles as novel therapeutics for airway inflammatory diseases. Int J Pharm 450(1–2):87–94

Zarogiannis S, Filippidis A, Fernandez S, Jurkuvenaite A, Ambalavanan N, Stanishevsky A, Vohra Y, Matalon S (2013) Nano-TiO_2 particles impair adhesion of airway epithelial cells to fibronectin. Respir Physiol Neurobiol 185:454–460

Zhang W, Yang H, Kong X, Mohapatra S, Juan-Vergara HS, Hellermann G, Behera S, Singam R, Lockey RF, Mohapatra SS (2005) Inhibition of respiratory syncytial virus infection with intranasal siRNA nanoparticles targeting the viral NS1 gene. Nat Med 11:56–62

Chapter 11
One-Carbon Metabolism Nutrients and Epigenetics: A Mechanistic Link Between Aberrant One-Carbon Metabolism and Cancer Risk?

Shannon Masih, Lesley A. Plumptre, and Young-In Kim

Abstract In relation to the genome, investigation of the epigenome is emerging as an equal, if not more influential factor in modulating human health and disease. Since epigenetic modifications are gradual in onset and potentially reversible, determining factors that modulate the epigenome is critical for possible preventive and therapeutic interventions. The development and progression of cancer is mechanistically linked to a number of epigenetic changes, including global DNA hypomethylation and gene-specific CpG promoter DNA hypermethylation. Environmental factors, including diet, have been shown to affect cancer risk, via epigenetic and non-epigenetic mechanisms. In this regard, one-carbon nutrients are prototypic dietary factors that may modulate cancer risk via epigenetic mechanisms. This chapter will discuss the role of nutrients involved in one-carbon metabolism and their effect on cancer risk via epigenetic modifications with a particular focus on DNA methylation.

Keywords One-carbon nutrients • Folate/folic acid • Vitamin B_6 • Vitamin B_{12} • Choline • Betaine • Epigenetics • DNA methylation • Cancer • Nutrition/vitamins

---

*Shannon Masih and Lesley A. Plumptre contributed equally to this work.

S. Masih, B.Sc., M.Sc. • L.A. Plumptre, B.Sc., M.Sc.
Department of Nutritional Sciences, University of Toronto, Toronto, ON, Canada

Keenan Research Center for Biomedical Science of St. Michael's Hospital, 16CC-038, 30 Bond Street, Toronto, ON, Canada M5B 1W8
e-mail: shannon.masih@mail.utoronto.ca; lesley.hoyt@mail.utoronto.ca

Y.-I. Kim, M.D., F.R.C.P.C. (✉)
Department of Nutritional Sciences, University of Toronto, Toronto, ON, Canada

Department of Medicine, University of Toronto, Toronto, ON, Canada

Keenan Research Center for Biomedical Science of St. Michael's Hospital, 16CC-038, 30 Bond Street, Toronto, ON, Canada M5B 1W8

Division of Gastroenterology, Department of Medicine, St. Michael's Hospital, 16CC-038, 30 Bond Street, Toronto, ON, Canada M5B 1W8
e-mail: youngin.kim@utoronto.ca

N. Maulik and T. Karagiannis (eds.), *Molecular Mechanisms and Physiology of Disease: Implications for Epigenetics and Health*, DOI 10.1007/978-1-4939-0706-9_11,

11.1 Introduction

A convincing body of evidence suggests that environmental factors, including diet, influence our susceptibility to disease including cancer. Individual variations in disease susceptibility to environmental factors cannot entirely be accounted for by conventional genetic mechanisms. In this regard, epigenetics modifications including DNA methylation, histone modifications, chromatin remodeling, and microRNA interference have emerged as a promising mechanism by which environmental factors including diet can influence disease susceptibility because they are potentially reversible. Several individual nutrients have been linked to epigenetic modifications and cancer susceptibility. Given that many nutrients are consumed simultaneously, however, investigating interrelated nutrients may be of more significance. Given this consideration, the aim of this chapter is to highlight the potential role of one-carbon metabolism nutrients in site-specific cancer risk modifications via its modulating effect on DNA methylation.

11.2 One-Carbon Metabolism Nutrients

One-carbon metabolism is a network of biochemical reactions resulting in the transfer of one-carbon units. This system of reactions is necessary for essential biological processes: de novo purine and thymidylate synthesis (i.e., the nucleotide synthesis pathway) and the remethylation of homocysteine to methionine (i.e., the biological methylation pathway). Methionine is the precursor to *S*-adenosyl methionine (SAM), the primary methyl donor for most biological methylation reactions including DNA. Some of the key nutrients involved in one-carbon metabolism, which will be further discussed in detail, include folate and synthetic folic acid, vitamins B_6 and B_{12}, and choline and its metabolite, betaine.

11.2.1 Folate and Folic Acid

Folate is the overarching term used to describe naturally occurring dietary folates and synthetically manufactured folic acid. The main structure of folate and folic acid consists of a pteridine moiety attached to P-aminobenzoic acid with one or more glutamate residues attached via γ-peptide bonds. Dietary folates have reduced pteridine rings and are characterized by their polyglutamylated tails whereas folic acid is monoglutamylated and fully oxidized. Due to their structural difference, folic acid is more stable and therefore, more bioavailable than naturally occurring folates (Shane 1995).

Although endogenous folate is synthesized by colonic bacteria (Kim et al. 2004) and can be absorbed across the large intestine (Aufreiter et al. 2009; Rong et al. 1991), mammals cannot synthesize folate and therefore must obtain folate from the

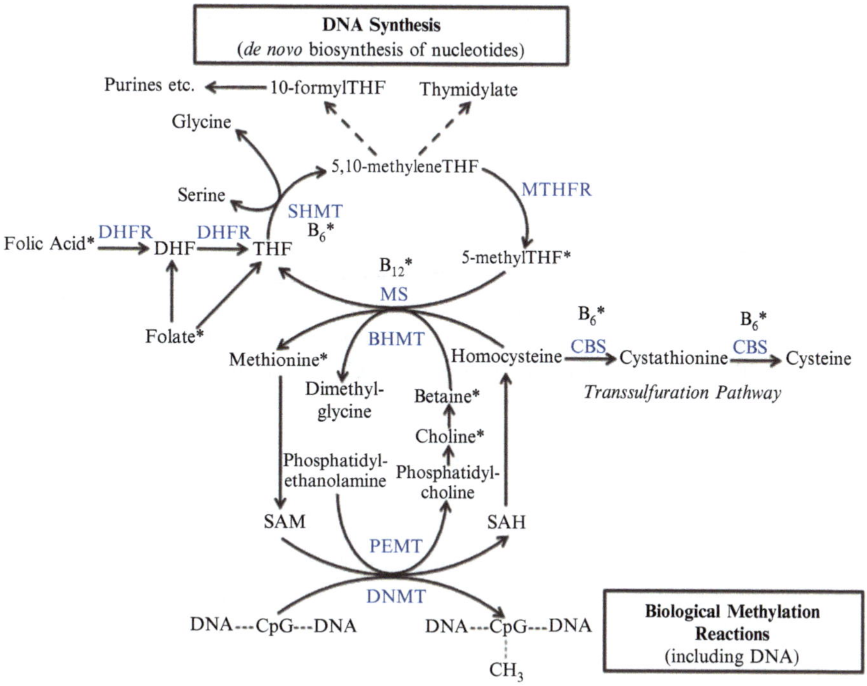

Fig. 11.1 Dietary factors, enzymes, and substrates in one-carbon metabolism involved in DNA synthesis and biological methylation reactions, including that of DNA. Enzymes are shown in bold. The *asterisk* denotes nutrients which are available from the diet. SAM is both an allosteric inhibitor of MTHFR and an activator of cystathionine β-synthase. B_{12} vitamin B_{12}, *BHMT* betaine-homocysteine methyltransferase, *CβS* cystathionine β-synthase, CH_3 methyl group, *CpG* cytosine-guanine dinucleotide sequence, *DHF* dihydrofolate, *DHFR* dihydrofolate reductase, *DNMT* DNA methyltransferase, *MS* methionine synthase, *SAH* *S*-adenosylhomocysteine, *SAM* S-adenosylmethionine, *SHMT* serine hydroxymethyltransferase, *THF* tetrahydrofolate, *TS* thymidylate synthase, *PEMT* phosphatidylethanolamine

diet for its sole biological function, which is mediating the transfer of one-carbon moieties (Shane 1995). The form of folate directly related to the formation of SAM is 5-methylTHF. 5-methylTHF is synthesized from 5,10-methyleneTHF by the enzyme methylenetetrahydrofolate reductase (MTHFR). This conversion is irreversible and plays a critical role in the maintenance of the 5-methylTHF pool available for methionine regeneration. 5-methylTHF and homocysteine are converted to THF and methionine, respectively via methionine synthase (MS), which requires vitamin B_{12} as a coenzyme (Fig. 11.1). Methionine, an amino acid, is then converted to SAM via methionine adenosyltransferases (MAT1A and MAT2A), which maintain the pool of SAM for biological methylation reactions.

Historically, folate was consumed primarily from dietary sources including leafy green vegetables, citrus fruits, organ meats, and legumes. More recently, however, with the use of supplements and fortified foods, synthetic folic acid is considered a significant contributor to total folate intake (Yeung et al. 2008). Due to its role in purine and thymidylate synthesis and methylation reactions, folate is essential for proper cell division, normal growth and development, and therefore plays a crucial role in human health and disease. Folate deficiency has been linked to a number of human diseases such as anemia, neural tube and other congenital defects, adverse pregnancy outcomes, coronary heart disease, stroke, neuropsychiatric disorders, cognitive impairments, osteoporosis, and certain cancers (1998). Due to the overwhelming body of evidence regarding folate's protective effect on neural tube defect risk (MRC Vitamin Study Research Group 1991; De Wals et al. 2007; Wilson et al. 2007), several countries have adopted voluntary or mandatory folic acid fortification programs (Crider et al. 2011), although this practice still remains controversial (Kim 2004b, 2007b). The United States and Canada introduced mandatory folic acid fortification in 1998, which led to significant increases in North American total folate intakes and serum and red blood cell (RBC) concentrations of folate (Shakur et al. 2010; Quinlivan and Gregory 2003). For example, nationally representative data in Canada found virtually no folate deficiency in the Canadian population as determined by RBC folate concentrations. In the same survey, 40 % of the Canadian population had concentrations above an established high cutoff (97th percentile from National Health and Nutrition Examination Survey (NHANES) 1999–2004) (Colapinto et al. 2011). Serum folate concentrations are indicative of short-term folate intake and can be influenced by recent folate and/or folic acid consumption prior to blood draw. RBC folate concentrations are indicative of long-term folate intake and its measurement is considered the gold standard for determining folate status (Yetley et al. 2011a).

11.2.2 Vitamin B_{12}

Vitamin B_{12} (cobalamin) is an essential water-soluble B vitamin, which is required for the production of RBCs and for optimal neurological function. Vitamin B_{12} is a very complex vitamin compound that exists in many different forms, all of which have the commonality of a central cobalt atom. Depending on the compounds attached to the cobalt atom, different types of cobalamins are formed such as methylcobalamin, cyanocobalamin, hydroxycobalamin, aquacobalamin, and 5′-deoxyadenosylcobalamin. The most stable form of this vitamin is cyanocobalamin, the pharmacological form (1998), which is readily converted to the coenzyme forms, methylcobalamin, and 5′-deoxyadenosylcobalamin (Watanabe 2007). These coenzyme forms are biologically active in human metabolism.

The methylcobalamin form of vitamin B_{12} serves as one of the key enzymatic cofactors in the one-carbon metabolism cycle. Vitamin B_{12} functions as the coenzyme for MS, which catalyzes the transfer of a methyl group from 5-methylTHF to homocysteine, forming methylcobalamin and regenerating THF. In this biochemical pathway, homocysteine is remethylated to methionine in what is considered to be the folate-dependent remethylation pathway of one-carbon metabolism

(Dominguez-Salas et al. 2012) (Fig. 11.1). The other metabolically active form of vitamin B_{12}, 5′-deoxyadenosylcobalamin, serves as the coenzyme for L-methylmalonyl-CoA (MMA) mutase, the enzyme which catalyzes the isomerization of L-MMA to succinyl-CoA, an enzymatic reaction which is involved in amino acid and fatty acid metabolism (1998; Watanabe 2007). Vitamin B_{12} is naturally found in foods of animal origin such as fish, shellfish, chicken, red meat, liver, and dairy products. The bioavailability of natural dietary B_{12} is approximately 50 % of synthetic (crystalline) B_{12} (Watanabe 2007).

The metabolic pathways of vitamin B_{12} and folate interact significantly, thereby leading to a correlation between folate and vitamin B_{12} status. High folic acid intake has the potential to mask the hematological symptoms and signs associated with vitamin B_{12} deficiency considering that both these vitamins play a role in normal blood cell formation. However, folic acid cannot correct vitamin B_{12} deficiency-associated neurological symptoms. Therefore, when initial hematological symptoms are masked through the corrective action of folic acid, the untreated vitamin B_{12} deficiency can lead to severe neurological symptoms such as irreversible sensory neuropathy (1998).

The main clinical manifestations of vitamin B_{12} deficiency are hyperhomocysteinemia, megaloblastic anemia, myelopathy, neurodegeneration, depression, and cognitive decline (1998; MacFarlane et al. 2011; Ryan-Harshman and Aldoori 2008). Inadequate B_{12} status during pregnancy has been shown to be an independent risk factor for neural tube defects, which may be in part due to its close metabolic relationship to folate (Ray et al. 2008; Molloy et al. 2009).

Healthy adults, with normal gastrointestinal function, are believed to absorb 50 % of dietary vitamin B_{12} (Watanabe 2007). With aging, however, malabsorption of food-bound vitamin B_{12} increases due to gastrointestinal impairments, such as reduced gastric acidity, which results in a higher prevalence of vitamin B_{12} deficiency in the elderly (1998; MacFarlane et al. 2011). Vitamin B_{12} status can be measured directly in the blood or determined by measuring functional or metabolic biomarkers. Serum B_{12} and plasma holotranscobalamin (holoTC), both measure circulating concentrations of B_{12}. Of the two B_{12} transport proteins (haptocorrin and transcobalamin), B_{12} bound to transcobalamin represents the fraction of circulating B_{12} available for cellular uptake (Yetley et al. 2011b).

MMA and homocysteine, both functional indicators, increase when there is an inadequate concentration of vitamin B_{12}. MMA increases when there is a decrease in the activity of the enzyme L-MMA mutase due to an inadequate source of the coenzyme form of vitamin B_{12}, 5′-deoxyadenosylcobalamin. Therefore, MMA is an accurate inverse indicator of serum B_{12} concentrations. Homocysteine also serves as a functional indicator of B_{12} status but is considered a nonspecific inverse indicator considering that increased levels of homocysteine are also associated with inadequate folate concentrations and to lesser extent, inadequate vitamin B_6 and riboflavin (vitamin B_2). The results of a recent roundtable summary of the NHANES reported that the most definitive assessment of B_{12} status requires the measurement of a combination of at least one biomarker of circulating vitamin B_{12} (serum vitamin B_{12} or holoTC) and one functional biomarker (MMA or homocysteine) (Yetley et al. 2011b).

11.2.3 Vitamin B_6

Vitamin B_6 is a water-soluble B vitamin comprised of several vitamers: pyridoxine (PN), pyridoxal (PL), and pyridoxamine (PM). These three compounds also exist in the 5′ phosphate forms (PLP, PNP, and PMP). The B_6 vitamers act as coenzymes in over 100 biological reactions involving metabolism of carbohydrates, lipids, and proteins (1998). With respect to one-carbon metabolism, vitamin B_6 is related to folate, B_{12}, and choline in several biochemical pathways including condensing homocysteine with serine to form the amino acid cysteine in the two-step transsulfuration pathway which requires the biologically active form of vitamin B_6, PLP, as a cofactor. Vitamin B_6 is also involved in the conversion of serine to glycine by acting as a coenzyme for serine hydroxymethyltransferase (SHMT), which catalyzes the conversion of THF to 5,10-methyleneTHF (Fig. 11.1). Rich sources of vitamin B_6 include fish, beef liver, chickpeas, and non-citrus fruits. However, the US population consumes vitamin B_6 primarily from fortified cereals, beef, poultry, and starchy vegetables (1998).

Vitamin B_6 status is most commonly measured by assessing biologically active PLP in plasma since it most closely reflects tissue stores. However, status can also be measured in RBCs or urine by measuring other individual B_6 vitamers or total vitamin B_6 forms. PLP concentrations above 20 nmol/L in plasma are considered adequate B_6 concentrations in adults. Deficiency of vitamin B_6 is rarely seen in the general population, although certain subgroups such as those with gastrointestinal disorders, renal dysfunction, autoimmune diseases, or alcohol dependence may be at higher risk of B_6 inadequacy. Vitamin B_6 deficiency can lead to microcytic anemia, irregularities in electroencephalography readings, depression and confusion, and compromised immune function. Deficiency symptoms include cracked corners of the mouth and swollen tongue (1998).

Vitamin B_6 status has been linked to a number of diseases, most notably coronary heart disease, stroke, and cognitive dysfunction. Although theoretically plausible due to the potential homocysteine-lowering effects of vitamin B_6, previous randomized control trials and combined analyses have failed to show vitamin B_6, alone or in combination with other B vitamins reduce the risk of coronary heart disease and stroke (Ebbing et al. 2010; Albert et al. 2008; Toole et al. 2004). With regard to cognitive function, low vitamin B_6 status has been hypothesized to lead to cognitive decline in the elderly. However, a systematic review and a Cochrane review concluded there is not enough evidence to determine if vitamin B_6 supplementation would improve cognitive function in the elderly (Balk et al. 2007; Malouf and Grimley Evans 2003).

11.2.4 Choline and Betaine

Choline is an essential nutrient which functions as a precursor for the neurotransmitter acetylcholine and along with its derivatives, also serves as the component of structural lipoproteins (such as phospholipids), blood and membrane lipids.

Although an essential nutrient, choline can also be synthesized de novo by the endogenous pathway via methylation of phosphatidylethanolamine. This reaction is catalyzed by phosphatidylethanolamine *N*-methyltransferase (PEMT) with SAM serving as the methyl donor. Although PEMT-catalyzed de novo biosynthesis of phosphatidylcholine provides a significant source of choline, this does not provide an adequate amount to meet the nutrient requirements when concentrations of other one-carbon metabolism nutrients, namely methionine, folate, vitamin B_6, and B_{12}, are not available in sufficient amounts to sustain normal growth and function (Fig. 11.1). This interrelationship confirms the importance of studying one-carbon nutrients together and adds the complexity to this field of research, as it is often very difficult to tease apart effects of individual one-carbon nutrients. Natural dietary sources of choline are egg yolks, beef, chicken, liver, and soybeans (1998; Ueland 2011; Zeisel and da Costa 2009).

Choline is naturally found in the food supply as free choline, or bound as esters such as phosphocholine, glycerophosphocholine, sphingomyelin, or phosphatidylcholine. Choline plays important roles in structural integrity of cell membranes, methyl metabolism, cholinergic neurotransmission, transmembrane signaling, and lipid and cholesterol transport, and metabolism (1998; Ueland 2011). Furthermore, choline is associated with the neurodevelopment and cell proliferation of the hippocampus, differentiation, and apoptosis in animals and is also believed to affect brain development and cognitive functions in humans (Jun Ying et al. 2013).

Most choline is irreversibly oxidized to betaine in the liver and kidney. Betaine serves as a methyl donor by transferring a methyl group to homocysteine for the conversion to methionine catalyzed by BHMT, which is the focal point where betaine and choline are linked to one-carbon metabolism (Dominguez-Salas et al. 2012). On an intracellular level, betaine functions as an osmolyte that regulates cell volume and tissue integrity, thereby protecting cells and proteins from environmental stresses such as ionic stress and elevated temperature. Betaine is abundantly found in animal foods, seafood, and plant foods such as wheat bran and spinach (Ying et al. 2013).

Inadequate choline intakes lead to fatty liver which is a result of the lack of phosphatidylcholine, thereby limiting the export of excess triacylglycerols from the liver (Zeisel 2005) and therefore both choline and betaine are considered lipotropes based on their ability to prevent fatty liver (Rosenfeld 2010). Other signs of choline deficiency are liver and muscle damage (Ueland 2011; Zeisel 2005; Zeisel and da Costa 2009). Although both men and postmenopausal women experience fatty liver or muscle damage during choline deprivation, sex, and menopausal status have been shown to influence dietary requirements through the manifestation of a significantly lower frequency of symptoms of fatty liver and muscle damage, in premenopausal women which can be explained by the upregulation of the de novo synthesis of choline mediated by estrogen (Fischer et al. 2007). Furthermore, because choline is involved in the remethylation pathway of homocysteine, choline deficiency can result in increased levels of plasma homocysteine (Ueland 2011). Although the homocysteine-lowering effect of choline and betaine is well documented, there is conflicting evidence concerning the potential of these two nutrients to modulate the risk of coronary heart disease and stroke via the homocysteine-lowering effect (1998).

Finally, inadequate choline has also been shown to be a risk factor for neural tube defects, likely due to its involvement in the remethylation of homocysteine, independent of folate (Shaw et al. 2004, 2009).

11.3 Epigenetics: DNA Methylation

Epigenetic modification refers to changes in compounds attached to or related to the DNA sequence independent of the genome itself. Epigenetic mechanisms include DNA methylation, covalent modifications of histones, chromatin remodeling, and RNA interference, all of which alter gene expression and function without changing the nucleotide sequence (Egger et al. 2004). In contrast to genetic changes, epigenetic changes are gradual in onset and progressive, their effects are dose-dependent, and are potentially reversible by dietary and pharmacologic manipulations (Kim 2005; Ballestar and Esteller 2002).

The most widely studied of all the epigenetic modifications in mammals is DNA methylation. DNA methylation is characterized by methyl groups attached at cytosine base pairs located within cytosine-guanine (CpG) sequences, and the pattern of these methyl groups is heritable, and tissue- and species-specific (Ballestar and Esteller 2002; Jones and Baylin 2002). DNA methylation is an important epigenetic inverse determinant in gene expression, in the maintenance of DNA integrity and stability, and in chromatin modifications (Jones and Baylin 2002; Esteller 2007). CpG sites are unevenly distributed throughout the genome. Up to 80 % of all CpG sites in human DNA are normally methylated. However, this methylation occurs primarily in the bulk of the genome where CpG density is low, including exons, noncoding regions, and repeat DNA sites, and allows correct organization of chromatin in active and inactive states (Klose and Bird 2006; Robertson and Wolffe 2000). Most CpG rich areas are clustered in small stretches of DNA termed "CpG islands", which span the 5′ end of approximately 50 % of human genes including the promoter, untranslated region, and exon 1, are unmethylated in normal cells, thereby allowing transcription (Robertson and Wolffe 2000). When methylated, CpG islands cause stable heritable transcriptional silencing, which is mediated by the transcriptional repressor, methyl-CpG-binding protein-2 (MeCP2), which binds methylated CpG islands and recruits a complex containing a transcriptional co-repressor and a histone deacetylase (HDAC) (Jones et al. 1998; Canman et al. 1998). HDACs suppress transcription by tightly packing the DNA structure, thus rendering an inactive chromatin conformation (Bird and Wolffe 1999). Not surprisingly, HDACs have been shown to be overexpressed in certain cancers and lead to gene silencing (Ropero and Esteller 2007). DNA methylation is a dynamic process between active methylation, mediated by CpG methyltransferases (DNMT1, 3a, 3b) (Li and Jaenisch 2000) using SAM (Tajima and Suetake 1998), and removal of methyl groups from 5-methylcytosine residues by several mechanisms, including active demethylation by a purported demethylase, methyl DNA-binding domain protein 2 (MBD2). SAM is converted to *S*-adenosylhomocysteine (SAH) by DNMTs, and accumulation of SAH can act as an inhibitor of DNMTs (James et al. 2002).

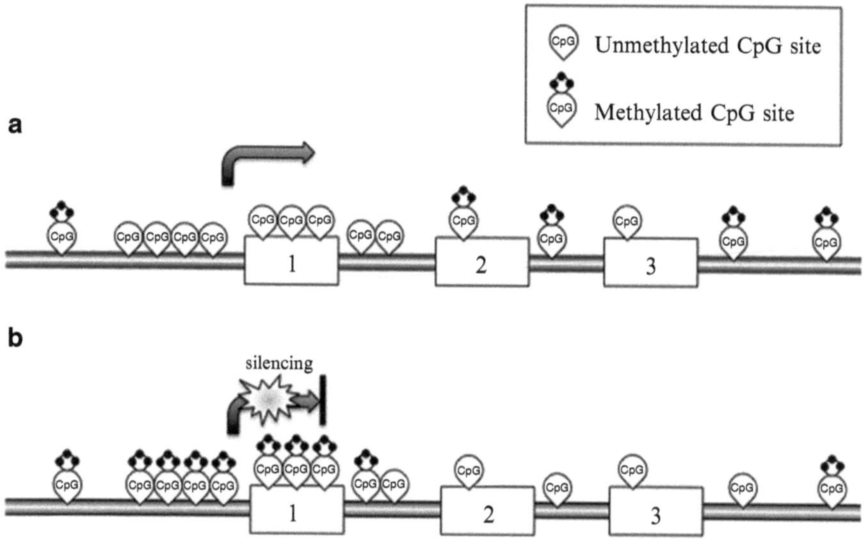

Fig. 11.2 Global and gene-specific methylation patterns in normal and cancer cells. (**a**) In normal cells, methylation generally occurs globally and CpG islands in promoter regions of genes are unmethylated, allowing for transcription of genes. (**b**) In cancer, global hypomethylation and CpG promoter region hypermethylation lead to blockage of transcription and silencing of gene expression. Functional ramifications of global hypomethylation include DNA strand breakage and chromosome instability

Aberrant DNA methylation patterns are mechanistically related to the development of several human diseases including cancer (Jones and Baylin 2002; Esteller 2012). Cancer, in particular, has been shown to have distinct epigenetic characteristics that appear early on in cancer initiation and progression. Cancer is characterized by global DNA hypomethylation as well as gene-specific CpG DNA hypermethylation of generally unmethylated gene promoter regions leading to transcriptional gene silencing (Kim 2005) (Fig. 11.2). Global hypomethylation contributes to the development of cancer through a number of mechanisms including chromosomal instability, increased mutations, reactivation of intragenomic parasitic sequences, loss of heterozygosity, rearrangements, aneuploidy, loss of imprinting, and upregulation of proto-oncogenes (Esteller 2012). DNA methylation at promoter CpG islands silences transcription, thereby inactivating the function of a wide array of genes that have classic tumor suppressor function or play critical roles in cancer development and progression (Herman and Baylin 2003).

11.3.1 Involvement of One-Carbon Nutrients

Folate, vitamins B_{12} and B_6, and choline, which are metabolically interrelated in one-carbon transfer reactions leading to biological methylation reactions, may play a role in cancer development and progression through their effects on DNA methylation (Fig. 11.1) (Davis and Uthus 2004). The conversion of SAM to SAH via DNMTs and other methyltransferases is the key step in methylating DNA and other bioactive molecules, respectively. One-carbon nutrients are involved in this crucial step by contributing to the regeneration of SAM. Another crucial enzyme involved in a number of pathways leading to SAM regeneration is MS. MS requires vitamin B_{12} as a coenzyme for optimal function. MS converts the main form of circulating folate, 5-methylTHF, to THF while simultaneously converting homocysteine to methionine. Homocysteine can also be converted to methionine independent of MS by BHMT, which demethylates betaine to form dimethylglycine. De novo synthesis of the endogenous form of choline (phosphatidylcholine) occurs through the conversion of phosphatidylethanolamine via PEMT. Phosphatidylcholine is further converted to choline, which is the precursor of betaine. Methionine is converted to SAM to be used for methyl group donation. To avoid buildup of toxic levels of homocysteine, vitamin B_6 aids in the clearance of homocysteine to cysteine via the transsulfuration pathway mediated by cystathionine beta synthase (CβS). Vitamin B_6 is also involved in the conversion of serine to glycine by acting as a coenzyme for SHMT, which catalyzes the conversion of THF to 5,10-methyleneTHF. Given that one-carbon nutrients can directly affect the generation of SAM, deficiencies or imbalances of these nutrients can potentially increase cancer risk by impairing DNA methylation capacity.

Experimental Studies

A number of in vitro studies utilizing various human cell lines have demonstrated the potential of folic acid to modulate DNA methylation changes and subsequently, altered gene expression. The results of these studies show that the effects of folate concentrations in in vitro systems on global and gene-specific DNA methylation vary vastly by cell line. A bulk of the literature has investigated these effects in the colonic epithelial cells. Supraphysiological concentrations of folic acid, but not physiologically relevant doses, have been shown to induce Long Interspersed Nuclear Element-1 (LINE-1) hypomethylation and gene-specific CpG island hypermethylation in normal WI-38 fibroblasts and FHC colon epithelial cells (Charles et al. 2012). Another study demonstrated increased global DNA hypomethylation, higher levels of uracil misincorporation, and inhibition of DNA excision repair in normal colonocytes in response to folate deficiency (Duthie et al. 2000). These findings demonstrate that both supraphysiological concentrations and depletion of folic acid can increase global DNA hypomethylation in normal colonic epithelial cells. Collectively, these studies provide proof-of-principle of the ability of folate to modulate DNA methylation.

Rodent models have been widely used to help elucidate the relationship between diet-mediated DNA methylation changes. Overall, studies have demonstrated that diets deficient in one-carbon nutrients in rodent models result in global DNA hypomethylation in the liver (Wainfan et al. 1989; Christman et al. 1993; Wilson et al. 1984; Pogribny et al. 2007, 2009b) and brain (Pogribny et al. 2008).

Furthermore, alterations of gene-specific DNA methylation patterns in various tissues, some with subsequent changes in gene expression, were shown to be sensitive to folate levels (Kim et al. 1997; Qin et al. 2013) and one-carbon nutrient intakes (Waterland et al. 2006b). Although these studies may not be entirely applicable to humans, they provide strong experimental evidence to support that dietary intakes of one-carbon nutrients may influence global and gene-specific DNA methylation. Finally, the effect of dietary intake of one-carbon nutrients on specific tissues and organs in animal models will be discussed in more detail in subsequent specific cancer site sections.

Human Studies

Several studies in humans have corroborated the evidence in in vitro and in vivo systems of the ability of one-carbon nutrients to modulate DNA methylation. Epidemiological studies have investigated peripheral blood mononuclear cell (PBMC) DNA methylation changes in adult women. A nonsignificant decreased trend in leukocyte global DNA methylation in young women was observed with low intake of dietary folate for 7 weeks using a controlled feeding protocol (Shelnutt et al. 2004). After an additional 7 weeks of repletion, global methylation increased nonsignificantly, but only in the *MTHFR* 677CC genotypes. Another study found varying intakes of choline, betaine, and/or folate of healthy women did not influence global DNA methylation in leukocytes after a 12 week follow-up, even when accounting for *MTHFR* C677T genotype (Abratte et al. 2009). Significant decreases in lymphocyte global DNA methylation were observed after intervention with a low folate diet for 3 months in postmenopausal (Jacob et al. 1998) and 7 weeks in elderly women (Rampersaud et al. 2000). However, after folate-repletion, global DNA methylation levels were restored in postmenopausal (Jacob et al. 1998) but not in elderly (Rampersaud et al. 2000) women. In a cross-sectional study, postmenopausal women exhibited an inverse relationship between lymphocyte LINE-1 methylation and sex hormone concentrations when stratified by serum folate levels (Ulrich et al. 2012). LINE-1 methylation was positively associated with immune markers in those with high serum folate levels (Ulrich et al. 2012). Furthermore, there was a nonsignificant inverse trend with supplemental vitamin B_6 intakes and LINE-1 methylation (Jin et al. 2009). In men and women, low folate status measured in blood directly correlated with lower PBMC global DNA methylation, which was specific for individuals with the *MTHFR* 677TT genotype (Friso et al. 2002).

Few studies have investigated the effects of one-carbon nutrients on global and gene-specific DNA methylation patterns at a tissue-specific level in healthy individuals. In normal colonic biopsies, serum and RBC folate concentrations were significantly inversely correlated with DNA hypomethylation (Pufulete et al. 2005b).

Al-Ghnaneim et al. investigated gene-specific DNA methylation in the promoter regions of *mutL homolog 1* (*hMLH1*) and *Estrogen Receptor α* (*ERα*) in healthy colonic mucosa. *ERα* promoter DNA methylation was significantly inversely correlated with serum vitamin B_{12} concentrations but not with serum and RBC folate concentrations. *hMLH1* promoter DNA methylation was however not significantly correlated with vitamin B_{12} or folate status in this study (Al-Ghnaniem et al. 2007). Although the functional ramifications of these changes in DNA methylation resulting from alterations in B vitamin and choline status have yet to be explained, these studies offer insight into how one-carbon nutrient diet can play a potentially significant role in modulating DNA methylation in PBMC and specific tissues/organs in humans. Moreover, the mosaic of results presented herein point to the fact that the direction of DNA methylation changes may be specific to cell type and target organ and may not always be consistent between global and gene- or site-specific DNA methylation. Contrary to permanent genetic mutations that elicit "programmed" disease risks, alterations of the epigenome are potentially reversible and thus provide novel prospects on the forefront of disease prevention and treatment.

11.4 The Link Between One-Carbon Metabolism Nutrients and Cancer

In both normal and cancer cells, the diet has been shown to modify a myriad of fundamental processes including DNA repair, cell proliferation and apoptosis, gene expression, and inflammatory and immunological reactions. Furthermore, dietary components are important modulators of DNA methylation and cancer risk. As demonstrated through varying levels of evidence in both experimental models and in human studies, a relationship between one-carbon metabolism nutrients and DNA methylation has been established. Furthermore, the most convincing evidence regarding the interaction of dietary components and cancer risk via DNA methylation changes is from investigation of one-carbon nutrients and DNA methylation changes since they directly contribute to the methyl donor supply available for biological methylation reactions (Davis and Uthus 2004; Marmot et al. 2007).

Of all the one-carbon nutrients, folate has been the most widely studied in its ability to modulate cancer risk. Although the traditional role of folate in cancer is evident with use of antifolate drugs in cancer chemotherapy (Robien 2005), epidemiologic studies have suggested folate insufficiency may increase the risk of several human cancers and that folic acid supplementation may reduce this risk (Kim 2003, 2004b, 2007a). However, animal studies from the author's lab suggest folate possesses dual modulatory effects on cancer development and progression depending on the dose and stage of cell transformation at the time of intervention (Kim 2003, 2004b, 2007a, 2008). Cancer develops over decades, if not a lifetime, through different stages of premalignant lesions. Deficiency of folate in normal tissues predisposes cells to neoplastic transformation, and modest supplemental levels suppress transformation, whereas supraphysiologic doses enhance, the development of tumors in normal tissues. In contrast, folate deficiency has an inhibitory effect whereas folate supplementation has a promoting effect on the progression of established intraepithelial neoplasms (Fig. 11.2).

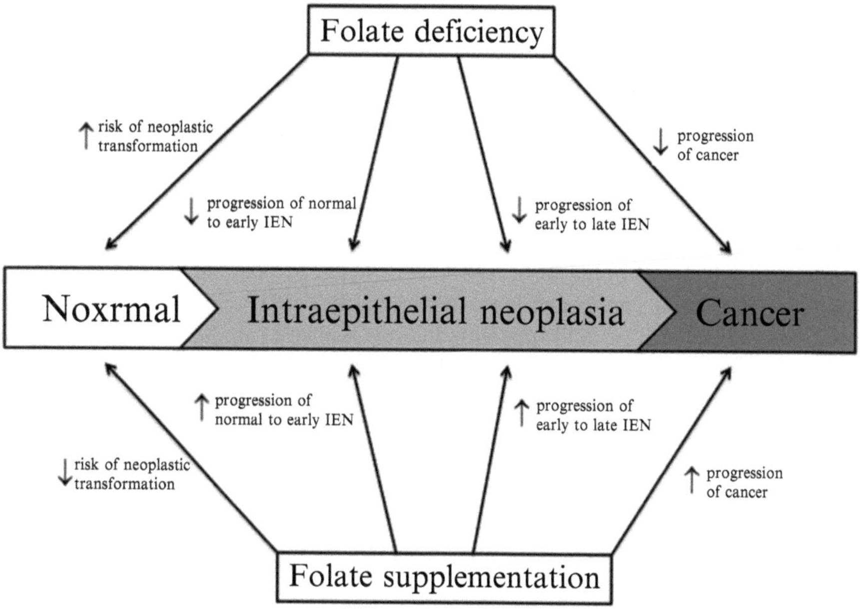

Fig. 11.3 Dual modulatory role of folate in carcinogenesis. Folate deficiency in normal tissues predisposes them to neoplastic transformation, and modest supplemental levels suppress, whereas supraphysiologic doses of supplementation enhance, the development of tumors in normal tissues. In contrast, folate deficiency has an inhibitory effect whereas folate supplementation has a promoting effect on the progression of established neoplasms. The mechanisms by which folate exerts dual modulatory effects on carcinogenesis depending on the timing and dose of folate intervention relate to its essential role in one-carbon transfer reactions involved in DNA synthesis and biological methylation reactions. *ACF* aberrant crypt foci

The mechanisms by which folate exerts dual modulatory effects on carcinogenesis depending on the timing and dose of folate intervention relate to its essential role in one-carbon transfer reactions involved in DNA synthesis and biological methylation reactions. The tumor promoting mechanisms of folate deficiency in normal, untransformed tissue include DNA strand breaks, impaired DNA repair, increased mutagenesis and global DNA hypomethylation, whereas tumor inhibitory mechanisms in transformed tissue include impaired DNA synthesis therefore inhibiting tumorigenesis and reversal of aberrant CpG hypermethylation (Fig. 11.3). On the contrary, tumor inhibitory mechanisms of folate supplementation in normal, untransformed tissue include: DNA stability and integrity, optimal DNA repair, decreased mutagenesis, and prevention of aberrant DNA methylation. Finally, tumor promoting mechanisms of folate supplementation in transformed tissue include provision of nucleotide precursors for proliferation and growth of neoplastic cells, inactivation of tumor suppressor genes induced by de novo methylation of promoter CpG islands, and hypermutability of methylated cytosine in CpG dinucleotides (Smith et al. 2008) (Fig. 11.2).

Although the evidence is stronger in animal models, the potential tumor promoting effect of high folic acid is corroborated in two human studies. The Aspirin/Folate Polyp Prevention Study (Cole et al. 2007) reported that daily folic acid supplementation for up to 10 years in individuals likely harboring (pre)neoplastic lesions significantly increased the risk of advanced and multiple colorectal adenomas (Cole et al. 2007) and of prostate cancer (Figueiredo et al. 2009a). A Norwegian study found treatment with folic acid and vitamin B_{12} for a median follow-up of 36 months significantly increased overall cancer incidence and mortality by 21 % and 38 %, respectively (Ebbing et al. 2009). However, a meta-analysis investigating folate's effects on all cancer incidence and mortality failed to demonstrate the tumor promoting effect of folic acid supplementation (Vollset et al. 2013). At present, the potential tumor promoting effect on the progression of (pre)neoplastic lesions remains the most controversial and potentially most serious adverse effect of folic acid supplementation and fortification.

This relationship has recently been expanded to other one-carbon nutrients, particularly choline and prostate cancer. Data from the Health Professionals Follow-up Study cohort reported a positive association between dietary choline intake and increased lethal prostate cancer risk after a 22-year follow-up (Richman et al. 2012). Prostate cancer has been shown to alter choline metabolism and large amounts of choline are taken up by prostate cells (Ackerstaff et al. 2001). However it is still unknown how choline may adversely affect prostate cancer risk. To complicate these results, synergistic effects of one-carbon nutrients have been demonstrated. For example, changes in other B vitamin intakes have been shown to modulate choline status in rats since choline can be generated de novo and the increase in related B vitamin nutrients could spare use of choline or increase endogenous choline synthesis (van Wijk et al. 2012). Therefore, it may be difficult to attribute adverse effects to one nutrient in particular, considering one-carbon nutrients are integrally linked and regulated.

In the following sections, an in-depth discussion of the complex relationships between one-carbon nutrients and cancer risks for several different cancers of the gastrointestinal system (esophageal, gastric, liver, pancreatic, and colorectal) and reproductive cancers (breast, cervical, endometrial, ovarian, and prostate) will be presented. Trans-generational studies will also be discussed, focusing on maternal one-carbon nutrient intakes and its effects on cancer risk and DNA methylation changes in the offspring. Evidence from in vitro, animal, and epidemiological studies as well as interventional trials in human will be reviewed. Furthermore, the role of one-carbon nutrients in the modulation of DNA methylation within each cancer site will be reviewed.

11.4.1 Cancers of the Gastrointestinal System

Esophageal Cancer

Esophageal cancer incidence rates vary geographically demonstrating a significantly higher frequency in Southern and Eastern Africa and Eastern Asian countries. This type of cancer has been shown to affect males 3–4 times more commonly than females (Jemal et al. 2011). Cancer of the esophagus exists in two forms: esophageal squamous cell carcinoma (ESCC), which affects the upper-third of the esophagus and esophageal adenocarcinoma (EAC), which affects the distal esophagus and the junction of the esophagus and stomach. Only environmental risk factors of esophageal cancer have been presently identified including poor nutritional status, low fruit and vegetable intake and drinking hot beverages which pertain to high-risk areas whereas in the low-risk areas (i.e., Western countries), smoking and excessive alcohol consumption have been characterized as the most common risk factors. Barrett's esophagus (BE), a major predisposing risk factor of EAC, is associated with smoking, overweight or obesity, and chronic gastroesophageal reflux disease (Jemal et al. 2011).

There is a paucity of animal studies that examined the effect of one-carbon nutrients on esophageal carcinogenesis. One preliminary study presented as an abstract reported that folic acid supplementation reduced gastroesophageal reflux-induced hyperproliferative esophagitis in rats (Menezes et al. 2008).

A meta-analysis of four case–control studies for ESCC and three case–controls for EAC has found dietary folate to be protective against these esophageal cancers (Larsson et al. 2006a). Two recent case–control studies have also found dietary folate to be protective against all forms of esophageal cancer (Bravi et al. 2012; Jessri et al. 2011). However, an interesting finding from a recent case–control study demonstrated that dietary folate, but not supplemental folic acid, was protective against BE (Ibiebele et al. 2011). Supplemental folic acid was associated with a twofold increased BE risk with dysplasia and supplemental vitamin B_6 and B_{12} were associated with increased EAC risk. The direct relationship between esophageal cancer progression and high folic acid is consistent with the tumor promoting effect of folic acid supplementation on established (pre)neoplastic foci observed primarily in colorectal cancer (CRC) animal studies (Kim 2006). Although this study captured dietary changes and supplemental use prior to diagnosis, there is still a possibility that supplemental B vitamin use may have increased post-diagnosis, which is in agreement with previous research indicating that supplement and multivitamin use increases after cancer diagnosis (Velicer and Ulrich 2008).

A few case–control studies have found dietary vitamin B_6 to be protective against ESCC (Bravi et al. 2012; Mayne et al. 2001) and EAC (Mayne et al. 2001). Interestingly, dietary vitamin B_{12} was associated with an increased risk of ESCC and EAC (Mayne et al. 2001). It has been suggested that plant-based nutrients (folate, vitamin B_6) may be protective and animal-based nutrients (vitamin B_{12}) may be detrimental to cancer risk of the esophagus and stomach (Mayne et al. 2001). Furthermore, another case–control study performed in Iranian middle aged to elderly adults used factor analysis to investigate nutrients and dietary patterns associated with ESCC. Two dietary nutrient patterns were identified and factor 2, but not factor 1, had an inverse relationship with ESCC risk (Hajizadeh et al. 2012). Interestingly, dietary folate and vitamin B_6 were loaded in factor 1 which saw no effect whereas vitamin B_{12} was loaded in factor 2 which was protective. Identifying dietary patterns can be beneficial when determining how whole diets affect cancer risk. However, it is difficult to tease out the effects of individual nutrients given each dietary nutrient pattern consisted of over 15 nutrients and therefore, could have synergistic effects. A Chinese intervention trial with over 3,000 subjects examined whether multivitamin use over a 6-year period improved esophageal dysplasia. The multivitamin treatment contained 26 nutrients including folic acid, vitamin B_6, and vitamin B_{12}. Overall, multivitamin supplementation had no significant protective effect on total or esophageal/gastric cardia cancer incidence. However, when analyzed separately, there was a nonsignificant trend of decreased incidence of esophageal but increased gastric cancer in the supplemented group (Li et al. 1993).

One case–control study that considered dietary intake of methyl donors including betaine, choline, and methionine found a significant inverse association between dietary betaine, but not choline or methionine, and BE risk (Ibiebele et al. 2011).

A limited number of studies have investigated the effects of one-carbon nutrients on DNA methylation as related to esophageal cancer. Two epidemiological studies found no association between dietary folate intake and CpG promoter DNA methylation of the tumor suppressor *p16* and DNA mismatch repair *hMLH1* genes in esophageal cancer tissues. However, CpG promoter DNA methylation of the *MGMT* gene was positively associated with dietary folate intake. Furthermore, CpG promoter DNA hypermethylation of the *MGMT* gene was associated with increased ESCC risk (Wang et al. 2008; Chen et al. 2012). *MGMT* is an important DNA repair gene and has been previously implicated in carcinogenesis across various tissues (Chen et al. 2012; Nakai et al. 2012). On the contrary, Lu et al. did not find any associations between global or *p16*, *MGMT* and *hMLH1* gene-specific CpG promoter DNA methylation and ESCC risk; this study, however, did not investigate whether dietary folate affected DNA methylation patterns in esophageal cancer (Lu et al. 2011). Nonetheless, dietary folate intake appeared to decrease the risk of death after esophagectomy in ESCC patients (Lu et al. 2011).

Esophageal Cancer Summary. Collectively, human studies have shown dietary folate and vitamin B_6 to be protective against esophageal cancer. However, there are inconsistent results regarding the effects of vitamin B_{12}. Furthermore, the source of nutrient (dietary or supplemental) is an important factor to consider since supplemental B vitamins were shown to have increased risk of esophageal cancer. More research is needed regarding choline's effects on esophageal cancer risk.

At present, given the paucity of data, it is difficult to conclude whether one-carbon nutrients might play a role in esophageal carcinogenesis via alterations in DNA methylation. Dietary folate has been shown to modulate CpG promoter DNA methylation of the *MGMT* gene, but not the tumor suppressor *p16* or DNA mismatch repair *hMLH1* genes (Wang et al. 2008; Chen et al. 2012). CpG promoter DNA hypermethylation of *MGMT* was a consistent characteristic observed in various cancer tissues (Chen et al. 2012; Nakai et al. 2012). Studies are warranted to investigate the effect of one-carbon nutrients on global DNA methylation and CpG promoter DNA methylation of other target genes involved in esophageal cancer. For example, genome-wide DNA hypomethylation was observed in early BE progression, both globally and in CpG islands located in gene promoter regions (Alvarez et al. 2011). Interestingly, DNA hypomethylation at CpG dinucleotides outside of CpG islands in the promoter region has been observed in precancerous esophageal tissues. DNA methylation of *Deleted in Malignant Brain Tumor 1* (*DMBT1*) gene, which does not contain a CpG island gene promoter region, has been shown to be hypomethylated in CpG sites, leading to transcriptional upregulation in early BE progression (Alvarez et al. 2011). More studies are needed to determine whether one-carbon nutrients are able to modulate global and gene-specific DNA methylation in precancerous esophageal lesions such as BE as it could help identify early epigenetic biomarkers of and potential therapeutic target for esophageal cancer development and progression.

Gastric Cancer

Gastric cancer (GC) is the fourth most common cancer worldwide and the second leading cause of cancer mortality (Gonda et al. 2012). Similar to esophageal cancer, GC rates vary widely around the world, with the highest rates in Japan, Korea, and China (Forman and Burley 2006). The *Helicobacter pylori* (*H. pylori*) infection is a recognized independent risk factor for development of GC (Forman and Burley 2006). Dietary factors linked to increased GC risk include high salt intake, excess alcohol, and low fruit and vegetable consumption (Forman and Burley 2006).

Studies investigating the chemopreventive effects of nutrients on GC have generally found one-carbon nutrients to be protective against GC (Mayne et al. 2001). Only few animal studies have investigated this link. High folic acid supplementation was found to be protective against *N*-ethyl-*N*-nitrosoguanidine-induced GC in beagles (Xiao et al. 2002). In a rodent model, folic acid supplementation was able to decrease mucosal inflammation and dysplasia induced in hypergastrinemic mice infected with *Helicobacter felis* (Gonda et al. 2012).

In case–control and cohort studies, low folate status has been linked to an increased risk of GC (Weng et al. 2006; Fang et al. 1997), and higher folate status, particularly in individuals with the *MTHFR* 677TT genotype (Mu et al. 2007; Galvan-Portillo et al. 2009, 2010; Gao et al. 2013), was inversely related to GC risk, although two case–control studies found a null association with plasma folate concentrations (Vollset et al. 2007) and dietary folate and vitamin B_6 (Lazarevic et al. 2011). Higher intakes of choline and vitamin B_6 have also demonstrated a protective effect against GC in *MTHFR* 677TT carriers in a Mexican case–control study (Galvan-Portillo et al. 2009). A nested case–control study in the prospective European Prospective Investigation into Cancer and Nutrition (EPIC) cohort found plasma PLP concentrations (vitamin B_6) to have a protective effect on GC risk (Eussen et al. 2010b). Previous studies have shown that low concentrations of plasma PLP (vitamin B_6) can act adversely on SHMT enzyme activity, thereby interfering with one-carbon nutrient metabolism, leading to impaired DNA repair and global DNA hypomethylation (Choi et al. 1998; Jones and Buckley 1990). With regard to vitamin B_{12} status and GC risk, one case–control study found vitamin B_{12} intakes prior to diagnosis decreased GC mortality risk (Galvan-Portillo et al. 2010), while plasma vitamin B_{12} concentrations in the EPIC study were significantly associated with decreased GC risk (Vollset et al. 2007). At present, no intervention trials have investigated the effects of one-carbon nutrients on GC risk and progression.

Decreased global DNA methylation, hypomethylation of oncogenes, and hypermethylation of tumor suppressor genes are frequently observed epigenetic alterations associated with GC (Fang et al. 1997). In the previously mentioned hypergastrinemic transgenic mouse model infected with *Helicobacter felis*, high doses of folic acid provided post-infection mitigated the loss of global DNA methylation in gastric tissue, which was associated with dysplasia progression (Gonda et al. 2012). In humans, global DNA hypomethylation has been associated with low plasma folate concentrations in those with GC (Fang et al. 1997). Additionally, global LINE-1 DNA methylation levels in blood leukocytes in a Polish population were lower in those with GC compared to control subjects (Hou et al. 2010). However, it is unclear whether global DNA hypomethylation in blood leukocytes reflect that in the gastric tissue and its reliability as a potential biomarker of GC risk requires further investigation.

Several studies have investigated gene-specific DNA methylation patterns related to GC. CpG promoter DNA hypomethylation of *c-myc* with overexpression of c-myc protein in cancerous tissues has been demonstrated in gastric biopsies from GC patients compared to noncancerous tissues from controls (Weng et al. 2006). Furthermore, tissue folate concentrations were lower in biopsies exhibiting hypomethylated *c-myc* compared to biopsies with normally methylated *c-myc*. Fang and colleagues demonstrated similar associations with plasma folate concentrations and *c-myc* hypomethylation in 21 patients with advanced GC (Fang et al. 1997). *C-myc* is a proto-oncogene known to promote cell growth and proliferation, and regulates apoptosis and therefore, dysregulation of *c-myc* transcription could impact downstream pathways such as uncontrolled cell proliferation (Pelengaris et al. 2002; Calcagno et al. 2008). Overexpression of *c-myc* has been described in over 40 % of GC and is associated with poor cancer prognosis (Pelengaris et al. 2002; Calcagno et al. 2008). Activation or loss of silencing of *c-myc* via DNA hypomethylation is thought to be the mechanism behind overexpression of *c-myc* mRNA. Furthermore, synergistic epigenetic mechanisms could be associated with increased expression of *c-myc*. *C-myc* is known to recruit histone acetyl transferases (HATs) to the chromatin structure (Pelengaris et al. 2002). Opposite to the action of HDACs, HATs loosen the DNA structure, leading to enhanced transcription. DNA hypomethylation of the proto-oncogene *c-Ha-ras* has also been implicated in GC risk. In advanced gastric cancer patients, 40 % of precancerous and 50 % of cancerous biopsies exhibited *c-Ha-ras* DNA hypomethylation (Fang et al. 1997). Similar to the consequences of *c-myc* hypomethylation, hypomethylation of *c-Ha-ras* is thought to promote gastric carcinogenesis by activation of gene transcription and uncontrolled cell proliferation (Fang et al. 1997). A recent case–control study investigated the effect of folate intake on DNA methylation of *COX-2*, *MGMT* and *hMLH1* genes as related to GC risk and a potential modifying effect of the *MTHFR* C677T polymorphism (Gao et al. 2013). Individuals with the *MTHFR* 677TT genotype and low folate intake had the highest GC risk. Similarly, the *MTHFR* 677TT genotype combined with hypermethylated *MGMT* had the highest GC risk compared to the *MTHFR* CC genotype. *COX-2* and *hMLH2* DNA methylation patterns were not associated with folate status or the *MTHFR* C677T genotype (Gao et al. 2013).

Gastric Cancer Summary. Overall, both animal and human studies are generally suggestive of a protective effect of one-carbon nutrients on GC risk. However, this relationship has been shown to be dependent on genetic variants in several enzymes in the folate-metabolic pathway, particularly the *MTHFR* C677T polymorphism.

With respect to the role of DNA methylation in gastric carcinogenesis, global DNA hypomethylation was observed in animal and human studies, whereas CpG promoter DNA hypermethylation of several tumor suppressor genes was exhibited in several human studies. DNA hypomethylation and subsequent activation of *c-myc* and *c-Ha-ras* proto-oncogenes have been implicated in increased GC risk. Dietary folate, in particular, was also shown to influence DNA hypomethylation of these genes, indicating that folate could play a significant role in GC risk modulation. Furthermore, CpG promoter DNA hypermethylation of the tumor suppressor *MGMT* gene was also linked to increased GC risk. DNA methylation changes in several other genes have been associated with increased GC risk (Calcagno et al. 2013), although they are yet to be linked to one-carbon nutrient exposure. Nevertheless, further investigation is needed to explore the role of one-carbon nutrients in modulating DNA methylation and consequent GC development and progression.

Liver Cancer

Liver cancer, one of the most common and fatal cancers in the world, is often diagnosed in advanced stages and is associated with poor prognosis (Song et al. 2013). The most common risk factors of liver cancer are chronic infection with hepatitis B and hepatitis C virus (HBV and HCV), alcohol abuse, environmental toxins, and certain metabolic and immune disorders (Song et al. 2013).

An overwhelming body of evidence exists in rodent models that supports the role of folate and other one-carbon nutrients in liver carcinogenesis via aberrant DNA methylation both at the global (Wainfan et al. 1989; Christman et al. 1993; Wilson et al. 1984; Wainfan and Poirier 1992; Pogribny et al. 2006, 2009a; Asada et al. 2006; Dizik et al. 1991) and gene-specific levels (Christman et al. 1993; Kim et al. 1997; Wainfan and Poirier 1992; Dizik et al. 1991; Du et al. 2009; Chagas et al. 2011; Okabe et al. 2011; Shimizu et al. 2007). More specifically, these studies have demonstrated that diets deficient in one-carbon nutrients (such as folic acid, methionine, choline, and vitamin B_{12}) were able to induce an increase in global DNA hypomethylation compared to a diet sufficient in these nutrients in the liver (Christman et al. 1993; Wilson et al. 1984; Pogribny et al. 2009b; Kim et al. 1997; Wainfan and Poirier 1992; Asada et al. 2006; Dizik et al. 1991; Okabe et al. 2011; Shimizu et al. 2007), in some cases leading to neoplastic transformation (Pogribny et al. 2006; Asada et al. 2006).

On a gene-specific scale, diets deficient in these one-carbon nutrients led to an increase in expression of the proto-oncogenes *c-myc*, *c-fos*, and *c-Ha-ras* (Christman et al. 1993; Wainfan and Poirier 1992; Dizik et al. 1991), which was primarily mediated through DNA hypomethylation of these genes in rat liver tissue (Christman et al. 1993; Dizik et al. 1991). Similarly, another study demonstrated that a choline-deficient diet resulted in DNA hypomethylation of *c-fos*, translating into an increase in gene expression, and in DNA hypermethylation of *E-cadherin*, *connexin 26*, and *Rassf1a*, which however did not result in altered gene expression (Shimizu et al. 2007). The relationship of folate with liver carcinogenesis was examined in a study, which demonstrated that folic acid supplementation decreased the expression of *c-myc* in liver tumor of rats and led to a return of the preneoplastic phenotype and inhibition of DNA damage compared to the control basal folic acid diet (Chagas et al. 2011). Collectively, these studies demonstrate that one-carbon nutrients have the potential to mediate DNA methylation at the global and gene-specific levels with functional ramifications, and that this epigenetic mechanism may in part account for the role of these one-carbon nutrients in liver carcinogenesis.

Other aspects that play a role in DNA methylation, such as timing of intervention and the effect it has on the components of DNA methylation machinery, have been well characterized in rodent models. For example, sensitivity to folate deficiency in early life has been demonstrated in a study in which global DNA hypomethylation resulting from exposure to a diet deficient in folic acid from post weaning to puberty persisted even after returning to the control diet at puberty (Kotsopoulos et al. 2008). This study also demonstrated that these early life epigenetic alterations were permanent and may have downstream effects on liver carcinogenesis, given that DNA hypomethylation is a hallmark characteristic of the early stages of carcinogenesis including that of the liver. The importance of the timing of deficiency of one-carbon nutrients on DNA methylation and cancer was highlighted in another study that demonstrated liver tumors of rats fed a choline-deficient diet exhibited increasing DNA methylation and a decrease in expression of *lysophosphatidic acid* (*LPA*) *receptor-3* gene, which mediates several cellular effects including cell proliferation, with increasing age (Okabe et al. 2011). The effect of one-carbon nutrients on DNMTs has also been studied in the liver in order to elucidate the mechanics of DNA methylation in response to one-carbon nutrients. Increased DNMT1 and DNMT3a expression in liver tumor tissue was demonstrated in response to a diet deficient in methionine, folic acid, and choline, compared to controls (Ghoshal et al. 2006). This observation may be explained in part by the compensatory upregulation of DNMT in order to maintain optimal DNA methylation status in the state of substrate deficiency. Another study also showed an increase in DNMT activity in liver tissue in rats on a folate-deficient diet as well as an upregulation of DNA repair proteins, which may have been in response to an increase in DNA damage resulting from folate deficiency (Duthie et al. 2010).

There is a paucity of human studies investigating the effects of one-carbon nutrients on the risk of liver cancer, despite the large body of evidence that clearly demonstrates the mechanistic link between one-carbon-mediated DNA methylation changes and liver cancer. Two prospective studies revealed a relationship between folate and the risk of liver cancer. In a cohort of high-risk individuals (those with chronic HBV infection), RBC folate concentration was inversely associated with the development of liver cancer (Welzel et al. 2007). Similarly, a protective effect was observed in a prospective study that examined the combined effects of dietary folate intake and alcohol consumption on liver cancer risk. Although a significant effect on liver cancer was not observed when comparing the highest intake of both folate and alcohol compared to the lowest folate and alcohol intake level, this study revealed that high folate was able to protect against the effect of high alcohol consumption on the development of liver cancer (Persson et al. 2013). These results render some support for the protective effects of folate in individuals who are at high risk of liver damage and/or cancer, by ameliorating the impact of HBV infection and alcohol, two known risk factors of liver carcinogenesis.

Furthermore, there is evidence to support the relationship between perturbed folate metabolism and liver cancer risk. A recent meta-analysis demonstrated a significantly reduced risk associated with the variant CC genotype of the *MTHFR* A1298C polymorphism compared to the wild-type AA genotype (Qin et al. 2013). Another meta-analysis revealed a significantly increased risk of liver cancer in individuals with the variant TT genotype of the *MTHFR* C677T polymorphism compared to the wild-type CC genotype, a risk which was even more pronounced in patients with chronic liver disease (Jin et al. 2009).

Aberrant DNA methylation patterns, both global and gene-specific, have been identified through a study that performed genome-wide methylation profiling of liver cancer and normal tissues in humans. This study revealed a significant decrease in genome-wide DNA methylation in liver cancer tissue compared to normal. Furthermore, the most significant differentially methylated promoter CpG island changes were observed in four genes (*BMP4*, *CDKN2A*, *GSTP1*, and *NFATC1*) that play a role in cellular development, gene expression, and cell death (Song et al. 2013). Aside from these candidate genes, which provide a viable starting point for the investigation of one-carbon nutrient-mediated DNA methylation alterations in liver carcinogenesis, another study found that *insulin-like growth factor 2* (*IGF2*) gene-specific DNA hypomethylation in peripheral mononuclear blood cells is significantly higher in liver cancer patients compared to controls in the setting of chronic HCV-related liver cirrhosis (Couvert et al. 2012). Therefore, a decrease in *IGF2* DNA methylation in the setting of liver cirrhosis is another postulated risk factor and its relationship to one-carbon nutrients is worthy of further investigation.

Liver Cancer Summary. Studies in the rodent model have clearly demonstrated that a diet deficient in folate and other one-carbon nutrients increase liver cancer risk and this process may be mediated in part through aberrant DNA methylation of the genome and of specific proto-oncogenes and tumor suppressor genes. There is also evidence to support the protective role of folate sufficiency or supplementation in the early stages of liver carcinogenesis. These results were mirrored in human studies which showed that the increased risk of liver cancer associated with alcohol and chronic HBV infection was ameliorated by higher folate intakes. Furthermore, there is evidence to support that disturbed folate metabolism (e.g., perturbed folate homeostasis resulting from *MTHFR* polymorphisms) may induce aberrant global and gene-specific DNA methylation patterns and influence the risk of liver cancer risks in humans. Given the direct relationship between one-carbon nutrients and

DNA methylation, the role of these nutrients in liver cancer risk mediated through this epigenetic mechanism should be investigated in concert. Additionally, there is a need to determine if there is an association between other one-carbon nutrients (vitamin B_{12}, vitamin B_6 and choline), and liver cancer risk in addition to folate.

Pancreatic Cancer

Pancreatic cancer is the fourth most common cause of cancer-related mortality worldwide, which can largely be attributed to diagnosis at an advanced stage due to the lack of effective screening (Chuang et al. 2011; Hariharan et al. 2008). Symptoms are often vague and nonspecific making it exceptionally difficult to diagnose pancreatic cancer at an early stage when it is potentially treatable (Gong et al. 2009). Currently, the most common identified risk factors are cigarette smoking, family history, obesity, diabetes mellitus, and chronic pancreatitis (Chuang et al. 2011; Gong et al. 2009; Oaks et al. 2010).

A few case–control studies have evaluated the association between one-carbon nutrients and risk of pancreatic cancer. A protective effect has been associated with folate intakes from dietary and supplemental sources combined (Gong et al. 2009; Baghurst et al. 1991). Although a protective effect for pancreatic cancer was mirrored in a study investigating circulating levels of folate in male smokers (Stolzenberg-Solomon et al. 1999), a lack of significant association between plasma folate and risk of pancreatic cancer has more commonly been found (Chuang et al. 2011; Schernhammer et al. 2007). Several large population-based prospective studies found a protective effect of dietary folate intake on pancreatic cancer risk confined to women (Oaks et al. 2010), in both men and women (Larsson et al. 2006b) or in male smokers (Stolzenberg-Solomon et al. 2001). However, results from prospective studies have not been uniformly consistent as there have been several studies demonstrating a null association between dietary and supplemental folate intakes and the risk of pancreatic cancer (Skinner et al. 2004; Keszei et al. 2009).

With respect to case–control analyses of vitamin B_6, one study found a significant risk reduction with higher vitamin B_6 intakes compared to lower (Baghurst et al. 1991). However, the majority of the published case–control studies found a lack of an association between intakes and/or blood levels of vitamin B_6 and pancreatic cancer risk (Chuang et al. 2011; Gong et al. 2009; Stolzenberg-Solomon et al. 1999; Schernhammer et al. 2007). There is a lack of prospective studies investigating the relationship between vitamins B_6 and pancreatic cancer risk with only one study reporting a null association (Stolzenberg-Solomon et al. 2001).

Higher vitamin B_{12} levels in non-supplement users (Schernhammer et al. 2007) and in male Finnish smokers (Stolzenberg-Solomon et al. 1999) have been shown to have a protective effect on pancreatic cancer. However, other studies have found no association between the risk of pancreatic cancer and plasma vitamin B_{12} levels (Chuang et al. 2011) or dietary vitamin B_{12} intakes (Baghurst et al. 1991). Most surprising was the significantly increased risk of pancreatic cancer, by almost two-fold, observed with higher intakes of dietary vitamin B_{12} (Gong et al. 2009).

The influence of epigenetics on the development and progression of pancreatic cancer has been well characterized in the most common pancreatic neoplasm (accounting for greater than 85 % of pancreatic tumor cases), known as pancreatic ductal adenocarcinoma (PDAC) (Hezel et al. 2006; McCleary-Wheeler et al. 2013). In particular, the inactivation of the tumor suppressor gene *p16*, via CpG promoter DNA hypermethylation, has been implicated in PDAC (McCleary-Wheeler et al. 2013; Heichman and Warren 2012). From a broad list of genes that are commonly hypermethylated in various cancers, Heichman and Warren identified 21 candidate hypermethylated genes that are specific to pancreatic carcinogenesis, (Heichman and Warren 2012). A comprehensive review by McCleary-Wheeler and colleagues has elucidated other unique DNA methylation events which occur in pancreatic cancer including increasing aberrant gene-specific DNA methylation patterns in the progression of intraepithelial neoplastic changes in PDAC (*NPTX2*, *SARP2*, *Remprimo*, and *LHX1*) and promoter DNA hypomethylation which results in over-expression of certain genes that are normally silenced in normal epithelium (McCleary-Wheeler et al. 2013). This defined list of genes that are regulated by DNA methylation in pancreatic carcinogenesis is particularly useful for identifying possible biomarkers of pancreatic cancer risk and/or potential therapeutic targets for prevention and treatment using dietary, including one-carbon nutrients, or pharmacologic strategies (Heichman and Warren 2012).

Pancreatic Cancer Summary. Due to the small number of animal and epidemiologic studies and the inconsistent results from the small number of published studies, it is difficult to determine whether one-carbon nutrients may play a role in pancreatic cancer development. Some of the published epidemiological studies were confined to smokers whose requirements and intakes of folate may be different than the general population. Genetic polymorphisms in one-carbon nutrient metabolizing genes have been shown to affect various cancer risks including pancreatic cancer and therefore taking into account genetic variation of these genes may help to elucidate the observed disparities in cancer risk outcomes of the pancreas. Furthermore, considering that DNA methylation is an epigenetic mechanism purportedly implicated in various cancer types including pancreatic cancer, it would be valuable to assess global and gene-specific DNA methylation to use them as diagnostic and prognostic markers and potential prevention and treatment targets. A mechanistic link between one-carbon nutrients and epigenetic changes in pancreatic cancer development should be interrogated to determine whether these nutrients are related to pancreatic cancer risk through epigenetic mechanisms and whether these nutrients can be used to modulate pancreatic cancer risk.

Colorectal Cancer

CRC is the third most common cancer worldwide in men and women. Ecological observations and epidemiologic studies have suggested that environmental factors, including diet, significantly influence the development of CRC. Some of the dietary risk factors associated with increased CRC risk include red and processed meat consumption and excessive alcohol intake and probable protective factors are increased dietary fiber, and calcium supplementation (Vargas and Thompson 2012).

Animal studies have generally demonstrated a causal relationship between folate depletion and CRC risk. On the other hand, modest levels of folate supplementation resulted in an inhibitory effect on colorectal carcinogenesis (Kim 2003; Kadaveru et al. 2012). As previously discussed, however, animals studies have also shown that folic acid supplementation may increase CRC risk and accelerate CRC progression if supraphysiological levels of folic acid are supplemented or if it is provided after neoplastic foci are established in the colorectum (Kim 2003, 2004a) (Fig. 11.3).

Epidemiological studies collectively suggest a ~20–40 % reduction in the risk of CRC or its precursor, adenoma in subjects with the highest dietary intake or blood levels of folate compared with those with the lowest intake or blood levels (Kim 1999, 2007a, 2008; Kennedy et al. 2011). Pooled and meta-analytical approaches of case–control and cohort studies of dietary and/or supplemental folate intake on CRC risk have found a very modest protective effect of dietary folate (Sanjoaquin et al. 2005) and the combination of dietary and supplemental folate (Kim et al. 2010) on CRC risk. The role of folate in colorectal carcinogenesis has been further strengthened by the observations that genetic polymorphisms in the folate-metabolic pathway (e.g., *MTHFR* C677T polymorphism) modify CRC risk (Potter 2002; Bailey 2003; Kim 2009). Although there is no definitive evidence supporting the protective effect of folic acid supplementation on colorectal carcinogenesis from human experiments at present, several small intervention studies have demonstrated that folic acid supplementation can improve or reverse surrogate endpoint biomarkers of CRC (Cravo et al. 1994, 1998b; Kim et al. 2001; Paspatis and Karamanolis 1994; Biasco et al. 1997; Khosraviani et al. 2002; Lashner et al. 1999) and some epidemiologic studies have shown a beneficial effect of multivitamin supplements containing ≥0.4 mg folic acid on CRC risk and mortality (Giovannucci et al. 1995, 1998; Jacobs et al. 2001). As previously mentioned, the Aspirin/Folate Polyp Prevention Study (Cole et al. 2007) reported that daily folic acid supplementation for up to 10 years in individuals likely harboring (pre)neoplastic lesions significantly increased the risk of advanced and multiple colorectal adenomas (Cole et al. 2007) and of prostate cancer (Figueiredo et al. 2009a). However, several meta-analyses of large intervention trials of folic acid supplementation in patients with a history of colorectal adenomas did not find folic acid supplementation to have a significant effect as a chemopreventive measure against recurrent adenomas (Figueiredo et al. 2011; Carroll et al. 2010; Fife et al. 2011; Ibrahim and Zekri 2010). Furthermore, a recent large meta-analysis on folic acid supplementation on all cancer risk from studies with cardiovascular disease and recurrent colorectal adenoma outcomes found folic acid supplementation to have no effect on site-specific cancer risk during the first 5 years of treatment (Vollset et al. 2013). In contrast, a marginally significant increase in overall cancer risk with folic acid supplementation was observed in a meta-analysis of ten randomized-controlled trials all with cancer incidence or mortality as the primary or secondary outcome (Wien et al.

2012). Two recent ecological studies that examined a temporal post-fortification trend of CRC incidence in the United States, Canada, and Chile reported increased CRC rates in these countries following fortification, suggesting that folic acid fortification may have been wholly or partly responsible for this disturbing trend (Mason et al. 2007; Hirsch et al. 2009). However, two large prospective studies conducted after folic acid fortification in the United States have suggested a CRC-protective effect of consuming adequate amounts of folate and have not demonstrated a CRC-promoting effect of folic acid supplementation (Stevens et al. 2011; Gibson et al. 2011).

Some studies have suggested that folate and the synthetic folic acid might have differential effects on colorectal carcinogenesis. For example, epidemiological studies have shown dietary folate, not synthetic folic acid, to be protective against CRC (Kennedy et al. 2011). Furthermore, combined randomized-controlled trials and a meta-analysis found no beneficial effect of folic acid supplementation on CRC risk (Figueiredo et al. 2011; Carroll et al. 2010). Although folate and folic acid ultimately are metabolized to the same metabolite (5-methylTHF) in the body, there is probable cause to believe folate and folic acid may exert different effects on tissues. Unlike reduced folate, folic acid requires additional reduction steps by the enzyme dihydrofolate reductase (DHFR) to be converted to tetrahydrofolate (THF) so it can then enter the folate metabolism pathway. DHFR in humans has been shown to be easily saturated and enzyme activity varies widely among individuals (Bailey and Ayling 2009). With this said, there is concern high concentrations of unmetabolized folic acid may compete with active forms of folate for folate-binding proteins and transporters, and folate-dependent enzymes, potentially creating an intracellular folate-deficient environment as well as perturbed intracellular folate metabolism (Bailey and Ayling 2009; Lucock 2004). This intracellular deficiency could potentially lead to altered methylation capacity. In addition, high intracellular folate concentrations have been shown in mathematical models to increase thymidylate synthesis leading to accumulation of dUMP, potentially leading to increased tumorigenesis (Neuhouser et al. 2011). Furthermore, excess dihydrofolate (DHF) resulting from folic acid supplementation inhibits MTHFR, thereby reducing the formation of 5-methylTHF and consequent DNA hypomethylation (Matthews and Haywood 1979).

A few animal studies have shown dietary vitamin B_6 reduces colorectal epithelial proliferation and tumorigenesis (Jeronimo et al. 2011; Lin et al. 2013). There is also an emerging body of evidence that suggests a potential association between vitamin B_6 and CRC in humans. The majority of case–control studies have concluded that high dietary vitamin B_6 intakes are modestly protective against CRC compared to the low intakes (Zhang et al. 2013b), although not all have found a protective effect (Key et al. 2012; Razzak et al. 2012). Nested case–control studies have suggested that plasma PLP (vitamin B_6) is associated with an approximately 30–50 % reduction in CRC risk when comparing the highest to lowest concentrations (Zhang et al. 2013b). However, one recent case–control study has found that plasma PLP concentrations had no effect on the development of colorectal adenomas (Chen et al. 2013). Several population-based prospective studies have investigated the relationship between vitamin B_6 and CRC risk. The Women's Health Initiative Study found a protective effect of dietary and supplemental vitamin B_6 on CRC risk in postmenopausal women (Zschabitz et al. 2013). The Iowa Women's Health Study initial analysis of vitamin B_6 intake on CRC found an increased risk of rectal cancer in women (Harnack et al. 2002). However, an extension of this study with a longer follow-up period, found no effect (Razzak et al. 2012). The prospective Netherlands Cohort Study found high total vitamin B_6 intake was related to an increased rectal cancer risk in women (de Vogel et al. 2008b). Other large prospective cohorts such as the EPIC Study, the Physicians' Health Study, and the Multiethnic Cohort Study have all reported slight decreases in CRC risk with the highest quartile PLP concentrations compared with the lowest (Eussen et al. 2010a; Le Marchand et al. 2011; Lee et al. 2009). A meta-analysis including these studies observed a lower CRC risk in the highest PLP compared to the lowest PLP concentrations but only a nonsignificant protective trend for dietary vitamin B_6 intakes (Larsson et al. 2010).

The relationship between CRC risk and vitamin B_{12} is less clear. Several epidemiological studies have found no association between dietary (Key et al. 2012; Razzak et al. 2012; Le Marchand et al. 2011) or plasma concentrations of vitamin B_{12} (Chen et al. 2013; Fujimori et al. 2011; de Vogel et al. 2011b; Weinstein et al. 2008; Eussen et al. 2010a) and CRC risk. Of the studies that did find an effect, there has generally been an inverse association between vitamin B_{12} and CRC risk (Williams et al. 2010; Dahlin et al. 2008). In a nested case–control study from the United States, high dietary vitamin B_{12} was inversely related to distal CRC risk, but only in Caucasians (Williams et al. 2010). The prospective Northern Swedish Health and Disease Study found high plasma vitamin B_{12} concentrations were associated with lower rectal cancer risk when comparing highest quintile to lowest after adjusting for BMI, smoking, physical activity, and alcohol (Dahlin et al. 2008).

There is limited evidence regarding choline intake/levels on CRC risk. In one animal study, postweaning dietary choline administered did not affect colorectal tumorigenesis in mice (Yang et al. 2008). In humans, the prospective Health Professionals Follow-up Study found no association with dietary choline or betaine intakes and CRC risk, even when stratifying choline by its metabolic derivatives (Lee et al. 2010). However, in the Norwegian Colorectal Cancer Prevention Trial, the occurrence of high-risk distal colorectal adenomas was inversely associated with plasma betaine concentrations but not with choline (de Vogel et al. 2011a). Interestingly, in the Nurses Health Study, dietary choline was positively associated, yet betaine was inversely associated with risk of colorectal adenoma when comparing the highest quintile of concentrations to the lowest (Cho et al. 2007a). Furthermore, phosphatidylcholine and sphingomyelin, two sources of choline, had the strongest association to increased adenoma risk (Cho et al. 2007b).

Studies looking at the combined effects of B vitamins and choline have not found significant effects on CRC risk. Although cardiovascular disease, and not cancer, was the primary outcome, pooled analysis of two Norwegian intervention trials found the combination of folic acid, vitamin B_6, vitamin B_{12} supplementation, folic acid and vitamin B_{12} together, or vitamin B_6 supplementation alone to have no effect on CRC risk or mortality (Ebbing et al. 2009). Similarly, the Women's Antioxidant and Folic Acid Cardiovascular Study found combined supplementation of folic acid, vitamins B_{12} and B_6 to have no increased risk of colorectal adenoma development in women at high risk for developing cardiovascular disease (Song et al. 2012). Furthermore, in a meta-analysis of eight randomized trials of B vitamin supplementation involving 37,485 individuals at increased risk of developing cardiovascular disease, folic acid supplementation of a median duration of 5 years had no significant effects on vascular outcomes (the primary endpoint), overall cancer incidence, cancer mortality, or all-cause mortality (Clarke et al. 2010).

Neoplastic cells simultaneously harbor widespread global DNA hypomethylation and more specific regional areas of DNA hypermethylation (Jones and Baylin 2002). Global DNA hypomethylation is an early, and consistent, event in colorectal carcinogenesis (Jones and Baylin 2002) and is associated with genomic instability, increased mutations, and upregulation of proto-oncogenes (Esteller 2003). In addition, site-specific DNA hypermethylation at promoter CpG islands of tumor suppressor and mismatch repair genes is an important mechanism in gene silencing in colorectal carcinogenesis (Herman and Baylin 2003). Increased frequencies of promoter region CpG methylation in specific groups of genes gave rise to the CpG island methylator phenotype (CIMP) in CRC (Issa 2004). One common characteristic of tumors in CIMP+ phenotypes is the consistent CpG promoter DNA hypermethylation and subsequent silencing of the mismatch repair gene *hMLH1* (Kane et al. 1997). Therefore, because of the potential modulatory effect of folate and other one-carbon nutrients on DNA methylation, studies have been conducted to elucidate a potential link between one-carbon nutrients and CRC risk via alterations in DNA methylation.

The link between aberrant DNA methylation patterns and CRC risk has almost exclusively been researched with varying degrees of folate exposure in in vitro models. In human colon SW620 adenocarcinoma cells, folate depletion demonstrated a significant increase in global DNA hypomethylation and an increase in region-specific DNA hypomethylation of the *p53* gene; global and *p53*-specific DNA hypomethylation were reverted by folic acid supplementation (Wasson et al. 2006). In contrast, global DNA methylation in HCT116 and Caco-2 colon cancer cell lines have been shown to be resistant to the effect of folate depletion (Stempak et al. 2005), thereby demonstrating the differential effects by cancer cell line.

A closer look at gene-specific changes further demonstrates the complexity of aberrant DNA methylation in response to variations in intracellular folate concentration. Although CpG site-specific DNA hypermethylation in the promoter region of the tumor suppressor gene, *ER*, was detected in folate-depleted HCT116 and Caco-2 cells, these aberrant promoter DNA methylation patterns did not lead to significant functional ramifications, as demonstrated by a lack of change in *ER* gene expression in both cell lines (Stempak et al. 2005). However, other studies found that increased DNA methylation in promoter regions of specific genes in response to folate concentrations had functional consequences as evidenced by altered expression of genes that play a role in tumor suppression and aggressiveness in colon cancer cells. For example, high concentrations of folic acid increased CpG promoter DNA methylation of tumor suppressor genes (*ESR1*, $p16^{INK4a}$, and $p15^{INK4b}$) and led to a significant reduction in *ESR1* gene expression in Caco-2 colon cancer cells (Berner et al. 2010). Another study found that HCT116 colon cancer cells cultured in a folate-depleted environment exhibited enhanced invasiveness as a result of CpG promoter DNA hypomethylation of the *sonic hedgehog* (*Shh*) gene which corresponded to an increase in *Shh* gene expression and increased activation of the Shh signal coupled with increased binding with nuclear factor-kappa B (nf-kB), both crucial events for increased cancer aggressiveness (Zhao et al. 2012). Overall, it is evident that both folate deficiency and high concentrations have the potential to affect the expression of genes involved in colorectal carcinogenesis via altered global and/or promoter DNA methylation patterns. In addition, the effects of folate deficiency and supplementation on DNA methylation appear to be cell-specific, consistent with prior studies that have indicated cell, tissue, and organ-specific effects of folate manipulation on DNA methylation (Kim 2004b; Crowell et al. 2011; Ly et al. 2012).

The effect of isolated folic acid supplementation on DNA methylation in rodent colon has not yet been clearly elucidated. The majority of animal studies have demonstrated a resistance of the colorectum to altered SAM and SAH levels resulting from folate deficiency and equivocally support an association between folate deficiency and global DNA hypomethylation in the colorectum (Kim 2004b, 2005; Crowell et al. 2011). One study found that a colon carcinogen, dimethylhydrazine administration in conjunction with folic acid supplementation of at least 8 mg/kg for 20 weeks in weanling rats did not alter concentrations of SAM, SAH, SAM to SAH concentration ratios, and global DNA methylation (Kim et al. 1996). In addition, weanling rats fed a folic acid-supplemented diet for 5 weeks showed no change in concentrations of SAM, SAH, and SAM to SAH ratios, global DNA methylation, and *p53* gene-specific DNA methylation in the colon (Sohn et al. 2003). Interestingly, dietary folic acid supplementation in both young and elder rats for 8 and 20 weeks resulted in a decrease in colonic SAH concentrations, although global DNA methylation in the colon was not altered (Choi et al. 2003). However, a recent study reported that chronic, severe folate deficiency in older adult mice induced significant global DNA hypomethylation in the colon (Linhart et al. 2009). Contrary to this observation, a study by Sohn et al. (Sohn et al. 2003) demonstrated that folate deficiency of a short duration and severe degree may induce global DNA hypermethylation in the colon (Sohn et al. 2003). The paradoxical effect of folate deficiency on increasing global methylation in colonic tissue is likely due to compensatory upregulation of *DNMT* and of the choline and betaine-dependent transmethylation pathway. Generally, the results from animal studies suggest that DNA methylation patterns are gene and site-specific and depend on cell type, target organ, and stage of transformation as well as on the timing, degree, and duration of folate intervention (Crowell et al. 2011; Ly et al. 2012).

In human studies, colonic global DNA methylation has been shown to be positively correlated with serum and RBC folate concentrations and negatively with plasma homocysteine concentrations in individuals with colonic adenomas and adenocarcinomas (Pufulete et al. 2003; Al-Ghnaniem et al. 2007) and in those without these lesions (Pufulete et al. 2005b). Folate levels and SAM to SAH concentration ratios have also been reported to be lower (by 28 %) in malignant tissue compared with normal-appearing adjacent colon mucosa in subjects with CRC

(Alonso-Aperte et al. 2008); it has been well established that colorectal neoplasms have lower global DNA methylation compared with nonneoplastic colonic tissue. Additional evidence lends support to a positive relationship between folate status and global DNA methylation. In a combined analysis of CRCs from participants from the Nurses' Health Study and the Health Professionals Follow-up Study, the risk of global DNA hypomethylation (determined by <55 % LINE-1 methylation) was 43 % lower in subjects with a high compared with a low total daily folate intake (Schernhammer et al. 2009). In a study that stratified folate intake according to pre- and post-fortification levels, the observed inverse association between leukocyte global DNA methylation and adenoma was stronger among subjects with low (<0.317 μg/1000 kcal pre- and <0.413 μg/1000 kcal post-fortification) as compared to high (≥0.317 μg/1000 kcal pre- and ≥0.413 μg/1000 kcal post-fortification) total folate intake in either fortification periods (Lim et al. 2008).

In some human intervention studies, folic acid supplementation at 12.5–25 times the daily requirement for 3–12 months significantly increased the extent of colonic genomic DNA methylation in subjects with resected colorectal adenoma or cancer (Cravo et al. 1994, 1998b; Kim et al. 2001), whereas no such effect was observed in patients with chronic ulcerative colitis who were given folate supplementation at 12.5 times the daily requirement for 6 months (Cravo et al. 1998a) or in participants from the Aspirin/Folate Polyp Prevention Study, as determined by methylation of long interspersed nucleotide elements (LINE-1) (Figueiredo et al. 2009b). Folic acid supplementation at three and five times the daily requirement, which was sufficient to improve and correct a marker of DNA damage, also failed to modulate genomic DNA methylation in lymphocytes of healthy volunteers (Basten et al. 2006; Fenech et al. 1998). In another study, daily folic acid supplementation with 15 mg 5-methylTHF for 8 weeks has been observed to restore genomic DNA methylation in lymphocytes to normal levels in 32 men with uremia, hyperhomocysteinemia, and preexisting genomic DNA hypomethylation (Ingrosso et al. 2003). In patients with colorectal adenomas, a physiological dose of folic acid (0.4 mg/day) for 10 weeks has been demonstrated to significantly increase both genomic DNA methylation in lymphocytes (by 31 %) and in colonic mucosa (by 25 %) compared with placebo (Pufulete et al. 2005a).

Aberrant CpG island methylation is characteristic of tumor development and specific promoter CpG islands are frequently and simultaneously methylated in sporadic CRC, leading to transcriptional silencing (Markowitz and Bertagnolli 2009; Kawakami et al. 2003; Curtin et al. 2007; Toyota et al. 1999). In the Netherlands Cohort Study on Diet and Cancer, the prevalence of CpG island promoter hypermethylation was higher, albeit nonsignificantly, in CRCs derived from patients with low folate/high alcohol intake compared with CRCs from patients with high folate/low alcohol intake for each of the six tested genes (*APC*, *p14*, *p16*, *hMLH1*, *O^6-MGMT*, and *RASSF1A*) (van Engeland et al. 2003). The number of CRCs with at least one gene methylated was higher (84 %) in the low folate intake/high alcohol intake group compared with the high folate intake/low alcohol intake group (van Engeland et al. 2003). A later follow-up analysis in a sub-cohort of this population did not report any effect of isolated dietary folate intake on risk of CRCs specifically presenting with *hMLH1* hypermethylation (de Vogel et al. 2008a). Furthermore, a recent trial found 400 μg/d folic acid supplementation had no effect on *ESR1* and *hMLH1* methylation in colonic mucosa (Abbadi et al. 2012). Al-Ghnaniem et al. also examined gene-specific methylation in biopsies of normal-appearing colorectal mucosa from subjects with and without colorectal neoplasia (Al-Ghnaniem et al. 2007). In general, patients with neoplasia were reported to have lower serum folate and promoter CpG hypermethylation of the *ERα* and *hMLH1* genes compared with disease-free patients. *ERα* methylation was also positively correlated with plasma homocysteine in all subjects but significant inverse correlations between promoter CpG methylation and folate status were not observed (Al-Ghnaniem et al. 2007). In contrast, a modest degree of colorectal CpG hypermethylation of the *ERα* and *SFRP1* genes was significantly associated with higher RBC folate levels in participants from the Aspirin/Folate Polyp Prevention Study (Wallace et al. 2010). The odds of promoter methylation of *CDKN2A*, *MLH1*, *CACNA1G*, *NEUROG1*, *RUNX3*, *SOCS1*, *IGF2*, and *CRABP1* in colorectal tumors are also greater in patients with high circulating levels of plasma folate (Van Guelpen et al. 2010). CRCs with frequent promoter methylation have been shown to have higher tumor concentrations of different folate metabolites, including 5,10-methyleneTHF and THF (Kawakami et al. 2003).

Epigenetic alterations in the Wnt-signaling pathway have been implicated in development of human CRC. The Wnt-signaling pathway is a series of proteins which participate and regulate stages of development, including embryogenesis and cell differentiation and tissue regeneration (Logan and Nusse 2004). Studies on gene aberrations in the Wnt-signaling pathway found that mutations in the *Axin2* predispose individuals to CRC (Lammi et al. 2004) and *Apc* gene mutations lead to increased Wnt-signaling pathway activation and consequent cell proliferation in the colorectum, thereby increasing the risk of CRC (Nishisho et al. 1991). For example, a few in vitro studies have shown folate depletion leads to changes in the Wnt-signaling pathway (Crott et al. 2008; Huang and Chen 2008; Morillon and Katula 2008). Furthermore, in cross-bred BAT-LacZxApc1638N mice, mild depletion of methyl donors including folate, B_6, and B_{12}, resulted in a fourfold increase in Wnt-signaling pathways in the colonic mucosa while simultaneously observing activation of genes related to the Wnt-signaling pathway (Liu et al. 2011). The authors suggested increased gene expression of *Jun* and *Pitx2*, two genes related to proliferation, transformation, and apoptosis regulation, to be responsible for upregulated Wnt-signaling and subsequent colonic tumorigenesis (Liu et al. 2011).

The direction and magnitude of effect due to dietary and blood folate concentrations on gene-specific DNA methylation remain unclear. Some studies demonstrate a greater prevalence or risk of aberrant DNA hypermethylation of certain genes involved in colorectal carcinogenesis in subjects with low folate while others have reported this in subjects with high folate. The discrepancies in identifying a clear association between folate status and gene-specific DNA methylation may be explained in part by the different methods of stratifying folate levels for comparison and the use of different markers to evaluate folate status. Dietary intake and serum levels of folate may not necessarily be reflective of folate concentrations in the target organ. Moreover, blood as a surrogate marker of methylation is not always representative of tissue-specific methylation (McKay et al. 2011d). These studies are also complicated by the lack of consistency in the specific genes investigated and sampling of different CpG sites in different tissues.

There is evidence that folate status influences DNA methylation through an interaction with the *MTHFR* C677T polymorphism. The *MTHFR* C677T polymorphism causes thermolability and reduced MTHFR activity, leading to lower levels of 5-methylTHF, an accumulation of 5,10-methyleneTHF, increased plasma homocysteine levels, and changes in cellular composition of one-carbon folate derivatives (Friso et al. 2002; Kim 1999). Studies have also indicated that another polymorphism in the *MTHFR* gene (A1298C) may modulate genomic DNA methylation in human lymphocytes, although the degree and direction of change have not been clearly established (Castro et al. 2004; Friso et al. 2005). More recent investigations of folate status and DNA methylation in humans include analysis of common *MTHFR* polymorphisms. In a study that investigated the combined effects of folic acid and vitamin B_{12} supplementation for 6 months on CpG promoter DNA methylation of six tumor suppressor and DNA repair genes frequently reported to be aberrantly methylated in CRC, a trend towards a 67 % increase in promoter DNA hypermethylation was reported in the rectal mucosa of patients with resected colorectal adenomas, although this did not reach statistical significance (van den Donk et al. 2007a). However, further investigation of the six genes revealed that folate intake interacted with the *MTHFR* C677T polymorphism to influence CpG promoter DNA methylation in colorectal adenomas such that among individuals homozygous for this variant, the risk of promoter DNA methylation was inversely related to dietary folate intake, but statistical significance was only observed for the *O6-MGMT* DNA-methyltransferase gene (van den Donk et al. 2007b). The results from this research group suggest higher folate intakes may increase *DNMT* expression and subsequent DNA methylation activity, particularly in individuals with adenomas and reduced MTHFR enzyme activity (van den Donk et al. 2007b). Furthermore, Slattery et al. initially failed to identify a significant association between dietary folate and colon tumor CpG island methylation of *p16, hMLH1, and MINT-1, -2, and -3 loci* (Slattery et al. 2006), but in their follow-up analysis, subjects heterozygous or homozygous for the *MTHFR* A1298C genotype with low folate/low methionine/high alcohol intake had an over twofold greater odds of developing tumors presenting CpG island DNA hypermethylation compared with subjects with the wild-type genotype and high folate/high methionine/low alcohol intake (Curtin et al. 2007). A greater risk of *p16* DNA hypermethylation in head and neck squamous cell carcinomas was also observed in subjects with low dietary folate intakes compared with those with high dietary folate, which was further exacerbated in subjects with the *MTHFR* 677TT genotype (Kraunz et al. 2006). In contrast, one study reported that the prevalence of CpG promoter DNA methylation of *p16*, but not *hMLH1* or *hMSH2*, was significantly higher in CRCs from patients with high serum folate concentrations, that the odds of tumor promoter DNA methylation were significantly higher in patients with high circulating folate levels, and that this positive association was further modified by the

MTHFR C677T polymorphism, reaching significance only in subjects heterozygous or homozygous for the *MTHFR* C677T polymorphism (Mokarram et al. 2008).

CRC Summary. Overall animal and epidemiological evidence suggests dietary folate and vitamin B_6 are protective against CRC. Supplemental folic acid, on the other hand, may have the potential to promote colorectal carcinogenesis if given at very high doses or if provided after preneoplastic lesions are established. There is limited data on the effects of vitamin B_{12}, choline, and betaine on CRC risk, although there is some evidence suggesting that choline may increase the risk of CRC. These observed differential effects of one-carbon nutrients on CRC risk may be in part related to the dietary sources of these nutrients. Folate and vitamin B_6 are generally considered plant-based nutrients; intakes are mainly derived from consumption of vegetables and fruits. However, vitamin B_{12} and choline are more exclusively found in animal products. It is possible that dietary patterns, and nutrients found in those foods, may also interact with one-carbon nutrients in modulating CRC risk.

DNA methylation alterations, global DNA hypomethylation and CpG promoter DNA hypermethylation of tumor suppressor and other cancer-related genes, are mechanistically related to the development of CRC. Due to the potential modulatory role of one-carbon nutrients in biological methylation reactions including DNA methylation, there has been intense interest in elucidating the effect of one-carbon nutrients on global and gene-specific DNA methylation implicated in colorectal carcinogenesis. The bulk of studies have interrogated the role of folate in DNA methylation in the colorectum and its link to colorectal carcinogenesis. At present, there is a paucity of studies investigation the effects of vitamins B_6, B_{12}, and choline on DNA methylation in the colorectum.

With respect to folate, the results from in vitro and animal studies suggest that the effects of folate deficiency and supplementation on DNA methylation are gene- and site-specific and depend on cell type, target organ, and stage of transformation as well as on the timing, degree, and duration of folate intervention. In human studies, there is inconsistent data concerning the effect of folate deficiency of a physiologically and clinical relevant degree on global DNA methylation in the colorectum. In contrast, folic acid supplementation, even at the modest supplemental levels, appears to be able to increase global DNA methylation in the colorectum in certain situations. The majority of observational studies have described a direct correlation between dietary and blood levels of folate and global DNA methylation in both lymphocytes and colonic tissues such that a low folate status is associated with global DNA hypomethylation. This positive association is more consistent in individuals with colorectal adenomas, adenocarcinomas, or previously resected neoplastic tumors as well as in those at a greater risk of health complications compared with normal subjects.

It is important to consider the common *MTHFR* polymorphisms associated with impaired enzyme activity and how they interact with folate in a manner to modulate both global and gene-specific DNA methylation. Human observational studies provide evidence that the *MTHFR* C677T polymorphism is associated with global

DNA hypomethylation in leukocytes, which may be mediated in part, by a low status in folate or other methyl donors. For the colorectum and other tissue sites, whether or not the *MTHFR* C677T and A1298C polymorphism in conjunction with marginal folate status affects DNA methylation needs to be further characterized. These studies emphasize the importance of taking into consideration interactions between folate status and critical genes in the folate and one-carbon metabolic pathways when investigating the effect of folate nutrition on DNA methylation.

In terms of gene-specific DNA methylation, the direction and magnitude of effect due to dietary and blood folate concentrations remain unclear. Some studies demonstrate a greater prevalence or risk of aberrant DNA hypermethylation of certain genes involved in colorectal carcinogenesis in subjects with low folate while others have reported this in subjects with high folate. The discrepancies in identifying a clear association between folate status and gene-specific DNA methylation may be explained in part by the different methods of stratifying folate levels for comparison and the use of different markers to evaluate folate status. Dietary intake and serum levels of folate may not necessarily be reflective of folate concentrations in the target organ.

The potential role of one-carbon nutrients in modulating DNA methylation in the colorectum and their contribution to colorectal carcinogenesis via this epigenetic mechanism is worthy of further investigation, given that these nutrients are integrally involved in biological methylation reactions including that of DNA.

11.4.2 Reproductive Cancers

Breast Cancer

Breast cancer accounts for 23 % of all cancers diagnosed across the globe and ranks as the fifth cause of death from cancer (Teegarden et al. 2012). Of growing concern is the increased incidence among premenopausal women with increasing aggressive neoplastic manifestations that may be less responsive to therapy (Teegarden et al. 2012). Breast cancer is characterized by abnormal genetic and epigenetic alterations, notably promoter CpG island DNA hypermethylation and global DNA hypomethylation (Xu and Chen 2009; Xu et al. 2009b).

The effects of folate deficiency and folic acid supplementation on mammary tumorigenesis have been investigated in the methylnitrosourea (MNU)-induced rat model. Collectively, these animal studies showed that a moderate folate-deficient diet suppressed mammary tumorigenesis, whereas modest supplemental levels of folic acid had no significant effect on mammary tumorigenesis (Baggott et al. 1992; Kotsopoulos et al. 2003). It appears that folate deficiency had no effect during the initiation stage but significantly suppressed the progression of or caused regression of the established mammary neoplastic foci during the promotion phase of the tumorigenesis (Kotsopoulos et al. 2005).

Case–control studies have generally demonstrated that higher dietary folate intakes significantly reduce the risk of breast cancer (Lajous et al. 2006a, 2006b; Shrubsole et al. 2001; Zhang et al. 2011) and a much more prominent effect was observed in women with higher plasma folate who consumed higher alcohol (Zhang et al. 2003). In contrast, a significant increase in breast cancer risk was observed in premenopausal women with higher dietary folate intakes (Ma et al. 2009b) and higher plasma folate concentrations (Lin et al. 2008). The majority of prospective studies examining the risk of breast cancer in relation to intakes of one-carbon nutrients demonstrated a lack of association with dietary folate (Cho et al. 2007b; Kabat et al. 2008; Feigelson et al. 2003; Shrubsole et al. 2011; Stevens et al. 2010). However, there is no clear pattern of association considering that some studies have also demonstrated a protective effect while others report an increased risk (Kotsopoulos et al. 2012). However, subgroup analyses have shown that high folate intakes may have a significant protective effect in women who also had higher dietary intakes of vitamin B_{12} (Lajous et al. 2006a, 2006b), in women with ER-hormonal status (Maruti et al. 2009) and in premenopausal women (Shrubsole et al. 2011).

The joint effect of genetic polymorphisms of *MTHFR* and one-carbon nutrient intakes on breast cancer risk has also been investigated; one study found a null association with dietary folate intakes and three *MTHFR* polymorphisms (C677T, A1298C, and A2756G) (Ma et al. 2009a), whereas another demonstrated a significant reduction in risk in women with the variant genotypes *MTHFR* 677CT and 677TT consuming low dietary folate (Ma et al. 2009b). In contrast, an increased risk associated with low dietary folate intake and the variant *MTHFR* 677TT compared to the wild-type 677CC genotype with higher intakes was observed (Chen et al. 2005). Finally, an increased risk was demonstrated in women with 1298AC and 1298CC genotypes with lower intakes compared to those with the wild-type 1298AA genotype (Ma et al. 2009b).

Inconsistencies exist in the case–control studies investigating the relationship between vitamin B_6 and breast cancer risk. Some studies have reported a significant decrease in the risk associated with higher dietary intakes (Zhang et al. 2011), higher circulating concentrations (Lin et al. 2008), and higher dietary intakes combined with lower alcohol intakes (Zhang et al. 2003), while other studies have reported a lack of association between higher vitamin B_6 intakes and breast cancer risk (Lajous et al. 2006a, 2006b; Shrubsole et al. 2001). The majority of prospective studies examining the risk of breast cancer in relation to dietary intakes of vitamin B_6 demonstrated a lack of effect (Cho et al. 2007b; Shrubsole et al. 2011; Stevens et al. 2010; Maruti et al. 2009).

Similarly, inconsistencies exist in regard to vitamin B_{12} intakes and breast cancer risk. Although some case–control studies have reported that higher intakes exhibited a protective effect (Lajous et al. 2006a, 2006b) and higher plasma concentrations exerted a protective effect only in premenopausal women (Zhang et al. 2003), a number of studies have reported a null association (Shrubsole et al. 2001; Zhang et al. 2011; Lin et al. 2008). Finally, some studies have shown an increase in breast cancer risk associated with higher dietary intakes of vitamin B_{12} (Lin et al. 2008) and higher serum vitamin B_{12} concentrations (Wu et al. 1999). The majority of prospective studies examining the risk of breast cancer in relation to dietary intakes of vitamin B_{12} demonstrated a lack of effect (Cho et al. 2007b; Shrubsole et al. 2011; Stevens et al. 2010; Maruti et al. 2009).

Only two case–control studies investigated the effects of choline and betaine on breast cancer risk and both found a significant protective effect associated with higher dietary choline intakes (Zhang et al. 2013a; Xu et al. 2009a), higher betaine intakes (Zhang et al. 2013a), and a combination of higher intakes of both choline and betaine (Zhang et al. 2013a). High intakes of choline and betaine were associated with a significant decrease in breast cancer-related mortality (Xu et al. 2009a).

The role of folate in breast carcinogenesis has been demonstrated in an in vitro system with respect to the potential it has to modulate gene expression via DNA methylation. In MCF-7 and MDA-MB-231 breast cancer cells, high folic acid concentrations led to an increase in CpG promoter DNA methylation and dose-dependent inverse regulation of the tumor suppressor genes, *PTEN*, *APC*, and *RAR*β_2 in both cell lines (Lubecka-Pietruszewska et al. 2013). Furthermore, a significant increase in the expression of DNMT1, which facilitates the addition of methyl groups to the DNA sequence, at the highest folic acid concentration in the MDA-MB-231 cell line was observed (Lubecka-Pietruszewska et al. 2013).

A few human studies have been conducted to elucidate the influence of one-carbon nutrient intakes on aberrant DNA methylation in breast cancer. One study confirmed that there is an increased risk of breast cancer associated with increasing levels of global DNA methylation measured in peripheral blood by the LUminometric Methylation Assay (LUMA) (Xu et al. 2012). However, there was no evidence of a significant association between LUMA levels and dietary intakes of folate, vitamin B_6, vitamin B_{12}, choline, or betaine (Xu et al. 2012). An investigation of gene-specific DNA methylation patterns in breast tumor tissue revealed that *BRCA1* CpG promoter DNA methylation is significantly more frequent in invasive cancers compared to in situ cancer and in postmenopausal compared to premenopausal women (Xu et al. 2009b) and moreover, CpG promoter DNA methylation of this gene resulted in a significant increase in breast cancer-specific mortality. However, *BRCA1* methylation was not modulated by dietary intakes of folate, vitamin B_6, vitamin B_{12}, choline, or betaine (Xu et al. 2009b). Similarly, another study found that CpG promoter DNA methylation of *CDH1* (*E-cadherin*), *p16*, and *RAR*-β_2 genes in breast tumor tissue was not modulated by dietary intakes of folate and vitamins B_6 and B_{12} (Tao et al. 2011). In contrast, another study reported a significant decrease in CpG promoter DNA methylation of *E-cadherin* was associated with higher levels of dietary vitamin B_6 and vitamin B_{12} intakes, and a significant decrease in CpG promoter DNA methylation of *CCDN2* (*cyclin D2*) was associated with higher levels of dietary folate and vitamin B_6 intakes (Xu et al. 2011). Finally, higher levels of dietary betaine and vitamin B_6 intakes were significantly associated with an increase in CpG promoter DNA methylation of *APC* and *HIN1*, respectively (Xu et al. 2011). Based on these findings, three genes found to be most sensitive to dietary intakes of one-carbon nutrients were *cyclin D2*, *HIN1*, *and E-cadherin*, genes with tumor suppressing and oncogenic properties, and vitamin B_6 possessed the greatest potential among all the one-carbon nutrients investigated to modulate CpG promoter DNA methylation (Xu et al. 2011). Overall, these results provide preliminary evidence that one-carbon nutrients may modulate breast cancer risk by potentially altering the expression of genes implicated in breast carcinogenesis via changes in gene-specific CpG promoter DNA methylation. Future research is warranted to investigate how these diet-mediated changes in gene-specific DNA methylation can affect gene expression and subsequently, the potential to modulate breast cancer risks.

Breast Cancer Summary. In conclusion, there are inconsistencies in the evidence to support a relationship between dietary one-carbon nutrient intakes and blood levels on breast carcinogenesis. In some cases, stratification by menopausal status, hormonal receptor status, and genetic predisposition (e.g., *MTHFR* polymorphisms) resulted in significant findings with respect to the modulatory effect on breast cancer. Altogether, the complexity and heterogeneity of breast cancer can lead to differential outcomes and responses observed in the studies discussed herein (Holm et al. 2010).

Distinct aberrant DNA methylation profiles in a large number of tumor suppressor genes with consequent inactivation of these genes have been observed in familial breast cancer (Xiang et al. 2013). Distinct methylation profiles specific to molecular subtype of breast cancer have also been detected, once again demonstrating prognostic and screening merit (Holm et al. 2010). A comprehensive review by Heichman and Warren identified commonly hypermethylated genes with functions other than tumor suppression in breast cancer and of these, 54 candidate hypermethylated genes specific to breast cancer were identified (Heichman and Warren 2012). The identification of differentially methylated genes implicated in breast carcinogenesis, which are sensitive to dietary intakes of one-carbon nutrients, can increase the potential for targeted prevention through dietary intervention.

Cervical Cancer

Cervical cancer is ranked as the second most common malignancy across the globe with the highest incidence and mortality in developing countries due to the lack of cervical screening programs. Two biomarkers of cervical dysplasia are identified either through routine pap smears known as squamous intraepithelial neoplastic lesions (SILs) or through biopsy of the cervix known as cervical intraepithelial neoplasia (CIN). Both of these cervical dysplastic events are graded based on the severity of neoplastic transformation. SIL is classified as low- or high-grade and CIN is classified as CIN1 (mild dysplasia), CIN2 (moderate or marked dysplasia), or CIN3 (ranging from severe dysplasia to carcinoma in situ).

Infection with the human papillomavirus (HPV) is a well-established risk factor for cervical cancer. Although infection with HPV is common, only 5 % of women infected with HPV eventually develop cervical cancer, thereby suggesting the role of genetic and epigenetic mechanisms in modulating the HPV-mediated cervical carcinogenesis (Pathak et al. 2012).

A number of case–control studies have been conducted to elucidate the role of one-carbon nutrients in cervical carcinogenesis with conflicting results. Some case–control studies have demonstrated a lack of an association between circulating folate levels and risk of cervical dysplasia or cervical cancer (Alberg et al. 2000) and between dietary intakes of folate and the risk of invasive cervical cancer (Ziegler et al. 1991). A modest nonsignificant protective trend in cervical cancer risks associated with higher circulating levels of folate has been observed (Alberg et al. 2000). In contrast, some case–control studies have demonstrated a significant reduction in cervical cancer risk associated with high dietary and total folate (including supplements) intake (Ghosh et al. 2008; Hernandez et al. 2003). Circulating concentrations of vitamin B_{12}, HPV infection and genetic polymorphisms have been shown to modulate folate-mediated cervical carcinogenesis. For example, a significant reduction in the risk of CIN $\geq$2 in women with higher plasma folate and sufficient plasma vitamin B_{12} compared to women with lower plasma folate and sufficient vitamin B_{12} has been reported, suggesting that folate has the potential to decrease the risk of cervical dysplasia in the presence of adequate stores of vitamin B_{12} (Piyathilake et al. 2009). RBC folate was incrementally lower in women with high-risk HPV across increasing grades of cervical dysplasia and cancer (Flatley et al. 2009). Furthermore, a dramatically increased risk of CIN $\geq$2 is associated with lower RBC folate concentrations in women infected with HPV-16 (one of the high-risk oncogenic types of HPV) compared to those with higher RBC folate concentrations and HPV-negative status (Piyathilake et al. 2007). Finally, three studies investigating the modulating potential of the *MTHFR* variant genotypes on cervical cancer risk demonstrated that women with lower dietary folate intakes and serum folate concentrations with the variant *MTHFR* 677T allele had a significantly higher risk of cervical dysplasia (Goodman et al. 2001) and cancer (Tong et al. 2011) compared to women with the wild-type 677C allele and higher concentrations of serum folate. Additionally, there was a significant association between cervical dysplasia/cancer and the *MTHFR* A1298C variant genotypes in conjunction with lower circulating folate concentrations compared to the wild-type and higher plasma folate concentrations (Tong et al. 2011) and this effect was also seen in combination with adequate circulating concentrations of vitamin B_{12} (Ragasudha et al. 2012).

The literature concerning the effects of vitamin B_6 and cervical cancer risk is sparse and demonstrates a lack of association with total dietary intakes of vitamin B_6 (Hernandez et al. 2003) and a lack of an association between total vitamin B_6 dietary intakes combined with the *MTHFR* 677T allele and cervical dysplasia (Goodman et al. 2001).

Conflicting results have also been reported concerning the association between vitamin B_{12} and cervical cancer risk. One case–control study reported a lack of association between plasma vitamin B_{12} concentration and risk of cervical dysplasia (Goodman et al. 2000), whereas another study demonstrated a potentially protective, albeit nonsignificant, effect of dietary vitamin B_{12} intakes on cervical cancer (Alberg et al. 2000). However, evidence of a protective effect was also found in case–control studies which demonstrated a significantly decreased risk in cervical dysplasia in women who used vitamin B_{12}-containing supplements compared to those who did not (Hernandez et al. 2003) and similarly, a significant increase in the risk of cervical dysplasia was noted in women with lower serum concentrations of vitamin B_{12} compared to those with higher concentrations (Kwanbunjan et al. 2006). The modulatory effect of *MTHFR* polymorphisms was also explored with respect to dietary vitamin B_{12} intakes which showed either no association in women with the variant *MTHFR* C677T T allele (Goodman et al. 2001) or a significantly increased risk associated with low circulating vitamin B_{12} and the variant T allele (Tong et al. 2011). Finally, a joint interactive effect of low serum folate and vitamin B_{12} concentrations and the variant *MTHFR* 1298C allele has been associated with an increased risk of cervical cancer as mentioned above (Ragasudha et al. 2012).

Previous research has demonstrated that there is a significant positive correlation between the extent of global DNA hypomethylation and increasing grade of cervical dysplasia and cancer (Flatley et al. 2009; Shuangshoti et al. 2007; Missaoui et al. 2010; Kim et al. 1994). Furthermore, a significantly increased frequency of CpG promoter DNA methylation in four tumor suppressor genes, *E-cadherin* (Pathak et al. 2012; Flatley et al. 2009), *DAPK* (Flatley et al. 2009), *HIC1* (Pathak et al. 2012; Flatley et al. 2009), and *RARβ* (Pathak et al. 2012), all of which play an important role in the early stages of cervical carcinogenesis, was correlated with an increasing severity of cervical neoplasia. Although these studies also found that lower folate (Pathak et al. 2012; Flatley et al. 2009) and vitamin B_{12} status (Pathak et al. 2012) was associated with cervical dysplasia, the direct effect of these one-carbon nutrients on DNA methylation at the global or gene-specific level was not investigated. One case–control study, however, did investigate this relationship; women with supraphysiological plasma folate concentrations (>19.8 ng/mL) and adequate serum vitamin B_{12} ($\geq$200.6 pg/mL) exhibited the highest level of global DNA methylation compared to those with lower folate and vitamin B_{12} status (Piyathilake et al. 2011). Furthermore, this increased level of global DNA methylation translated to a 56 % lower chance of being diagnosed with CIN $\geq$2. The population-wide increase in the intake of folic acid after mandatory fortification took place in the United States has caused a concern due to the potential cancer-promoting effects associated with high intakes. Piyathilake et al. demonstrated the lack of association between mandatory folic acid fortification and global DNA methylation in cervical tissue. Although there was a significantly higher degree of methylation in CIN $\geq$2 compared to CIN $\leq$1, this increase did not differ between pre- and post-fortification (Persson et al. 2013). However, another study conducted shortly after by the same authors demonstrated that folic acid fortification was associated with a significant increase in DNMT expression in the cervix (Liu et al. 2012).

Cervical Cancer Summary. Although the current body of literature suggests an equivocal association between folate and cervical cancer risk, there is some evidence of the protective potential of this one-carbon nutrient. Key modulators of the relationship between folate and cervical cancer risk have been identified including high-risk HPV, vitamin B_{12}, and *MTHFR* gene polymorphisms. Although findings are also inconsistent for vitamin B_{12}, there is a suggestive role of protection conferred with supplemental vitamin B_{12} intakes and adequate vitamin B_{12} status. The evidence regarding other one-carbon nutrients (vitamin B_6, choline, and betaine) is sparse or lacking. Future research should take a more comprehensive approach to elucidating the potential of one-carbon nutrient-mediated risk of cervical cancer by assessing a complete nutrient profile of these metabolically related nutrient intakes and circulating levels.

On the mechanistic forefront, both global and gene-specific DNA methylation has been implicated in cervical cancer. Transcriptional silencing of tumor suppressor genes and activation of oncogenes, due to aberrant CpG promoter DNA methylation, appear to play a significant role in cervical carcinogenesis (Saavedra et al. 2012). Although supraphysiological folate combined with adequate vitamin B_{12} status has been shown to modulate global DNA methylation, the role of one-carbon nutrients in the modulation of gene-specific DNA methylation of tumor suppressor genes implicated in cervical cancer (*E-cadherin*, *DAPK*, *HIC1*, and *RARβ*) is yet to be investigated. As reviewed by Saavedra et al., CpG promoter DNA hypermethylation-induced silencing of various genes involved in different regulatory pathways including apoptosis, cell cycle control, and DNA repair has been implicated in cervical cancer (Saavedra et al. 2012). The identification of the DNA methylation profile and expression of these genes specific to cervical cancer may provide novel early detection biomarkers of cervical cancer risk and may allow for more effective screening in conjunction with other screening modalities (Saavedra et al. 2012). If gene-specific CpG promoter DNA methylation in cervical carcinogenesis is potentially linked to one-carbon nutrients, it would present an opportunity to reduce risk through dietary strategies using one-carbon nutrients and would provide further mechanistic insights into cervical cancer etiology.

Endometrial Cancer

The most common type of uterine cancer occurs in the lining of the uterus, known as the endometrium. Endometrial cancer is one of the most common malignancies of the female reproductive system in western countries (Key et al. 2004). Several risk factors have been identified with the most prominent being obesity; obese women have a threefold increased risk compared to lean women (Key et al. 2004). Other risk factors include increased exposure to estradiol in postmenopausal women and diabetes (Key et al. 2004; Uccella et al. 2011a). There are two histological subtypes of endometrial cancer, type I and type II, which exhibit different molecular and clinical characteristics. Type I endometrial cancer comprises majority of all sporadic endometrial cancers (80–90 %) which can be either adenocarcinoma with or without squamous differentiation, often is well defined and is more common in pre- and perimenopausal women (Uccella et al. 2011a; Felix et al. 2010). Type II endometrial cancer, which is much less common than type I, is often associated with *p53* mutations and is not well characterized. Overall, type II endometrial cancer is generally associated with a higher incidence of advanced-stage cancer and a worse prognosis compared to type I and is most common in older women (Uccella et al. 2011a; Felix et al. 2010).

A few case–control studies have investigated the effect of folate on endometrial cancer risk. The majority of these studies demonstrated a lack of association between dietary folate intakes and endometrial cancer risk (Biel et al. 2011; Jain et al. 2000; Negri et al. 1996), even after taking into account the *MTHFR* C677T and A1298C genetic polymorphisms (Liu et al. 2013). Only one case–control study demonstrated a moderately protective effect of dietary folate when comparing high intakes to low (Xu et al. 2007). Also, this study showed a significant interaction between dietary folate intake and the *MTHFR* G1793A polymorphism in modulating endometrial cancer risk; non-supplement users with the variant genotypes 1793AG or 1793AA who consumed lower levels of dietary folate had a higher risk of endometrial cancer compared to those who consumed higher levels (Xu et al. 2007). Most of the published prospective studies showed no association between dietary folate and endometrial cancer risk. Kabat et al. demonstrated a lack of association between dietary folate and endometrial cancer risk (Kabat et al. 2008). Given the biological and clinical distinction between the two types of endometrial cancer, some prospective studies interrogated the potential association according to the types of endometrial cancer. One study which was restricted to women with type I endometrial cancer demonstrated a null association (Liu et al. 2013). Another large prospective study demonstrated a similar lack of effect of both dietary and total folate intakes on both type I and type II endometrial cancer (Uccella et al. 2011b). Interestingly, however, this study reported a significantly increased risk of type II endometrial cancer associated with supplemental intakes of folic acid (Uccella et al. 2011b).

In an even smaller pool of case–control studies, one study demonstrated that higher dietary vitamin B_6 intake compared to lower was associated with an increased risk of endometrial cancer but this association was nonlinear (Biel et al. 2011). Another study also found a lack of an association even after taking into consideration the *MTHFR* C677T, A1298C, and G1793A polymorphisms (Xu et al. 2007). Contrary to these findings, however, another study found a joint interactive effect between vitamin B_6 intakes from diet and supplements and the *MTHFR* C677T polymorphism on endometrial cancer risk despite a lack of an association between dietary and supplemental intake of vitamin B_6 and endometrial cancer. The variant 677TT genotype combined with higher intakes of dietary and supplemental vitamin B_6 was associated with a protective effect compared to the heterozygous and wild-type genotypes (677CT/CC) and lower intakes (Liu et al. 2013). The joint effect of vitamin B_6 and the *MTHFR* 677TT genotype demonstrates the intricate gene–nutrient interactions between one-carbon nutrients and genes involved in the one-carbon transfer reaction pathway. Two prospective studies revealed that vitamin B_6 intakes from diet and supplements were not associated with type I endometrial cancer risk (Liu et al. 2013) and supplemental intakes, similar to folic acid, was associated with an increased risk of type II endometrial cancer only (Uccella et al. 2011b).

Case–control studies investigating the relationship between dietary and/or dietary plus supplemental vitamin B_{12} intakes and endometrial cancer risk revealed that there was no association (Biel et al. 2011) and furthermore, there was no evidence of a joint interaction with the *MTHFR* C677T and A1298C (Liu et al. 2013; Xu et al. 2007) or

G1793A (Xu et al. 2007) polymorphisms. Similarly, two prospective studies revealed that vitamin B_{12} intakes from diet and supplements were not associated with type I endometrial cancer risk (Liu et al. 2013) whereas the second study demonstrated that supplemental intake, similar to folic acid, has been found to be associated with an increased risk only in type II endometrial cancer (Uccella et al. 2011b).

There are currently no case–control studies that have investigated the relationships between choline and/or betaine and endometrial cancer risk. However, one prospective study found a lack of association between choline and endometrial cancer risk (Liu et al. 2013). It would be interesting to investigate whether these results differ by type I or type II status, which was the trend observed with folate, vitamin B_6, and vitamin B_{12}.

Altered DNA methylation and resulting aberrant expression of multiple genes have been more frequently associated with endometrial cancer than genetic alterations and DNA hypomethylation (Balch et al. 2010). A consistent gene-specific DNA methylation change observed in endometrial cancer is the breakdown of the DNA mismatch repair mechanism mediated through CpG promoter DNA methylation-induced changes in the expression of *hMLH1* (Banno et al. 2012; Tao and Freudenheim 2010), an event which occurs in the early stages of endometrial carcinogenesis (Tao and Freudenheim 2010). This gene, among several others, has been frequently silenced via DNA methylation in endometrial cancer. Hypermethylation of the tumor suppressor gene, *RASSF1A*, is one of the most frequently methylated genes in endometrial tumors in both early and late stages of carcinogenesis (Balch et al. 2010) among many others which have been found to be commonly hypermethylated in endometrial cancer (Balch et al. 2010). There are, however, no studies that interrogated the role of one-carbon nutrients in modulating these DNA methylation changes implicated in endometrial carcinogenesis.

Endometrial Cancer Summary. Overall, these studies demonstrate that supplemental, not dietary, intakes of folic acid, vitamin B_6, and vitamin B_{12}, may increase the risk of type II endometrial cancer. The lack of association in the bulk of the studies discussed may be attributed to the fact that stratification by the type of endometrial cancer was not performed, considering that a significant effect emerged when type I and type II cases were considered individually. The body of evidence investigating the relationship between one-carbon nutrients and endometrial cancer risk has been restricted to intakes whereas circulating concentrations of these one-carbon nutrients may reveal different results. One-carbon nutrients available in the circulation, which is not always reflected by dietary intakes, may be more directly correlated with disease risks, especially considering the observed interactive effect of vitamin B_6 with the *MTHFR* C677T polymorphism on endometrial cancer risk.

Finally, few studies have investigated the relationship between aberrant DNA methylation and endometrial carcinogenesis. Type II endometrial cancer is characterized by a high frequency of *p53* mutations and chromosomal instability (Uccella et al. 2011b). Therefore, investigation of global DNA hypomethylation, a contributor to DNA instability, and gene-specific DNA methylation of the *p53* gene as well as other genes which have been found to be hypermethylated in endometrial cancer may further elucidate the mechanistic underpinnings of endometrial cancer. Finally, interrogation of the potential influence of one-carbon nutrients on DNA methylation alterations would provide mechanistic insights into endometrial carcinogenesis as well as potential targets for endometrial cancer prevention and treatment using one-carbon nutrients.

Ovarian Cancer

Ovarian cancer is the fifth leading cause of death from female gynecological malignancy (Hanna and Adams 2006). There is a need to improve prognosis and preventative strategies that may be made possible through improved understanding of the pathogenesis of ovarian cancer. Two types have been identified which may help to determine effective screening and treatment: type I is characterized by slow growth, is confined to the ovaries and is genetically stable and type II is highly aggressive displaying metastatic potential to organs in close proximity such as the Fallopian tube (Westgren 2012). Furthermore, type II tumors are much more common and due to its aggressive nature, it is responsible for 90 % of ovarian cancer-related deaths (Westgren 2012).

A number of studies have been conducted to investigate the relationship between one-carbon nutrients and ovarian cancer risk. The majority of case–control studies have found no association between dietary and/or total folate intakes and ovarian cancer risk (Harris et al. 2012; Pelucchi et al. 2005b; Salazar-Martinez et al. 2002) with the exception of one study which revealed that there was a reduced risk associated with increased intakes of dietary folate in current smokers (Webb et al. 2011). Interestingly, an increased ovarian cancer risk was observed with two MTR polymorphisms (rs736502 and rs7526063). However, there was a very low frequency of these polymorphisms in the study population and neither of these polymorphisms nor any of the other polymorphisms investigated, had any interaction with dietary intakes of folate, vitamin B_6, vitamin B_{12}, choline, or betaine (Webb et al. 2011). No significant association was observed between the *MTHFR* C677T and A1298C polymorphisms and dietary intakes of folate in modifying ovarian cancer risk in another study (Harris et al. 2012). Prospective studies investigating the relationship between dietary folate and ovarian cancer risk have generally found a lack of association (Kabat et al. 2008; Kelemen et al. 2004). However, subtype analysis revealed that there was a significant reduction of the risk of serous ovarian cancer associated with high dietary folate intake (Tworoger et al. 2006). Some studies have reported nonsignificant reductions in the risk of ovarian cancer associated with dietary intakes of folate (Larsson et al. 2004; Silvera et al. 2006) and a combination of alcohol and folate intakes when comparing higher folate and alcohol intakes to lower intakes of both (Kelemen et al. 2004; Silvera et al. 2006).

Conflicting results were found with vitamin B_6 intakes as one case–control study found that there was no association between total vitamin B_6 intakes and ovarian cancer risk even after stratification by histologic subtype (Webb et al. 2011) whereas another case–control study reported a significant risk reduction associated with dietary vitamin B_6 intakes, with the most prominent effect in women with the serous borderline subtype (Harris et al. 2012). Also, in this study, dietary vitamin B_6 intake was not significantly associated with the *MTHFR* C677T and A1298C polymorphisms (Harris et al. 2012).

Similar results were found with vitamin B_{12} intakes; generally, there was a lack of association between dietary and/or total vitamin B_{12} and ovarian cancer risk in case–control studies (Harris et al. 2012; Salazar-Martinez et al. 2002). Similarly, a lack of association was observed between the *MTHFR* C677T and A1298C polymorphisms, dietary intakes of vitamin B_{12} and ovarian cancer risk (Harris et al. 2012).

Only one case–control study investigated the association between choline and betaine and ovarian cancer risk and reported null findings for both nutrients (Webb et al. 2011). A prospective study investigating the relationship among choline, betaine, and ovarian cancer risk similarly demonstrated a lack of association (Kotsopoulos et al. 2010).

An increase in satellite DNA hypomethylation has been commonly detected in the advanced stages of ovarian cancer and a number of tumor suppressor gene silencing has been shown to be mediated via CpG promoter DNA hypermethylation (Welzel et al. 2007). For example, the promoter of *BRCA1* is commonly hypermethylated and silenced in sporadic cases of ovarian cancer (Welzel et al. 2007). Although *BRCA1* is one of the most studied genes in ovarian carcinogenesis, several hypermethylated and hypomethylated genes have been identified which can lead to gene silencing and overexpression, respectively (Seeber and van Diest 2012). Moreover, distinct DNA methylation profiles specific to histological subtypes of ovarian cancer have been identified (Seeber and van Diest 2012). Furthermore, Heichman and Warren reviewed commonly hypermethylated genes in ovarian cancer and found that 19 genes were ovarian cancer-specific (Heichman and Warren 2012). At present, however, the role of one-carbon nutrients in modulating these established DNA methylation alterations implicated in ovarian cancer has not been established.

Ovarian Cancer Summary. Overall, there is no strong evidence of an association between dietary intakes of one-carbon nutrients and ovarian cancer risk. However, there appears to be a protective effect of higher dietary intakes of folate on the histologic subtype of serous ovarian cancer and this effect may be further modified by alcohol intake. Of the other one-carbon nutrients investigated, there is a potential protective role of vitamin B_6 intakes in ovarian carcinogenesis. It is important to note that all of the studies discussed within this section only assessed the effects of dietary intakes, which is more prone to methodological errors such as misclassification of exposure. If circulating concentrations had been measured, in addition to genetic polymorphisms in one-carbon metabolizing genes, perhaps a significant effect would have emerged. Given the paucity of data, the potential effect of one-carbon nutrients on CpG promoter DNA methylation of several tumor suppressor and proto-oncogenes that exhibit aberrant DNA methylation patterns in ovarian cancer as well as global DNA methylation is worthy of further investigation. Provided a link exists, this may improve prevention, treatment, and prognosis of ovarian cancer through a DNA methylation-targeted approach using one-carbon nutrients.

Prostate Cancer

Prostate cancer is one of the most common cancers in the western world, with an increasing prevalence in developing countries, and one of the leading causes of cancer-related death in the United States (Haas et al. 2008). Prostate cancer progression is very slow and therefore, neoplastic transformation occurs many years before diagnosis (Collin et al. 2010). Established risk factors include age, ethnicity, and family history; a few dietary risk factors have also been identified including increased consumption of meat and animal fat (associated with an increased risk) and a diet rich in certain vegetables and other nutrients (associated with a decreased risk) (Stevens et al. 2006). Although traditionally thought of as an increased risk factor, endogenous sex hormone levels are not considered a risk factor for prostate cancer (Roddam et al. 2008).

An animal study investigating folate's role in prostate cancer risk found mild folate depletion to be protective against tumor progression and aberrant DNA methylation patterns (Bistulfi et al. 2011). Using the transgenic adenoma of mouse prostate (TRAMP) model, this study demonstrated that mild folate depletion was able to strongly inhibit prostate cancer progression in 96 % of TRAMP mice (25 of 26 animals) (Bistulfi et al. 2011). This was measured through various parameters including reduced size and grade of tumor and marked reduction in proliferative and apoptotic indices. Furthermore, aberrant DNA methylation patterns in the prostate tumors were less frequent in the folate-deficient TRAMP mice compared with the folate-sufficient controls (Bistulfi et al. 2011). However, supplying high folate levels did not significantly accelerate tumor growth or disease progression. With respect to aberrant DNA methylation patterns, a folate-depleted diet prevented methylation of four specific CpG sites and conversely, supplementation increased methylation at these four sites (Bistulfi et al. 2011). In an athymic nu/nu mouse model using established DU-145 and PC-3 prostate cancer xenografts, the importance of the timing of folate exposure and its relationship with prostate cancer risk was demonstrated. The growth of established DU-145 and PC-3 prostate cancer xenografts was not affected by a diet fortified with one-carbon nutrients (methionine, choline chloride, betaine, folic acid, vitamin B_{12}), nor did these high levels of one-carbon nutrients demonstrate the ability to influence the demethylation capacity of 5-aza-2′-deoxycytidine. However, the administration of this fortified diet in utero was able to exert a protective effect in the offspring against the development of high-grade prostate tumors (Shabbeer et al. 2012). Collectively, these animal studies illustrate that the timing of exposure to folate/one-carbon nutrients plays an important role in prostate carcinogenesis.

In general, the majority of case–control studies have demonstrated a lack of effect of dietary and/or supplemental intakes (Shannon et al. 2009; Verhage et al. 2012; Vlajinac et al. 1997; Weinstein et al. 2006) and circulating levels of folate (Collin et al. 2010; Beilby et al. 2010; Johansson et al. 2008) on prostate cancer risk. Likewise, a null association was observed between serum folate and prostate cancer risk when the study population included only male smokers, which may partially be explained by the high prevalence of inadequate serum folate status in 92 % of this population (Weinstein et al. 2003). Although the majority of these studies were adjusted for smoking habits and alcohol consumption, two risk factors postulated to affect prostate cancer (Collin et al. 2010; Beilby et al. 2010; Johansson et al. 2008; Shannon et al. 2009; Verhage et al. 2012), no effect of modification was found by either of these factors. Similarly, a number of studies stratified results by stage of cancer development (i.e., localized versus advanced stage) (Collin et al. 2010; Verhage et al. 2012; Weinstein et al. 2006; Johansson et al. 2008) and found no differences in prostate cancer risk with the exception of one study demonstrating that higher dietary folate intakes (from natural and fortified sources) significantly reduced the risk of high-grade prostate cancer when compared to controls and high-risk individuals (Shannon et al. 2009). Only a small number of case–control studies demonstrated the potential of folate to modulate prostate cancer risk. Of these, one showed a nonsignificant trend of an increasing risk associated with higher plasma folate concentrations (Hultdin et al. 2005) and another showed evidence of a significant increased risk as demonstrated by a multiplicative interaction of the variant *MTHFR* 677CT/TT genotype, higher folate intakes, and greater than five alcoholic drinks per week (Kobayashi et al. 2012). A modest effect of risk reduction associated with higher intakes of dietary folate has been observed and this effect was more prominent when coupled with higher intakes of vitamin B_6 (Pelucchi et al. 2005a). Furthermore, the modulating effects of alcohol have also been observed such that higher intakes of folate and lower intakes of alcohol was more protective (Pelucchi et al. 2005a) and conversely, lower intakes of folate and higher alcohol intakes were associated with a significantly increased risk of prostate cancer (Kobayashi et al. 2012). The effect of dietary folate on prostate cancer risk modulation ascertained through prospective studies is tenuous at best. In one study, the association between dietary folate intake and increased risk irrespective of disease stage did not exhibit a significant linear relationship (Bassett et al. 2012). Similarly, another prospective study demonstrated a weak reduction, albeit not statistically significant, in prostate cancer risk in advanced stage prostate cancer cases with higher dietary and total folate intakes (Stevens et al. 2006). Finally, a null association has been reported between dietary folate and prostate cancer risk, even after stratifying for disease stage in another study (Kasperzyk et al. 2009).

The Aspirin/Folate Polyp Prevention Study, designed to investigate the effects of folic acid and/or aspirin supplementation on colorectal neoplasms, found that aspirin had a protective effect whereas folic acid supplementation at 1 mg/day appeared to increase CRC risk (Cole et al. 2007). An investigation of secondary outcomes revealed that folic acid supplementation was associated with an increased risk of prostate cancer compared to placebo. However, high dietary folate intake at baseline was protective against prostate cancer (Figueiredo et al. 2009a). This study also revealed that dietary and plasma vitamin B_6 and dietary vitamin B_{12} were not associated with prostate cancer risk (Figueiredo et al. 2009a).

Vitamin B_6 demonstrated a lack of effect on prostate cancer risk with respect to dietary and/or supplemental intakes (Vlajinac et al. 1997; Pelucchi et al. 2005a; Key et al. 1997). Furthermore, no significant associations between vitamin B_6 intakes and prostate cancer emerged even after taking into consideration the *MTHFR* C677T polymorphism (Kobayashi et al. 2012). Stratification by histological tumor grade (Kobayashi et al. 2012) and adjustment for smoking habits (Pelucchi et al. 2005a; Key et al. 1997) did not result in any significant findings. These null findings, after adjustment for stage of disease progression, were mirrored in a study which investigated the relationship between circulating vitamin B_6 concentrations and prostate cancer risk (Johansson et al. 2009). A similar null effect was noted in male smokers of which 55 % of the cases had an inadequate vitamin B_6 status (Weinstein et al. 2003). On the contrary, one case–control study found a significant, albeit weak, reduction in prostate cancer risk in male smokers (Weinstein et al. 2006). Inconsistent results are also evident regarding dietary vitamin B_6 intakes and prostate cancer risk in prospective studies; a null association was reported after stratification for disease stage in one study (Bassett et al. 2012) whereas another prospective study found a significant protective effect of dietary vitamin B_6 intakes on localized prostate cancer (Kasperzyk et al. 2009).

A very different pattern of association was observed in case–control studies investigating the relationship between vitamin B_{12} and prostate cancer risk. Whereas only two studies reported a lack of association between prostate cancer and serum vitamin B_{12} concentrations in smokers (Weinstein et al. 2003) and dietary intakes after adjusting for the *MTHFR* C677T polymorphism (Kobayashi et al. 2012), the majority of the studies revealed a surprising increase in the risk of prostate cancer (Collin et al. 2010; Vlajinac et al. 1997; Weinstein et al. 2006; Johansson et al. 2008; Hultdin et al. 2005). Higher dietary intakes of vitamin B_{12} have been associated with an increased risk of prostate cancer (Vlajinac et al. 1997). This was also observed in smokers (Weinstein et al. 2006). Moreover, studies investigating the circulating levels of vitamin B_{12} also corroborate these findings of an increased prostate cancer risk associated with higher circulating concentrations of vitamin B_{12} (Collin et al. 2010; Johansson et al. 2008; Hultdin et al. 2005). Only two prospective studies investigated the relationship between dietary vitamin B_{12} and prostate cancer risks, both of which showed a null association (Bassett et al. 2012; Kasperzyk et al. 2009).

The relationship between choline and betaine and prostate cancer risk has only been investigated in one case–control study to date, revealing surprising results of a significant increasing trend in prostate cancer risk associated with plasma choline concentrations and a null association with plasma betaine concentrations (Johansson et al. 2009). Additionally, one prospective study investigated the effects of dietary choline on prostate cancer risk and found a significant 70 % increased risk of lethal prostate cancer with individuals consuming high choline intake compared to those consuming less choline (Richman et al. 2012). Collectively, these studies provide evidence to support the role of both dietary and circulating concentrations of choline in prostate carcinogenesis.

DNA hypo- and hypermethylation is associated with prostate carcinogenesis; global DNA hypomethylation has been implicated in metastatic prostate cancer and disease progression (Ho et al. 2011). On a gene-specific level, a few hypomethylated promoter regions have been identified in prostate cancer compared to over 50 hypermethylated and inactivated genes, all of which commonly occur in the early stages of prostate carcinogenesis (Ho et al. 2011). These genes include tumor suppressor genes in addition to genes involved in many different cellular pathways such as hormonal responses, cell invasion and architecture, cell cycle control, and

DNA damage repair (Heichman and Warren 2012). A comprehensive review by Heichman and Warren identified 14 cancer-specific hypermethylated genes in prostate cancer (Heichman and Warren 2012).

Aberrant DNA hypermethylation is the best characterized epigenetic alteration in prostate cancer, which contributes to genomic instability and aberrant gene expression (Shabbeer et al. 2012; Jeronimo et al. 2011). Global hypomethylation, also detected as a contributing epigenetic mechanism to prostate cancer, has been linked to metastatic prostate cancer (Jeronimo et al. 2011). A recent study aimed at characterizing the differences in the methylome of indolent versus aggressive prostate cancer led to the discovery of 13 CpG islands which are progressively hypermethylated with increasing disease severity (Lin et al. 2013). The identification of these specific CpG islands can lead to improved diagnosis and hence prognosis by enabling the early identification of more aggressive forms of prostate cancer (Lin et al. 2013). Of great value on a prognostic forefront, these DNA methylation alterations occur in the early stages of prostate cancer enabling early detection and prevention. Several factors have been shown to influence promoter DNA hypermethylation in the prostate, such as age, diet, and environmental factors, with age having the most significant impact. Therefore, this provides the rationale for investigating the relationship between one-carbon nutrients and aberrant DNA methylation in prostate carcinogenesis. The discovery of commonly hypermethylated CpG islands and specific genes in prostate carcinogenesis provides a basis for the further investigation of the potential of one-carbon nutrients to modulate these epigenetic alterations. This would provide a much more insightful understanding of the modifiable risk factors of prostate cancer.

Prostate Cancer Summary. Collectively, these studies demonstrate a weak or null association between folate, vitamin B_6, vitamin $B_{12,}$ and prostate carcinogenesis. Although a very limited amount of data is available to support the relationship between choline and prostate cancer risk, strong associations were found which were more conclusive than any of the other one-carbon nutrients discussed. This may be partly due to the fact that choline has been found to be highly concentrated in prostate cancer cells (Ackerstaff et al. 2001). In order to further elucidate the role of one-carbon nutrients on prostate cancer risk, the joint interactive effect between one-carbon metabolizing genes and intakes and/or status of these nutrients needs to be investigated. Furthermore, the additive effects of these metabolically related nutrients should be investigated, considering that a significant reduction in prostate cancer risk was observed comparing highest folate and vitamin B_6 intakes to the lowest intakes of both nutrients (Pelucchi et al. 2005a). A recent meta-analysis has revealed that there is a significant reduction in prostate cancer risk conferred by the *MTHFR* 677T allele (Li and Xu 2012) and therefore, the joint interactive effect of nutrient intakes/status and the *MTHFR* C677T polymorphism should be investigated to elucidate if the relationship between the one-carbon nutrients and prostate cancer risk has been confounded by this risk-modifying factor. Similar to the dual modulatory effects of folate observed in CRC, the dose, timing of administration, and stage of cancer may also play an important role in the preventive and cancer-promoting effects of folate and

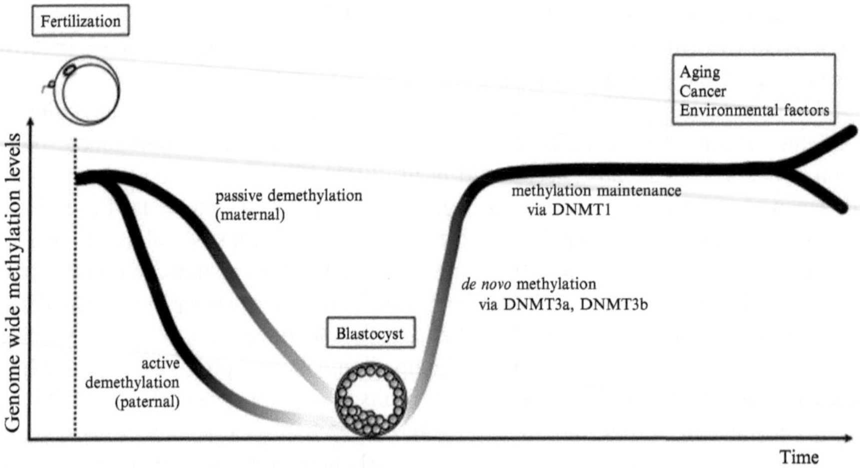

Fig. 11.4 DNA methylation throughout the life cycle. DNA methylation patterns are reprogrammed during embryogenesis by genome-wide demethylation after fertilization. Active and passive demethylation erases significant parts of the parental DNA methylation patterns, followed by de novo methylation, which establishes a new DNA methylation pattern soon after implantation when the blastocyst is formed. Maintenance methyltransferase (DNMT1) uses hemimethylated sites to ensure DNA methylation patterns, whereas de novo methyltransferases (DNMT3a, 3b) do not require preexisting methylation patterns. Later in life, factors such as aging, cancer, and environmental exposures can cause epigenetic divergence with increased or decreased methylation levels

some of other one-carbon nutrients in prostate cancer (Ho et al. 2011). At present, there is no evidence that one-carbon nutrients may influence prostate carcinogenesis via their modulatory effects on DNA methylation implicated in prostate carcinogenesis. Future studies are warranted to clarify the role of one-carbon nutrients in DNA methylation changes and prostate cancer risk modification.

11.4.3 Trans-generational Studies

Recently, the potential role of maternal and early life nutrition on cancer risk in the offspring has been investigated. Definitive evidence linking perinatal folic acid supplementation and the prevention of neural tube defects confirmed maternal diet has a significant effect on the health of the fetus (MRC Vitamin Study Research Group 1991). Epigenetic modifications in utero and in early postnatal period have been postulated as one of the underlying mechanisms by which maternal and early postnatal nutrition may modulate cancer risk in the offspring. In this regard, DNA methylation patterns are reprogrammed during embryogenesis by active and passive genome-wide demethylation, which erases large parts of the parental DNA methylation pattern (Fig. 11.4). After implantation of the blastocyst into the uterus, de novo methylation occurs, which establishes a new DNA methylation pattern for the offspring, with methylation mostly limited to non-CpG island areas, which is maintained postnatally (Fig. 11.4) (Nafee et al. 2008). Therefore, DNA methylation during the embryogenic stage, when DNA methylation patterns are reprogrammed, may be highly susceptible to the modulating effects of nutritional, hormonal, and metabolic factors. In particular, the one-carbon nutrient pool available during this critical time period of development can have a significant effect on the DNA methylation patterns in the offspring, and subsequently on chronic disease risk, including cancer, later in life.

Evidence for the potential cancer modulating effects of maternal environment in the offspring is largely demonstrated in animal studies, since cancer is an age-related disease and requires a lengthy follow-up period for diagnosis. Two animal studies have demonstrated the protective effects of maternal folic acid supplementation alone (Sie et al. 2011) or in combination with vitamins B_2, B_6, and B_{12} (Ciappio et al. 2011a) on colorectal and small intestinal tumorigenesis, respectively, in the offspring. In contrast, another animal study demonstrated that a maternal diet deficient in folic acid had a protective effect on small intestinal cancer in the offspring (Sie et al. 2011), whereas another study found no effect of maternal folate depletion on small intestinal tumorigenesis (McKay et al. 2011c). Two animal studies have demonstrated conflicting effects of maternal folic acid supplementation on mammary tumorigenesis; maternal folic acid supplementation significantly reduced a well-established intermediary biomarker of mammary tumors (i.e., terminal end buds) in the offspring at puberty in one study (Sie et al. 2009) but it significantly increased the incidence of a carcinogen-induced mammary tumors and accelerated the tumor development in the offspring in another study (Ly et al. 2011). Kovacheva et al. demonstrated the relationship between maternal choline and mammary tumors in the offspring in a carcinogen rat model; choline supplementation during early embryogenesis resulted in slower tumor growth in the pups compared to pups born to dams fed a choline-deficient diet (Kovacheva et al. 2009). It is important to pay heed to the fact that the experimental design and rodent models used in these studies may have contributed to disparate results observed in these animal models. For example, the proclivity and severity of the tumor phenotype of the rodent model may affect the protective ability of folate in carcinogenesis; folate has been shown to confer a protective effect in models with a relatively weak tumorigenic phenotype and this effect may be further bolstered in combination with other one-carbon nutrients (vitamins B_2, B_6, and B_{12}) whereas in models with an aggressive tumorigenic capacity, the protective potential of one-carbon nutrients is inundated and supplementation with these nutrients may in fact promote the progression of established (pre)neoplastic foci (Ciappio et al. 2011b).

Epidemiological studies have generally shown an inverse relationship between periconceptional supplementation with a folic acid-containing multivitamin and a number of pediatric cancers including acute lymphoblastic leukemia (ALL) (Thompson et al. 2001; Wen et al. 2002; Ross et al. 2005), brain tumors (Bunin et al. 1993, 2006; Michalek et al. 1996; Olshan et al. 2002) and retinoblastoma (Bunin et al. 1989). A recent case–control study demonstrated with high dietary, not supplemental, folate, and vitamin B_{12} intakes during pregnancy was associated with lower ALL risk in offspring (Bailey et al. 2012). An unexpected outcome was an increased, albeit weak, ALL risk with high maternal vitamin B_6 intakes (Bailey et al. 2012). Another case–control study by Schuz et al. reported an increase in neuroblastoma incidence in children associated with maternal periconceptional folic acid supplementation (Schuz et al. 2007). Finally, some studies also reported a lack of a protective effect on ALL associated with maternal folic acid supplementation (Dockerty et al. 2007; Milne et al. 2010).

Proof-of-principle studies have shown that maternal methyl donor supplementation can significantly change DNA methylation and associated phenotypes in the offspring (Waterland et al. 2006a; Wolff et al. 1998; Yen et al. 1994; Sinclair et al. 2007). For example, the pioneering research using the agouti viable yellow (A^{vy}/a) mouse showed that maternal supplementation of high levels of one-carbon nutrients, including folic acid, B_{12}, choline, and methionine, during pregnancy led to CpG DNA hypermethylation at the regulatory region of the A^{vy} gene which resulted in a change in offspring's coat color from yellow to brown (Wolff et al. 1998; Waterland and Jirtle 2003). Furthermore, the brown coat mice were found to have lower obesity, hypertension, and insulin resistance (Yen et al. 1994), and when challenged with a chemical carcinogen, tumor multiplicity in the liver and tumor incidence in the lungs were reduced (Wolff et al. 1987) compared to their yellow coat color siblings. In addition, supplementation with one-carbon nutrients during pregnancy led to a "straight tail" phenotype compared to the "kinky tailed" phenotype in the offspring in the *Axin Fused* (*Axin*^{FU}) mouse model. The "straight tail" phenotype was correlated with increased CpG promoter DNA methylation at the *Axin* gene (Waterland et al. 2006a). These observations demonstrate that the maternal supplementation of one-carbon nutrients can influence the phenotype and disease susceptibility in the offspring via DNA methylation changes.

Several animal studies have investigated the effect of maternal and early life folate manipulation on DNA methylation in the offspring with conflicting results. In a mouse model, daily administration of folinic acid by gavage during the periconceptional period until day 15.5 of gestation significantly decreased global DNA methylation in both the liver and the brain of the embryo (Finnell et al. 2002). Generally, folate depletion imposed during the in utero and lactation stages, but not during the postweaning stage, appears to decrease global DNA methylation in the offspring in an organ-specific manner (McKay et al. 2011a, 2011c; Maloney et al. 2007; Lawrance et al. 2009). These studies used different protocols of folate manipulation in several different models and hence, definitive conclusions cannot be drawn (McKay et al. 2011a, 2011c; Wolff et al. 1998; Finnell et al. 2002; Maloney et al. 2007; Lillycrop et al. 2005).

Sie et al. have recently reported that maternal folic acid supplementation can significantly decrease hepatic global DNA methylation in the offspring at weaning whereas at 14 weeks of age, a significant decrease in global DNA methylation was associated with postweaning, but not maternal, folic acid supplementation (Sie et al. 2013). In contrast, a previous animal study by the same authors reported that maternal folic acid supplementation of the same magnitude significantly increased global DNA methylation of the colon in the weanling rat offspring (Sie et al. 2011). This prior study found that postweaning, but not maternal, folic acid supplementation significantly decreased global DNA methylation of the colon in the rat at 14 weeks of age (Sie et al. 2011). However, the magnitude of change in global DNA methylation of the colon in the prior study (3–6.4 %) was quite modest, albeit significant, compared with that of the liver (10–25 %). In another study of a similar design, maternal, but not postweaning, folic acid supplementation of the same magnitude significantly decreased, by 7 %, global DNA methylation of the mammary glands in the offspring at 28 weeks of age (Ly et al. 2011). These observations collectively suggest that the effect of maternal and postweaning folic acid supplementation on global DNA methylation may be organ-specific.

Gene-specific DNA methylation has also been investigated in animal studies. Lillycrop et al. demonstrated that maternal folic acid supplementation (five times the control) prevented the promoter DNA hypomethylation of the *Ppar-* and *Gr* genes and the decreased Dnmt1 expression induced by protein restriction in maternal diet in the liver of adult rat offspring (Lillycrop et al. 2005, 2007). A recent study using pregnant CD1 mice found maternal folic acid supplementation followed by gestational arsenic exposure altered DNA methylation in CpG islands located in nearly 3,000 genes, including differentially methylated regions (DMRs) of imprinted genes involved in embryogenesis and cancer (Klaus and Birchmeier 2008). After identification of various gene networks, the most affected pathway was the Wnt-signaling pathway, which plays a role in fetal development and regeneration of tissues in adults (Klaus and Birchmeier 2008). Over 50 genes associated with this pathway were found to have altered DNA methylation patterns (e.g., *Wnt3*, *Fzd8*, *Fzd10*, *Dvl1*, *Dvl3*, *Axin2*, *Ctnnb1*, *Cdkn2a*, *Rara*, *Tgf-β*) (Tsang et al. 2012). As previously mentioned, alterations in the Wnt-signaling pathway have been implicated in the development of CRC; therefore these DNA methylation changes could have implications for the development of CRC in the offspring. A maternal diet deplete of folic acid fed to $Apc^{+/Min}$ and wild-type mice had no effect on gene-specific DNA methylation in weaning mice from both groups. However, decreased gene-specific DNA methylation at *p53* in the small intestine of adult mice was seen with the depleted maternal folic acid diet (McKay et al. 2011b). Furthermore, a low folic acid diet in the postweaning period increased DNA methylation at the *Apc* gene in the $Apc^{+/Min}$ mice but not in the wild-type mice (McKay et al. 2011b). Sie et al. have recently reported significant decreases in DNA methylation of the promoter regions of the *Pparγ* and *ERα* genes and in specific exons of the *p53* and *Apc* genes in the liver of offspring at weaning due to maternal folic acid supplementation (Sie et al. 2013). However, at 14 weeks of age, increased DNA methylation of the *Pparγ*, *p53*, and *p16* genes was observed due to maternal and postweaning folic acid supplementation (Sie et al. 2013). In contrast, postweaning, but not maternal, folic acid supplementation significantly increased DNA methylation of the *ERα* and *Apc* genes in the offspring at 14 weeks of age (Sie et al. 2013). *Apc*, *p16*, and *p53* are tumor suppressor genes and mutations and/or aberrant DNA methylation of these genes have been implicated in CRC (Fearon 2011). Therefore, aberrant DNA methylation of the *Apc*, *p16*, and *p53* could contribute to colorectal carcinogenesis via gene silencing as well as gene instability.

Maternal dietary choline intake has been directly associated with CpG promoter DNA methylation and decreased mRNA expression of the *Stratifin* (*Sfn*) gene in mammary tumor tissues of rat offspring (Kovacheva et al. 2009). *Sfn*, a tumor suppressor gene with cell cycle regulator function, has previously been shown to be hypermethylated in the CpG promoter region, and silenced, in breast cancer (Kovacheva et al. 2007).

Although not specifically tied to cancer risk, several human studies have investigated changes in LINE-1 DNA methylation patterns, a surrogate for global DNA methylation, with maternal folic acid. A study from the UK found LINE-1 methylation in 24 offspring was inversely associated with plasma homocysteine in umbilical cord blood; however, cord blood LINE-1 methylation was not significantly associated with maternal or cord serum folate concentrations (Fryer et al. 2009). As highlighted in Fig. 11.1, homocysteine is considered a nonspecific inverse indicator of folate status since 5-methylTHF is required for remethylation of homocysteine to methionine. A recent prospective study of 913 pregnancies did find LINE-1 methylation was lower in children born to mothers who reported folic acid supplementation after 12 weeks gestation but not with mothers who stopped folic acid supplementation at 12 weeks (Haggarty et al. 2013). Finally, Boeke et al. found a sex-specific effect where LINE-1 methylation was inversely associated with maternal choline intake in early pregnancy in male offspring only (Boeke et al. 2012). Although these studies do highlight how one-carbon nutrients can modulate LINE-1 methylation patterns in the offspring, functional ramifications of LINE-1 methylation changes and how they correlate with global DNA methylation status in specific target organs remain elusive. Furthermore, how changes in LINE-1 methylation can affect the offspring's chronic disease risk including cancer is still unclear. McKay et al. have recently reported maternal vitamin B_{12} status to be inversely correlated with global DNA methylation in the offspring determined by another surrogate measure of global DNA methylation, LUMA (McKay et al. 2012).

Human studies investigating maternal folic acid supplementation on DNA methylation patterns at specific genes in the offspring have either taken a candidate gene approach or an epigenomic approach using CpG island DNA methylation microarray. Fryer et al. found CpG DNA methylation patterns at over 27,000 loci in cord blood, using a CpG island DNA methylation microarray, that were significantly correlated with cord plasma homocysteine concentrations and infant birth weight percentile in 12 mother–child pairs; however, no association was found with maternal or cord serum folate concentrations (Fryer et al. 2011). Several studies that have investigated DNA methylation patterns of specific candidate genes have focused on the *IGF2* gene, likely because IGF2 acts as a growth-promoting hormone during fetal development. *IGF2* is an imprinted gene requiring DNA methylation to suppress either paternal or maternal allelic expression. Aberrant biallelic *IGF2* expression has been linked to the development of certain cancers including CRC (Cheng et al. 2010; Honda et al. 2008). Four studies investigating the effect of maternal folic acid supplementation on *IGF2*-specific DNA methylation in the infant found conflicting results; *IGF2* DNA methylation was increased with maternal folic acid supplementation (Steegers-Theunissen et al. 2009; Haggarty et al. 2013) and was positively associated with RBC folate concentrations in cord blood (Haggarty et al. 2013). The other two studies, however, demonstrated that DNA methylation of *IGF2* was decreased (Hoyo et al. 2011) or had no association (Ba et al. 2011) with maternal folic acid supplementation. Interestingly, Ba et al. found an inverse association between *IGF2* DNA methylation and maternal B_{12} levels (Ba et al. 2011). Furthermore, maternal *MTHFR* 677T variant allele was associated with an increase in *IGF2* methylation in the infant (McKay et al. 2012). There is also some evidence that one-carbon nutrients can affect DNA methylation of another related gene in the IGF pathway. For example, McKay et al. found that infant vitamin B_{12} concentrations were inversely correlated with Insulin-like Growth Factor-binding Protein 3 (*IGFBP3*) DNA methylation (McKay et al. 2012). Both *IGF2* and *IGFBP3* are members of the IGF system and therefore play a role in intrauterine growth (McKay et al. 2012).

Another candidate gene of interest is the imprinted gene paternally expressed 3 (*PEG3*). Haggarty et al. found that women reporting folic acid supplementation at 12 weeks of gestation had offspring with lower *PEG3* DNA methylation compared to those stopping folic acid supplementation at 12 weeks gestation (Haggarty et al. 2013). Similar to *p53* and *APC*, *PEG3* is a tumor suppressor gene, which inhibits the Wnt-signaling pathway and regulates tumor growth (Otsuka et al. 2009). Therefore, the lower *PEG3* DNA methylation observed with folic acid supplementation at 12 weeks gestation could cause improper gene imprinting required for normal *PEG3* function and tumor growth regulation.

A gap in the knowledge still exists with the lack of information on how changes in the maternal diet can influence epigenetics and/or disease risk in the mothers themselves. Studies to date have generally focused on the health outcomes of the offspring but we need to still consider these effects on the mother. For example, in mice fed folate-deficient diets from preconception through lactation, average postpartum DNA methylation across three tissues at *Igf2*, DMR1, and *Slc39a4* was lower compared to mice fed folate-adequate diets (McKay et al. 2011a). In humans, Krishnaveni et al. found at 5 years postpartum, gestational vitamin B_{12} deficiency was positively associated with greater insulin resistance and skinfold thickness measurements in mothers at follow-up (Krishnaveni et al. 2009). Investigating the effects of the perinatal diet on the mother herself is especially important considering diseases such as diabetes and cancer are age-related diseases and exposure to various levels of one-carbon nutrients, even for a select period of time (preconception, pregnancy, and lactation), may have a long-lasting and potentially, adverse effect later in life. Furthermore, the effect of paternal dietary status on epigenetic changes in the offspring is an emerging important area of research as some studies have suggested that paternal diet may modulate health outcomes of the offspring, independent of the maternal diet. Kim et al. reported fetuses from male mice fed folate-deficient diets prior to and during mating were associated with a lower percentage of fetal whole brain DNA methylation and Igf2 protein expression compared with fetuses from folic acid-supplemented males (Kim et al. 2013). Therefore, in future studies, it may be beneficial to investigate both parties involved in fetal development, as well as the offspring.

Whether or not DNA methylation changes are transmitted to future generations (i.e., trans-generational epigenetic inheritance) is a highly controversial topic (Cropley et al. 2006; Waterland et al. 2007). Overall interpretation of the available evidence does not provide unequivocal support for the transmission of epigenetic changes in the F1 generation induced by maternal or paternal diets to the F3-4 generations. Elucidating the mechanisms of epigenetic inheritance is particularly challenging, especially due to the phenomenon of passive genome-wide demethylation that occurs post-conception in mammalian embryogenesis. This is an important factor that needs to be taken into consideration when determining the heritability of parental DNA methylation patterns (Fisher and Brockdorff 2012). Epigenetic inheritance of histone modification, chromatin remodeling, and DNA methylation is an area of growing interest. To date, preliminary research has been conducted to establish proof-of-principle of the mechanisms behind trans-generational epigenetic inheritance in plants, worms, and flies (Fisher and Brockdorff 2012).

Trans-generational Studies Summary. Overall, the preconceptional to postweaning environment, including maternal supplementation of one-carbon nutrients, has been shown to have an effect on cancer risk in the offspring. Some of these studies have investigated epigenetic modulation as the underlying mechanism for this relationship. The body of evidence for trans-generational studies is found in animal models and therefore translation of outcomes to humans is questionable. More investigation is needed to clarify whether maternal one-carbon nutrient exposure has an effect on adult cancers in the offspring (Ciappio et al. 2011b). Studies differ quite significantly in the duration and amount of one-carbon nutrient exposure during pregnancy. Since it has been demonstrated that timing and dose are important factors in modulating cancer risk (Kim 2005), this may explain some of the differences seen in cancer risk in the offspring (Ciappio et al. 2011b). It has been suggested that readdressing recommendations of perinatal nutrient intakes, including folate, vitamin B_6 and B_{12}, and choline, could help minimize chronic disease, including cancer, in the offspring (Ciappio et al. 2011b). Overall, the available body of evidence from animal and human studies, although not uniformly consistent, provides considerable evidence that global and gene-specific DNA methylation can be modulated by maternal dietary and supplemental intakes of one-carbon nutrients, in particular folate/folic acid, as well as maternal genotype of critical enzymes involved in the folate pathway. Whether these DNA methylation changes resulting from maternal one-carbon nutrition play a mechanistic role in cancer development and progression in the offspring needs further exploration.

11.5 Summary and Conclusions

Genetic changes in cancer are abrupt in onset, their effects are often all-or-nothing, the loss of function occurs at a fixed level, and they are not reversible in most cases. In contrast, epigenetic changes are gradual in onset and progressive, their effects are dose-dependent, and are potentially reversible. These observations present new opportunities in cancer risk modification and prevention using dietary and lifestyle factors and potential chemopreventive drugs. In this regard, folate and related one-carbon metabolism nutrients have been a focus of intense interest because of their potential modulatory effects on cancer development observed in animal studies, epidemiological observations, and intervention trials in humans and their potential ability to modulate DNA methylation.

We have exclusively focused on DNA methylation in this chapter. However, one-carbon nutrients will likely influence other epigenetic mechanisms such as histone modifications, chromatin remodeling, and microRNA. Indeed, studies investigating these complex interactions have begun to emerge. Furthermore, investigating the effects of one-carbon nutrients on different methylation products beyond 5-methylcytosine (5-mC) such as 5-hydroxymethylcytosine (5-hmC) are of great interest. While 5-mC in the CpG promoter region of genes is generally linked to gene silencing, 5-hmC has been shown to be associated with gene activation (Mellen et al. 2012).

The portfolio of evidence from in vitro, animal, and human studies collectively suggests that the effects of deficiency and supplementation of one-carbon nutrients on DNA methylation are highly complex and appear to depend on cell type, target organ and stage of transformation and furthermore, these effects appear to be gene- and site-specific. These studies also suggest that changes in DNA methylation depend on the magnitude and duration of dietary manipulations of one-carbon nutrients, on interactions with one another and with other dietary factors, and on genetic variants of critical genes in the folate-metabolic and one-carbon transfer pathways. Consequently, linking DNA methylation changes resulting from dietary manipulations of one-carbon nutrients to cancer development and progression in specific organ sites remains a very difficult task. Nevertheless, a considerable amount of evidence supports that one-carbon nutrients may modulate the risk of cancer in certain sites via aberrant global and gene-specific DNA methylation. An emerging body of evidence suggests that the status of one-carbon nutrients during pregnancy and early postnatal life may play a critical role in the development of cancer of the offspring later in adult life because of their potential modulatory effects on DNA methylation reprogramming during this highly critical and susceptible period of epigenetic programming. Although the jury is still out, the potential for one-carbon metabolism nutrients to modulate DNA methylation and thus modify the risk of chronic diseases including cancer in humans across the lifespan remains provocative and is worthy of further studies.

References

(1998) Dietary reference intakes for thiamin, riboflavin, niacin, vitamin B6, folate, vitamin B12, pantothenic acid, biotin, and choline. The National Academies Press, Washington, DC

Abbadi RA-G, Emery P, Pufulete M (2012) Short-term folate supplementation in physiological doses has no effect on ESR1 and MLH1 methylation in colonic mucosa of individuals with adenoma. J Nutrigenet Nutrigenomics 5:327–338

Abratte CM, Wang W, Li R, Axume J, Moriarty DJ, Caudill MA (2009) Choline status is not a reliable indicator of moderate changes in dietary choline consumption in premenopausal women. J Nutr Biochem 20:62–69

Ackerstaff E, Pflug BR, Nelson JB, Bhujwalla ZM (2001) Detection of increased choline compounds with proton nuclear magnetic resonance spectroscopy subsequent to malignant transformation of human prostatic epithelial cells. Cancer Res 61:3599–3603

Alberg AJ, Selhub J, Shah KV, Viscidi RP, Comstock GW, Helzlsouer KJ (2000) The risk of cervical cancer in relation to serum concentrations of folate, vitamin B-12, and homocysteine. Cancer Epidemiol Biomarkers Prev 9:761–764

Albert CM, Cook NR, Gaziano JM, Zaharris E, Macfadyen J, Danielson E, Buring JE, Manson JE (2008) Effect of folic acid and B vitamins on risk of cardiovascular events and total mortality among women at high risk for cardiovascular disease: a randomized trial. JAMA 299:2027–2036

Al-Ghnaniem R, Peters J, Foresti R, Heaton N, Pufulete M (2007) Methylation of estrogen receptor alpha and mutL homolog 1 in normal colonic mucosa: association with folate and vitamin B-12 status in subjects with and without colorectal neoplasia. Am J Clin Nutr 86:1064–1072

Alonso-Aperte E, Gonzalez MP, Poo-Prieto R, Varela-Moreiras G (2008) Folate status and S-adenosylmethionine/S-adenosylhomocysteine ratio in colorectal adenocarcinoma in humans. Eur J Clin Nutr 62:295–298

Alvarez H, Opalinska J, Zhou L, Sohal D, Fazzari MJ, Yu YT, Montagna C, Montgomery EA, Canto M, Dunbar KB, Wang J, Roa JC, Mo YK, Bhagat T, Ramesh KH, Cannizzaro L, Mollenhauer J, Thompson RF, Suzuki M, Meltzer SJ, Melnick A, Greally JM, Maitra A, Verma A (2011) Widespread hypomethylation occurs early and synergizes with gene amplification during esophageal carcinogenesis. PLoS Genet 7

Asada K, Kotake Y, Asada R, Saunders D, Broyles RH, Towner RA, Fukui H, Floyd RA (2006) LINE-1 hypomethylation in a choline-deficiency-induced liver cancer in rats: dependence on feeding period. J Biomed Biotechnol 2006:17142

Aufreiter S, Gregory JF, Pfeiffer CM, Fazili Z, Kim YI, Marcon N, Kamalaporn P, Pencharz PB, O'Connor DL (2009) Folate is absorbed across the colon of adults: evidence from cecal infusion of C-13-labeled 6S–5-formyltetrahydrofolic acid. Am J Clin Nutr 90:116–123

Ba Y, Yu H, Liu F, Geng X, Zhu C, Zhu Q, Zheng T, Ma S, Wang G, Li Z, Zhang Y (2011) Relationship of folate, vitamin B12 and methylation of insulin-like growth factor-II in maternal and cord blood. Eur J Clin Nutr 65:480–485

Baggott JE, Vaughn WH, Juliana MM, Eto I, Krumdieck CL, Grubbs CJ (1992) Effects of folate-deficiency and supplementation on methylnitrosourea-induced RAT mammary-tumors. J Natl Cancer Inst 84:1740–1744

Baghurst PA, McMichael AJ, Slavotinek AH, Baghurst KI, Boyle P, Walker AM (1991) A case-control study of diet and cancer of the pancreas. Am J Epidemiol 134:167–179

Bailey LB (2003) Folate, methyl-related nutrients, alcohol, and the MTHFR 677C → T polymorphism affect cancer risk: intake recommendations. J Nutr 133:3748S–3753S

Bailey SW, Ayling JE (2009) The extremely slow and variable activity of dihydrofolate reductase in human liver and its implications for high folic acid intake. Proc Natl Acad Sci U S A 106:15424–15429

Bailey HD, Miller M, Langridge A, De Klerk NH, Van Bockxmeer FM, Attia J, Scott RJ, Armstrong BK, Milne E (2012) Maternal dietary intake of folate and vitamins B6 and B12 during pregnancy and the risk of childhood acute lymphoblastic leukemia. Nutr Cancer 64:1122–1130

Balch C, Matei DE, Huang TH, Nephew KP (2010) Role of epigenomics in ovarian and endometrial cancers. Epigenomics 2:419–447

Balk EM, Raman G, Tatsioni A, Chung M, Lau J, Rosenberg IH (2007) Vitamin B6, B12, and folic acid supplementation and cognitive function: a systematic review of randomized trials. Arch Intern Med 167:21–30

Ballestar E, Esteller M (2002) The impact of chromatin in human cancer: linking DNA methylation to gene silencing. Carcinogenesis 23:1103–1109

Banno K, Kisu I, Yanokura M, Masuda K, Kobayashi Y, Ueki A, Tsuji K, Yamagami W, Nomura H, Susumu N, Aoki D (2012) Endometrial cancer and hypermethylation: regulation of DNA and MicroRNA by epigenetics. Biochem Res Int 2012:738274

Bassett JK, Severi G, Hodge AM, Baglietto L, Hopper JL, English DR, Giles GG (2012) Dietary intake of B vitamins and methionine and prostate cancer incidence and mortality. Cancer Causes Control 23:855–863

Basten GP, Duthie SJ, Pirie L, Vaughan N, Hill MH, Powers HJ (2006) Sensitivity of markers of DNA stability and DNA repair activity to folate supplementation in healthy volunteers. Br J Cancer 94:1942–1947

Beilby J, Ambrosini GL, Rossi E, De Klerk NH, Musk AW (2010) Serum levels of folate, lycopene, beta-carotene, retinol and vitamin E and prostate cancer risk. Eur J Clin Nutr 64:1235–1238

Berner C, Aumuller E, Gnauck A, Nestelberger M, Just A, Haslberger AG (2010) Epigenetic control of estrogen receptor expression and tumor suppressor genes is modulated by bioactive food compounds. Ann Nutr Metab 57:183–189

Biasco G, Zannoni U, Paganelli GM, Santucci R, Gionchetti P, Rivolta G, Miniero R, Pironi L, Calabrese C, Di Febo G, Miglioli M (1997) Folic acid supplementation and cell kinetics of rectal mucosa in patients with ulcerative colitis. Cancer Epidemiol Biomarkers Prev 6:469–471

Biel RK, Csizmadi I, Cook LS, Courneya KS, Magliocco AM, Friedenreich CM (2011) Risk of endometrial cancer in relation to individual nutrients from diet and supplements. Public Health Nutr 14:1948–1960

Bird AP, Wolffe AP (1999) Methylation-induced repression–belts, braces, and chromatin. Cell 99:451–454

Bistulfi G, Foster BA, Karasik E, Gillard B, Miecznikowski J, Dhiman VK, Smiraglia DJ (2011) Dietary folate deficiency blocks prostate cancer progression in the TRAMP model. Cancer Prev Res 4:1825–1834

Boeke CE, Baccarelli A, Kleinman KP, Burris HH, Litonjua AA, Rifas-Shiman SL, Tarantini L, Gillman MW (2012) Gestational intake of methyl donors and global LINE-1 DNA methylation in maternal and cord blood Prospective results from a folate-replete population. Epigenetics 7:253–260

Bravi F, Edefonti V, Randi G, Garavello W, La Vecchia C, Ferraroni M, Talamini R, Franceschi S, Decarli A (2012) Dietary patterns and the risk of esophageal cancer. Ann Oncol 23:765–770

Bunin GR, Meadows AT, Emanuel BS, Buckley JD, Woods WG, Hammond GD (1989) Pre- and postconception factors associated with sporadic heritable and nonheritable retinoblastoma. Cancer Res 49:5730–5735

Bunin GR, Kuijten RR, Buckley JD, Rorke LB, Meadows AT (1993) Relation between maternal diet and subsequent primitive neuroectodermal brain tumors in young children. N Engl J Med 329:536–541

Bunin GR, Gallagher PR, Rorke-Adams LB, Robison LL, Cnaan A (2006) Maternal supplement, micronutrient, and cured meat intake during pregnancy and risk of medulloblastoma during childhood: a children's oncology group study. Cancer Epidemiol Biomarkers Prev 15:1660–1667

Calcagno DQ, Leal MF, Assumpcao PP, Smith MA, Burbano RR (2008) MYC and gastric adenocarcinoma carcinogenesis. World J Gastroenterol 14:5962–5968

Calcagno DQ, Gigek CO, Chen ES, Burbano RR, Smith MA (2013) DNA and histone methylation in gastric carcinogenesis. World J Gastroenterol 19:1182–1192

Canman CE, Lim DS, Cimprich KA, Taya Y, Tamai K, Sakaguchi K, Appella E, Kastan MB, Siliciano JD (1998) Activation of the ATM kinase by ionizing radiation and phosphorylation of p53. Science 281:1677–1679

Carroll C, Cooper K, Papaioannou D, Hind D, Tappenden P, Pilgrim H, Booth A (2010) Meta-analysis: folic acid in the chemoprevention of colorectal adenomas and colorectal cancer. Aliment Pharmacol Ther 31:708–718

Castro R, Rivera I, Ravasco P, Camilo ME, Jakobs C, Blom HJ, De Almeida IT (2004) 5,10-methylenetetrahydrofolate reductase (MTHFR) 677C→T and 1298A→C mutations are associated with DNA hypomethylation. J Med Genet 41:454–458

Chagas CE, Bassoli BK, De Souza CA, Deminice R, Jordao Junior AA, Paiva SA, Dagli ML, Ong TP, Moreno FS (2011) Folic acid supplementation during early hepatocarcinogenesis: cellular and molecular effects. Int J Cancer 129:2073–2082

Charles MA, Johnson IT, Belshaw NJ (2012) Supra-physiological folic acid concentrations induce aberrant DNA methylation in normal human cells *in vitro*. Epigenetics 7:689–694

Chen J, Gammon MD, Chan W, Palomeque C, Wetmur JG, Kabat GC, Teitelbaum SL, Britton JA, Terry MB, Neugut AI, Santella RM (2005) One-carbon metabolism, MTHFR polymorphisms, and risk of breast cancer. Cancer Res 65:1606–1614

Chen J, Huang Z-J, Duan Y-Q, Xiao X-R, Jiang J-Q, Zhang R (2012) Aberrant DNA methylation of P16, MGMT, and hMLH1 genes in combination with MTHFR C677T genetic polymorphism and folate intake in esophageal squamous cell carcinoma. Asian Pac J Cancer Prev 13:5303–5306

Chen F-P, Lin C-C, Chen T-H, Tsai M-C, Huang Y-C (2013) Higher plasma homocysteine is associated with increased risk of developing colorectal polyps. Nutr Cancer 65:195–201

Cheng YW, Idrees K, Shattock R, Khan SA, Zeng Z, Brennan CW, Paty P, Barany F (2010) Loss of imprinting and marked gene elevation are 2 forms of aberrant IGF2 expression in colorectal cancer. Int J Cancer 127:568–577

Cho E, Willett WC, Colditz GA, Fuchs CS, Wu K, Chan AT, Zeisel SH, Giovannucci EL (2007a) Dietary choline and betaine and the risk of distal colorectal adenoma in women. J Natl Cancer Inst 99:1224–1231

Cho EY, Holmes M, Hankinson SE, Willett WC (2007b) Nutrients involved in one-carbon metabolism and risk of breast cancer among premenopausal women. Cancer Epidemiol Biomarkers Prev 16:2787–2790

Choi SW, Kim YI, Weitzel JN, Mason JB (1998) Folate depletion impairs DNA excision repair in the colon of the rat. Gut 43:93–99

Choi SW, Friso S, Dolnikowski GG, Bagley PJ, Edmondson AN, Smith DE, Mason JB (2003) Biochemical and molecular aberrations in the rat colon due to folate depletion are age-specific. J Nutr 133:1206–1212

Christman JK, Sheikhnejad G, Dizik M, Abileah S, Wainfan E (1993) Reversibility of changes in nucleic-acid methylation and gene-expression induced in rat-liver by severe dietary methyl deficiency. Carcinogenesis 14:551–557

Chuang SC, Stolzenberg-Solomon R, Ueland PM, Vollset SE, Midttun O, Olsen A, Tjonneland A, Overvad K, Boutron-Ruault MC, Morois S, Clavel-Chapelon F, Teucher B, Kaaks R, Weikert C, Boeing H, Trichopoulou A, Benetou V, Naska A, Jenab M, Slimani N, Romieu I, Michaud DS, Palli D, Sieri S, Panico S, Sacerdote C, Tumino R, Skeie G, Duell EJ, Rodriguez L, Molina-Montes E, Huerta JM, Larranaga N, Gurrea AB, Johansen D, Manjer J, Ye WM, Sund M, Peeters PHM, Jeurnink S, Wareham N, Khaw KT, Crowe F, Riboli E, Bueno-de-Mesquita B, Vineis P (2011) A U-shaped relationship between plasma folate and pancreatic cancer risk in the European Prospective Investigation into Cancer and Nutrition. Eur J Cancer 47:1808–1816

Ciappio ED, Liu Z, Brooks RS, Mason JB, Bronson RT, Crott JW (2011a) Maternal B vitamin supplementation from preconception through weaning suppresses intestinal tumorigenesis in Apc(1638N) mouse offspring. Gut 60:1695–1702

Ciappio ED, Mason JB, Crott JW (2011b) Maternal one-carbon nutrient intake and cancer risk in offspring. Nutr Rev 69:561–571

Clarke R, Halsey J, Lewington S, Lonn E, Armitage J, Manson JE, Bonaa KH, Spence JD, Nygard O, Jamison R, Gaziano JM, Guarino P, Bennett D, Mir F, Peto R, Collins R (2010) Effects of lowering homocysteine levels with B vitamins on cardiovascular disease, cancer, and cause-specific mortality: meta-analysis of 8 randomized trials involving 37 485 individuals. Arch Intern Med 170:1622–1631

Colapinto CK, O'Connor DL, Tremblay MS (2011) Folate status of the population in the Canadian Health Measures Survey. CMAJ 183:E100–E106

Cole BF, Baron JA, Sandler RS, Haile RW, Ahnen DJ, Bresalier RS, McKeown-Eyssen G, Summers RW, Rothstein RI, Burke CA, Snover DC, Church TR, Allen JI, Robertson DJ, Beck GJ, Bond JH, Byers T, Mandel JS, Mott LA, Pearson LH, Barry EL, Rees JR, Marcon N, Saibil F, Ueland PM, Greenberg ER (2007) Folic acid for the prevention of colorectal adenomas: a randomized clinical trial. JAMA 297:2351–2359

Collin SM, Metcalfe C, Refsum H, Lewis SJ, Zuccolo L, Smith GD, Chen L, Harris R, Davis M, Marsden G, Johnston C, Lane JA, Ebbing M, Bonaa KH, Nygard O, Ueland PM, Grau MV, Baron JA, Donovan JL, Neal DE, Hamdy FC, Smith AD, Martin RM (2010) Circulating folate, vitamin B12, homocysteine, vitamin B12 transport proteins, and risk of prostate cancer: a case-control study, systematic review, and meta-analysis. Cancer Epidemiol Biomarkers Prev 19:1632–1642

Couvert P, Carrie A, Du Montcel ST, Vaysse J, Sutton A, Barget N, Trinchet J-C, Beaugrand M, Ganne N, Giral P, Chelly J (2012) Insulin-like growth factor 2 gene methylation in peripheral blood mononuclear cells of patients with hepatitis C related cirrhosis or hepatocellular carcinoma. Clin Res Hepatol Gastroenterol 36:345–351

Cravo M, Fidalgo P, Pereira AD, Gouveia-Oliveira A, Chaves P, Selhub J, Mason JB, Mira FC, Leitao CN (1994) DNA methylation as an intermediate biomarker in colorectal cancer: modulation by folic acid supplementation. Eur J Cancer Prev 3:473–479

Cravo ML, Albuquerque CM, Salazar De Sousa L, Gloria LM, Chaves P, Dias Pereira A, Nobre Leitao C, Quina MG, Costa Mira F (1998a) Microsatellite instability in non-neoplastic mucosa of patients with ulcerative colitis: effect of folate supplementation. Am J Gastroenterol 93:2060–2064

Cravo ML, Pinto AG, Chaves P, Cruz JA, Lage P, Nobre Leitao C, Costa Mira F (1998b) Effect of folate supplementation on DNA methylation of rectal mucosa in patients with colonic adenomas: correlation with nutrient intake. Clin Nutr 17:45–49

Crider KS, Bailey LB, Berry RJ (2011) Folic acid food fortification-its history, effect, concerns, and future directions. Nutrients 3

Cropley JE, Suter CM, Beckman KB, Martin DIK (2006) Germ-line epigenetic modification of the murine A(vy) allele by nutritional supplementation. Proc Natl Acad Sci U S A 103:17308–17312

Crott JW, Liu Z, Keyes MK, Choi S-W, Jang H, Moyer MP, Mason JB (2008) Moderate folate depletion modulates the expression of selected genes involved in cell cycle, intracellular signaling and folate uptake in human colonic epithelial cell lines. J Nutr Biochem 19:328–335

Crowell J, Ly A, Kim YI (2011) Folate and DNA methylation. In: Maulik N, Maulik G (eds) Nutrition, epigenetic mechanisms, and human disease. CRC, Boca Raton

Curtin K, Slattery ML, Ulrich CM, Bigler J, Levin TR, Wolff RK, Albertsen H, Potter JD, Samowitz WS (2007) Genetic polymorphisms in one-carbon metabolism: associations with CpG island methylator phenotype (CIMP) in colon cancer and the modifying effects of diet. Carcinogenesis 28:1672–1679

Dahlin AM, Van Guelpen B, Hultdin J, Johansson I, Hallmans G, Palmqvist R (2008) Plasma vitamin B12 concentrations and the risk of colorectal cancer: a nested case-referent study. Int J Cancer 122:2057–2061

Davis CD, Uthus EO (2004) DNA methylation, cancer susceptibility, and nutrient interactions. Exp Biol Med 229:988–995

De Vogel S, Bongaerts BW, Wouters KA, Kester AD, Schouten LJ, De Goeij AF, De Bruine AP, Goldbohm RA, Van Den Brandt PA, Van Engeland M, Weijenberg MP (2008a) Associations of dietary methyl donor intake with MLH1 promoter hypermethylation and related molecular phenotypes in sporadic colorectal cancer. Carcinogenesis 29:1765–1773

De Vogel S, Dindore V, Van Engeland M, Goldbohm RA, Van Den Brandt PA, Weijenberg MP (2008b) Dietary folate, methionine, riboflavin, and vitamin B-6 and risk of sporadic colorectal cancer. J Nutr 138:2372–2378

De Vogel S, Schneede J, Ueland PM, Vollset SE, Meyer K, Fredriksen A, Midttun O, Bjorge T, Kampman E, Bretthauer M, Hoff G (2011a) Biomarkers related to one-carbon metabolism as

potential risk factors for distal colorectal adenomas. Cancer Epidemiol Biomarkers Prev 20:1726–1735
De Vogel S, Wouters KAD, Gottschalk RWH, Van Schooten FJ, De Goeij AFPM, De Bruine AP, Goldbohm RA, Van Den Brandt PA, Van Engeland M, Weijenberg MP (2011b) Dietary methyl donors, methyl metabolizing enzymes, and epigenetic regulators: diet-gene interactions and promoter CpG island hypermethylation in colorectal cancer. Cancer Causes Control 22:1–12
De Wals P, Tairou F, Van Allen MI, Uh SH, Lowry RB, Sibbald B, Evans JA, Van Den Hof MC, Zimmer P, Crowley M, Fernandez B, Lee NS, Niyonsenga T (2007) Reduction in neural-tube defects after folic acid fortification in Canada. N Engl J Med 357:135–142
Dizik M, Christman JK, Wainfan E (1991) Alterations in expression and methylation of specific genes in livers of rats fed a cancer promoting methyl-deficient diet. Carcinogenesis 12:1307–1312
Dockerty JD, Herbison P, Skegg DC, Elwood M (2007) Vitamin and mineral supplements in pregnancy and the risk of childhood acute lymphoblastic leukaemia: a case-control study. BMC Public Health 7:136
Dominguez-Salas P, Cox SE, Prentice AM, Hennig BJ, Moore SE (2012) Maternal nutritional status, C-1 metabolism and offspring DNA methylation: a review of current evidence in human subjects. Proc Nutr Soc 71:154–165
Du YP, Peng JS, Sun A, Tang ZH, Ling WH, Zhu HL (2009) Assessment of the effect of betaine on p16 and c-myc DNA methylation and mRNA expression in a chemical induced rat liver cancer model. BMC Cancer 9
Duthie SJ, Narayanan S, Blum S, Pirie L, Brand GM (2000) Folate deficiency *in vitro* induces uracil misincorporation and DNA hypomethylation and inhibits DNA excision repair in immortalized normal human colon epithelial cells. Nutr Cancer 37:245–251
Duthie SJ, Grant G, Pirie LP, Watson AJ, Margison GP (2010) Folate deficiency alters hepatic and colon MGMT and OGG-1 DNA repair protein expression in rats but has no effect on genome-wide DNA methylation. Cancer Prev Res (Phila) 3:92–100
Ebbing M, Bonaa KH, Nygard O, Arnesen E, Ueland PM, Nordrehaug JE, Rasmussen K, Njolstad I, Refsum H, Nilsen DW, Tverdal A, Meyer K, Vollset SE (2009) Cancer incidence and mortality after treatment with folic acid and vitamin B12. JAMA 302:2119–2126
Ebbing M, Bonaa KH, Arnesen E, Ueland PM, Nordrehaug JE, Rasmussen K, Njolstad I, Nilsen DW, Refsum H, Tverdal A, Vollset SE, Schirmer H, Bleie O, Steigen T, Midttun O, Fredriksen A, Pedersen ER, Nygard O (2010) Combined analyses and extended follow-up of two randomized controlled homocysteine-lowering B-vitamin trials. J Intern Med 268:367–382
Egger G, Liang G, Aparicio A, Jones PA (2004) Epigenetics in human disease and prospects for epigenetic therapy. Nature 429:457–463
Esteller M (2003) Relevance of DNA methylation in the management of cancer. Lancet Oncol 4:351–358
Esteller M (2007) Cancer epigenomics: DNA methylomes and histone-modification maps. Nat Rev Genet 8:286–298
Esteller M (2012) Human cancer epigenetics. Eur J Cancer 48:S6
Eussen SJPM, Vollset SE, Hustad S, Midttun O, Meyer K, Fredriksen A, Ueland PM, Jenab M, Slimani N, Boffetta P, Overvad K, Thorlacius-Ussing O, Tjonneland A, Olsen A, Clavel-Chapelon F, Boutron-Ruault M-C, Morois S, Weikert C, Pischon T, Linseisen J, Kaaks R, Trichopoulou A, Zilis D, Katsoulis M, Palli D, Pala V, Vineis P, Tumino R, Panico S, Peeters PHM, Bueno-De-Mesquita HB, Van Duijnhoven FJB, Skeie G, Munoz X, Martinez C, Dorronsoro M, Ardanaz E, Navarro C, Rodriguez L, Vanguelpen B, Palmqvist R, Manjer J, Ericson U, Bingham S, Khaw K-T, Norat T, Riboli E (2010a) Plasma vitamins B2, B6, and B12, and related genetic variants as predictors of colorectal cancer risk. Cancer Epidemiol Biomarkers Prev 19:2549–2561
Eussen SJPM, Vollset SE, Hustad S, Midttun O, Meyer K, Fredriksen A, Ueland PM, Jenab M, Slimani N, Ferrari P, Agudo A, Sala N, Capella G, Del Giudice G, Palli D, Boeing H, Weikert C, Bueno-De-Mesquita HB, Buechner FL, Carneiro F, Berrino F, Vineis P, Tumino R, Panico S, Berglund G, Manjer J, Stenling R, Hallmans G, Martinez C, Arrizola L, Barricarte A,

Navarro C, Rodriguez L, Bingham S, Linseisen J, Kaaks R, Overvad K, Tjonneland A, Peeters PHM, Numans ME, Clavel-Chapelon F, Boutron-Ruault M-C, Morois S, Trichopoulou A, Lund E, Plebani M, Riboli E, Gonzalez CA (2010b) Vitamins B2 and B6 and genetic polymorphisms related to one-carbon metabolism as risk factors for gastric adenocarcinoma in the European prospective investigation into cancer and nutrition. Cancer Epidemiol Biomarkers Prev 19:28–38

Fang JY, Xiao SD, Zhu SS, Yuan JM, Qiu DK, Jiang SJ (1997) Relationship of plasma folic acid and status of DNA methylation in human gastric cancer. J Gastroenterol 32:171–175

Fearon ER (2011) Molecular genetics of colorectal cancer. Annu Rev Pathol 6:479–507

Feigelson HS, Jonas CR, Robertson AS, McCollough ML, Thun MJ, Calle EE (2003) Alcohol, folate, methionine, and risk of incident breast cancer in the American Cancer Society Cancer Prevention Study II Nutrition Cohort. Cancer Epidemiol Biomarkers Prev 12:161–164

Felix AS, Weissfeld JL, Stone RA, Bowser R, Chivukula M, Edwards RP, Linkov F (2010) Factors associated with Type I and Type II endometrial cancer. Cancer Causes Control 21:1851–1856

Fenech M, Aitken C, Rinaldi J (1998) Folate, vitamin B12, homocysteine status and DNA damage in young Australian adults. Carcinogenesis 19:1163–1171

Fife J, Raniga S, Hider PN, Frizelle FA (2011) Folic acid supplementation and colorectal cancer risk: a meta-analysis. Colorectal Dis 13:132–137

Figueiredo JC, Grau MV, Haile RW, Sandler RS, Summers RW, Bresalier RS, Burke CA, McKeown-Eyssen GE, Baron JA (2009a) Folic acid and risk of prostate cancer: results from a randomized clinical trial. J Natl Cancer Inst 101:432–435

Figueiredo JC, Grau MV, Wallace K, Levine AJ, Shen L, Hamdan R, Chen X, Bresalier RS, McKeown-Eyssen G, Haile RW, Baron JA, Issa J-PJ (2009b) Global DNA hypomethylation (LINE-1) in the normal colon and lifestyle characteristics and dietary and genetic factors. Cancer Epidemiol Biomarkers Prev 18:1041–1049

Figueiredo JC, Mott LA, Giovannucci E, Wu K, Cole B, Grainge MJ, Logan RF, Baron JA (2011) Folic acid and prevention of colorectal adenomas: a combined analysis of randomized clinical trials. Int J Cancer 129:192–203

Finnell RH, Spiegelstein O, Wlodarczyk B, Triplett A, Pogribny IP, Melnyk S, James JS (2002) DNA methylation in Folbp1 knockout mice supplemented with folic acid during gestation. J Nutr 132:2457S–2461S

Fischer LM, Dacosta KA, Kwock L, Stewart PW, Lu TS, Stabler SP, Allen RH, Zeisel SH (2007) Sex and menopausal status influence human dietary requirements for the nutrient cholin. Am J Clin Nutr 85:1275–1285

Fisher AG, Brockdorff N (2012) Epigenetic memory and parliamentary privilege combine to evoke discussions on inheritance. Development 139:3891–3896

Flatley JE, McNeir K, Balasubramani L, Tidy J, Stuart EL, Young TA, Powers HJ (2009) Folate status and aberrant DNA methylation are associated with HPV infection and cervical pathogenesis. Cancer Epidemiol Biomarkers Prev 18:2782–2789

Forman D, Burley VJ (2006) Gastric cancer: global pattern of the disease and an overview of environmental risk factors. Best Pract Res Clin Gastroenterol 20:633–649

Friso S, Choi SW, Girelli D, Mason JB, Dolnikowski GG, Bagley PJ, Olivieri O, Jacques PF, Rosenberg IH, Corrocher R, Selhub J (2002) A common mutation in the 5,10-methylenetetrahydrofolate reductase gene affects genomic DNA methylation through an interaction with folate status. Proc Natl Acad Sci U S A 99:5606–5611

Friso S, Girelli D, Trabetti E, Olivieri O, Guarini P, Pignatti PF, Corrocher R, Choi SW (2005) The MTHFR 1298A>C polymorphism and genomic DNA methylation in human lymphocytes. Cancer Epidemiol Biomarkers Prev 14:938–943

Fryer AA, Nafee TM, Ismail KM, Carroll WD, Emes RD, Farrell WE (2009) LINE-1 DNA methylation is inversely correlated with cord plasma homocysteine in man: a preliminary study. Epigenetics 4:394–398

Fryer AA, Emes RD, Ismail KM, Haworth KE, Mein C, Carroll WD, Farrell WE (2011) Quantitative, high-resolution epigenetic profiling of CpG loci identifies associations with cord blood plasma homocysteine and birth weight in humans. Epigenetics 6

Fujimori S, Gudis K, Takahashi Y, Kotoyori M, Tatsuguchi A, Ohaki Y, Sakamoto C (2011) Determination of the minimal essential serum folate concentration for reduced risk of colorectal adenoma. Clin Nutr 30:653–658

Galvan-Portillo MV, Cantoral A, Onate-Ocana LF, Chen J, Herrera-Goepfert R, Torres-Sanchez L, Hernandez-Ramirez RU, Palma-Coca O, Lopez-Carrillo L (2009) Gastric cancer in relation to the intake of nutrients involved in one-carbon metabolism among MTHFR 677 TT carriers. Eur J Nutr 48:269–276

Galvan-Portillo MV, Onate-Ocana LF, Perez-Perez GI, Chen J, Herrera-Goepfert R, Chihu-Amparan L, Flores-Luna L, Mohar-Betancourt A, Lopez-Carrillo L (2010) Dietary folate and vitamin B12 intake before diagnosis decreases gastric cancer mortality risk among susceptible MTHFR 677TT carriers. Nutrition 26:201–208

Gao S, Ding LH, Wang JW, Li CB, Wang ZY (2013) Diet folate, DNA methylation and polymorphisms in methylenetetrahydrofolate reductase in association with the susceptibility to gastric cancer. Asian Pac J Cancer Prev 14:299–302

Ghosh C, Baker JA, Moysich KB, Rivera R, Brasure JR, McCann SE (2008) Dietary intakes of selected nutrients and food groups and risk of cervical cancer. Nutr Cancer 60:331–341

Ghoshal K, Li X, Datta J, Bai S, Pogribny I, Pogribny M, Huang Y, Young D, Jacob ST (2006) A folate- and methyl-deficient diet alters the expression of DNA methyltransferases and methyl CpG binding proteins involved in epigenetic gene silencing in livers of F344 rats. J Nutr 136:1522–1527

Gibson TM, Weinstein SJ, Pfeiffer RM, Hollenbeck AR, Subar AF, Schatzkin A, Mayne ST, Stolzenberg-Solomon R (2011) Pre- and postfortification intake of folate and risk of colorectal cancer in a large prospective cohort study in the United States. Am J Clin Nutr 94:1053–1062

Giovannucci E, Rimm EB, Ascherio A, Stampfer MJ, Colditz GA, Willett WC (1995) Alcohol, low-methionine–low-folate diets, and risk of colon cancer in men. J Natl Cancer Inst 87:265–273

Giovannucci E, Stampfer MJ, Colditz GA, Hunter DJ, Fuchs C, Rosner BA, Speizer FE, Willett WC (1998) Multivitamin use, folate, and colon cancer in women in the Nurses' Health Study. Ann Intern Med 129:517–524

Gonda TA, Kim YI, Salas MC, Gamble MV, Shibata W, Muthupalani S, Sohn KJ, Abrams JA, Fox JG, Wang TC, Tycko B (2012) Folic acid increases global DNA methylation and reduces inflammation to prevent helicobacter-associated gastric cancer in mice. Gastroenterology 142:824–833

Gong Z, Holly EA, Bracci PM (2009) Intake of folate, vitamins B6, B12 and methionine and risk of pancreatic cancer in a large population-based case-control study. Cancer Causes Control 20:1317–1325

Goodman MT, McDuffie K, Hernandez B, Wilkens LR, Selhub J (2000) Case-control study of plasma folate, homocysteine, vitamin B-12, and cysteine as markers of cervical dysplasia. Cancer 89:376–382

Goodman MT, McDuffie K, Hernandez B, Wilkens LR, Bertram CC, Killeen J, Le Marchand L, Selhub J, Murphy S, Donlon TA (2001) Association of methylenetetrahydrofolate reductase polymorphism C677T and dietary folate with the risk of cervical dysplasia. Cancer Epidemiol Biomarkers Prev 10:1275–1280

Haas GP, Delongchamps N, Brawley OW, Wang CY, De La Roza G (2008) The worldwide epidemiology of prostate cancer: perspectives from autopsy studies. Can J Urol 15:3866–3871

Haggarty P, Hoad G, Campbell DM, Horgan GW, Piyathilake C, McNeill G (2013) Folate in pregnancy and imprinted gene and repeat element methylation in the offspring. Am J Clin Nutr 97:94–99

Hajizadeh B, Jessri M, Akhoondan M, Moasheri SM, Rashidkhani B (2012) Nutrient patterns and risk of esophageal squamous cell carcinoma: a case-control study. Dis Esophagus 25:442–448

Hanna L, Adams M (2006) Prevention of ovarian cancer. Best Prac Res Clin Obstet Gynaecol 20:339–362

Hariharan D, Saied A, Kocher HM (2008) Analysis of mortality rates for pancreatic cancer across the world. HPB 10:58–62

Harnack L, Jacobs DR, Nicodemus K, Lazovich D, Anderson K, Folsom AR (2002) Relationship of folate, vitamin B-6, vitamin B-12, and methionine intake to incidence of colorectal cancers. Nutr Cancer 43:152–158

Harris HR, Cramer DW, Vitonis AF, Depari M, Terry KL (2012) Folate, vitamin B6, vitamin B12, methionine and alcohol intake in relation to ovarian cancer risk. Int J Cancer 131:E518–E529

Heichman KA, Warren JD (2012) DNA methylation biomarkers and their utility for solid cancer diagnostics. Clin Chem Lab Med 50:1707–1721

Herman JG, Baylin SB (2003) Gene silencing in cancer in association with promoter hypermethylation. N Engl J Med 349:2042–2054

Hernandez BY, McDuffie K, Wilkens LR, Kamemoto L, Goodman MT (2003) Diet and premalignant lesions of the cervix: evidence of a protective role for folate, riboflavin, thiamin, and vitamin B-12. Cancer Causes Control 14:859–870

Hezel AF, Kimmelman AC, Stanger BZ, Bardeesy N, Depinho RA (2006) Genetics and biology of pancreatic ductal adenocarcinoma. Genes Dev 20:1218–1249

Hirsch S, Sanchez H, Albala C, De La Maza MP, Barrera G, Leiva L, Bunout D (2009) Colon cancer in Chile before and after the start of the flour fortification program with folic acid. Eur J Gastroenterol Hepatol 21:436–439

Ho E, Beaver LM, Williams DE, Dashwood RH (2011) Dietary factors and epigenetic regulation for prostate cancer prevention. Adv Nutr 2:497–510

Holm K, Hegardt C, Staaf J, Vallon-Christersson J, Jonsson G, Olsson H, Borg A, Ringner M (2010) Molecular subtypes of breast cancer are associated with characteristic DNA methylation patterns. Breast Cancer Res 12:R36

Honda S, Arai Y, Haruta M, Sasaki F, Ohira M, Yamaoka H, Horie H, Nakagawara A, Hiyama E, Todo S, Kaneko Y (2008) Loss of imprinting of IGF2 correlates with hypermethylation of the H19 differentially methylated region in hepatoblastoma. Br J Cancer 99:1891–1899

Hou L, Wang H, Sartori S, Gawron A, Lissowska J, Bollati V, Tarantini L, Zhang FF, Zatonski W, Chow W-H, Baccarelli A (2010) Blood leukocyte DNA hypomethylation and gastric cancer risk in a high-risk Polish population. Int J Cancer 127:1866–1874

Hoyo C, Murtha AP, Schildkraut JM, Jirtle RL, Demark-Wahnefried W, Forman MR, Iversen ES, Kurtzberg J, Overcash F, Huang Z, Murphy SK (2011) Methylation variation at IGF2 differentially methylated regions and maternal folic acid use before and during pregnancy. Epigenetics 6:928–936

Huang M, Chen H (2008) Effect of folate deprivation on WNT signaling pathway in human colon cell lines. FASEB J 22

Hultdin J, Van Guelpen B, Bergh A, Hallmans G, Stattin M (2005) Plasma folate, vitamin B12, and homocysteine and prostate cancer risk: a prospective study. Int J Cancer 113:819–824

Ibiebele TI, Hughes MC, Pandeya N, Zhao Z, Montgomery G, Hayward N, Green AC, Whiteman DC, Webb PM, Study of Digestive Health, Australian Cancer Study (2011) High intake of folate from food sources is associated with reduced risk of esophageal cancer in an Australian population. J Nutr 141:274–283

Ibrahim EM, Zekri JM (2010) Folic acid supplementation for the prevention of recurrence of colorectal adenomas: metaanalysis of interventional trials. Med Oncol 27:915–918

Ingrosso D, Cimmino A, Perna AF, Masella L, De Santo NG, De Bonis ML, Vacca M, D'esposito M, D'urso M, Galletti P, Zappia V (2003) Folate treatment and unbalanced methylation and changes of allelic expression induced by hyperhomocysteinaemia in patients with uraemia. Lancet 361:1693–1699

Issa JP (2004) CpG island methylator phenotype in cancer. Nat Rev Cancer 4:988–993

Jacob RA, Gretz DM, Taylor PC, James SJ, Pogribny IP, Miller BJ, Henning SM, Swendseid ME (1998) Moderate folate depletion increases plasma homocysteine and decreases lymphocyte DNA methylation in postmenopausal women. J Nutr 128:1204–1212

Jacobs EJ, Connell CJ, Patel AV, Chao A, Rodriguez C, Seymour J, McCullough ML, Calle EE, Thun MJ (2001) Multivitamin use and colon cancer mortality in the Cancer Prevention Study II cohort (United States). Cancer Causes Control 12:927–934

Jain MG, Rohan TE, Howe GR, Miller AB (2000) A cohort study of nutritional factors and endometrial cancer. Eur J Epidemiol 16:899–905

James SJ, Melnyk S, Pogribna M, Pogribny IP, Caudill MA (2002) Elevation in S-adenosylhomocysteine and DNA hypomethylation: potential epigenetic mechanism for homocysteine-related pathology. J Nutr 132:2361S–2366S

Jemal A, Bray F, Center MM, Ferlay J, Ward E, Forman D (2011) Global cancer statistics. CA Cancer J Clin 61:69–90

Jeronimo C, Bastian PJ, Bjartell A, Carbone GM, Catto JW, Clark SJ, Henrique R, Nelson WG, Shariat SF (2011) Epigenetics in prostate cancer: biologic and clinical relevance. Eur Urol 60:753–766

Jessri M, Rashidkhani B, Hajizadeh B, Jessri M, Gotay C (2011) Macronutrients, vitamins and minerals intake and risk of esophageal squamous cell carcinoma: a case-control study in Iran. Nutr J 10

Jin F, Qu LS, Shen XZ (2009) Association between the methylenetetrahydrofolate reductase C677T polymorphism and hepatocellular carcinoma risk: a meta-analysis. Diagn Pathol 4:39

Johansson M, Appleby PN, Allen NE, Travis RC, Roddam AW, Egevad L, Jenab M, Rinaldi S, Kiemeney LA, Bueno-De-Mesquita HB, Vollset SE, Ueland PM, Sanchez MJ, Quiros JR, Gonzalez CA, Larranaga N, Chirlaque MD, Ardanaz E, Sieri S, Palli D, Vineis P, Tumino R, Linseisen J, Kaaks R, Boeing H, Pischon T, Psaltopoulou T, Trichopoulou A, Trichopoulos D, Khaw KT, Bingham S, Hallmans G, Riboli E, Stattin P, Key TJ (2008) Circulating concentrations of folate and vitamin B12 in relation to prostate cancer risk: results from the European Prospective Investigation into Cancer and Nutrition study. Cancer Epidemiol Biomarkers Prev 17:279–285

Johansson M, Van Guelpen B, Vollset SE, Hultdin J, Bergh A, Key T, Midttun O, Hallmans G, Ueland PM, Stattin P (2009) One-carbon metabolism and prostate cancer risk: prospective investigation of seven circulating B vitamins and metabolites. Cancer Epidemiol Biomarkers Prev 18:1538–1543

Jones PA, Baylin SB (2002) The fundamental role of epigenetic events in cancer. Nat Rev Genet 3:415–428

Jones PA, Buckley JD (1990) The role of DNA methylation in cancer. Adv Cancer Res 54:1–23

Jones PL, Veenstra GJ, Wade PA, Vermaak D, Kass SU, Landsberger N, Strouboulis J, Wolffe AP (1998) Methylated DNA and MeCP2 recruit histone deacetylase to repress transcription. Nat Genet 19:187–191

Jun Ying MHR, Michael Hallman D, Hernandez LM, Spitz MR, Forman MR, Gorlova OY (2013) Associations between dietary intake of choline and betaine and lung cancer risk. PLoS One 8:e54561

Kabat GC, Miller AB, Jain M, Rohan TE (2008) Dietary intake of selected B vitamins in relation to risk of major cancers in women. Br J Cancer 99:816–821

Kadaveru K, Protiva P, Greenspan EJ, Kim YI, Rosenberg DW (2012) Dietary methyl donor depletion protects against intestinal tumorigenesis in ApcMin/+ mice. Cancer Prev Res 5:911–920

Kane MF, Loda M, Gaida GM, Lipman J, Mishra R, Goldman H, Jessup JM, Kolodner R (1997) Methylation of the hMLH1 promoter correlates with lack of expression of hMLH1 in sporadic colon tumors and mismatch repair-defective human tumor cell lines. Cancer Res 57:808–811

Kasperzyk JL, Fall K, Mucci LA, Hakansson N, Wolk A, Johansson JE, Andersson SO, Andren O (2009) One-carbon metabolism-related nutrients and prostate cancer survival. Am J Clin Nutr 90:561–569

Kawakami K, Ruszkiewicz A, Bennett G, Moore J, Watanabe G, Iacopetta B (2003) The folate pool in colorectal cancers is associated with DNA hypermethylation and with a polymorphism in methylenetetrahydrofolate reductase. Clin Cancer Res 9:5860–5865

Kelemen LE, Sellers TA, Vierkant RA, Harnack L, Cerhan JR (2004) Association of folate and alcohol with risk of ovarian cancer in a prospective study of postmenopausal women. Cancer Causes Control 15:1085–1093

Kennedy DA, Stern SJ, Moretti M, Matok I, Sarkar M, Nickel C, Koren G (2011) Folate intake and the risk of colorectal cancer: a systematic review and meta-analysis. Cancer Epidemiol 35:2–10

Keszei AP, Verhage BA, Heinen MM, Goldbohm RA, Van Den Brandt PA (2009) Dietary folate and folate vitamers and the risk of pancreatic cancer in the Netherlands cohort study. Cancer Epidemiol Biomarkers Prev 18:1785–1791

Key TJA, Silcocks PB, Davey GK, Appleby PN, Bishop DT (1997) A case-control study of diet and prostate cancer. Br J Cancer 76:678–687

Key T, Schatzkin A, Willett WC, Allen NE, Spencer EA, Travis RC (2004) Diet, nutrition and the prevention of cancer. Public Health Nutr 7:187–200

Key TJ, Appleby PN, Masset G, Brunner EJ, Cade JE, Greenwood DC, Stephen AM, Kuh D, Bhaniani A, Powell N, Khaw K-T (2012) Vitamins, minerals, essential fatty acids and colorectal cancer risk in the United Kingdom Dietary Cohort Consortium. Int J Cancer 131:E320–E325

Khosraviani K, Weir HP, Hamilton P, Moorehead J, Williamson K (2002) Effect of folate supplementation on mucosal cell proliferation in high risk patients for colon cancer. Gut 51:195–199

Kim YI (1999) Folate and carcinogenesis: evidence, mechanisms, and implications. J Nutr Biochem 10:66–88

Kim YI (2003) Role of folate in colon cancer development and progression. J Nutr 133:3731S–3739S

Kim YI (2004a) Folate and DNA methylation: a mechanistic link between folate deficiency and colorectal cancer? Cancer Epidemiol Biomarkers Prev 13:511–519

Kim YI (2004b) Will mandatory folic acid fortification prevent or promote cancer? Am J Clin Nutr 80:1123–1128

Kim YI (2005) Nutritional epigenetics: impact of folate deficiency on DNA methylation and colon cancer susceptibility. J Nutr 135:2703–2709

Kim YI (2006) Folate: a magic bullet or a double edged sword for colorectal cancer prevention? Gut 55:1387–1389

Kim YI (2007a) Folate and colorectal cancer: an evidence-based critical review. Mol Nutr Food Res 51:267–292

Kim YI (2007b) Folic acid fortification and supplementation—good for some but not so good for others. Nutr Rev 65:504–511

Kim YI (2008) Folic acid supplementation and cancer risk: point. Cancer Epidemiol Biomarkers Prev 17:2220–2225

Kim YI (2009) Role of the MTHFR polymorphisms in cancer risk modification and treatment. Future Oncol 5:523–542

Kim YI, Giuliano A, Hatch KD, Schneider A, Nour MA, Dallal GE, Selhub J, Mason JB (1994) Global DNA hypomethylation increases progressively in cervical dysplasia and carcinoma. Cancer 74:893–899

Kim YI, Salomon RN, Graeme-Cook F, Choi SW, Smith DE, Dallal GE, Mason JB (1996) Dietary folate protects against the development of macroscopic colonic neoplasia in a dose responsive manner in rats. Gut 39:732–740

Kim YI, Pogribny IP, Basnakian AG, Miller JW, Selhub J, James SJ, Mason JB (1997) Folate deficiency in rats induces DNA strand breaks and hypomethylation within the p53 tumor suppressor gene. Am J Clin Nutr 65:46–52

Kim YI, Baik HW, Fawaz K, Knox T, Lee YM, Norton R, Libby E, Mason JB (2001) Effects of folate supplementation on two provisional molecular markers of colon cancer: a prospective, randomized trial. Am J Gastroenterol 96:184–195

Kim TH, Yang J, Darling PB, O'Connor DL (2004) A large pool of available folate exists in the large intestine of human infants and piglets. J Nutr 134:1389–1394

Kim DH, Smith-Warner SA, Spiegelman D, Yaun SS, Colditz GA, Freudenheim JL, Giovannucci E, Goldbohm RA, Graham S, Harnack L, Jacobs EJ, Leitzmann M, Mannisto S, Miller AB, Potter JD, Rohan TE, Schatzkin A, Speizer FE, Stevens VL, Stolzenberg-Solomon R, Terry P,

Toniolo P, Weijenberg MP, Willett WC, Wolk A, Zeleniuch-Jacquotte A, Hunter DJ (2010) Pooled analyses of 13 prospective cohort studies on folate intake and colon cancer. Cancer Causes Control 21:1919–1930

Kim HW, Kim KN, Choi YJ, Chang N (2013) Effects of paternal folate deficiency on the expression of insulin-like growth factor-2 and global DNA methylation in the fetal brain. Mol Nutr Food Res 57(4):671–676

Klaus A, Birchmeier W (2008) Wnt signalling and its impact on development and cancer. Nat Rev Cancer 8:387–398

Klose RJ, Bird AP (2006) Genomic DNA methylation: the mark and its mediators. Trends Biochem Sci 31:89–97

Kobayashi LC, Limburg H, Miao Q, Woolcott C, Bedard LL, Massey TE, Aronson KJ (2012) Folate intake, alcohol consumption, and the methylenetetrahydrofolate reductase (MTHFR) C677T gene polymorphism: influence on prostate cancer risk and interactions. Front Oncol 2:100

Kotsopoulos J, Sohn KJ, Martin R, Choi M, Renlund R, McKerlie C, Hwang SW, Medline A, Kim YIJ (2003) Dietary folate deficiency suppresses N-methyl-N-nitrosourea-induced mammary tumorigenesis in rats. Carcinogenesis 24:937–944

Kotsopoulos J, Medline A, Renlund R, Sohn KJ, Martin R, Hwang SW, Lu S, Archer MC, Kim YI (2005) Effects of dietary folate on the development and progression of mammary tumors in rats. Carcinogenesis 26:1603–1612

Kotsopoulos J, Sohn KJ, Kim YI (2008) Postweaning dietary folate deficiency provided through childhood to puberty permanently increases genomic DNA methylation in adult rat liver. J Nutr 138:703–709

Kotsopoulos J, Hankinson SE, Tworoger SS (2010) Dietary betaine and choline intake are not associated with risk of epithelial ovarian cancer. Eur J Clin Nutr 64:111–114

Kotsopoulos J, Kim YI, Narod SA (2012) Folate and breast cancer: what about high-risk women? Cancer Causes Control 23:1405–1420

Kovacheva VP, Mellott TJ, Davison JM, Wagner N, Lopez-Coviella I, Schnitzler AC, Blusztajn JK (2007) Gestational choline deficiency causes global and Igf2 gene DNA hypermethylation by up-regulation of dnmt1 expression. J Biol Chem 282:31777–31788

Kovacheva VP, Davison JM, Mellott TJ, Rogers AE, Yang S, O'Brien MJ, Blusztajn JK (2009) Raising gestational choline intake alters gene expression in DMBA-evoked mammary tumors and prolongs survival. FASEB J 23:1054–1063

Kraunz KS, Hsiung D, McClean MD, Liu M, Osanyingbemi J, Nelson HH, Kelsey KT (2006) Dietary folate is associated with p16(INK4A) methylation in head and neck squamous cell carcinoma. Int J Cancer 119:1553–1557

Krishnaveni GV, Hill JC, Veena SR, Bhat DS, Wills AK, Karat CLS, Yajnik CS, Fall CHD (2009) Low plasma vitamin B-12 in pregnancy is associated with gestational 'diabesity' and later diabetes. Diabetologia 52:2350–2358

Kwanbunjan K, Saengkar P, Cheeramakara C, Tangjitgamol S, Chitcharoenrung K (2006) Vitamin B12 status of Thai women with neoplasia of the cervix uteri. Southeast Asian J Trop Med Public Health 37:178–183

Lajous M, Lazcano-Ponce E, Hernandez-Avila M, Willett W, Romieu I (2006a) Folate, vitamin B-6, and vitamin B-12 intake and the risk of breast cancer among Mexican women. Cancer Epidemiol Biomarkers Prev 15:443–448

Lajous M, Romieu I, Sabia S, Boutron-Ruault MC, Clavel-Chapelon F (2006b) Folate, vitamin B12 and postmenopausal breast cancer in a prospective study of French women. Cancer Causes Control 17:1209–1213

Lammi L, Arte S, Somer M, Jarvinen H, Lahermo P, Thesleff I, Pirinen S, Nieminen P (2004) Mutations in AXIN2 cause familial tooth agenesis and predispose to colorectal cancer. Am J Hum Genet 74:1043–1050

Larsson SC, Giovannucci E, Wolk A (2004) Dietary folate intake and incidence of ovarian cancer: The Swedish Mammography Cohort. J Natl Cancer Inst 96:396–402

Larsson SC, Giovannucci E, Wolk A (2006a) Folate intake, MTHFR polymorphisms, and risk of esophageal, gastric, and pancreatic cancer: a meta-analysis. Gastroenterology 131:1271–1283
Larsson SC, Hakansson N, Giovannucci E, Wolk A (2006b) Folate intake and pancreatic cancer incidence: a prospective study of Swedish women and men. J Natl Cancer Inst 98:407–413
Larsson SC, Orsini N, Wolk A (2010) Vitamin B6 and risk of colorectal cancer: a meta-analysis of prospective studies. JAMA 303:1077–1083
Lashner BA, Shapiro BD, Husain A, Goldblum JR (1999) Evaluation of the usefulness of testing for p53 mutations in colorectal cancer surveillance for ulcerative colitis. Am J Gastroenterol 94:456–462
Lawrance AK, Deng L, Rozen R (2009) Methylenetetrahydrofolate reductase deficiency and low dietary folate reduce tumorigenesis in Apcmin/+ mice. Gut 58:805–811
Lazarevic K, Nagorni A, Bogdanovic D, Rancic N, Stosic L, Milutinovic S (2011) Dietary micronutrients and gastric cancer: hospital based study. Central Eur J Med 6:783–787
Le Marchand L, Wang H, Selhub J, Vogt TM, Yokochi L, Decker R (2011) Association of plasma vitamin B6 with risk of colorectal adenoma in a multiethnic case-control study. Cancer Causes Control 22:929–936
Lee JE, Li H, Giovannucci E, Lee IM, Selhub J, Stampfer M, Ma J (2009) Prospective study of plasma vitamin B(6) and risk of colorectal cancer in men. Cancer Epidemiol Biomarkers Prev 18:1197–1202
Lee JE, Giovannucci E, Fuchs CS, Willett WC, Zeisel SH, Cho E (2010) Choline and betaine intake and the risk of colorectal cancer in men. Cancer Epidemiol Biomarkers Prev 19:884–887
Li E, Jaenisch R (2000) DNA methylation and methyltransferases. In: Ehrlich M (ed) DNA alterations in cancer: genetic and epigenetic changes. Eaton, Natick
Li XL, Xu JH (2012) MTHFR polymorphism and the risk of prostate cancer: a meta-analysis of case-control studies. Prostate Cancer Prostatic Dis 15:244–249
Li JY, Taylor PR, Li B, Dawsey S, Wang GQ, Ershow AG, Guo WD, Liu SF, Yang CS, Shen Q, Wang W, Mark SD, Zou XN, Greenwald P, Wu YP, Blot WJ (1993) Nutrition intervention trials in Linxian, China—multiple vitamin mineral supplementation, cancer incidence, and disease-specific mortality among adults with esophageal dysplasia. J Natl Cancer Inst 85:1492–1498
Lillycrop KA, Phillips ES, Jackson AA, Hanson MA, Burdge GC (2005) Dietary protein restriction of pregnant rats induces and folic acid supplementation prevents epigenetic modification of hepatic gene expression in the offspring. J Nutr 135:1382–1386
Lillycrop KA, Slater-Jefferies JL, Hanson MA, Godfrey KM, Jackson AA, Burdge GC (2007) Induction of altered epigenetic regulation of the hepatic glucocorticoid receptor in the offspring of rats fed a protein-restricted diet during pregnancy suggests that reduced DNA methyltransferase-1 expression is involved in impaired DNA methylation and changes in histone modifications. Br J Nutr 97:1064–1073
Lim U, Flood A, Choi SW, Albanes D, Cross AJ, Schatzkin A, Sinha R, Katki HA, Cash B, Schoenfeld P, Stolzenberg-Solomon R (2008) Genomic methylation of leukocyte DNA in relation to colorectal adenoma among asymptomatic women. Gastroenterology 134:47–55
Lin J, Lee IM, Cook NR, Selhub J, Manson JE, Buring JE, Zhang SM (2008) Plasma folate, vitamin B-6, vitamin B-12, and risk of breast cancer in women. Am J Clin Nutr 87:734–743
Lin PC, Giannopoulou EG, Park K, Mosquera JM, Sboner A, Tewari AK, Garraway LA, Beltran H, Rubin MA, Elemento O (2013) Epigenomic alterations in localized and advanced prostate cancer. Neoplasia 15:373–383
Linhart HG, Troen AM, Bell GW, Cantu E, Chao WH, Moran E, Steine E, He T, Jaenisch R (2009) Folate deficiency induces genomic uracil misincorporation and hypomethylation but does not increase DNA point mutations. Gastroenterology 136:227–235
Liu ZH, Ciappio ED, Crott JW, Brooks RS, Nesvet J, Smith DE, Choi SW, Mason JB (2011) Combined inadequacies of multiple B vitamins amplify colonic Wnt signaling and promote intestinal tumorigenesis in BAT-LacZ X Apc1638N mice. FASEB J 25:3136–3145
Liu WR, Shi YH, Peng YF, Fan J (2012) Epigenetics of hepatocellular carcinoma: a new horizon. Chin Med J (Engl) 125:2349–2360

Liu JJ, Hazra A, Giovannucci E, Hankinson SE, Rosner B, De Vivo I (2013) One-carbon metabolism factors and endometrial cancer risk. Br J Cancer 108:183–187

Logan CY, Nusse R (2004) The Wnt signaling pathway in development and disease. Annu Rev Cell Dev Biol 20:781–810

Lu C, Xie H, Wang F, Shen H, Wang J (2011) Diet folate, DNA methylation and genetic polymorphisms of MTHFR C677T in association with the prognosis of esophageal squamous cell carcinoma. BMC Cancer 11

Lubecka-Pietruszewska K, Kaufman-Szymczyk A, Stefanska B, Fabianowska-Majewska K (2013) Folic acid enforces DNA methylation-mediated transcriptional silencing of PTEN, APC and RARbeta2 tumour suppressor genes in breast cancer. Biochem Biophys Res Commun 430:623–628

Lucock M (2004) Is folic acid the ultimate functional food component for disease prevention? BMJ 328:211–214

Ly A, Lee H, Chen J, Sie KKY, Renlund R, Medline A, Sohn K-J, Croxford R, Thompson LU, Kim Y-I (2011) Effect of maternal and postweaning folic acid supplementation on mammary tumor risk in the offspring. Cancer Res 71:988–997

Ly A, Hoyt L, Crowell J, Kim Y-I (2012) Folate and DNA methylation. Antioxid Redox Signal 17:302–326

Ma E, Iwasaki M, Junko I, Hamada GS, Nishimoto IN, Carvalho SM, Motola J Jr, Laginha FM, Tsugane S (2009a) Dietary intake of folate, vitamin B6, and vitamin B12, genetic polymorphism of related enzymes, and risk of breast cancer: a case-control study in Brazilian women. BMC Cancer 9:122

Ma E, Iwasaki M, Kobayashi M, Kasuga Y, Yokoyama S, Onuma H, Nishimura H, Kusama R, Tsugane S (2009b) Dietary intake of folate, vitamin B2, vitamin B6, vitamin B12, genetic polymorphism of related enzymes, and risk of breast cancer: a case-control study in Japan. Nutr Cancer 61:447–456

Macfarlane AJ, Greene-Finestone LS, Shi YP (2011) Vitamin B-12 and homocysteine status in a folate-replete population: results from the Canadian Health Measures Survey. Am J Clin Nutr 94:1079–1087

Maloney CA, Hay SM, Rees WD (2007) Folate deficiency during pregnancy impacts on methyl metabolism without affecting global DNA methylation in the rat fetus. Br J Nutr 97:1090–1098

Malouf R, Grimley Evans J (2003) The effect of vitamin B6 on cognition. Cochrane Database Syst Rev CD004393

Markowitz SD, Bertagnolli MM (2009) Molecular origins of cancer: molecular basis of colorectal cancer. N Engl J Med 361:2449–2460

Marmot M, Atinmo T, Byers T, Chen J, Hirohata T, Jackson A, James W, Kolonel L, Kumanyika S, Leitzmann C (2007) Food, nutrition, physical activity, and the prevention of cancer: a global perspective. AICR, Washington, DC

Maruti SS, Ulrich CM, White E (2009) Folate and one-carbon metabolism nutrients from supplements and diet in relation to breast cancer risk. Am J Clin Nutr 89:624–633

Mason JB, Dickstein A, Jacques PF, Haggarty P, Selhub J, Dallal G, Rosenberg IH (2007) A temporal association between folic acid fortification and an increase in colorectal cancer rates may be illuminating important biological principles: a hypothesis. Cancer Epidemiol Biomarkers Prev 16:1325–1329

Matthews RG, Haywood BJ (1979) Inhibition of pig liver methylenetetrahydrofolate reductase by dihydrofolate: some mechanistic and regulatory implications. Biochemistry 18:4845–4851

Mayne ST, Risch HA, Dubrow R, Chow WH, Gammon MD, Vaughan TL, Farrow DC, Schoenberg JB, Stanford JL, Ahsan H, West AB, Rotterdam H, Blot WJ, Fraumeni JF (2001) Nutrient intake and risk of subtypes of esophageal and gastric cancer. Cancer Epidemiol Biomarkers Prev 10:1055–1062

McCleary-Wheeler AL, Lomberk GA, Weiss FU, Schneider G, Fabbri M, Poshusta TL, Dusetti NJ, Baumgart S, Iovanna JL, Ellenrieder V, Urrutia R, Fernandez-Zapico ME (2013) Insights

into the epigenetic mechanisms controlling pancreatic carcinogenesis. Cancer Lett 328:212–221
McKay JA, Waltham KJ, Williams EA, Mathers JC (2011a) Folate depletion during pregnancy and lactation reduces genomic DNA methylation in murine adult offspring. Genes Nutr 6:189–196
Mckay JA, Williams EA, Mathers JC (2011b) Effect of maternal and post-weaning folate supply on gene-specific DNA methylation in the small intestine of weaning and adult apc and wild type mice. Front Genet 2:23
McKay JA, Wong YK, Relton CL, Ford D, Mathers JC (2011c) Maternal folate supply and sex influence gene-specific DNA methylation in the fetal gut. Mol Nutr Food Res 55:1717–1723
McKay JA, Xie L, Harris S, Wong YK, Ford D, Mathers JC (2011d) Blood as a surrogate marker for tissue-specific DNA methylation and changes due to folate depletion in post-partum female mice. Mol Nutr Food Res 55:1026–1035
McKay JA, Groom A, Potter C, Coneyworth LJ, Ford D, Mathers JC, Relton CL (2012) Genetic and non-genetic influences during pregnancy on infant global and site specific DNA methylation: role for folate gene variants and vitamin B-12. PLoS One 7
Mellen M, Ayata P, Dewell S, Kriaucionis S, Heintz N (2012) MeCP2 binds to 5hmC enriched within active genes and accessible chromatin in the nervous system. Cell 151:1417–1430
Menezes CA, Bonde P, Bell Z, Swarbrick C, O'Hara G, Hoper M, Khosraviani K, Campbell FC, McGuigan J (2008) Folate supplementation in rats reduces gastro-esophageal reflux induced hyperproliferative esophagitis. Gastroenterology 134:A634
Michalek AM, Buck GM, Nasca PC, Freedman AN, Baptiste MS, Mahoney MC (1996) Gravid health status, medication use, and risk of neuroblastoma. Am J Epidemiol 143:996–1001
Milne E, Royle JA, Miller M, Bower C, De Klerk NH, Bailey HD, Van Bockxmeer F, Attia J, Scott RJ, Norris MD, Haber M, Thompson JR, Fritschi L, Marshall GM, Armstrong BK (2010) Maternal folate and other vitamin supplementation during pregnancy and risk of acute lymphoblastic leukemia in the offspring. Int J Cancer 126:2690–2699
Missaoui N, Hmissa S, Dante R, Frappart L (2010) Global DNA methylation in precancerous and cancerous lesions of the uterine cervix. Asian Pac J Cancer Prev 11:1741–1744
Mokarram P, Naghibalhossaini F, Saberi Firoozi M, Hosseini SV, Izadpanah A, Salahi H, Malek-Hosseini SA, Talei A, Mojallal M (2008) Methylenetetrahydrofolate reductase C677T genotype affects promoter methylation of tumor-specific genes in sporadic colorectal cancer through an interaction with folate/vitamin B12 status. World J Gastroenterol 14:3662–3671
Molloy AM, Kirke PN, Troendle JF, Burke H, Sutton M, Brody LC, Scott JM, Mills JM (2009) Maternal vitamin B12 status and risk of neural tube defects in a population with high neural tube defect prevalence and no folic acid fortification. Pediatrics 123:917–923
Morillon M, Katula K (2008) Alterations in Wnt signaling in folate deficient cells. FASEB J 22
MRC Vitamin Study Research Group (1991) Prevention of neural tube defects: results of the Medical Research Council Vitamin Study. Lancet 338:131–137
Mu LN, Cao W, Zhang ZF, Yu SZ, Jiang QW, You NC, Lu QY, Zhou XF, Ding BG, Chang J, Chen CW, Wei GR, Cai L (2007) Polymorphisms of 5,10-methylenetetralydrofolate reductase (MTHFR), fruit and vegetable intake, and the risk of stomach cancer. Biomarkers 12:61–75
Nafee TM, Farrell WE, Carroll WD, Fryer AA, Ismail KMK (2008) Epigenetic control of fetal gene expression. BJOG 115:158–168
Nakai K, Mitomi H, Alkam Y, Arakawa A, Yao T, Tokuda E, Saito M, Kasumi F (2012) Predictive value of MGMT, hMLH1, hMSH2 and BRCA1 protein expression for pathological complete response to neoadjuvant chemotherapy in basal-like breast cancer patients. Cancer Chemother Pharmacol 69:923–930
Negri E, Lavecchia C, Franceschi S, Levi F, Parazzini F (1996) Intake of selected micronutrients and the risk of endometrial carcinoma. Cancer 77:917–923
Neuhouser ML, Nijhout HF, Gregory JF III, Reed MC, James SJ, Liu A, Shane B, Ulrich CM (2011) Mathematical modeling predicts the effect of folate deficiency and excess on cancer-related biomarkers. Cancer Epidemiol Biomarkers Prev 20:1912–1917
Nishisho I, Nakamura Y, Miyoshi Y, Miki Y, Ando H, Horii A, Koyama K, Utsunomiya J, Baba S, Hedge P, Markham A, Krush AJ, Petersen G, Hamilton SR, Nilbert MC, Levy DB, Bryan TM,

Preisinger AC, Smith KJ, Su LK, Kinzler KW, Vogelstein B (1991) Mutations of chromosome-5Q21 genes in FAP and colorectal-cancer patients. Science 253:665–669

Oaks BM, Dodd KW, Meinhold CL, Jiao L, Church TR, Stolzenberg-Solomon RZ (2010) Folate intake, post-folic acid grain fortification, and pancreatic cancer risk in the Prostate, Lung, Colorectal, and Ovarian Cancer Screening Trial. Am J Clin Nutr 91:449–455

Okabe K, Hayashi M, Yoshida I, Nishimura K, Fukushima N, Tsujiuchi T (2011) Distinct DNA methylation patterns of lysophosphatidic acid receptor genes during rat hepatocarcinogenesis induced by a choline-deficient L-amino acid-defined diet. Arch Toxicol 85:1303–1310

Olshan AF, Smith JC, Bondy ML, Neglia JP, Pollock BH (2002) Maternal vitamin use and reduced risk of neuroblastoma. Epidemiology 13:575–580

Otsuka S, Maegawa S, Takamura A, Kamitani H, Watanabe T, Oshimura M, Nanba E (2009) Aberrant promoter methylation and expression of the imprinted PEG3 gene in glioma. Proc Jpn Acad Ser B Phys Biol Sci 85:157–165

Paspatis GA, Karamanolis DG (1994) Folate supplementation and adenomatous colonic polyps. Dis Colon Rectum 37:1340–1341

Pathak S, Bhatla N, Singh N (2012) Cervical cancer pathogenesis is associated with one-carbon metabolism. Mol Cell Biochem 369:1–7

Pelengaris S, Khan M, Evan G (2002) c-MYC: more than just a matter of life and death. Nat Rev Cancer 2:764–776

Pelucchi C, Galeone C, Talamini R, Negri E, Parpinel M, Franceschi S, Montella M, La Vecchia C (2005a) Dietary folate and risk of prostate cancer in Italy. Cancer Epidemiol Biomarkers Prev 14:944–948

Pelucchi C, Mereghetti M, Talamini R, Negri E, Montella M, Ramazzotti V, Franceschi S, La Vecchia C (2005b) Dietary folate, alcohol consumption, and risk of ovarian cancer in an Italian case-control study. Cancer Epidemiol Biomarkers Prev 14:2056–2058

Persson EC, Schwartz LM, Park Y, Trabert B, Hollenbeck AR, Graubard BI, Freedman ND, McGlynn KA (2013) Alcohol consumption, folate intake, hepatocellular carcinoma, and liver disease mortality. Cancer Epidemiol Biomarkers Prev 22:415–421

Piyathilake CJ, Macaluso M, Brill I, Heimburger DC, Partridge EE (2007) Lower red blood cell folate enhances the HPV-16-associated risk of cervical intraepithelial neoplasia. Nutrition 23:203–210

Piyathilake CJ, Macaluso M, Alvarez RD, Bell WC, Heimburger DC, Partridge EE (2009) Lower risk of cervical intraepithelial neoplasia in women with high plasma folate and sufficient vitamin B12 in the post-folic acid fortification era. Cancer Prev Res 2:658–664

Piyathilake CJ, Macaluso M, Alvarez RD, Chen M, Badiga S, Siddiqui NR, Edberg JC, Partridge EE, Johanning GL (2011) A higher degree of LINE-1 methylation in peripheral blood mononuclear cells, a one-carbon nutrient related epigenetic alteration, is associated with a lower risk of developing cervical intraepithelial neoplasia. Nutrition 27:513–519

Pogribny IP, Ross SA, Wise C, Pogribna M, Jones EA, Tryndyak VP, James SJ, Dragan YP, Poirier LA (2006) Irreversible global DNA hypomethylation as a key step in hepatocarcinogenesis induced by dietary methyl deficiency. Mutat Res 593:80–87

Pogribny IP, Tryndyak VP, Muskhelishvili L, Rusyn I, Ross SA (2007) Methyl deficiency, alterations in global histone modifications, and carcinogenesis. J Nutr 137:216S–222S

Pogribny IP, Karpf AR, James SR, Melnyk S, Han T, Tryndyak VP (2008) Epigenetic alterations in the brains of Fisher 344 rats induced by long-term administration of folate/methyl-deficient diet. Brain Res 1237:25–34

Pogribny IP, Shpyleva SI, Muskhelishvili L, Bagnyukova TV, James SJ, Beland FA (2009a) Role of DNA damage and alterations in cytosine DNA methylation in rat liver carcinogenesis induced by a methyl-deficient diet. Mutat Res 669:56–62

Pogribny IP, Tryndyak VP, Bagnyukova TV, Melnyk S, Montgomery B, Ross SA, Latendresse JR, Rusyn I, Beland FA (2009b) Hepatic epigenetic phenotype predetermines individual susceptibility to hepatic steatosis in mice fed a lipogenic methyl-deficient diet. J Hepatol 51:176–186

Potter JD (2002) Methyl supply, methyl metabolizing enzymes and colorectal neoplasia. J Nutr 132:2410S–2412S

Pufulete M, Al-Ghnaniem R, Leather AJ, Appleby P, Gout S, Terry C, Emery PW, Sanders TA (2003) Folate status, genomic DNA hypomethylation, and risk of colorectal adenoma and cancer: a case control study. Gastroenterology 124:1240–1248

Pufulete M, Al-Ghnaniem R, Khushal A, Appleby P, Harris N, Gout S, Emery PW, Sanders TA (2005a) Effect of folic acid supplementation on genomic DNA methylation in patients with colorectal adenoma. Gut 54:648–653

Pufulete M, Al-Ghnaniem R, Rennie JA, Appleby P, Harris N, Gout S, Emery PW, Sanders TA (2005b) Influence of folate status on genomic DNA methylation in colonic mucosa of subjects without colorectal adenoma or cancer. Br J Cancer 92:838–842

Qin X, Peng Q, Chen Z, Deng Y, Huang S, Xu J, Li H, Li S, Zhao J (2013) The association between MTHFR gene polymorphisms and hepatocellular carcinoma risk: a meta-analysis. PLoS One 8:e56070

Quinlivan EP, Gregory JF 3rd (2003) The impact of food fortification on folic acid intake in Canada. Can J Public Health 94:154

Ragasudha PN, Thulaseedharan JV, Wesley R, Jayaprakash PG, Lalitha P, Pillai MR (2012) A case-control nutrigenomic study on the synergistic activity of folate and vitamin B12 in cervical cancer progression. Nutr Cancer 64:550–558

Rampersaud GC, Kauwell GP, Hutson AD, Cerda JJ, Bailey LB (2000) Genomic DNA methylation decreases in response to moderate folate depletion in elderly women. Am J Clin Nutr 72:998–1003

Ray JG, Goodman J, O'Mahoney PRA, Mamdani MM, Jiang D (2008) High rate of maternal vitamin B12 deficiency nearly a decade after Canadian folic acid flour fortification. QJM 101:475–477

Razzak AA, Oxentenko AS, Vierkant RA, Tillmans LS, Wang AH, Weisenberger DJ, Laird PW, Lynch CF, Anderson KE, French AJ, Haile RW, Harnack LJ, Potter JD, Slager SL, Smyrk TC, Thibodeau SN, Cerhan JR, Limburg PJ (2012) Associations between intake of folate and related micronutrients with molecularly defined colorectal cancer risks in the Iowa Women's Health Study. Nutr Cancer 64:899–910

Richman EL, Kenfield SA, Stampfer MJ, Giovannucci EL, Zeisel SH, Willett WC, Chan JM (2012) Choline intake and risk of lethal prostate cancer: incidence and survival. Am J Clin Nutr 96:855–863

Robertson KD, Wolffe AP (2000) DNA methylation in health and disease. Nat Rev Genet 1:11–19

Robien K (2005) Folate during antifolate chemotherapy: what we know... and do not know. Nutr Clin Pract 20:411–422

Roddam A, Allen N, Appleby P, Key T (2008) Endogenous sex hormones and prostate cancer: a collaborative analysis of 18 prospective studies. J Natl Cancer Inst 100:170–183

Rong N, Selhub J, Goldin BR, Rosenberg IH (1991) Bacterially synthesized folate in rat large-intestine is incorporated into host tissue folyl polyglutamates. J Nutr 121:1955–1959

Ropero S, Esteller M (2007) The role of histone deacetylases (HDACs) in human cancer. Mol Oncol 1:19–25

Rosenfeld CS (2010) Animal models to study environmental epigenetics. Biol Reprod 82:473–488

Ross JA, Blair CK, Olshan AF, Robison LL, Smith FO, Heerema NA, Roesler M (2005) Periconceptional vitamin use and leukemia risk in children with Down syndrome: a Children's Oncology Group study. Cancer 104:405–410

Ryan-Harshman M, Aldoori W (2008) Vitamin B12 and health. Can Fam Physician 54:536–541

Saavedra KP, Brebi PM, Roa JC (2012) Epigenetic alterations in preneoplastic and neoplastic lesions of the cervix. Clin Epigenetics 4:13

Salazar-Martinez E, Lazcano-Ponce EC, Lira-Lira GG, Escudero-De Los Rios P, Hernandez-Avila M (2002) Nutritional determinants of epithelial ovarian cancer risk: a case-control study in Mexico. Oncology 63:151–157

Sanjoaquin MA, Allen N, Couto E, Roddam AW, Key TJ (2005) Folate intake and colorectal cancer risk: a meta-analytical approach. Int J Cancer 113:825–828

Schernhammer E, Wolpin B, Rifai N, Cochrane B, Manson JA, Ma J, Giovannucci E, Thomson C, Stampfer MJ, Fuchs C (2007) Plasma folate, vitamin B6, vitamin B12, and homocysteine and pancreatic cancer risk in four large cohorts. Cancer Res 67:5553–5560

Schernhammer ES, Giovannucci E, Kawasaki T, Rosner B, Fuchs C, Ogino S (2009) Dietary folate, alcohol, and B vitamins in relation to LINE-1 hypomethylation in colon cancer. Gut 59(6):794–799

Schuz J, Weihkopf T, Kaatsch P (2007) Medication use during pregnancy and the risk of childhood cancer in the offspring. Eur J Pediatr 166:433–441

Seeber LM, Van Diest PJ (2012) Epigenetics in ovarian cancer. Methods Mol Biol 863:253–269

Shabbeer S, Williams SA, Simons BW, Herman JG, Carducci MA (2012) Progression of prostate carcinogenesis and dietary methyl donors: temporal dependence. Cancer Prev Res (Phila) 5:229–239

Shakur YA, Garriguet D, Corey P, O'Connor DL (2010) Folic acid fortification above mandated levels results in a low prevalence of folate inadequacy among Canadians. Am J Clin Nutr 92:818–825

Shane B (1995) Folate chemistry and metabolism. In: Bailey LB (ed) Folate in health and disease. Marcel Dekker, New York

Shannon J, Phoutrides E, Palma A, Farris P, Peters L, Forester A, Tillotson CJ, Garzotto M (2009) Folate intake and prostate cancer risk: a case-control study. Nutr Cancer 61:617–628

Shaw GM, Carmichael SL, Yang W, Selvin S, Schaffer DM (2004) Periconceptional dietary intake of choline and betaine and neural tube defects in offspring. Am J Epidemiol 160:102–109

Shaw GM, Finnell RH, Blom HJ, Carmichael SL, Vollset SE, Yang W, Ueland PM (2009) Choline and risk of neural tube defects in a folate-fortified population. Epidemiology 20:714–719

Shelnutt KP, Kauwell GP, Gregory JF 3rd, Maneval DR, Quinlivan EP, Theriaque DW, Henderson GN, Bailey LB (2004) Methylenetetrahydrofolate reductase 677C→T polymorphism affects DNA methylation in response to controlled folate intake in young women. J Nutr Biochem 15:554–560

Shimizu K, Onishi M, Sugata E, Sokuza Y, Mori C, Nishikawa T, Honoki K, Tsujiuchi T (2007) Disturbance of DNA methylation patterns in the early phase of hepatocarcinogenesis induced by a choline-deficient L-amino acid-defined diet in rats. Cancer Sci 98:1318–1322

Shrubsole MJ, Jin F, Dai Q, Shu XO, Potter JD, Hebert JR, Gao YT, Zheng W (2001) Dietary folate intake and breast cancer risk: results from the Shanghai Breast Cancer Study. Cancer Res 61:7136–7141

Shrubsole MJ, Shu XO, Li HL, Cai H, Yang G, Gao YT, Gao J, Zheng W (2011) Dietary B vitamin and methionine intakes and breast cancer risk among Chinese women. Am J Epidemiol 173:1171–1182

Shuangshoti S, Hourpai N, Pulmsuk U, Mutirangura A (2007) Line-1 hypomethylation in multistage carcinogenesis of the uterine cervix. Asian Pac J Cancer Prev 8:307–309

Sie KKY, Chen JM, Sohn KJ, Croxford R, Thompson LU, Kim YI (2009) Folic acid supplementation provided in utero and during lactation reduces the number of terminal end buds of the developing mammary glands in the offspring. Cancer Lett 280:72–77

Sie KK, Medline A, Van Weel J, Sohn KJ, Choi SW, Croxford R, Kim YI (2011) Effect of maternal and postweaning folic acid supplementation on colorectal cancer risk in the offspring. Gut 60(12):1687–1694

Sie KK, Li J, Ly A, Sohn KJ, Croxford R, Kim YI (2013) Effect of maternal and postweaning folic acid supplementation on global and gene-specific DNA methylation in the liver of the rat offspring. Mol Nutr Food Res 57(4):677–685

Silvera SAN, Jain M, Howe GR, Miller AB, Rohan TE (2006) Dietary folate consumption and risk of ovarian cancer: a prospective cohort study. Eur J Cancer Prev 15:511–515

Sinclair KD, Allegrucci C, Singh R, Gardner DS, Sebastian S, Bispham J, Thurston A, Huntley JF, Rees WD, Maloney CA, Lea RG, Craigon J, McEvoy TG, Young LE (2007) DNA methylation, insulin resistance, and blood pressure in offspring determined by maternal periconceptional B vitamin and methionine status. Proc Natl Acad Sci U S A 104:19351–19356

Skinner HG, Michaud DS, Giovannucci EL, Rimm EB, Stampfer MJ, Willett WC, Colditz GA, Fuchs CS (2004) A prospective study of folate intake and the risk of pancreatic cancer in men and women. Am J Epidemiol 160:248–258

Slattery ML, Curtin K, Sweeny C, Levin TR, Potter JD, Wolff RK, Albertsen H, Samowitz WS (2006) Diet and lifestyle factor associations with CpG island methylator phenotype and BRAF mutations in colon cancer. Int J Cancer 120:656–663

Smith AD, Kim YI, Refsum H (2008) Is folic acid good for everyone? Am J Clin Nutr 87:517–533

Sohn KJ, Stempak JM, Reid S, Shirwadkar S, Mason JB, Kim YI (2003) The effect of dietary folate on genomic and p53-specific DNA methylation in rat colon. Carcinogenesis 24:81–90

Song YQ, Manson JE, Lee IM, Cook NR, Paul L, Selhub J, Giovannucci E, Zhang SMM (2012) Effect of combined folic acid, vitamin B-6, and vitamin B-12 on colorectal adenoma. J Natl Cancer Inst 104:1562–1575

Song MA, Tiirikainen M, Kwee S, Okimoto G, Yu H, Wong LL (2013) Elucidating the landscape of aberrant DNA methylation in hepatocellular carcinoma. PLoS One 8:e55761

Steegers-Theunissen RP, Obermann-Borst SA, Kremer D, Lindemans J, Siebel C, Steegers EA, Slagboom PE, Heijmans BT (2009) Periconceptional maternal folic acid use of 400 microg per day is related to increased methylation of the IGF2 gene in the very young child. PLoS One 4:e7845

Stempak JM, Sohn KJ, Chiang EP, Shane B, Kim YI (2005) Cell and stage of transformation-specific effects of folate deficiency on methionine cycle intermediates and DNA methylation in an *in vitro* model. Carcinogenesis 26:981–990

Stevens VL, Rodriguez C, Pavluck AL, McCullough ML, Thun MJ, Calle EE (2006) Folate nutrition and prostate cancer incidence in a large cohort of US men. Am J Epidemiol 163:989–996

Stevens VL, McCullough ML, Sun JZ, Gapstur SM (2010) Folate and other one-carbon metabolism-related nutrients and risk of postmenopausal breast cancer in the Cancer Prevention Study II Nutrition Cohort. Am J Clin Nutr 91:1708–1715

Stevens VL, McCullough ML, Sun J, Jacobs EJ, Campbell PT, Gapstur SM (2011) High levels of folate from supplements and fortification are not associated with increased risk of colorectal cancer. Gastroenterology 141:98–105, 105.e1

Stolzenberg-Solomon RZ, Albanes D, Nieto FJ, Hartman TJ, Tangrea JA, Rautalahti M, Sehlub J, Virtamo J, Taylor PR (1999) Pancreatic cancer risk and nutrition-related methyl-group availability indicators in male smokers. J Natl Cancer Inst 91:535–541

Stolzenberg-Solomon RZ, Pietinen P, Barrett MJ, Taylor PR, Virtamo J, Albanes D (2001) Dietary and other methyl-group availability factors and pancreatic cancer risk in a cohort of male smokers. Am J Epidemiol 153:680–687

Tajima S, Suetake I (1998) Regulation and function of DNA methylation in vertebrates. J Biochem 123:993–999

Tao MH, Freudenheim JL (2010) DNA methylation in endometrial cancer. Epigenetics 5:491–498

Tao MH, Mason JB, Marian C, McCann SE, Platek ME, Millen A, Ambrosone C, Edge SB, Krishnan SS, Trevisan M, Shields PG, Freudenheim JL (2011) Promoter methylation of E-Cadherin, p16, and RAR-beta(2) genes in breast tumors and dietary intake of nutrients important in one-carbon metabolism. Nutr Cancer 63:1143–1150

Teegarden D, Romieu I, Lelievre SA (2012) Redefining the impact of nutrition on breast cancer incidence: is epigenetics involved? Nutr Res Rev 25:68–95

Thompson JR, Gerald PF, Willoughby ML, Armstrong BK (2001) Maternal folate supplementation in pregnancy and protection against acute lymphoblastic leukaemia in childhood: a case-control study. Lancet 358:1935–1940

Tong SY, Kim MK, Lee JK, Lee JM, Choi SW, Friso S, Song ES, Lee KB, Lee JP (2011) Common polymorphisms in methylenetetrahydrofolate reductase gene are associated with risks of cervical intraepithelial neoplasia and cervical cancer in women with low serum folate and vitamin B12. Cancer Causes Control 22:63–72

Toole JF, Malinow MR, Chambless LE, Spence JD, Pettigrew LC, Howard VJ, Sides EG, Wang CH, Stampfer M (2004) Lowering homocysteine in patients with ischemic stroke to prevent recurrent stroke, myocardial infarction, and death: the Vitamin Intervention for Stroke Prevention (VISP) randomized controlled trial. JAMA 291:565–575

Toyota M, Ahuja N, Ohe-Toyota M, Herman JG, Baylin SB, Issa JP (1999) CpG island methylator phenotype in colorectal cancer. Proc Natl Acad Sci U S A 96:8681–8686

Tsang V, Fry RC, Niculescu MD, Rager JE, Saunders J, Paul DS, Zeisel SH, Waalkes MP, Styblo M, Drobna Z (2012) The epigenetic effects of a high prenatal folate intake in male mouse fetuses exposed in utero to arsenic. Toxicol Appl Pharmacol 264:439–450

Tworoger SS, Hecht JL, Giovannucci E, Hankinson SE (2006) Intake of folate and related nutrients in relation to risk of epithelial ovarian cancer. Am J Epidemiol 163:1101–1111

Uccella S, Cha SS, Melton LJ 3rd, Bergstralh EJ, Boardman LA, Keeney GL, Podratz KC, Ciancio FF, Mariani A (2011a) Risk factors for developing multiple malignancies in patients with endometrial cancer. Int J Gynecol Cancer 21:896–901

Uccella S, Mariani A, Wang AH, Vierkant RA, Robien K, Anderson KE, Cerhan JR (2011b) Dietary and supplemental intake of one-carbon nutrients and the risk of type I and type II endometrial cancer: a prospective cohort study. Ann Oncol 22:2129–2136

Ueland PM (2011) Choline and betaine in health and disease. J Inherit Metab Dis 34:3–15

Ulrich CM, Toriola AT, Koepl LM, Sandifer T, Poole EM, Duggan C, McTiernan A, Issa JP (2012) Metabolic, hormonal and immunological associations with global DNA methylation among postmenopausal women. Epigenetics 7:1020–1028

Van Den Donk M, Pellis L, Crott JW, Van Engeland M, Friederich P, Nagengast FM, Van Bergeijk JD, De Boer SY, Mason JB, Kok FJ, Keijer J, Kampman E (2007a) Folic acid and vitamin B-12 supplementation does not favorably influence uracil incorporation and promoter methylation in rectal mucosa DNA of subjects with previous colorectal adenomas. J Nutr 137:2114–2120

Van Den Donk M, Van Engeland M, Pellis L, Witteman BJ, Kok FJ, Keijer J, Kampman E (2007b) Dietary folate intake in combination with MTHFR C677T genotype and promoter methylation of tumor suppressor and DNA repair genes in sporadic colorectal adenomas. Cancer Epidemiol Biomarkers Prev 16:327–333

Van Engeland M, Weijenberg MP, Roemen GM, Brink M, De Bruine AP, Goldbohm RA, Van Den Brandt PA, Baylin SB, De Goeij AF, Herman JG (2003) Effects of dietary folate and alcohol intake on promoter methylation in sporadic colorectal cancer: the Netherlands cohort study on diet and cancer. Cancer Res 63:3133–3137

Van Guelpen B, Dahlin AM, Hultdin J, Eklof V, Johansson I, Henriksson ML, Cullman I, Hallmans G, Palmqvist R (2010) One-carbon metabolism and CpG island methylator phenotype status in incident colorectal cancer: a nested case-referent study. Cancer Causes Control 21:557–566

Van Wijk N, Watkins CJ, Bohlke M, Maher TJ, Hageman RJJ, Kamphuis P, Broersen LM, Wurtman RJ (2012) Plasma choline concentration varies with different dietary levels of vitamins B-6, B-12 and folic acid in rats maintained on choline-adequate diets. Br J Nutr 107:1408–1412

Vargas AJ, Thompson PA (2012) Diet and nutrient factors in colorectal cancer risk. Nutr Clin Pract 27:613–623

Velicer CM, Ulrich CM (2008) Vitamin and mineral supplement use among US adults after cancer diagnosis: a systematic review. J Clin Oncol 26:665–673

Verhage BAJ, Cremers P, Schouten LJ, Goldbohm RA, Van Den Brandt PA (2012) Dietary folate and folate vitamers and the risk of prostate cancer in The Netherlands Cohort Study. Cancer Causes Control 23:2003–2011

Vlajinac HD, Marinkovic JM, Ilic MD, Kocev NI (1997) Diet and prostate cancer: a case-control study. Eur J Cancer 33:101–107

Vollset SE, Igland J, Jenab M, Fredriksen A, Meyer K, Eussen S, Gjessing HK, Ueland PM, Pera G, Sala N, Agudo A, Capella G, Del Giudice G, Palli D, Boeing H, Weikert C, Bueno-De-Mesquita HB, Carneiro F, Pala V, Vineis P, Tumino R, Panico S, Berglund G, Manjer J, Stenling R, Hallmans G, Martinez C, Dorronsoro M, Barricarte A, Navarro C, Quiros JR, Allen N, Key TJ, Bingham S, Linseisen J, Kaaks R, Overvad K, Tjonneland A, Buchner FL, Peeters PHM, Numans ME, Clavel-Chapelon F, Boutron-Ruault M-C, Trichopoulou A, Lund E, Slimani N, Ferrari P, Riboli E, Gonzalez CA (2007) The association of gastric cancer risk with plasma folate, cobalamin, and Methylenetetrahydrofolate reductase polymorphisms in the European prospective investigation into cancer and nutrition. Cancer Epidemiol Biomarkers Prev 16:2416–2424

Vollset SE, Clarke R, Lewington S, Ebbing M, Halsey J, Lonn E, Armitage J, Manson JE, Hankey GJ, Spence JD, Galan P, Bonaa KH, Jamison R, Gaziano JM, Guarino P, Baron JA, Logan RF, Giovannucci EL, Den Heijer M, Ueland PM, Bennett D, Collins R, Peto R (2013) Effects of folic acid supplementation on overall and site-specific cancer incidence during the randomised trials: meta-analyses of data on 50 000 individuals. Lancet 381(9871):1029–1036

Wainfan E, Poirier LA (1992) Methyl-groups in carcinogenesis—effects on DNA methylation and gene-expression. Cancer Res 52:S2071–S2077

Wainfan E, Dizik M, Stender M, Christman JK (1989) Rapid appearance of hypomethylated DNA in livers of rats fed cancer-promoting, methyl-deficient diets. Cancer Res 49:4094–4097

Wallace K, Grau MV, Levine AJ, Shen L, Hamdan R, Chen X, Gui J, Haile RW, Barry EL, Ahnen D, McKeown-Eyssen G, Baron JA, Issa JP (2010) Association between folate levels and CpG Island hypermethylation in normal colorectal mucosa. Cancer Prev Res (Phila) 3:1552–1564

Wang J, Sasco AJ, Fu C, Xue H, Guo G, Hua Z, Zhou Q, Jiang Q, Xu B (2008) Aberrant DNA methylation of P16, MGMT, and hMLH1 genes in combination with MTHFR C677T genetic polymorphism in esophageal squamous cell carcinoma. Cancer Epidemiol Biomarkers Prev 17:118–125

Wasson GR, McGlynn AP, McNulty H, O'Reilly SL, McKelvey-Martin VJ, McKerr G, Strain JJ, Scott J, Downes CS (2006) Global DNA and p53 region-specific hypomethylation in human colonic cells is induced by folate depletion and reversed by folate supplementation. J Nutr 136:2748–2753

Watanabe F (2007) Vitamin B12 sources and bioavailability. Exp Biol Med 232:1266–1274

Waterland RA, Jirtle RL (2003) Transposable elements: targets for early nutritional effects on epigenetic gene regulation. Mol Cell Biol 23:5293–5300

Waterland RA, Dolinoy DC, Lin JR, Smith CA, Shi X, Tahiliani KG (2006a) Maternal methyl supplements increase offspring DNA methylation at Axin Fused. Genesis 44:401–406

Waterland RA, Lin JR, Smith CA, Jirtle RL (2006b) Post-weaning diet affects genomic imprinting at the insulin-like growth factor 2 (Igf2) locus. Hum Mol Genet 15:705–716

Waterland RA, Travisano M, Tahiliani KG (2007) Diet-induced hypermethylation at agouti viable yellow is not inherited transgenerationally through the female. FASEB J 21:3380–3385

Webb PM, Ibiebele TI, Hughes MC, Beesley J, Van Der Pols JC, Chen X, Nagle CM, Bain CJ, Chenevix-Trench G, Australian Cancer Study (Ovarian Cancer), Australian Ovarian Cancer Study Group (2011) Folate and related micronutrients, folate-metabolising genes and risk of ovarian cancer. Eur J Clin Nutr 65:1133–1140

Weinstein SJ, Hartman TJ, Stolzenberg-Solomon R, Pietinen P, Barrett MJ, Taylor PR, Virtamo J, Albanes D (2003) Null association between prostate cancer and serum folate, vitamin B(6), vitamin B(12), and homocysteine. Cancer Epidemiol Biomarkers Prev 12:1271–1272

Weinstein SJ, Stolzenberg-Solomon R, Pietinen P, Taylor PR, Virtamo J, Albanes D (2006) Dietary factors of one-carbon metabolism and prostate cancer risk. Am J Clin Nutr 84:929–935

Weinstein SJ, Albanes D, Selhub J, Graubard B, Lim U, Taylor PR, Virtamo J, Stolzenberg-Solomon R (2008) One-carbon metabolism biomarkers and risk of colon and rectal cancers. Cancer Epidemiol Biomarkers Prev 17:3233–3240

Welzel TM, Katki HA, Sakoda LC, Evans AA, London WT, Chen G, O'Broin S, Shen FM, Lin WY, McGlynn KA (2007) Blood folate levels and risk of liver damage and hepatocellular carcinoma in a prospective high-risk cohort. Cancer Epidemiol Biomarkers Prev 16:1279–1282

Wen W, Shu XO, Potter JD, Severson RK, Buckley JD, Reaman GH, Robison LL (2002) Parental medication use and risk of childhood acute lymphoblastic leukemia. Cancer 95:1786–1794

Weng Y-R, Sun D-F, Fang J-Y, Gu W-Q, Zhu H-Y (2006) Folate levels in mucosal tissue but not methylenetetrahydrofolate reductase polymorphisms are associated with gastric carcinogenesis. World J Gastroenterol 12:7591–7597

Westgren M (2012) Prevention of ovarian cancer—let's do something. Acta Obstet Gynecol Scand 91:1009–1010

Wien TN, Pike E, Wisløff T, Staff A, Smeland S, Klemp M (2012) Cancer risk with folic acid supplements: a systematic review and meta-analysis. BMJ Open 2

Williams CD, Satia JA, Adair LS, Stevens J, Galanko J, Keku TO, Sandler RS (2010) Antioxidant and DNA methylation-related nutrients and risk of distal colorectal cancer. Cancer Causes Control 21:1171–1181

Wilson MJ, Shivapurkar N, Poirier LA (1984) Hypomethylation of hepatic nuclear-DNA in rats fed with a carcinogenic methyl-deficient diet. Biochem J 218:987–990

Wilson RD, Johnson JA, Wyatt P, Allen V, Gagnon A, Langlois S, Blight C, Audibert F, Desilets V, Brock JA, Koren G, Goh YI, Nguyen P, Kapur B (2007) Pre-conceptional vitamin/folic acid supplementation 2007: the use of folic acid in combination with a multivitamin supplement for the prevention of neural tube defects and other congenital anomalies. J Obstet Gynaecol Can 29:1003–1026

Wolff GL, Roberts DW, Morrissey RL, Greenman DL, Allen RR, Campbell WL, Bergman H, Nesnow S, Frith CH (1987) Tumorigenic responses to lindane in mice—potentiation by a dominant mutation. Carcinogenesis 8:1889–1897

Wolff GL, Kodell RL, Moore SR, Cooney CA (1998) Maternal epigenetics and methyl supplements affect agouti gene expression in Avy/a mice. FASEB J 12:949–957

Wu KN, Helzlsouer KJ, Comstock GW, Hoffman SC, Nadeau MR, Selhub J (1999) A prospective study on folate, B12, and pyridoxal 5′-phosphate (B6) and breast cancer. Cancer Epidemiol Biomarkers Prev 8:209–217

Xiang TX, Yuan Y, Li LL, Wang ZH, Dan LY, Chen Y, Ren GS, Tao Q (2013) Aberrant promoter CpG methylation and its translational applications in breast cancer. Chin J Cancer 32:12–20

Xiao SD, Meng XJ, Shi Y, Hu YB, Zhu SS, Wang CW (2002) Interventional study of high dose folic acid in gastric carcinogenesis in beagles. Gut 50:61–64

Xu XR, Chen J (2009) One-carbon metabolism and breast cancer: an epidemiological perspective. J Genet Genomics 36:203–214

Xu WH, Shrubsole MJ, Xiang YB, Cai QY, Zhao GM, Ruan ZX, Cheng JR, Zheng W, Shu XO (2007) Dietary folate intake, MTHFR genetic polymorphisms, and the risk of endometrial cancer among Chinese women. Cancer Epidemiol Biomarkers Prev 16:281–287

Xu XR, Gammon MD, Zeisel SH, Bradshaw PT, Wetmur JG, Teitelbaum SL, Neugut AI, Santella RM, Chen J (2009a) High intakes of choline and betaine reduce breast cancer mortality in a population-based study. FASEB J 23:4022–4028

Xu XR, Gammon MD, Zhang YJ, Bestor TH, Zeisel SH, Wetmur JG, Wallenstein S, Bradshaw PT, Garbowski G, Teitelbaum SL, Neugut AI, Santella RM, Chen J (2009b) BRCA1 promoter methylation is associated with increased mortality among women with breast cancer. Breast Cancer Res Treat 115:397–404

Xu X, Gammon MD, Jefferson E, Zhang Y, Cho YH, Wetmur JG, Teitelbaum SL, Bradshaw PT, Terry MB, Garbowski G, Hibshoosh H, Neugut AI, Santella RM, Chen J (2011) The influence of one-carbon metabolism on gene promoter methylation in a population-based breast cancer study. Epigenetics 6:1276–1283

Xu XR, Gammon MD, Hernandez-Vargas H, Herceg Z, Wetmur JG, Teitelbaum SL, Bradshaw PT, Neugut AI, Santella RM, Chen J (2012) DNA methylation in peripheral blood measured by LUMA is associated with breast cancer in a population-based study. FASEB J 26:2657–2666

Yang K, Kurihara N, Fan K, Newmark H, Rigas B, Bancroft L, Corner G, Livote E, Lesser M, Edelmann W, Velcich A, Lipkin M, Augenlicht L (2008) Dietary induction of colonic tumors in a mouse model of sporadic colon cancer. Cancer Res 68:7803–7810

Yen TT, Gill AM, Frigeri LG, Barsh GS, Wolff GL (1994) Obesity, diabetes, and neoplasia in yellow A(vy)/-mice—ectopic expression of the agouti gene. FASEB J 8:479–488

Yetley EA, Coates PM, Johnson CL (2011a) Overview of a roundtable on NHANES monitoring of biomarkers of folate and vitamin B-12 status: measurement procedure issues. Am J Clin Nutr 94:297S–302S

Yetley EA, Pfeiffer CM, Phinney KW, Bailey RL, Blackmore S, Bock JL, Brody LC, Carmel R, Curtin LR, Durazo-Arvizu R, Eckfeldt JH, Green R, Gregory JF, Hoofnagle AN, Jacobsen DW, Jacques PF, Lacher DA, Molloy AM, Massaro J, Mills JL, Nexo E, Rader JI, Selhub J, Sempos C, Shane B, Stabler S, Stover P, Tamura T, Tedstone A, Thorpe SJ, Coates PM, Johnson CL, Picciano MF (2011b) Biomarkers of vitamin B-12 status in NHANES: a roundtable summary. Am J Clin Nutr 94:313S–321S

Yeung L, Yang QH, Berry RJ (2008) Contributions of Total Daily Intake of Folic Acid to Serum Folate Concentrations. JAMA 300:2486–2487

Ying J, Rahbar MH, Hallman DM, Hernandez LM, Spitz MR, Forman MR, Gorlova OY (2013) Associations between dietary intake of choline and betaine and lung cancer risk. PLoS One 8

Zeisel SH (2005) Choline, homocysteine, and pregnancy. Am J Clin Nutr 82:719–720

Zeisel SH, Da Costa KA (2009) Choline: an essential nutrient for public health. Nutr Rev 67:615–623

Zhang SM, Willett WC, Selhub J, Hunter DJ, Giovannucci EL, Holmes MD, Colditz GA, Hankinson SE (2003) Plasma folate, vitamin B6, vitamin B12, homocysteine, and risk of breast cancer. J Natl Cancer Inst 95:373–380

Zhang C-X, Ho SC, Chen Y-M, Lin F-Y, Fu J-H, Cheng S-Z (2011) Dietary folate, vitamin B6, vitamin B12 and methionine intake and the risk of breast cancer by oestrogen and progesterone receptor status. Br J Nutr 106:936–943

Zhang CX, Pan MX, Li B, Wang L, Mo XF, Chen YM, Lin FY, Ho SC (2013a) Choline and betaine intake is inversely associated with breast cancer risk: a two-stage case-control study in China. Cancer Sci 104:250–258

Zhang X-H, Ma J, Smith-Warner SA, Lee JE, Giovannucci E (2013b) Vitamin B6 and colorectal cancer: current evidence and future directions. World J Gastroenterol 19:1005

Zhao H, Li Q, Wang J, Su X, Ng KM, Qiu T, Shan L, Ling Y, Wang L, Cai J, Ying J (2012) Frequent epigenetic silencing of the folate-metabolising gene cystathionine-beta-synthase in gastrointestinal cancer. PLoS One 7

Ziegler RG, Jones CJ, Brinton LA, Norman SA, Mallin K, Levine RS, Lehman HF, Hamman RF, Trumble AC, Rosenthal JF et al (1991) Diet and the risk of in situ cervical cancer among white women in the United States. Cancer Causes Control 2:17–29

Zschabitz S, Cheng TYD, Neuhouser ML, Zheng YY, Ray RM, Miller JW, Song XL, Maneval DR, Beresford SAA, Lane D, Shikany JM, Ulrich CM (2013) B vitamin intakes and incidence of colorectal cancer: results from the Women's Health Initiative Observational Study cohort. Am J Clin Nutr 97:332–343

Chapter 12
Principles of the Warburg Effect and Cancer Cell Metabolism

Natalie Molino, K. Ververis, and Tom C. Karagiannis

Abstract The implication of cancer metabolism is gaining recent interest in cancer research after nearly nine decades since Dr. Otto Warburg first discovered the differing metabolic pathway of cancer cells. His early observations established that in contrast to normal cellular metabolism, most cancer cells rely on aerobic glycolysis. Although aerobic glycolysis is inefficient with respect to production of ATP it may provide a selective advantage for cancer cells producing glycolytic intermediates to support cell growth and division. It is becoming evident that genetic alterations associated with cancer have a role to play in aberrant cellular metabolism. In this chapter we discuss the current concepts of cancer metabolism and the relationship to tumor suppressor genes and oncogenes. The widespread recognition of the complex interplay between genetic alterations, cellular metabolism, and the tumor microenvironment could establish a framework for exploitable cancer therapies and potential targets of therapeutic intervention. In this chapter we outline these prospects.

Keywords Warburg effect • Cancer metabolism • Cellular proliferation • Aerobic glycolysis • Oncogenes • Tumor suppressor genes

Abbreviations

AKT	A serine/threonine protein kinase named AKT
AMPK	5′ AMP-activated protein kinase
ATP	Adenosine triphosphate

N. Molino, B.Sc. • K. Ververis, B.Sc. (Hons.) • T.C. Karagiannis, B.Sc. (Hons.), Ph.D. (✉)
Epigenomic Medicine, Baker IDI Heart and Diabetes Institute, The Alfred Medical Research and Education Precinct, 75 Commercial Road, Melbourne 3004, VIC, Australia

Department of Pathology, The University of Melbourne, Parkville, VIC, Australia
e-mail: n.molino@student.unimelb.edu.au; katherine.ververis@bakeridi.edu.au; tom.karagiannis@bakeridi.edu.au

N. Maulik and T. Karagiannis (eds.), *Molecular Mechanisms and Physiology of Disease: Implications for Epigenetics and Health*, DOI 10.1007/978-1-4939-0706-9_12,

EPO	Erythropoietin
GADD45A	Growth arrest and DNA-damage-inducible protein
GsH	Glutathione
HIF-1	Hypoxia-inducible factor 1
IGF-IBP-3	Insulin-like growth factor binding protein 3
LDHA	Lactate dehydrogenase A
mTOR	Mammalian target of rapamycin
NADPH	Nicotinamide adenine dinucleotide phosphate
p53	Tumor protein 53
PDGFB	Platelet-derived growth factor subunit B
PI3K	Phosphoinositide-3-kinase
PKM2	Pyruvate kinase isozyme M2
PTEN	Phosphatase and tensin homolog
RAS	Reticular activating system
SCO2	Synthesis of cytochrome c oxidase subunit 2
TCA	Tricarboxylic acid
TSC-2	Tuberous sclerosis complex 2
VEGF-A	Vascular endothelial growth factor A

12.1 Introduction

In the early 1920s Dr. Otto Warburg pioneered an investigation of cancer cell metabolism and made a prominent discovery. Warburg found that unlike normal adult differentiated cells that catabolize glucose through glycolysis and the citric acid cycle, most cancer cells, even in the presence of ample oxygen, rely on aerobic glycolysis (Warburg 1924, 1956). Whilst Warburg's observations were revolutionary, his findings were largely disregarded (Koppenol 2011). This was primarily due to inadequacies in molecular biology at the time. It was unclear why it would be advantageous for cancer cells to mediate a switch from high-energy mitochondrial oxidative phosphorylation to an energetically insufficient pathway (Hanahan and Weinberg 2011). Further, the biological mechanisms and signalling pathways to enable this switch were poorly understood (Levine and Puzio Kuter 2010). With our increased understanding of the link between genetic mutations and metabolic processes, attention has been revisited to the Warburg phenomenon.

Cancer is a multifaceted and complex disease with each tumor composed of heterogeneous cell types that have acquired biological capabilities to enable growth and metastatic dissemination. The hallmarks of cancer comprise six biological capabilities, namely to sustain proliferation, evade tumor suppressors, resist cell death, enable replicative immortality, induce angiogenesis, and activate invasion and metastasis (Hanahan and Weinberg 2011). Whilst these hallmarks of cancer continue to provide a solid framework for cancer biology, the recurrence of the

Warburg effect has added the reprogramming of energy metabolism as an emerging hallmark to this list (Hanahan and Weinberg 2011). The uncontrolled growth and proliferation of cancer cells represents the essence of neoplastic disease (Hanahan and Weinberg 2011). As this proliferation is facilitated by changes in metabolism, the reliance of cancer cells on specific metabolic pathways represents an ideal set of possible targets for cancer therapy. Cancer involves not only deregulated control of oncogenes and tumor suppressor genes but also adjustments of energy metabolism to fuel cell proliferation and tumorigenesis.

12.2 Metabolism of Glucose by Normal and Cancer Cells

Normal differentiated cells display a low division rate and predominantly metabolize glucose through glycolysis and oxidative phosphorylation (Levine and Puzio Kuter 2010). During glycolysis, glucose is transported into the cell by a facilitative transporter. Glycolysis then involves a ten-step reaction, which produces a series of intermediates catalyzed by different enzymes. At its completion, a glucose molecule is converted into two molecules of pyruvate, with a net gain of two molecules of adenosine triphosphate (ATP) and two molecules of NADH (Boyle 2008). Oxidative phosphorylation then involves an oxygen-dependent process coupling the oxidation of macromolecules and the electron transport chain with ATP synthesis (Cairns 2011) generating 36 ATP.

According to Schrödinger (1992) proliferating cells have a high requirement for energy and mass to replicate and hence must be adapted to facilitate the uptake and incorporation of nutrients into biomass. This metabolic switch places emphasis on producing intermediates for cell growth and division such as synthesis of substrates for membranes, nucleic acids, and proteins (Vander Heiden et al. 2009) instead of excess ATP production. Glucose and glutamine supply most of the carbon skeleton, nitrogen, free energy, and nicotinamide adenine dinucleotide phosphate (NADPH) to support cell growth and division of new cancer cells. Hence, the bulk of glucose cannot be committed to carbon catabolism for ATP production when the cell must instead accumulate biomass, produce glycolytic intermediates, and replicate DNA (Vander Heiden et al. 2009). Increased glycolysis to allow for the conversion of glycolytic intermediates into biosynthetic pathways has been seen in many rapidly dividing embryonic tissues and in the retina of the frog *Xenopus laevis*. Proliferating cells of the *Xenopus laevis* retina in vivo depend on glycogen to fuel aerobic glycolysis. These proliferating cells exhibited high lactate levels and decreased oxygen consumption regardless of adequate oxygen. It was only after terminal differentiation to non-proliferating cells that the transition from glycolysis to oxidative phosphorylation occurred (Fiske and Vander Heiden 2012; Agathocleous et al. 2012). This once again supports the hypothesis that large-scale biosynthetic programs and conversion to aerobic glycolysis are required for active cell proliferation (Hanahan and Weinberg 2011). Indeed, the marked increase of glucose uptake in cancer cells has been documented in many tumor types. This is most readily visualized by non-invasive positron emission tomography (PET) scans with a radiolabeled analog of glucose (^{18}F-fluorodeoxyglucose, FDG) (Hanahan and Weinberg 2011).

12.3 Building Blocks for Growing Cells

It is clear that there are fundamental differences in the central metabolic pathways operating in malignant cancer cells. The proliferating cancer cell provides increased intermediates for cell growth and accumulated biomass by favoring alternative pathways of glucose metabolism. This is accomplished, in part, by slowing the entry of pyruvate into the mitochondria, inhibiting the conversion of pyruvate to acetyl-CoA to begin the tricarboxylic acid (TCA) cycle, redirecting glucose metabolites into the pentose phosphate shunt, as well as promoting nucleic acid and lipid synthesis.

The delayed entry of pyruvate into mitochondria is facilitated by tyrosine kinase signalling which negatively regulates flux of the late step of glycolysis (Vander Heiden et al. 2009). This delay enables phosphorylated glycolytic intermediates to be used in anabolic synthesis as well as supports NADPH production (Christofk et al. 2008). The normal conversion of pyruvate to acetyl-CoA by the enzyme pyruvate dehydrogenase is inhibited by the activation of hypoxia-inducible factor 1 (HIF-1). HIF-1 activates pyruvate dehydrogenase kinase which subsequently shunts pyruvate to lactate formation (Kim 2006) and causes the redirection of glucose metabolites into the pentose phosphate shunt (Vander Heiden et al. 2009). The pentose phosphate pathway generates ribose for nucleic acid synthesis as well as supports NADPH production (Dang 2012). Whilst citrate is a key component in the TCA cycle, during proliferation of cancer cells it can be extruded into the cytosol to form the intermediates acetyl-CoA and oxaloacetate. Acetyl-CoA is utilized in the formation of lipid synthesis and oxaloacetate is used for amino acid synthesis (Dang 2012). However, it is not only glucose that provides the major substrates for the proliferating cell. Glutamine serves as a major bioenergetic substrate (Deberardinis and DeBerardinis 2010) and is a nitrogen donor for the synthesis of purines, pyrimidines, and nonessential amino acids (Levine and Puzio Kuter 2010) whilst also supporting NADPH production.

12.4 Key Regulators of Altered Metabolism in Cancer Cells

Activated oncogenes and loss of tumor suppressor genes have been found to be key regulators in altered metabolism of cancer cells. Detailed bioinformatics and high-throughput sequencing data suggest cancer-related driver mutations—oncogenes and downregulated tumor suppressor genes—largely affect the central metabolic pathways responsible for cancer cell proliferation (Parsons et al. 2008). Whilst key oncogenes have been identified as driver mutations, we still know relatively little regarding the signalling controls of these pathways. This is largely due to the complex nature of the signalling. Oncogenic signalling strength may be the product of a concession between maximum proliferative signals and avoidance of the antiproliferative tumor suppressor defenses. Alternatively, as seen in many human cancers, the cancer may adapt to high levels of oncogenic signalling by disabling apoptotic mechanisms (Hanahan and Weinberg 2011). Oncogenic mutations have diverse effects on multiple hallmark capabilities. Concerning the reprogramming of cell metabolism, oncogenic mutations result in the facilitated uptake of glucose and glucose metabolism. These mutations allow uptake of nutrients that may exceed the bioenergetic demands of the proliferating cancer cell (Vander Heiden et al. 2009). Conversely, loss of tumor suppressor genes results in loss of inhibition of cell growth, an increase in the Warburg effect, and an increase in the cancer phenotype (Levine and Puzio Kuter 2010). These findings present strong evidence that metabolic transformation may be a key step in the development of cancer. The mechanism of action of each oncogene and tumor suppressor gene driving cancer cell metabolism will be examined in more detail in this chapter.

12.4.1 PI3K

The phosphoinositide-3-kinase (PI3K) pathway is one of the most commonly altered signalling pathways in human cancers. The pathway can be altered by several mechanisms including a defect in the PI3K component, deregulation of the tumor suppressor gene that acts on PI3K, phosphatase and tensin homolog (PTEN), or aberrant signalling from tyrosine kinases (Wong et al. 2010; Cairns 2011). Mutations in the catalytic subunit of PI3K or in PTEN are now being detected in an array of human cancers and these mutations have been found to hyperactivate the PI3K circuitry. Once expressed PI3K activates many growth signals for the proliferating cell and its metabolism (Cairns 2011). PIK3 not only regulates glucose transporter expression but also enhances glucose capture by hexokinase, thereby rendering cells dependent on high levels of glucose flux (Vander Heiden et al. 2009).

12.4.2 AKT

PI3K is the upstream activator of AKT, a known driver of the glycolytic phenotype. The effect of constitutively active AKT is a common oncogenic perturbation in cancer cells and is often due to the hyper-amplification of PI3K or deletion of PTEN (Vivanco and Sawyers 2002). AKT has been found to increase the expression and membrane translocation of GLUT transporters and phosphorylates key glycolytic enzymes (Cairns 2011; Elstrom et al. 2004). The stimulation of key glycolytic enzymes, hexokinase and phosphofructokinase 2, renders cells dependent on aerobic glycolysis for continued growth and survival (Elstrom et al. 2004) and enhances glucose capture. AKT also stimulates ATP generation (Elstrom et al. 2004) ensuring that the cells have the bioenergetic scope to utilize the increased glycolytic flux. Although a major stimulator of glucose metabolism, AKT may also contribute to tumor cell malignancy. Recent studies suggest that AKT may be able to generate apoptotic resistance in vitro (Elstrom et al. 2004), driving an antiapoptotic phenotype.

12.4.3 mTOR

A well-studied downstream effector of AKT is mammalian target of rapamycin (mTOR). Regulating cellular metabolism mTOR directs available amino acids into protein synthesis via mRNA translation and ribosome biogenesis, whilst also coupling nutrient availability with growth signals (Elstrom et al. 2004). This aids to balance biomass growth with proliferation. Activated mTOR not only stimulates protein synthesis but also promotes lipid biosynthesis, thereby ensuring adequate nutrients to accumulate biomass and facilitate the proliferation of cells during tumorigenesis (Guertin and Sabatini 2007).

12.4.4 AMPK

5′ AMP-activated protein kinase (AMPK) is a crucial energy sensor in cellular homeostasis and responds to metabolic stress. As a potent inhibitor of both AKT and mTOR, AMPK couples energy status with growth signalling thereby reducing the proliferative potential of the cell to respond to growth signals. Several oncogenic mutations have been identified that suppress the AMPK signalling pathway, enabling cells to divide under abnormal energy status, supporting a shift towards glycolytic metabolism (Shackelford and Shaw 2009; Cairns 2011).

12.4.5 RAS

The oncogenic potential of reticular activating system (RAS) is complex, as mutations in RAS compromise its intrinsic negative feedback mechanism in cellular metabolism (Hanahan and Weinberg 2011). As a consequence, the mechanism that normally ensures transitory signal transmission is undermined. Subsequently, RAS upregulates the GLUT 1 transporter, increasing both glucose uptake and its retention inside the cell (Levine and Puzio Kuter 2010).

12.4.6 Myc

Myc is an essential oncogene for the activation of glutaminolysis as well as glycolysis and affects a number of diverse cellular processes involved in cell growth, proliferation, apoptosis, and metabolism (Boxer and Dang 2001). The primary role of Myc in cellular metabolism is to regulate the expression of several components of the protein synthetic machinery (Barna 2008). Its induction to stimulate genes involved in ribosomal biogenesis, including ribosomal proteins and ribosomal DNA, has proven to be essential for proliferating cancer cells (Barna 2008). According to Barna (2008), the ability of a cancer cell to deregulate Myc and increase protein synthesis directly augments cell size and accelerates cell cycle production. In fact, the ability to deregulate Myc activity is one of the most frequent oncogenic mutations underlying human cancers (Boxer and Dang 2001). These results confer that the oncogenic potential of Myc is dependent on its ability to regulate protein synthesis (Barna 2008). Molecular developments have implicated downstream Myc signalling events—decreased IRES-dependent translation of $Cdk11^{p58}$—to be an early sign of tumorigenesis (Barna 2008), supporting the oncogenic potential of Myc. Further, studies have identified an unrecognized molecular connection between aberrant control of protein synthesis downstream of Myc and chromosomal abnormalities (Barna 2008; Boxer and Dang 2001). This suggests that increased downstream protein synthesis of Myc is a rate-limiting determinant in cancer.

Myc not only increases the rate of protein synthesis but is also involved in parallel pathways that involve all phases of cell growth and metabolism. Myc has been implicated in the transition of cells from G0 into G1 phase, in which cells accumulate mass in preparation for DNA synthesis (Boxer and Dang 2001). Myc has also been found to interact with transcription factors driving proliferating cells to S phase for DNA replication and nucleotide synthesis (Dang 2012). In both in vivo and in vitro studies over-expression of Myc has resulted in increased lymphocytosis (Iritani and Eisenman 1999). With respect to glycolysis, highly expressed Myc has been found to increase the rate of some of the GLUT transporters, increasing glycolytic metabolism of the cell. Myc not only activates the expression of GLUT transporters but also collaboratively with HIF-1 regulates lactate dehydrogenase A (LDHA), phosphofructokinase, and enolase A (Boxer and Dang 2001; Cairns 2011).

Myc has also been proven to have a major role in driving glutamine metabolism. Not only does Myc-induced glutaminolysis support NADPH production, but it also produces antioxidants that are required for cell growth (Cairns 2011). Myc increases glutamine uptake by inducing the sIC5A1 and sIC7A1, glutamine transporters. After glutamine enters the cell it is converted into glutamate and has several fates. One fate is that it is converted into glutathione (GsH), a vital antioxidant controlling reactive oxygen species (ROS) levels and cellular homeostasis (Vaughn and Deshmukh 2008). It is becoming clear that perturbations in translational control of protein synthesis by Myc have a profound effect on gene expression, genomic stability, and cancer initiation.

12.5 The Tumor Microenvironment and Hypoxia

It has also been postulated that glycolytic metabolism arises as an adaptation to hypoxic conditions in the tumor microenvironment (Huang et al. 1998). The tumor microenvironment is highly irregular comprising abnormal tumor vasculature, excess lactate production, and modified metabolism. These cell conditions create both spatial and temporal heterogeneity (Cairns 2011). This heterogeneity across the tumor was explored by Thomlinson and Gray in a study in 1955 (Thomlinson and Gray 1955). Examining histological specimens of human lung tumors Thomlinson and Gray found that tumors were exposed to varying oxygen concentrations. From near-efficient oxygenation at the stroma, the oxygen gradient gradually decreased after 170 μm to near anoxia at the necrotic regions (Dang and Semenza 1999). It has since been shown that the hypoxic environment has significant effects on cellular processes (Bertout 2008). These hypoxic regions within solid tumors are largely due to atypical vasculature. Blood vessels in tumors are often the product of chronic angiogenesis driven by vascular endothelial growth factor A (VEGF-A). Unbalanced angiogenic signalling and upregulated VEGF-A expression orchestrate aberrant blood vessel formation, anomalous blood flow, leakiness, and an increased homeostatic survival of endothelial cells (Hanahan and Weinberg 2011). Other genes regulating the vasculature and haematopoiesis induced by hypoxia include IL1A, PDGFB (platelet-derived growth factor subunit B), EPO (erythropoietin), and GADD45A (growth arrest and DNA-damage-inducible protein) (Bertout 2008). Cancer cells must respond to this tumor cell stress and subsequently those that adapt to hypoxia convey a selective advantage for survival. Not only the hypoxic regions are associated with altered vasculature and metabolism, but they also display an increased resistance to radiotherapy and chemotherapy (Cairns 2011). Interestingly the multidrug resistance (MDR1) gene product P-glycoprotein is induced by hypoxia and is associated with increased tumor resistance to chemotherapeutics (Comerford et al. 2002). Upon examination, studies have identified a binding site on the MDR1 gene for HIF-1, the primary catalyst of hypoxia (Comerford et al. 2002). This indicates that adaptations to hypoxic conditions may represent a crucial step in tumorigenesis (Dang and Semenza 1999).

Adaptation to excess lactate production and an acidic environment may further drive the development of the glycolytic phenotype (Cairns 2011; Gatenby and Gillies 2004). Consequently, hypoxia and the tumor microenvironment pleiotropically act to enhance the glycolytic pathway of cancer cells to accumulate biomass and proliferate. The discovery of the transcription factor, HIF-1, was a landmark discovery that enabled scientists to understand how the cell sensed oxygen deprivation and consequently how it regulated gene transcription under hypoxia (Bertout 2008). Semenza and Wang were the first to biochemically purify and clone HIF-1 in 1993, heralding a new era of interest in hypoxia and the tumor microenvironment (Wang and Semenza 1993).

12.5.1 HIF-1

HIF-1 is a heterodimeric transcription factor that is rapidly induced by hypoxia. Oxygen-deprived cells inhibit the hydroxylation of HIF-1, leading to accumulation of HIF-1, translocation to the nucleus, and target gene activation (Bertout 2008). Hence, it is the absence of oxygen that stabilizes HIF-1. HIF-1 can activate the transcription of >100 genes, influencing key components of proliferating cancer cells such as cell survival, energy metabolism, angiogenesis, and cell migration (Levine and Puzio Kuter 2010). Increasing energy metabolism and cell proliferation, HIF-1 encodes aldolase A, enolase 1, phosphofructokinase L, phosphoglycerate kinase 1, and pyruvate kinase M (Bertout 2008). All of these stimulate the glycolytic pathway and foster cell proliferation. Increasing the capacity for the cell to carry out glycolysis, HIF-1 has been found to regulate nine out of the ten enzymes that function in glycolysis (Semenza 2003; Kim 2006). Glycolysis is further promoted as high AKT and mTOR activities have downstream effects on HIF-1, upregulating its activity (Levine and Puzio Kuter 2010). HIF-1 further influences cell metabolism by transcriptionally regulating LDHA and pyruvate dehydrogenase which both shunt glucose from entering the mitochondria and thereby reduce the flow of glucose into oxidative phosphorylation (Semenza 2003).

HIF-1 is also associated with targeted expression of genes that are crucial to hypoxic responses, such as angiogenesis and erythropoiesis. For example, VEGF is a target gene of HIF-1. Alternate systemic responses to hypoxia include autophagy. It has been substantiated that hypoxia and induction of HIF-1 may be one of the key mechanisms that activate autophagy. Autophagy is a catabolic mechanism that enables cells to break down organelles, allowing catabolites to be recycled and thus used for energy metabolism and biomass synthesis and accumulation (Hanahan and Weinberg 2011). A recent study (Liao et al. 2007) has found that in addition to HIF-1's role in cellular metabolism and proliferation, HIF-1 may have a pivotal role in promoting metastatic potential. When HIF-1 was ablated in the mammary epithelium of mice there was decreased tumor growth, delayed tumor onset, and a significant decrease in metastases (Liao et al. 2007).

12.5.2 Pyruvate Kinase

The M2 isoform of pyruvate kinase is selectively advantageous for the metabolic requirements of a cancer cell. As the cancer cell must obtain biomass accumulation for the synthesis of lipids, proteins, and nucleotides, pyruvate kinase isozyme M2 (PKM2) aids this development by diverting pyruvate away from the mitochondria and creating a build-up of glycolytic intermediates (Cairns 2011). In cancer cells PKM1 may be replaced by the splice variant PKM2. The induction of PKM2 interacts with HIF-1 inducing an inactivated state (Vander Heiden et al. 2009; Christofk et al. 2008). Inhibition of PKM2 slows glycolysis, allowing a build-up of carbohydrate metabolites to enter the pentose phosphate pathway, the hexosamine pathway, uridine diphosphate (uDP)-glucose synthesis, and glycerol synthesis pathways (Cairns 2011). These pathways generate macromolecular biosynthetic precursors necessary to support tumorigenesis (Vander Heiden et al. 2009).

In addition to stimulating macromolecular biosynthesis, PKM2 also bolsters redox control that is pivotal in a cancer cell that has eliminated many of its apoptotic mechanisms regulating cellular ROS homeostasis as described below. A consequence of cellular metabolism is the production of ROS, which evoke an abundance of destructive downstream effects. Cells oppose the effects of excessive ROS by producing antioxidant molecules. Key antioxidant mechanisms include the molecules GsH and thioredoxin (TRX), both of which depend on the redox buffering capacity of NADPH (Hamanaka and Chandel 2011). NADPH contributes to proper redox control and detoxifies ROS. NADPH is a crucial source of reducing equivalents for both lipid synthesis and GsH protecting damage mediated by ROS. NADPH is produced as a result of the inhibition of PKM2. As PKM2 shunts glucose to the pentose phosphate pathway, the result is increased NADPH levels.

12.5.3 p53

Inactivation of p53 is a hallmark of many human cancers (Vousden and Lu 2002). Deregulated or eliminated p53 in many cancers leads to an increase in the Warburg effect and glycolysis (Levine and Puzio Kuter 2010). p53 possesses many crucial roles for cellular homeostasis. One such role is to regulate cell cycle damage. p53 receives inputs from stress and abnormality sensors within the cell's intracellular machinery, managing levels of nucleotide pools, degree of genomic damage, double-stranded DNA breaks, growth-promoting signals, and glucose and oxygenation levels (Hanahan and Weinberg 2011). If any of these are suboptimal for cellular homeostasis, p53 can mediate arrest of the cell cycle or induce apoptosis until the erroneous conditions have been stabilized. As p53 is deregulated in many cancer cells this critical damage sensor control pathway is obliterated from the apoptosis-circuitry. In addition to the role of p53 in cell cycle damage control, it also directly stimulates genes that inhibit glycolysis and induce senescence and apoptosis.

It is now increasingly clear that p53 is an important regulator of metabolism (Cairns 2011). Expression of TIGAR is activated by p53. TIGAR lowers fructose-2,6-biphosphate levels in cells, shunting glucose into the pentose phosphate pathway and thus decreasing glycolysis (Vander Heiden et al. 2009). The decrease of intracellular ROS levels is thought to decrease sensitivity to p53 and aid the ability of p53 to protect cells from genomic instability (Bensaad 2006). p53 also regulates metabolism by enhancing utilization of mitochondrial oxidative phosphorylation, transcribing the gene SCO2 (synthesis of cytochrome c oxidase subunit 2) which assembles into oxidative phosphorylation complexes (Levine and Puzio Kuter 2010). A key function of p53 is as a transcription factor (Vogelstein et al. 2000). Stress induction leads to the activation of a large number of target genes, four of which are PTEN, insulin-like growth factor binding protein 3 (IGF-1BP-3), tuberous sclerosis complex 2 (TSC-2), and AMPK, which function to inhibit the activities of PI3K, AKT, and mTOR. p53 fundamentally protects cells from the potential outgrowth of malignant alterations (Bensaad 2006) and the activities of p53 are designed to regulate and control aberrant cell growth and induce apoptosis if necessary.

12.6 Potential Therapeutics

In the metabolic milieu of the tumor environment many metabolic pathways have been identified as potential therapeutic targets and targeting cancer metabolism has emerged as an area of drug discovery (Jones 2012). Blocking or restoring these altered pathways could ultimately lead to new drug targets and novel approaches to antitumor treatments. Whilst there is an in-depth understanding of the biological mechanisms of cellular metabolism, the knowledge of how pathways are affiliated and regulated to facilitate cell proliferation is incomplete (Vander Heiden 2011). Hence research to discover the extent to which integrated signalling pathways play a role in metabolic tumorigenesis is a key therapeutic avenue.

One potential therapeutic target of tumor cell metabolism is related to the excessive nutrient demand that cancer cells exhibit. This increased requirement of cancer cells addicts them to downstream metabolic signalling events (Vander Heiden 2011). Glucose withdrawal is seen to induce cell death in the same manner as growth factor signalling withdrawal. As a result, blood glucose lowering drugs, such as the antidiabetic drug Metformin, are being explored as potential antitumorigenic drugs (Vander Heiden 2011). In a study conducted in diabetic patients, those that were treated with Metformin were less likely to develop cancer than those on other treatment regimes (Tennant et al. 2010).

One major metabolic target that has become a key interest for antitumor drugs is HIF-1. Targeting the actions of HIF-1 is an important research direction as HIF-1 regulates many factors that are pivotal for tumorigenesis, particularly, angiogenesis, cellular metabolism, and metastasis (Bertout 2008). A hypoxia-activated drug, tirapazamine, has been found to be promising, with its reduction in a hypoxic environment inducing double-stranded breaks in the DNA of cancer cells (Bertout 2008). However, as direct targeting of transcription factors and signalling molecules proves to be challenging for drug development, scientists are now looking towards strategies that may inhibit or block the target signalling pathways. Strategies to oppose HIF-1 accumulation include addressing transcription and translation of the gene. This approach could reduce the potent downstream effects of HIF-1 subsequently curbing elaborate angiogenesis, reducing the build-up of carbohydrate metabolites, and depreciating the high levels of glucose entering the cancer cell.

An additional strategy has been to target expression of key genes that are central for the oncogenic potential of HIF-1. A drug that has already been trialled and approved for clinical use is Bevacizumab. Bevacizumab is a monoclonal antibody against VEGF-A, the potent angiogenic gene involved in aberrant tumor angiogenesis (Bertout 2008; Ferrara et al. 2004). As mutations in RAS, PI3K, AKT, and mTOR have been found to upregulate HIF-1, drugs focusing on these metabolic pathways could also diminish the overall net effect of irregular cellular metabolism in cancer cells. Targeting the key areas of metabolism is thought to slow growth and ultimately alter the Warburg effect and its consequences. As a major effector downstream of PI3K, mTOR is a target of new drug treatments. mTOR inhibitors are increasingly being studied in preclinical and clinical trials and have been shown to have a promising effect on renal cell carcinoma (Tennant et al. 2010). However, more effective advances could focus on combining mTOR inhibitors with other signalling pathway inhibitors to maximize their antitumorigenic potential (Tennant et al. 2010).

Glutamine targeting is also becoming a key drug target designed to capitalize on the cells' dependence on cellular metabolism. Glutamine is the amino acid found at highest concentrations in the plasma and several studies have identified the dependence of the cancer cell on this bioenergetic substrate (Tennant et al. 2010). Phenylacetate is a drug that reduces the availability of glutamine in the blood and thereby has been thought to reduce the proliferative potential of cancer cells. Inhibiting lactate production or transport has also been proposed as a potential drug therapeutic. Inhibiting LDHA has been shown to slow the growth of xenograft tumors in mice (Vander Heiden 2011). Inhibiting the removal of lactate from the cell could also acidify the intracellular environmental, killing the tumor cells (Jones 2012). Whilst this represents a highly attractive drug target, potential negative side effects on muscle metabolism must also be avoided (Jones 2012).

Other metabolic stress conditions that are altered in cancer such as the importance of NADPH to control cellular redox homeostasis have been implicated as potential drug targets. Inhibiting NADPH production through the mitigation of the pentose phosphate pathway could lead to increased ROS levels. With decreased antioxidant reducing molecules, increasing ROS could lead to radical-mediated damage (Cairns 2011). Preclinical studies have shown that this could be a potential therapeutic avenue with drugs inhibiting G6PD, the enzyme that initiates the pentose phosphate pathway, demonstrating antitumorigenic effects (Cairns 2011).

12.7 Therapeutic Challenges

Whilst the direct targeting of specific molecular signalling pathways, oncogenes, and their downstream effects has been fundamental ambition of cancer research, it has been put forward that such a specificity of action may not always be beneficial (Hanahan and Weinberg 2011). In principle, a selective drug target is ideal to target cancer, producing only low side effects and lessened toxicity. However, in reality such a selective drug target may not be the exemplar solution. Many principle mechanisms of cancer, such as cellular metabolism, angiogenesis, cellular invasion, and metastasis, are regulated by partially redundant signalling pathways (Hanahan and Weinberg 2011). Consequently, inhibiting just one aspect of the signalling pathway may not completely oppose a principle mechanism. Switching off one aspect of a pathway may potentially allow cancer cells to strengthen their selective adaptability and oppose the therapeutic being treated. In certain preclinical studies, potent anti-angiogenic inhibitors were taken to oppose the crucial role of aberrant angiogenesis in the cancer cell (Azam et al. 2010; Ebos et al. 2009; Bergers and Hanahan 2008). Instead of seeing reduced tumor growth, the researchers identified a shift in dependence from angiogenesis to heightened invasion and metastasis. This evasive response of the cancer provided a selective and aggressive advantage, as invasion of nearby tissue enabled access to preexisting vasculature (Hanahan and Weinberg 2011). This adaptive response illustrates the complex pathways regulating cancer. In addition to adopting new selective pressures under specific targeted therapeutics, cancer cells can also develop acquired drug resistance to selective drug targets. In a study on the effects of B-RAF inhibition in human melanomas, it was demonstrated that there was an unprecedented 80 % reduction in tumor response following treatment (Nazarian et al. 2010). However, acquired drug resistance and recovery of ERK phosphorylation were observed in many patients. It was later observed that the melanoma escaped B-RAF targeting not through secondary drug mutations, but through receptor tyrosine kinase-mediated activation of alternative survival pathways (Nazarian et al. 2010). Highlighting that clinical responses to specific drug targets have often been transitory in nature suggests that additional combination therapeutics must be involved in targeting cancer (Hanahan and Weinberg 2011). Given that the network of signalling pathways must be limited for principle capabilities of tumorigenesis, it may become possible to target all of these additional pathways, thereby reducing the cancer cells' ability to adapt and confer resistance to treatment (Hanahan and Weinberg 2011).

12.8 Conclusion

As the fruitions of cancer research continue to provide new, exciting directions for drug developments it is clear that incorporating the principal mechanisms of cancer hallmarks and their biochemical networks in an integrated approach is fundamental. Whilst these advances in the regulation of signalling pathways and cellular metabolism remain at the forefront of cancer research development, epigenetic alterations have also been identified as significant for tumorigenesis. The components of these epigenetic mechanisms and their influence on cellular metabolism and the Warburg effect may provide new insights into the regulatory circuitry defining the cancer cell (Hanahan and Weinberg 2011). Furthermore the discovery of microRNAs may also provide unknown advances in the genetic sphere regarding cancer phenotypes. Nearly a century after Otto Warburg's pioneering studies, the relationship between aerobic glycolysis and cellular metabolism in cancer is as contemporary as ever before. Specifically the complex network regulating the hypoxic tumor environment and subsequent interaction with deregulated oncogene and tumor suppressor genes will definitely be a focal aspect for further research.

Acknowledgments The support of the Australian Institute of Nuclear Science and Engineering is acknowledged. T.C.K. was the recipient of AINSE awards. T.C.K. is a Future Fellow and Epigenomic Medicine Laboratory is supported by the Australian Research Council. This work was supported in part by the Victorian Government's Operational Infrastructure Support Program.

References

Agathocleous M, Love N, Randlett O, Harris J, Liu J, Murray A, Harris W (2012) Metabolic differentiation in the embryonic retina. Nat Cell Biol 14:859–864

Azam F, Mehta S, Harris AL (2010) Mechanisms of resistance to antiangiogenesis therapy. Eur J Cancer 46:1323–1332

Barna M (2008) Suppression of Myc oncogenic activity by ribosomal protein haploinsufficiency. Nature 456:971–975

Bensaad K (2006) TIGAR, a p53-inducible regulator of glycolysis and apoptosis. Cell 126: 107–120

Bergers G, Hanahan D (2008) Modes of resistance to anti-angiogenic therapy. Nat Rev Cancer 8:592–603

Bertout JA (2008) The impact of O2 availability on human cancer. Nat Rev Cancer 8:967–975

Boxer LM, Dang CV (2001) Translocations involving c-myc and c-myc function. Oncogene 20:5595–5610

Cairns RA (2011) Regulation of cancer cell metabolism. Nat Rev Cancer 11:85–95

Christofk HR, Vander Heiden MG, Wu N, Asara JM, Cantley LC (2008) Pyruvate kinase M2 is a phosphotyrosine-binding protein. Nature 452:181–186

Comerford K, Wallace T, Karhausen J, Louis N, Montalto M, Colgan S (2002) Hypoxia-inducible factor-1-dependent regulation of the multidrug resistance (MDR1) gene. Cancer Res 62:3387–3394

Dang C (2012) Links between metabolism and cancer. Genes Dev 26:877–890

Dang CV, Semenza GL (1999) Oncogenic alterations of metabolism. Trends Biochem Sci 24: 68–72

Deberardinis RJ, Cheng T (2010) Q's next: the diverse functions of glutamine in metabolism, cell biology and cancer. Oncogene 29:313–324

Ebos JM, Lee CR, Kerbel RS (2009) Tumor and host-mediated pathways of resistance and disease progression in response to antiangiogenic therapy. Clin Cancer Res 15:5020–5025

Elstrom R, Bauer D, Buzzai M, Karnauskas R, Harris M, Plas D, Zhuang H, Cinalli R, Alavi A, Rudin C, Thompson C (2004) Akt stimulates aerobic glycolysis in cancer cells. Cancer Res 64:3892–3899

Ferrara N, Hillan K, Gerber H-P, Novotny W (2004) Discovery and development of bevacizumab, an anti-VEGF antibody for treating cancer. Nat Rev Drug Discov 3:391–400

Fiske B, Vander Heiden M (2012) Seeing the Warburg effect in the developing retina. Nat Cell Biol 14:790–791

Gatenby R, Gillies R (2004) Why do cancers have high aerobic glycolysis? Nat Rev Cancer 4: 891–899

Guertin D, Sabatini D (2007) Defining the role of mTOR in cancer. Cancer Cell 12:9–22

Hamanaka R, Chandel N (2011) Cell biology. Warburg effect and redox balance. Science 334:1219–1220

Hanahan D, Weinberg R (2011) Hallmarks of cancer: the next generation. Cell 144:646–674

Huang LE, Gu J, Schau M, Bunn HF (1998) Regulation of hypoxia-inducible factor 1alpha is mediated by an O2-dependent degradation domain via the ubiquitin-proteasome pathway. Proc Natl Acad Sci U S A 95:7987–7992

Iritani BM, Eisenman RN (1999) c-Myc enhances protein synthesis and cell size during B lymphocyte development. Proc Natl Acad Sci U S A 96:13180–13185

Jones NP (2012) Targeting cancer metabolism—aiming at a tumour's sweet-spot. Drug Discov Today 17:232–241
Kim JW (2006) HIF-1-mediated expression of pyruvate dehydrogenase kinase: a metabolic switch required for cellular adaptation to hypoxia. Cell Metab 3:177–185
Koppenol WH (2011) Otto Warburg's contributions to current concepts of cancer metabolism. Nat Rev Cancer 11:325–337
Levine A, Puzio Kuter A (2010) The control of the metabolic switch in cancers by oncogenes and tumor suppressor genes. Science 330:1340–1344
Liao D, Corle C, Seagroves T, Johnson R (2007) Hypoxia-inducible factor-1alpha is a key regulator of metastasis in a transgenic model of cancer initiation and progression. Cancer Res 67:563–572
Nazarian R, Shi H, Wang Q, Kong X, Koya R, Lee H, Chen Z, Lee M-K, Attar N, Sazegar H, Chodon T, Nelson S, Mcarthur G, Sosman J, Ribas A, Lo R (2010) Melanomas acquire resistance to B-RAF(V600E) inhibition by RTK or N-RAS upregulation. Nature 468:973–977
Parsons DW, Jones S, Zhang X, Lin J, Leary R, Angenendt P, Mankoo P, Carter H, Siu IM, Gallia G, Olivi A, Mclendon R, Rasheed BA, Keir S, Nikolskaya T, Nikolsky Y, Busam D, Tekleab H, Diaz L, Hartigan J, Smith D, Strausberg R, Marie SKN, Shinjo SMO, Yan H, Riggins G, Bigner D, Karchin R, Papadopoulos N, Parmigiani G, Vogelstein B, Velculescu V, Kinzler K (2008) An integrated genomic analysis of human glioblastoma multiforme. Science 321: 1807–1812
Schrödinger E (1992) What is life? The physical aspect of the living cell; with, mind and matter; & autobiographical sketches/Erwin Schrödinger. Cambridge University Press, Cambridge
Semenza GL (2003) Targeting HIF-1 for cancer therapy. Nat Rev Cancer 3:721–732
Shackelford D, Shaw R (2009) The LKB1-AMPK pathway: metabolism and growth control in tumour suppression. Nat Rev Cancer 9:563–575
Tennant D, Durn R, Gottlieb E (2010) Targeting metabolic transformation for cancer therapy. Nat Rev Cancer 10:267–277
Thomlinson RH, Gray LH (1955) The histological structure of some human lung cancers and the possible implications for radiotherapy. Br J Cancer 9:539–549
Vander Heiden MG (2011) Targeting cancer metabolism: a therapeutic window opens. Nat Rev Drug Discov 10:671–684
Vander Heiden M, Cantley L, Thompson C (2009) Understanding the Warburg effect: the metabolic requirements of cell proliferation. Science 324:1029–1033
Vaughn A, Deshmukh M (2008) Glucose metabolism inhibits apoptosis in neurons and cancer cells by redox inactivation of cytochrome c. Nat Cell Biol 10:1477–1483
Vivanco I, Sawyers C (2002) The phosphatidylinositol 3-kinase AKT pathway in human cancer. Nat Rev Cancer 2:489–501
Vogelstein B, Lane D, Levine AJ (2000) Surfing the p53 network. Nature 408:307–310
Vousden K, Lu X (2002) Live or let die: the cell's response to p53. Nat Rev Cancer 2:594–604
Wang GL, Semenza GL (1993) Characterization of hypoxia-inducible factor 1 and regulation of DNA binding activity by hypoxia. J Biol Chem 268:21513–21518
Warburg O (1924) Über den Stoffwechsel der Carcinomzelle. Naturwissenschaften 12:1131
Warburg O (1956) On the origin of cancer cells. Science 123:309–314
Wong K-K, Engelman J, Cantley L (2010) Targeting the PI3K signaling pathway in cancer. Curr Opin Genet Dev 20:87–90

Chapter 13
Molecular Aspects of the Warburg Effect

Elba Balding, Katherine Ververis, and Tom C. Karagiannis

Abstract The Warburg effect is a quality of cancer cells which is so defining of them that it is considered an important emerging hallmark of disease. Discovered by Dr. Otto Warburg in the 1920s, it was not until the last decade that the importance of this phenomenon was more widely realised and exploited in medical research, and its future possibilities conceived, largely due to an increase in our understanding of cellular metabolism. The Warburg effect itself is an observed change in the metabolism of cancer cells, where they metabolise a much larger amount of glucose than normal cells, utilising aerobic glycolysis rather than oxidative phosphorylation. While aerobic glycolysis creates less ATP energy for the cell it creates metabolic products, which allows the tumour to increase its biomass, important for the growth of the tumour and its ability to metastasize. Although it is still unknown why this metabolic change occurs, it is driven, at least in part, by the actions of activated oncogenes, in particular HIF-1α, and suppression of tumour suppressor proteins, such as p53. This quality provides the basis for the cancer monitoring technique of positron emission tomography, and there are emerging drugs which take advantage of this change from normal cells for potential therapeutic benefits. For example, metabolic drugs such as the type II diabetes drug metformin are being investigated and trialled as tools to starve cancer cells of their large energy requirements. In this chapter we provide an outline of the molecular characteristics of the Warburg effect and discuss related potential therapeutic developments.

Keywords Warburg effect • Cancer metabolism • Mitochondria • Lactic acid • Aerobic glycolysis

---

E. Balding, B.Sc. • K. Ververis, B.Sc. (Hons.) • T.C. Karagiannis, B.Sc. (Hons.), Ph.D. (✉)
Epigenomic Medicine, Baker IDI Heart and Diabetes Institute, The Alfred Medical Research and Education Precinct, 75 Commercial Road, Melbourne 3004, VIC, Australia

Department of Pathology, The University of Melbourne, Parkville, VIC, Australia
e-mail: elba.balding@bakeridi.edu.au; katherine.ververis@bakeridi.edu.au; tom.karagiannis@bakeridi.edu.au

N. Maulik and T. Karagiannis (eds.), *Molecular Mechanisms and Physiology of Disease: Implications for Epigenetics and Health*, DOI 10.1007/978-1-4939-0706-9_13,

13.1 Introduction to the Warburg Effect

The Warburg effect is a metabolic condition observed in cancerous cells, which is so defining of their physiology that now it can be classified as a hallmark of cancer (Resendis-Antonio et al. 2010; Hsu and Sabatini 2008). Originally the findings were published in 1924 by the German biochemist Otto Warburg, when he noticed that cancer cells metabolised glucose at a far more rapid rate than healthy cells, and produced lactate as a metabolic by-product (Warburg 1924). This led him to discover that instead of using the typical method of metabolism, mitochondrial oxidative phosphorylation, cancer cells largely used aerobic glycolysis, even in the presence of abundant oxygen (Warburg 1924). Since early in his career, Otto Warburg had been interested in metabolism, respiration and the role of the consumption of oxygen in cellular expansion. Early in his career he investigated respiration in sea urchin eggs, where he noticed that at fertilisation, the cells had a rapid increase in the intake of oxygen and then rapid increase in division. He believed that this scenario may be similar in cancer cells (Koppenol et al. 2011). Later in his career, Otto Warburg published a paper in 1924 entitled "Über den Stoffwechsel der Carcinomzelle". In this paper he outlined the changes in metabolism he noted in tumour cells compared to healthy cells, importantly the change from relying mainly on oxidative phosphorylation to aerobic glycolysis (Warburg 1924). Warburg's research has only recently become of interest again, attributed largely to the discovery of the relationship between oncogenes and metabolism (Pedersen 2007). Apart from a few dedicated research groups, this discovery was largely ignored for many decades due to lack of knowledge in areas of metabolism and cancer cell biology (Levine and Puzio-Kuter 2010), but has recently been the focus of considerable work. This is especially so with the return of metabolism to the forefront of cancer research, for example, for its potential in tumour growth inhibition (Bayley and Devilee 2012), cancer prevention and treatment as well as its importance for combating multiple drug resistance (MDR) (Xu et al. 2005).

There is accumulating evidence that the Warburg effect may provide advantageous features for cancer cells. The first factor, which is theorised to be an advantageous quality, involves energy utilisation. Aerobic glycolysis produces far less energy compared to oxidative phosphorylation; however, instead of producing ATP energy, aerobic glycolysis produces metabolic intermediates for the production of essential cell products, such as nucleotides, fatty acids and lipids (Vander Heiden et al. 2009; DeBerardinis 2008). Therefore, the change in metabolic pathway leads to an increase in metabolic intermediates, which are used to increase the biomass, necessary for the tumour's expansion. This is important for the rapid growth of the cancer biomass (Vander Heiden et al. 2009). Proliferating cancer cells which have an adequate blood supply, as is often the case, have a continual supply of oxygen and nutrients, perhaps making it not necessary to have such an efficient ATP production method (Vander Heiden et al. 2009). The second hypothesis for an advantage of the Warburg effect for cancer is also related to detoxification from reactive oxygen species (ROS). ROS, such as superoxide, are important mediators in key cellular processes in cancerous (and other) cells such as adhesion, immune response, cell growth and differentiation (Boonstra and Post 2004; Droge 2002). The synthesis of many NADPH molecules through the change in metabolism may operate to detoxify cancer cells from ROS production.

There is no known reason for the change from oxidative phosphorylation to aerobic glycolysis, but there are some theories. One theory, which is contentious, is that aerobic glycolysis is, selected for, due to tumour hypoxia (Vander Heiden et al. 2009). This has been rebutted by two main arguments. The first being that even before they experience hypoxic conditions, tumour cells seem to use aerobic glycolytic metabolism (Vander Heiden et al. 2009). The second major argument is that even tumours which have an abundant access to oxygen, such as those in lung cancers, still exhibit aerobic glycolysis (Vander Heiden et al. 2009). It is more likely, then, that the hypothesis that the change in metabolism is caused by mutations in the oncogenes and tumour suppressor genes which drive the metabolic pathways to increase proliferation. Whether this is due to the reversion to embryonic modes, or due to evolutionary advances in cell growth for the cancer, is still unknown (Vander Heiden et al. 2009). It is noteworthy that the metabolic change to oxidative glycolysis utilised by cancer cells is not displayed by them alone, but also by developing cells. This may lead to an explanation of how the Warburg effect originated, and it has been hypothesised by some that it may be due, at least in part, to the reversion to foetal forms of metabolism.

13.2 Altered Metabolism and Biochemical Pathways

Instead of following the usual mitochondrial oxidative phosphorylation pathway from glucose, as in normal cells, cancer cells metabolise via aerobic glycolysis, even in the presence of abundant oxygen (Vander Heiden et al. 2009). This seems detrimental for the cell because it leads to the production of only four ATP molecules compared to the 36 molecules if undergoing oxidative phosphorylation. As discussed earlier, cancer cells have much larger macromolecule requirements than healthy cells, due to their rapid growth and frequent divisions (Vander Heiden et al. 2009). Aerobic glycolysis leads to the production of nucleotides, fatty acids, membrane lipids and proteins (Hsu and Sabatini 2008), as well as NADPH molecules to assist in detoxifying the cell from the large ROS levels caused by such rapid expansion (Hamanaka and Chandel 2011). This change in metabolism is stimulated by signalling pathways controlled by oncogenes and tumour suppressor genes, which promote the use of other metabolic side shunts (Levine and Puzio-Kuter 2010). This includes changes to GLUT transporters, the use of the pentose phosphate pathway (PPP) and glutaminolysis. One of the ways that cancer cells satisfy their energy needs is through the uptake of much larger amounts of glucose than healthy cells (Ganapathy et al. 2009). It is through the facilitated GLUT transporters that glucose is taken up into the cell (Locasale et al. 2009). There are 14 different GLUT transporters, of which GLUT 1, 2, 3, 4 and, the more recently discovered, GLUT 12 are related to the high uptake of glucose in cancer (Macheda 2005; Rogers et al. 2003). These transporters are over-expressed in cancer cells compared to healthy epithelial cells (Barron et al. 2012; Krzeslak et al. 2012b). However, some cancers express GLUT transporters which would not normally be expressed in that tissue type (Medina and Owen 2002). This may be due to the large energy requirements of the cancer leading to changes in GLUT expression, and expression may be impacted by other molecules such as lactate, which is known to increase GLUT 1 and GLUT 4 expression (Medina et al. 2002). Further, GLUT transporters are regulated by p53, which is frequently mutated in many cancers. Loss of functional p53 leads to unregulated glucose uptake into the cell (Levine and Puzio-Kuter 2010). Also the oncogenes, MYC and HIF-1, increase the transcription of certain GLUT transporters, facilitating glucose uptake and retention in the cancer cell (Levine and Puzio-Kuter 2010). Modifiers of GLUT transporters have been explored as anticancer agents because of the hypothesis that they may represent the rate controlling step of glycolysis, as indicated in hepatocarcinoma and HeLa cell lines (Rodríguez-Enríquez et al. 2009; Diaz-Ruiz et al. 2011).

An important metabolic pathway related to the Warburg effect, that is being actively explored, is the PPP. This pathway involves partially shunting glucose-6-phosphate for use in nucleotide synthesis, and NADPH for redox control and fatty acid synthesis (Vander Heiden et al. 2009). Production of nucleotides, fatty acids and amino acids is also due to the increased utilisation of glutamine by cancer cells. Similar to glucose, cancer cells uptake much larger levels of glutamine, which is involved in fuelling the tricarboxylic acid (TCA) cycle for energy, NADPH molecules and glycolytic intermediates (Dang 2012; Vander Heiden et al. 2009). The TCA cycle produces malate and citrate, both producing more NADPH as part of their conversions. Citrate is also converted to acetyl-CoA and oxaloacetate for fatty acid and amino acid synthesis. The pyruvate generated from these reactions may undergo reverse glycolysis and shunted to the PPP to produce more NADPH and nucleotides. Excess glutamate is converted to aspartate, which also contributes to nucleotide synthesis (Levine and Puzio-Kuter 2010). Further, there is support for the idea that the production of excess carbons as the by-product of lactate allows for the carbon to be taken up more rapidly into the biomass (Vander Heiden et al. 2009). It has also been observed that excess lactate enters the Cori cycle in the liver, where it is converted into glucose. There is a similar pathway for the conversion of the by-product alanine, from glutamine metabolism (Vander Heiden et al. 2009). TCA cycle enzymes have also been linked with the development of cancer. Mutated genes encoding for the enzymes IDH1 (isocitrate dehydrogenase 1), SDH (succinate dehydrogenase) and FH (fumarate hydratase) can lead to the development of cancer (Chen and Russo 2012). The enzyme lactate dehydrogenase (LDH) is involved in the conversion of both glutamine and glucose into lactate, and is under the effect of HIF-1α and *c*-MYC. The oncogene *c*-MYC induces LDH-A expression (Shim et al. 1997; Yeung et al. 2008). If LDH activity is inhibited it impairs the cells' ability to proliferate (Vander Heiden et al. 2009).

13.3 Oncogenes and Tumour Suppressor Genes

In most cancer cells, sustained aerobic glycolysis is associated with the activation of oncogenes and deactivation or mutation of tumour suppressors (Dang 2012), which in turn directly control the metabolic pathways and rate being followed in the proliferating cells (Vander Heiden et al. 2009). The rate of glycolysis can vary by more than a 100-fold (Levine and Puzio-Kuter 2010), and is usually linked to the action of HIF-1α and *c*-MYC, as they are key drivers in energy producing pathways (Zawacka-Pankau et al. 2011). The oncogenic pathways involving phosphoinositide 3-kinase (PI3K), AKT, mammalian target of rapamycin (mTor), hexokinase, *c*-MYC and HIF-1α all influence the metabolic change, as well as promote cell proliferation and growth (Zawacka-Pankau et al. 2011). The p53 tumour suppressor gene is also influential in the Warburg effect, and the VHL tumour suppressor gene is also important. HIF is a transcription factor which is stabilised under situations of hypoxia, activating numerous pathways by upregulating target genes which promote tumour growth. HIF is stabilised in conditions of hypoxia aided by the inhibition of the tumour suppressor VHL to bind with HIF-1α. When there is adequate oxygen HIF subunits are unstable and are swiftly degraded in the ubiquitin–proteasome pathway (Maxwell et al. 1999; Kaelin 2002; Harris 2002). The activation of HIF is quite different in areas of a tumour which differ in oxygen content (Harris 2002; Yijun et al. 2009). HIF-1α is most commonly found in human tumours comparative to normal healthy tissue (Harris 2002; Zhong et al. 1999). HIF activation is clinically significant and is associated with poor outcome, although whether HIF activation is the causal factor or a result of tumour aggression is as yet unknown. High AKT and mTOR activity leads to high HIF-1 activity, and HIF-1 is activated and stabilised in cancer by mitochondrial dysfunction (Levine and Puzio-Kuter 2010). In support of this data, multiple studies have shown that the knockdown of HIF-1α in tumour cells suppresses tumour growth (Gao et al. 2007; Li et al. 2006; Mendez et al. 2010).

HIF-1 upregulates the transcription of many genes; in fact 9 of the 10 enzymes that function in glycolysis are regulated by HIF-1 (Macheda 2005; Semenza 2003). HIF was first identified as a regulator of erythropoietin, a haemopoietic growth factor. One of the relevant roles of HIF-1 is its role in increasing the transcription of GLUT transporters 1 and 3 (Stubbs and Griffiths 2010; Levine and Puzio-Kuter 2010). As previously mentioned, increasing the transcription of GLUT transporters is critical in the high glucose uptake seen in cancer cells (Levine and Puzio-Kuter 2010; Adekola et al. 2012). Another key role HIF-1 plays in cancer is via encoding glycolytic enzymes. In fact HIF-1 encodes nearly all of the key glycolytic enzymes involved in the Warburg effect (Stubbs and Griffiths 2010; Semenza 2001; Maxwell et al. 2001). HIF has been shown to affect mitochondrial respiration through the action of PDK-1 and COX-4, which are downstream targets. There are two main forms of hexokinase that we are interested in for the Warburg effect, hexokinase 1 (HK1) and hexokinase 2 (HK2). HK2 expression is linked with increased malignancy, and worse overall survival rates, for example, in human glioblastoma multiforme (Wolf et al. 2011). In terms of the Warburg effect, it is linked with increased aerobic glycolysis rates and decreased sensitivity to radiation, temozolomide and other inducers of cell death. HK2 levels are often reflected in HIF-1α levels (Wolf et al. 2011).

The oncogene MYC encodes for an important transcription factor, *c*-MYC (Dang 2010). *c*-MYC is very important for the Warburg effect, and cancer proliferation, because of its role in regulating glucose metabolism and the cell cycle, as well as promoting glutamine catabolism, HK2 and PDK1 (Kroemer and Pouyssegur 2008; Dang 2010; Dang et al. 2008). The expression of the MYC oncogene is dysregulated in approximately 30 % of human cancers (Dang et al. 2008). It controls these functions through the regulation and importantly repression of many microRNAs (Dang 2010). *c*-MYC also induces LDH-A expression (Shim et al. 1997; Yeung et al. 2008). However, sustained elevated levels of *c*-MYC are directly linked to increased mitochondrial ROS, which can cause mitochondrial DNA mutations, contributing to mitochondrial dysfunction (Kim and Dang 2006; Vander Heiden et al. 2009). The AKT (protein kinase B) pathway has been shown to promote cell growth as well as have a role in the glycolytic pathways promoting glycolysis, and promoting glucose uptake by the cell (Yeung et al. 2008). It activates mTor which regulates functions such as cell growth, cycle progression and autophagy (Guertin and Sabatini 2007; Falasca 2010).

PI3K is one of the most important intracellular signalling pathway regulators (Falasca 2010). Important for the Warburg effect, it regulates the AKT/mTOR pathway, which is key in the increased rate of metabolism (Bayley and Devilee 2012). If PI3K is disrupted, it leads to the tumour having decreased glucose uptake, and then glucose withdrawal (Levine and Puzio-Kuter 2010). Glucose withdrawal shows similar physiological conditions to growth factor withdrawal in healthy tissues, and induces cell death in the tumour (Vander Heiden et al. 2009). There are two important isoforms of pyruvate kinase (which are oncogenic tyrosine kinases), PKM1 and PKM2 (Vander Heiden et al. 2009). The M2 isoform is present in rapidly proliferating foetal cells, and during development it is slowly replaced by the M1 isoform (Bayley and Devilee 2012; Chaneton and Gottlieb 2012; Yacovan et al. 2012). It has been discovered however that tumour cells re-express PKM2 (Hamanaka and Chandel 2011; Ferguson and Rathmell 2009; Luo and Semenza 2012). This supports the idea that the Warburg effect is essentially a reversion back to foetal rates of metabolism. The activated Ras oncogene promotes cell growth and differentiation, as well as the cancer cell's survival from apoptosis (Diaz-Ruiz et al. 2011). Ras promotes glycolysis by controlling MAPK and PI3K regulation of HIF-1α. It also promotes angiogenesis through its interaction with host-mediated immune responses. Ras also promotes invasive and metastatic behaviour in cancer cells through its role in the alteration of motility and cellular adhesion (Pylayeva-Gupta et al. 2011).

p53 is a prominent tumour suppressor in the Warburg effect due to its vast regulatory effect on the cell (Yeung et al. 2008). p53 is suppressed in many cancer cells, which generally has a large impact on the rapid growth of the cancer (Dang 2012). With respect to the Warburg effect, p53 inhibits glycolysis (Zawacka-Pankau et al. 2011). p53 directly activates the TIGAR gene, which shunts glycolysis through the PPP (Bayley and Devilee 2012). It also inhibits PGM, inhibiting the conversion of fructose 1,6-bisphosphate into pyruvate and the travel of glycolysis (Levine and

Puzio-Kuter 2010). p53 also represses GLUT transporters 1 and 4, inhibiting glycolysis through the inhibition of glucose uptake into the cell (Vander Heiden et al. 2009). p53 also activates SCO2 and GLS2, which enhances mitochondrial respiration efficiency via increasing the use of the TCA cycle (Dang 2012; Zhang et al. 2010). p53 also inhibits the action of many genes (such as PI3K, AKT/mTOR pathways and IGF) which results in the shutdown of cell growth, decrease in the Warburg effect and lower HIF levels, and thus a reversal of the development of cancer (Vander Heiden et al. 2009; Levine 1997). If the cell is too damaged, then p53 can also direct apoptosis (Levine and Puzio-Kuter 2010). This is evidence supporting the idea that inactivation or suppression of p53 may, either independently or cooperatively with other tumour suppressors or oncogenes, directly contribute to the Warburg effect (Kim and Dang 2006). The von Hippel–Lindau tumour suppressor protein interacts closely with HIF-1α. In environments with abundant oxygen, VHL binds to HIF-1α and targets it for protein degradation. However, in hypoxic environments HIF-1α and VHL protein binding is inhibited due to the absence of prolyl hydroxylation. This leads to its stabilisation, allowing it to increase the transcription of many genes as outlined in the above section (Stubbs and Griffiths 2010).

13.4 Mitochondrial Dysfunction

The mitochondria are not only the major place of cellular ATP production but also have an important role in regulation through their role in apoptotic regulation (Alirol and Martinou 2006). This role involves their position as a central checkpoint of apoptosis via integrating endogenous and exogenous cellular signals (Chiaradonna et al. 2012). It is believed that mutations in mitochondrial DNA can also contribute to tumourigenesis (Vander Heiden et al. 2009). An important mitochondrial alteration involves metabolic genes which change to work as cancer genes (Dang 2012). Mutations in FH and SDH disrupt the TCA cycle in the mitochondria, leading to the accumulation of fumarate or succinate, which can inhibit enzymes such as dioxygenases, which control the degradation of HIF proteins (Dang 2012). HIF is also elevated due to the activation of other oncogenes and enzymes in the dysfunctional mitochondria (Dang 2012). Elevation of HIF proteins is pro-oncogenic (Dang 2012). Until the identification of mutant TCA cycle enzymes associated with familial cancer syndromes, alterations of metabolic genes that could provide a direct genetic link to altered metabolism were not known (Dang 2012).

13.5 Practical Applications, Future Directions and Conclusions

The Warburg effect has been an underestimated discovery until the last 10 years, when its importance and potential for cancer treatment have been recognised. It may hold the key for utilising the link between metabolism and proliferation, in the context of tumour suppression. New advances in treatment methods can be forged by targeting the physiological changes exhibited in the Warburg effect to create new therapeutic and diagnostic tools. Some important areas of research include drugs targeting key points in the control of cellular metabolism, especially for aerobic glycolysis. This includes PKM2 and LDH, as well as the PI3K/AKT pathway (Vander Heiden et al. 2009). There is also research into the link between drugs already used to treat metabolic diseases and their potential role in cancer treatment. There have been clinical studies into the use of metformin, used currently in the treatment of diabetes type II, which have found links with positive benefits for cancer patient outcomes, as well as help with cancer prevention (Vander Heiden et al. 2009). Most of these drugs or strategies are used or have the potential to be used as combination drugs with currently used anticancer drugs, as is the case with metformin (Milane et al. 2011b; Wenger et al. 2011). Another area is in combating MDR, with evidence emerging that cancer treatment strategies may be less susceptible to MDR if they take advantage of the cancer cell's distinctive metabolic needs (Locasale et al. 2009).

One of the most important and widely used technologies that is directly linked to the Warburg effect is PET imaging. Radioactively tagged glucose analogues are being used to track the rate of glucose utilisation. The high rate of glucose intake in cancers leads to this imaging being successfully used to diagnose, monitor and determine the stage of many cancers. The most common glucose tracer used in the clinic is ^{18}F-fluoro-2-deoxy-D-glucose (FDG) (Kwee et al. 2011; Gallagher et al. 1978; Lucignani and Larson 2010). It is taken into the cell through GLUT transporters, and while it can be phosphorylated by hexokinase as with glucose, it cannot travel any further through the metabolic pathway, trapping it inside the cell. This allows the measurement of the accumulated FDG in tissues, with higher amounts being observed in areas of higher metabolism (Yanagawa et al. 2010; van Ginkel et al. 1996; Busk et al. 2008).

Glucose restriction has been explored in the recent years due to its key role in the Warburg effect, with the same information as that used to develop PET imaging but used in an entirely different way. Glucose restriction during chemotherapy has been shown to promote survival free from cancer in experiments with mice with certain cancer types (breast cancer, melanoma and neuroblastoma) by reducing the toxicity to healthy tissues and by promoting tumour growth inhibition (Adekola et al. 2012; Lee et al. 2012). The over-expression of GLUT transporters has also been explored as a possible diagnostic indicator for cancer. One study has found that the expression of GLUT 3 in the urine of postmenopausal women with bladder cancer can potentially be used diagnostically (Krzeslak et al. 2012a).

MDR occurs when a cancer becomes resilient to certain drugs, usually chemotherapeutic agents, and it is a significant challenge for patients who are treated for recurrent disease, such as many cancers, because the dosage levels and changes in usual chemotherapy courses cultivate acquired MDR (Milane et al. 2011b). One important factor, however, is that even MDR cancer cells are susceptible to the inhibition of aerobic glycolysis (Xu et al. 2005; Milane et al. 2011a). This is a very important area for therapeutic target for aggressive and resistant cancers. Due to this information, therapeutics are being trialled to target the metabolic requirements of cancer, to combat and potentially reverse the Warburg effect (Fiske 2012). In fact, many new drugs are at clinical trial stage, culminating 20 years of research in this area, such as agents that target PI3K and other downstream signalling molecules and targeting HIF and hypoxia-related changes in cancer cells (Dang 2012; Yeoa et al. 2004). An example of the Warburg effect's role in combating MDR is Taxol resistance. Taxol (paclitaxel) is one of the most successful treatments of breast cancer currently used; however, it has a high rate of patients developing drug resistance. It was discovered that LDH-A activity was increased in cancers with this Taxol resistance (Zhou et al. 2010). Oxamate is known to decrease LDH activity leading to re-sensitisation of the breast cancer to Taxol (Zhou et al. 2010).

Overall, the Warburg effect is a metabolic characteristic of cancer cells, which although has been known for nearly 90 years, many molecular aspects are still poorly understood. Further characterisation of the aetiology of the metabolic switch and the molecular characteristics will aid in the development of new anti-metabolic therapy for cancer. Following the resurrection of research in the field in the past decade, there are already new compounds undergoing preclinical and early phase clinical trials. Further, it is emerging that various nutrients may influence the metabolic predisposition of cancer and this may provide another means for intervention.

Acknowledgments The support of the Australian Institute of Nuclear Science and Engineering is acknowledged. T.C.K. was the recipient of AINSE awards. T.C.K. is a Future Fellow and Epigenomic Medicine Laboratory is supported by the Australian Research Council. This work was supported in part by the Victorian Government's Operational Infrastructure Support Program.

References

Adekola K, Rosen ST, Shanmugam M (2012) Glucose transporters in cancer metabolism. Curr Opin Oncol 24:650–654

Alirol E, Martinou JC (2006) Mitochondria and cancer: is there a morphological connection? Oncogene 25:4706–4716

Barron C, Tsiani E, Tsakiridis T (2012) Expression of the glucose transporters GLUT1, GLUT3, GLUT4 and GLUT12 in human cancer cells. BMC Proc 6:1

Bayley JP, Devilee P (2012) The Warburg effect in 2012. Curr Opin Oncol 24:62–67

Boonstra J, Post JA (2004) Molecular events associated with reactive oxygen species and cell cycle progression in mammalian cells. Gene 337:1–13

Busk M, Horsman MR, Jakobsen S, Bussink J, Van Der Kogel A, Overgaard J (2008) Cellular uptake of PET tracers of glucose metabolism and hypoxia and their linkage. Eur J Nucl Med Mol Imaging 35:2294–2303

Chaneton B, Gottlieb E (2012) Rocking cell metabolism: revised functions of the key glycolytic regulator PKM2 in cancer. Trends Biochem Sci 37:309–316

Chen J-Q, Russo J (2012) Dysregulation of glucose transport, glycolysis, TCA cycle and glutaminolysis by oncogenes and tumor suppressors in cancer cells. Biochim Biophys Acta 1826: 370–384

Chiaradonna F, Moresco RM, Airoldi C, Gaglio D, Palorini R, Nicotra F, Messa C, Alberghina L (2012) From cancer metabolism to new biomarkers and drug targets. Biotechnol Adv 30:30–51

Dang CV (2010) Rethinking the Warburg effect with Myc micromanaging glutamine metabolism. Cancer Res 70:859–862

Dang C (2012) Links between metabolism and cancer. Genes Dev 26:877–890
Dang CV, Kim J-W, Gao P, Yustein J (2008) The interplay between MYC and HIF in cancer. Nat Rev Cancer 8:51–56
Deberardinis RJ (2008) Is cancer a disease of abnormal cellular metabolism? New angles on an old idea. Genet Med 10:767–777
Diaz-Ruiz R, Rigoulet M, Devin A (2011) The Warburg and Crabtree effects: on the origin of cancer cell energy metabolism and of yeast glucose repression. Biochim Biophys Acta 1807:568–576
Droge W (2002) Free radicals in the physiological control of cell function. Physiol Rev 82:47–95
Falasca M (2010) PI3K/Akt signalling pathway specific inhibitors: a novel strategy to sensitize cancer cells to anti-cancer drugs. Curr Pharm Des 16:1410–1416
Ferguson E, Rathmell J (2009) New roles for pyruvate kinase M2: working out the Warburg effect. Trends Biochem Sci 33:359–362
Fiske VH (2012) Seeing the Warburg effect in the developing retina. Nat Cell Biol 14:790–791
Gallagher BM, Fowler JS, Gutterson NI, Macgregor RR, Wan CN, Wolf AP (1978) Metabolic trapping as a principle of oradiopharmaceutical design: some factors responsible for the biodistribution of [18F] 2-deoxy-2-fluoro-D-glucose. J Nucl Med 19:1154–1161
Ganapathy V, Thangaraju M, Prasad PD (2009) Nutrient transporters in cancer: relevance to Warburg hypothesis and beyond. Pharmacol Ther 121:29–40
Gao P, Zhang H, Dinavahi R, Li F, Xiang Y, Raman V et al (2007) HIF dependent antitumorigenic effect of antioxidants in vivo. Cancer Cell 12:230–238
Guertin DA, Sabatini DM (2007) Defining the role of mTOR in cancer. Cancer Cell 12:9–22
Hamanaka R, Chandel N (2011) Warburg effect and redox balance. Science 334:1219–1220
Harris A (2002) Hypoxia—a key regulatory factor in tumour growth. Nat Rev Cancer 2:38–47
Hsu PP, Sabatini DM (2008) Cancer cell metabolism: Warburg and beyond. Cell 134:703–707
Kaelin W (2002) Molecular basis of the VHL hereditary cancer syndrome. Nat Rev Cancer 2:673–682
Kim J, Dang C (2006) Cancer's molecular sweet tooth and the Warburg effect. Cancer Res 66:8927–8930
Koppenol WH, Bounds PL, Dang CV (2011) Otto Warburg's contributions to current concepts of cancer metabolism. Nat Rev Cancer 11:325–337
Kroemer G, Pouyssegur J (2008) Tumor cell metabolism: cancer's Achilles' heel. Cancer Cell 13:472–482
Krzeslak A, Jozwiak P, Forma E, Brys M, Wozniak P, Wikosz J, Lipinski M, Rozanski W (2012a) Diagnostic value of glucose transporter 1 and 3 (GLUT1 and GLUT3) mRNA level in postmenopausal women with urinary bladder cancer. Przegl Menopauzalny 3:178–182
Krzeslak A, Wojcik-Krowiranda K, Forma E, Jozwiak P, Romanowicz H, Bienkiewicz A, Brys M (2012b) Expression of GLUT1 and GLUT3 glucose transporters in endometrial and breast cancers. Pathol Oncol Res 18:721–728
Kwee T, Basu S, Saboury B, Ambrosini V, Torigian D, Alavi A (2011) A new dimension of FDG-PET interpretation: assessment of tumor biology. Eur J Nucl Med Mol Imaging 38: 1158–1170
Lee C, Raffaghello L, Brandhorst S, Safdie FM, Bianchi G, Martin-Montalvo A, Pistoia V, Wei M, Hwang S, Merlino A, Emionite L, De Cabo R, Longo VD (2012) Fasting cycles retard growth of tumors and sensitize a range of cancer cell types to chemotherapy. Sci Transl Med 4:124ra27
Levine A (1997) p53, the cellular gatekeeper for growth and division. Cell 88:323–331
Levine A, Puzio-Kuter A (2010) The control of the metabolic switch in cancers by oncogenes and tumour suppressor genes. Science 330:1340–1344
Li J, Shi M, Cao Y, Yuan W, Pang T, Li B, Sun Z, Chen L, Zhao RC (2006) Knockdown of hypoxia-inducible factor-1alpha in breast carcinoma MCF-7 cells results in reduced tumor growth and increased sensitivity to methotrexate. Biochem Biophys Res Commun 342: 1341–1351
Locasale J, Cantley L, Vander Heiden M (2009) Cancer's insatiable appetite. Nature 27:916–917

Lucignani G, Larson SM (2010) Doctor, what does my future hold? The prognostic value of FDG-PET in solid tumours. Eur J Nucl Med Mol Imaging 37:1032–1038
Luo W, Semenza G (2012) Emerging roles of PKM2 in cell metabolism and cancer progression. Trends Endocrinol Metab 23:560–566
Macheda MEA (2005) Molecular and cellular regulation of glucose transporter (GLUT) proteins in cancer. J Cell Physiol 202:654–662
Maxwell PH, Wiesener MS, Chang GW, Clifford SC, Vaux EC, Cockman ME et al (1999) The tumour suppressor protein VHL targets hypoxia-inducible factors for oxygen-dependent proteolysis. Nature 399:271–275
Maxwell PH, Pugh CW, Ratcliffe PJ (2001) Activation of the HIF pathway in cancer. Curr Opin Genet Dev 11:293–299
Medina RA, Owen GI (2002) Glucose transporters: expression, regulation and cancer. Biol Res 35:9–26
Medina RA, Southworth R, Fuller W, Garlick PB (2002) Lactate-induced translocation of GLUT1 and GLUT4 is not mediated by phosphatidylinositol-3-kinase pathway in the rat heart. Basic Res Cardiol 97:168–176
Mendez O, Zavadil J, Esencay M, Lukyanov Y, Santovasi D, Wang SC, Newcomb EW, Zagzag D (2010) Knock down of HIF-1alpha in glioma cells reduces migration in vitro and invasion in vivo and impairs their ability to form tumor spheres. Mol Cancer 9:133
Milane L, Duan Z, Amiji M (2011a) Role of hypoxia and glycolysis in the development of multi-drug resistance in human tumor cells and the establishment of an orthotopic multi-drug resistant tumor model in nude mice using hypoxic pre-conditioning. Cancer Cell Int 11:3
Milane L, Ganesh S, Shah S, Duan Z-F, Amiji M (2011b) Multi-modal strategies for overcoming tumor drug resistance: hypoxia, the Warburg effect, stem cells, and multifunctional nanotechnology. J Control Release 155:237–247
Pedersen PL (2007) The cancer cell's "power plants" as promising therapeutic targets: an overview. J Bioenerg Biomembr 39:1–12
Pylayeva-Gupta Y, Grabocka E, Bar-Sagi D (2011) RAS oncogenes: weaving a tumorigenic web. Nat Rev Cancer 11:761–774
Resendis-Antonio O, Checa A, Encarnación S (2010) Modeling core metabolism in cancer cells: surveying the topology underlying the Warburg effect. PLoS One 5:1–11
Rodríguez-Enríquez S, Marín-Hernández A, Gallardo-Pérez JC, Moreno-Sánchez R (2009) Kinetics of transport and phosphorylation of glucose in cancer cells. J Cell Physiol 221: 552–559
Rogers S, Docherty SE, Slavin JL, Henderson MA, Best JD (2003) Differential expression of GLUT12 in breast cancer and normal breast tissue. Cancer Lett 193:225–233
Semenza GL (2001) Hypoxia-inducible factor 1: oxygen homeostasis and disease pathophysiology. Trends Mol Med 7:345–350
Semenza G (2003) Targeting HIF-1 for cancer therapy. Nat Rev Cancer 3:721–732
Shim H, Dolde C, Lewis BC, Wu C-S, Dang G, Jungmann RA, Dalla-Favera R, Dang CV (1997) c-Myc transactivation of LDH-A: implications for tumor metabolism and growth. Proc Natl Acad Sci U S A 94:6658–6653
Stubbs M, Griffiths JR (2010) The altered metabolism of tumors: HIF-1 and its role in the Warburg effect. Adv Enzyme Regul 50:44–55
Van Ginkel RJ, Hoekstra HJ, Pruim J, Nieweg OE, Molenaar WM, Paans AM, Willemsen AT, Vaalburg W, Koops HS (1996) FDG-PET to evaluate response to hyperthermic isolated limb perfusion for locally advanced soft-tissue sarcoma. J Nucl Med 37:984–990
Vander Heiden MG, Cantley LC, Thompson CB (2009) Understanding the Warburg effect: the metabolic requirements of cell proliferation. Science 324:1029–1033
Warburg O (1924) Uber den Stoffwechsel der Carcinomzelle. Naturwissenschaften 12: 1131–1137
Wenger JB, Chun SY, Dang DYT, Luesch H, Dang LH (2011) Combination therapy targeting cancer metabolism. Med Hypotheses 76:169–172

Wolf A, Agnihotri S, Micallef J, Mukherjee J, Sabha N, Cairns R, Hawkins C, Guha A (2011) Hexokinase 2 is a key mediator of aerobic glycolysis and promotes tumor growth in human glioblastoma multiforme. J Exp Med 208:313–326

Xu R, Pelicano H, Zhou Y, Carew J, Feng L, Bhalla K, Keating M, Huang P (2005) Inhibition of glycolysis in cancer cells: a novel strategy to overcome drug resistance associated with mitochondrial respiratory defect and hypoxia. Cancer Res 65:613–621

Yacovan A, Ozeri R, Kehat T, Mirilashvili S, Sherman D, Aizikovich A, Shitrit A, Ben-Zeev E, Schutz N, Bohana-Kashtan O, Konson A, Behar V, Becker OM (2012) 1-(Sulfonyl)-5-(arylsulfonyl)indoline as activators of the tumor cell specific M2 isoform of pyruvate kinase. Bioorg Med Chem Lett 22:6460–6468

Yanagawa T, Watanabe H, Shinozaki T, Takagishi K (2010) Usefulness of FDG PET in primary bone tumors. Open Bone J 2:19–23

Yeoa E-J, Chunb Y-S, Parka J-W (2004) New anticancer strategies targeting HIF-1. Biochem Pharmacol 68:1061–1069

Yeung SJ, Pan J, Lee MH (2008) Roles of p53, MYC and HIF-1 in regulating glycolysis—the seventh hallmark of cancer. Cell Mol Life Sci 65:3981–3999

Yijun C, Cairns R, Papandreou I, Koong A, Denko NC (2009) Oxygen consumption can regulate the growth of tumors, a new perspective on the Warburg effect. PLoS One 4:1–9

Zawacka-Pankau J, Grinkevich VV, Hünten S, Nikulenkov F, Gluch A, Li H, Enge M, Kel A, Selivanova G (2011) Inhibition of glycolytic enzymes mediated by pharmacologically activated p53: targeting Warburg effect to fight cancer. J Biol Chem 286:41600–41615

Zhang X-D, Qin Z-H, Wang J (2010) The role of p53 in cell metabolism. Acta Pharmacol Sin 31:1208–1212

Zhong H, De Marzo AM, Laughner E, Lim M, Hilton DA, Zagzag D, Buechler P, Isaacs WB, Semenza GL, Simons JW (1999) Overexpression of hypoxia-inducible factor 1α in common human cancers and their metastases. Cancer Res 59:5830–5835

Zhou M, Zhao Y, Ding Y, Liu H, Liu Z, Fodstad O, Riker AI, Kamarajugadda S, Lu J, Owen LB, Ledoux SP, Tan M (2010) Warburg effect in chemosensitivity: targeting lactate dehydrogenase-A re-sensitizes taxol-resistant cancer cells to taxol. Mol Cancer 9:33

Chapter 14
Epigenetic Perturbations in the Context of the Multi-hit Hypothesis of Carcinogenesis

Francesca Migheli and Lucia Migliore

Abstract Since 1950, a large number of models tried to represent the tumorigenesis process; numerous theories have overlapped and have been changed or improved along the way. Recently the significant role of epigenetics has emerged. These dynamic modifications are strongly linked to genetic and environmental factors and are involved in cancer development. The chapter highlights the contribution of the epigenetics in the multi-hit process of carcinogenesis, focusing on tumors with the highest incidence and mortality. Moreover the complex interplay between genetics and epigenetics will be discussed, such as the link between *DNMT* mutations and aberrant promoter methylation in several critical genes. Lifestyle factors such as smoking, alcohol and fat consumption, and stress have definitely an important role in carcinogenesis, and may act through epigenetic mechanisms. However numerous questions remain still unresolved, and continuous experimental researches are required.

Keywords Epigenetics • Carcinogenesis • Multi-hit process

14.1 Carcinogenesis Models Toward Epigenetics

Several observations during the past years permitted to divide carcinogenesis in three main steps: initiation, promotion, and progression. However to identify the temporal sequence of the molecular events involved in tumor development could be very hard, considering the large amount of possibilities that the different tumor types could undertake. In 2010 Vineis et al. reported five models that, since 1950, "have roughly represented" tumorigenesis; however, these theories have overlapped during time and have been changed or ameliorated along the way (Vineis et al. 2010).

F. Migheli • L. Migliore (✉)
Department of Translational Research and New Technologies in Medicine and Surgery, Division of Medical Genetics, University of Pisa, Via S. Maria 55, 56126 Pisa, Italy
e-mail: lucia.migliore@med.unipi.it

N. Maulik and T. Karagiannis (eds.), *Molecular Mechanisms and Physiology of Disease: Implications for Epigenetics and Health*, DOI 10.1007/978-1-4939-0706-9_14,

The first models intended the multistage hypothesis as "few changes, each heritable when somatic cells divide in the tissues, needed to alter an ordinary epithelial cell into the progenitor of a carcinoma." Multistage models that gave importance on aging were developing, as well as the Armitage and Doll model (Armitage and Doll 1954). Moreover researchers were studying the effect of smoke on tumor progression and its probable involvement in the early stages (Doll and Peto 1978); other papers dealt with the interaction between some compounds such as polycyclic aromatic hydrocarbons (PAH) and macromolecules (DNA adducts), to evaluate irreversible alterations in the differentiation capacity of the target cells (Slaga 1984). Another model regarded the "two-hit" theory hypothesis developed by Alfred and Knudson (1971). Analyzing 48 cases of retinoblastoma and basing on published reports, Knudson formulated the hypothesis that retinoblastoma is caused by two mutations, each of which occurs at a rate of the order of 2×10^{-7} per year. "In the inherited form, one mutational event is inherited via the germinal cells and the second occurs in somatic cells. In the nonhereditary form, both mutations occur in somatic cells" (Alfred and Knudson 1971). The *RB1* gene was the first hereditary cancer gene and the first tumor suppressor gene (TSG) to be cloned; following this gene several hereditary cancer genes have been investigated including familial adenomatous polyposis (FAP) gene (*APC*) (Knudson 2005). The investigation about familial colorectal cancer has then led to the study of DNA mismatch repair (MMR) genes focusing on genomic instability. In effect mutations, e.g., in h*MLH1*, *MSH2* genes may cause the acceleration of tumor development. HNPCC or Lynch Syndrome is a common hereditary disorder caused by germline mutations of MMR genes. Together with the somatic loss of the other normal allele, MMR mutations lead to a "mutator phenotype," that means a high percent of mutations in repetitive sequences (Whitehouse et al. 1998). Fearon and Vogelstein (1990) proposed a model of colorectal carcinogenesis according to which cancer arises from the activation of oncogenes and the inactivation of TSGs; mutations in four or five genes need to form a malignant tumor and the total accumulation of changes, rather than their order, is important to determine the tumor biologic properties. According to their model, patients with FAP inherit a mutation on chromosome 5q, where resides the *APC* gene, that may be responsible of the hyperproliferative epithelium; the same might happen in patients without polyposis. Hypomethylation is present in adenomas and may cause aneuploidy with consequent loss of TSGs correlated with the colorectal cancer (CRC) progression (Fearon and Vogelstein 1990). Beyond the concept of global hypomethylation linked to genomic instability, the importance of the different possible roles of epigenetics in the multistage carcinogenesis was emerging. It was postulated that the inactivation of suppressor genes or activation of oncogenes could be caused by altered methylation.

It was becoming clear that methylation of cytosines in CpGs or CpNpGs was associated with a "gene-off" signal repression with implications in gene regulation and in the aetiology of the diseases (Clark et al. 1995). Moreover the researchers referred to chemical compounds which can influence the epigenotype of eukaryotic cells (MacPhee 1998). In carcinogenesis a great role to clonal expansion (selection) of cells was always given; that is, their capacity to acquire selective advantage over cells not having a particular mutation, causing loss of cell-cycle control, lack of response to external signals, and uncontrolled proliferation. This Darwinian model of carcinogenesis is not new, having been proposed by several authors since the 1970s (see review of Vineis et al. 2010). Darwinian paradigm could become a unifying theory that explains several biologic phenomena, giving also a great importance to the environment (both macro and micro) in selecting cells that have some acquired advantage.

Vineis and co-workers (2010) reported some examples of environmental factors that can influence clonal selection such as the influence of beta-carotene supplementation in smokers on lung cancer; in effect the mutated cells (after tobacco smoke exposure) could have a greater advantage because the agent beta-carotene suppresses the replication of normal cells but not of cells with "specific" mutations. A good example to unifying the different theories developed could be, according to our opinion, the relationship between folic acid and colon cancer. Folic acid metabolism is involved in both DNA synthesis and methylation. Restricted folate diet or SNPs in one-carbon metabolism genes might result in hypomethylation of repetitive DNA sequences, contributing to the origin of cancer cells by generating chromosomal instability, reactivation of transposable elements, and loss of imprinting; moreover, hypomethylation may activate proto-oncogenes. The misincorporation of uracil into human DNA, favored when thymidylate availability is restricted, might also increase the frequency of chromosome breakage. On the other hand, TSGs could undergo CpG island methylation, resulting in the inactivation of these protecting proteins (Beetstra et al. 2005; Coppedè 2014; Crider et al. 2012; Kim 2005).

In this way epigenetics contributes to integrate the previous models about carcinogenesis consisting in several dynamic modifications strongly linked to genetic and environmental factors, affecting together the clonal cell selection.

14.2 Focusing on Epigenetics

14.2.1 Epigenetic Marks and Their Role in Carcinogenesis

Epigenetics is "the transmitted inherited genome activity that does not depend on the naked DNA sequence" and consists of different chemical modifications as well as DNA methylation, histone modifications, chromatin remodelling, nucleosome positioning, and noncoding RNA (ncRNA) modulation. They in turn affect transcriptional gene activation/silencing (Baylin and Jones 2011; Esteller 2012; Sandoval and Esteller 2012). To date no cancer or other pathologies have been identified with only genetic or epigenetic background; but rather genetics and epigenetics could interact and influence themselves during carcinogenesis (van Engeland et al. 2011).

A good example of the interplay between genetic and epigenetic alterations is the early involvement of an epigenetic event such as the altered methylation of a DNA repair gene. MGMT removes carcinogen-induced O6-methylguanine adducts from DNA, resulting in G to A transition. Cancers with the *MGMT* gene precociously hypermethylated/silenced are associated in the late stages to mutations in genes such as p53 or KRAS (Baylin and Jones 2011; Ogino et al. 2007). Moreover loss of *MGMT* expression seems to interfere with the development of *PIK3CA* G>A mutation, interfering with cell growth in various cancers, such as CRC (Nosho et al. 2008). At the same time genetic alterations (mutations) of one of the DNA methyltransferases (DNMTs) or histone deacetylases (HDACs), etc. may cause epigenetic changes such as DNA methylation and histone modifications. *DNMT3A* mutations have been effectively associated to methylation changes in acute myeloid leukemia (Ley et al. 2010). In the same way, mutations of *CREBBP*, *EP300*, or *MEF2B* may affect expression of *MEF2* target genes influencing the acetylation of nucleosomes near these genes, and then playing a role in the development of non-Hodgkin lymphoma (Morin et al. 2011).

Hence both these genetic and epigenetic alterations can determine abnormal gene expression and/or genomic instability, influencing tumor development (Rodríguez-Paredes and Esteller 2011; You and Jones 2012).

Gene-specific methylation in h*MLH1* and *MGMT* has been associated with aging in some type of cancers (Menigatti et al. 2009; Tserga et al. 2012). Both genes are involved in DNA repair processes and their methylation might reflect an age-related decline of DNA repair capabilities. It is well known that global hypomethylation and CpG island hypermethylation are accumulated during aging and could contribute to tumor progression (Fraga et al. 2007). These changes might then depend on genetic, environmental, or "stochastic" factors occurring during senescence (Calvanese et al. 2009). During aging also oxidative stress and damage are enhanced due to an increase of reactive oxygen species (ROS) generation and tissue susceptibility to oxidative damage. By inducing cellular oxidative stress O'Hagan et al. showed that DNMTs and HDAC complexes are recruited at damaged DNA. The authors suggest that the delocalization of epigenetic key enzymes as consequence of cellular stress might be one cause of global or local epigenetic alterations of cancer cells (O'Hagan et al. 2011).

In cancer the increased epigenetic heterogeneity/expression variability could represent the capacity of tumor cells to adapt rapidly to changing environments, such as augmented oxygen with neovascularization or decreased oxygen with necrosis (Hansen et al. 2011).

14.2.2 DNA Methylation and Human Malignancies

Below is a list of tumors with the highest incidence and mortality, for which methylation studies have been reported.

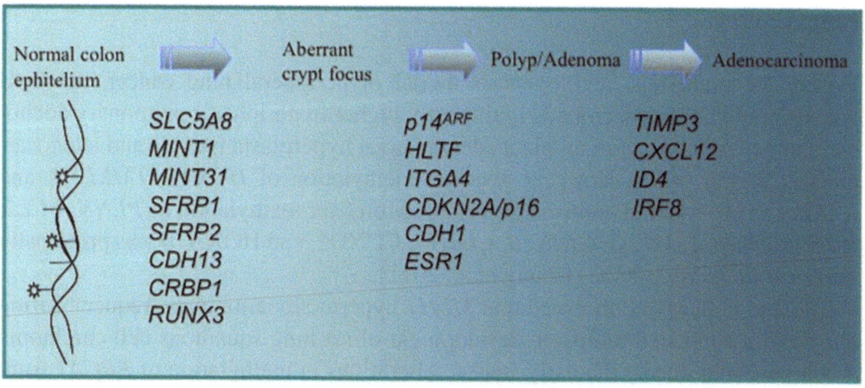

Fig. 14.1 Methylated genes during the colorectal cancer polyp–adenoma–carcinoma sequence. Adapted from Lao and Grady (2011)

Colon Cancer

In contrast to the widely accepted genetic model of CRC development, few CRCs were found to embody mutations in all of the *APC*, *K-ras*, and *p53* genes (Imai and Yamamoto 2008). Issa revised the model of CRC development and proposed three distinct multiple pathways, each based on different molecular mechanisms, instead of a linear progression of single events (Issa 2008; Pancione et al. 2012). In effect it is not so simple to define the way that the tumor undertakes since three molecular subtypes have been reported: MIN (or MSI, for "microsatellite instability"), CIN (for "chromosomal instability"), and CIMP (for "CpG island methylator phenotype"). The CIN phenotype is found in 85 % of sporadic CRCs and is characterized by aneuploidy, chromosomal rearrangements, and accumulation of mutations in oncogenes and TSGs such as *K-ras*, *APC*, and *p53*. The age-dependent accumulation of DNA hypomethylation is also associated to genomic instability in several types of cancers. The MSI phenotype is found in 15 % of sporadic CRCs and is associated with insertions and deletions particularly in repetitive sequences (microsatellites). Therefore CIMP consists in a gene silencing due to hypermethylation of CpG islands involving several CRC key genes (Centelles 2012; Grady and Carethers 2008; Imai and Yamamoto 2008). An important overlap between CIMP and MSI-H has been observed, since CIMP could explain silencing of the h*MLH1* gene in cancers with MSI-High. In the review of Imai and Yamamoto (2008) it is clarified the extensive interaction between CIMP-H and MSI in colorectal carcinogenesis. Numerous papers are investigating the DNA methylation's contribution in colorectal carcinogenesis; until now a large amount of genes, including cell-cycle control, DNA repair, growth and differentiation, and inflammation genes, has been identified during the multistage process of colorectal tumorigenesis (Migheli and Migliore 2012).

Figure 14.1 is a scheme, adapted from Lao and Grady (2011), representing the most common genes (and loci) found methylated during the steps of the colorectal cancer polyp–adenoma–carcinoma sequence.

Lung Cancer

During the multistep carcinogenesis model of peripheral lung cancer, the CpG islands methylation was found significantly higher in peripheral pulmonary adenocarcinoma (ADC) than in atypical adenomatous hyperplasia (AAH) and adenocarcinoma in situ (AIS). However aberrant methylation of *HOXA1*, *TMEFF2*, and *RARB* was observed in preinvasive lesions. Moreover methylation of *PENK*, *BCL2*, *RUNX3*, *DLEC1*, *MT1G*, *GRIN2B*, *CDH13*, CCND2, and HOXA10 was principally observed in invasive ADC (Chung et al. 2011).

Liu and colleagues observed that *MLH1* hypermethylation was frequent during the chemical-induced multistep development of rat lung squamous cell carcinoma and it appeared during the early stages. Alterations in methylation of *BRCA1* were found in the infiltrating carcinoma, showing lacked expression of the *BRCA1* protein. However several tumors with unmethylated *BRCA1* did not express *BRCA1*; that means other mechanisms, such as somatic mutations, may reflect the reduced expression of *BRCA1* (Liu et al. 2011a). A large number of genes found methylated in lung cancer are well documented in Liloglou and co-workers review (Liloglou et al. 2014).

Prostate Cancer

PTEN, *RB1*, and *TP53* genes, frequently altered in several human cancers, are not generally hypermethylated in prostate cancer (PC); however, point mutations or allelic loss is observed in the late stages. A large variety of genes involved in the regulation of the cell cycle and cell adhesion or DNA repair and hormone signalling genes present an altered methylation during the development of PC. For example, CpG islands hypermethylation of *GSTP1* or *MGMT* plays an important role in development of prostate carcinoma. Moreover *ERα*, *ERβ*, and *RASSF1A* genes such as other relevant TSGs were found hypermethylated in PC (see review of Albany et al. 2011). Methylation of *ER1* and *ER2* promoters was also observed in benign prostatic hyperplasia (BPH); however, this epigenetic modification was significantly higher in prostatic tumors than in BPH (Majumdar et al. 2011). It could mean that prostate carcinogenesis induces ER gene hypermethylation or vice versa. Hypomethylation might contribute to carcinogenesis involving the activation of oncogenes such as *c-MYC*, *K-ras*, or latent retrotransposons and in turn to affect chromosome instability. A strong association between *MYC* over-expression in PC and clinical progression was observed. At the same time the *PLAU* gene was found over-expressed in several PC tissues and able to promote metastasis in most human tumors (Albany et al. 2011; Bernard et al. 2003; Helenius et al. 2001; Li et al. 2005).

Moreover DNMTs activity has been found higher in primary cell lines derived from PC compared with the nonneoplastic counterparts, and was associated with altered methylation levels (Gravina et al. 2012).

Breast Cancer

Some alterations in DNA methylation do not occur in promoters, nor in CpG islands but in sequences up to 2 kb distant termed "CpG island shores" that strongly correlate with gene expression. In breast cancer, with respect to hypomethylation of Cav1 promoter CpG island, hypermethylation of CpG island shores was associated with low expression of the same gene (Rao et al. 2012). Some of the most frequently methylated genes involving cell-cycle regulation, apoptosis, DNA repair, cellular homeostasis, cell adhesion and invasion in breast cancer (BC) are summarized in Jovanovic and co-workers review (2010).

Esteller and co-workers, analyzing samples from BC tissues, observed *BRCA1* inactivation as consequence of one allele loss (generally deletions) and of the hypermethylation of the other, resulting in a decreased capacity of DNA repair (Catteau and Morris 2002; Esteller et al. 2000). Catteau and colleagues observed that since BRCA1 hypermethylation occurs preferentially in the same tumor types of BRCA1 families, it could not be a random event, but might result by the selection in specific tissues, as well as it happens for genetic mutations (Catteau and Morris 2002). However contrasting results have been observed; in effect a small study did not report *BRCA1* and *BRCA2* promoter hypermethylation as a frequent "second-hit" in tumors in *BRCA1* or *BRCA2* carriers (Dworkin et al. 2009).

Methylation of the *CST6* (cystatin M) gene occurs in noninvasive ductal carcinoma in situ (DCIS), suggesting its role in affecting disease progression and contributing to the invasive cellular phenotype of breast carcinoma (IBC) (Ai et al. 2006).

As previously described, global DNA hypomethylation was reported in the first studies about epigenetics and cancer; thereafter, authors focused their attention principally on gene-specific hypermethylation. Nevertheless genome-wide hypomethylation, generally identified as the loss of methylation at normally heavily methylated repeat elements including long interspersed nuclear element (LINE-1), has also a critical role during carcinogenesis. Global hypomethylation phenomenon has been observed at the earliest stages of epithelial carcinogenesis (Alvarez et al. 2011). Generally genome-wide hypomethylation occurs in the first stages of most cancers and might induce a cascade effect with direct implications in the determination of tumor progression (Kitkumthorn and Mutirangura 2011).

NcRNAs and Cancer

MicroRNAs (miRNAs) are small ncRNAs that regulate up to 30 % of human genes (Liloglou et al. 2014), by affecting mRNA stability. Several studies show evidence that altered miRNAs expression contributes to the initiation and progression of human cancers (Croce 2009; Esquela-Kerscher and Slack 2006).

Tumor cells reprogramming provides significant changes in miRNA profiles (Lopez-Serra and Esteller 2012). Chromosomal loci can be lost, giving rise to miRNA down-regulation such as in case of miR-204; it acts as a potent tumor

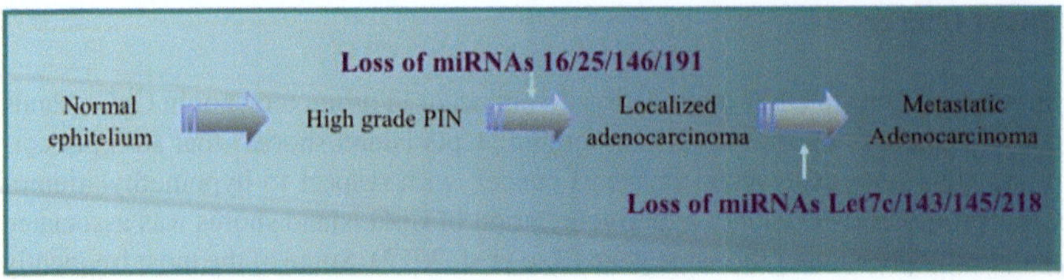

Fig. 14.2 Representation of microRNAs involved in the progression of prostate carcinogenesis. Adapted from Leite et al. (2013)

growth and metastasis suppressor and its lower expression appears in multiple cancers (Imam et al. 2012).

Moreover epigenetic mechanisms could play a critical role in the regulation of miRNA expression. Hypermethylation of miRNAs and their functional consequences such as cell proliferation, gene transcription alteration, and invasive capacity are discussed in Lopez-Serra and Esteller (2012) and Suzuki et al. (2012) reviews. Therefore some of the most deregulated miRNAs which have a central role in carcinogenesis are well documented in Harquail et al. review (2012).

miR-10b is highly expressed and correlates with metastatic breast cancer, inducing migration and invasion targeting HOXD10, a repressor of genes involved in cell migration and extracellular matrix remodelling (Ma et al. 2007). It was observed that the miR-204 over-expression significantly reduced the migratory and invasive capabilities of BS and ovarian cancer cells in vitro (Imam et al. 2012). Figure 14.2 (adapted from Leite et al. 2013) is an example of the PC progression focused on the contribution of miRNA in the multistep process from normal epithelium to metastasis.

Several papers reported a critical involvement of miRNAs in colorectal carcinogenesis. A pivotal role of miR-499-5p, whose expression frequently increases in CRC, has been noticed; in fact it seems to promote cell migration and invasion in human CRC cell lines, by affecting FOXO4 and PDCD4 mRNA expression (Liu et al. 2011b).

miR-31, miR-1, miR-9, miR-99a, miR-137, and miR-135b, that are involved in critical pathways of CRC, including APC/WNT signalling or cMYC, presented a differential expression in adenomas compared to normal colon tissue. Moreover miRNA changes observed in the early stages of tumor seem to have a critical role in both proficient MMR and defective MMR tumors. Since these two types of tumors present molecular differences between them, these results underline the involvement of shared pathways as well as differences at the miRNA level as consequence of altered methylation (Oberg et al. 2011). Therefore miR-218 seems to have an important role in CRC development through inhibiting cell proliferation and cycle progression and promoting apoptosis by down-regulating BMI-1 (He et al. 2012).

14.3 Histone Tail Modifications and Examples in Tumors

Histones are subject to several posttranslational covalent modifications such as phosphorylation, acetylation, methylation, ubiquitination, and citrulation (Nair and Kumar 2012). Histone modifications may change the accessibility of chromatin; for example, acetylation of histone tail consents the access of transcription factors, resulting in gene activation; instead deacetylation of histones causes a repressive chromatin state (Tyler and Kadonaga 1999). Aberrant patterns of histone modifications are another hallmark of cancer. The histone acetyltransferases (HATs) and the counterpart called HDACs are the enzymes involved in the process of acetylation. Alterations of HDAC and HAT proteins result in aberrant expression of TSGs (Glozak and Seto 2007). An exhaustive table with the most histone modification genes found altered in several types of cancer is available in Kanwal and Gupta (2012) review.

Expression of histone lysine demethylases (KDMs) is elevated in BC. Amplification/mutations of KDMs have been associated to histone methyl modifications and to many types of cancer. KDM1, through the demethylation of H3K4, seems to affect the expression of several genes involved in early-stage breast carcinogenesis (Paolicchi et al. 2013). The expression of histone demethylase JMJD2C is correlated with the expression of *GLUT1*, *LDHA*, *PDK1*, *LOX*, *LOXL2*, and *L1CAM* in BS tissues; the JMJD2C knockdown on the other hand seems to inhibit BS growth and metastasis to the lungs of mice (Luo et al. 2012). An aberrant pattern of H4K20 modifications was found during PC progression, compared to normal prostate tissue; particularly a general hypomethylation of H4K20me1 and H4K20me2 was observed (Behbahani et al. 2012).

Another paper reported that an increased H3K4me3 in PS tissues, compared with normal tissues, correlated with the expression of genes involved in cell proliferation and survival, therefore affecting tumorigenesis (Chen et al. 2010). The HATs CREBBP and EP300, the histone methyltransferase CARM1, and several HDACs involved in PC were expressed at significantly higher levels in metastatic lesions with respect to primary tumors suggesting their role in PC tumorigenesis (Bianco-Miotto et al. 2010). Aberrant histone modification marks implicated in CRC pathogenesis are well documented in Gargalionis et al. (2012).

14.4 Epigenetics and Environment Interplay

Besides genetic background, which confers susceptibility to certain diseases, epigenetic changes might represent the contributions of environmental insults occurring during life, starting even from prenatal period. Aging is one of the major risk factors for the development of cancers, since the environment can exert its influence during all the life-span. This also reflects the well-founded correlation between age and tumor incidence, and the consequent cancer heterogeneity strongly influenced by environmental factors.

Lifestyle factors such as smoking, alcohol consumption, excess exposure to sunlight, fat consumption, and stress may contribute to cancer development. Exposure to environmental agents therefore to carcinogens, infectious agents, and lifestyle causes epigenetic changes leading to transformation and consequently to cancer (Herceg and Vaissière 2011). Diet has a great influence on human health and numerous studies take in account its important role. An interesting work reported that a severe energy restriction during puberty is inversely correlated with CIMP tumor in older age, suggesting that the adolescence is a critical period involving epigenetic changes with consequent increased CRC risk in adults (Hughes et al. 2009).

Folate metabolism is fundamental for the synthesis of DNA and RNA precursors or for the conversion of homocysteine to methionine, which is then used to form the main DNA methylating agent or SAM. Low dietary folate intake (<200 μg/day) has also been associated with an increased frequency of hypomethylated long interspersed nucleotide element/LINE (a marker of genome-wide DNA methylation) repeats in human colon tumors (Schernhammer et al. 2010). Folic acid supplementation may protect from the development of colorectal cancer because of its critical role in maintaining DNA stability. However in the normal colorectum, folate deficiency appears to enhance, whereas folic acid supplementation suppresses, the development of CRC. In contrast, once aberrant crypt foci are established, folate deficiency inhibits the progression and induces regression of these established preneoplastic foci (Kim 2007, 2008). An interesting study showed that folate deficiency significantly increased hepatic OGG-1 and MGMT repair activity. Probably the up-regulation of these two proteins indicates the occurrence of DNA damage corroborating the hypothesis that increased DNA damage (including DNA strand breaks, uracil misincorporation, and oxidized bases) is a consequence of folate deficiency. Low folate intake significantly increased 8-oxo-7,8-dihydroguanine levels in DNA in lymphocytes from rats fed the folate-deficient diet. Although there were highly significant changes in *OGG-1* and *MGMT* expression in rat liver in response to folate depletion, no such effects were seen in colon, indicating that the ability of the liver to respond to folate deficiency is not shared by the colon. The colon cannot respond as the liver to damage and would therefore be more susceptible to the genotoxic effects induced by folate deficiency (Duthie et al. 2010). It was also observed a trend for the association between serum folate/vitamin B12 levels and gene promoter methylation: higher serum folate/vitamin B12 levels were strongly associated with promoter methylation of p16 and had an association trend with promoter methylation of *MLH1* and *MLH2* genes (Mokarram et al. 2009). The effect of intervention with folic acid on DNA methylation is thereby conflicting and highly dependent on initial folate status, level and duration of supplementation, tissues examined, stage of malignant transformation, and polymorphisms in folate metabolizing genes (Sie et al. 2011).

Air pollutants can influence epigenetic changes, including DNA methylation as well as up- or down-regulation of miRNAs (Jardim 2011). The effects of particulate matter (PM) exposure on Alu, LINE-1, and gene-specific methylation were examined in steel plant workers. Long-term exposure to PM10 was negatively associated with methylation in both Alu and LINE-1. Exposure to black carbon, a marker of traffic particles, was also associated with decreased DNA methylation in LINE-1. NOS2 promoter methylation was significantly lower in post-exposure blood samples compared to baseline (Madrigano et al. 2011; Tarantini et al. 2009).

Fetal exposure to maternal smoking during pregnancy is associated with increased DNA methylation in *AXL* and *PTPRO* and *IGF2* genes and, in adult lung cancer patients, quantity and duration of smoking are correlated with increased DNA methylation of *p16*, *MGMT*, and *DAPK* genes. Tobacco smoke might also induce epigenetic changes in esophageal squamous cell carcinoma (see review by Cortessis et al. 2012). Modifications due to exposures to smoking have been widely investigated and include DNA adducts, gene mutations, micronuclei, chromosome aberrations, and DNA strand breaks. DNMTs seem to bind DNA damage sites, giving rise to altered methylation patterns on these regions; it suggests a molecular mechanism for the generation of aberrant DNA methylation by exposure to chemicals such as those present in cigarettes (Terry et al. 2011). Alcohol consumption was found to be a risk factor for cancer, including CRC. In an in vivo study (male rats), it was observed that a decrease in RFC1 (reduced folate carrier) mRNA and protein expression correlates with alcoholism. That is a possible reason of lower blood folate levels commonly found in chronic alcoholics (Hamid et al. 2009). Alcohol in murine studies appears to reduce methionine synthase (MTR) levels; MTR is an enzyme involved in folate pathway; thus, its aberrant expression can interfere with SAM availability, and in turn induce DNA hypomethylation (Arasaradnam et al. 2008).

Exposure to metals such as cadmium and chromium is associated with altered miRNAs expression (Bollati et al. 2010); moreover, occupational exposure to nickel is correlated with H3K4me3 and H3K9me2 modifications (Arita et al. 2011). Other examples of the role of environmental agents in inducing epigenetic modifications are reported in the exhaustive review by Cortessis et al. (2012).

14.5 Concluding Remarks

For several years cancer has been defined just as a "genetic disease," and the multi-hit hypothesis of carcinogenesis was based on the accumulation of genetic mutations after a "first event," spontaneous or induced by environmental mutagenic carcinogens. Now it is quite clear that epigenetic modifications play a critical role in tumorigenesis, acting as the dynamic link between the environment and the genome. At present an extensive network, having as "engine" the environment and including both the influence of genetics on epigenetics and the epigenetic effects on genetics, might better represent a plausible carcinogenesis model. The effect of both types of factors depends probably on the cell type and on the specific locus (Huidobro et al. 2012).

Figure 14.3 represents some of the numerous and complex interactions occurring during the carcinogenesis processes.

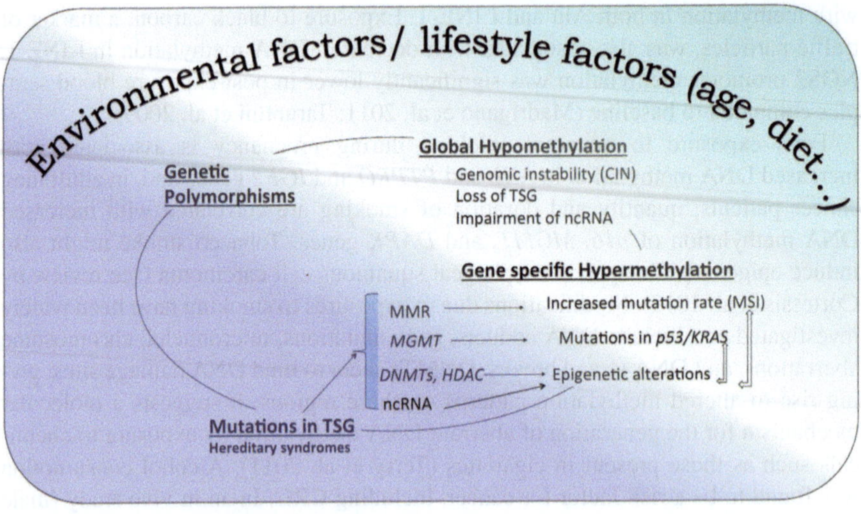

Fig. 14.3 Interplay between genetics and epigenetics in the multi-hit carcinogenesis processes

The coevolution of genetic and epigenetic changes may determine tumor heterogeneity, giving selective advantages for the tumor evolution (Sadikovic et al. 2008).

To partially explain or understand the tumor complexity and the individual susceptibility, an interesting tumorigenesis theory was recently proposed by Berger and co-workers (2011). These authors suggest a model that arises from a continuum variation of TSG expression, rather than from discrete changes in DNA copy number. In fact in the classical "discrete model," tumorigenesis is induced either by complete loss of a TSG or after single-copy loss of a TSG (haploinsufficiency). Instead differences in gene expression may be caused by polymorphisms in TSG promoter regions or in miRNA binding sites and may also be induced by environmental factors through epigenetic regulation of critical genes (Berger et al. 2011). Analysis of numerous human cancer cell lines, by whole exome sequencing, showed a great amount of potential mutations in genes of enzymes involved in epigenetic modifications (Barretina et al. 2012). Germline mutations in these "epigenetic enzymes" may contribute to development of several human diseases. Therefore altered epigenetic profiles seem to be the consequence of a defective epigenetic machinery, generally observed in cancer initiation and progression (Berdasco and Esteller 2013).

It could be interesting to identify networks of interaction among different types of cancer and multiple genes to clarify the involvement of specific genes in cancer progression. Data showed by Frey and colleagues demonstrated that causal genetic mutations correlate with high multiplicity. Consequently other mutations with high multiplicity might have a causal role in the multistep process of carcinogenesis. The same concept of multiplicity can also be generalized to epigenetic modifications (Frey et al. 2011). Recent experiments highlighted the importance of 5-hydroxymethylcytosine modifications and the existence of critical regulatory regions outside the promoter, such as CpG island shores and enhancer that might also influence gene expression. Thus epigenetic field is constantly in expansion, and continuous investigations are needed to clarify all the possible complex interactions among all these molecular mechanisms; indeed since epigenetic marks are erasable it is possible to intervene, they being good targets for anticancer drugs (Esteller 2012). Standardization of terminology and testing protocols might help to enlarge knowledge and to develop useful networks to better understand the cancer complexity.

Numerous theories have followed one another until now, without a specific, representative, unique cause–effect relationship between genetics, epigenetics, and environmental factors. Only by the contribution of numerous questions, formulation of hypotheses, and continuous experimental researches, new interesting theories may emerge, each of which will try to get closer to a process that is constantly evolving.

References

Ai L, Kim WJ, Kim TY, Fields CR, Massoll NA, Robertson KD, Brown KD (2006) Epigenetic silencing of the tumor suppressor cystatin M occurs during breast cancer progression. Cancer Res 66:7899–7909

Albany C, Alva AS, Aparicio AM, Singal R, Yellapragada S, Sonpavde G, Hahn NM (2011) Epigenetics in prostate cancer. Prostate Cancer 2011:580318

Alfred G, Knudson JR (1971) Mutation and cancer: statistical study of retinoblastoma. Proc Natl Acad Sci U S A 68:820–823

Alvarez H, Opalinska J, Zhou L, Sohal D, Fazzari MJ, Yu Y, Montagna C, Montgomery EA, Canto M, Dunbar KB, Wang J, Roa JC, Mo Y, Bhagat T, Ramesh KH, Cannizzaro L, Mollenhauer J, Thompson RF, Suzuki M, Meltzer SJ, Melnick A, Greally JM, Maitra A, Verma A (2011) Widespread hypomethylation occurs early and synergizes with gene amplification during esophageal carcinogenesis. PLoS Genet 7:e1001356

Arasaradnam RP, Commane DM, Bradburn D, Mathers JC (2008) A review of dietary factors and its influence on DNA methylation in colorectal carcinogenesis. Epigenetics 3:193–198

Arita A, Niu J, Qu Q, Zhao N, Ruan Y, Nadas A, Chervona Y, Wu F, Sun H, Hayes RB, Costa M (2011) Global levels of histone modifications in peripheral blood mononuclear cells of subjects with exposure to nickel. Environ Health Perspect 120:198–203

Armitage P, Doll R (1954) The age distribution of cancer and a multi-stage theory of carcinogenesis. Br J Cancer 8:1–12

Barretina J, Caponigro G, Stransky N, Venkatesan K, Margolina A, Kim S, Wilson CJ, Leha RJ, Kryukov GV, Sonkin D et al (2012) The Cancer Cell Line Encyclopedia enables predictive modelling of anticancer drug sensitivity. Nature 483:603–607

Baylin SB, Jones PA (2011) A decade of exploring the cancer epigenome—biological and translational implications. Nat Rev Cancer 11:726–734

Beetstra S, Thomas P, Salisbury C, Turner J, Fenech M (2005) Folic acid deficiency increases chromosomal instability, chromosome 21 aneuploidy and sensitivity to radiation-induced micronuclei. Mutat Res 578:317–326

Behbahani TE, Kahl P, von der Gathen J, Heukamp LC, Baumann C, Gütgemann I, Walter B, Hofstädter F, Bastian PJ, von Ruecker A, Müller SC, Rogenhofer S, Ellinger J (2012) Alterations of global histone H4K20 methylation during prostate carcinogenesis. BMC Urol 13:12–15

Berdasco M, Esteller M (2013) Genetic syndromes caused by mutations in epigenetic genes. Hum Genet 132:359–383
Berger AH, Knudson AG, Pandolfi PP (2011) A continuum model for tumour suppression. Nature 476:163–169
Bernard D, Pourtier-Manzanedo A, Gil J, Beach DH (2003) Myc confers androgen-independent prostate cancer cell growth. J Clin Invest 112:1724–1731
Bianco-Miotto T, Chiam K, Buchanan G, Jindal S, Day TK, Thomas M, Pickering MA, O'Loughlin MA, Ryan NK, Raymond WA, Horvath LG, Kench JG, Stricker PD, Marshall VR, Sutherland RL, Henshall SM, Gerald WL, Scher HI, Risbridger GP, Clements JA, Butler LM, Tilley WD, Horsfall DJ, Ricciardelli C, Australian Prostate Cancer Bioresource (2010) Global levels of specific histone modifications and an epigenetic gene signature predict prostate cancer progression and development. Cancer Epidemiol Biomarkers Prev 19:2611–2622
Bollati V, Marinelli B, Apostoli P, Bonzini M, Nordio F, Hoxha M, Pegoraro V, Motta V, Tarantini L, Cantone L, Schwartz J, Bertazzi PA, Baccarelli A (2010) Exposure to metal-rich particulate matter modifies the expression of candidate microRNAs in peripheral blood leukocytes. Environ Health Perspect 118:763–768
Calvanese V, Lara E, Kahn A, Fraga MF (2009) The role of epigenetics in aging and age-related diseases. Ageing Res Rev 8:268–276
Catteau A, Morris JR (2002) BRCA1 methylation: a significant role in tumour development? Semin Cancer Biol 12:359–371
Centelles JJ (2012) General aspects of colorectal cancer. ISRN Oncol 2012:139268. doi:10.5402/2012/139268
Chen Z, Wang L, Wang Q, Li W (2010) Histone modifications and chromatin organization in prostate cancer. Epigenomics 2:551–560
Chung JH, Lee HJ, Kim BH, Cho NY, Kang GH (2011) DNA methylation profile during multistage progression of pulmonary adenocarcinomas. Virchows Arch 459:201–211
Clark SJ, Harrison J, Frommer M (1995) CpNpG methylation in mammalian cells. Nat Genet 10:20–27
Coppedè F (2014) Epigenetic biomarkers of colorectal cancer: focus on DNA methylation. Cancer Lett 342:238–247
Cortessis VK, Thomas DC, Levine AJ, Breton CV, Mack TM, Siegmund KD, Haile RW, Laird PW (2012) Environmental epigenetics: prospects for studying epigenetic mediation of exposure-response relationships. Hum Genet 131:1565–1589
Crider KS, Yang TP, Berry RJ, Bailey LB (2012) Folate and DNA methylation: a review of molecular mechanisms and the evidence for folate's role. Adv Nutr 3:21–38
Croce CM (2009) Causes and consequences of microRNA dysregulation in cancer. Nat Rev Genet 10:704–714
Doll R, Peto R (1978) Cigarette smoking and bronchial carcinoma: dose and time relationships among regular smokers and lifelong non-smokers. J Epidemiol Community health 32:303–313
Duthie SJ, Pirie LP, Grant G, Watson AJ, Margison GP (2010) Long term folate deficiency differentially alters hepatic and colon MGMT and OGG-1 DNA repair protein expression in rats but has no impact on genome-wide DNA methylation. Cancer Prev Res (Phila) 3:92–100
Dworkin AM, Spearman AD, Tseng SY, Sweet K, Toland AE (2009) Methylation not a frequent "second hit" in tumors with germline BRCA mutations. Fam Cancer 8:339–346
Esquela-Kerscher A, Slack FJ (2006) Oncomirs—microRNAs with a role in cancer. Nat Rev Cancer 6:259–269
Esteller M (2012) Cancer, epigenetics and the Nobel Prizes. Mol Oncol 6:565–566
Esteller M, Silva JM, Dominguez G, Bonilla F, Matias-Guiu X, Lerma E, Bussaglia E, Prat J, Harkes C, Repasky EA et al (2000) Promoter hypermethylation and BRCA1 inactivation in sporadic breast and ovarian tumors. J Natl Cancer Inst 92:564–569
Fearon ER, Vogelstein B (1990) A genetic model for colorectal tumorigenesis. Cell 61:759–767
Fraga MF, Agrelo R, Esteller M (2007) Cross-talk between aging and cancer: the epigenetic language. Ann N Y Acad Sci 1100:60–74

Frey LJ, Piccolo SR, Edgerton ME (2011) Multiplicity: an organizing principle for cancers and somatic mutations. BMC Med Genomics 4:52

Gargalionis AN, Piperi C, Adamopoulos C, Papavassiliou AG (2012) Histone modifications as a pathogenic mechanism of colorectal tumorigenesis. Int J Biochem Cell Biol 44:1276–1289

Glozak MA, Seto E (2007) Histone deacetylases and cancer. Oncogene 26:5420–5432

Grady WM, Carethers JM (2008) Genomic and epigenetic instability in colorectal cancer pathogenesis. Gastroenterology 135:1079–1099

Gravina GL, Ranieri G, Muzi P, Marampon F, Mancini A, Di Pasquale B, Di Clemente L, Dolo V, D'Alessandro AM, Festuccia C (2012) Increased levels of DNA methyltransferases are associated with the tumorigenic capacity of prostate cancer cells. Oncol Rep 29:1189–1195

Hamid A, Kiran M, Rana S, KAUR J (2009) Low folate transport across intestinal basolateral surface is associated with down-regulation of reduced folate carrier in in vivo model of folate malabsorption. IUBMB Life 61:236–243

Hansen KD, Timp W, Bravo HC, Sabunciyan S, Langmead B, Mcdonald OG, Wen B, Wu H, Liu Y, Diep D, Briem E, Zhang K, Irizarry RA, Feinberg AP (2011) Increased methylation variation in epigenetic domains across cancer types. Nat Genet 43:768–775

Harquail J, Benzina S, Robichaud GA (2012) MicroRNAs and breast cancer malignancy: an overview of miRNA-regulated cancer processes leading to metastasis. Cancer Biomark 11: 269–280

He X, Dong Y, Wu CW, Zhao Z, Ng SS, Chan FK, Sung JJ, Yu J (2012) MicroRNA-218 inhibits cell cycle progression and promotes apoptosis in colon cancer by downregulating oncogene BMI-1. Mol Med 18:1491–1498

Helenius MA, Saramäki OR, Linja MJ, Tammela TLJ, Visakorpi T (2001) Amplification of urokinase gene in prostate cancer. Cancer Res 61:5340–5344

Herceg Z, Vaissière T (2011) Epigenetic mechanisms and cancer: an interface between the environment and the genome. Epigenetics 6:804–819

Hughes LA, van den Brandt PA, de Bruïne AP, Wouters KA, Hulsmans S, Spiertz A, Goldbohm RA, de Goeij AF, Herman JG, Weijenberg MP, van Engeland M (2009) Early life exposure to famine and colorectal cancer risk: a role for epigenetic mechanisms. PLoS One 4:e7951

Huidobro C, Fernandez AF, Fraga MF (2012) Aging epigenetics: causes and consequences. Mol Aspects Med 34:765–781

Imai K, Yamamoto H (2008) Carcinogenesis and microsatellite instability: the interrelationship between genetics and epigenetics. Carcinogenesis 29:673–680

Imam JS, Plyler JR, Bansal H, Prajapati S, Bansal S, Rebeles J, Chen HI, Chang YF, Panneerdoss S, Zoghi B, Buddavarapu KC, Broaddus R, Hornsby P, Tomlinson G, Dome J, Vadlamudi RK, Pertsemlidis A, Chen Y, Rao MK (2012) Genomic loss of tumor suppressor miRNA-204 promotes cancer cell migration and invasion by activating AKT/mTOR/Rac1 signaling and actin reorganization. PLoS One 7:e52397. doi:10.1371/journal.pone.0052397

Issa JP (2008) Colon cancer: it's CIN or CIMP. Clin Cancer Res 14:5939–5940

Jardim MJ (2011) microRNAs: implications for air pollution research. Mutat Res 717:38–45

Jovanovic J, Rønneberg JA, Tost J, Kristensen V (2010) The epigenetics of breast cancer. Mol Oncol 4:242–254

Kanwal R, Gupta S (2012) Epigenetic modifications in cancer. Clin Genet 81:303–311

Kim YI (2005) Nutritional epigenetics: impact of folate deficiency on DNA methylation and colon cancer susceptibility. J Nutr 1:35,2703–9

Kim YI (2007) Folate and colorectal cancer: an evidence-based critical review. Mol Nutr Food Res 51:267–292

Kim YI (2008) Folic acid supplementation and cancer risk: point. Cancer Epidemiol Biomarkers Prev 17:2220–2225

Kitkumthorn N, Mutirangura A (2011) Long interspersed nuclear element 1 hypomethylation in cancer: biology and clinical applications. Clin Epigenet 2:315–330

Knudson A (2005) Retinoblastoma: teacher of cancer biology and medicine. PLoS Med 2:e349

Lao VV, Grady WM (2011) Epigenetics and colorectal cancer. Nat Rev Gastroenterol Hepatol 8:686–700

Leite KR, Tomiyama A, Reis ST, Sousa-Canavez JM, Sañudo A, Camara Lopes LH, Srougi M (2013) MicroRNA expression profiles in the progression of prostate cancer-from high-grade prostate intraepithelial neoplasia to metastasis. Urol Oncol 31:796–801
Ley TJ, Ding L, Walter MJ, McLellan MD, Lamprecht T, Larson DE, Kandoth C, Payton JE, Baty J, Welch J et al (2010) DNMT3A mutations in acute myeloid leukemia. N Engl J Med 363:2424–2433
Li LC, Carroll PR, Dahiya R (2005) Epigenetic changes in prostate cancer: implication for diagnosis and treatment. J Natl Cancer Inst 97:103–115
Liloglou T, Bediaga NG, Brown BR, Field JK, Davies MP (2014) Epigenetic biomarkers in lung cancer. Cancer Lett 342:200–212
Liu WB, Ao L, Cui ZH, Zhou ZY, Zhou YH, Yuan XY, Xiang YL, Cao J, Liu JY (2011a) Molecular analysis of DNA repair gene methylation and protein expression during chemical-induced rat lung carcinogenesis. Biochem Biophys Res Commun 408:595–601
Liu X, Zhang Z, Sun L, Chai N, Tang S, Jin J, Hu H, Nie Y, Wang X, Wu K, Jin H, Fan D (2011b) MicroRNA-499-5p promotes cellular invasion and tumor metastasis in colorectal cancer by targeting FOXO4 and PDCD4. Carcinogenesis 32:1798–1805
Lopez-Serra P, Esteller M (2012) DNA methylation-associated silencing of tumor-suppressor microRNAs in cancer. Oncogene 31:1609–1622
Luo W, Chang R, Zhong J, Pandey A, Semenza GL (2012) Histone demethylase JMJD2C is a coactivator for hypoxia-inducible factor 1 that is required for breast cancer progression. Proc Natl Acad Sci U S A 109:E3367–E3376
Ma L, Teruya-Feldstein J, Weinberg RA (2007) Tumour invasion and metastasis initiated by microRNA-10b in breast cancer. Nature 449:682–688
Macphee DG (1998) Epigenetics and epimutagens: some new perspectives on cancer, germ line effects and endocrine disrupters. Mutat Res 400:369–379
Madrigano J, Baccarelli A, Mittleman MA, Wright RO, Sparrow D, Vokonas PS, Tarantini L, Schwartz J (2011) Prolonged exposure to particulate pollution, genes associated with glutathione pathways and DNA methylation in a cohort of older men. Environ Health Perspect 119:977–982
Majumdar S, Buckles E, Estrada J, Koochekpour S (2011) Aberrant DNA methylation and prostate cancer. Curr Genomics 12:486–505
Menigatti M, Truninger K, Gebbers JO, Marbet U, Marra G, Schär P (2009) Normal colorectal mucosa exhibits sex- and segment-specific susceptibility to DNA methylation at the hMLH1 and MGMT promoters. Oncogene 28:899–909
Migheli F, Migliore L (2012) Epigenetics of colorectal cancer. Clin Genet 81:312–318
Mokarram P, Kumar K, Brim H, Naghibalhossaini F, Saberifiroozi M, Nouraie M, Green R, Lee E, Smoot DT, Ashktorab H (2009) Distinct high-profile methylated genes in colorectal cancer. PLoS One 4:e7012
Morin RD, Mendez-Lago M, Mungall AJ, Goya R, Mungall KL, Corbett RD, Johnson NA, Severson TM, Chiu R, Field M, Jackman S, Krzywinski M, Scott DW, Trinh DL, Tamura-Wells J, Li S, Firme MR, Rogic S, Griffith M, Chan S, Yakovenko O, Meyer IM, Zhao EY, Smailus D, Moksa M, Chittaranjan S, Rimsza L, Brooks-Wilson A, Spinelli JJ, Ben-Neriah S, Meissner B, Woolcock B, Boyle M, Mcdonald H, Tam A, Zhao Y, Delaney A, Zeng T, Tse K, Butterfield Y, Birol I, Holt R, Schein J, Horsman DE, Moore R, Jones SJ, Connors JM, Hirst M, Gascoyne RD, Marra MA (2011) Frequent mutation of histone-modifying genes in non-Hodgkin lymphoma. Nature 476:298–303
Nair SS, Kumar R (2012) Chromatin remodelling in cancer: a gateway to regulate gene transcription. Mol Oncol 6:611–619
Nosho K, Kawasaki T, Ohnishi M, Suemoto Y, Kirkner GJ, Zepf D, Yan L, Longtine JA, Fuchs CS, Ogino S (2008) PIK3CA mutation in colorectal cancer: relationship with genetic and epigenetic alterations. Neoplasia 10:534–541
Oberg AL, French AJ, Sarver AL, Subramanian S, Morlan BW, Riska SM, Borralho PM, Cunningham JM, Boardman LA, Wang L, Smyrk TC, Asmann Y, Steer CJ, Thibodeau SN

(2011) miRNA expression in colon polyps provides evidence for a multihit model of colon cancer. PLoS One 6:e20465

Ogino S, Kawasaki T, Kirkner GJ, Suemoto Y, Meyerhardt JA, Fuchs CS (2007) Molecular correlates with MGMT promoter methylation and silencing support CpG island methylator phenotype-low (CIMP-low) in colorectal cancer. Gut 56:1564–1571

O'Hagan HM, Wang W, Sen S, Desthhefano Shields C, Lee SS, Zhang YW, Clements EG, Cai Y, Van Neste L, Easwaran H, Casero RA, Sears CL, Baylin SB (2011) Oxidative damage targets complexes containing DNA methyltransferases, SIRT1, and polycomb members to promoter CpG Islands. Cancer Cell 20:606–619

Pancione M, Remo A, Colantuoni V (2012) Genetic and epigenetic events generate multiple pathways in colorectal cancer progression. Pathol Res Int 2012:509348

Paolicchi E, Crea F, Farrar WL, Green JE, Danesi R (2013) Histone lysine demethylases in breast cancer. Crit Rev Oncol Hematol 86:97–103

Rao X, Evans J, Chae H, Pilrose J, Kim S, Yan P, Huang RL, Lai HC, Lin H, Liu Y, Miller D, Rhee JK, Huang YW, Gu F, Gray JW, Huang TM, Nephew KP (2012) CpG island shore methylation regulates caveolin-1 expression in breast cancer. Oncogene. doi:10.1038/onc.2012.474

Rodríguez-Paredes M, Esteller M (2011) Cancer epigenetics reaches mainstream oncology. Nat Med 17:330–339

Sadikovic B, Al-Romaih K, Squire JA, Zielenska M (2008) Cause and consequences of genetic and epigenetic alterations in human cancer. Curr Genomics 9:394–408

Sandoval J, Esteller M (2012) Cancer epigenomics: beyond genomics. Curr Opin Genet Dev 22:50–55

Schernhammer ES, Giovannucci E, Kawasaki T, Rosner B, Fuchs CS, Ogino S (2010) Dietary folate, alcohol and B vitamins in relation to LINE-1 hypomethylation in colon cancer. Gut 59:794–799

Sie KK, Medline A, van Weel J, Sohn KJ, Choi SW, Croxford R, Kim YI (2011) Effect of maternal and postweaning folic acid supplementation on colorectal cancer risk in the offspring. Gut 60:1687–1694

Slaga TJ (1984) Multistage skin carcinogenesis: a useful model for the study of the chemoprevention of cancer. Acta Pharmacol Toxicol (Copenh) 55:107–124

Suzuki H, Maruyama R, Yamamoto E, Kai M (2012) DNA methylation and microRNA dysregulation in cancer. Mol Oncol 6:567–578

Tarantini L, Bonzini M, Apostoli P, Pegoraro V, Bollati V, Marinelli B, Cantone L, Rizzo G, Hou L, Schwartz J, Bertazzi PA, Baccarelli A (2009) Effects of particulate matter on genomic DNA methylation content and iNOS promoter methylation. Environ Health Perspect 117:217–222

Terry MB, Delgado-Cruzata L, Vin-Raviv N, Wu HC, Santella RM (2011) DNA methylation in white blood cells: association with risk factors in epidemiologic studies. Epigenetics 6: 828–837

Tserga A, Michalopoulos NV, Levidou G, Korkolopoulou P, Zografos G, Patsouris E, Saetta AA (2012) Association of aberrant DNA methylation with clinicopathological features in breast cancer. Oncol Rep 27:1630–1638

Tyler JK, Kadonaga JT (1999) The "dark side" of chromatin remodeling: repressive effects on transcription. Cell 99:443–446

van Engeland M, Derks S, Smits KM, Meijer GA, Herman JG (2011) Colorectal cancer epigenetics: complex simplicity. J Clin Oncol 29:1382–1391

Vineis P, Schatzkin A, Potter JD (2010) Models of carcinogenesis: an overview. Carcinogenesis 31:1703–1709

Whitehouse A, Meredith DM, Markham AF (1998) DNA mismatch repair genes and their association with colorectal cancer (Review). Int J Mol Med 1:469–474

You JS, Jones PA (2012) Cancer genetics and epigenetics: two sides of the same coin? Cancer Cell 22:9–20

Chapter 15
Epigenetic Mechanisms of Colon Cancer Prevention: What Can Nutrition Do?

Yuan-Xiang Pan, Yukun Zhang, and Hong Chen

Abstract Colon cancer is the fourth most common cause of death from cancer worldwide. Colon cancer occurs as a consequence of the accumulation of abnormal DNA methylation and the disruption of the histone code. The present review summarizes etiology and risk factors of colon cancer and the potential of nutrition to counteract these cancer-related epigenetic alterations. Unlike the genome, epigenetic structure can be reshaped, which has the potential to offer access for the prevention and treatment of cancer by any compounds that directly target the epigenome

Y.-X. Pan (✉)
Nutrient-gene and Epigenetics Group (NEG), University of Illinois at Urbana-Champaign, 905 South Goodwin Avenue, Urbana, IL 61801, USA

Food Science and Human Nutrition (FSHN), University of Illinois at Urbana-Champaign, 461 Bevier Hall, MC-182, 905 South Goodwin Avenue, Urbana, IL 61801, USA

Division of Nutritional Sciences (DNS), University of Illinois at Urbana-Champaign, 905 South Goodwin Avenue, Urbana, IL 61801, USA

Illinois Informatics Institute (I^3), University of Illinois at Urbana-Champaign, 905 South Goodwin Avenue, Urbana, IL 61801, USA
e-mail: yxpan@illinois.edu

Y. Zhang
Nutrient-gene and Epigenetics Group (NEG), University of Illinois at Urbana-Champaign, 905 South Goodwin Avenue, Urbana, IL 61801, USA

Food Science and Human Nutrition (FSHN), University of Illinois at Urbana-Champaign, 461 Bevier Hall, MC-182, 905 South Goodwin Avenue, Urbana, IL 61801, USA

H. Chen
Nutrient-gene and Epigenetics Group (NEG), University of Illinois at Urbana-Champaign, 905 South Goodwin Avenue, Urbana, IL 61801, USA

Food Science and Human Nutrition (FSHN), University of Illinois at Urbana-Champaign, 461 Bevier Hall, MC-182, 905 South Goodwin Avenue, Urbana, IL 61801, USA

Division of Nutritional Sciences (DNS), University of Illinois at Urbana-Champaign, 905 South Goodwin Avenue, Urbana, IL 61801, USA

N. Maulik and T. Karagiannis (eds.), *Molecular Mechanisms and Physiology of Disease: Implications for Epigenetics and Health*, DOI 10.1007/978-1-4939-0706-9_15,

by influencing the activity or expression of DNA methyltransferases (DNMTs) and histone modifying enzymes. Epigenetic processes are detailed thoroughly and the molecular details of carcinogenesis and colon cancer prevention are reviewed comprehensively. The review also includes specific dietary components, including chromatin modifiers and anti-inflammatory agents, that can directly influence cancer cell growth. Overall, information is still mainly derived from in vitro investigations, and data from animal models or human intervention studies that demonstrate the functional relevance of epigenetic mechanisms for health promoting or cancer preventive efficacy are limited. With the emergence of novel technologies, future research exploring epigenomics will help to better understand the importance of epigenetic mechanisms for nutrition in cancer prevention.

Keywords Diet • Chromatin • Histone modifications • miRNA • Transcription factor • Bioactive • Inflammation • Signal pathway

15.1 Introduction

15.1.1 Prevalence and Severity of Colon Cancer

Colon cancer is the fourth most common cause of death from cancer worldwide (Ferlay et al. 2010). On top of an individual's genetic background, environmental factors, including diet, are additional risk factors for cancer development. More importantly, unlike the former, the latter can be modified, and is thus of great interest for the study of cancer prevention. It is estimated by the American Institute for Cancer Research (AICR) that 45 % of colon cancer cases could be preventable (Colorectal Cancer Report 2010 Summary, 2011, AICR). This recent report emphasized the importance of dietary alterations, such as decreasing the intake of red and processed meat and increasing the intake of dietary fiber and calcium, in the prevention of colon cancer. As the importance of diet in cancer prevention becomes increasingly evident, it is critical to understand the underlying mechanisms by which dietary factors contribute to colon cancer prevention.

15.1.2 What Is Epigenetics?

The term "epigenetics" refers to heritable changes that alter gene expression and cell phenotype without changing the DNA sequence itself, and was coined by C.H. Waddington in 1942. Since then, tremendous work has been done to illustrate the fascinating role of this "layer above the genome" in controlling cellular gene expression in different life stages and environments. Epigenetic regulations, including DNA methylation, histone modifications, and micro-RNA (miR) regulation, closely orchestrate the transcriptional activity of genes by switching gene on and off without altering the DNA sequence. Furthermore, epigenome is essential for keeping genetic activity stabilized and organized, and consequentially maintaining normal physiological functions.

15.1.3 Epigenetic Shifts During Colon Cancer Development

An aberrant epigenome structure activates genes that promote cancer development and progression, while suppressing tumor suppressor genes, leading to the development of various cancers (Esteller 2007). Colon cancer occurs as a consequence of the accumulation of abnormal DNA methylation and the disruption of the histone code (Kim et al. 2010). Dysregulated DNA methylation occurs at the earliest stages of colon cancer development in normal-appearing colon mucosa. Global DNA hypomethylation is a general indicator of a cancer epigenome, and causes chromosome instability and activates cancer-promoting genes (Salhia et al. 2010; Strathdee and Brown 2002). However, increased methylation intensity at regulatory regions of genes, such as promoter CpG-rich regions, which are normally unmethylated and facilitate transcription activation, results in the silencing of tumor suppressor genes (Jones and Takai 2001). In addition to the DNA methylation status, the histone code is another regulator of gene expression. The majority of histone modifications occur at specific residues on the tails of histone H3 and H4, including acetylation, biotinylation, mono-, di-, and tri-methylation, phosphorylation, ubiquitination, and sumoylation (Jenuwein and Allis 2001). In general, histone acetylation creates a loose, active chromatin structure, which allows the transcription machinery to bind to the DNA sequence, promoting gene expression. In contrast, histone methylation leads to a tight, repressive chromatin structure, which obstructs the initiation of transcription, thus suppressing gene expression. miRs are 21- to 23-nucleotide RNAs that negatively regulate gene expression by binding to the 3′ untranslated regions (3UTRs) of target transcripts, leading to mRNA degradation or the inhibition of protein translation. In normal tissues, miR-induced regulation contributes to maintaining a normal state of cell growth, proliferation, differentiation, and apoptosis. The deregulation of miRs' expression leads to abnormal activity of miR target genes. For instance, overexpression of an oncogenic miR may cause excessive repression of a targeted tumor suppressor gene, and conversely, silencing of a tumor-suppressive miR may give rise to ectopic levels of a target onco-protein. Consequently, deregulation of certain miRs may result in cells with selective advantages such as increased proliferation or survival (Iorio and Croce 2012). In addition to their individual impact on the genetic accessibility, DNA methylation, histone modifications, and miRs cross talk with each other, thereby determining the overall chromatin structure and transcriptional activity.

15.1.4 How Diet Affects Epigenetic Structure of Gene in Relation to Disease Prevention or Attenuation

As one of the primary environmental factors that modify cancer risk, dietary preferences and intakes and their associations with cancer are constantly evaluated in epidemiologic investigations. Through numerous cohort studies, researchers have reached the consensus that a diet high in fat increases colon cancer risk, while a diet rich in fruits and vegetables helps to lower the risk of developing colon cancer. Unlike genetic predisposition, one's diet is modifiable. Therefore, more and more efforts have been devoted to elucidate the impact of dietary factor on attenuating colon cancer development and progression. Food-derived natural compounds have been reported to prevent epigenetic dysregulation and facilitate in reestablishing an organized epigenetic network. However, the other side of the coin is that cytotoxicity and other side effects were also observed in response to natural compounds. In order to take the most advantage of the anticancer properties of natural products, it is necessary to understand the mechanisms of how these different bioactive dietary components affect the epigenome in different conditions. Here, we discuss the etiology of colon cancer, summarize the diet-induced epigenetic modifications and molecular signaling pathways involved in colon cancer development, review what is known about the colon cancer preventive diet, and propose strategies for optimizing our diet to promote colon health.

15.2 Etiology and Risk Factors of Colon Cancer

Colon cancer is a disease that develops over a relatively long period of time compared to other acute diseases (Kelloff et al. 2004). It can take up to 20 years for a normal colon epithelium to transform into an adenoma, and it takes an additional 10 years to develop carcinoma. The development of colon cancer can be attributed to numerous factors (Fig. 15.1). Uncontrollable factors include gender, ethnicity, and family medical history. Specifically, it was recently reported that colon cancer occurs more often in men than in women (Colorectal Cancer Report 2010 Summary, 2011, AICR), is less common in Africa and the majority of Asia, and the incidence and death rate climb with age. Only 20 % of colon cancer cases are categorized as inherited, while the other 80 % are considered sporadic cancers, which means that they develop naturally, probably caused by factors other than genetic susceptibility. Besides the above-mentioned factors that an individual cannot control, there are modifiable factors that can lower the chance of developing colon cancer. Convincing evidence shows that increased tobacco use strongly correlates with an elevation of colon cancer (Secretan et al. 2009). Furthermore, the type and amount of ingested food and drink impact greatly on colon health (Milner 2006; Berlau et al. 2004). At the end of the gastrointestinal tract, the luminal side

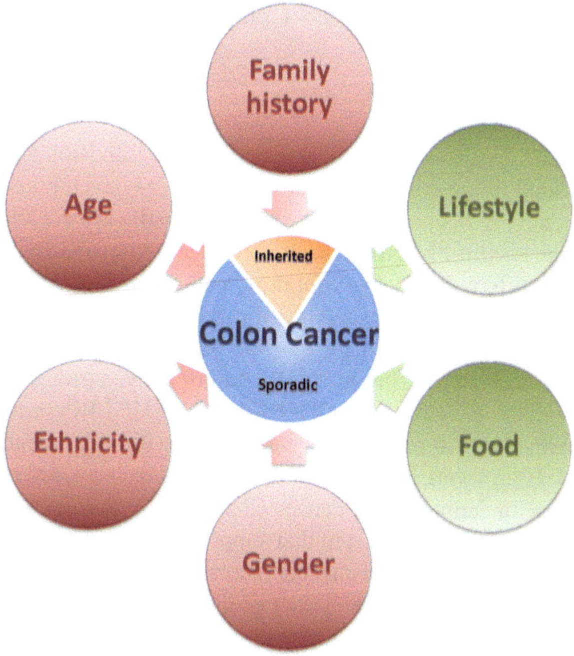

Fig. 15.1 Driving forces of colon cancer. There are various contributing factors of colon cancer development. Some of them are uncontrollable factors (in *red circles*), such as one's family medical history, ethnicity, gender, and ages. Some of them are controllable factors (in *green circles*), such as one's food choices and lifestyles, including physical activity, smoking, and drinking status

of the colon is directly exposed to and therefore is easily disturbed by pathogenic remnant or metabolites from the daily diet. Keeping this in mind, experts in the field and medical professionals agree that it is much easier and less painful to change one's lifestyle to include a healthier diet and prevent colon cancer than to treat the disease after its development. Therefore, elucidating the colon cancer preventive potential of different diets and dietary compounds becomes critical. To do so, we first must understand the molecular mechanisms that drive colon cancer development and progression.

As was discussed in the above introduction, epigenetic instability greatly dysregulates gene expression networks and causes pathogenic activity, and therefore has the ability to initiate and promote cancer development. Like genetic information, the epigenetic code is heritable. However, unlike the genome, epigenetic structure can be reshaped, which has the potential to offer access for the prevention and treatment of cancer by any compounds that directly target the epigenome. Therefore, for the past 15 years, the involvement of epigenetics in cancer has been intensively investigated. The three major epigenetic regulations that attract cancer researchers the most are DNA methylation, histone modifications, and miR-mediated regulations, and dietary component could interfere with all of these, thereby preventing or attenuating cancer development.

15.2.1 *Dysregulated Cell Growth*

Growth Factors

Cells that have naturally enhanced rates of division and proliferation, such as those found in the colon, are predisposed to the development of cancer (Preston-Martin et al. 1990), and several key signaling pathways are known to regulate this process. The insulin-like growth factor (IGF) system plays a vital role in controlling apoptosis, differentiation, proliferation, and transformation by working together with the IGF receptors on the cell membrane. The interaction between IGFs and their receptors is also controlled by IGF binding proteins (i.e., IGFBP-1 to IGFBP-6) (Marshman and Streuli 2002). Additionally, a group of IGFBP proteases indirectly regulate the action of IGFs by degrading IGFBPs, causing the discharge of bound IGFs that will regain their interactions with IGF receptors (Lelbach et al. 2005).

High levels of IGFs are associated with increased risk for several common cancers, including colon cancer (Park 2008). The level of IGFBP-3, a major IGF-I-binding protein in serum that suppresses the action of IGF-I, is inversely associated with the risk of these cancers (Kansra et al. 2000). IGF-I has both instant and enduring effects on cellular activities mediated through IGF-I receptor. IGF-I increases cellular uptake of amino acids and glucose and stimulates glycogen and protein synthesis (Jones and Clemmons 1995). IGF-I also has a lasting effect on apoptosis, differentiation, and cell proliferation (Clemmons et al. 1995) through inducing DNA synthesis and increasing the expression of cyclin D1, which in turn speed up progression of the cell cycle from G1 to S phase (Ren et al. 2009). In addition to controlling cell cycle advancement, IGF-I also affects apoptosis by reducing the expression of Bax and inducing the expression of Bcl proteins. This change causes an increase in the Bcl/Bax heterodimer, which stops the initiation of the apoptotic pathway (Yang et al. 2008).

Energy restriction in animals can reduce the risk of cancer and inhibit tumor growth (Kritchevsky 2002) and this effect can be attributed in part to IGFs (Berrigan et al. 2005). In humans, dietary energy intake and nutritional status are key controllers of IGF level. Malnutrition causes reduction of IGF-I levels, and increase in energy intake reverts the level back to normal (Soliman et al. 1986). Over-nutrition results in an increase in IGF-I level (Shimizu et al. 2006), while fasting decreases IGF-I level, but the effect is less evident in obese subjects who are less dependent on energy intake to maintain IGF-I levels (Svensson et al. 1998). Studies in adults have demonstrated a positive correlation between protein intake and serum IGF-I levels (Castaneda et al. 2000), and a 50 % reduction in calorie intake or a 30 % reduction in protein intake results in a decline in serum IGF-I and IGFBP-3 levels and an increase in IGFBP-2 level (Smith et al. 1995). The finding is consistent with animal experiments (Chujo et al. 2013) showing that restriction of nutrients has diverse effects on IGF-I gene transcription (Nogueira et al. 2012). Although all of these studies show a significant influence of food intake on the IGF family, the precise nature of the relationships between energy and protein intake and levels of IGFs in circulation and colon cancer remains to be determined (Qu et al. 1997).

Wnt Signaling: Dysregulated Motility, Cytostasis and Differentiation

During normal development, the colon epithelium is organized to maintain a homeostatic renewal status. Many signaling pathways are required to maintain this complexity, including the Wnt signaling pathway. The importance of Wnt in colon cancer development did not become clear until the discovery of mutated adenomatous polyposis coli (APC) in many sporadic colon cancer cases 20 years ago (Jenuwein and Allis 2001). Suppression of APC was shown to lead to constitutive activation of Wnt signaling, which was later found to be a robust stimulator of cancerous growth in the colon (Morin et al. 1997). Wnt signaling is categorized as either canonical, which is Tcf/β-catenin dependent, or noncanonical, which is calcium related but Tcf/β-catenin independent (Groden et al. 1991). In general, the former one has been of more interest in the field of colon cancer research. Wnt signaling is initiated when a ligand protein, such as WNT1, binds to its frizzle receptor (FZ), and forms a complex with co-receptor LRP on the cell membrane. After associating and disassociating with several cytoplasmid components, including Dishevelled (Dvl), GSK3β, and Axin1, the signal is passed down to β-catenin, which then dissociates from APC, and translocates into the nucleus. After forming a complex with TCF/LEF in the nucleus, β-catenin then stimulates the expression of Wnt target genes, including those that are critical for cell cycle control, such as *Cyclin D1* and *c-Myc* (Moon 2005). Wnt signaling also interplays with other cellular signaling pathways, thereby supervising cancer cell growth (FGF and BMP) (Klapholz-Brown et al. 2007), migration (matrix metalloproteases, MMPs) (Arozarena et al. 2011), viability (Bcl9) (Deka et al. 2010), and adhesion (CD44 and LGR5) (Saigusa et al. 2012). As the colon develops during normal physiological conditions, Wnt signaling is supervised and maintained at a normal active level by its antagonists, such as secreted frizzled-related protein sFRP1, sFRP2, and sFRP5 (Bovolenta et al. 2008). By restraining Wnt proteins, such as WNT1, in the cytoplasm, Sfrps obstruct the initial binding of Wnt ligand protein to FZ, thus preventing the signal transduction to the nucleus. Another Wnt antagonist, Dikkopf-1 (DKK-1), is a secretary protein that competitively binds to LRP with Wnt ligand and prevents downstream signaling (Gonzalez-Sancho et al. 2005) (Fig. 15.2).

Silencing of Wnt antagonist leads to uncontrolled Wnt signaling, which is closely associated with pathogenic events, thus promoting colon carcinogenesis (Gregorieff and Clevers 2005), and this process is likely strongly regulated by epigenetic events. Abnormal Wnt signaling was observed in approximately 90 % of colorectal cancer cases (Fevr et al. 2007), and an increasing body of evidence has shown that silencing of Wnt antagonists results from epigenetic dysregulation. Hypermethylation at the promoter region of Sfrp1, Sfrp2, Sfrp4, Sfrp5, and Wif was reported in various colon cancer cell lines and occurs from the earliest stages through the whole span of colon caner development. After treating colon cancer cells with a demethylating agent, such as 5-aza-cytidine (5-aza-C) and 5-aza-2′-deoxycytidine (5-aza-2dC), the gene expression of Sfrps and Wif was restored, resulting in attenuation of constitutively activated Wnt signaling (Suzuki et al. 2004; Taniguchi et al. 2005; Zhang and

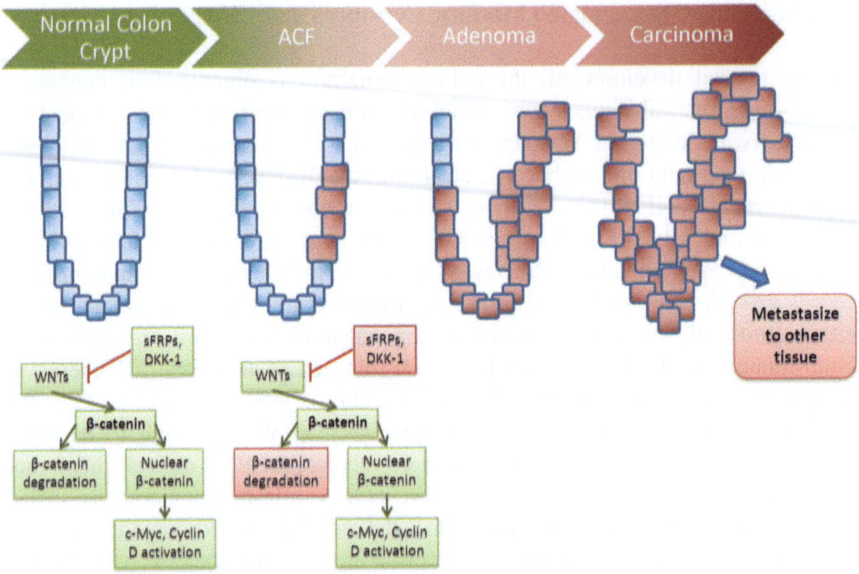

Fig. 15.2 Wnt signaling during colon cancer development. Colon cancer is a chronic disease, which composes of several steps. When normal crypts are disturbed, they may develop into aberrant crypt foci (ACF), a group of pre-neoplastic lesions in colon. Over a period of time, adenoma will form and deteriorate, and eventually becomes carcinoma. In normal colon crypts, Wnt signaling is closely controlled by its antagonists and keeps a balanced activity. β-Catenin, the hallmarker of Wnt, either translocates into nucleus and triggers downstream gene activation, or enters depredation process initiated by phosphorylation. During the early stage of colon cancer development, when ACF initially occur, the balance of Wnt signaling is often disturbed. The degradation process of β-catenin is impaired; thereby, β-catenin accumulates excessively in nucleus, accelerating cell cycle progression. Constitutively activated Wnt signaling drives colon cancer progression. Wnt antagonists are normally silenced in colon tumor tissues, and the repression of Wnt antagonist may occur at the initiation part of colon cancer. *Green text boxes* indicate active steps. *Red text boxes* indicate inactive steps

Chen 2011a). Treating with a histone deacetylase (HDAC) inhibitor, which promotes histone acetylation, restored the expression of sFRP1 and sFRP2 in gastric cancer cells (Shin et al. 2012). To date, however, there is little evidence showing the impact of histone modifications on reestablishing sFRPs expression in colon cancers. Another Wnt inhibitor, DKK-1, was not as effective as Sfrps in attenuating aberrant Wnt signaling, but the re-expression of this gene suppressed colon cancer growth (Aguilera et al. 2006; Sato et al. 2007). Decreased DNA methylation was shown to upregulate the expression of DKKs (Maehata et al. 2008), and methylation at the promoter CpG island of DKK-1 was identified in multiple colon cancer cell lines and it is proposed that epigenetic silencing of DKK-1 happens more often in higher stage colon carcinomas. Besides DNA methylation, histone modification also plays a role in regulating DKK-1 expression. It has been reported that DKK-1 expression was restored by histone acetylation at histone H3, and this was associated with decreased cell proliferation in colon cancer cell lines SW480 and HCT15

(Wang et al. 2012). In addition to the regulation that occurs at the initial steps of Wnt signaling, the pathway is also mediated at other levels. DACT3, which interferes with Dvl during signal transduction and thereby antagonizes Wnt signaling, has been reported to be repressed in colorectal cancer through histone alteration but not promoter methylation. Using a combination of inhibitors of histone methylation and acetylation, the depression on DACT3 expression was rescued (Jiang et al. 2008).

In addition to its control by DNA methylation and histone modifications, emerging studies have shown that Wnt signaling is also regulated through miR-mediated mechanisms in colon cancer. miR-34, a downstream target of p53, was reported to be a Wnt suppressor by interacting with the UTR of Wnt1, LRP6, and β-catenin genes (Kim et al. 2011). Tumor-suppressive miR-101 showed a gradient decrease from normal mucosa, to the cancerous epithelium, and to the invasive front of cancerous tissues (Strillacci et al. 2013). Silencing of miR-101 increased Wnt signaling by upregulating β-catenin nuclear accumulation. On the contrary, over-expression of miR-101 resulted in a decrease of β-catenin in the nucleus. miR-1290 was increasingly expressed in colon cancer tissues, and has been proposed to promote tumor development (Wu et al. 2013). Stabilized expression of miR-1290 promoted colon cancer cell SW620 reprogramming by silencing DKK-3, upregulating Wnt signaling, and thereby activating c-Myc. Overall, appropriate Wnt signaling has been shown to be critical for maintaining colonic homeostasis, making it critical to unravel the epigenetic modifications that occur when this pathway is dysregulated during colon cancer development.

Cell Cycle Control

The expression of cyclin-dependent kinase (CDK) inhibitors is the primary molecular indicator of aging and senescence in cells and tissues (Campisi and Di Dda 2007). $p16^{INK4a}$ (inhibitor of CDK4a) is a cell cycle control gene that inhibits the formation of the CDK4/cyclin D complex (Serrano et al. 1993), which controls the arrest of the progression from the first gap phase (G1) to DNA synthesis (S) (Sherr 1994; Hunter and Pines 1994; Morgan 1995), while $p21^{Cip1}$ (cell cycle inhibitory protein 1) affects the activities of cyclin D-, E-, and A-dependent kinases (Sherr and Roberts 1999). Both $p16^{INK4a}$ and $p21^{Cip1}$ inhibit the formation of CDK/cyclin complexes, thereby decreasing the phosphorylation of downstream Rb (Mittnacht 1998). Hypophosphorylation of Rb increases its binding to transcription factor E2F, which causes a decrease of E2F-regulated transcription of downstream DNA synthesis-related genes responsible for cell progression from G1 to S phase (Sherr and Roberts 1999; Dyson 1998). Long-term changes in $p21^{Cip1}$ expression were reported in response to DNA damage, and in association with change in p53 and Rb protein (Almasan et al. 1995). Senescence stress was also shown to affect $p16^{INK4a}$ expression, which further led to cellular senescence in cells (Campisi and Di Dda 2007; Krishnamurthy et al. 2004). $p16^{INK4a}$ and $p21^{Cip1}$ can bind to CDK and inhibit the phosphorylation of retinoblastoma protein (Rb protein) by CDK (Mittnacht 1998; Liggett and Sidransky 1998). Oncogenic Ras/Raf may induce $p16^{INK4a}$ expression by activating ETS family transcription factors (Graves and Petersen 1998), which interact with ID transcription factors (Ohtani et al. 2001), and these are repressed during senescence (Hara et al. 1994). Other studies have shown that p16^{INK4a} was regulated by chromatin state (Jung et al. 2010; Feng et al. 2009).

As a critical regulator in the cell cycle control system, p16^{INK4a} has been proposed to be a tumor suppressor gene. Inactivation of the p16^{INK4a} gene, which often occurs through promoter hypermethylation of CpG islands, results in carcinogenesis in many types of cancers (Taghavi et al. 2010; Hinshelwood et al. 2009). In a study of 326 sporadic colorectal cancer patients, it was found that the p16^{INK4a} gene was methylated in about 25 % of tumor tissues and this occurrence depended on the differentiation grade of tumors and the location of the cancerous tissue. Promoter hypermethylation was observed more often in poorly differentiated tumors, and the distal colon had a low incidence of p16^{INK4a} hypermethylation (Veganzones-de-Castro et al. 2012). Contrary to this study, in a recent report, it was shown that both the gene expression and the promoter methylation level of p16^{INK4a} were higher in tumor tissues compared with the normal counterparts, but there was no correlation between these two (Yoruker et al. 2012). Furthermore, the level of methylation was not associated with the stage or tumor location in these patients. The methylation level of histone H3 lysine9 (H3K9) was similar in tumor and normal tissues, and there was no association between mono-, di-, and tri-methylation at histone H3K9 and the expression of p16^{INK4a}. This controversial observation might be due to the small sample size in each colon cancer stage category. While studies in humans are limited, the role of cell cycle control genes in colon cancer progression warrants further investigation, with a focus on the epigenetic marks that regulate the dysregulated expression of these genes.

15.2.2 *Promotion of Cell Viability*

Tumor-Promoting Inflammation

Inflammation is associated with all types of malignant tumors, and has been shown to promote tumor proliferation, angiogenesis, and metastasis, and increase resistance to hormonal- or chemotherapies (Wagner et al. 2012). Inflammation is driven by cytokines and chemokines, which are produced by tumor cells themselves and by the cells recruited to the tumor microenvironment, such as macrophages and mast cells (Fridman et al. 2011). Monocyte/macrophage cells are usually the major components of the inflammatory infiltrate in the microenvironment of most malignant tumors. The infiltration begins early in the noninvasive stage of the tumor, and continues progressively, with an eventual switch from the M1 pro-inflammatory phenotype to the M2 cancer-promoting phenotype (McClellan et al. 2012; Vendramini-Costa and Carvalho 2012). These changes directly influence tumor cells, and stimulate tissue remodeling, immunomodulation, angiogenesis, and tumor progression (Mantovani and Sica 2010). In colorectal cancer cells, stimulation by preexisting inflammation activates the local stromal cells to secrete a variety of cytokines (Oshima and Oshima 2012). Among the cytokines involved in inflammation-associated tumorigenesis, interleukin-6 (IL-6) and interleukin-8 (IL-8) play an important role. Both IL-6 and IL-8 were over-expressed in colon cancer with increased angiogenic and metastatic potential (Wilkening et al. 2008; Gunter et al. 2006). Tumor-associated macrophages are the most likely source of these cytokines, but colon cancer cells themselves were reported to synthesize IL-8, under inflammatory conditions (Rubie et al. 2007). Therefore, the IL-6- and IL-8-associated inflammatory networks are thought to significantly contribute to oncogene-induced cellular senescence with tumor progression (Kuilman et al. 2008).

Transforming growth factor b1 (TGFb1), MMPs, and vascular endothelial growth factor (VEGF) are inflammation-related molecules that become predominant during advanced tumor stages (Mantovani and Sica 2010). TGFb1 plays a major role in the relationship between inflammation and carcinogenesis. Under physiological conditions, TGFb1 is involved in regulating cell processes, including apoptosis, differentiation, proliferation, and survival. TGFb1 inhibits the growth of intestinal epithelial cells by increasing differentiation and apoptotic pathways, while promoting the proliferation of fibroblasts and myofibroblasts, and deposition of the extracellular matrix (Hawinkels et al. 2014). TGFb1 is an immunosuppressive factor that inhibits activation and differentiation of immune effector cells; thus, its over-expression might contribute to promoting the invasive and metastatic properties of tumor cells (Bellam and Pasche 2010). Disruption of the TGFb1 signal transduction pathway by mutations or polymorphisms of its receptors and/or of the transduction molecules (SMADs) contributes to the development and progression of several types of cancers, including colon cancer (Bellam and Pasche 2010). Disorder of TGFb1 receptors (TGFb1RI or TGFb1RII) does not only contribute to colon cancer initiation but also increase metastasis (Munoz et al. 2006; Zhang et al. 2009), and mutations in the TGFb1RII are estimated to occur in approximately 30 % of colon cancer cases (Biasi et al. 2002). Inflammatory cytokines are also activators of MMPs, which are enzymes that demolish the extracellular matrix and are involved in all steps of colorectal carcinogenesis (Lee et al. 2012). However, the actual pattern and relevance of their serum levels during both benign and malignant phases of colorectal carcinogenesis are less clear.

miRNAs as Survival Factors

The involvement of oncogenes and tumor suppressor genes in the control of cell proliferation and survival pathways is well established for protein-coding genes (Corvinus et al. 2005). miRs are important regulators of cancer survival genes because approximately half of the protein-coding genes are related to miR-mediated regulation (Chekulaeva and Filipowicz 2009). miRs are reported to play an important role in the pathogenesis of human cancers with dysregulated genomic signatures (Galasso et al. 2012). In colon cancer, a subset of miRs was aberrantly expressed in colon cancer, and most of the miRs are related to cell proliferation, apoptosis, and tumor metastasis (Wu et al. 2011; Schetter et al. 2012; Vickers et al. 2012). Because individual miRs can regulate multiple targets, their function may differ between cell types, depending on which of their target genes are being expressed (Ross and Davis 2011). Several miRs are described to be dysregulated in colon cancer, correlated with poor patient prognosis, and are proposed to be cancer biomarkers or important factors in cancer development and progression (Lu et al. 2005). Importantly, miR expression profiling of colon cancer cell lines revealed that fairly low number of the miRs were expressed, which is consistent with the hypothesis that the miR expression is tissue specific (Cummins et al. 2006). Investigating the involvement of miRs will reveal new information related to multistage colon cancer development, which may help us to identify novel potential biomarkers and therapeutic targets.

15.3 Dietary Adjustment to Prevent and Repress Colon Cancer: What We Do Not Know

15.3.1 Growth Suppressors

Uncontrolled cell growth is a major characteristic of cancer, and overconsumption of dietary energy boosts excessive cell growth. Additionally, loss of cell cycle control results in constitutive cell proliferation. Therefore, dietary modifications that could attenuate either of these two events may have roles in cancer prevention or anticancer treatment.

Reducing Excessive Energy Intake

Caloric restriction may reduce cancer risk by decreasing levels of plasma insulin and IGF-I, inhibiting cell proliferation, increasing cell death due to apoptosis, increasing activities of antioxidant enzymes, enhancing DNA repair, reducing oncogene expression, and influencing the levels of immunological responsiveness (Hursting et al. 2003). These studies suggest the importance of energy balance as a determinant for cancer risk. Although animal models clearly demonstrated a protective effect of energy restriction on cancer risk, it is less clear that such a protective effect exists in humans. A study in normal-weight humans found that a 20 % energy restriction for 10 weeks did not reduce oxidative DNA damage (Loft et al. 1995). Study of the 1944–1945 Dutch famine and subsequent overall cancer incidence found no evidence that the short famine affected overall cancer risk (Elias et al. 2005). However, another study found that higher energy intake in childhood may increase the risk of developing cancer in adult life (Uauy and Solomons 2005). Case–control studies suggested an increased risk of colon cancer associated with high energy intake (Howe et al. 1997), but the relationship could not be confirmed in a later study (Dahm et al. 2010), indicating that the precise relationships between energy intake and risk of colon cancer remains to be determined.

Obesity increases the risk of colorectal cancer in men and women (Vanamala et al. 2008), and the association is generally more consistent and stronger for men than for women (Robsahm et al. 2013; Xiao et al. 2014; Renehan et al. 2012), and for cancer of the distal colon than the proximal colon (Laake et al. 2010; MacInnis et al. 2004). Age, menopausal status, and HRT use may modify the association between BMI and colorectal cancer (Hou et al. 2006). Some studies reported a nearly twofold increase in colorectal cancer risk in people with a BMI $\geq$ 30 kg/m^2 compared with those with a BMI < 23 kg/m^2 (Calle and Kaaks 2004). Approximately 35.4 and 20.8 % of colorectal cancer cases in the US men and women as well as 27.5 and 14.2 % in European men and women could be attributed to overweight and obesity (Hull and Lagergren 2013; Bardou et al. 2013; Ma et al. 2013; Aleksandrova et al. 2013; Gribovskaja-Rupp et al. 2011). Waist circumference and waist–hip ratio are strong indicators of colorectal cancer risk in both sexes (Moore et al. 2004), suggesting that fat distribution may be more important than BMI for colorectal cancer risk. Furthermore, a positive relationship between body fatness and colorectal adenomas, and the association with colorectal cancer may imply that obesity may affect progression from adenoma to cancer (Chung et al. 2006; Sedjo et al. 2007).

Food Components That Inhibit Cancer Cell Growth

Dietary compounds have long been of interest for their anticancer potentials. As established epigenome modifiers, numerous dietary components can switch gene expression on and off, inducing or suppressing carcinogenesis. Functional dietary components are categorized into DNA methylation modulators, histone code modulators, and miR modulators (Chen and Xu 2010; Parasramka et al. 2012). Inside the nucleus, DNA methylation is regulated by a group of enzymes called DNMTs. DNMTs play different roles in the DNA methylation process. DNMT1 mainly plays a role in maintaining methylation, while DNMT3a and 3b are responsible for de novo DNA methylation (Okano et al. 1999; Rhee et al. 2002). In cancer cells, tumor suppressor genes are commonly repressed by DNA methylation, and DNMT inhibitors such as 5-aza-C and 5-aza-2dC suppress DNA methylation, thereby restoring the expression of genes that are silenced by promoter methylation in cancer cells. HDAC inhibitors such as trichostatin A (TSA) and Vorinostat (suberoylanilide hydroxamic acid, SAHA) repress histone acetylation at lysine residues, and natural dietary products have been shown to be epigenetic modifiers, acting as DNMT inhibitors, HDAC inhibitors, or both (Table 15.1).

Sulforaphane (SFN), one of the isothiocyanates extracted from cruciferous vegetables, such as broccoli, was reported to reduce tumor formation in colon of APC^{min} mice (Shen et al. 2007). It was also shown to suppress DNMT1 expression in colon cancer cell CaCo2 (Traka et al. 2005). Treatment of these cells with a physiologically relevant dose of SFN increased the gene expression pattern of genes that were

Table 15.1 Example of dietary epigenetic modifiers

Bioactive compound	Food source	Epigenetic impacts	References
Sulforaphane	Broccoli	Inhibits DNMT1; inhibits HDAC	Shen et al. (2007) and Traka et al. (2005)
Epigallocatechin gallate	Green tea	Inhibits DNMT; affects SAM; promotes miR-16, miR-210	Fang et al. (2007), Park et al. (2012), Tsang and Kwok (2010), and Wang et al. (2011)
Genistein	Soybean	Inhibits DNMT; promotes histone acetylation; promotes miR-200, miRNA	Li et al. (2009a), Wang et al. (2012), Wang and Chen (2010), and Zhang and Chen (2011a)

critical in increasing cell differentiation and decreasing cell proliferation (KLF4, a gut transcription factor), and regulating cell cycle arrest (p21). Additionally, SFN was shown to impact HDAC activity (Nian et al. 2009), and colon cancer cells treated with SFN underwent G2/M arrest, which coincided with depletion of HDACs, including HDAC1, selected HDAC2, and HDAC3 (Rajendran et al. 2011). Removal of SFN after SFN treatment resulted in recovery of HDAC expression and activity and a rescue of cell arrest from G2/M phase. The effect of SFN on miRs, however, currently remains unclear.

Epigallocatechin gallate (EGCG) is a natural polyphenol from green tea. It has been shown that EGCG was able to inhibit colon cancer cell growth by suppressing the activation of endothelial growth factor-mediated signaling (Shimizu et al. 2010). EGCG has also been reported as a DNMT inhibitor (Fang et al. 2007). By decreasing DNA methylation and elevating histone acetylation, EGCG treatment induced the re-expression of silenced tumor suppressor genes, p16 and p21, in human skin cancer cells (Nandakumar et al. 2011). In colon cancer cells, EGCG restored the expression of tumor suppressive p15 and p16 by decreasing their promoter methylation, and repressed colon cancer cell proliferation (Berner et al. 2010). In addition to regulating DNMTs and causing direct changes in DNA methylation, EGCG has been reported to impact the level of *S*-adenosylmethionine (SAM), the universal methyl donor, thereby indirectly mediating DNA methylation (Park et al. 2012). After treatment with EGCG, pancreatic cancer cell showed decreased invasive metastatic activity (Kim and Kim 2013), and this was potentially due to EGCG's action as an HDAC inhibitor, thereby inducing the expression of a Raf kinase inhibitor protein, which was important for repressing tumor invasion. EGCG was also shown to regulate miR expression in several cancers. It was shown to increase the expression of miR-16, which played a critical role in apoptotic activity of hepatoma cells (Tsang and Kwok 2010). Another study suggested that EGCG reduced the growth of lung cancer cells by promoting the expression of miR-210 (Wang et al. 2011). More studies have to be conducted to illustrate the specific mechanisms by which EGCG modulates miRs in colon cancer.

Genistein, a major isoflavone derived from soybeans, has been reported to be an epigenome modifier, and exerts anticancer effects (Zhang and Chen 2011b). Recently, Zhang et al. reported that feeding genistein-containing diets decreased the formation and severity of aberrant crypt foci, which were pre-neoplastic lesions after carcinogen-induction in rat descending colons (Zhang et al. 2013). Wnt signaling was protected from carcinogen-induced over-activation by the genistein diets. Furthermore, in vitro studies suggest that after genistein treatment, colon cancer cells show increased apoptotic activity and decreased cell proliferation. Genistein treatment restored the expression of Wnt5a and sFRP2 by increasing promoter demethylation and acting as a DNMT inhibitor, thereby attenuating aberrant Wnt signaling in colon cancer cells (Zhang and Chen 2011a; Wang and Chen 2010). Genistein has also been reported as a histone modifier. It induced the expression of DKK-1 through upregulating histone H3 acetylation at gene promoter (Wang et al. 2012). Although genistein had been reported to regulate miRs in several types of cancers (Li et al. 2009b), little is known about miR action in colon cancer.

15.3.2 Anti-inflammatory Dietary Agent

Numerous natural products have received significant interest over the years for their health promoting properties, especially those characterized as having anti-inflammatory (Poeckel et al. 2008), neuroprotective (Kim et al. 2006), antioxidant (Satoh et al. 2008), and anticancer properties (Johnson et al. 2010). Inflammation has been shown to occur throughout the multi-step tumorigenic cascade, which has led to the search for dietary agents that decrease inflammatory signaling pathways.

Curcumin is a major constituent of curry powder, which has been used in traditional Asian and African medicine to treat a wide variety of ailments for over 4,000 years (Epstein et al. 2010). Curcumin's molecular mechanisms of action, including its anti-inflammatory, antioxidant, and anticancer properties, have been extensively investigated (Mosieniak et al. 2012; Johnson and Mukhtar 2007; Wang et al. 2006; Rao et al. 1995).

Carnosol, another botanical compound, was isolated from sage (*Salvia carnosa*) and rosemary, and sage has been shown to contain a variety of polyphenols, including carnosol, carnosic acid, rosmanol, rosmarinic acid, as well as other active ingredients (Chang et al. 2008; Ngo et al. 2011). Carnosol was shown to reduce LPS-stimulated NO production in a mouse monocyte macrophage cell line, which led to an inhibition of the NF-κB, p38, and p44/42 mitogen-activated protein kinase (MAPK) (Lo et al. 2002). Additionally, carnosol was shown to activate the peroxisome proliferator-activated receptor gamma, reduce the pro-inflammatory mediators leukotrienes, inhibit lipoxygenase and the secretion of leukocyte elastase, and antagonize the intracellular Ca^{2+} mobilization (Poeckel et al. 2008). Carnosol also inhibited protein kinase C signaling and the binding of AP-1 to the COX-2 promoter, which was different than the synthetic COX-2 inhibitor celecoxib (Subbaramaiah et al. 2002). In vitro, rosemary extract was shown to inhibit the proliferation of ovarian cancer cell lines by affecting the cell cycle at multiple phases and modifying the expression of multiple genes regulating apoptosis (Tai et al. 2012). Carnosol also decreased intestinal multiplicity by 46 % and restored E-cadherin and b-catenin to the enterocyte membranes, thus producing a phenotype similar to the APC+/+ wild-type (WT) littermate in a mouse model (Moran et al. 2005). As previously discussed, inherited mutations in the APC tumor suppressor gene result in the generation of familial APC with somatic mutations in >80 % of sporadic colon cancers (Sancho et al. 2004). There is growing evidence that carnosol from rosemary extract can suppress the development of tumors in several organs including the colon, breast, liver, and stomach, as well as melanoma and leukemia cells, which could make it a potential candidate as a chemopreventive agent.

Lupeol is a triterpene (member of phytosterol family) derived from vegetable oils, cereals, fruits, and vegetables (Liby et al. 2007). Lupeol has been shown to possess many pharmacological properties, including anticancer and anti-inflammatory effects (Siddique et al. 2011). Topical application of Lupeol decreases neutrophil specific marker myeloperoxidase levels thus causing reduction in cell infiltration into inflamed tissues in mice (Fernandez et al. 2001). Lupeol pretreatment significantly reduced prostaglandin E2 production in stimulated macrophages (Fernandez et al. 2001), and in an animal model of carrageenan-induced inflammation, Lupeol treatment exhibits anti-inflammatory activity with a maximum inhibition of 57.14 % while α-mangostin, another well-known anti-inflammatory agent, showed inhibition activity of only 38.70 % at similar dose (Nguemfo et al. 2009). Lupeol further decreased the generation of pro-inflammatory cytokines, such as tumor necrosis factor α (TNFα) and interleukin b (ILb), in lipopolysaccharide-treated macrophages (Nguemfo et al. 2009). Lupeol was also shown to modulate the phagocytic activity of macrophages and T lymphocytes, and suppress $CD4^+$ T cell-mediated cytokine generation by reducing $CD4^+$ T and $CD8^+$ T cell counts and the level of cytokines (IL-2, IFN-gamma, and IL-4) in a mouse model (Bani et al. 2006). Additionally, Lupeol inhibited the activity of lipoxygenase with relatively low IC50 (Gutierrez-Lugo et al. 2004), and Lupeol treatment significantly reduced prostate cancer cell viability in a dose-dependent manner and caused apoptosis through degradation of acinus protein and poly(ADP-ribose) polymerase protein, and increase in the expression of FADD protein (Saleem et al. 2005). Among the targets studied, Lupeol caused a specific increase in the expression of Fas receptor in apoptotic pathway (Saleem et al. 2009a; Prasad et al. 2008), decrease in the protein levels of cyclins-A, -B1, -D1, -D2, and -E2, and CDK2, and increase in the protein level of CDK-inhibitor p21, which induce G2/M cell cycle arrest (Saleem et al. 2009b). Lupeol was also shown to exhibit multitarget efficacy within the beta-catenin signaling network resulting in the inhibition of prostate cancer cell proliferation (Saleem et al. 2009a). In a mouse skin cancer model, Lupeol treatment inhibited TPA-induced activation of PI3K, activation of NF-kappaB and IKKalpha, and degradation and phosphorylation of IkappaBalpha, which showed significantly reduced tumor incidence, lower tumor body burden, and a significant delay in the latency period for tumor appearance (Saleem et al. 2004). Taken together, these studies suggest that the therapeutic potency of natural compounds for inflammatory conditions in colon cancer warrants further investigation.

15.4 Future Direction

Colon cancer is a primary cause of cancer-related disease and death globally. Although colon cancer has been strongly associated with a Western lifestyle, increasing fiber, fruits, or vegetable in our diet is not enough to ward off a large percentage of colon cancer (Chan and Giovannucci 2010; Qasim and O'Morain 2010). Tumor progression in colon involves multiple stages associated by accumulation of inflammation, deregulated signaling pathways, and modified metabolic pathways, which provides excellent opportunities for the evaluation of chemopreventive agents, especially epigenetic modifiers from natural food compounds. Epigenetic intervention can help to reduce colon cancer incidence by intervening key development pathways that promote growth and metastases of colon cancer.

Due to its importance for regulating transcriptional activity of genes, the promoter region of genes has historically been of particular interest to researchers investigating the epigenetic impact of food compound on the activation or suppression of gene expression. In early studies, most analyses were focused and conducted only in a few target genes, primarily due to the limitations of research technology for DNA methylation. Recently, with the application of high-throughput sequencing technology combined with advanced computational analysis, the discoveries of epigenetic phenomena have been brought up to a genome-wide scale. Using methylated DNA immunoprecipitation sequencing (MeDIP-seq), and Infinium HumanMethylation450 BeadChip, researchers can now map out even the slightest shifts in DNA methylation within the human genome. Moreover, in addition to looking at the 20,000 protein-coding genes, which only composes 1.5 % of the human genome, these new technologies allow us to screen any changes occur throughout the rest of genome, including intra- and intergenic regions. It has been clearly demonstrated that the changes in DNA methylation that lead to changes in gene expression occur not only within the promoter region CpG islands but also within other gene regions. For example, intragenic methylation, or the DNA methylation that occurs on CGIs within gene bodies rather than promoters, greatly affects the regulation of gene transcription, especially the tissue-specific gene expression. Moreover, the change of intragenic methylation was shown to be inversely correlated with that of tri-methylation of histone H3K4 (Maunakea et al. 2010). Diet has the ability to induce genome-wide DNA methylation, and one study reported that a short-term overfeeding of a high-fat diet affected the methylation status of about 45 % of genes within human skeletal muscle (Jacobsen et al. 2012), and the changed genes were primarily associated with inflammation, reproduction, and cancer. However, despite the drastic change in DNA methylation, few genes were found to have altered expression levels that corresponded to their changed methylation profiles. More efforts have to be made to clearly elucidate the mechanisms underlying the impact of diet on molecular signaling pathways, and on a genome-wide scale. By doing so, we will have a comprehensive view of how diet can affect our body functions and health through a network of inter- and intracellular signaling pathways, which will lead to further dietary interventions for preventing diseases such as colon cancer.

Acknowledgments This publication or project was made possible by Grant Number P50AT006268 from the National Center for Complementary and Alternative Medicines (NCCAM), the Office of Dietary Supplements (ODS), and the National Cancer Institute (NCI). Its contents are solely the responsibility of the authors and do not necessarily represent the official views of the NCCAM, ODS, NCI, or the National Institutes of Health.

The authors wish to acknowledge Dr. Rita S. Strakovsky (University of Illinois) for her critical reading and helpful comments on the manuscript.

References

Aguilera O, Fraga MF, Ballestar E, Paz MF, Herranz M, Espada J, Garcia JM, Munoz A, Esteller M, Gonzalez-Sancho JM (2006) Epigenetic inactivation of the Wnt antagonist DICKKOPF-1 (DKK-1) gene in human colorectal cancer. Oncogene 25:4116–4121

Aleksandrova K, Nimptsch K, Pischon T (2013) Obesity and colorectal cancer. Front Biosci (Elite Ed) 5:61–77

Almasan A, Linke SP, Paulson TG, Huang LC, Wahl GM (1995) Genetic instability as a consequence of inappropriate entry into and progression through S-phase. Cancer Metastasis Rev 14:59–73

Arozarena I, Bischof H, Gilby D, Belloni B, Dummer R, Wellbrock C (2011) In melanoma, beta-catenin is a suppressor of invasion. Oncogene 30:4531–4543

Bani S, Kaul A, Khan B, Ahmad SF, Suri KA, Gupta BD, Satti NK, Qazi GN (2006) Suppression of T lymphocyte activity by lupeol isolated from Crataeva religiosa. Phytother Res 20:279–287

Bardou M, Barkun AN, Martel M (2013) Obesity and colorectal cancer. Gut 62:933–947

Bellam N, Pasche B (2010) Tgf-beta signaling alterations and colon cancer. Cancer Treat Res 155:85–103

Berlau J, Glei M, Pool-Zobel BL (2004) Colon cancer risk factors from nutrition. Anal Bioanal Chem 378:737–743

Berner C, Aumuller E, Gnauck A, Nestelberger M, Just A, Haslberger AG (2010) Epigenetic control of estrogen receptor expression and tumor suppressor genes is modulated by bioactive food compounds. Ann Nutr Metab 57:183–189

Berrigan D, Lavigne JA, Perkins SN, Nagy TR, Barrett JC, Hursting SD (2005) Phenotypic effects of calorie restriction and insulin-like growth factor-1 treatment on body composition and bone mineral density of C57BL/6 mice: implications for cancer prevention. In Vivo 19:667–674

Biasi F, Tessitore L, Zanetti D, Cutrin JC, Zingaro B, Chiarpotto E, Zarkovic N, Serviddio G, Poli G (2002) Associated changes of lipid peroxidation and transforming growth factor beta1 levels in human colon cancer during tumour progression. Gut 50:361–367

Bovolenta P, Esteve P, Ruiz JM, Cisneros E, Lopez-Rios J (2008) Beyond Wnt inhibition: new functions of secreted Frizzled-related proteins in development and disease. J Cell Sci 121:737–746

Calle EE, Kaaks R (2004) Overweight, obesity and cancer: epidemiological evidence and proposed mechanisms. Nat Rev Cancer 4:579–591

Campisi J, Di Dda FF (2007) Cellular senescence: when bad things happen to good cells. Nat Rev Mol Cell Biol 8:729–740

Castaneda C, Gordon PL, Fielding RA, Evans WJ, Crim MC (2000) Marginal protein intake results in reduced plasma IGF-I levels and skeletal muscle fiber atrophy in elderly women. J Nutr Health Aging 4:85–90

Chan AT, Giovannucci EL (2010) Primary prevention of colorectal cancer. Gastroenterology 138:2029–2043.e10

Chang CH, Chyau CC, Hsieh CL, Wu YY, Ker YB, Tsen HY, Peng RY (2008) Relevance of phenolic diterpene constituents to antioxidant activity of supercritical CO(2) extract from the leaves of rosemary. Nat Prod Res 22:76–90

Chekulaeva M, Filipowicz W (2009) Mechanisms of miRNA-mediated post-transcriptional regulation in animal cells. Curr Opin Cell Biol 21:452–460

Chen J, Xu X (2010) Diet, epigenetic, and cancer prevention. Adv Genet 71:237–255

Chujo Y, Fujii N, Okita N, Konishi T, Narita T, Yamada A, Haruyama Y, Tashiro K, Chiba T, Shimokawa I, Higami Y (2013) Caloric restriction-associated remodeling of rat white adipose tissue: effects on the growth hormone/insulin-like growth factor-1 axis, sterol regulatory element binding protein-1, and macrophage infiltration. Age (Dordr) 35:1143–1156

Chung YW, Han DS, Park YK, Son BK, Paik CH, Lee HL, Jeon YC, Sohn JH (2006) Association of obesity, serum glucose and lipids with the risk of advanced colorectal adenoma and cancer: a case-control study in Korea. Dig Liver Dis 38:668–672

Clemmons DR, Busby WH, Arai T, Nam TJ, Clarke JB, Jones JI, Ankrapp DK (1995) Role of insulin-like growth factor binding proteins in the control of IGF actions. Prog Growth Factor Res 6:357–366

Corvinus FM, Orth C, Moriggl R, Tsareva SA, Wagner S, Pfitzner EB, Baus D, Kaufmann R, Huber LA, Zatloukal K, Beug H, Ohlschlager P, Schutz A, Halbhuber KJ, Friedrich K (2005) Persistent STAT3 activation in colon cancer is associated with enhanced cell proliferation and tumor growth. Neoplasia 7:545–555

Cummins JM, He Y, Leary RJ, Pagliarini R, Diaz LA Jr, Sjoblom T, Barad O, Bentwich Z, Szafranska AE, Labourier E, Raymond CK, Roberts BS, Juhl H, Kinzler KW, Vogelstein B, Velculescu VE (2006) The colorectal microRNAome. Proc Natl Acad Sci U S A 103: 3687–3692

Dahm CC, Keogh RH, Lentjes MA, Spencer EA, Key TJ, Greenwood DC, Cade JE, Burley VJ, Shipley MJ, Brunner EJ, Stephen AM, Mishra G, Kuh D, Fentiman IS, White IR, Luben R, Khaw KT, Rodwell Bingham SA (2010) Intake of dietary fats and colorectal cancer risk: prospective findings from the UK Dietary Cohort Consortium. Cancer Epidemiol 34:562–567

Deka J, Wiedemann N, Anderle P, Murphy-Seiler F, Bultinck J, Eyckerman S, Stehle JC, Andre S, Vilain N, Zilian O, Robine S, Delorenzi M, Basler K, Aguet M (2010) Bcl9/Bcl9l are critical for Wnt-mediated regulation of stem cell traits in colon epithelium and adenocarcinomas. Cancer Res 70:6619–6628

Dyson N (1998) The regulation of E2F by pRB-family proteins. Genes Dev 12:2245–2262

Elias SG, Peeters PH, Grobbee DE, Van Noord PA (2005) The 1944-1945 Dutch famine and subsequent overall cancer incidence. Cancer Epidemiol Biomarkers Prev 14:1981–1985

Epstein J, Sanderson IR, Macdonald TT (2010) Curcumin as a therapeutic agent: the evidence from in vitro, animal and human studies. Br J Nutr 103:1545–1557

Esteller M (2007) Epigenetics provides a new generation of oncogenes and tumour-suppressor genes. Br J Cancer 96(suppl):R26–R30

Fang M, Chen D, Yang CS (2007) Dietary polyphenols may affect DNA methylation. J Nutr 137:223S–228S

Feng Y, Wang X, Xu L, Pan H, Zhu S, Liang Q, Huang B, Lu J (2009) The transcription factor ZBP-89 suppresses p16 expression through a histone modification mechanism to affect cell senescence. FEBS J 276:4197–4206

Ferlay J, Shin HR, Bray F, Forman D, Mathers C, Parkin DM (2010) Estimates of worldwide burden of cancer in 2008: GLOBOCAN 2008. Int J Cancer 127:2893–2917

Fernandez MA, De Las Heras B, Garcia MD, Saenz MT, Villar A (2001) New insights into the mechanism of action of the anti-inflammatory triterpene lupeol. J Pharm Pharmacol 53:1533–1539

Fevr T, Robine S, Louvard D, Huelsken J (2007) Wnt/beta-catenin is essential for intestinal homeostasis and maintenance of intestinal stem cells. Mol Cell Biol 27:7551–7559

Fridman WH, Galon J, Dieu-Nosjean MC, Cremer I, Fisson S, Damotte D, Pages F, Tartour E, Sautes-Fridman C (2011) Immune infiltration in human cancer: prognostic significance and disease control. Curr Top Microbiol Immunol 344:1–24

Galasso M, Sandhu SK, Volinia S (2012) MicroRNA expression signatures in solid malignancies. Cancer J 18:238–243

Gonzalez-Sancho JM, Aguilera O, Garcia JM, Pendas-Franco N, Pena C, Cal S, De Garcia HA, Bonilla F, Munoz A (2005) The Wnt antagonist DICKKOPF-1 gene is a downstream target of beta-catenin/TCF and is downregulated in human colon cancer. Oncogene 24:1098–1103

Graves BJ, Petersen JM (1998) Specificity within the ets family of transcription factors. Adv Cancer Res 75:1–55

Gregorieff A, Clevers H (2005) Wnt signaling in the intestinal epithelium: from endoderm to cancer. Genes Dev 19:877–890

Gribovskaja-Rupp I, Kosinski L, Ludwig KA (2011) Obesity and colorectal cancer. Clin Colon Rectal Surg 24:229–243

Groden J, Thliveris A, Samowitz W, Carlson M, Gelbert L, Albertsen H, Joslyn G, Stevens J, Spirio L, Robertson M et al (1991) Identification and characterization of the familial adenomatous polyposis coli gene. Cell 66:589–600

Gunter MJ, Canzian F, Landi S, Chanock SJ, Sinha R, Rothman N (2006) Inflammation-related gene polymorphisms and colorectal adenoma. Cancer Epidemiol Biomarkers Prev 15: 1126–1131

Gutierrez-Lugo MT, Deschamps JD, Holman TR, Suarez E, Timmermann BN (2004) Lipoxygenase inhibition by anadanthoflavone, a new flavonoid from the aerial parts of Anadenanthera colubrina. Planta Med 70:263–265

Hara E, Yamaguchi T, Nojima H, Ide T, Campisi J, Okayama H, Oda K (1994) Id-related genes encoding helix-loop-helix proteins are required for G1 progression and are repressed in senescent human fibroblasts. J Biol Chem 269:2139–2145

Hawinkels LJ, Paauwe M, Verspaget HW, Wiercinska E, Van Der Zon JM, Van Der Ploeg K, Koelink PJ, Lindeman JH, Mesker W, Ten Dijke P, Sier CF (2014) Interaction with colon cancer cells hyperactivates TGF-beta signaling in cancer-associated fibroblasts. Oncogene 33:97–107

Hinshelwood RA, Melki JR, Huschtscha LI, Paul C, Song JZ, Stirzaker C, Reddel RR, Clark SJ (2009) Aberrant de novo methylation of the p16INK4A CpG island is initiated post gene silencing in association with chromatin remodelling and mimics nucleosome positioning. Hum Mol Genet 18:3098–3109

Hou L, Ji BT, Blair A, Dai Q, Gao YT, Potter JD, Chow WH (2006) Body mass index and colon cancer risk in Chinese people: menopause as an effect modifier. Eur J Cancer 42:84–90

Howe GR, Aronson KJ, Benito E, Castelleto R, Cornee J, Duffy S, Gallagher RP, Iscovich JM, Deng-Ao J, Kaaks R, Kune GA, Kune S, Lee HP, Lee M, Miller AB, Peters RK, Potter JD, Riboli E, Slattery ML, Trichopoulos D, Tuyns A, Tzonou A, Watson LF, Whittemore AS, Shu Z et al (1997) The relationship between dietary fat intake and risk of colorectal cancer: evidence from the combined analysis of 13 case-control studies. Cancer Causes Control 8: 215–228

Hull M, Lagergren J (2013) Obesity and colorectal cancer. Gut 62:933–947

Hunter T, Pines J (1994) Cyclins and cancer. II: Cyclin D and CDK inhibitors come of age. Cell 79:573–582

Hursting SD, Lavigne JA, Berrigan D, Perkins SN, Barrett JC (2003) Calorie restriction, aging, and cancer prevention: mechanisms of action and applicability to humans. Annu Rev Med 54: 131–152

Iorio MV, Croce CM (2012) MicroRNA dysregulation in cancer: diagnostics, monitoring and therapeutics. A comprehensive review. EMBO Mol Med 4:143–159

Jacobsen SC, Brons C, Bork-Jensen J, Ribel-Madsen R, Yang B, Lara E, Hall E, Calvanese V, Nilsson E, Jorgensen SW, Mandrup S, Ling C, Fernandez AF, Fraga MF, Poulsen P, Vaag A (2012) Effects of short-term high-fat overfeeding on genome-wide DNA methylation in the skeletal muscle of healthy young men. Diabetologia 55:3341–3349

Jenuwein T, Allis CD (2001) Translating the histone code. Science 293:1074–1080

Jiang X, Tan J, Li J, Kivimae S, Yang X, Zhuang L, Lee PL, Chan MT, Stanton LW, Liu ET, Cheyette BN, Yu Q (2008) DACT3 is an epigenetic regulator of Wnt/beta-catenin signaling in colorectal cancer and is a therapeutic target of histone modifications. Cancer Cell 13:529–541

Johnson JJ, Mukhtar H (2007) Curcumin for chemoprevention of colon cancer. Cancer Lett 255:170–181
Johnson JJ, Syed DN, Suh Y, Heren CR, Saleem M, Siddiqui IA, Mukhtar H (2010) Disruption of androgen and estrogen receptor activity in prostate cancer by a novel dietary diterpene carnosol: implications for chemoprevention. Cancer Prev Res (Phila) 3:1112–1123
Jones JI, Clemmons DR (1995) Insulin-like growth factors and their binding proteins: biological actions. Endocr Rev 16:3–34
Jones PA, Takai D (2001) The role of DNA methylation in mammalian epigenetics. Science 293:1068–1070
Jung JW, Lee S, Seo MS, Park SB, Kurtz A, Kang SK, Kang KS (2010) Histone deacetylase controls adult stem cell aging by balancing the expression of polycomb genes and jumonji domain containing 3. Cell Mol Life Sci 67:1165–1176
Kansra S, Ewton DZ, Wang J, Friedman E (2000) IGFBP-3 mediates TGF beta 1 proliferative response in colon cancer cells. Int J Cancer 87:373–378
Kelloff GJ, Schilsky RL, Alberts DS, Day RW, Guyton KZ, Pearce HL, Peck JC, Phillips R, Sigman CC (2004) Colorectal adenomas: a prototype for the use of surrogate end points in the development of cancer prevention drugs. Clin Cancer Res 10:3908–3918
Kim SO, Kim MR (2013) (-)-Epigallocatechin 3-gallate inhibits invasion by inducing the expression of Raf kinase inhibitor protein in AsPC1 human pancreatic adenocarcinoma cells through the modulation of histone deacetylase activity. Int J Oncol 42:349–358
Kim SJ, Kim JS, Cho HS, Lee HJ, Kim SY, Kim S, Lee SY, Chun HS (2006) Carnosol, a component of rosemary (Rosmarinus officinalis L.) protects nigral dopaminergic neuronal cells. Neuroreport 17:1729–1733
Kim MS, Lee J, Sidransky D (2010) DNA methylation markers in colorectal cancer. Cancer Metastasis Rev 29:181–206
Kim YH, Lee HC, Kim SY, Yeom YI, Ryu KJ, Min BH, Kim DH, Son HJ, Rhee PL, Kim JJ, Rhee JC, Kim HC, Chun HK, Grady WM, Kim YS (2011) Epigenomic analysis of aberrantly methylated genes in colorectal cancer identifies genes commonly affected by epigenetic alterations. Ann Surg Oncol 18:2338–2347
Klapholz-Brown Z, Walmsley GG, Nusse YM, Nusse R, Brown PO (2007) Transcriptional program induced by Wnt protein in human fibroblasts suggests mechanisms for cell cooperativity in defining tissue microenvironments. PLoS One 2:e945
Krishnamurthy J, Torrice C, Ramsey MR, Kovalev GI, Al-Regaiey K, Su L, Sharpless NE (2004) Ink4a/Arf expression is a biomarker of aging. J Clin Invest 114:1299–1307
Kritchevsky D (2002) Caloric restriction and experimental carcinogenesis. Hybrid Hybridomics 21:147–151
Kuilman T, Michaloglou C, Vredeveld LC, Douma S, Van Doorn R, Desmet CJ, Aarden LA, Mooi WJ, Peeper DS (2008) Oncogene-induced senescence relayed by an interleukin-dependent inflammatory network. Cell 133:1019–1031
Laake I, Thune I, Selmer R, Tretli S, Slattery ML, Veierod MB (2010) A prospective study of body mass index, weight change, and risk of cancer in the proximal and distal colon. Cancer Epidemiol Biomarkers Prev 19:1511–1522
Lee SW, Ahn YY, Kim YS, Kang SB, Nam SW, Lee DS, Jeong HY, Kim JM (2012) The immunohistochemical expression of STAT3, Bcl-xL, and MMP-2 proteins in colon adenoma and adenocarcinoma. Gut Liver 6:45–51
Lelbach A, Muzes G, Feher J (2005) The insulin-like growth factor system: IGFs, IGF-binding proteins and IGFBP-proteases. Acta Physiol Hung 92:97–107
Li Y, Liu L, Andrews LG, Tollefsbol TO (2009a) Genistein depletes telomerase activity through cross-talk between genetic and epigenetic mechanisms. Int J Cancer 125:286–296
Li Y, Vandenboom TG 2nd, Kong D, Wang Z, Ali S, Philip PA, Sarkar FH (2009b) Up-regulation of miR-200 and let-7 by natural agents leads to the reversal of epithelial-to-mesenchymal transition in gemcitabine-resistant pancreatic cancer cells. Cancer Res 69:6704–6712
Liby KT, Yore MM, Sporn MB (2007) Triterpenoids and rexinoids as multifunctional agents for the prevention and treatment of cancer. Nat Rev Cancer 7:357–369

Liggett WH Jr, Sidransky D (1998) Role of the p16 tumor suppressor gene in cancer. J Clin Oncol 16:1197–1206
Lo AH, Liang YC, Lin-Shiau SY, Ho CT, Lin JK (2002) Carnosol, an antioxidant in rosemary, suppresses inducible nitric oxide synthase through down-regulating nuclear factor-kappaB in mouse macrophages. Carcinogenesis 23:983–991
Loft S, Velthuis-Te Wierik EJ, Van Den Berg H, Poulsen HE (1995) Energy restriction and oxidative DNA damage in humans. Cancer Epidemiol Biomarkers Prev 4:515–519
Lu J, Getz G, Miska EA, Alvarez-Saavedra E, Lamb J, Peck D, Sweet-Cordero A, Ebert BL, Mak RH, Ferrando AA, Downing JR, Jacks T, Horvitz HR, Golub TR (2005) MicroRNA expression profiles classify human cancers. Nature 435:834–838
Ma Y, Yang Y, Wang F, Zhang P, Shi C, Zou Y, Qin H (2013) Obesity and risk of colorectal cancer: a systematic review of prospective studies. PLoS One 8:e53916
Macinnis RJ, English DR, Hopper JL, Haydon AM, Gertig DM, Giles GG (2004) Body size and composition and colon cancer risk in men. Cancer Epidemiol Biomarkers Prev 13:553–559
Maehata T, Taniguchi H, Yamamoto H, Nosho K, Adachi Y, Miyamoto N, Miyamoto C, Akutsu N, Yamaoka S, Itoh F (2008) Transcriptional silencing of Dickkopf gene family by CpG island hypermethylation in human gastrointestinal cancer. World J Gastroenterol 14:2702–2714
Mantovani A, Sica A (2010) Macrophages, innate immunity and cancer: balance, tolerance, and diversity. Curr Opin Immunol 22:231–237
Marshman E, Streuli CH (2002) Insulin-like growth factors and insulin-like growth factor binding proteins in mammary gland function. Breast Cancer Res 4:231–239
Maunakea AK, Chepelev I, Zhao K (2010) Epigenome mapping in normal and disease States. Circ Res 107:327–339
Mcclellan JL, Davis JM, Steiner JL, Enos RT, Jung SH, Carson JA, Pena MM, Carnevale KA, Berger FG, Murphy EA (2012) Linking tumor-associated macrophages, inflammation, and intestinal tumorigenesis: role of MCP-1. Am J Physiol Gastrointest Liver Physiol 303: G1087–G1095
Milner JA (2006) Diet and cancer: facts and controversies. Nutr Cancer 56:216–224
Mittnacht S (1998) Control of pRB phosphorylation. Curr Opin Genet Dev 8:21–27
Moon RT (2005) Wnt/beta-catenin pathway. Sci STKE 2005:cm1
Moore LL, Bradlee ML, Singer MR, Splansky GL, Proctor MH, Ellison RC, Kreger BE (2004) BMI and waist circumference as predictors of lifetime colon cancer risk in Framingham Study adults. Int J Obes Relat Metab Disord 28:559–567
Moran AE, Carothers AM, Weyant MJ, Redston M, Bertagnolli MM (2005) Carnosol inhibits beta-catenin tyrosine phosphorylation and prevents adenoma formation in the C57BL/6J/Min/+ (Min/+) mouse. Cancer Res 65:1097–1104
Morgan DO (1995) Principles of CDK regulation. Nature 374:131–134
Morin PJ, Sparks AB, Korinek V, Barker N, Clevers H, Vogelstein B, Kinzler KW (1997) Activation of beta-catenin-Tcf signaling in colon cancer by mutations in beta-catenin or APC. Science 275:1787–1790
Mosieniak G, Adamowicz M, Alster O, Jaskowiak H, Szczepankiewicz AA, Wilczynski GM, Ciechomska IA, Sikora E (2012) Curcumin induces permanent growth arrest of human colon cancer cells: link between senescence and autophagy. Mech Ageing Dev 133:444–455
Munoz NM, Upton M, Rojas A, Washington MK, Lin L, Chytil A, Sozmen EG, Madison BB, Pozzi A, Moon RT, Moses HL, Grady WM (2006) Transforming growth factor beta receptor type II inactivation induces the malignant transformation of intestinal neoplasms initiated by Apc mutation. Cancer Res 66:9837–9844
Nandakumar V, Vaid M, Katiyar SK (2011) (-)-Epigallocatechin-3-gallate reactivates silenced tumor suppressor genes, Cip1/p21 and p16INK4a, by reducing DNA methylation and increasing histones acetylation in human skin cancer cells. Carcinogenesis 32:537–544
Ngo SN, Williams DB, Head RJ (2011) Rosemary and cancer prevention: preclinical perspectives. Crit Rev Food Sci Nutr 51:946–954
Nguemfo EL, Dimo T, Dongmo AB, Azebaze AG, Alaoui K, Asongalem AE, Cherrah Y, Kamtchouing P (2009) Anti-oxidative and anti-inflammatory activities of some isolated

constituents from the stem bark of Allanblackia monticola Staner L.C (Guttiferae). Inflammopharmacology 17:37–41
Nian H, Delage B, Ho E, Dashwood RH (2009) Modulation of histone deacetylase activity by dietary isothiocyanates and allyl sulfides: studies with sulforaphane and garlic organosulfur compounds. Environ Mol Mutagen 50:213–221
Nogueira LM, Lavigne JA, Chandramouli GV, Lui H, Barrett JC, Hursting SD (2012) Dose-dependent effects of calorie restriction on gene expression, metabolism, and tumor progression are partially mediated by insulin-like growth factor-1. Cancer Med 1:275–288
Ohtani N, Zebedee Z, Huot TJ, Stinson JA, Sugimoto M, Ohashi Y, Sharrocks AD, Peters G, Hara E (2001) Opposing effects of Ets and Id proteins on p16INK4a expression during cellular senescence. Nature 409:1067–1070
Okano M, Bell DW, Haber DA, Li E (1999) DNA methyltransferases Dnmt3a and Dnmt3b are essential for de novo methylation and mammalian development. Cell 99:247–257
Oshima H, Oshima M (2012) The inflammatory network in the gastrointestinal tumor microenvironment: lessons from mouse models. J Gastroenterol 47:97–106
Parasramka MA, Ho E, Williams DE, Dashwood RH (2012) MicroRNAs, diet, and cancer: new mechanistic insights on the epigenetic actions of phytochemicals. Mol Carcinog 51:213–230
Park JH (2008) Inhibition of colon cancer cell growth by dietary components: role of the insulin-like growth factor (IGF) system. Asia Pac J Clin Nutr 17(suppl 1):257–260
Park LK, Friso S, Choi SW (2012) Nutritional influences on epigenetics and age-related disease. Proc Nutr Soc 71:75–83
Poeckel D, Greiner C, Verhoff M, Rau O, Tausch L, Hornig C, Steinhilber D, Schubert-Zsilavecz M, Werz O (2008) Carnosic acid and carnosol potently inhibit human 5-lipoxygenase and suppress pro-inflammatory responses of stimulated human polymorphonuclear leukocytes. Biochem Pharmacol 76:91–97
Prasad S, Nigam N, Kalra N, Shukla Y (2008) Regulation of signaling pathways involved in lupeol induced inhibition of proliferation and induction of apoptosis in human prostate cancer cells. Mol Carcinog 47:916–924
Preston-Martin S, Pike MC, Ross RK, Jones PA, Henderson BE (1990) Increased cell division as a cause of human cancer. Cancer Res 50:7415–7421
Qasim A, O'morain C (2010) Primary prevention of colorectal cancer: are we closer to reality? Eur J Gastroenterol Hepatol 22:9–17
Qu Z, Chow JC, Ling PR, Ziegler TR, Bistrian BR, Smith RJ (1997) Tissue-specific effects of chronic dietary protein restriction and gastrostomy on the insulin-like growth factor-I pathway in the liver and colon of adult rats. Metabolism 46:691–697
Rajendran P, Delage B, Dashwood WM, Yu TW, Wuth B, Williams DE, Ho E, Dashwood RH (2011) Histone deacetylase turnover and recovery in sulforaphane-treated colon cancer cells: competing actions of 14-3-3 and Pin1 in HDAC3/SMRT corepressor complex dissociation/reassembly. Mol Cancer 10:68
Rao CV, Rivenson A, Simi B, Reddy BS (1995) Chemoprevention of colon cancer by dietary curcumin. Ann N Y Acad Sci 768:201–204
Ren M, Zhong X, Ma CY, Sun Y, Guan QB, Cui B, Guo J, Wang H, Gao L, Zhao JJ (2009) Insulin-like growth factor-1 promotes cell cycle progression via upregulation of cyclin D1 expression through the phosphatidylinositol 3-kinase/nuclear factor-kappaB signaling pathway in FRTL thyroid cells. Acta Pharmacol Sin 30:113–119
Renehan AG, Flood A, Adams KF, Olden M, Hollenbeck AR, Cross AJ, Leitzmann MF (2012) Body mass index at different adult ages, weight change, and colorectal cancer risk in the National Institutes of Health-AARP Cohort. Am J Epidemiol 176:1130–1140
Rhee I, Bachman KE, Park BH, Jair KW, Yen RW, Schuebel KE, Cui H, Feinberg AP, Lengauer C, Kinzler KW, Baylin SB, Vogelstein B (2002) DNMT1 and DNMT3b cooperate to silence genes in human cancer cells. Nature 416:552–556
Robsahm TE, Aagnes B, Hjartaker A, Langseth H, Bray FI, Larsen IK (2013) Body mass index, physical activity, and colorectal cancer by anatomical subsites: a systematic review and meta-analysis of cohort studies. Eur J Cancer Prev 22:492–505

Ross SA, Davis CD (2011) MicroRNA, nutrition, and cancer prevention. Adv Nutr 2:472–485
Rubie C, Frick VO, Pfeil S, Wagner M, Kollmar O, Kopp B, Graber S, Rau BM, Schilling MK (2007) Correlation of IL-8 with induction, progression and metastatic potential of colorectal cancer. World J Gastroenterol 13:4996–5002
Saigusa S, Inoue Y, Tanaka K, Toiyama Y, Matsushita K, Kawamura M, Okugawa Y, Hiro J, Uchida K, Mohri Y, Kusunoki M (2012) Clinical significance of LGR5 and CD44 expression in locally advanced rectal cancer after preoperative chemoradiotherapy. Int J Oncol 41: 1643–1652
Saleem M, Afaq F, Adhami VM, Mukhtar H (2004) Lupeol modulates NF-kappaB and PI3K/Akt pathways and inhibits skin cancer in CD-1 mice. Oncogene 23:5203–5214
Saleem M, Kweon MH, Yun JM, Adhami VM, Khan N, Syed DN, Mukhtar H (2005) A novel dietary triterpene Lupeol induces fas-mediated apoptotic death of androgen-sensitive prostate cancer cells and inhibits tumor growth in a xenograft model. Cancer Res 65:11203–11213
Saleem M, Murtaza I, Tarapore RS, Suh Y, Adhami VM, Johnson JJ, Siddiqui IA, Khan N, Asim M, Hafeez BB, Shekhani MT, Li B, Mukhtar H (2009a) Lupeol inhibits proliferation of human prostate cancer cells by targeting beta-catenin signaling. Carcinogenesis 30:808–817
Saleem M, Murtaza I, Witkowsky O, Kohl AM, Maddodi N (2009b) Lupeol triterpene, a novel diet-based microtubule targeting agent: disrupts survivin/cFLIP activation in prostate cancer cells. Biochem Biophys Res Commun 388:576–582
Salhia B, Baker A, Ahmann G, Auclair D, Fonseca R, Carpten J (2010) DNA methylation analysis determines the high frequency of genic hypomethylation and low frequency of hypermethylation events in plasma cell tumors. Cancer Res 70:6934–6944
Sancho E, Batlle E, Clevers H (2004) Signaling pathways in intestinal development and cancer. Annu Rev Cell Dev Biol 20:695–723
Sato H, Suzuki H, Toyota M, Nojima M, Maruyama R, Sasaki S, Takagi H, Sogabe Y, Sasaki Y, Idogawa M, Sonoda T, Mori M, Imai K, Tokino T, Shinomura Y (2007) Frequent epigenetic inactivation of DICKKOPF family genes in human gastrointestinal tumors. Carcinogenesis 28:2459–2466
Satoh T, Izumi M, Inukai Y, Tsutsumi Y, Nakayama N, Kosaka K, Shimojo Y, Kitajima C, Itoh K, Yokoi T, Shirasawa T (2008) Carnosic acid protects neuronal HT22 Cells through activation of the antioxidant-responsive element in free carboxylic acid- and catechol hydroxyl moieties-dependent manners. Neurosci Lett 434:260–265
Schetter AJ, Okayama H, Harris CC (2012) The role of microRNAs in colorectal cancer. Cancer J 18:244–252
Secretan B, Straif K, Baan R, Grosse Y, El Ghissassi F, Bouvard V, Benbrahim-Tallaa L, Guha N, Freeman C, Galichet L, Cogliano V, WHO International Agency for Research on Cancer Monograph Working Group (2009) A review of human carcinogens—Part E: tobacco, areca nut, alcohol, coal smoke, and salted fish. Lancet Oncol 10:1033–1044
Sedjo RL, Byers T, Levin TR, Haffner SM, Saad MF, Tooze JA, D'agostino RB Jr (2007) Change in body size and the risk of colorectal adenomas. Cancer Epidemiol Biomarkers Prev 16:526–531
Serrano M, Hannon GJ, Beach D (1993) A new regulatory motif in cell-cycle control causing specific inhibition of cyclin D/CDK4. Nature 366:704–707
Shen G, Khor TO, Hu R, Yu S, Nair S, Ho C-T, Reddy BS, Huang M-T, Newmark HL, Kong A-NT (2007) Chemoprevention of familial adenomatous polyposis by natural dietary compounds sulforaphane and dibenzoylmethane alone and in combination in ApcMin/+ mouse. Cancer Res 67:9937–9944
Sherr CJ (1994) G1 phase progression: cycling on cue. Cell 79:551–555
Sherr CJ, Roberts JM (1999) CDK inhibitors: positive and negative regulators of G1-phase progression. Genes Dev 13:1501–1512
Shimizu M, Beckman BR, Hara A, Dickhoff WW (2006) Measurement of circulating salmon IGF binding protein-1: assay development, response to feeding ration and temperature, and relation to growth parameters. J Endocrinol 188:101–110

Shimizu M, Shirakami Y, Sakai H, Yasuda Y, Kubota M, Adachi S, Tsurumi H, Hara Y, Moriwaki H (2010) (-)-Epigallocatechin gallate inhibits growth and activation of the VEGF/VEGFR axis in human colorectal cancer cells. Chem Biol Interact 185:247–252

Shin H, Kim JH, Lee YS, Lee YC (2012) Change in gene expression profiles of secreted frizzled-related proteins (SFRPs) by sodium butyrate in gastric cancers: induction of promoter demethylation and histone modification causing inhibition of Wnt signaling. Int J Oncol 40: 1533–1542

Siddique HR, Mishra SK, Karnes RJ, Saleem M (2011) Lupeol, a novel androgen receptor inhibitor: implications in prostate cancer therapy. Clin Cancer Res 17:5379–5391

Smith WJ, Underwood LE, Clemmons DR (1995) Effects of caloric or protein restriction on insulin-like growth factor-I (IGF-I) and IGF-binding proteins in children and adults. J Clin Endocrinol Metab 80:443–449

Soliman AT, Hassan AE, Aref MK, Hintz RL, Rosenfeld RG, Rogol AD (1986) Serum insulin-like growth factors I and II concentrations and growth hormone and insulin responses to arginine infusion in children with protein-energy malnutrition before and after nutritional rehabilitation. Pediatr Res 20:1122–1130

Strathdee G, Brown R (2002) Aberrant DNA methylation in cancer: potential clinical interventions. Expert Rev Mol Med 4:1–17

Strillacci A, Valerii MC, Sansone P, Caggiano C, Sgromo A, Vittori L, Fiorentino M, Poggioli G, Rizzello F, Campieri M, Spisni E (2013) Loss of miR-101 expression promotes Wnt/beta-catenin signalling pathway activation and malignancy in colon cancer cells. J Pathol 229:379–389

Subbaramaiah K, Cole PA, Dannenberg AJ (2002) Retinoids and carnosol suppress cyclooxygenase-2 transcription by CREB-binding protein/p300-dependent and -independent mechanisms. Cancer Res 62:2522–2530

Suzuki H, Watkins DN, Jair KW, Schuebel KE, Markowitz SD, Chen WD, Pretlow TP, Yang B, Akiyama Y, Van EM, Toyota M, Tokino T, Hinoda Y, Imai K, Herman JG, Baylin SB (2004) Epigenetic inactivation of SFRP genes allows constitutive WNT signaling in colorectal cancer. Nat Genet 36:417–422

Svensson J, Lonn L, Jansson JO, Murphy G, Wyss D, Krupa D, Cerchio K, Polvino W, Gertz B, Boseaus I, Sjostrom L, Bengtsson BA (1998) Two-month treatment of obese subjects with the oral growth hormone (GH) secretagogue MK-677 increases GH secretion, fat-free mass, and energy expenditure. J Clin Endocrinol Metab 83:362–369

Taghavi N, Biramijamal F, Sotoudeh M, Khademi H, Malekzadeh R, Moaven O, Memar B, A'rabi A, Abbaszadegan MR (2010) p16INK4a hypermethylation and p53, p16 and MDM2 protein expression in esophageal squamous cell carcinoma. BMC Cancer 10:138

Tai J, Cheung S, Wu M, Hasman D (2012) Antiproliferation effect of Rosemary (Rosmarinus officinalis) on human ovarian cancer cells in vitro. Phytomedicine 19:436–443

Taniguchi H, Yamamoto H, Hirata T, Miyamoto N, Oki M, Nosho K, Adachi Y, Endo T, Imai K, Shinomura Y (2005) Frequent epigenetic inactivation of Wnt inhibitory factor-1 in human gastrointestinal cancers. Oncogene 24:7946–7952

Traka M, Gasper AV, Smith JA, Hawkey CJ, Bao Y, Mithen RF (2005) Transcriptome analysis of human colon Caco-2 cells exposed to sulforaphane. J Nutr 135:1865–1872

Tsang WP, Kwok TT (2010) Epigallocatechin gallate up-regulation of miR-16 and induction of apoptosis in human cancer cells. J Nutr Biochem 21:140–146

Uauy R, Solomons N (2005) Diet, nutrition, and the life-course approach to cancer prevention. J Nutr 135:2934S–2945S

Vanamala J, Tarver CC, Murano PS (2008) Obesity-enhanced colon cancer: functional food compounds and their mechanisms of action. Curr Cancer Drug Targets 8:611–633

Veganzones-De-Castro S, Rafael-Fernandez S, Vidaurreta-Lazaro M, De-La-Orden V, Mediero-Valeros B, Fernandez C, Maestro-De Las Casas ML (2012) p16 gene methylation in colorectal cancer patients with long-term follow-up. Rev Esp Enferm Dig 104:111–117

Vendramini-Costa DB, Carvalho JE (2012) Molecular link mechanisms between inflammation and cancer. Curr Pharm Des 18:3831–3852

Vickers MM, Bar J, Gorn-Hondermann I, Yarom N, Daneshmand M, Hanson JE, Addison CL, Asmis TR, Jonker DJ, Maroun J, Lorimer IA, Goss GD, Dimitroulakos J (2012) Stage-dependent differential expression of microRNAs in colorectal cancer: potential role as markers of metastatic disease. Clin Exp Metastasis 29:123–132

Wagner M, Bjerkvig R, Wiig H, Melero-Martin JM, Lin RZ, Klagsbrun M, Dudley AC (2012) Inflamed tumor-associated adipose tissue is a depot for macrophages that stimulate tumor growth and angiogenesis. Angiogenesis 15:481–495

Wang Z, Chen H (2010) Genistein increases gene expression by demethylation of WNT5a promoter in colon cancer cell line SW1116. Anticancer Res 30:4537–4545

Wang X, Wang Q, Ives KL, Evers BM (2006) Curcumin inhibits neurotensin-mediated interleukin-8 production and migration of HCT116 human colon cancer cells. Clin Cancer Res 12: 5346–5355

Wang H, Bian S, Yang CS (2011) Green tea polyphenol EGCG suppresses lung cancer cell growth through upregulating miR-210 expression caused by stabilizing HIF-1alpha. Carcinogenesis 32:1881–1889

Wang H, Li Q, Chen H (2012) Genistein affects histone modifications on Dickkopf-related protein 1 (DKK1) gene in SW480 human colon cancer cell line. PLoS One 7:e40955

Wilkening S, Tavelin B, Canzian F, Enquist K, Palmqvist R, Altieri A, Hallmans G, Hemminki K, Lenner P, Forsti A (2008) Interleukin promoter polymorphisms and prognosis in colorectal cancer. Carcinogenesis 29:1202–1206

Wu WK, Law PT, Lee CW, Cho CH, Fan D, Wu K, Yu J, Sung JJ (2011) MicroRNA in colorectal cancer: from benchtop to bedside. Carcinogenesis 32:247–253

Wu J, Ji X, Zhu L, Jiang Q, Wen Z, Xu S, Shao W, Cai J, Du Q, Zhu Y, Mao J (2013) Up-regulation of microRNA-1290 impairs cytokinesis and affects the reprogramming of colon cancer cells. Cancer Lett 329:155–163

Xiao Q, Hsing AW, Park Y, Moore SC, Matthews CE, De Gonzalez AB, Kitahara CM (2014) Body mass index and mortality among blacks and whites in the prostate, lung, colorectal and ovarian (PLCO) cancer screening trial. Obesity (Silver Spring) 22:260–268

Yang SY, Sales KM, Fuller BJ, Seifalian AM, Winslet MC (2008) Inducing apoptosis of human colon cancer cells by an IGF-I D domain analogue peptide. Mol Cancer 7:17

Yoruker EE, Mert U, Bugra D, Yamaner S, Dalay N (2012) Promoter and histone methylation and p16(INK4A) gene expression in colon cancer. Exp Ther Med 4:865–870

Zhang Y, Chen H (2011a) Genistein attenuates WNT signaling by up-regulating sFRP2 in a human colon cancer cell line. Exp Biol Med (Maywood) 236:714–722

Zhang Y, Chen H (2011b) Genistein, an epigenome modifier during cancer prevention. Epigenetics 6:888–891

Zhang B, Halder SK, Zhang S, Datta PK (2009) Targeting transforming growth factor-beta signaling in liver metastasis of colon cancer. Cancer Lett 277:114–120

Zhang Y, Li Q, Zhou D, Chen H (2013) Genistein, a soya isoflavone, prevents azoxymethane-induced up-regulation of WNT/beta-catenin signalling and reduces colon pre-neoplasia in rats. Br J Nutr 109:33–42

Chapter 16
Dietary Antioxidants and Chromatin Modifying Compounds as Potential Anti-cancer Therapies

Nadia Mazarakis and Tom C. Karagiannis

Abstract Despite the efficacy of novel chemotherapeutic agents and radiation therapy, considerable developments are still necessary to improve the tolerance and reduce the toxicity in healthy cells of cancer patients. Dietary polyphenols have received increasing interest as an alternative approach, particularly in cancer treatment, as they display strong antioxidant properties and reduced toxicity profiles in normal cells. For decades, the Mediterranean diet but particularly olive oil has been linked with increased health benefits and has been associated with decreased risks in cardiovascular diseases and cancers. The minor constituents of olives are its phenolic compounds including oleuropein, tyrosol, hydroxytyrosol and homovanillic alcohol. These main phenolic compounds all possess antioxidant activity, with particular potency exhibited in hydroxytyrosol (HT). An imbalance of reactive oxygen species cause oxidative stress that can damage cells, and consequently lead to the formation of cancer or various diseases. Here, we provide evidence for the dietary antioxidant and polyphenolic compound hydroxytyrosol for its potential application as both a chemopreventive and anti-cancer agent for the treatment of haematological and solid malignancies.

Aberrant gene expression caused by histone acetylation has also been associated with cancer and represents a potentially useful therapeutic target for dietary compounds. The use of histone deacetylase inhibitors derived naturally from the diet, such as butyrate, dially disulphide and sulforaphane, in the inhibition of carcinogenesis has

N. Mazarakis, B.Sc.
Epigenomic Medicine, Baker IDI Heart and Diabetes Institute, The Alfred Medical Research and Education Precinct, 75 Commercial Road, Melbourne, VIC 3004, Australia
e-mail: nadia.mazarakis@bakeridi.edu.au

T.C. Karagiannis, B.Sc. (Hons). Ph.D. (✉)
Epigenomic Medicine, Baker IDI Heart and Diabetes Institute, The Alfred Medical Research and Education Precinct, 75 Commercial Road, Melbourne, VIC 3004, Australia

Department of Pathology, The University of Melbourne, Parkville, VIC, Australia
e-mail: tom.karagiannis@bakeridi.edu.au

N. Maulik and T. Karagiannis (eds.), *Molecular Mechanisms and Physiology of Disease: Implications for Epigenetics and Health*, DOI 10.1007/978-1-4939-0706-9_16,

been supported by numerous studies. Similarly probiotics have been reported to have positive health effects. We describe the underlying mechanisms and discuss the possible limitations and benefits of using dietary antioxidants and chromatin modifying compounds as potential anti-cancer agents.

Keywords Dietary antioxidants • Histone acetylation • Epigenetics • Hydroxytyrosol • Histone deacetylase inhibitors

Abbreviations

ADP	Adenosine diphosphate
ADR	Adriamycin
ARE	Antioxidant response element
DADS	Dially disulfide
DNA	Deoxyribonucleic acid
H_2O_2	Hydrogen peroxide
HDAC	Histone deacetylases
HDACi	Histone deacetylase inhibitor
HT	Hydroxytyrosol
IEC	Intestine epithelial
Keap1	Kelch-like ECH-associating protein 1
LDL	Low-density lipoprotein
MUFA	Monosaturated fatty acids
NAD+	Nicotinamide adenine dinucleotide
NO·	Nitric oxide
Nrf2	Nuclear factor E2-related factor 2
$O_2^{\cdot-}$	Superoxide anion radical
$^1O_2$	Singlet oxygen
·OH	Hydroxyl radical
–OH	Hydroxyl group
PUFA	Polyunsaturated fatty acids
RNS	Reactive nitrogen species
ROO	Peroxyl radical
RT-PCR	Real time-polymerase chain reaction
SAHA	Suberoylanilide hydroxamic acid
SCFA	Short chain fatty acid
SFN	Sulforaphane
SFN-CYS	Sulforaphane-cysteine
SFN-NAC	Sulforaphane-*N*-acetylase
Sfrp4	Secreted fizzled-related protein 4
Sir2	Silent information regulator 2
SIRT	Sirtuin
UV	Ultra violet

16.1 Introduction

There has been a long-standing interest in dietary compounds as chemopreventatives and potential cancer therapeutics. Current cancer therapies involving surgery, chemotherapy and radiation are limited and have numerous side-effects in healthy cells (Chen and Kong 2005) as research has predominantly focused on developing drugs to specifically interrupt the carcinogenesis process (Manson 2003). The importance of diet on the other hand is often overlooked, although recent research has found that poor diet could contribute to as much as one-third of all human cancers (Doll 1981). As a result, various epidemiological studies support dietary compounds and the long-term health benefits associated with them (Gazzani 2012).

Chemoprevention was a term coined in 1976 by Michael Sporn which demonstrates the advances in the understanding of the molecular mechanisms involved in cancer prevention (Weinstein 1991). Carcinogenesis involves three stages: initiation, promotion and progression (Chen and Kong 2005). Food components are believed to have the capability to inhibit these carcinogenic processes, and hence can act as chemopreventatives. This can be achieved by activation of the phase I enzymes in the two phase drug-metabolizing enzyme system, scavenging of the reactive deoxyribonucleic acid (DNA) agents, suppression of carcinogenic proliferation in the early formation of neoplasms or by intervening within the cancer cell itself (Wargovich 1997).

A seven country study conducted by Keys, over a span of 15 years was a vital study to determine the positive correlation between healthy life and the increased consumption of olive oil (Keys 1986; Cicerale 2012). The study recruited 11,579 'healthy' men aged between 40 and 59 years old who were monitored over a course of 15 years. By the completion of the study 2,288 men had died, and from this outcome they examined multivariates of deaths and coronary heart disease in different populations: Yugoslavia, Italy, Greece, Finland, Netherlands, United States and Japan. The results of study found an increase in life expectancy and overall decrease in chronic diseases in the countries that followed the Mediterranean diet (Keys 1986). Keys' study emphasized the importance of diet, and hence led to further research into the breakdown of the components within the Mediterranean diet that provides its significant health effects.

The Mediterranean diet is rich in vitamins, antioxidants and high in monounsaturated fatty acids, comprised of predominantly, fruits and vegetables, with low meat consumption (Khymenets 2011), whereby, the main source of fat is olive oil, making up approximately 50 % of the diet (Trichopoulou and Lagiou 1997). In 1990, a study involving Australians with Greek heritage who consume the traditional Mediterranean diet, rich in olive oils, was conducted in order to identify whether it was in fact the diet or other lifestyle factors that provided those results found by Keys in 1986 (Halliwell and Gutteridge 1990). The results from this study were found to be similar to that of Keys, reinforcing the importance of olives in the diet. Additional studies found that a change in diet from the high red meat and low fruit and vegetable consumption (commonly found in the Scandinavian countries, the

United Kingdom and the Unites States) to the Mediterranean diet could reduce the incidence of colorectal, breast, prostate, endometrial and pancreatic cancer by an estimated of 25 %, 15 %, 10 %, 10 % and 10 % respectively (Trichopoulou 2000). Initially it was thought that oleic acid, the major constituent of olive oil, contributed to providing the health benefits seen by the consumption of the Mediterranean diet. However, foods which contribute largely to a Western diet, such as pork or chicken, are also rich in oleic acid (Granados-Principal 2010). This initiated investigations into the other constituents of olive oil that could be providing the increased health benefits in humans (Kyrtopoulos et al. 2003).

In this chapter, we investigate the importance of our diet, with particular focus on olives, other dietary antioxidants and dietary chromatin modifying agents in the process of carcinogenesis and explore their potential as chemopreventatives and other health benefits.

16.2 Dietary Antioxidants

Numerous epidemiological studies have offered a greater insight to the impact of dietary antioxidants as anti-cancer and anti-atherogenic agents and their therapeutic potential in cardiovascular diseases. For example, cinnamon and its antioxidant catechin significantly reduce the fasting blood glucose levels, providing a positive impact for type 2 diabetes or pre-diabetes patients (Davis and Yokoyama 2011; Akilen et al. 2012). Moreover, broccoli is abundant in the antioxidant L-SULFORAPHANE, a key potential therapeutic in asthma patients (Dye 2009; Park et al. 2012). Similarly, the polyphenols compound, resveratrol, found in red grape skin, berries and peanuts, has been found to decrease low-density lipoprotein (LDL) cholesterol levels (Ramprasath and Jones 2010). Hence, the role of resveratrol in reducing the risk of cardiovascular disease plays a salient role in prevention therapies (Tome-Carneiro et al. 2012). Consequently, there has been an increased interest in the advantageous health effects of dietary antioxidants (Gordon 2001).

Antioxidants are either hydrophilic, whereby the inhibition or retardation of oxidation of other molecules occur in the cytosol or hydrophobic where they protect the cell membrane from lipid peroxidation (Halliwell 1996). However, in order for antioxidants to work effectively they must possess three necessary mechanisms. Firstly the propagation rate must be faster than the oxidation rate (Gordon 2001), secondly the product must be a less reactive oxygen, and thirdly antioxidants must be bioavailable (Gulcin 2012). Antioxidants work by direct scavenging these free radical species, binding with metal ions, deactivating singlet oxygen, absorbing UV radiation or by converting hydroperoxides to form a less reactive product (Kwak and Kensler 2010; Gulcin 2012). Furthermore, antioxidants structure steams from a catechol group where the catechol structure form hydrogen bonds with the free radicals, stabilizing the compounds (Visioli et al. 2002). The phenolic compounds in olives are strong antioxidants, as they possess this orthodiphenolic/catecholic structure (Tuck 2002). Hence membranes rich in monosaturated fatty acids (MUFA) are less susceptible to oxidation by free radicals than polyunsaturated fatty acids (PUFA), as the hydrogen atoms between the double bonds within the carbon chain are highly reactive (Fitó 2005).

The importance of antioxidants is generally associated with the inhibition of reactive oxygenated species (ROS) (Tome-Carneiro et al. 2012) and/or reactive nitrogen species (RNS) which can accumulate intracellularly due to endogenous or exogenous factors (Silva and Coutinho 2010). Under normal conditions these reactive species are maintained in their state of equilibrium via the cells' antioxidant defence systems. However under stressful conditions, the balance may be disrupted causing excess ROS or RNS—this is commonly known as oxidative stress (Silva and Coutinho 2010).

Oxidative stress is often associated with carcinogenesis due to the damage it creates in biomolecules, which consequently leads to DNA damage and cell death (Gulcin 2012; Halliwell and Gutteridge 1990; Aruoma 1994). There are several endogenous sources for the formation of ROS, the major source which occurs in the mitochondria (Cadenas 1989). It has been estimated that 1–3 % of the oxygen molecules in the electron transport chain (ETC) are released forming these highly reactive superperoxides (Wojcik et al. 2010; Fridovich 1986). ROS consist of a peroxyl radical (ROO·), hydroxyl radical (·OH) and a superoxide anion radical ($O_2^{\cdot -}$). Non-radical compounds, for instance, hydrogen peroxide (H_2O_2) or the singlet oxygen ($^1O_2$) can be converted into radical species under the Fenton reaction, initiated by a metal ions, either copper or iron (Wojcik et al. 2010). On the other hand, RNS are formed through a Ca^{2+}-sensitive mitochondrial NO synthase, such as nitric oxide (NO·) (Ghafourifar and Sen 2007). Therefore dietary antioxidants effectively can clear these free radical species and reduce the risk of oxidative stress (Gulcin 2012). However, dietary antioxidants can activate the body's natural cellular defence system that induces detoxifying enzymes that can prevent cellular damage.

The natural cellular defence mechanism is linked to the antioxidant response element (ARE) (Manson) and its binding property to nuclear factor E2-related factor 2, (Nrf2) which was found to be a regulator in gene transcription of detoxifying genes (Venugopal and Jaiswal 1998). Detoxifying enzymes are part of phase II, of the two phase drug-metabolizing enzyme system, and monofunctional inducers of phase II have proven to be an effective promoter in cytoprotective mechanisms against carcinogens and ROS (Chen and Kong 2005). The specific importance of the Nrf2 protein was found in a study of Nrf2 knockout mice, whereby the Nrf2 knockout mice had decreased these levels of detoxifying enzymes, compared to the wild-type mice examined, and hence was at a greater risk of xenobiotics (Enomoto et al. 2001).

The ARE-Nrf2 pathway (Fig. 16.1) is initiated via dietary detoxifying enzyme inducers which initiate the release of Nrf2 protein in the cytoplasm that binds to the Kelch-like ECH-associating protein 1 (Keap1) (Kobayashi et al. 2004). Keap1 protein is an cysteine enriched protein (Kwak and Kensler 2010) that binds to the Nrf2 protein at two binding sites in the Neh2 domain, ETGE and DLG motifs (McMahon et al. 2004). Following the chemical signal released from the inducer, the cysteine residues in the Keap1 complex are modified and release Nrf2 to the nucleus causing

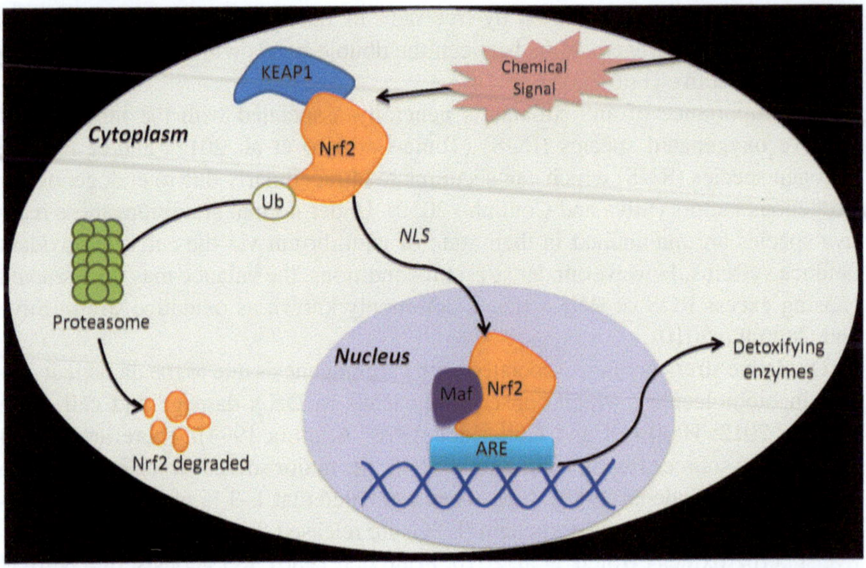

Fig. 16.1 Dietary antioxidants activate the cells natural defence mechanism via the ARE-Nrf2 pathway. Dietary antioxidants induce the release of the Nrf2 protein from the Keap1 protein in the cytoplasm, by an unknown chemical signal. Nrf2 complex binds to Maf proteins in the nucleus to form heterodimers, and from here transcription of detoxifying enzymes can begin. Without the dietary antioxidant induction, the Nrf2 complex would remain in the cytosol and conjoin with Ub, inducing proteasome-mediated rapid degradation

its rapid degradation (Kwak and Kensler 2010). Once the Nrf2 complex is in the nucleus it can bind to Maf proteins to form heterodimers where transcription of the detoxifying enzymes can begin (Kwak and Kensler 2010).

Ultimately, detoxifying the carcinogenic compounds, via the antioxidant properties of certain compounds is a vital area of research for developing chemopreventative agents. Olives are an abundant source of phenolic compounds but more specifically ortho-phenolic structures, which provide the dietary source with its antioxidant properties (Tuck 2002). This gives rise to olive constituents as a potential and promising approach for anti-cancer therapeutics.

16.3 Phenolic Compounds in Olives

Olea europaea is a native olive tree in most Mediterranean regions, allowing the production of olives and olive oil to become a major component of Mediterranean diet for over 5,000 years (Omar 2010; Rafehi 2012; Stark and Madar 2002; Alarcón de la Lastra et al. 2001). The interest of the medicinal properties of olives and olive oil has intensified since epidemiological studies have linked the Mediterranean diet to decrease health risks (Simopoulos 2001; Gill et al. 2005).

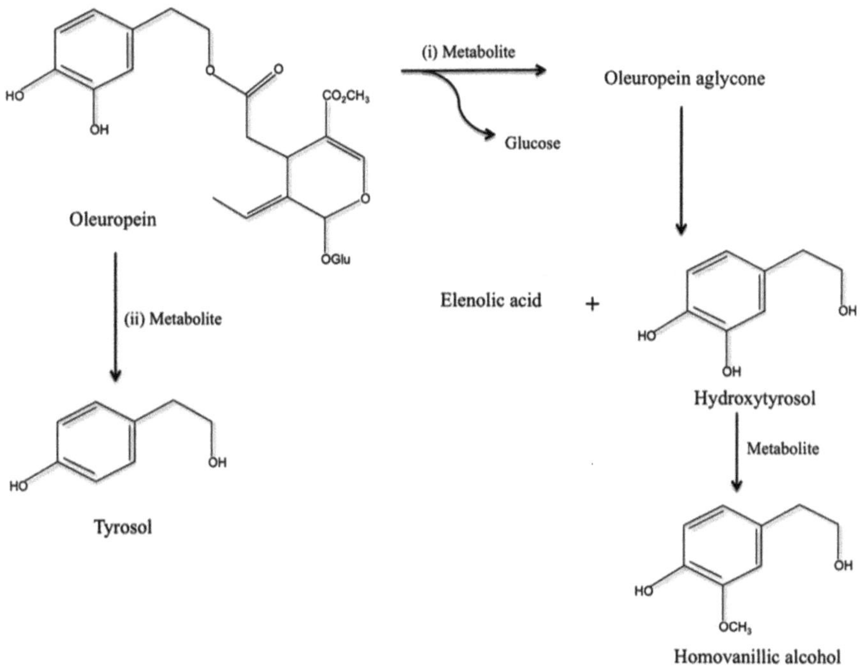

Fig. 16.2 The origin of the relevant phenolic compounds found in olives, from the main compound oleuropein. Oleuropein hydrolyses to form HT and tyrosol, and homovanillic alcohol is formed as a metabolite of HT

Initially these studies believed that the major component of olive oil, oleic acid which contributes to 68–81.5 % (Tuck 2002) of the saponifiable fraction, was the main reason for the health benefits witnessed across these regions. However, the minor constituents of the unsaponifiable fraction primarily made up of phenolic compounds and their antioxidant properties are now deemed of greater importance (Granados-Principal 2010).

The concentration of phenolic compounds in olives and virgin olive oil can vary considerably dependent on agronomical and technological factors, such as cultivator, climate, season or the production process (Fernández Arroyo 2012). These alterations in the concentration of phenolic compounds determine the olive colour, taste and texture (Malheiro 2011). For example, green olives have the highest phenolic content in comparison to the other types (Othman 2009). There are at least 30 phenolic compounds present in olives, with the most prominent being oleuropein (Boskou 1996). Oleuropein can reach concentrations of 60–90 mg g^{-1} in the dry matter of olive leaves, and up to 140 mg g^{-1} in young olives (Omar 2010). As the maturation process progresses, the concentration of oleuropein in olives decreases as it hydrolysis into polyphenolic constituents: tyrosol, HT (Kobayashi et al. 2004) and consequently the metabolite of HT, homovanillic alcohol (Granados-Principal 2010) (Fig. 16.2). These four phenolic compounds and minor constituents of olives are the focus in biomedical research investigating their therapeutic potential.

HT has been found to be the most potent phenolic compound and in comparison to oleuropein, homovanillic alcohol and tyrosol, it has been shown that homovanillic alcohol and tyrosol lack the ability to induce H_2O_2 accumulation, a pro-apoptotic mechanism (Fabiani et al. 2011), in comparison to HT where pro-apoptotic activity is achieved. Furthermore, it was reported that although oleuropein produced an abundance of H_2O_2 accumulation, apoptotic activity was not observed. It is proposed that the biological functions of the polyphenols are due to their differing structure (Fig. 16.2), and the catechol moiety in the phenol molecule is required to produce H_2O_2 (Fabiani et al. 2011). There is a general consensus that both HT and oleuropein are efficient scavengers of superoxide radicals (Kyrtopoulos et al. 2003; de la Puerta et al. 2001; Visioli et al. 1998). In in vivo studies comparing tyrosol with HT, tyrosol was found to have no effects, where HT inhibited endothelial cell proliferation, migration and anti-apoptotic activity in vivo (Fortes et al. 2012). These findings were attributed to the additional hydroxyl group (–OH) found in HT (Warleta et al. 2011) (Fig. 16.2). Homovanillic alcohol, the methylated metabolite of HT, on the other hand, has the ability to inhibit hydrogen peroxide induction in kidney cell injuries. However, a much higher concentration than HT is necessary in order to produce the same effect, such that 0.3 and 10 μM are required respectively (Incani et al. 2010). The phenolic compounds are concentration-dependent, as observed in the urinary excretion of HT and homovanillic alcohol, whereby the excretion of 8-iso-$PGF_2\alpha$ decreased when the concentration of the phenolic doses increased (Visioli et al. 2002). It is considered that the biological function of these phenols are via passive transfusion, and that the phenols absorbed are quantitatively absorbed in the intestine (Manna et al. 2000).

HT is a powerful antioxidant (Della Ragione et al. 2000) with possible prospects of chemoprevention. HT has been reported to upregulate heme-oxygenase, an enzyme that catalyses the degradation of heme to an antioxidant biliverdin, which further degrades in bilirubin, a singlet oxygen scavenger (Rafehi 2012). Additionally, HT has the ability to induce self-antioxidant defences against oxidative stress (Rafehi 2012). It was shown that HT displays dose-dependent anti-tumour effects by inhibiting proliferation and inducing apoptosis in several cancer cell lines via increased expression of the secreted fizzled-related protein 4 (Sfrp4), a protein modulator for Wnt-signalling pathway (Granados-Principal 2010; Granados-Principal et al. 2011). In a rat in vivo model of breast cancer, HT decreased the mammary tumour volume by inhibiting its growth to a similar degree of the positive control, a current chemotherapeutic, adriamycin (Granados-Principal et al. 2011). In several other cancer lines, it has shown that HT inhibits cell proliferation by blocking the G1 phase of the cell cycle (Fabiani et al. 2006; Corona et al. 2007; Granados-Principal 2010). Conversely, studies performed in human mammary epithelial cells (MCF10A) or breast cancer cells (MCF7 and MDA-MB-231) showed no noticeable effect on cell proliferation, the cell cycle or apoptosis following treatments of HT (Warleta et al. 2011). Although, numerous studies support HT as a possible anticancer agent, further research is required to understand its full therapeutic potential in the treatment of cancer and other pathologies.

16.4 Histone Acetylation and Histone Deacetylases

Epigenetics is a relatively new field of research involving the reversible alterations in gene expression without changing the specific sequences of DNA (Santini et al. 2007). Although cancer is known for an abundance of epigenetic alterations, the ability to modify the expression of DNA could also be the key in cancer therapy (Bolden et al. 2006). Histone acetylation is one of the most widely investigated post-translational modifications regulating gene transcription (Miller et al. 2011). Acetylation is regulated by the balance of the opposing activities of histone acetyl-transferases (HATs) and histone deacetylases (HDACs). HATs catalyse the acetylation of the lysine residues on the N-terminal tail of histones, neutralizing the positive charge and relaxing the DNA increasing its accessibility for transcription to occur (Spiegel et al. 2012). HDACs, on the other hand, remove any acetyl groups from the histones, restoring its positive charge (Tsukamoto et al. 1997), inducing chromatin condensation, or tightening the DNA wrap around the histones, silencing transcription (Spiegel et al. 2012). The dynamic control of the opposing actions of the enzymes, HATs and HDACs, maintains the cells homeostasis (Tsukamoto et al. 1997). Changes in expression of HDACs have been associated with cancer, for example HDAC1 is overexpressed in gastric cancer, HDAC2 and 3 expression are increased in colon cancer (Bolden et al. 2006), and a decreased expression of HDAC5 and 10 are observed in lung cancer, and often correlates to poor prognosis (Noureen et al. 2010). To date, 18 mammalian HDACs have been identified based on their homology to yeasts and can be categorized into four classes: class I (HDAC1, 2, 3 and 8), class II-a (HDAC4, 5, 7 and 9), class II-b (HDAC6 and 10) and class IV (HDAC11). These 11 HDAC enzymes are zinc-dependent (Spiegel et al. 2012). Finally, class III comprises the sirtuin deacetylases or nicotinamide adenine dinucleotide (NAD+)-dependent enzymes (Santini et al. 2007).

Sirtuins have been conserved throughout eukaryotes, maintaining two functionalities; firstly, the deacetylase activity which requires NAD+ as an essential cofactor (Saunders and Verdin 2007) and secondly, the adenosine diphosphate (ADP) ribosyl transferase (Schapira 2011). The sirtuins contain a catalytic domain of approximately 275 amino acids (Schapira 2011) and are comprised of seven members, with SIRT1, 6 and 7 located in the nucleus, SIRT2 found in the cytoplasm and, SIRT3, 4 and 5 in the mitochondria of the cell (Saunders and Verdin 2007). SIRT1 and 2 have been the most extensively researched of all sirtuins and there has been conflicting views of the effects as either is a promoter or inhibitor for cancer (Assam El-Osta and Karagiannis 2010). The prototypical member of sirtuins is the silent information regulator 2 (Sir2) (Raghavan and Shah 2012). In yeasts Sir2 contributes to the regulation of chromatin silencing, and repair double-stranded breaks (Tsukamoto et al. 1997). Sirtuins are well known for their role in longevity of prokaryote organisms and anti-ageing effects, yet their anti-cancer effects in humans are still questionable (Saunders and Verdin 2007). For instance, a study showed that SIRT1 can act as a tumour suppressor gene, in the case of colon cancer cells in rodent models (Fridovich 1986). On the contrary, it is believed that tumour cells can become 'addicted' to sirtuin regulation and induce rapid proliferation, eventuating to the loss of accuracy of check points and subsequently lead to increased mutations which give rise for tumourigenesis to occur (Saunders and Verdin 2007). SIRT1 has also been found to be overexpressed in prostate cancer cells (Fridovich 1974). Moreover, p53 is a tumour suppressor protein (Aruoma 1994) and a substrate for SIRT1 (Halliwell 1996), and has been suggested to be activated via the inhibition of the SIRT1 gene (Assam El-Osta and Karagiannis 2010). Therefore there is much controversy over the molecular mechanisms of sirtuins and their biological effects. Future research is required in order to determine the efficacy of these enzymes in relation to anti-cancer modalities.

Class I, II and IV of the metal-dependent HDAC enzymes often referred to as 'classical' HDACs, and they require metal ions, such as zinc, to catalyse their reaction (Schapira 2011). The zinc ion (Zn^{2+}) polarizes the carbonyl group on the departing acetyl group, via chelating compounds such as hydroxamic acids, improving the electrophilicity of the carbonyl carbon (Tsukamoto et al. 1997). These metal-dependent HDAC enzymes differ in their function and location in the cell. Class I is expressed in all tissue, and primarily located in the nucleus (Spiegel et al. 2012). Their roles include cell survival and cell proliferation, which is more esteemed than class II's role in cell proliferation. Furthermore class I HDAC2 plays a key role in the suppression of apoptosis (Noureen et al. 2010). Class II HDACs are phosphoryl-dependent and very tissue-specific, transporting to and from the nucleus and cytoplasm (Noureen et al. 2010). HDACs can alter the phenotypic gene expression, and hence have been found in a number of cancer cells (Ho et al. 2009). Classical HDACs are known to be associated with oncogenesis, due to their ability to deacetylate numerous transcriptional factors and proteins, including tumour suppressors, such as p53, TFIIE, TFIIF, GATA-1 and ER-α (Assam El-Osta and Karagiannis 2010; Juan et al. 2000), resulting in excessive cell proliferation and tumourigenesis (Cress and Seto 2000). Therefore, HDAC inhibitors (HDACi) have been recently developed to inhibit the classical HDACs and prevent aberrant expression of oncogenes caused by HDAC enzymes (Seidel 2012). Through the use of cDNA arrays, it was determined that the expression of genes exposed to HDACi can change between 2 and 20 % (Bolden et al. 2006; Dokmanovic et al. 2007).

HDACi induce cell differentiation, cell cycle arrest, ROS production and apoptosis in specific cancer cell lines (Tho and Jiyoung 2012). The great interest of HDACi is their behaviour to maintain a greater resistance within normal cell lines than compared to cancer cell lines, whereby their anti-cancer effect is more prominent (Cadenas 1989; Stark and Madar 2002; Tsukamoto et al. 1997). The anti-cancer properties of HDACi are due to the accumulation of acetylated nuclear core histones, hence altering the transcription of certain genes, which subsequently result in cancer or various other pathologies (Marks and Xu 2009). Several HDACi have been identified and are classified into structurally different classes: short chain fatty acid (SCFA), benzamides, hydroxamic acids, electrophilic ketones and cyclic tetrapeptides (Miller et al. 2011). These isoform-specific HDACi have three similar characteristics, which include (1) zinc moiety, (2) an opposite capping group, and (3) a straight chain alkyl, vinyl/aryl to linker (Marks and Xu 2009). The first of the

HDACi to be approved by the US Federal Drug Administration was suberoylanilide hydroxamic acid (SAHA) (Mateos et al. 2011), now commonly referred to as vorinostat (Tho and Jiyoung 2012) for the treatment of cutaneous T-cell lymphoma. Moreover, HDACi are known to increase ROS production, a pro-apoptotic effect (Marks and Xu 2009). This mechanism of action is still not yet understood; however, possible theories include mitochondrial injury or affecting antioxidant levels (Miller et al. 2011; Spiegel et al. 2012). However, there are a few generally conceived processes, such as HDACi alter the acetylation of gene promoters, neutralizing the acidic chargers of the histone tails (Bolden et al. 2006). Understanding the molecular mechanisms of HDACi will aid in the development of novel HDACi as anti-cancer agents (Marks and Xu 2009). Furthermore, the efficacy of naturally derived HDACi obtained from the diet can be determined.

16.5 Dietary Histone Deacetylase Inhibitors

Numerous naturally occurring HDACi found in the diet have been identified and include butyrate, dially disulfide (DADS), sulforaphane (SFN) and probiotics (Myzak and Dashwood 2006; Kumar et al. 2013) (Fig. 16.3). Dietary HDACi, however, are weaker ligands compared to the synthetic compounds, such as SAHA, required at a higher concentrations in order for them to be effective (Dashwood et al. 2006).

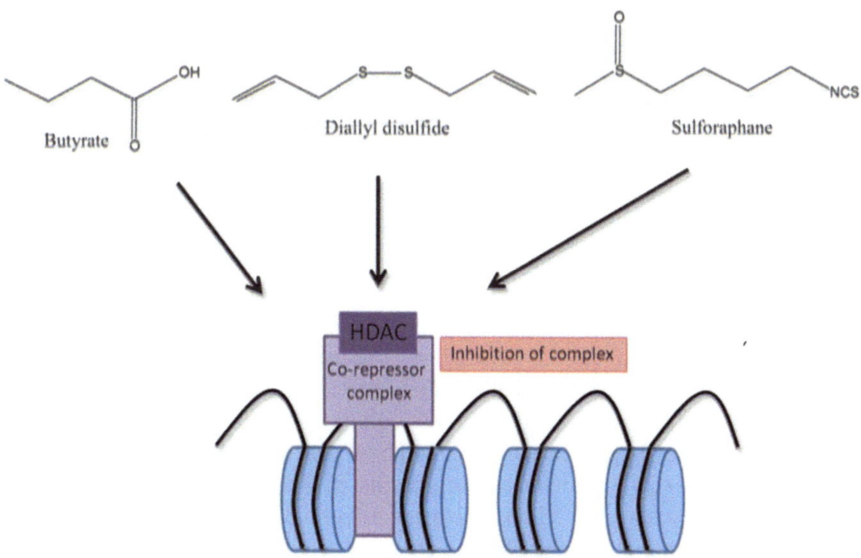

Fig. 16.3 Chemical structure of the dietary HDAC inhibitors; butyrate, dially disulfide and sulforaphane. Dietary HDAC inhibitors inhibit the HDAC co-repressor complex to allow transcriptional activation

Although, the effective concentration range of many dietary compounds is still unknown and further investigations are necessary to determine their efficacy in inhibiting HDAC activity.

Butyrate, an organoselenium compound, was the smallest (Dashwood et al. 2006) and first dietary HDACi discovered in 1978 (Rajendran et al. 2011; Song et al. 2009). Butyrate, a SCFA is formed in the digestive tract via the fermentation of dietary fibres (Tho and Jiyoung 2012). Its structure contains a 'spacer' made up of three carbons attached to carboxylic acid group, and with the combination of a buried zinc atom creates a bidenate ligand, allowing it to enter the active site of the enzyme (Dashwood et al. 2006). The primary research involving SCFA are associated with gastrointestinal diseases, however there is much research interest of butyrate as a HDACi and its potential as a anti-cancer (Tho and Jiyoung 2012). Butyrate inhibits most HDAC enzymes except the class II-b enzymes, HDAC6 and 10 (Licciardi et al. 2010). In the small intestine epithelial cell line (IEC-6 cells) and in human colonic adenocarcinoma cell line (HT-29 cells), butyrate was found to induce the Nrf2 pathway via the inhibition of HDAC enzymes (Nakagawa et al. 2001). Similarly, butyrate has been reported to have chemopreventative effects against colorectal cancer (Zhang et al. 1992) and acute myelogenous leukaemia (Gibbs et al. 2009), although the results seem to be showing some discrepancies, thus questioning the authenticity of the compound itself. These discrepancies could be due to poor bioavailability within the target site (Myzak and Dashwood 2006) or alternatively as butyrate can be used as an energy source by nutrient-deficient cells for proliferation, the tested cells could have had different energy statuses and hence altered their HDACi activity efficiency (Zhang et al. 1992). Research into the optimization of butyrate via the route of administration and chronic exposure through daily consumption will provide evidence of whether there may be a significant chemopreventative affect (Myzak and Dashwood 2006).

DADS is another dietary HDACi originating from organosulfur compounds, found in the *Allium* family, which includes foods such as onions, garlic or shallots, and contains small thiol molecules (Rajendran et al. 2011). DADS resembles butyrate after the metabolite conversion to *S*-allylmercaptocystiene, where it has obtained a 'spacer' carboxylic acid functional group ending (Dashwood et al. 2006). Its chemopreventative properties are recognized through its ability to inhibit the carcinogenic activation, by inducing both phase I and II detoxification pathways (Myzak and Dashwood 2006), as well as altering the cell differentiation and apoptosis pathways (Rajendran et al. 2011). Additionally Song et al. (2009) suggests that the mitochondrial ROS also plays a role in the induction of DADS apoptosis, with the activation of p53, in HCT-116 colon cancer cells. This study observed p53-independent cell cycle arrest at G2/M phase following treatment with DADS and apoptosis after 24 h, which was triggered by ROS production, hence activating the p53 protein (Song et al. 2009). The same effect has also occurred within other cancer cell lines, including lung (Myzak et al. 2004), leukaemia, breast (Nakagawa et al. 2001) and neuroblastoma (Filomeni et al. 2003). DADS has proven to be an effective HDACi in several cancer cell lines and is a promising candidate as a dietary chemopreventative.

Similar to DADS, SFN also originates from organosulfur compounds, however part of the isothiocynate group, which provides its anti-cancer effects. Cruciferous vegetables, such as broccoli, cabbage, cauliflower and kale are a rich source of SFN, and it is activated when the plant is damaged releasing the enzyme myrosinase, which hydrolyzes glucosinolate to activate the isothiocynate, releasing the chemopreventative effects of SFN (Tho and Jiyoung 2012; Rajendran et al. 2011). SFN was first recognized for its ability to induce phase II of the detoxification process (Zhang et al. 1992) and the Keap1/Nrf2 pathway, activating the antioxidant response element (Myzak and Dashwood 2006). Gibbs et al. conducted a study and reported the results of SFN in prostate cancer to inhibit HDAC6 specifically. Furthermore this study explains how SFN downregulates HDAC6 which is usually upregulated in the formation of cancer, and interferes with key mechanisms that initiate the carcinogenesis process, such as the enhancement of HSP96 acetylation which inhibits the association with the androgen receptor, hence preventing the formation of the prostate cancer (Gibbs et al. 2009). Other studies have also confirmed its anti-cancer effects in colon (Myzak et al. 2004) and breast cancer cells (Pledgie-Tracy et al. 2007). However, recent studies have found SFN role in reducing the risk of cancer as a HDACi to be due to its metabolites that are generated through the mercapturic acid pathway, rather than acting as a parent compound in the detoxification process (Rajendran et al. 2011). Through the mercapturic acid pathway it has been found that all effective mechanisms of SFN are due to its metabolites, such as SFN-cysteine (SFN-CYS) and SFN-*N*-acetylase (SFN-NAC) that can fit into the HDAC binding domain, and hence as a spacer, inhibit HDAC activity (Somoza et al. 2004). In summary, numerous studies have confirmed the effectiveness of SFN as a dietary HDACi and chemopreventative in several cancer cell lines, although the mechanism of action is slowly being elucidated. Further research into mechanisms of action of SFN is required to provide accurate advances in the effects SFN has as a chemopreventative in humans.

Fermented dairy products are a very rich source for probiotics and have consequently been marketed for their digestive health benefits (Kumar et al. 2013; Roberfroid et al. 2010). Probiotics are living microbes that are consumed through one's diet for the use of their health benefits in gut microbiome (Kumar et al. 2013). Specifically, there are two commonly used probiotics that are proven to be beneficial for human consumption, the first is a lactic acid producing microorganisms (e.g. *Bifidobacterium* spp.) and the second is low-GC-content lactic acid bacteria (e.g. *Lacctobacillus* spp., *Streptococcus* spp., *Lactococcus* spp. and *Pediococcus* spp.) (Kumar et al. 2013; Kleerebezem and Vaughan 2009). The purpose of probiotics is to restore the microbiota profile balance, a key in preventing immune-mediated diseases, although the precise mechanisms of actions are not yet clearly understood (Licciardi et al. 2010). What is known is that the probiotics and its metabolites are involved in the formation of SCFA in the intestinal tract. SCFA is known to produce butyrate, as mentioned earlier, and is a clearly established HDACi. This suggests the potential avenues of an alternative method with the use of probiotics in the prevention of chronic immune-mediated diseases, carcinogenesis and more specifically colon cancer cell lines. A study conducted by Sokol et al. (2008) provided evidence of the effects of probiotics with the use of *Faecalibacterium prausnitzii,* whereby *F. pranusnitzi* is a lactic acid producing probiotic and one of the most abundant bacterium found in the human digestive tract, with butyrate as its by-product (Sokol et al. 2008). The positive effects of this bacterium were demonstrated in the treatment of inflammatory bowel disease, with reduced levels of interleukin-12 levels and increased interleukin-10 levels (Berni Canani et al. 2011). These alterations in the interleukin levels reduce the swelling and inflammation in colon, therefore suggesting the benefits of butyrate in the reduction of inflammatory bowel disease (Berni Canani et al. 2011). Moreover, another probiotic *Propionibacterium freudenreichii* was found to display apoptotic effects in colorectal adenocarcinoma cells (Visioli et al. 1998).

Although dietary sources of HDACi are found to be less potent in comparison to the synthetically derived HDACi's, strong evidence exists for their potential in preventing the initiation of carcinogenesis in multiple cancer cell lines. Given the low toxicity profiles of the diet-derived compounds and their lifetime consumption, accumulative effects over time may be achieved to strengthen their capacity as chemopreventatives provide multiple health benefits across several pathologies.

16.6 Conclusion

The increasing interest in dietary compounds as potential chemopreventatives has been highlighted as promising future prospectives within this chapter. Dietary antioxidants have been shown to be advantageous across multiple health disciplines including the reduction of fasting blood glucose levels, diabetics, asthma and cardiovascular diseases. Although the mechanisms of antioxidants are still being elucidated, improvements have been made involving the cells natural defence system and the role of Nrf2 in the detoxifying pathways to remove oxidative stress in cells which has lead to tumourigenesis. We focus on the minor constituents and phenolic compounds found in olives and their potential as salient anti-cancer agents. Although, most studies have explored the potent antioxidant effects of HT, we also discussed the therapeutic potential of tyrosol, homovanillic alcohol and oleuropein.

Finally, the aberrant overexpression of HDACs in various cancers has been identified, and the prospects of synthetic and dietary HDACi as plausible chemopreventative and anti-cancer agents have been explored. The mechanisms of action of these compounds are still not clearly understood; however, we highlight in vivo and in vitro investigations into chromatin modifying compounds as potential anti-cancer therapies. Although, dietary HDACi's are less potent compared to the synthetically derived compounds, the safety and efficacy of the lifetime consumption of dietary HDACi's may provide an overall greater effect.

Acknowledgements The support of the Australian Institute of Nuclear Science and Engineering is acknowledged. TCK was the recipient of AINSE awards. TCK is a Future Fellow and Epigenomic Medicine Laboratory supported by the Australian Research Council. Also supported in part by the Victorian Government's Operational Infrastructure Support Program.

References

Akilen R, Tsiami A, Devendra D, Robinson N (2012) Cinnamon in glycaemic control: systematic review and meta analysis. Clin Nutr 31:609–615

Alarcón De La Lastra C, Barranco MD, Motilva V, Herrerías JM (2001) Mediterranean diet and health: biological importance of olive oil. Curr Pharm Des 7:933–950

Aruoma OI (1994) Nutrition and health aspects of free radicals and antioxidants. Food Chem Toxicol 32:671

Assam El-Osta, Karagiannis TC (2010) Histone deacetylase inhibitors in medicine. Transworld Research Network, Kerala

Berni Canani R, Di Costanzo M, Leone L, Bedogni G, Brambilla P, Cianfarani S, Nobili V, Pietrobelli A, Agostoni C (2011) Epigenetic mechanisms elicited by nutrition in early life. Nutr Res Rev 24:198

Bolden JE, Peart MJ, Johnstone RW (2006) Anticancer activities of histone deacetylase inhibitors. Nat Rev Drug Discov 5:769–784

Boskou D (1996) In: Boskou D (ed) Olive oil: chemistry and technology. AOCS, Champaign, IL

Cadenas E (1989) Biochemistry of oxygen toxicity. Annu Rev Biochem 58:79–110

Chen C, Kong A-NT (2005) Dietary cancer-chemopreventive compounds: from signaling and gene expression to pharmacological effects. Trends Pharmacol Sci 26:318–326

Cicerale S (2012) Antimicrobial, antioxidant and anti-inflammatory phenolic activities in extra virgin olive oil. Curr Opin Biotechnol 23:129–135

Corona G, Deiana M, Incani A, Vauzour D, Dessi MA, Spencer JPE (2007) Inhibition of p38/CREB phosphorylation and COX-2 expression by olive oil polyphenols underlies their anti-proliferative effects. Biochem Biophys Res Commun 362:606–611

Cress WD, Seto E (2000) Histone deacetylases, transcriptional control, and cancer. J Cell Physiol 184:1–16

Dashwood RH, Myzak MC, Ho E (2006) Dietary HDAC inhibitors: time to rethink weak ligands in cancer chemoprevention? Carcinogenesis 27:344–349

Davis PA, Yokoyama W (2011) Cinnamon intake lowers fasting blood glucose: meta-analysis. J Med Food 14(9):884–889

De La Puerta R, Dominguez MEM, Ruiz-Gutierrez V, Flavill JA, Hoult JRS (2001) Effects of virgin olive oil phenolics on scavenging of reactive nitrogen species and upon nitrergic neurotransmission. Life Sci 69:1213–1222

Della Ragione F, Cucciolla V, Borriello A, Della Pietra V, Manna C, Galletti P, Zappia V (2000) Pyrrolidine dithiocarbamate induces apoptosis by a cytochrome c-dependent mechanism. Biochem Biophys Res Commun 268:942–946

Dokmanovic M, Clarke C, Marks PA (2007) Histone deacetylase inhibitors: overview and perspectives. Mol Cancer Res 5:981–989

Doll R (1981) The causes of cancer: quantitative estimates of avoidable risks of cancer in the United States today. J Natl Cancer Inst 66:1191

Dye D (2009) Compound in broccoli may help protect against asthma and other respiratory diseases. Life Extension 15:18

Enomoto A, Itoh K, Nagayoshi E, Haruta J, Kimura T, O'connor T, Harada T, Yamamoto M (2001) High sensitivity of Nrf2 knockout mice to acetaminophen hepatotoxicity associated with decreased expression of ARE-regulated drug metabolizing enzymes and antioxidant genes. Toxicol Sci 59:169–177

Fabiani R, Bartolomeo AD, Rosignoli P, Servili M, Selvaggini R, Montedoro GF, Saverio CD, Morozzi G (2006) Virgin olive oil phenols inhibit proliferation of human promyelocytic leukemia cells (HL6O) by inducing apoptosis and differentiation. J Nutr 136:614–619

Fabiani R, Rosignoli P, de Bartolomeo A, Fuccelli R, Servili M, Morozzi G (2011) The production of hydrogen peroxide is not a common mechanism by which olive oil phenols induce apoptosis on HL60 cells. Food Chem 125:1249–1255

Fernández Arroyo S (2012) Application of nanoLC-ESI-TOF-MS for the metabolomic analysis of phenolic compounds from extra-virgin olive oil in treated colon-cancer cells. J Pharm Biomed Anal 63:128–134

Filomeni G, Aquilano K, Rotilio G, Ciriolo MR (2003) Reactive oxygen species-dependent c-Jun NH2-terminal kinase/c-Jun signaling cascade mediates neuroblastoma cell death induced by diallyl disulfide. Cancer Res 63:5940–5949

Fitó M (2005) Antioxidant effect of virgin olive oil in patients with stable coronary heart disease: a randomized, crossover, controlled, clinical trial. Atherosclerosis 181:149–158

Fortes C, Garcia-Vilas JA, Quesada AR, Medina MA (2012) Evaluation of the anti-angiogenic potential of hydroxytyrosol and tyrosol, two bio-active phenolic compounds of extra virgin olive oil, in endothelial cell cultures. Food Chem 134:134–140

Fridovich I (1974) Superoxide dismutases. Adv Enzymol Relat Areas Mol Biol 41:35–97

Fridovich I (1986) Superoxide dismutases. Adv Enzymol Relat Areas Mol Biol 58:61–97

Gazzani G (2012) Functional foods and their expanding applications in the improvement of human health. Curr Opin Biotechnol 23:127–128

Ghafourifar P, Sen CK (2007) Mitochondrial nitric oxide synthase. Front Biosci-Landmark 12:1072–1078

Gibbs A, Schwartzman J, Deng V, Alumkal J (2009) Sulforaphane destabilizes the androgen receptor in prostate cancer cells by inactivating histone deacetylase 6. Proc Natl Acad Sci U S A 106:16663–16668

Gill CIR, Boyd A, Mcdermott E, Mccann M, Servili M, Selvaggini R, Taticchi A, Esposto S, Montedoro G, Mcglynn H, Rowland I (2005) Potential anti-cancer effects of virgin olive oil phenols on colorectal carcinogenesis models in vitro. Int J Cancer 117:1–7

Gordon MH (2001) Antioxidant activity of hydroxytyrosol acetate compared with that of other olive oil polyphenols. J Agric Food Chem 49:2480–2485

Granados-Principal S (2010) Hydroxytyrosol: from laboratory investigations to future clinical trials. Nutr Rev 68:191–206

Granados-Principal S, Quiles JL, Ramirez-Tortosa C, Camacho-Corencia P, Sanchez-Rovira P, Vera-Ramirez L, Ramirez-Tortosa M (2011) Hydroxytyrosol inhibits growth and cell proliferation and promotes high expression of sfrp4 in rat mammary tumours. Mol Nutr Food Res 55:S117–S126

Gulcin I (2012) Antioxidant activity of food constituents: an overview. Arch Toxicol 86:345–391

Halliwell B (1996) Antioxidants in human health and disease. Annu Rev Nutr 16:33–50

Halliwell B, Gutteridge JM (1990) Role of free radicals and catalytic metal ions in human disease: an overview. Methods Enzymol 186:1–85

Ho E, Clarke JD, Dashwood RH (2009) Dietary sulforaphane, a histone deacetylase inhibitor for cancer prevention. J Nutr 139:2393–2396

Incani A, Deiana M, Corona G, Vafeiadou K, Vauzour D, Dessì MA, Spencer JPE (2010) Involvement of ERK, Akt and JNK signalling in H2O2-induced cell injury and protection by hydroxytyrosol and its metabolite homovanillic alcohol. Mol Nutr Food Res 54:788–796

Juan LJ, Shia WJ, Chen MH, Yang WM, Seto E, Lin YS, WU CW (2000) Histone deacetylases specifically down-regulate p53-dependent gene activation. J Biol Chem 275:20436–20443

Keys A (1986) The diet and 15-year death rate in the seven countries study. Am J Epidemiol 124:903

Khymenets O (2011) Direct analysis of glucuronidated metabolites of main olive oil phenols in human urine after dietary consumption of virgin olive oil. Food Chem 126:306–314

Kleerebezem M, Vaughan EF (2009) Probiotic and gut lactobacilli and bifidobacteria: molecular approaches to study diversity and activity. Annu Rev Microbiol 63:269–290

Kobayashi A, Kang MI, Okawa H, Ohtsuji M, Zenke Y, Chiba T, Igarashi K, Yamamoto M (2004) Oxidative stress sensor Keap1 functions as an adaptor for Cul3-based E3 ligase to regulate for proteasomal degradation of Nrf2. Mol Cell Biol 24:7130–7139

Kumar M, Nagpal R, Verma V, Kumar A, Kaur N, Hemalatha R, Gautam SK, Singh B (2013) Probiotic metabolites as epigenetic targets in the prevention of colon cancer. Nutr Rev 71:23–34

Kwak M-K, Kensler TW (2010) Targeting NRF2 signaling for cancer chemoprevention. Toxicol Appl Pharmacol 244:66–76

Kyrtopoulos SA, Xenakis A, Sotiroudis GT, Sotiroudis TG (2003) Chemopreventative potential of minor components of olive oil against cancer. Italian J Food Sci 15:169

Licciardi PV, Wong S-S, Tang ML, Karagiannis TC (2010) Epigenome targeting by probiotic metabolites. Gut Pathog 2:24–28

Malheiro R (2011) Cultivar effect on the phenolic composition and antioxidant potential of stoned table olives. Food Chem Toxicol 49:450–457

Manna C, Galletti P, Maisto G, Cucciolla V, D'angelo S, Zappia V (2000) Transport mechanism and metabolism of olive oil hydroxytyrosol in Caco-2 cells. FEBS Lett 470:341–344

Manson MM (2003) Cancer prevention—the potential for diet to modulate molecular signalling. Trends Mol Med 9:11–18

Marks PA, Xu WS (2009) Histone deacetylase inhibitors: potential in cancer therapy. J Cell Biochem 107:600–608

Mateos R, Pereira-Caro G, Saha S, Cert R, Redondo-Horcajo M, Bravo L, Kroon PA (2011) Acetylation of hydroxytyrosol enhances its transport across differentiated Caco-2 cell monolayers. Food Chem 125:865–872

McMahon M, Thomas N, Itoh K, Yamamoto M, Hayes JD (2004) Redox-regulated turnover of Nrf2 is determined by at least two separate protein domains, the redox-sensitive Neh2 degron and the redox-insensitive Neh6 degron. J Biol Chem 279(30):31556–31567

Miller CP, Singh MM, Rivera-Del Valle N, Manton CA, Chandra J (2011) Therapeutic strategies to enhance the anticancer efficacy of histone deacetylase inhibitors. J Biomed Biotechnol 2011:514261

Myzak MC, Dashwood RH (2006) Histone deacetylases as targets for dietary cancer preventive agents: lessons learned with butyrate, diallyl disulfide, and sulforaphane. Curr Drug Targets 7:443–452

Myzak MC, Karplus PA, Chung FL, Dashwood RH (2004) A novel mechanism of chemoprotection by sulforaphane: inhibition of histone deacetylase. Cancer Res 64:5767–5774

Nakagawa H, Tsuta K, Kiuchi K, Senzaki H, Tanaka K, Hioki K, Tsubura A (2001) Growth inhibitory effects of diallyl disulfide on human breast cancer cell lines. Carcinogenesis 22(6): 891–897

Noureen N, Rashid H, Kalsoom S (2010) Identification of type-specific anticancer histone deacetylase inhibitors: road to success. Cancer Chemother Pharmacol 66:625–633

Omar SH (2010) Cardioprotective and neuroprotective roles of oleuropein in olive. Saudi Pharm J 18:111–121

Othman NB (2009) Antioxidant phenolic compounds loss during the fermentation of Chétoui olives. Food Chem 116:662–669

Park JH, Kim JW, Lee C-M, Kim YD, Chung SW, Jung ID, Noh KT, Park JW, Heo DR, Shin YK, Seo JK, Park Y-M (2012) Sulforaphane inhibits the Th2 immune response in ovalbumin-induced asthma. BMB Rep 45:311–316

Pledgie-Tracy A, Sobolewski MD, Davidson NE (2007) Sulforaphane induces cell type-specific apoptosis in human breast cancer cell lines. Mol Cancer Ther 6:1013–1021

Rafehi H (2012) Investigation into the biological properties of the olive polyphenol, hydroxytyrosol: mechanistic insights by genome-wide mRNA-Seq analysis. Genes Nutr 7:343

Raghavan A, Shah ZA (2012) Sirtuins in neurodegenerative diseases: a biological-chemical perspective. Neurodegener Dis 9:1–10

Rajendran P, Williams DE, Ho E, Dashwood RH (2011) Metabolism as a key to histone deacetylase inhibition. Crit Rev Biochem Mol Biol 46:181–199

Ramprasath VR, Jones PJH (2010) Anti-atherogenic effects of resveratrol. Eur J Clin Nutr 64:660–668

Roberfroid M, Gibson GR, Hoyles L, Mccartney AL, Rastall R, Rowland I, Wolvers D, Watzl B, Szajewska H, Stahl B, Guarner F, Respondek F, Whelan K, Coxam V, Davicco M-J, Léotoing L, Wittrant Y, Delzenne NM, Cani PD, Neyrinck AM, Meheust A (2010) Prebiotic effects: metabolic and health benefits. Br J Nutr 104(Suppl 2):S1–S63

Santini V, Gozzini A, Ferrari G (2007) Histone deacetylase inhibitors: molecular and biological activity as a premise to clinical application. Curr Drug Metab 8:383–393

Saunders LR, Verdin E (2007) Sirtuins: critical regulators at the crossroads between cancer and aging. Oncogene 26:5489–5504

Schapira M (2011) Structural biology of human metal-dependent histone deacetylases. Handb Exp Pharmacol 206:225–240

Seidel C (2012) Chromatin-modifying agents in anti-cancer therapy. Biochimie 94:2264

Silva JP, Coutinho OP (2010) Free radicals in the regulation of damage and cell death—basic mechanisms and prevention. Drug Discov Ther 4:144–167

Simopoulos AP (2001) The Mediterranean diets: what is so special about the diet of Greece? The scientific evidence. J Nutr 131:3065S–3073S

Sokol H, Pigneur B, Watterlot L, Lakhdari O, Bermúdez-Humarán LG, Gratadoux J-J, Blugeon S, Bridonneau C, Furet J-P, Corthier G, Grangette C, Vasquez N, Pochart P, Trugnan G, Thomas G, Blottière HM, Doré J, Marteau P, Seksik P, Langella P (2008) *Faecalibacterium prausnitzii* is an anti-inflammatory commensal bacterium identified by gut microbiota analysis of Crohn disease patients. Proc Natl Acad Sci U S A 105:16731–16736

Somoza JR, Skene RJ, Katz BA, Mol C, Ho JD, Jennings AJ, Luong C, Arvai A, Buggy JJ, Chi E, Tang J, Sang B-C, Verner E, Wynands R, Leahy EM, Dougan DR, Snell G, Navre M, Knuth MW, Swanson RV, Mcree DE, Tari LW (2004) Structural snapshots of human HDAC8 provide insights into the class I histone deacetylases. Structure (London, England: 1993) 12:1325–1334

Song J-D, Lee SK, Kim KM, Park SE, Park S-J, Kim KH, Ahn SC, Park YC (2009) Molecular mechanism of diallyl disulfide in cell cycle arrest and apoptosis in HCT-116 colon cancer cells. J Biochem Mol Toxicol 23:71

Spiegel S, Milstien S, Grant S (2012) Endogenous modulators and pharmacological inhibitors of histone deacetylases in cancer therapy. Oncogene 31:537–551

Stark AH, Madar Z (2002) Olive oil as a functional food: epidemiology and nutritional approaches. Nutr Rev 60:170

Tho XP, Jiyoung L (2012) Dietary regulation of histone acetylases and deacetylases for the prevention of metabolic diseases. Nutrients 4:1868

Tome-Carneiro J, Gonzalvez M, Larrosa M, Yanez-Gascon MJ, Garcia-Almagro FJ, Ruiz-Ros JA, Garcia-Conesa MT, Tomas-Barberan FA, Espin JC (2012) One-year consumption of a grape nutraceutical containing resveratrol improves the inflammatory and fibrinolytic status of patients in primary prevention of cardiovascular disease. Am J Cardiol 110:356–363

Trichopoulou A (2000) Cancer and Mediterranean dietary traditions. Cancer Epidemiol Biomarkers Prev 9:869

Trichopoulou A, Lagiou P (1997) Healthy traditional Mediterranean diet: an expression of culture, history, and lifestyle. Nutr Rev 55:383–389

Tsukamoto Y, Kato J, Ikeda H (1997) Silencing factors participate in DNA repair and recombination in *Saccharomyces cerevisiae*. Nature 388:900–903

Tuck KL (2002) Major phenolic compounds in olive oil: metabolism and health effects. J Nutr Biochem 13:636

Venugopal R, Jaiswal AK (1998) Nrf2 and Nrf1 in association with Jun proteins regulate antioxidant response element-mediated expression and coordinated induction of genes encoding detoxifying enzymes. Oncogene 17:3145–3156

Visioli F, Bellosta S, Galli C (1998) Oleuropein, the bitter principle of olives, enhances nitric oxide production by mouse macrophages. Life Sci 62:541–546

Visioli F, Poli A, Galli C (2002) Antioxidant and other biological activities of phenols from olives and olive oil. Med Res Rev 22:65–75

Wargovich MJ (1997) Experimental evidence for cancer preventive elements in foods. Cancer Lett 114:11

Warleta F, Quesada CS, Campos M, Allouche Y, Beltran G, Gaforio JJ (2011) Hydroxytyrosol protects against oxidative DNA damage in human breast cells. Nutrients 3:839–857

Weinstein IB (1991) Cancer prevention: recent progress and future opportunities. Cancer Res 51:5080s–5085s

Wojcik M, Burzynska-Pedziwiatr I, Wozniak LA (2010) A review of natural and synthetic antioxidants important for health and longevity. Curr Med Chem 17:3262–3288

Zhang Y, Talalay P, Cho C-G, Posner GH (1992) A major inducer of anticarcinogenic protective enzymes from broccoli: isolation and elucidation of structure. Proc Natl Acad Sci U S A 89(6):2399–2403

Chapter 17
Combination Therapy for Cancer: Phototherapy and HDAC Inhibition

Jane Jisun Sung and Tom C. Karagiannis

Abstract Cutaneous T-cell lymphomas (CTCLs) are rare extranodal non-Hodgkin's characterised by pleomorphic skin lesions and distinct T-cell markers. The annual incidence of these non-Hodgkin's lymphomas are approximately 0.2–0.8/100,000 and mycosis fungoides (MF) or its leukemic variant, Sézary syndrome (SS), account for the majority of cases. CTCL is a relatively benign disease in its early stages, but survival rates decrease significantly as it progresses. As curative therapy remains elusive, the goal of therapy is preventing or slowing progression from early stages while minimising long-term toxicity associated with the treatments. Early-stage CTCL can often be controlled with skin-directed therapies including topical steroids, topical retinoids and phototherapy, while patients with late-stage or refractory MF and SS are given systemic therapies including extracorporeal photopheresis (ECP), interferon (IFN), histone deacetylase inhibitors (HDACi) and denileukin diftitox. Since no single therapy can control disease progression fully, combination therapy is employed to enhance response rates. A novel combination treatment using ultraviolet light phototherapy and HDACi has shown to be a potent radiosensitiser, allowing the use of lower radiation doses and minimising the adverse effects of phototherapy. Such combination reduces the carcinogenic risks associated with the long-term use of phototherapy. Studies have shown that HDACi, such as suberoylanilide hydroxamic acid (Vorinostat, Zolinza®), Romidepsin (Istodax®) and sodium butyrate, induce increased radiosensitivity and decreased double-strand break repair capacity. This is due to the action of HDACi modifying the chromatin compaction and thus changing DNA accessibility. By blocking deacetylation of

J.J. Sung, B.Sc. • T.C. Karagiannis, B.Sc. (Hons), Ph.D. (✉)
Epigenomic Medicine, Baker IDI Heart and Diabetes Institute, The Alfred Medical Research and Education Precinct, 75 Commercial Road, Melbourne, VIC, Australia

Department of Pathology, The University of Melbourne, Parkville, VIC, Australia
e-mail: tom.karagiannis@bakeridi.edu.au

N. Maulik and T. Karagiannis (eds.), *Molecular Mechanisms and Physiology of Disease: Implications for Epigenetics and Health*, DOI 10.1007/978-1-4939-0706-9_17,

histones, they promote an open chromatin structure altering the expression of genes involved in cell survival, proliferation, differentiation and apoptosis. As a result, there is a significant increase in the level of ultraviolet-induced apoptosis.

Keywords Phototherapy • Photochemotherapy • Iodinated DNA ligand • Histone deacetylases • Histone deacetylase inhibitors • Vorinostat • Cutaneous T-cell lymphoma

17.1 Disease Overview and Cells of Origin

Cutaneous T-cell lymphomas (CTCLs) refer to a spectrum of non-Hodgkin's T-cell lymphomas (Willemze et al. 2005), which are characterised by the malignant proliferation of T lymphocytes, which primarily manifest in the skin (Wollina 2012). The subtypes include mycosis fungoides (MF), Sézary syndrome (SS), cutaneous $CD30^+$ T-cell lymphoproliferative disorders, subcutaneous panniculitis-like T-cell lymphoma, $CD8^+$ T-cell lymphoma and gamma-delta T-cell lymphoma (Bradford et al. 2009). The annual incidence of CTCL has been increasing and is currently 6.4 per million persons (Criscione and Weinstock 2007) with a higher prevalence amongst males and in patients over 70. In 2009, the prevalence in the United States ranged from 16,000 to 20,000 cases (Horwitz 2011). The two most common subtypes of CTCL worldwide are MF and SS (Weinstock and Gardstein 1999). MF accounts for 54 % of CTCLs with an annual incidence of approximately four new cases per million people, and SS accounts for approximately 5 % of CTCLs (Duhovic et al. 2012).

CTCL is characterised by clonal proliferations of mature $CD4^+$ $CD45RO^+$ T cells, which have a marked homing capacity for the papillary dermis and epidermis (Whittaker and Foss 2007). The skin contains approximately one million T cells per cubic centimetre, which is twice as many T cells compared to peripheral blood (Clark et al. 2006). Following antigenic activation, naïve T cells differentiate into effector and memory cells, which then undergo migration according to their expression of tissue-specific homing factors (Clark et al. 2006). The homing to skin is mediated by expression of skin-homing addressins such as the cutaneous lymphoid antigen (CLA) and chemokine receptors such as CCR4, CCR6 and CCR10, which promote binding to E-selectin (Whittaker and Foss 2007; Reiss et al. 2001; Homey et al. 2002). E-selectin is expressed on post-capillary venules in the skin and is required for lymphocyte tethering and rolling (Schaerli et al. 2006). Such ligand–receptor interaction facilitates the exit of T cells from the circulation and into the skin. Hence, early clinical stages of CTCL are limited to the epidermis with infiltrates building up particularly along the basal layer and adjacent to Langerhans cells (Wollina 2012).

17.2 Diagnosis

17.2.1 Mycosis Fungoides

Mycosis fungoides (MF) is considered as the prototype of CTCL and is characterised by formation of three distinct cutaneous lesions, which are classified as patches, plaques and tumours. Patches are highly variable and there are islands of uninvolved skin between the patches. They can give rise to plaques forming palpable reddish-brown skin lesions (Wollina 2012). Patches and plaques have an asymmetric distribution involving the breasts, axillae, groin, lower trunk and the buttocks (Diamandidou et al. 1996). In the later stages of MF, there is deeper infiltration of malignant $CD4^+$ T helper cells into the dermis and plaques can be followed by the development of tumours characterised by protruding, ulcerative lesions (Diamandidou et al. 1996). However, patients commonly have patch, plaque and tumour lesions on different parts of the body simultaneously (Diamandidou et al. 1996). Such skin lesions disrupt the normal skin barrier leading to opportunistic infections (Beyer et al. 2011). Infectious complications are considered to be a significant cause of morbidity and mortality (Dummer et al. 2000). These skin lesions are accompanied by physical symptoms such as pain and intense pruritus and as a result, MF patients often experience sleep deprivation and limitation in daily tasks causing detrimental effects on the patient's social and emotional well-being (Demierre et al. 2006).

17.2.2 Sézary Syndrome

Sézary syndrome (SS) is a more aggressive form of CTCL that is characterised by circulating, atypical, malignant T lymphocytes with cerebriform nuclei called Sézary cells (Willemze et al. 2005). It can occur as a progression of MF but can also occur de novo (Kavanaugh et al. 2010). It leads to lymphadenopathy, erythroderma, and the presence of malignantly proliferating T lymphocytes in the skin, lymph nodes and peripheral blood (Willemze et al. 2005). The fundamental difference between MF and SS lies in the putative cell of origin. T cells in MF resemble effector memory T cells (T_{EM}), which form a persistent population of skin-resident T cells (Clark et al. 2006). In contrast, T cells in SS resemble central memory T cells (T_{CM}) expressing CCR7 and L-selectin, which are required for lymph node homing and circulation in the peripheral blood (Campbell 2010). Such difference is responsible for the distinct clinical behaviour of the two subtypes, where T_{EM} derived MF has skin-resident T cells that are unable to circulate in peripheral blood, remaining fixed within the skin (Campbell 2010). T_{CM}-derived SS has long-lived T cells resistant to apoptosis in all of the peripheral blood, lymph node and skin (Campbell 2010).

17.3 Clinical Staging and Prognosis of MF and SS

MF and SS are commonly staged based on a tumour–node–metastasis–blood (TNMB) classification, which was recently modified by the International Society for Cutaneous Lymphomas/EORTC (Table 17.1) (Olsen et al. 2008; Bunn and Lamberg 1979). TNMB classification involves examining the level of skin involvement and the level of lymph node or visceral involvement (de Coninck et al. 2001). In stage IA CTCL, less than 10 % of the skin's surface is covered with plaques or patches. When it advances into stage IB, 10 % or more of the skin's surface is covered with plaques or patches. Stage IIA CTCL is characterised by the enlargement of lymph nodes. Stage IIB differs only in the presence of cutaneous tumours. Stage III CTCL is characterised by reddening of the majority of the skin and the presence of cutaneous lesions, which have not yet spread to lymph nodes. Cancer is spread to the lymph nodes and to other organs in stages IVA and IVB, respectively (Kavanaugh et al. 2010). The most commonly affected organs in metastasis are the lungs, the spleen, and the liver (Lansigan et al. 2008). MF patients with patch-plaque stage (stages IA, IB and IIA) have a survival of >12 years and those with mid-grade tumours or erythroderma have a median survival of approximately 4 years (Sausville et al. 1988). However, for patients with stage IV, who have lymph node or visceral involvement, the median survival is less than 3 years (Dummer et al. 2000). Similarly, SS patients have a median survival of less than 3 years (Demierre et al. 2003; Vonderheid et al. 2002).

Table 17.1 Different stages of cutaneous T cell lymphoma based on a tumour–node–metastasis–blood classification (Olsen et al. 2008)

<table>
<tr><th></th><th>T1: Limited patch and plaque</th><th>T2: Generalised patch and plaque</th><th>T3: Mid-grade Tumour</th><th>T4: Erythroderma</th></tr>
<tr><td>N0: No clinical involvement of lymph nodes</td><td>IA</td><td>IB</td><td rowspan="2">IIB</td><td>IIIA</td></tr>
<tr><td>N1: Enlargement of lymph nodes, but no histological involvement (N1)</td><td colspan="2">IIA</td><td>IIIB</td></tr>
<tr><td>N2-3: Enlargement and histological involvement of lymph nodes</td><td colspan="4">IVA</td></tr>
<tr><td>M1: Visceral involvement</td><td colspan="4">IVB</td></tr>
</table>

N: node; **T:** tumour; **M:** metastases.

17.4 Treatment Overview

Currently, there are no curative therapies available for CTCL. This has severe consequences for patients with advanced-stage disease as CTCL is a relatively benign disease in its early stages but once it progresses to tumour and lymph-node involvement stages, survival rates decrease dramatically (Kim et al. 2003). The current therapeutic goals are, therefore, to prevent progression from early-stage disease, in which survival rates are good, to advanced-stage disease, which has a poor prognosis (Stadler 2007). Also, therapies should aim to minimise cytotoxic effects against normal lymphocytes and hence, to maintain long-term treatment tolerability (Hymes 2010; Stadler 2007). Such goals are considered to be best achieved using stage-adapted therapy on preventing disease progression while minimising treatment-related toxicity (Stadler 2007). Classification of different stages of CTCL has allowed for the development of stage-adapted therapy, which is crucial in minimising the adverse effects of various treatment options. Decisions on the treatment offered to each patient are determined by their clinical stage of the disease, the use of previous therapies, availability and side-effect profile of the treatment, and patient quality of life (Lansigan and Foss 2010; Zain et al. 2010).

The aim of this chapter is to outline treatment algorithms according to disease stage (Table 17.2) and to discuss in detail various skin-directed (Table 17.3) and systemic therapies (Table 17.4) currently available for the treatment of CTCL. Early-stage CTCL has limited skin involvement and the use of early aggressive therapy for CTCL was shown to provide no survival advantage. Hence, patients with early-stage CTCL are given skin-directed therapies as the initial management (Lansigan et al. 2008). Currently available skin-directed therapies include topical corticosteroids (Lansigan and Foss 2010), nitrogen mustard (Apisarnthanarax et al. 2002), psoralen and ultraviolet A (PUVA) radiation therapy (Breuckmann et al. 2004) and broad and narrowband ultraviolet B (NB-UVB) radiation therapy

Table 17.2 Treatment options according to disease stage (Lansigan and Foss 2010)

Disease stage	Manifestation	Treatment algorithms
Early	IA: <10 % BSA patch-plaque stage (T1)	Skin-directed therapy
	IB: ≥10 % BSA patch-plaque stage (T2)	Skin-directed therapy ± biological therapy
	IIA: T1–T2, enlargement of nodes but no histological involvement	Skin-directed therapy ± biological therapy
Intermediate	IIB: Cutaneous tumours (T3)	Skin-directed therapy + biological therapy
	III: Erythroderma (T4)	Systemic therapy ± skin-directed therapy
	IVA: T1–T4, enlargement of nodes with histological involvement	Systemic therapy + skin-directed therapy
Advanced	IVB: T1–T4, visceral involvement	Systemic therapy + skin-directed therapy

BSA body surface area, *T* tumour

Table 17.3 Skin-directed therapies for the treatment of early and intermediate stage CTCL

Agent	Mechanism of action	References
Topical corticosteroids	Induction of direct apoptosis of T cells Decreases the number of Langerhan's cells	Kaye et al. (1989), Zackheim et al. (1998), Pitzalis et al. (1997), Diederen et al. (2003)
Topical methchlorethamine (nitrogen mustard)	0.01 or 0.02 % aqueous solution or ointment induce direct cytotoxic effect	Apisarnthanarax et al. (2002), Huber et al. (2006)
Topical bexarotene	Induction of apoptosis of malignant T cells	Apisarnthanarax et al. (2002), Breneman et al. (2002)
Phototherapy	Orally administered psoralen is absorbed by epidermal cells, which then gets photo-activated upon exposure to UVA radiation forming DNA adducts	Pothiawala et al. (2010), Arbiser et al. (2006), Clark et al. (2000), Knobler (2004), Wollina (2012), Whittaker and Foss (2007)
Psoralen + UVA	More effective for thicker plaques than UVB	
Broadband UVB	Does not require ingestion of psoralen	Cimino et al. (1985), Duhovic et al. (2012), Zarebska (1994),
Narrowband UVB	Used for thinner plaques	Osella-Abate et al. (2001), Hönigsmann et al. (1984), Introcaso et al. (2008)
External beam radiation	Dose ranging from 10 to 30 cGy Provide effective palliative therapy	Akilov et al. (2012)

UVA ultraviolet A, *UVB* ultraviolet B, *DNA* deoxyribonucleic acid

Table 17.4 Systemic therapies for the treatment of late-stage CTCL

Agent	Mechanism of action	References
Extracorporeal photopheresis	Induction of apoptosis of malignant T cells; conversion of blood monocytes to dendritic cells; induction of anti-tumour $CD8^+$ T-cell response	Introcaso et al. (2008), Knobler and Girardi (2001), Chiesa Fuxench (2010), Berger et al. (2001)
Retinoids (e.g. bexarotene)	Induction of apoptosis by decreasing surviving and activating caspase-3	Vittorio et al. (2001)
Denileukin diftitox	Infusional toxin combined with diphtheria toxin that binds to IL-2R (CD25)	Foss et al. (2001), Olsen et al. (2001)
Vorinostat (SAHA)	Histone deacetylase inhibitor	Kavanaugh et al. (2010), Zain and O'Connor (2010), Grant et al. (2007)
Romidepsin	Histone deacetylase inhibitor	Kavanaugh et al. (2010)
Alematuzemab (monoclonal antibodies)	Target CD52 on malignant T cells	Lundin et al. (2003)
IFN-α, -γ	Modulation of immune response	

SAHA suberoylanilide hydroxamic acid, *IL* interleukin, *IFN* interferon

(Resnik and Vonderheid 1993; Clark et al. 2000). Skin-directed therapies produce long-term responses and complete response rates of approximately 60 %, together with limited toxicity (Zackheim et al. 1998).

In contrast, although systemic therapies produce rapid responses and are associated with high response rates, they are reserved for patients with advanced disease or progressive disease due to the associated toxicities (Hymes 2007). Currently available systemic therapies include retinoids such as bexarotene (Breneman et al. 2002), interferons (IFNs) (Vittorio et al. 2001), histone deacetylase (HDAC) inhibitors (Kavanaugh et al. 2010), ECP (Zic 2012) and denileukin diftitox (Foss et al. 2001).

Various treatment modalities are employed as combination therapy, such as ECP, IFN and bexarotene for improved response rates (Zain et al. 2010). A new treatment modality is commonly added to an existing one if the patient shows evidence of disease progression. Due to synergistic effects of combination therapies, they are often reserved for more advanced cases but the responses to current combination therapies are short-lived and most patients ultimately relapse (Rosen and Foss 1995).

As a result, novel drugs that target epigenetic abnormalities associated with CTCL have been developed (Kavanaugh et al. 2010). Effects of chromatin-modifying compounds and their potential synergy with the most phototherapy in particular will be the main focus of this chapter. Such novel combination therapy allows specific targeting of biological pathways involved in the pathogenesis of CTCL broadening the scope of therapeutic options for the patients with late-stage CTCL (Zain et al. 2010).

17.5 Treatment of Early-Stage CTCL

17.5.1 Skin-Directed Therapy

Topical Chemotherapy

For early stages of CTCL (IA, IB and IIA), local treatment approaches including topical chemotherapy and phototherapy are highly recommended, as they present with disease limited to the skin without systemic involvement (Kaye et al. 1989). One example of primary therapy in stage IA disease is topical corticosteroids. They have demonstrated high efficacy in the treatment of patch-stage CTCL achieving complete clinical remission in 25–63 % of the patients (Lansigan and Foss 2010; Zackheim et al. 1998). Corticosteroids block intercellular adhesion and lymphocyte binding to endothelium, inducing apoptosis of lymphocytes (Diederen et al. 2003; Pitzalis et al. 1997), they are associated with short duration of benefit and prolonged administration could cause cutaneous atrophy (Zackheim et al. 1998). Hence, in the more advanced stages (clinical stage IB), corticosteroids should be applied adjuvantly (Dummer 2003).

Another example of topical chemotherapy commonly used for early-stage CTCL is mechlorethamine (nitrogen mustard), which is an alkylating agent (Apisarnthanarax et al. 2002). Nitrogen mustard is anti-cancer drug that induces cytotoxic damage to deoxyribonucleic acid (DNA) in malignant cells (Broch et al. 1991). However, nitrogen mustard is associated with hypersensitivity reactions but such adverse effects could be minimised when it is compounded into an ointment form (Huber et al. 2006). In addition, topical retinoids, such as bexarotene, is a therapeutic alternative for early-stage patients who have refractory or persistent disease after other therapies (Breneman et al. 2002). Bexarotene decreases proliferation of lymphocytes, increases differentiation and induces apoptosis. Hence, histological observations include a marked decrease of T-cell infiltrates in skin lesions seen in the early stage of disease (Apisarnthanarax et al. 2002).

Phototherapy as Monotherapy

UV phototherapy is employed for more widespread patches and plaques (Pothiawala et al. 2010). Examples include PUVA therapy, broadband ultraviolet B (BB-UVB) and narrowband UVB (NB-UVB) (Pothiawala et al. 2010; Whittaker and Foss 2007). Ultraviolet B phototherapy was the first form of phototherapy used for the treatment of early-stage CTCL. It interrupts the proliferation of T cells via DNA damage and, consequently, causes apoptosis of malignant cutaneous T cells (Pothiawala et al. 2010). BB-UVB (with a wavelength of 290–320 nm) had been utilised for the treatment of psoriasis since 1920s using crude coal tar (CCT) as a photosensitiser (Goekerman 1925). Combination of BB-UVB and CCT inhibits hyperproliferation of keratinocytes, modulates inflammatory cytokines, eliminates T-lymphocytes in psoriatic skin and finally, inhibits angiogenesis (Lowe et al. 1982; Finch et al. 1997; Fiala et al. 2006; Arbiser et al. 2006).

NB-UVB with a wavelength of 311–312 nm is an alternative treatment option that has been shown effective for the treatment of early-stage CTCL in the recent years. Increased complete remission rates of up to 83 % have been reported with NB-UVB phototherapy (Clark et al. 2000). Also, there is evidence that NB-UVB is less carcinogenic than BB-UVB (Knobler 2004).

However, the efficacy of UVB is limited to the patch stage because UVB is unable to penetrate the skin into the deeper dermis. UVA penetrates deeper into the layer than UVB and hence, PUVA is a better treatment option for patients in the plaque stage (Wollina 2012). Therefore, PUVA therapy is considered as the cornerstone of treatment in early-stage CTCL. It involves oral administration of 8-methoxypsoralen (8-MOP), which sensitises skin to UVA irradiation. It has a planar aromatic structure and hydrophobic nature that allows it to intercalate into DNA at alternating pyrimidine-purine sites (Cimino et al. 1985). When the localised psoralen molecules are exposed to UVA radiation (320–400 nm), they form covalent interstrand cross-links with pyramidines, which are also termed as photoadducts (Cimino et al. 1985). PUVA-induced interstrand cross-links in chromosomal DNA induce an anti-proliferative effect (Yoo et al. 1996). Also, interstrand cross-links induce apoptosis as they interfere with DNA synthesis at higher doses (Duhovic et al. 2012). In addition, they also react with molecular oxygen producing a reactive oxygen singlet, which can damage cell membranes by lipid peroxidation (Zarebska 1994).

Damage in cell membranes results in various structural and functional modifications such as altered fluidity, increased permeability and inactivation of cellular enzymes and transport proteins (Breuckmann et al. 2004). Such changes in the cell membranes have detrimental effects on the survival of the cell eventually leading to cell death (Pothiawala et al. 2010). In addition, apoptosis is induced via bcl-2 family members and various extrinsic cell death pathways (Bladon and Taylor 2006; Osella-Abate et al. 2001). Several studies have confirmed that PUVA produced high remission rates in early stage MF and have reported complete response in up to 71 % of patients (Hönigsmann et al. 1984). PUVA therapy had been surveyed to be the most effective treatment option for stage IA and IB patients (Introcaso et al. 2008).

Consequently, extensive research has been done to extend the PUVA concept and a new class of UV photosensitisers has emerged. Although it is still at its preclinical stage, its phototoxicity has shown to be significantly higher than that of PUVA therapy. The initial discovery was an analogue of the bisbenzimidazole, Hoechst 33258 (Martin et al. 1990). It binds non-covalently to A-T rich sites of the minor groove of DNA (Adhikary et al. 2003). *Ortho*-iodoHoechst, which was produced by iodination of Hoechst 33258 in the *ortho* position, exhibited the highest UV_A-induced cytotoxicity in human chronic myelogenous leukaemia cells and human epidermoid carcinoma cells (Karagiannis et al. 2006b). It was denoted the name UV_ASens and further studies proved that its photopotency was approximately 1,000-fold that of psoralens (Karagiannis et al. 2006a). With 1 μm of UV_ASens, the UV_A fluence required to kill 90 % of the cell culture was approximately 2 J/m^2, whereas for PUVA therapy, the fluences required for comparable cell kill was approximately three orders of magnitude higher (Karagiannis et al. 2006a; Procaccini et al. 1996).

17.6 Treatment of Late-Stage or Refractory CTCL

In late-stage CTCL, especially in patients with Sézary syndrome, T-cell immunity is considerably weakened (Poligone and Heald 2012). With the advancing of the disease, there is a decline in the production of cytokines necessary for the activation and differentiation of cell-mediated immunity such as interferon-γ (IFNγ) and interleukin-2 (IL-2), which are produced by T_H1 cells (Rook et al. 1993). The increased tumour burden in the advanced CTCL as well as the imbalance between T_H1 and T_H2 cytokines result in a significant impairment of cell-mediated immunity (Vittorio et al. 2001). Therefore, immunosuppressive and cytotoxic therapies are used as backup therapies as they can lead to various complications and to increased morbidity (Poligone and Heald 2012).

17.6.1 Extracorporeal Photopheresis

For the advanced leukemic forms of CTCL (stage III), ECP is considered most effective (Introcaso et al. 2008). It is an extension of PUVA therapy that involves the extracorporeal exposure of peripheral mononuclear cells mixed with 8-methoxypsoralen (8-MOP) to 1 or 2 J of UVA light (Vittorio et al. 2001). When the photosensitised cells are exposed to UVA irradiation ex vivo, 8-MOP is activated causing cross-linking of DNA in leukocytes and subsequently, they are re-infused (Knobler and Girardi 2001). Cross-linking of DNA induces apoptosis-releasing tumour antigens that lead to a systemic anti-tumour response against the malignant T-cell clone in the patient (Chiesa Fuxench 2010; Vittorio et al. 2001). In addition, ECP also leads to dendritic cell differentiation, which is considered to further enhance antigen presentation and persistence of the host immune response (Berger et al. 2001).

17.6.2 Histone Deacetylase Inhibitors

Most genomic DNA in eukaryotic cells is packaged around histone proteins to allow it to be accommodated within the nucleus (New et al. 2012). Approximately 146 base pairs of DNA are wrapped around an octamer of histone proteins termed as a nucleosome, which consists of one H3–H4 tetramer and two H2A–H2B dimmers (Bentley et al. 1984). Nucleosomes interact with the linker histone protein H1 and other chromatin-associated proteins to be further compacted into chromatin (Inche and La Thangue 2006). The N-terminal tail of histones undergoes various epigenetic modifications including acetylation, methylation, phosphorylation and ubiquitination (Moniot et al. 2012). These post-translational modifications of histones form the "histone code", which is read by transcription factors resulting in specific gene expression pattern required for a particular phenotype (Santini et al. 2007).

Epigenetic modifications, in contrast to genetic modifications, are reversible changes that are transmitted from a cell to its progeny (Santini et al. 2007). Abnormal epigenetic control appears as an early mechanism in the tumoural transformation of cells, which results in alterations in normal gene expression, particularly in tumour suppressor genes and oncogenes (Blanquart et al. 2011; New et al. 2012). These events are independent of alterations in the DNA sequence as they modulate gene expression without changing DNA sequence and without introducing any new genetic information (Santini et al. 2007). Therefore, maintaining the equilibrium of these reversible chromatin rearrangements is critical as many studies have shown that alterations of this equilibrium are frequently involved in the genesis of cancer.

The two best characterised chromatin-remodelling mechanisms are histone acetylation and DNA methylation. These two modifications are interdependent and their equilibrium contributes to overall regulation of gene expression (Santini et al. 2001; Das and Singal 2004). Human DNA contains small regions of DNA called

"CpG islands" where CpGs are clustered together at higher frequency (Das and Singal 2004). Nearly half of all human genes have CpG islands in their 5′-promoter regions, which are usually unmethylated in normal tissues, regardless of the transcriptional status of the gene (Santini et al. 2007). DNA methylation results in changes in chromatin structure and the consequent repression of gene transcription have fundamental roles during embryogenesis, differentiation, but also in cancer genesis (Santini et al. 2001; Das and Singal 2004). In normal cells, methylation of CpG islands is mediated by methyl-binding proteins, which interact with transcription repressors, HDACs and DNA methyltransferases (DNMTs) (Santini et al. 2001; Das and Singal 2004). This event leads to inaccessibility of DNA to RNA polymerase. In cancerous cells, DNMTs are over-expressed hypermethylating CpG islands, which results in permanent repression of gene transcription (Santini et al. 2001; Das and Singal 2004).

Another essential epigenetic modification that contributes to the equilibrium of gene expression is histone acetylation. Histone acetylation is regulated by the equilibrium of two enzymes: HDACs and histone acetyltransferases (HATs), which are recruited locally by sequence-specific DNA-binding proteins, attracted to the site by CpG-methylated islands (Santini et al. 2007). They reversibly and dynamically alter the acetylation status of histones at multiple lysine residues in their N-terminal tails (Lansigan and Foss 2010; Khan and La Thangue 2008). HATs transfer acetyl groups to lysine residues, which leads to expansion of chromatin. This results in increased accessibility of regulatory proteins to the structure of DNA resulting in a transcriptionally active state (Bolden et al. 2006). In contrast, HDAC removes acetyl modification from lysine residues resulting in chromatin condensation and limited access of the transcription factors to DNA leading to transcriptional repression (Marks et al. 2000; Johnstone 2002). Chromatin condensation occurs via elimination of the charge-neutralising acetyl groups resulting in a closed chromatin structure (Fig. 17.1) (Kuo 1998).

There has been increasing evidence of alterations in histone acetylation regulatory enzymes and the subsequent aberrant acetylation in cancers (Kavanaugh et al. 2010). Particularly, cancers of hematologic and epithelial origin (e.g. CTCL) have been associated with hypoacetylation of histones. Hypoacetylation of histones results in a significant decrease in the expression of anti-tumour genes including those responsible for cell differentiation, cell-cycle control, apoptosis and tumour suppression (Jenuwein and Allis 2001; Zain and O'Connor 2010; Kavanaugh et al. 2010). Therefore, HDAC inhibitors were developed to reverse the epigenetic modifications in order to correct uncontrolled proliferation and aberrant apoptotic pathways in cancer cells (Jain and Odenike 2010; Kwa et al. 2011).

HDAC inhibitors cause the accumulation of hyperacetylated histones and this induces various anti-cancer effects including apoptosis, cytostasis, differentiation and inhibition of tumour angiogenesis (Di Gennaro et al. 2004; Dokmanovic and Marks 2005; Liu et al. 2006; Johnstone and Licht 2003) (Fig. 17.2). HDAC inhibitor are able to block cell proliferation and cause apoptosis by inducing cell cycle arrest in G1 or G2/M phase through dysregulation of proteins that mediate cell cycle progression and coordinate G1/S and G2/M transition such as cyclins and

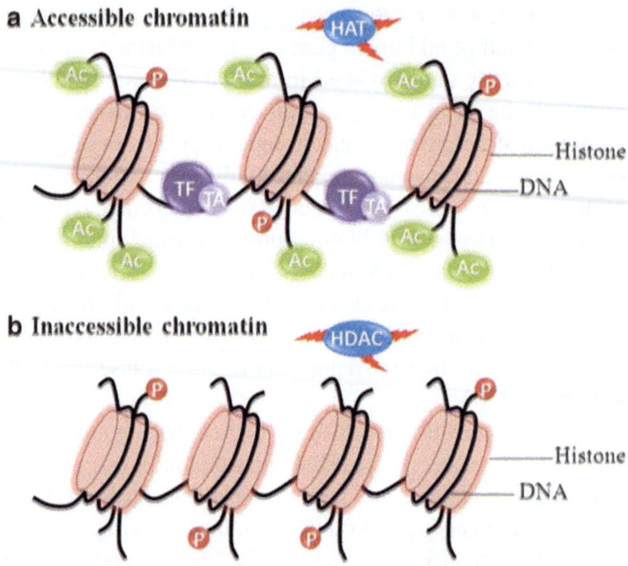

Fig. 17.1 Structure of chromatin and the effect of histone acetyltransferases and histone deacetylases on the transcriptional activity. (**a**) Transcriptional activation through histone acetylation (Ac) catalysed by HATs and demethylation of DNA resulting in a relaxed configuration of chromatin, which is accessible to transcription factors (TFs) and transcriptional activators (TAs). (**b**) Transcriptional repression through histone deacetylation catalysed by histone deacetylases and DNA methylation inducing a closed configuration of chromatin, which is not accessible to transcription factors and activators

cyclin-dependent kinases (Sandor et al. 2000; Sambucetti et al. 1999; Classon and Harlow 2002). In addition, HDAC inhibitors cause apoptosis by upregulating the expression of genes that encode for death receptors and their ligands such as Fas and the Apo 2L/TRAIL receptors, death receptor 4 (DR4) and DR5, and also by generating reactive oxygen species (ROS) (Peart et al. 2003; Ruefli et al. 2001; Jones and Saha 2002; Nebbioso et al. 2005). Furthermore, HDAC inhibitors have been associated with antiangiogenic effects. They have been shown to upregulate angiogenesis inhibitors such as thrombospondin and von-Hippel Lindau factor and downregulate vasculogenesis promoting factors such as vascular endothelial growth factor (VEGF) and hypoxia-induced protein (Kim et al. 2001).

HDAC inhibitors bind to the active site of specific classes of HDACs inhibiting the enzymes from removing the acetyl groups from lysine residues (Kavanaugh et al. 2010). This induces hyperacetylation of both histone and non-histone targets promoting a more open chromatin structure as acetyl groups reduce affinity for DNA (Miller et al. 2003; Santini et al. 2007). The loosening of the histone complex from the DNA exposes more DNA regions to the transcriptional machinery (Santini et al. 2007). This results in an increase in the expression of genes for preventing carcinogenesis and tumourigenesis (Codd et al. 2009). HDAC inhibitors promote cell cycle arrest, terminal differentiation, apoptosis and/or autophagic cell death

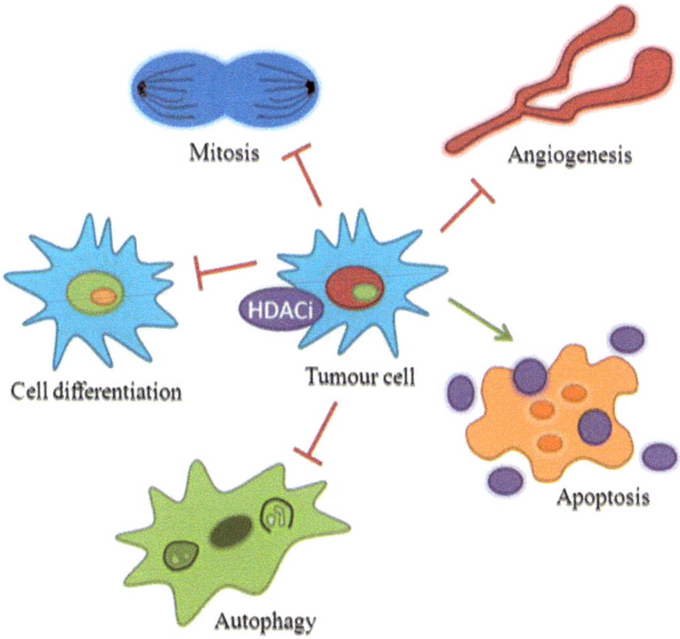

Fig. 17.2 Effects of histone deacetylase inhibition. Histone deacetylase inhibition blocks cell proliferation, angiogenesis, cell differentiation and development and autophagy, indicated by *red lines*. In contrast, it favours apoptosis, indicated by the *green arrow*

(Minucci and Pelicci 2006; Richon et al. 2009; Vrana et al. 1999). They achieve this through upregulating the expression of pro-apoptotic factors such as Bak, Bax and Bim and downregulating the expression of anti-apoptotic factors such as Bcl-2, Bcl-XI, XIAP and Mcl-1, which were shown to play important roles in their anti-tumour activity (Zhang et al. 2004). The overriding effect of such epigenetic alterations in any tumour may be cell-context dependent (Jain and Odenike 2010).

In vivo studies demonstrated that HDAC inhibitors induce tumour apoptosis at concentrations to which normal cells are resistant and such selective induction of apoptosis makes them well suited for cancer therapy (Marks and Jiang 2005). They are a structurally diverse group of compounds, which include both natural and synthetic compounds (Karagiannis and El Osta 2006). HDAC inhibitors include the cyclic and non-cyclic hydroxamates (e.g. Trichostatin A (TSA), vorinostat (Zolinza®, SAHA)) and cyclic peptides (depsipeptide, apicidin), benzamides and electrophilic ketones (Karagiannis and El Osta 2006). Currently, the two most potent HDAC inhibitors are intravenous romidepsin and oral vorinostat. They are the only two approved monotherapies used for the treatment of relapsed and refractory CTCLs (Fig. 17.3) (Kavanaugh et al. 2010). Several other HDACs are currently in clinical trials for CTCLs (Table 17.5) (Zain et al. 2010).

Vorinostat increases expression of genes governing growth arrest, differentiation; and both caspase-dependent and caspase-independent apoptosis (Rasheed et al. 2008).

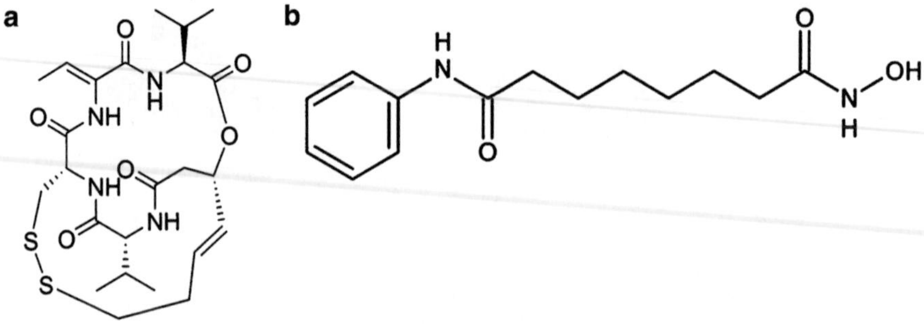

Fig. 17.3 Chemical structures of the two approved histone deacetylase inhibitors: (**a**) Romidepsin; (**b**) Vorinostat

Table 17.5 Histone deacetylase inhibitors currently being evaluated for cutaneous T cell lymphoma

Chemical structure	Compound	Phase of clinical trials	FDA approval status for CTCL	References
Hydroxamic acid	Vorinostat	Phase I, II, III	Approved in 2006	Olsen et al. (2007)
	Belinostat	Phase I, II	Not approved	Carew et al. (2008)
	Panobinostat	Phase I, II	Not approved	Duvic et al. (2013)
	Givinostat	Phase I, II	Not approved	Galli et al. (2010)
Cyclic peptide	Romidepsin	Phase I, II	Approved in 2009	Piekarz et al. (2009)
Benzamide	Entinostat	Phase I, II	Not approved	Egger et al. (2004)
Short-chain fatty acids	Valproic acid	Phase I, II	Not approved	Dokmanovic et al. (2007)
	Phenyl butyrate	Phase I, II	Not approved	Xu et al. (2007)

CTCL cutaneous T cell lymphoma, *FDA* Food and Drug Administration

In vitro studies have an accumulation of acetylated histones H2B, H3 and H4 in both normal and tumour cells, inducing increased expression of death receptor 5, tumour necrosis factor and the cell cycle arrest inducer, p21(Zain and O'Connor 2010; Rasheed et al. 2008). This results in inhibition of cell cycle progression is inhibited and selective apoptosis in the malignant CTCL cell lines (Grant et al. 2007). Cell cycle is arrested in G2 phase as p21 binds to and inactivates cyclin-dependent kinase 2. Vorinostat treatment was associated with a significant decrease in the level of dermal and epidermal lymphocytes (Duvic et al. 2007). In addition, vorinostat was shown to downregulate expression of VEGF, which controls angiogenesis, a critical component for cancer growth and metastasis (Heider et al. 2006). Furthermore, studies have shown that CTCL cells have an abnormally high expression of anti-apoptotic protein Bcl-2 possibly increasing their survival and resistance against various therapies (Osella Abate et al. 2001; Zhang et al. 2002; Breuckmann et al. 2002). Vorinostat has shown to correct this imbalance between expression of Bcl-2 and pro-apoptotic protein Bax and promote activation of caspase-3 pathway resulting in increased apoptosis (Zhang et al. 2005). Response to vorinostat was also associated with a significant increase in the level of the antiangiogenic protein TSP-1 and a decrease in microvessel density in patients' skin lesions (Duvic et al. 2007). Vorinostat produced objective clinical and symptomatic relief with meaningful duration in patients with advanced, refractory CTCLs (Duvic et al. 2007). Consequently, vorinostat was approved by the United States Food and Drug Administration (FDA) in 2006 for patients who have progressive, persistent or recurrent CTCL following two prior systemic treatments (Marks and Breslow 2007; Mann et al. 2007).

Romidepsin (Istodax®) is a cyclic peptide, which is a natural product obtained from *Chromobacterium violaceum* (Shiozawa et al. 2009). It has been reported to show anti-proliferative and apoptotic effects in various malignant cells and hence, is used as a broad-spectrum, intravenously administered HDAC inhibitor (Jain and Odenike 2010). Its mechanism of action includes the activation of caspase 3 and 9 and the downregulation of Bcl-2 and Bcl-xl (Shiozawa et al. 2009). It is a prodrug that is reduced to the active compound when it enters cells (Furumai et al. 2002). Early in vitro studies demonstrated that romidepsin induces significant apoptosis in the HUT78 human CTCL cell line. Romidepsin also causes accumulation of highly acetylated histones within cells, which results in cell cycle arrest in G1 and G2/M, differentiation, morphologic reversion and/or apoptosis of transformed cells (Ueda et al. 1994; Aron et al. 2003; Klisovic et al. 2003; Yang et al. 2007) Like vorinostat, romidepsin is associated with increased expression of p21 (Piekarz et al. 2001). Phase I trials revealed that romidepsin is particularly effective for patients who had refractory Sézary syndrome (Piekarz et al. 2001). A rapid decline in the number of Sézary cells was reported as well as significant improvement in skin erythema and oedema (Piekarz et al. 2001). Romidepsin was also approved by FDA in 2009 for CTCL patients who have received at least one prior systemic therapy (Lyseng Williamson and Yang 2012). One potential advantage with vorinostat over romidepsin may be the convenience associated with oral administration as vorinostat is orally bioavailable (Jain and Odenike 2010).

17.6.3 Combination Therapy

Currently, there is no single agent available that is potent enough to control CTCL (Stadler 2007). Overall response rates to monotherapies are only about 50–60 % (Stadler 2007). Even PUVA therapy, which is the most commonly employed monotherapy for MF, had the recurrence rate of 32 % over a median follow-up of 94 months (Querfeld et al. 2005). Combinations of monotherapies with different mechanisms may increase patients' response rates (Stadler 2007). Also, combining therapies may reduce doses that the each monotherapy is given at, minimising the toxicity of each individual treatment (Stadler 2007). Furthermore, decreased risks of the progression of disease to extracutaneous T-cell lymphoma involvement have been associated with combination therapy (Duvic 2007). Therefore, combination therapy has potential to consolidate remission and allow a long-term treatment with an acceptable side-effect profile (Stadler 2007). As there are no curative therapies for CTCL available currently, long-term treatment with appropriate treatment combinations is mandatory to maintain recurrence-free survival (Stadler 2007).

The most common treatment combinations for the treatment of CTCL in the widespread patch-plaque phase are PUVA with either bexarotene gel (retinoid) or IFNα, although PUVA is also used with other retinoid compounds and denileukin diftitox (Stadler 2007). Bexarotene gel is a retinoid that are known to increase Th1 cytokines and IFNα (Vittorio et al. 2001). CTCL is characterised by a shift in cytokine profiles from Th1- to Th2-dominating, and hence retinoids are able to correct the imbalance between Th1 and Th2 cytokines (Stadler and Kremer 2006; Stadler 1998). IFNα is known to inhibit proliferation of malignant T cells in response to growth stimulatory factors. The response rates were higher in patients who received the combination therapy compared with patients who received PUVA alone (Stadler et al. 1998). Also, the cumulative treatment dose of UVA was significantly lower for the combination therapy and consequently, it toxicity which enables it to be used as long-term treatment (Stadler et al. 1998). In addition, the recurrence-free time was about 60 weeks longer for PUVA plus IFNα compared with PUVA alone (Stadler et al. 1998). Another commonly used treatment combination for early-stage CTCL is bexarotene and denileukin diftitox (Gorgun and Foss 2002). They act synergistically as the immunomodulatory capacity of bexarotene (retinoid) upregulates the expression of the high-affinity IL-2 receptor and, hence, increase susceptibility to denileukin diftitox (Gorgun and Foss 2002).

17.6.4 Phototherapy in Combination with HDACi

HDAC inhibitors have the capacity to trigger the intrinsic and extrinsic apoptotic pathways that enables them to lower the apoptotic threshold in malignant cells. Thus, the malignant cells are made more susceptible to cytotoxic agents during combination therapy (Ma et al. 2009). Such combination therapy would be particularly effective against the cancer cells that are chemoresistant (Kwa et al. 2011). However, not all drug combinations involving HDAC inhibitors enhance anti-tumour activity. For example, the combination of vorinostat and the anti-leukaemic DNA-damaging drug cytarabine has shown to act antagonistically (Shiozawa et al. 2009). Vorinostat induces the G_1/G_2 cell cycle arrest reducing the availability of S phase cells for cytarabine to work, thereby limiting the DNA-damaging effects (Shiozawa et al. 2009).

Recently, the potential synergy of HDAC inhibitors and phototherapy has been attracting attention (Santini et al. 2007). In particular, HDAC inhibitor sodium butyrate (SB) was shown to augment radiosensitivity in cancer cells by downregulating the expression of double-strand breaks repair proteins, especially non-homologous end rejoining-related (NHEJ) proteins (Munshi et al. 2005; Munshi et al. 2006). In addition, SB was shown to enhance the cytotoxic effects of PUVA therapy in several cancer cell lines, including skin melanoma (Toyooka and Ibuki 2009). Compared to a single treatment with PUVA or SB, combined treatment with SB and PUVA induced much more apoptosis in cancer cells within 24 h (Toyooka and Ibuki 2009). Pretreatment with SB significantly reduced the number of double-strand breaks that normally form as intermediates during the repair of PUVA-induced interstrand cross-links in chromosomal DNA (Toyooka and Ibuki 2009). Disruption of the nucleotide excision repair resulted in the accumulation of interstrand cross-links in cancer cells, which blocked replication and eventually cell death (Toyooka and Ibuki 2009).

In addition, the ability of vorinostat to induce an open chromatin conformation allows UVA radiation to cause increased numbers of double-strand breaks (Rodd et al. 2012). The loosening of the chromatin structure leads to an increase in the number of susceptible sites for free radicals to cause DNA damage (Cadet et al. 2005). Recently, investigated the effect of pre-incubation with vorinostat on haematological cell lines, which were photosensitised by UV_ASens followed by exposure to UVA radiation (Rodd et al. 2012). The resultant DNA double-strand breaks were analysed with γH2AX immunofluorescence assays. While they exist as by-products of various normal endogenous processes, they are also formed as a consequence of exogenous insults including ionising radiation ultraviolet rays, oxidative stress and chemical agents (Mah et al. 2010). Upon induction of a double-strand break, the histone variant H2AX on the Ser-139 residue becomes phosphorylated, resulting in discrete foci at the site of damage. Thus, γH2AX formation is a rapid and accurate cellular marker to the presence of DNA double-strand breaks (Dickey et al. 2009). The accumulation of γH2AX foci was shown to be the greatest in cells that received both UV_ASens/UVA radiation and vorinostat treatment (Rodd et al. 2012). Such prolonged expression of γH2AX foci indicates a decrease in the rate of repair of radiation-induced DNA double-strand breaks (Munshi et al. 2006). In addition to the inhibition of the DNA repair pathway, vorinostat is known to induce a significant increase in apoptosis (Dong et al. 2008). It induces apoptosis via mechanisms such as the activation of the intrinsic death pathway resulting in caspase-dependent cell death (Dong et al. 2008; Amin et al. 2001; Rodd et al. 2012). Such findings indicate that vorinostat attenuates the cytotoxic effects of UV_ASens/UVA radiation phototherapy further highlighting the synergistic therapeutic potential for combination therapy involving HDAC inhibitor and phototherapy (Rodd et al. 2012).

In vivo treatment with romidepsin has been shown to upregulate the expression of IL-2 receptor, which is the target for denileukin diftitox (ONTAK®). Denileukin diftitox is a fusion molecule of IL-2 to diphtheria toxin, which targets the CD25 subunit of the IL-2 receptor resulting in a delivery of the toxin to CD-25 bearing T-cells (Foss et al. 2001; Olsen et al. 2001; Lansigan et al. 2010). By upregulating the expression of IL-2 receptors, romidepsin treatment increases the sensitivity of the cells to denileukin diftitox with a synergistic increase in apoptosis. Such combination of drugs is another potential combination therapy that may be developed for increased therapeutic efficacy.

17.7 Conclusion

Although there is no curative therapy for CTCL yet, there is a myriad of distinct therapeutic options available for different forms of CTCL. Hence, it is of crucial importance that the patient is diagnosed with the correct subset of CTCL commonly by using the TNM staging system. An accurate staging of the disease and stratifying risks are then to be followed by a careful selection of therapy tailored for the patient. Therapies are commonly divided into skin-directed and systemic therapies, which are used for the management of the localised and widespread disease, respectively. These treatment modalities may be used as a monotherapy or as a multi-modal combination therapy. Various therapies are combined in order to produce higher overall response rates. In addition, low-dose combination therapy can possibly minimise the toxicity of each individual treatment allowing good long-term treatment tolerability for the patients. In particular, synergistic actions of HDAC inhibitors and phototherapy have highlighted the potential for another novel combination therapy for the treatment of CTCL. HDAC inhibitors acetylate the core histone protein increasing DNA accessibility, which promotes the formation of UV-induced interstrand cross-links (photoadducts). These photoadducts accumulate causing increased DNA double-strand breaks and consequently, activating various apoptotic pathways. In addition, downregulation of the expression of various DNA repair proteins by HDAC inhibitors result in suppression of the DNA repair capacity of CTCL cells, exacerbating the DNA damage. Therefore, future studies should further explore the potential of this combination therapy. The potential of a combination therapy may be assessed by its capacity to minimise risks, maximise response and to improve the quality of life in CTCL patients.

Acknowledgments The support of the Australian Institute of Nuclear Science and Engineering is acknowledged. TCK was the recipient of AINSE awards. TCK is a Future Fellow and Epigenomic Medicine Laboratory is supported by the Australian Research Council. Supported in part by the Victorian Government's Operational Infrastructure Support Program.

References

Adhikary A, Buschmann V, Muller C, Sauer M (2003) Ensemble and single-molecule fluorescence spectroscopic study of the bidning modes of the bis-bensimidazole derivative Hoechst 33258 with DNA. Nucleic Acids Res 31:2178–2186

Akilov OE, Grant C, Frye R, Bates S, Piekarz R, Geskin LJ (2012) Low-dose electron beam radiation and romidepsin therapy for symptomatic cutaneous T-cell lymphoma lesions. Br J Dermatol 167:194–197

Amin HM, Saeed S, Alkan S (2001) Histone deacetylase inhibitors induce caspase-dependent apoptosis and downregulation of daxx in acute promyelocytic leukaemia with T(15;17). Br J Haematol 115:287–297

Apisarnthanarax N, Talpur R, Duvic M (2002) Treatment of cutaneous T cell lymphoma: current status and future directions. Am J Clin Dermatol 3:193–215

Arbiser J, Govindarajan B, Battle T, Lynch R, Frank D, Ushio Fukai M, Perry B, Stern D, Bowden GT, Liu A, Klein E, Kolodziejski P, Eissa NT, Hossain C, Nagle D (2006) Carbazole is a naturally occurring inhibitor of angiogenesis and inflammation isolated from antipsoriatic coal tar. J Investig Dermatol 126:1396–1402

Aron J, Parthun M, Marcucci G, Kitada S, Mone A, Davis M, Shen T, Murphy T, Wickham J, Kanakry C, Lucas D, Reed J, Grever M, Byrd J (2003) Depsipeptide (FR901228) Induces histone acetylation and inhibition of histone deacetylase in chronic lymphocytic leukemia cells concurrent with activation of caspase 8-mediated apoptosis and down-regulation of c-FLIP protein. Blood 102:652–658

Bentley GA, Finch JT, Lewit Bentley A, Roth M (1984) The crystal structure of the nucleosome core particle by contrast variation. Basic Life Sci 27:105–117

Berger C, Xu A, Hanlon D, Lee C, Schechner J, Glusac E, Christensen I, Snyder E, Holloway V, Tigelaar R, Edelson RL (2001) Induction of human tumor-loaded dendritic cells. Int J Cancer 9:438–447

Beyer M, Mobs M, Humme D, Sterry W (2011) Pathogenesis of mycosis fungoides. J German Soc Dermatol 9:594–598

Bladon J, Taylor PC (2006) Extracorporeal photopheresis: a focus on apoptosis and cytokines. J Dermatol Sci 43:85–94

Blanquart C, François M, Charrier C, Bertrand P, Gregoire M (2011) Pharmacological characterization of histone deacetylase inhibitor and tumor cell-growth inhibition properties of new benzofuranone compounds. Curr Cancer Drug Targets 11:919–928

Bolden J, Peart M, Johnstone R (2006) Anticancer activities of histone deacetylase inhibitors. Nat Rev Drug Discov 5:769–784

Bradford P, Devesa S, Anderson W, Toro J (2009) Cutaneous lymphoma incidence patterns in the United States: a population-based study of 3884 cases. Blood 113:5064–5073

Breneman D, Duvic M, Kuzel T, Yocum R, Truglia J, Stevens V (2002) Phase 1 and 2 trial of bexarotene gel for skin-directed treatment of patients with cutaneous T-cell lymphoma. Arch Dermatol 138:325–332

Breuckmann F, Von Kobyletzki G, Avermaete A, Kreuter A, Altmeyer P (2002) Efficacy of ultraviolet A1 phototherapy on the expression of bcl-2 in atopic dermatitis and cutaneous T-cell lymphoma in vivo: a comparison study. Photodermatol Photoimmunol Photomed 18:217–222

Breuckmann F, Gambichler T, Altmeyer P, Kreuter A (2004) UVA/UVA1 phototherapy and PUVA photochemotherapy in connective tissue diseases and related disorders: a research based review. BMC Dermatol 4:11

Broch H, Viani R, Vasilescu D (1991) Quantum molecular study of the alkylating agent mechlorethamine. Int J Quantum Chem 40:119–130

Bunn PA, Lamberg SI (1979) Report of the Committee on Staging and Classification of Cutaneous T-Cell Lymphomas. Cancer Treat Rep 63:725–728

Cadet J, Sage E, Douki T (2005) Ultraviolet radiation-mediated damage to cellular DNA. Mutat Res 571:3–17

Campbell JJ (2010) Sezary syndrome and mycosis fungoides arise from distinct T-cell subsets: a biologic rationale for their distinct clinical behaviors. Blood 116:767

Carew J, Giles F, Nawrocki S (2008) Histone deacetylase inhibitors: mechanisms of cell death and promise in combination cancer therapy. Cancer Lett 269:7–17

Chiesa Fuxench ZC (2010) Extracorporeal photopheresis: a review on the immunological aspects and clinical applications. P R Health Sci J 29:337

Cimino GD, Gamper HB, Isaacs ST, Hearst JE (1985) Psoralens as photoactive probes of nucleic acid structure and function: organic chemistry, photochemistry, and biochemistry. Annu Rev Biochem 54:1151–1193

Clark C, Dawe RS, Evans AT, Lowe G, Ferguson J (2000) Narrowband TL-01 phototherapy for patch-stage mycosis fungoides. Arch Dermatol 136:748–752

Clark RA, Chong B, Mirchandani D (2006) The vast majority of CLA+T cells are resident in normal skin. J Immunol 176:4431–4439

Classon M, Harlow E (2002) The retinoblastoma tumour suppressor in development and cancer. Nat Rev Cancer 2:910–917
Codd R, Braich N, Liu J, Soe C, Pakchung AAH (2009) Zn(II)-dependent histone deacetylase inhibitors: suberoylanilide hydroxamic acid and trichostatin A. Int J Biochem Cell Biol 41:736–739
Criscione VD, Weinstock MA (2007) Incidence of cutaneous T-cell lymphoma in the United States, 1973–2002. Arch Dermatol 143:854–859
Das P, Singal R (2004) DNA methylation and cancer. J Clin Oncol 22:4632–4642
De Coninck EC, Kim YH, Varghese A, Hoppe RT (2001) Clinical characteristics and outcome of patients with extracutaneous mycosis fungoides. J Clin Oncol 19:779–784
Demierre M-F, Kim Y, Zackheim H (2003) Prognosis, clinical outcomes and quality of life issues in cutaneous T-cell lymphoma. Hematol Oncol Clin North Am 17:1485–1507
Demierre M-F, Gan S, Jones J, Miller D (2006) Significant impact of cutaneous T-cell lymphoma on patients' quality of life: results of a 2005 National Cutaneous Lymphoma Foundation Survey. Cancer 107:2504–2511
Di Gennaro E, Bruzzese F, Caraglia M, Abruzzese A, Budillon A (2004) Acetylation of proteins as novel target for antitumor therapy: review article. Amino Acids 26:435–441
Diamandidou E, Cohen PR, Kurzrock R (1996) Mycosis fungoides and sezary syndrome. Blood 88:2385–2409
Dickey J, Redon C, Nakamura A, Baird B, Sedelnikova O, Bonner W (2009) H2AX: functional roles and potential applications. Chromosoma 118:683–692
Diederen PVMM, van Weelden H, Sanders CJG, Toonstra J, van Vloten W (2003) Narrowband UVB and psoralen-UVA in the treatment of early-stage mycosis fungoides: a retrospective study. J Am Acad Dermatol 48:215–219
Dokmanovic M, Marks P (2005) Prospects: histone deacetylase inhibitors. J Cell Biochem 96:293–304
Dokmanovic M, Clarke C, Marks P (2007) Histone deacetylase inhibitors: overview and perspectives. Mol Cancer Res 5:981–989
Dong G, Wang L, Wang C-Y, Yang T, Kumar MV, Dong Z (2008) Induction of apoptosis in renal tubular cells by histone deacetylase inhibitors, a family of anticancer agents. J Pharmacol Exp Ther 325:978–984
Duhovic C, Child F, Wain EM (2012) CME dermatology. Clin Med 12:160–164
Dummer R (2003) Therapy of cutaneous lymphoma–current practice and future developments. Onkologie 26:366
Dummer R, Hess Schmid M, Burg G (2000) Cutaneous T-cell lymphomas: prognosis and quality-of-life issues. Clin Lymphoma 1(suppl 1):S21–S25
Duvic M (2007) Systemic monotherapy vs combination therapy for CTCL: rationale and future strategies. Oncology 21:33–40
Duvic M, Talpur R, Ni X, Zhang C, Hazarika P, Kelly C, Chiao J, Reilly J, Ricker J, Richon V, Frankel S (2007) Phase 2 trial of oral vorinostat (suberoylanilide hydroxamic acid, SAHA) for refractory cutaneous T-cell lymphoma (CTCL). Blood 109:31–39
Duvic M, Dummer R, Becker JR, Poulalhon N, Ortiz Romero P, Grazia Bernengo M, Lebbé C, Assaf C, Squier M, Williams D, Marshood M, Tai F, Prince HM (2013) Panobinostat activity in both bexarotene-exposed and -naive patients with refractory cutaneous T-cell lymphoma: results of a phase II trial. Eur J Cancer 49:386–394
Egger G, Liang G, Aparicio A, Jones P (2004) Epigenetics in human disease and prospects for epigenetic therapy. Nature 429:457–463
Fiala Z, Borska L, Pastorkova A, Kremlacek J, Cerna M, Smejkalova J, Hamakova K (2006) Genotoxic effect of Goeckerman regimen of psoriasis. Arch Dermatol Res 298:243–251
Finch PW, Murphy F, Cardinale I, Krueger JG (1997) Altered expression of keratinocyte growth factor and its receptor in psoriasis. Am J Pathol 151:1619–1628
Foss FM, Bacha P, Osann KE, Demierre MF, Bell T, Kuzel T (2001) Biological correlates of acute hypersensitivity events with DAB(389)IL-2 (denileukin diftitox, ONTAK) in cutaneous T-cell

lymphoma: decreased frequency and severity with steroid premedication. Clin Lymphoma 1:298–302
Furumai R, Matsuyama A, Kobashi N, Lee K-H, Nishiyama M, Nakajima H, Tanaka A, Komatsu Y, Nishino N, Yoshida M, Horinouchi S (2002) FK228 (depsipeptide) as a natural prodrug that inhibits class I histone deacetylases. Cancer Res 62:4916–4921
Galli M, Salmoiraghi S, Golay J, Gozzini A, Crippa C, Pescosta N, Rambaldi A (2010) A phase II multiple dose clinical trial of histone deacetylase inhibitor ITF2357 in patients with relapsed or progressive multiple myeloma. Ann Hematol 89:185–190
Goekerman EH (1925) The treatment of psoriasis. Northwest Med 24:229–231
Gorgun G, Foss F (2002) Immunomodulatory effects of RXR rexinoids: modulation of high-affinity IL-2R expression enhances susceptibility to denileukin diftitox. Blood 100: 1399–1403
Grant S, Easley C, Kirkpatrick P (2007) Vorinostat. Nat Rev Drug Discov 6:21–22
Heider U, Kaiser M, Sterz J, Zavrski I, Jakob C, Fleissner C, Eucker J, Possinger K, Sezer O (2006) Histone deacetylase inhibitors reduce VEGF production and induce growth suppression and apoptosis in human mantle cell lymphoma. Eur J Haematol 76:42–50
Homey B, Alenius H, Müller A, Soto H, Bowman E, Yuan W, Mcevoy L, Lauerma A, Assmann T, Bünemann E, Lehto M, Wolff H, Yen D, Marxhausen H, To W, Sedgwick J, Ruzicka T, Lehmann P, Zlotnik A (2002) CCL27-CCR10 interactions regulate T cell-mediated skin inflammation. Nat Med 8:157–165
Hönigsmann H, Brenner W, Rauschmeier W, Konrad K, Wolff K (1984) Photochemotherapy for cutaneous T cell lymphoma: a follow-up study. J Am Acad Dermatol 10:238–245
Horwitz S (2011) CTCL-MF fast facts. Getting the facts. Available http://www.clfoundation.org. Accessed 21 Nov 2012
Huber MA, Staib G, Pehamberger H, Scharffetter-Kochanek K (2006) Management of refractory early-stage cutaneous T-cell lymphoma. Am J Clin Dermatol 7:155–169
Hymes K (2007) Choices in the treatment of cutaneous T-cell lymphoma. Oncology 21:18–23
Hymes K (2010) The role of histone deacetylase inhibitors in the treatment of patients with cutaneous T-cell lymphoma. Clin Lymphoma Myeloma Leuk 10:98–109
Inche A, La Thangue N (2006) Chromatin control and cancer-drug discovery: realizing the promise. Drug Discov Today 11:97–109
Introcaso C, Leber B, Greene K, Ubriani R, Rook A, Kim EJ (2008) Stem cell transplantation in advanced cutaneous T-cell lymphoma. J Am Acad Dermatol 58:645–649
Jain N, Odenike O (2010) Emerging role of the histone deacetylase inhibitor romidepsin in hematologic malignancies. Expert Opin Pharmacother 11:3073–3084
Jenuwein T, Allis CD (2001) Translating the histone code. Science 293:1074–1080
Johnstone RW (2002) Histone-deacetylase inhibitors: novel drugs for the treatment of cancer. Nat Rev Drug Discov 1:287–299
Johnstone R, Licht J (2003) Histone deacetylase inhibitors in cancer therapy: is transcription the primary target? Cancer Cell 4:13–18
Jones L, Saha V (2002) Chromatin modification, leukaemia and implications for therapy. Br J Haematol 118:714–727
Karagiannis T, El Osta A (2006) Clinical potential of histone deacetylase inhibitors as stand alone therapeutics and in combination with other chemotherapeutics or radiotherapy for cancer. Epigenetics 1:121–126
Karagiannis TC, Lobachevsky PN, Leung BKY, White JM, Martin RF (2006a) Receptor-mediated DNA-targeted photoimmunotherapy. Cancer Res 66:10548–10552
Karagiannis TC, Lobachevsky PN, Martin RF (2006b) DNA targeted UVA photosensitisation: characterisation of an extremely photopotent iodinated minor groove binding DNA ligand. J Photochem Photobiol B 83:195–204
Kavanaugh S, White L, Kolesar J (2010) Vorinostat: a novel therapy for the treatment of cutaneous T-cell lymphoma. Am J Health Syst Pharm 67:793–797
Kaye FJ, Bunn PA, Steinberg SM, Stocker JL, Ihde DC, Fischmann AB, Glatstein EJ, Schechter GP, Phelps RM, Foss FM (1989) A randomized trial comparing combination electron-beam

radiation and chemotherapy with topical therapy in the initial treatment of mycosis fungoides. N Engl J Med 321:1784–1790

Khan O, La Thangue N (2008) Drug insight: histone deacetylase inhibitor-based therapies for cutaneous T-cell lymphomas. Nat Clin Pract Oncol 5:714–726

Kim MS, Kwon HJ, Lee YM, Baek JH, Jang JE, Lee SW, Moon EJ, Kim HS, Chung HY, Kim CW, Kim KW (2001) Histone deacetylases induce angiogenesis by negative regulation of tumor suppressor genes. Nat Med 7:437–443

Kim Y, Liu H, Mraz Gernhard S, Varghese A, Hoppe R (2003) Long-term outcome of 525 patients with mycosis fungoides and sezary syndrome: clinical prognostic factors and risk for disease progression. Arch Dermatol 139:857–866

Klisovic MI, Maghraby EA, Parthun MR, Guimond M, Sklenar AR, Whitman SP, Chan KK, Murphy T, Anon J, Archer KJ, Rush LJ, Plass C, Grever MR, Byrd JC, Marcucci G (2003) Depsipeptide (FR 901228) promotes histone acetylation, gene transcription, apoptosis and its activity is enhanced by DNA methyltransferase inhibitors in AML1/ETO-positive leukemic cells. Leukemia 17:350–358

Knobler E (2004) Current management strategies for cutaneous T-cell lymphoma. Clin Dermatol 22:197–208

Knobler R, Girardi M (2001) Extracorporeal photochemoimmunotherapy in cutaneous T cell lymphomas. Ann N Y Acad Sci 941:123–138

Kuo MH (1998) Roles of histone acetyltransferases and deacetylases in gene regulation. Bioessays 20:615

Kwa FAA, Balcerczyk A, Licciardi P, El Osta A, Karagiannis T (2011) Chromatin modifying agents—the cutting edge of anticancer therapy. Drug Discov Today 16:543–547

Lansigan F, Foss F (2010) Current and emerging treatment strategies for cutaneous T-cell lymphoma. Drugs 70:273–286

Lansigan F, Choi J, Foss F (2008) Cutaneous T-cell lymphoma. Hematol Oncol Clin North Am 22:979–996

Lansigan F, Stearns D, Foss F (2010) Role of denileukin diftitox in the treatment of persistent or recurrent cutaneous T-cell lymphoma. Cancer Manag Res 2:53–59

Liu T, Kuljaca S, Tee A, Marshall G (2006) Histone deacetylase inhibitors: multifunctional anticancer agents. Cancer Treat Rev 32:157–165

Lowe NJ, Breeding JH, Wortzman MS (1982) New coal tar extract and coal tar shampoos. evaluation by epidermal cell DNA synthesis suppression assay. Arch Dermatol 118:487–489

Lundin J, Hagberg H, Repp R, Cavallin-Ståhl E, Fredén S, Juliusson G, Rosenblad E, Tjønnfjord G, Wiklund T, Osterborg A (2003) Phase 2 study of alemtuzumab (anti-CD52 monoclonal antibody) in patients with advanced mycosis fungoides/sezary syndrome. Blood 101: 4267–4272

Lyseng Williamson K, Yang LPH (2012) Romidepsin: a guide to its clinical use in cutaneous T-cell lymphoma. Am J Clin Dermatol 13:67–71

Ma X, Ezzeldin H, Diasio R (2009) Histone deacetylase inhibitors: current status and overview of recent clinical trials. Drugs 69:1911–1934

Mah LJ, El Osta A, Karagiannis TC (2010) GammaH2AX: a sensitive molecular marker of DNA damage and repair. Leukemia 24:679–686

Mann B, Johnson J, He K, Sridhara R, Abraham S, Booth B, Verbois L, Morse D, Jee J, Pope S, Harapanhalli R, Dagher R, Farrell A, Justice R, Pazdur R (2007) Vorinostat for treatment of cutaneous manifestations of advanced primary cutaneous T-cell lymphoma. Clin Cancer Res 13:2318–2322

Marks P, Breslow R (2007) Dimethyl sulfoxide to vorinostat: development of this histone deacetylase inhibitor as an anticancer drug. Nat Biotechnol 25:84–90

Marks P, Jiang X (2005) Histone deacetylase inhibitors in programmed cell death and cancer therapy. Cell Cycle 4:549–551

Marks PA, Richon VM, Rifkind RA (2000) Histone deacetylase inhibitors: inducers of differentiation or apoptosis of transformed cells. J Natl Cancer Inst 92:1210–1216

Martin RF, Murray V, D'cunha G, Pardee M, Kampouris E, Haigh A, Kelly DP, Hodgson GS (1990) Radiation sensitization by an iodine-labelled DNA ligand. Int J Radiat Biol 57:939–946
Miller T, Witter D, Belvedere S (2003) Histone deacetylase inhibitors. J Med Chem 46: 5097–5116
Minucci S, Pelicci P (2006) Histone deacetylase inhibitors and the promise of epigenetic (and more) treatments for cancer. Nat Rev Cancer 6:38–51
Moniot SB, Weyand M, Steegborn C (2012) Structures, substrates, and regulators of mammalian sirtuins—opportunities and challenges for drug development. Front Pharmacol 3:16
Munshi A, Kurland J, Nishikawa T, Tanaka T, Hobbs M, Tucker S, Ismail S, Stevens C, Meyn R (2005) Histone deacetylase inhibitors radiosensitize human melanoma cells by suppressing DNA repair activity. Clin Cancer Res 11:4912–4922
Munshi A, Tanaka T, Hobbs M, Tucker S, Richon V, Meyn R (2006) Vorinostat, a histone deacetylase inhibitor, enhances the response of human tumor cells to ionizing radiation through prolongation of gamma-H2AX foci. Mol Cancer Ther 5:1967–1974
Nebbioso A, Clarke N, Voltz E, Germain E, Ambrosino C, Bontempo P, Alvarez R, Schiavone E, Ferrara F, Bresciani F, Weisz A, De Lera A, Gronemeyer H, Altucci L (2005) Tumor-selective action of HDAC inhibitors involves trail induction in acute myeloid leukemia cells. Nat Med 11:77–84
New M, Olzscha H, La Thangue N (2012) HDAC inhibitor-based therapies: can we interpret the code? Mol Oncol 6:637–656
Olsen E, Duvic M, Frankel A, Kim Y, Martin A, Vonderheid E, Jegasothy B, Wood G, Gordon M, Heald P, Oseroff A, Pinter Brown L, Bowen G, Kuzel T, Fivenson D, Foss F, Glode M, Molina A, Knobler E, Stewart S, Cooper K, Stevens S, Craig F, Reuben J, Bacha P, Nichols J (2001) Pivotal phase III trial of two dose levels of denileukin diftitox for the treatment of cutaneous T-cell lymphoma. J Clin Oncol 19:376–388
Olsen E, Kim Y, Kuzel T, Pacheco T, Foss F, Parker S, Frankel S, Chen C, Ricker J, Arduino J, Duvic M (2007) Phase IIB multicenter trial of vorinostat in patients with persistent, progressive, or treatment refractory cutaneous T-cell lymphoma. J Clin Oncol 25:3109–3115
Olsen E, Vonderheid E, Pimpinelli N, Willemze R, Kim Y, Knobler R, Zackheim H, Duvic M, Estrach T, Lamberg S, Wood G, Dummer R, Ranki A, Burg G, Heald P, Pittelkow M, Bernengo M, Sterry W, Laroche L, Trautinger F, Whittaker S (2008) Revisions to the staging and classification of mycosis fungoides and sezary syndrome: a proposal of The International Society For Cutaneous Lymphomas (ISCL) and the Cutaneous Lymphoma Task Force of the European Organization of Research and Treatment of Cancer (EORTC). Blood 111:4830
Osella-Abate S, Zaccagna A, Savoia P, Quaglino P, Salomone B, Bernengo MG (2001) Expression of apoptosis markers on peripheral blood lymphocytes from patients with cutaneous T-cell lymphoma during extracorporeal photochemotherapy. J Am Acad Dermatol 44:40–47
Peart M, Tainton K, Ruefli A, Dear A, Sedelies K, O'reilly L, Waterhouse N, Trapani J, Johnstone R (2003) Novel mechanisms of apoptosis induced by histone deacetylase inhibitors. Cancer Res 63:4460–4471
Piekarz RL, Robey R, Sandor V, Bakke S, Wilson WH, Dahmoush L, Kingma DM, Turner ML, Altemus R, Bates SE (2001) Inhibitor of histone deacetylation, depsipeptide (FR901228), in the treatment of peripheral and cutaneous T-cell lymphoma: a case report. Blood 98: 2865–2868
Piekarz R, Frye R, Turner M, Wright J, Allen S, Kirschbaum M, Zain J, Prince HM, Leonard J, Geskin L, Reeder C, Joske D, Figg W, Gardner E, Steinberg S, Jaffe E, Stetler Stevenson M, Lade S, Fojo AT, Bates S (2009) Phase II multi-institutional trial of the histone deacetylase inhibitor romidepsin as monotherapy for patients with cutaneous T-cell lymphoma. J Clin Oncol 27:5410–5417
Pitzalis C, Pipitone N, Bajocchi G, Hall M, Goulding N, Lee A, Kingsley G, Lanchbury J, Panayi G (1997) Corticosteroids inhibit lymphocyte binding to endothelium and intercellular adhe-

sion: an additional mechanism for their anti-inflammatory and immunosuppressive effect. J Immunol 158:5007–5016
Poligone B, Heald P (2012) Menus for managing patients with cutaneous T-cell lymphoma. Semin Cutan Med Surg 31:25–32
Pothiawala S, Baldwin B, Cherpelis B, Lien M, Fenske N (2010) The role of phototherapy in cutaneous T-cell lymphoma. J Drugs Dermatol 9:764–772
Procaccini E, Selleri C, Monfrecola G (1996) In vitro photoinhibition by psoralen and ultraviolet A radiation of human hematopoietic progenitors. Photodermatol Photoimmunol Photomed 12:200–203
Querfeld C, Rosen S, Kuzel T, Kirby K, Roenigk H, Prinz B, Guitart J (2005) Long-term follow-up of patients with early-stage cutaneous T-cell lymphoma who achieved complete remission with psoralen plus UV-A monotherapy. Arch Dermatol 141:305–311
Rasheed W, Bishton M, Johnstone R, Prince HM (2008) Histone deacetylase inhibitors in lymphoma and solid malignancies. Expert Rev Anticancer Ther 8:413–432
Reiss Y, Proudfoot AE, Power CA, Campbell JJ, Butcher EC (2001) CC chemokine receptor (CCR)4 and the CCR10 ligand cutaneous T cell-attracting chemokine (CTACK) in lymphocyte trafficking to inflamed skin. J Exp Med 194:1541–1547
Resnik KS, Vonderheid EC (1993) Home UV phototherapy of early mycosis fungoides: long-term follow-up observations in thirty-one patients. J Am Acad Dermatol 29:73–77
Richon V, Garcia Vargas J, Hardwick J (2009) Development of vorinostat: current applications and future perspectives for cancer therapy. Cancer Lett 280:201–210
Rodd AL, Ververis K, Karagiannis TC (2012) Combination Phototherapy with a histone deacetylase inhibitor and a potent DNA-binding bibenzimidazole: effects in haematological cell lines. Lymphoma 2012:13
Rook AH, Vowels BR, Jaworsky C, Singh A, Lessin SR (1993) The immunopathogenesis of cutaneous T-cell lymphoma. abnormal cytokine production by Sézary T cells. Arch Dermatol 129: 486–489
Rosen S, Foss F (1995) Chemotherapy for mycosis fungoides and the sezary syndrome. Hematol Oncol Clin North Am 9:1109–1116
Ruefli AA, Ausserlechner MJ, Bernhard D, Sutton VR, Tainton KM, Kofler R, Smyth MJ, Johnstone RW (2001) The histone deacetylase inhibitor and chemotherapeutic agent suberoylanilide hydroxamic acid (SAHA) induces a cell-death pathway characterized by cleavage of bid and production of reactive oxygen species. Proc Natl Acad Sci U S A 98:10833–10838
Sambucetti LC, Fischer DD, Zabludoff S, Kwon PO, Chamberlin H, Trogani N, Xu H, Cohen D (1999) Histone deacetylase inhibition selectively alters the activity and expression of cell cycle proteins leading to specific chromatin acetylation and antiproliferative effects. J Biol Chem 274:34940–34947
Sandor V, Senderowicz A, Mertins S, Sackett D, Sausville E, Blagosklonny MV, Bates SE (2000) P21-dependent g(1)arrest with downregulation of cyclin D1 and upregulation of cyclin E by the histone deacetylase inhibitor FR901228. Br J Cancer 83:817–825
Santini V, Kantarjian HM, Issa JP (2001) Changes in DNA methylation in neoplasia: pathophysiology and therapeutic implications. Ann Intern Med 134:573–586
Santini V, Gozzini A, Ferrari G (2007) Histone deacetylase inhibitors: molecular and biological activity as a premise to clinical application. Curr Drug Metab 8:383–393
Sausville EA, Eddy JL, Makuch RW, Fischmann AB, Schechter GP, Matthews M, Glatstein E, Ihde DC, Kaye F, Veach SR (1988) Histopathologic staging at initial diagnosis of mycosis fungoides and the Sézary syndrome. definition of three distinctive prognostic groups. Ann Intern Med 109:372–382
Schaerli P, Ebert L, Moser B (2006) Comment on "the vast majority of CLA+T cells are resident in normal skin". J Immunol 177:1375–1376
Shiozawa K, Nakanishi T, Tan M, Fang H-B, Wang W-C, Edelman M, Carlton D, Gojo I, Sausville E, Ross D (2009) Preclinical studies of vorinostat (suberoylanilide hydroxamic acid) combined with cytosine arabinoside and etoposide for treatment of acute leukemias. Clin Cancer Res 15:1698–1707

Stadler R (1998) Interferons in dermatology. Present-day standard. Dermatol Clin 16:377–398
Stadler R (2007) Optimal combination with PUVA: rationale and clinical trial update. Oncology 21:29–32
Stadler R, Kremer A (2006) Therapeutic advances in cutaneous T-cell lymphoma (CTCL): from retinoids to rexinoids. Semin Oncol 33:S7–S10
Stadler R, Otte HG, Luger T, Henz BM, Kühl P, Zwingers T, Sterry W (1998) Prospective randomized multicenter clinical trial on the use of interferon-2a plus acitretin versus interferon-2a plus PUVA in patients with cutaneous T-cell lymphoma stages I and II. Blood 92:3578–3581
Toyooka T, Ibuki Y (2009) Histone deacetylase inhibitor sodium butyrate enhances the cell killing effect of psoralen plus UVA by attenuating nucleotide excision repair. Cancer Res 69: 3492–3500
Ueda H, Nakajima H, Hori Y, Goto T, Okuhara M (1994) Action of FR901228, a novel antitumor bicyclic depsipeptide produced by Chromobacterium violaceum no. 968, on Ha-ras transformed NIH3T3 cells. Biosci Biotechnol Biochem 58:1579–1583
Vittorio CC, Rook AH, French LE, Shapiro M, Lehrer MS, Junkins Hopkins JM (2001) Therapeutic advances in biological response modifiers in the treatment of cutaneous T-cell lymphoma. Biodrugs 15:431–437
Vonderheid E, Bernengo M, Burg GN, Duvic M, Heald P, Laroche L, Olsen E, Pittelkow M, Russell Jones R, Takigawa M, Willemze R (2002) Update on erythrodermic cutaneous T-cell lymphoma: report of the International Society For Cutaneous Lymphomas. J Am Acad Dermatol 46:95–106
Vrana JA, Decker RH, Johnson CR, Wang Z, Jarvis WD, Richon VM, Ehinger M, Fisher PB, Grant S (1999) Induction of apoptosis in U937 human leukemia cells by suberoylanilide hydroxamic acid (SAHA) proceeds through pathways that are regulated by Bcl-2/Bcl-XL, C-Jun, and p21CIP1, but independent of p53. Oncogene 18:7016–7025
Weinstock MA, Gardstein B (1999) Twenty-year trends in the reported incidence of mycosis fungoides and associated mortality. Am J Public Health 89:1240–1244
Whittaker S, Foss F (2007) Efficacy and tolerability of currently available therapies for the mycosis fungoides and sezary syndrome variants of cutaneous T-cell lymphoma. Cancer Treat Rev 33:146–160
Willemze R, Jaffe E, Burg G (2005) Who-EORTC classification for cutaneous lymphomas. Blood 105:3768–3785
Wollina U (2012) Cutaneous T cell lymphoma: update on treatment. Int J Dermatol 51: 1019–1036
Xu WS, Parmigiani RB, Marks PA (2007) Histone deacetylase inhibitors: molecular mechanisms of action. Oncogene 26:5541–5552
Yang G, Thompson MA, Brandt SJ, Hiebert SW (2007) Histone deacetylase inhibitors induce the degradation of the T(8;21) fusion oncoprotein. Oncogene 26:91–101
Yoo EK, Rook AH, Elenitsas R, Gasparro FP, Vowels BR (1996) Apoptosis induction of ultraviolet light A and photochemotherapy in cutaneous T-cell lymphoma: relevance to mechanism of therapeutic action. J Investig Dermatol 107:235–242
Zackheim HS, Kashani Sabet M, Amin S (1998) Topical corticosteroids for mycosis fungoides. Experience in 79 patients. Arch Dermatol 134:949–954
Zain J, O'connor O (2010) Targeting histone deacetyalses in the treatment of B- and T-cell malignancies. Invest New Drugs 28(suppl 1):S58–S78
Zain J, Kaminetzky D, O'connor O (2010) Emerging role of epigenetic therapies in cutaneous T-cell lymphomas. Expert Rev Hematol 3:187–203
Zarebska Z (1994) Cell membrane, a target for PUVA therapy. Cell Membr 23:101–109
Zhang C, Hazarika P, Ni X, Weidner D, Duvic M (2002) Induction of apoptosis by bexarotene in cutaneous T-cell lymphoma cells: relevance to mechanism of therapeutic action. Clin Cancer Res 8:1234–1240

Zhang X, Gillespie S, Borrow J, Hersey P (2004) The histone deacetylase inhibitor suberic bishydroxamate regulates the expression of multiple apoptotic mediators and induces mitochondria-dependent apoptosis of melanoma cells. Mol Cancer Ther 3:425–435
Zhang C, Richon V, Ni X, Talpur R, Duvic M (2005) Selective induction of apoptosis by histone deacetylase inhibitor SAHA in cutaneous T-cell lymphoma cells: relevance to mechanism of therapeutic action. J Investig Dermatol 125:1045–1052
Zic J (2012) Photopheresis in the treatment of cutaneous T-cell lymphoma: current status. Curr Opin Oncol 24(suppl 1):S1–S10

Chapter 18
Nano-Based Drug Delivery Modalities for the Treatment of Cancer: The Formulation of Tumour-Specific and -Targeted Nanoparticles

Li-Jeen Mah, Stephanie Tortorella, and Tom C. Karagiannis

Abstract The rapidly evolving and expanding discipline of nanotechnology, coupled with the continuous advancements in the understanding of cancer biology is a highly promising basis for cancer therapeutic research. Current clinical management strategies, including chemotherapy and radiotherapy fail to adequately treat malignancies in a subset of patients with advanced or severe forms of cancer. Multivariable dose-limiting factors, such as systemic toxicity and multi-drug resistance limit therapeutic benefit, quality of life, and complete long-term remission rates. The ability to deliver therapeutic compounds to the tumour site is thus an attractive area of research, with nanoparticle systems exhibiting the most promise. Formulation of a nanoparticle drug delivery platform that possesses the ideal properties for effective cancer-targeting requires the optimisation and characterisation of different materials at the nano-scale. Despite advancements, translation of this system to become clinically-relevant has proven to be difficult, with only six anticancer nanoparticle drug delivery systems FDA-approved. Numerous formulations are currently in clinical trials, demonstrating the theoretical relevance and potential in cancer treatment. The ability to modulate the surface chemistry of nanoparticles with relative ease provides rationale for their use in targeting strategies. Rapid clearance of untargeted nanoparticles also demonstrates the requirement to functionalise drug carriers in order to increase their time in systemic circulation, and enhance drug bioavailability. Cancer targeting strategies involve both passive and active mechanisms. The tumour microenvironment possesses distinct characteristics as

L.-J. Mah, B.Sc. • T.C. Karagiannis, B.Sc. (Hons), Ph.D. (✉)
Epigenomic Medicine, Baker IDI Heart and Diabetes Institute, The Alfred Medical Research and Education Precinct, 75 Commercial Road, Melbourne, VIC, Australia

Department of Pathology, The University of Melbourne, Parkville, VIC, Australia
e-mail: tom.karagiannis@bakeridi.edu.au

S. Tortorella, B.Sc. (Hons)
Epigenomic Medicine, Baker IDI Heart and Diabetes Institute, The Alfred Medical Research and Education Precinct, 75 Commercial Road, Melbourne, VIC, Australia

N. Maulik and T. Karagiannis (eds.), *Molecular Mechanisms and Physiology of Disease: Implications for Epigenetics and Health*, DOI 10.1007/978-1-4939-0706-9_18,

compared to normal tissue, which may be exploited for passive targeting. Utilisation of knowledge at the molecular level, including distinct differences in gene transcription and the expression of receptors between cancer and normal cells, allows for the active targeting of malignancies. Through the conjugation of a relevant moiety to the nanoparticle surface, cancer cells with an upregulation of the corresponding receptor exhibit high binding affinities allowing for cellular uptake, and subsequent drug release. Further research is required for such systems to become clinically-relevant; however with continued advancements in both nanoparticles theory and the understanding of cancer biology, the limitations observed in current management strategies may be overcome.

Keywords Nanoparticles • Nanomedicine • Drug-delivery • Liposomal drug carrier • Solid–lipid nanoparticles • Polymeric nanoparticles • Cancer therapy

18.1 Introduction

Cancer is a highly heterogeneous and complex disease that encompasses a group of disorders characterised by multiple genetic-, epigenetic- and environmentally-induced modifications to cellular components causing continuous indefinite growth (Feinberg et al. 2006). As the leading cause of death worldwide, cancer remains a heavy burden on health-care systems with incidence rates continuing to rise. By the year 2020, it is projected that over 15 million cancer cases will be reported (Jemal et al. 2011; Boyle and Levin 2008). Despite advances in the understanding of cancer biology, translation to clinically-relevant applications have proven to be difficult. A number of factors are attributable to this, including the difficulty in developing early diagnostic tools, the inability to effectively contain/and or treat metastatic growth, and the limitation of current treatment to selectively treat cancer cells with marginal effects on its surrounding healthy tissue (Tannock 1998; Peer et al. 2007).

The rapidly evolving and expanding discipline of nanotechnology is a science of engineering material and systems on a molecular scale. Due to their unique size-dependent physical and chemical properties (Whitesides 2003), and the exploitable nature of each property, development of functional nanoparticles may be designed for a range of applications. In cancer biology, nanoparticles theory is expected to be useful for various therapeutic and diagnostic strategies, with nanoparticles in the size range of 1–200 nm demonstrating important and often unique interactions with biological systems (Wen et al. 2008). Current limitations in the clinical management of cancer have caused an intense interest in refining treatment for the effective and targeted delivery of anti-cancer drugs, with nanoparticle drug delivery systems the most promising platform to date. Although progress has been made with the approval of six nano-based cancer medicines worldwide (Schütz et al. 2013), research requires the development of new materials by engineering molecules at the nanoscale which will interact with tumour cells and/or diseased tissues (Koutsopoulos 2012).

By focusing on the use of nanoparticles in the treatment of cancer, this chapter highlights the potential of nanomedicine, and provides rationale to continue research in this highly promising field.

18.2 Current Clinically-Relevant Cancer Therapy

Conventional treatment options, including surgical excision of cancerous tissue, radiotherapy, and chemotherapy have their own limitations (Jabir et al. 2012). Surgery may not be applied for all types of cancers, with the possibility of complete organ loss, healthy and functional tissue excision and cancer reoccurrence. Radiotherapy and chemotherapy aim to destroy cancerous tissue, however due to lack of specificity, insufficient concentrations of drug or radiation administered at tumour site, unwanted systemic distribution leading to nonspecific toxicity and other dose-related side effects, limitations in efficacy especially in advanced stages of cancer are observed (Peer et al. 2007). Although these therapeutic platforms confer with good disease-free survival, this is only for a limited period of time with systemic toxicity and drug resistance confining their utility to early-stage and non-metastatic cancers.

Current research emphasises the requirement to develop tumour-specific and -targeting therapeutics in order to enhance efficacy, limit unwanted adverse effects and overcome drug resistance. Utilising knowledge of cancer pathogenesis at the molecular scale, a number of targeted drugs are FDA-approved for clinical use, including rituximab, imatinib, lapatinib, and cetuximib (Sikora 2002). Although they have the capacity to target malignant cells or modulate the tumour microenvironment and limit toxic effects in healthy tissue, drug resistance is common and a positive response to treatment, based on cancer-type, -stage, and -molecular progression, may only be observed in a subset of patients. Thus, there is scope to develop new agents or site-specific delivery systems for the administration of cytotoxic and anti-cancer therapeutics in order to overcome current limitations.

18.3 The Significance of Nanotechnology in Medicine and Cancer

The use of nanotechnology in medicine, coupled with the continued advancement in the understanding of human disease, has become the proposed basis in a number of therapeutic and diagnostic applications. Immense potential in the enhanced delivery of anticancer therapeutics to cancer cells and/or to the tumour microenvironment provides rationale to continue such research, and has the capacity to transform its clinical management in the future.

Nanoparticle drug delivery systems have the ability to deliver high concentrations of therapeutic or diagnostic compound with up to a 100-fold increase, to the tumour site as compared to free drug (Peer and Margalit 2006). This property may be attributed to their small size, and surface functionalisation causing a decrease in systemic toxicity, maximised bioavailability, and modification of drug pharmacokinetics (Suri et al. 2007). A number of advantages for the use of such a system for the delivery of chemotherapeutic and other cancer therapeutics have been reported (Jabir et al. 2012). Drug-incorporated nanoparticles generally have a safety profile that is superior to that of free anticancer agents in healthy cells. They aim to overcome the lack of selectivity of anticancer drugs with the ability to target malignant cells and increase site-specific efficacy, while reducing toxicity in healthy tissue. Furthermore, limitations currently observed in current treatment options involving multiple drug resistance (MDR) in more advanced stages and/or serious cases of cancer and the inability to treat metastasis may be overcome with nanoparticle-based drug delivery systems (Ali et al. 2011). Due to their low aqueous solubility, administration of most anticancer drugs results in reduced bioavailability and efficacy, with higher interpatient variability (Jabir et al. 2012). Encapsulation of these drugs in nanoparticles engineered with soluble properties, allow for the administration of higher concentrations of compound to the site of disease limiting the pharmacokinetic influence of its innate insolubility.

18.4 Strategies for the Formulation of a Nanoparticle Drug Delivery System in Cancer

Formulation of a nanoparticle drug delivery system that possesses ideal properties, requires the optimisation and characterisation of different materials that may be manipulated at the nano-scale. The most researched for the application in cancer include polymeric, liposomal, solid–lipid, organic, and inorganic nanoparticles. Figure 18.1 provides an illustrated summary of these different nanoparticle systems, with a focus on those important in cancer therapeutic strategies.

18.4.1 Polymeric Nanoparticles as a Drug-Delivery System

Engineered from biocompatible and biodegradable polymers, such as chitosan (naturally-derived) and poly(lactic co-glycolic acid) (PLGA), polymeric nanoparticles are commonly formulated drug carriers (Calvo et al. 1997; Hrkach et al. 1997; El Samaligy and Rohdewald 1983). Delivery of anticancer agents with nanoparticles enhance efficacy, reduce toxicity, control release, prolong bioactivity, increase patient compliance, and co-deliver multiple drugs with synergistic effects at the same site (Brewer et al. 2011). The ability to manipulate their physicochemical and biological properties with relative ease makes them ideal for this application. Polymeric nanoparticles have been formulated to encapsulate either

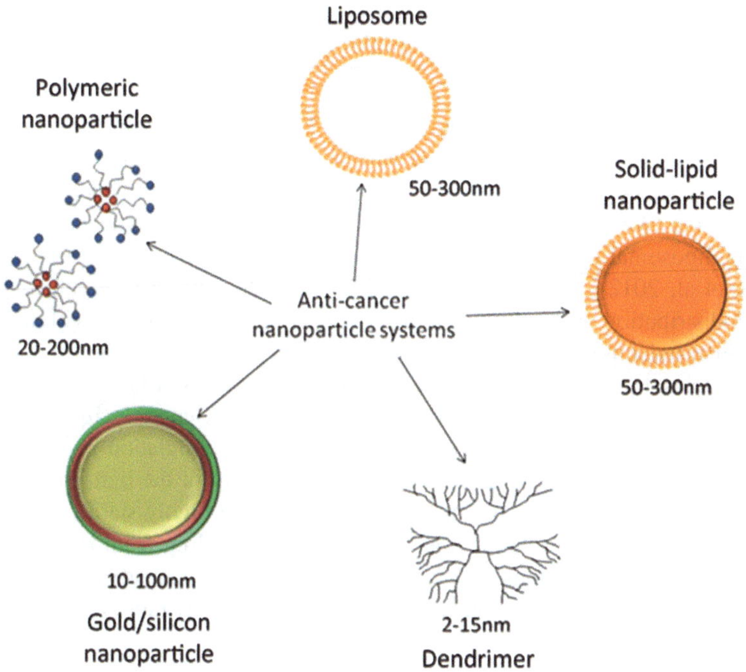

Fig. 18.1 Summary of general anti-cancer drug delivery systems approved for use in the treatment of cancer, and/or currently in human clinical trials. Each formulation is engineered from different material/s in order to encapsulate a chemotherapeutic drug. Design is determined by the nature of the intended drug, the nature of the nanomaterial and its subsequent characteristics, and the nature of the biological interaction between drug carrier and its cellular target. The following carriers possess properties unique to the material; liposome, solid–lipid nanoparticle, dendrimer, gold/silicon nanoparticle and polymeric nanoparticles

hydrophilic or hydrophobic small drug molecules, as well as proteins and nucleic acids (Nitta and Numata 2013). They are routinely prepared with the pairing of poly(ethylene glycol) (PEG) with the polymer in order to avoid immune recognition and premature opsonisation (Jain and Jain 2008). PEG inhibits binding of plasma proteins to the surface of the polymeric nanoparticles, which provides prolonged systemic circulation, and the opportunity to reach disease site (Gabizon 2001; Otsuka et al. 2003; Gref et al. 2000; Cheng et al. 2007; Veronese and Pasut 2005). The release of drug, through surface erosion or diffusion in response to the local environment may be controlled through the functionalisation of the nanoparticle surface (Moses et al. 2003). The key limitation in the formulation of such nanoparticles is the intrinsic structural heterogeneity of polymers (Peer et al. 2007). Numerous polymeric nanoparticles for the delivery of cancer therapeutics are currently in preclinical and clinical investigation (Moses et al. 2003; Farokhzad and Langer 2006; LaVan et al. 2003).

18.4.2 Liposomal Nanocarrier System for Drug Delivery

One of the first nanoparticle platforms to be applied in medicine, liposomes are spherical vesicles with an aqueous core and a bilayered membrane structure composed of natural and synthetic lipids (Hyodo et al. 2013). Due to their biocompatible and biodegradable composition, as well as their unique ability to encapsulate both hydrophobic (within their lamellae) and hydrophilic (in their aqueous core) compounds, liposomal nanocarriers have been extensively studied for this application (Wang et al. 2012). Liposomal formulations typically improve pharmacokinetics and biodistribution of the drug, achieving high drug concentrations within disease site while reducing drug concentration in normal tissue. Similar to polymeric nanoparticles, liposomes may be coated in polymers such as PEG to improve stability and increase the time of circulation (Torchilin 2005). Generally, liposomes have advantages over polymer-based nanoparticles for the formulation of cancer drug delivery systems (Fenske et al. 2001). Due to the ability of the lipid membrane to mimic common biological structures, these carriers are highly stable, providing a permeable barrier from degradation and protecting the drug from extraliposomal reactions. Liposomal nanomedicines represent one of the most advanced classes of drug-delivery systems, with several in clinical use and many more in trials (Fenske and Cullis 2008).

18.4.3 Drug Delivery via Solid–Lipid Nanoparticles

Solid–lipid nanoparticles are a lipid-based system that offers systemic stability, protection of drug from degradation, simple preparation, and low toxicity (Kang et al. 2010). Developed as an alternative to the existing carrier formulations, such as liposomes and polymer-based nanoparticles, solid–lipid nanoparticles have the capacity to provide unique properties for cancer-targeting drug delivery systems. Their small size, large surface area, high encapsulation yield, and interaction with disease cells are attractive for their potential to improve drug delivery system design and preparation (Uner and Yener 2007). Surface modification of solid–lipid nanoparticles is especially important in enhancing their stability in circulation and targeting diseased tissue.

18.4.4 Organic and Inorganic Nanocarriers

The formulation of nanoparticles with either organic or inorganic materials has been widely investigated for the use in drug delivery and diagnostic applications. Organic nanoparticles, such as branched macromolecule dendrimers are well defined and may be easily modified and conjugated to therapeutic compounds (Baker 2009).

They are characterised by enhanced biocompatibility and water solubility, well-defined chemical structures, rapid systemic clearance mechanisms and the ability to be conjugated to compounds with relative ease. Although production is costly and time-poor (Gillies and Fréchet 2005), dendrimers have been shown to efficiently deliver growth-inhibitory oligonucleotide to breast, ovarian, and prostate cell lines (Santhakumaran et al. 2004). Conversely, inorganic nanoparticles predominately metal-based including silicon and gold nanoparticles, are extensively studied for both diagnostic and therapeutic use through their ability to control drug release and possess optical resonance. Although metal nanoparticles are relatively inert and biocompatible, a significant amount may be retained in the body following administration, and substantial accumulation has been observed to lead to toxicity (Minelli et al. 2010).

18.5 Anticancer Nanoparticle Drug Delivery Platforms

Despite the recent advancements in nanoparticle design and development, difficulties in the translation from theoretical and laboratory research to clinically-relevant drug delivery platforms are a major limitation. There are currently six nano-based systems specifically formulated and approved for the treatment of cancer, which include Myocet (liposomal doxorubicin), DaunoXome (liposomal daunorubicin citrate), Abraxane (albumin-bound paclitaxel-loaded nanoparticle), and Doxil (PEGylated liposomal doxorubicin) (Haley and Frenkel 2008).

Chemotherapeutic drugs doxorubicin and paclitaxel are the most widely investigated in nanoparticle formulations due to their conventional use in the treatment of a wide range of cancers, and ensuing limitations. Doxorubicin, an anthracyclin antibiotic is used in the treatment of a variety of cancers including, breast, ovarian, sarcomas, lymphomas, and acute leukaemias (Speth et al. 1988). Despite the capacity to treat such a range of malignancies, the administration of doxorubicin is dose-limiting with the drug found to accumulate in the heart causing cardiotoxicity (Carvalho et al. 2013). Thus the ability to target its delivery to the tumour site becomes increasingly important in order to enhance its therapeutic index. Incorporation of doxorubicin in a nanoparticles drug carrier such as in the early case of Doxil, allows for drug diffusion within the leaky vasculature of the tumour micro-environment while limiting its capacity to escape normal blood vessels due to its size constraints (Gabizon and Martin 1997). Conversely paclitaxel also used as a broad spectrum anti-cancer drug (Hensley et al. 1999), promotes tubulin polymerisation, mitotic inhibition, and ultimately causes cell death. However, due to its poor solubility in aqueous solutions, its efficacy is significantly hindered by inefficient cellular uptake (Wang et al. 1996). Its encapsulation within a nano-based drug carrier has been extensively studied in order to overcome such a limitation. FDA-approved albumin-based Abraxane, used in the treatment of metastatic breast cancer is an example of a nanoparticle formulation for the enhanced delivery of paclitaxel to tumour cells (Damascelli et al. 2001).

In addition to these clinically approved nanoparticle therapeutics, a number of nano-based drug delivery systems are currently in preclinical and clinical investigation. Table 18.1 summarises the important nanoparticle formulations currently undergoing human clinical trials, with both untargeted and targeted nanoparticles drug delivery systems under development. These systems are significantly advanced when compared to those currently in clinical use, utilising the rapid progression and increasing knowledge in the field of nanotechnology and nanomedicine. A 2006 global survey conducted by the European Science and Technology Observatory (ESTO) found that over 150 companies were developing nano-based therapeutics (Zhang et al. 2007). In 2013, this number is sure to have risen significantly, with cancer nanomedicines the most widely investigated.

Trends in the development of nano-based drug delivery systems for cancer therapeutics provide insight into current research strategies. Most aim to encapsulate chemotherapeutics that are used in the treatment of a wide-range of cancers. For example, the most common strategy for the encapsulation of water-soluble platinum drugs such as cisplatin and carboplatin, utilise liposomal formulations. PEGylated liposomal carriers such as Lipoplatin demonstrate lower renal toxicity as compared to free drug, with therapeutic efficacy mixed in phase-II clinical trials mainly due to the key limitation of poor drug release from the liposome (Mylonakis et al. 2010). Thus, development of a nanocarrier which allows for the controlled release of drug, in addition to limiting the adverse systemic effects is of high importance. Continued efforts in the case of cisplatin have proved to be unsuccessful thus far. LiPlaCis, a more recent liposomal formulation with an internal drug release trigger allowed for controlled delivery, yet failed to limit toxicity with significant renal effects causing its cessation from further investigation (de Jonge et al. 2010). Polymeric nanoparticles have also been developed for the delivery of platinum drugs. NC-6004 is a polymeric cisplatin-encapsulated micelles, with Phase-I clinical trial data showing low but significant renal toxicities and hypersensitivity reactions (Plummer et al. 2011). Although the development of a cisplatin-incorporated nanoparticle is ongoing, the requirement for precise and efficacious delivery of drug to the site of disease is highlighted. Cancer targeting is thus shown to be crucial in the design of novel cancer therapeutics, with nanoparticle drug delivery systems well suited for customised drug delivery and imaging applications.

18.6 Nano-Based Drug Delivery Systems: Important Concepts for Cancer Targeting

The importance of targeted drug delivery for cancer treatment is immense with current cancer management strategies failing to be adequate, resulting in the development of unwanted adverse effects and a reduction in efficacy in long-term, advanced-staged, and metastatic cancers. Nanoparticle theory provides a basis for the development of novel drug delivery systems, with application in cancer significant.

Table 18.1 A sample of nanoparticle drug delivery systems undergoing clinical investigation and human trials

Agent	Formulation; therapeutic	Company	Indication	Status
Nontargeted nano-based cancer therapeutics				
S-CKD602	PEGylated liposome; topoisomerase inhibitor, CKD602	Alza Corporation	Various cancers	Phase I/II
CRLX101	Polymeric (cyclodextrin) nanoparticle; camptothecin	Cerulean Pharma	Various cancers	Phase II
CPX-1	Liposome; irinotecan	Celator Pharmaceuticals	Colorectal cancer	Phase II
LE-SN38	Liposome; active metabolite of irinotecan, SN38	NeoPharm	Colorectal cancer	Phase II
NC-6004	Polymeric (PEG-poly amino acid) nanoparticle; cisplatin	NanoCarrier Co.	Various cancers	Phase I
NK105	Polymeric (PEG-poly aspartate) nanoparticle; paclitaxel	Nippon Kayaku Co., Ltd.	Various cancers	Phase II
NK911	Polymeric (PEG-poly aspartate) nanoparticle; doxorubicin	Nippon Kayaku Co., Ltd.	Various cancers	Phase I
SP1049C	Glycoprotein micelle; doxorubicin	Supratek Pharma Inc.	Various cancers	Phase II
SPI-077	PEGylated liposome; cisplatin	Alza Corporation	Lung cancer	Phase II
NK102	Polymeric micelle; SN38	Nippon Kayaku Co., Ltd.	Various cancers	Phase II
ALN-VSP	Lipid nanoparticle; siRNA against vascular endothelial growth factor and kinesin spindle protein	Alnylam Pharmaceuticals	Liver cancer	Phase I
CPX-351	Liposome; cytarabine and doxorubicin (5:1)	Celator Pharmaceuticals	Acute myeloid leukemia	Phase I
OSI-7904L	Liposome; thymidylate synthase inhibitor	OSI Pharmaceuticals	Various cancers	Phase II
OSI-211	Liposome; lurtotecan	OSI Pharmaceuticals	Various cancers	Phase II
ABI-007	Albumin-stabilised nanoparticle; paclitaxel	National Cancer Institute	Various cancers	Phase II
Cellular and molecular targeting nanoparticle therapeutics				
BIND-014	Polymeric (PEG-PLGA) prostate-specific membrane antigen (PSMA) targeted nanoparticle; docetaxel	BIND Bioscience	Various cancers	Phase I

(continued)

Table 18.1 (continued)

Agent	Formulation; therapeutic	Company	Indication	Status
MCC-465	Human antibody fragment (GAH) targeted liposome; doxorubicin	National Cancer Centre, Japan	Gastric cancer	Phase I
MBP-426	Transferrin targeted liposome; oxaliplatin	Mebiopharm Co., Ltd.	Various cancers	Phase II
CALAA-01	Polymeric (cyclodextrin) transferrin targeted nanoparticle; siRNA against expression of M2 subunit of ribonucleotide reductase	Calando Pharmaceuticals	Solid tumours	Phase I
SGT53-01	Transferrin targeted liposome; p53 gene	SynerGene Therapeutics	Solid tumours	Phase I

Adapted from Wang et al. (2012)

18.6.1 Limitations of Unmodified and Non-targeting Nanoparticle for Drug Delivery in Cancer

Unmodified or 'naked' nanoparticles display very short circulation times, with rapid clearance from the bloodstream by the mononuclear phagocyte system (MPS) (Grislain et al. 1983). Phagocytic uptake and clearance favours hydrophobic nanoparticles above the 100-nm threshold (Storm et al. 1995). Drug distribution is highly dependent on the composition of the polymer used in nanofabrication (type, hydrophobicity, biodegradation profile) and the properties of the associated drug (molecular weight, charge, localisation in the nanocarrier—adsorbed or incorporated). Nanoparticle size, charge and surface properties determine their biological fate. Although it is demonstrated that all nanoparticles within the 2–100 nm size range facilitate a change in the signalling processes essential for cellular function, 40- and 50-nm nanoparticles are found to have the greatest effect (Jiang et al. 2008). Additionally, positively charged surfaces promote endocytosis (Byrne et al. 2008). The surface chemistry and its functionalisation, provides an effective way to control the interface between the nanoparticle and the biological system (Kim et al. 2013), through the capacity to determine and control receptor binding (Lynch et al. 2007).

The functionality of untargeted nanoparticles has been demonstrated using doxorubicin–polyisohexylcyanoaerylate (PIHCA) nanoparticles to treat hepatic metastases in a mouse model bearing reticulum cell sarcoma. Administration of doxorubicin–PIHCA nanoparticles results in a reduction of metastases as well as an increase in the life expectancy of mice compared to free doxorubicin, indicating greater anti-metastatic efficacy (Chiannilkulchai et al. 1989). The underlying mechanism responsible for the increased therapeutic efficacy of this nanoparticle formulation was observed to be the transfer of doxorubicin to healthy hepatic tissue, which acts as a drug reservoir to the malignant cells (Chiannilkulchai et al. 1990). High uptake concentration of nanoparticles via phagocytosis was detected in Kupffer cells in the liver. The release of doxorubicin from the nanoparticles within the Kupffer cells generates a concentration gradient, facilitating diffusion of the free active drug towards adjacent metastatic cells (Chiannilkulchai et al. 1990). Similarly, in vivo studies involving the administration of chitosan nanospheres loaded with dextran–doxorubicin conjugates and PIHCA nanospheres containing doxorubicin result in an increase in life expectancy due to higher nanoparticle accumulation and intratumoral drug release which surpassed the efficacy of free conjugates (Verdun et al. 1990).

However, the intrinsic constraints of using unmodified nanoparticles as drug carriers limit their application for clinical use in cancer treatment. Their inherent toxicity against cells of the MPS, result in a reduction of these cells due to cell death mechanisms, and exposes MPS organs to the risk of bacteriemia. Additionally, the unselective toxicity may enhance myelosuppressive effects in drug-resistant bone marrow cells and acute renal toxicity due to glomerular damage in the kidney (Brigger et al. 2002). Therefore, the use of unmodified anticancer drug-loaded nanoparticles is limited to targeting tumours associated with MPS organs, such as the liver, spleen, lungs, and bone marrow (Grislain et al. 1983).

18.6.2 Targeting Strategies: Passive Targeting Mechanism

Cancer therapy has progressively become more specific with the discovery of biologic targets that are either uniquely expressed or exhibit an upregulated expression on tumour cells. Cancer-cell targeting strategies aim to exploit the differences between malignant and normal cells. The two main mechanisms used for the formulation of targeted nanoparticles for the application in cancer are strategies based on either passively targeting the tumour's microenvironment or actively targeting the malignant cells directly. While passive targeting relies on the size and physical properties of the nanoparticle to target the site of the tumour, active targeting requires covalent linking of the nanoparticle with targeting moieties that bind to a specific antigen and/or receptor on the tumour cell surface.

Exploiting the abnormal tumour microenvironment including its vasculature for the selective and homogenous delivery of nanotherapeutics is a potential passive targeting strategy (Reynolds et al. 2003; Eatock et al. 2000). A major advantage of passively targeted nanoparticle drug delivery systems is that they have the ability to overcome physiological barriers that prevent effective nanoparticle distribution, restrained tumour growth and metastatic tendencies, low prospect of drug resistance, and adaptability to most cancer cell types (Niethammer et al. 2002; Kumar and Li 2001).

Angiogenesis, the formation of new blood vessels involves various factors and mediators that control the aggressive proliferation of endothelial and smooth muscle cells that are vital for rapid tumour growth through their capacity to supply adequate amounts of nutrients and oxygen to the site. Characteristics of angiogenesis including abnormalities in the basement membrane and a deficiency of pericytes lining the outer surface endothelial cells, result in the formation of leaky vessels with pores that lie between 380 and 780 nm depending upon the tumour type (Hosoda et al. 1995; Hobbs et al. 1998). Large pores, open gaps, and fenestrations in the vasculature are common in the process of rapid vessel formation, with regions of necrosis and/or haemorrhage also observed in some cases. Passive targeting allows for the accumulation of nanoparticles in the tumour due to the combination of increased endothelial fenestrations and compromised lymphatic drainage also known as the enhanced permeability and retention (EPR) effect (Byrne et al. 2008; Maeda et al. 2000). This biochemical difference may be exploited for nanoparticle delivery to the site of the tumour and is more favourable than non-specific free drug diffusion. Formulation of nanoparticles smaller than the size of the vasculature's pores, or fenestrations allows them to infiltrate into the interstitium, and subsequently into the lymphatic system for their eventual diffusion into the tumour site (Peer et al. 2007; Hawley et al. 1997; Nishioka and Yoshino 2001; Matsumura and Maeda 1986). Studies have shown that extravasation into tumours is limited to nanoparticles with diameters of 400 nm and lower (Couvreur and Vauthier 2006; Hobbs et al. 1998; Torchilin 2005).

The tumour interstitium is characterised by poor lymphatic drainage and high interstitial pressure that causes an outward flow of interstitial fluid (Jain 1987). Interstitial pressure is known to be higher at the internal regions of tumours compared to its periphery as they lack a well-defined lymphatic system (Jain and Baxter 1988). The outward convection of interstitial fluid decreases drug penetration into the interior of the tumour (Haley and Frenkel 2008). Nano-based drug delivery platforms however, allow for successful targeting as they have the capacity to successfully gain access to the tumour via the interstitium. These nanoparticles display higher retention rates than normal tissues. Passively targeted nanoparticles exploiting this cancer feature is beneficial against tumours of the lymphatic systems, particularly when administered subcutaneously or within the tumour itself (intratumoural) (Xie et al. 2009).

Other passive targeting strategies involve the manipulation of conditions expressed preferentially at tumour sites such as low pH. This is highlighted in a study which demonstrates that release of paclitaxel by biodegradable polymer nanoparticles may be triggered to release drug under conditions of low pH, allowing its activity to be concentrated at the tumour site (Potineni et al. 2003). Passive targeting may also involve charges such as those on cationic liposomes that undergo electrostatic interaction with negatively charged phospholipid headgroups that are expressed on tumour endothelial cells (Krasnici et al. 2003).

Despite the promise and key rationale towards the development of a passive targeting strategy, similar to all current nano-based platform limitations have been reported. Due to heterogeneity, not every tumour exhibits the EPR effect, with vascular permeability also observed to be diverse in different cancers and even in different regions within the same tumour tissue (Jain 1994). Unpredictable blood flow and altered osmotic pressure within cancer lesions causes an adverse force balance, hindering extravasation, and the diffusion of drug-loaded nanoparticles into the tumour (Jain 1994; Sarntinoranont et al. 2003; Netti et al. 1995). Furthermore, not all drugs diffuse efficiently within the cancer microenvironment and controlling their cellular uptake has proven to be a challenge due to random distribution. In turn, promotion of overexpression of drug transporter proteins have shown to contribute to multiple-drug resistance (MDR) where tumour cells are rendered resistant due to the expulsion of chronically-administered drug, and thereby reducing biodistribution and overall therapeutic efficacy (Peer and Margalit 2006; Couvreur and Vauthier 2006; Ferrari 2005). In order to provide an alternative route of diffusion into cancer cells, nanoparticles may be actively targeted to enable specific targeting to cellular components.

18.6.3 Active Targeting of Cancer Cells Utilising Molecular Biology and Nanoparticle Theory

Neoplastic transformation of a cell is frequently associated with a significant increase in the number of physiological-relevant surface receptors. These receptors generally have an established physiological function within the tissue but aberrant stimulation of signalling pathways lead to their overexpression in the development of cancer. Overexpression of receptors such as the estrogen receptor, human epidermal receptor-2 (HER2), vascular endothelial growth factor receptor (VEGFR), and epidermal growth factor receptor (EGFR) lead to cancers such as those of the breast, colon, and lung (Mendelsohn and Baselga 2000; Ross and Fletcher 1998; Hogemann-Savellano et al. 2003; Ferrara 2004). Given the significant role of tumour cell-surface receptors in tumour development and growth, any disruption in the pathway leading to their aberrant expression would theoretically result in the cessation of tumour growth and ultimately lead to malignant cell death (Mendelsohn and Baselga 2000; Krenning et al. 1993; Kersten et al. 2000; Ferrara et al. 2005; Hicklin and Ellis 2005).

Numerous studies have been conducted in order to actively target tumour cells which would enhance selectivity and efficacy, while decreasing systemic exposure. Active targeting may be achieved through molecular recognition mechanisms, including those which allow for the interaction between receptors with their relevant ligands or antigens and corresponding antibodies (Byrne et al. 2008). The ability to modify the nanoparticle surface with high ligand densities, due to an increase in the surface-area-to-volume ratio provides a basis for the continued development of nano-based cancer therapeutics (Banerjee et al. 2004).

The principle behind receptor-mediated targeting involves the ability to activate specific interactions between a ligand and its receptor to evoke downstream processes. Receptor-specific associations are characterised by high affinity binding and a low dissociation rate, with agonists resulting in physiologic changes upon receptor binding and antagonists inhibiting receptor activation through their ability to structurally mimic the naturally-occurring agonist (Krohn 2001). The majority of cancer therapeutics have antagonist properties, with examples such as tamoxifen, trastuzumab, and cetuximab. These compounds have the capacity to bind to the estrogen receptor, HER2, and EGFR respectively, limiting their activation by inhibiting the binding of such receptors with their naturally-derived agonist (Slamon et al. 2001; Shiau et al. 1998; Cunningham et al. 2004). To enhance efficacy and efficiency, actively targeted drug delivery strategies must utilise knowledge in cancer pathogenesis and molecular biology. It is of high importance that the receptor of interest is significantly overexpressed on cancer cells relative to healthy tissue (Park et al. 2002; Lopes de Menezes et al. 1998). The ideal antigen or receptor should be expressed exclusively on tumour cells.

Various conjugation chemistries may be utilised to link nanoparticles to targeting ligands that bind specifically to cellular membrane-bound receptors, in order to enhance efficiency of uptake into target cells (Torchilin 2005). Binding of the ligand-coated targeting nanoparticle with its corresponding receptor activates receptor-mediated internalisation and subsequently allows for intracellular drug release (Peer et al. 2007; Allen 2002; Pastan et al. 2007). Receptor-mediated internalisation, the most important interaction in the drug delivery process allows for the concentrated accumulation of therapeutic-encapsulating ligand-conjugated nanoparticles to selectively kill tumour cells (Koenig and Edwardson 1997; Schally and Nagy 1999; Ottaway 1992; Mantyh et al. 1995; Roettger et al. 1995).

Receptor-mediated endocytosis facilitates the cellular internalisation of targeted nanoparticles in a more efficient manner compared to the uptake of non-targeted complexes. Endocytosis involves the invagination of the cell membrane to encase molecules or large complexes within an intracellular vesicle, known as an endosome (Bareford and Swaan 2007). Depending on the mode of internalisation, drug molecules within the vesicle are either transported to early endosomes where they are recycled and exocytosed, or trafficked to organelles such as the lysosome, Golgi apparatus, or mitochondria (Steinman et al. 1983). It is important to note that clathrin-dependent endocytosis is linked to lysosomal fusion and enzymatic degradation; while, non-clathrin-dependent internalisation is more ideal in that it leads to drug accumulation in the endosome and involves the exposure of drug to acidic environments facilitating hydrolysis (Qaddoumi et al. 2003). Internalisation of the targeted complex and its drug load occurs rapidly (in minutes) and H^+ ions are pumped into early endosomes, creating an acidic internal environment between pH 5.5–6.0. Eventually these early endosomes mature into late endosomes that fuse with prelysosomal vesicles to produce lysosomes, within which enzymatic degradation ensues (Bareford and Swaan 2007). Avoiding the lysosomal pathway, shown in Fig. 18.2 is of high importance in the majority of targeting strategies to protect drug molecules from enzymatic degradation.

In addition, encapsulation of the drug complex within a vehicle provides protection from resistance associated with efflux transporters such as P-glycoprotein, multidrug related protein (MRP), and breast cancer resistance protein (BCRP) which are frequently upregulated in malignant cells. The conjugation of drug to a targeting agent, with studies involving drug conjugation with oligonucleotides, peptides, proteins, antibodies, and other high-affinity ligands, establish successful

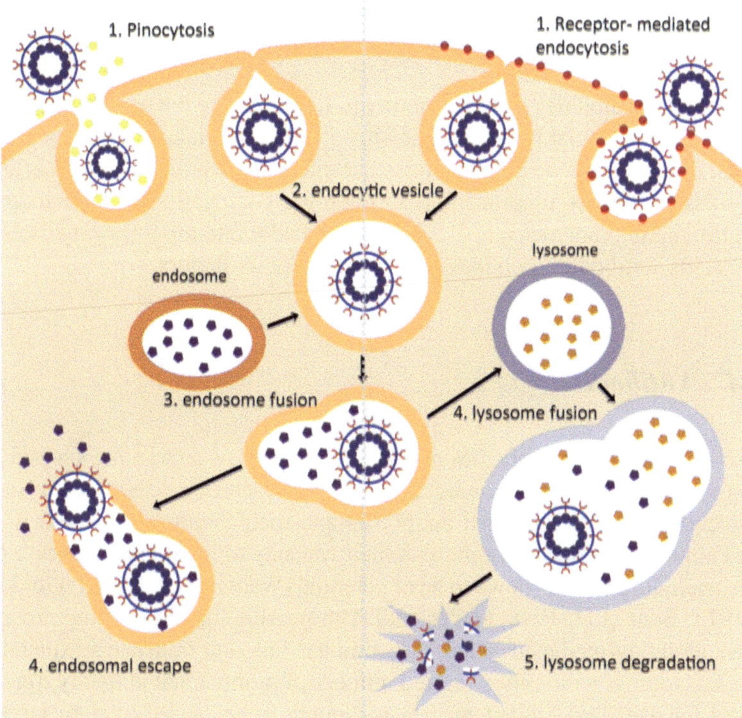

Fig. 18.2 Endosomal and lysosomal pathway following cellular uptake of nanoparticles. The major mechanisms of cellular uptake are receptor-mediated endocytosis and pinocytosis, which result in the internalisation of nanoparticles into an endocytic vesicle. Fusion with the endosome follows with two resultant pathways observed. The first and most important for drug delivery purposes is the ability to evade the lysosomal pathway and release the nanoparticle and its contents from the endosomal-fused vesicle. This is thought to increase the bioavailability of the drug in the cytoplasm. Avoidance of the lysosomal pathway is important in the delivery of therapeutics to the intended cellular compartment. Fusion of the lysosome with the vesicle indicates the initiation and activation of the degradation pathway, limiting drug bioavailability and neutralising active drug

facilitation of drug (above 1 kDa in size) uptake into the intracellular compartment via receptor-mediated internalisation and thus, circumventing the impervious cell membrane (Bareford and Swaan 2007; Reddy and Low 1998; Ishida et al. 2001; Citri et al. 2002).

Biological targeting may be improved by multivalent binding which involves the conjugation of a nanoparticle to several targeting ligands. In general, targeting efficacy increases with enhanced binding affinities. Only in very rare cases does excessive high-binding affinity hamper tumour penetration by impairing the diffusion of nanoparticles through the tumour (Adams et al. 2001). Although specificity is increased in active targeting drug delivery platforms, additional complexity in preparation with the possible increase in dimensions and the risk of adverse reactions to the targeting moiety have restricted its successful clinical application to date (Ray et al. 2011).

It is thus important to design, develop, and engineer a drug delivery system with cell-targeting moieties exhibiting optimal properties in an attempt to overcome such limitations.

Cell-targeting moieties and their ability to functionalise the surface properties of nanoparticles are crucial in the mechanisms of cellular uptake, increasing the probability of receptor-mediated endocytosis in target cells while limiting its effects on healthy tissue. Surface modifications through the conjugation of the nanoparticle with cell-targeting moieties may be classified broadly into the following categories: nucleic acids, antibodies, peptides, vitamins, and other factors.

18.6.4 Antibodies

The use of monoclonal antibodies and their fragments as a targeting surface modification through their direct conjugation to the surface of a nanoparticle has been extensively studied. Monoclonal antibodies and their therapeutic benefit was discovered almost three decades ago, with their capacity to intervene in disease pathogenesis, including cancer shown to be of potential (Warenius et al. 1981; Goldenberg et al. 1978; Behr et al. 1998; Miller et al. 1982). Although an exciting prospect at the time, antibody-mediated cancer targeting has failed to translate adequately into the clinical setting, with only a limited number of monoclonal antibody treatments approved for use. This limited success is thought to be caused by their large size, with the molecular mass of antibodies shown to be as high as 150 kDa (Weiner and Adams 2000; Reubi 2003; Ferrara et al. 2005; Bross et al. 2001; Glennie and Johnson 2000; Goldenberg 1999; Mehren et al. 2003).

Monoclonal antibody therapy may utilise native-state antibodies or their fragments as the targeting moiety. While whole antibodies are favourable due to their stability and higher binding affinity to their relevant antigen, there is potential for the nonspecific binding of antibody to normal cells, leading to an increase in immunogenicity. Antibody fragments, which consist of only the variable region of the antibody have been shown to be more selective in that they maintain target specificity, and limit unwanted immunogenic effects (Allen 2002; Carter 2001; Marks 2004; Chapman 2002). A murine study reported that PEGylated immunoliposomes conjugated to antibody fragments had nearly double the circulation time and twice the intratumoral accumulation compared to immunoliposomes decorated with whole monoclonal antibodies, indicating more efficacious tumour delivery and targeting with antibody fragments (Maruyama et al. 1997). Other studies comparing the targeting efficacy of whole monoclonal antibodies, its antibody fragments, and single chain variable fragments similarly found that immunoliposomes conjugated to monoclonal antibody fragments display the lengthiest circulation times (Cheng and Allen 2008; Sapra et al. 2004).

The tumour microenvironment, including its vasculature and increased interstitial hydrostatic pressure, coupled with antigen heterogeneity on the malignant cells themselves, are factors shown to influence the efficacy of monoclonal antibody therapeutics (Jain and Baxter 1988; Lobo et al. 2004). These factors have been established to limit the movement of monoclonal antibodies and their fragments within the tumour and its microenvironment. Studies report that intact antibody may traverse through the tumour site 1 mm in 48 h, while fragments traverse quicker at 1 mm in 24h. Upon successful delivery to the target cell, activation of antibody-dependent cellular cytotoxicity requires the presence of a sufficient number of immune effector cells. This may prove to be a limitation as immunosuppression and hypoxia are common in many tumour microenvironments (Mehren et al. 2003; Badger et al. 1987). It has also been reported that only 0.01 % of monoclonal antibodies administered intravenously are able to reach their target tissues in vivo (Li et al. 2004). Other limitations associated with this targeting strategy include non-specific binding, inadequate tumour penetration, and a possible reduction in receptor affinity depending on conjugation methods (Byrne et al. 2008). Immunogenicity and patient safety is another drawback as many early monoclonal antibodies were derived from mice which have been shown to trigger human antimouse antibody (HAMA) responses (Khazaeli et al. 1994).

The first FDA-approved anticancer monoclonal antibody, rituximab was developed for the treatment of low-grade B cell lymphoma in 1997 (James and Dubs 1997; Boyiadzis and Foon 2008). Following this, approval of several monoclonal antibody therapies were gained, with the development of treatments such as trastuzumab, gemtuzumab ozogamicin, and cetuximab for the clinical management of breast cancer, leukaemia, and colorectal cancer respectively (Lobo et al. 2004; Brannon-Peppas and Blanchette 2004). Despite these advances in the development of cancer targeted therapeutics as compared to conventional treatment options, specificity and efficacy of this type of modality are significantly limited in their clinical applicability (Sapra et al. 2005).

In order to enhance specificity, and overcome the limitations of conventional monoclonal antibody administration, including adverse immunogenic events, nanoparticle delivery systems have been studied. Conjugation of the antibody to the surface of the nanoparticle allows for the tolerability of drug in systemic circulation through its encapsulation within the nanoparticle itself, the enhancement of drug efficacy and for the concentrated biodistribution at the tumour site. Factors that influence the design of antibody-conjugated nanoparticle drug delivery systems include knowledge regarding the source of the antibody and its arrangement, as well as the method of antibody conjugation to the nanoparticle surface (Byrne et al. 2008).

The attachment of antibody or its fragments, whether random or specific, are facilitated by either linker molecules such as PEG or by direct conjugation to the nanoparticle surface (Byrne et al. 2008). Random conjugation relies on the formation of an amide bond between carboxylic acid groups in the nanocarrier and primary amine groups on the antibody or fragment through carbodiimide chemistry. However, there is a risk of activity reduction with this method of attachment as antibody binding sites may be inhibited due to the absence of specificity (Chapman 2002). Site-specific binding is a more favourable method of conjugation as it ensures the activity of the antibody or fragment is not compromised. This type of binding is achievable using maleimide chemistry at regions that do not coincide with known antigen-binding sites (Chapman et al. 1999).

Numerous attempts to develop nano-based antibody-conjugated targeted delivery systems for cancer treatment have resulted in positive preliminary data. For example, trastuzumab a monoclonal antibody antagonist for the HER-2 receptor has been used as a targeting moiety for various nanoparticle systems (Harries and Smith 2002). Promising results were published when nanoparticles coated with human serum albumin (HSA) and crosslinked to trastuzumab were administered to mice (Steinhauser et al. 2006). It was shown that trastuzumab-conjugated nanoparticles have the capacity to be specifically uptaken by HER2-overexpressing cells via receptor-mediated endocytosis. Similarly, humanised anti-HER-2 antibody fragments have been conjugated to carboxylic acid groups on the surface of PLGA nanoparticles loaded with a model protein toxin against HER-2-positive tumours (Cheng and Allen 2008). The therapeutic efficacy of targeted nanoparticles was demonstrated by a very significant reduction in the half maximal inhibitory concentration (IC_{50}) by a factor of 20 when compared to free toxin and non-targeted nanoparticles. In addition, cetuximab-conjugated nanoparticles have gained interest through the ability to target tumour cells. Cetuximab has a high affinity for the extracellular domain of EGFR, a receptor highly upregulated in a range of cancers including breast, lung colorectal, and brain (Laskin and Sandler 2004; Nicholson et al. 2001). The use of cetuximab antibody fragments conjugated to immunoliposomes exhibit an eightfold increase in immunoliposome delivery to EGFR-positive cells compared to its non-targeted counterparts (Harries and Smith 2002; Pan et al. 2007).

Despite these advancements, the rational design of antibodies as targeting moieties for successful site-specific drug delivery remains an important research platform. As the progression of knowledge on nano-based medicines, pharmacogenomics and cancer biology continues, its application in the development of novel treatment may be applied to create innovative antibody-based therapies with enhanced selectivity and reduced toxicity.

18.6.5 Nucleic Acids

As an emerging targeting moiety, nucleic acids such as aptamers are studied for their potential use in targeted therapeutic delivery. Aptamers are short single-stranded DNA or RNA oligonucleotide sequences that have the capacity to mimic the properties of antibodies, binding to a number of targets with high affinity and specificity (Jayasena 1999; Hermann and Patel 2000). Due to their favourable characteristics such as small size, low immunogenicity, and ease of isolation (Farokhzad and Langer 2006), aptamers have been used to functionalise the surface of nanoparticles for their targeted delivery. The surface of PEGylated biodegradable polymeric nanoparticles containing either docetaxel or paclitaxel has been modified with aptamers in order to target prostate specific membrane antigen (PSMA) on the surface of prostate cancer cells efficiency (Farokhzad and Langer 2006; Farokhzad et al. 2004). Results from this study showed a 3.77-fold increase in drug delivery.

Cellular uptake of these surface modified, targeted nanoparticles was significantly enhanced as demonstrated by the cytotoxic effect in vivo, leading to complete tumour reduction in mice (Farokhzad and Langer 2006).

18.6.6 Peptides

The molecular structure and defining characteristics of peptides support their use as cell targeting moieties in a number of active drug delivery strategies. As small molecules consisting of two or more amino acids, peptides may be synthesised for therapeutic use with the absence of the tertiary structural constraints of complete protein (Reubi 2003). Physiologically-relevant peptides, including neurotensin and somatostatin possess a regulatory role and are known to have diverse effects on numerous targets. Through their ability to bind to G protein-coupled membrane-bound receptors (Cattaneo et al. 1996), peptides are intimately involved in the function and modulation of important physiological and pathological processes (Reubi 1995).

Peptides are suitable for disease targeting due to their small size, low immunogenicity, and high avidity to its target receptor (Ruoslahti 2012). Those that are naturally occurring are usually hydrophilic, and may access the tumour site with efficiency due to their innate ability to permeate into tissue. Their low lipophilicity restricts their simple diffusion through the blood–brain barrier, thus limiting central nervous system toxicity unless the barrier function is compromised through the presence of inflammation or, in some cases glial tumours (Haldemann et al. 1995). Peptides are inherently non-toxic, and lack antigenicity which minimises the potential for systemic toxicity and unwanted side effects (Reubi 2003). Additionally, most including those involved in peptide modification and labelling possess the capacity to tolerate pathological conditions and highly unstable environments.

Peptide-based cancer targeting strategies encompass the exploration of suitable receptor targets for specific tumours, and the development and optimisation of radiolabeled and cytotoxic peptides. Functionalisation of nano-based drug delivery systems with targeting peptides has improved intracellular delivery and therapeutic outcome both in vitro and in vivo (Ishida et al. 2001; Kukowska-Latallo et al. 2005).

The first of such peptides for targeted cancer therapy was a small radiolabeled peptide somatostatin analogue, which was shown to bind with high affinity to the somatostatin receptor (Krenning et al. 1989). Highly dependent on the number of somatostatin receptors expressed on the surface of a cell, the application of this targeting strategy is limited to the diagnosis and treatment of a select range of cancers, including neuroendocrine tumours (Gibril et al. 1996; Jensen 2000; Otte et al. 1998; Waldherr et al. 2001). Radiolabelled peptides synthesised to bind to this receptor, and subsequently administered intravenously may be detected by γ-camera scintigraphy. Due to its sensitivity and the ability to identify tumour location and its metastases, successful reports indicate that it may be used in the diagnosis of a number of cancers, including small cell lung carcinomas (Brink et al. 2004), neuroendocrine malignancies, thyroid cancers, and gastro-entero-pancreatic tumours

(Waldherr et al. 2001; Waldherr et al. 2002; Kwekkeboom et al. 2003). In addition, scintigraphy information may be extrapolated in order to determine the course of treatment and cancer management for each individual patient (Krenning et al. 1993). This targeting strategy has prompted further studies of other peptide-receptor targeting systems, with a focus on those that are known to be highly relevant to human cancer (Reubi 2003).

Transferrin peptide is widely investigated and vastly relevant in cancer targeting strategies. An 80 kDa iron-binding glycoprotein, transferrin is internalised upon binding to cell surface transferrin receptors (Singh 1999). Due to the upregulation of transferrin receptors on metastatic and drug-resistant malignant cells, including those of pancreatic, colon, lung, and bladder cancers (Ekblom et al. 1983; Prost et al. 1998; Qian et al. 2002), drug targeting using this peptide-receptor system has become an important drug delivery platform. High concentrations of the membrane-bound receptor allows for the site-specific targeting and efficient cellular uptake of the transferrin peptide (Qian et al. 2002). This peptide has been shown to be non-toxic, non-immunogenic, and biodegradable with transferrin-mediated endocytosis of iron, the main uptake mechanism. As highly proliferating cells, tumour cells require increased levels of iron compared to normal cells (Singh 1999; Gatter et al. 1983). Furthermore, the efficiency of transferrin to bind to iron plays an important role in targeting, with low iron saturation resulting in decreased levels of peptide-receptor interaction and cellular uptake (Bellocq et al. 2003). Manipulation of this knowledge allows for the selective delivery of cancer therapeutics, proteins, and genes (Richardson and Ponka 1997). For example, transferrin-conjugated nanoparticle formulations designed for targeted gene delivery demonstrate a fourfold increase in cellular uptake compared to non-targeted nanoparticles in the established transferrin-overexpressing cell line, chronic myeloid leukemic-derived K562 (Bellocq et al. 2003). Similarly, administration of transferrin-conjugated liposomes encapsulating doxorubicin and verapamil to K562 cells enhance cellular uptake compared to non-targeted liposomes, successfully limiting drug resistance and increasing efficacy (Wu et al. 2007). A limitation of using naturally occurring peptides, including transferrin is its innate sensitivity to pH and enzymatic degradation, which has prompted studies investigating the design of more stable analogues, and synthesis of evasive nanoparticle delivery systems to prolong their circulating half-life in order to optimise formulations for clinically-relevant applications (Schally 1988).

18.6.7 Vitamins and Other Factors

The targeting of cancer cells may also be achieved through the ability to target vitamin and other growth factor receptors that are shown to be upregulated in some cancers. These cell-surface receptors are overexpressed in some cancers in order to supply the fast-metabolising tumour cells with excess energy and nutrition. The most important, and perhaps most extensively studied example of such are the folate receptors. Frequently overexpressed in a range of rapidly dividing malignant cells including those in the ovary, endometrium, and kidney (Wu et al. 2006; Sudimack and Lee 2000), they possess very high affinities for its corresponding ligand, folate or folic acid. The activation of these receptors following the binding of these ligands is necessary for essential cell function. As compared to other receptor-ligand systems that are either recycled or degraded in the lysosome, the folate receptor-folate/folic acid complex possesses the unique ability of remaining stable within the endosome or releasing into the cytoplasm following cellular uptake (Turek et al. 1993). As a nontoxic, nonimmunogenic, stable compound that may be conjugated to carriers with relative ease and cost effectiveness, the use of folate as a targeting moiety is practical for current research endeavours (Lu et al. 2004).

In vivo studies involving the administration of chemotherapeutic methotrexate-conjugated folate-linked dendrimers to mice, reported that higher levels of drug accumulation in tumours, inhibition of tumour growth, and significantly lower systemic toxicity may be observed as compared to free methotrexate at an equal cumulative dose (Kukowska-Latallo et al. 2005). Studies using dendrimer-based anticancer nanotherapeutics have demonstrated positive results both in vitro and in vivo in cancer cell targeting efficacy with the dramatic rise by up to 170,000-fold when conjugated to five or more folate molecules (Hong et al. 2007). Additionally, a heparin-folate-paclitaxel nanoparticle system coupled to folate was tested in a murine model. The potency of this novel nanoparticle system was obvious based on the reduction in the growth of paclitaxel-resistant tumour xenografts (Cho et al. 2008). Many in vitro studies using folate to functionalise the surface of nanoparticles have also observed improvements in cell internalisation such as folate receptor-targeted paclitaxel-loaded polymer-based nanoparticles, folate-conjugated dendritic polymers, and folic acid-conjugated PEGylated magnetite nanoparticles (Kukowska-Latallo et al. 2005; Pan and Feng 2008; Zhang and Zhang 2005). In the majority of studies, cancer cells treated with folic acid-conjugated nanoparticles demonstrate higher efficiency of intracellular uptake compared to their non-targeting counterparts.

Folate receptor targeting has also been studied as a strategy with the ability to limit resistance for the increased delivery of other chemotherapeutic agents such as paclitaxel and doxorubicin using liposomal formulations (Wu et al. 2006; Shmeeda et al. 2006). For example, folate-conjugated liposomes were found to successfully overcome multi-drug resistance in a model of acute myelogenous leukaemia by evading P-glycoprotein-mediated drug efflux (Ratnam et al. 2003).

Similar to any nanoparticle system, folate targeting is limited by non-specific targeting. Although the rationale for their development utilises the knowledge that cancer is characterised by the aberrantly increased proliferation of cells, folate receptors are also expressed on normal highly-proliferating cells such as fibroblasts and endothelial cells which inadvertently leads to decreased specificity (Peer et al. 2007).

The use of growth factors as targeting ligands is well documented, with their ability to repress tumour vascularity. This approach appears to be promising, as inhibiting the tumour blood and nutrient supply allows for the regulation of tumour growth, and perhaps most importantly prevent metastases (Folkman 1996).

The upregulation of proangiogenic factors such as vascular endothelial growth factor (VEGF), basic fibroblast growth factor (bFGF), matrix metalloproteinases, and tumour necrosis factor-α (TNF-α) are important in the development of the tumour microenvironment. Their cell-surface receptors, including VEGFR, matrix metalloproteinase receptor (MMPs), and $\alpha_v\beta_3$ integrin as well as other proteins such as vascular cell adhesion molecule-1 (VCAM-1) are suitable targets for the improvement of nanoparticle delivery (Byrne et al. 2008). Already clinically-relevant with the antiangiogenic Avastin for the treatment of metastatic colorectal cancer, this targeting strategy aims to inhibit the expression of growth factors important in angiogenesis. Avastin targets VEGF, an angiogenic-stimulating protein which is commonly overexpressed in solid cancers. In turn, this treatment has been shown to increase the permeability of tumour blood vessels, leading to their swelling and subsequent incapacitation as an adequate blood supply to the tumour. A nano-based delivery system actively targeted to VEGF may be important in order to deliver chemotherapeutic and other antiangiogenic drugs to the cancer site.

18.6.8 Other Targeting Strategies

The overexpression of extracellular matrix (ECM) receptors on tumour cells allows for their targeting with ECM ligands, including heparin sulphate, chondroitin sulphate, and hyaluronan (Peer and Margalit 2004). An example involves the formulation of hyaluronan-coated nanoparticles which were shown to increase drug circulation time in vivo, and increase their binding affinity to tumour cells expressing the hyaluronan receptor as compared to uncoated nanoparticles (Peer and Margalit 2004; Eliaz and Szoka 2001).

Development of targeted nanoparticles with the ability to distribute their therapeutic payload through remote activation and release by an external trigger has also been studied as a potential targeting strategy. The use of ultrasound scanning to rupture lipid-encapsulated microbubbles or different light wavelengths to stimulate thermal damage of cancer lesions has been investigated with success in preliminary studies (Hirsch et al. 2003; May et al. 2002; Roy et al. 2003; Yan and Kopelman 2003).

18.7 Conclusion

Limitations in the current clinical management of cancer have prompted further refinement of such strategies in order to increase the probability of long-term remission, limit adverse systemic toxicity, and reduce secondary metastatic disease (thus, increasing survival rates). Difficulties in the administration of chemotherapeutic drugs including those involved in limiting the biodistribution of an adequate amount of compound to the tumour site, restrict their effectiveness in severe forms and advanced stages of cancer. Development of a drug delivery platform for the targeted delivery of anticancer therapeutics to the site of malignancy is a major focus of current research efforts. Continued advancements in nanotechnology and nanomedicine provide the theoretical basis for the use of nano-sized particles as an anticancer drug carrier. The ability to define and modulate the properties and surface chemistry of nanoparticles in order to engineer a biologically-relevant delivery system is an important component for targeting. Formulation of such a nanoparticle system requires the optimisation and characterisation of different materials for the successful encapsulation of adequate concentrations of therapeutic compound. Despite continuous advancements in this field, translation into clinically-significant therapeutics has proven to be difficult, with currently six nano-based systems specifically formulated and approved for the treatment of cancer. The exploitation of knowledge from observations made in regard to their interactions with both cancerous and normal cellular components, allows for the development of scientifically designed and engineered nanoparticles with deliberate specificity to malignant cells. There are two main targeting mechanisms used for the formulation of such nanoparticles, with strategies based on either the passive targeting of the tumour's microenvironment or the active targeting of the cancer cells directly. Each mechanism aims to take advantage of the innate biochemical differences between tumour cells and normal tissue. Progression in the understanding of cancer biology in conjunction with the highly evolving expansion of nanoparticle theory to medical applications provides the basis for future anticancer research, and the greatest potential in the investigation to improve treatment strategies.

References

Adams GP, Schier R, Mccall AM, Simmons HH, Horak EM, Alpaugh RK, Marks JD, Weiner LM (2001) High affinity restricts the localization and tumor penetration of single-chain fv antibody molecules. Cancer Res 61:4750–4755

Ali I, Rahis U, Salim K, Rather MA, Wani W, Haque A (2011) Advances in nano drugs for cancer chemotherapy. Curr Cancer Drug Targets 11:135–146

Allen TM (2002) Ligand-targeted therapeutics in anticancer therapy. Nat Rev Cancer 2:750–763

Badger CC, Anasetti C, Davis J, Bernstein ID (1987) Treatment of malignancy with unmodified antibody. Pathol Immunopathol Res 6:419–434

Baker JR (2009) Dendrimer-based nanoparticles for cancer therapy. Hematology Am Soc Hematol Educ Program 2009:708–719

Banerjee R, Tyagi P, Li S, Huang L (2004) Anisamide-targeted stealth liposomes: a potent carrier for targeting doxorubicin to human prostate cancer cells. Int J Cancer 112:693–700

Bareford LM, Swaan PW (2007) Endocytic mechanisms for targeted drug delivery. Adv Drug Deliv Rev 59:748–758

Behr TM, Memtsoudis S, Sharkey RM, Blumenthal RD, Dunn RM, Gratz S, Wieland E, Nebendahl K, Schmidberger H, Goldenberg DM, Becker W (1998) Experimental studies on the role of antibody fragments in cancer radio-immunotherapy: influence of radiation dose and dose rate on toxicity and anti-tumor efficacy. Int J Cancer 77:787–795

Bellocq N, Pun S, Jensen G, Davis M (2003) Transferrin-containing, cyclodextrin polymer-based particles for tumor-targeted gene delivery. Bioconjug Chem 14:1122–1132

Boyiadzis M, Foon KA (2008) Approved monoclonal antibodies for cancer therapy. Expert Opin Biol Ther 8:1151–1158

Boyle P, Levin B (2008) World cancer report. IARC, World Health Organization Press, Lyon

Brannon-Peppas L, Blanchette JO (2004) Nanoparticle and targeted systems for cancer therapy. Adv Drug Deliv Rev 56:1649–1659

Brewer E, Coleman J, Lowman A (2011) Emerging technologies of polymeric nanoparticles in cancer drug delivery. J Nanomater 2011:1–10

Brigger I, Dubernet C, Couvreur P (2002) Nanoparticles in cancer therapy and diagnosis. Adv Drug Deliv Rev 54:631–651

Brink I, Schumacher T, Mix M, Ruhland S, Stoelben E, Digel W, Henke M, Ghanem N, Moser E, Nitzsche EU (2004) Impact of [18F]FDG-PET on the primary staging of small-cell lung cancer. Eur J Nucl Med Mol Imaging 31:1614–1620

Bross PF, Beitz J, Chen G, Chen XH, Duffy E, Kieffer L, Roy S, Sridhara R, Rahman A, Williams G, Pazdur R (2001) Approval summary: gemtuzumab ozogamicin in relapsed acute myeloid leukemia. Clin Cancer Res 7:1490–1496

Byrne JD, Betancourt T, Brannon-Peppas L (2008) Active targeting schemes for nanoparticle systems in cancer therapeutics. Adv Drug Deliv Rev 60:1615–1626

Calvo P, Remuñan-López C, Vila Jato JL, Alonso MJ (1997) Chitosan and chitosan/ethylene oxide-propylene oxide block copolymer nanoparticles as novel carriers for proteins and vaccines. Pharm Res 14:1431–1436

Carter P (2001) Improving the efficacy of antibody-based cancer therapies. Nat Rev Cancer 1: 118–129

Carvalho FS, Burgeiro A, Garcia R, Moreno AJ, Carvalho RA, Oliveira PJ (2013) Doxorubicin-induced cardiotoxicity: from bioenergetic failure and cell death to cardiomyopathy. Med Res Rev 34:106–135

Cattaneo MG, Amoroso D, Gussoni G, Sanguini AM, Vicentini LM (1996) A somatostatin analogue inhibits MAP kinase activation and cell proliferation in human neuroblastoma and in human small cell lung carcinoma cell lines. FEBS Lett 397:164–168

Chapman AP (2002) Pegylated antibodies and antibody fragments for improved therapy: a review. Adv Drug Deliv Rev 54:531–545

Chapman AP, Antoniw P, Spitali M, West S, Stephens S, King DJ (1999) Therapeutic antibody fragments with prolonged in vivo half-lives. Nat Biotechnol 17:780–783

Cheng WWK, Allen TM (2008) Targeted delivery of anti-CD19 liposomal doxorubicin in B-cell lymphoma: a comparison of whole monoclonal antibody, Fab′ fragments and single chain Fv. J Control Release 126:50–58

Cheng J, Teply BA, Sherifi I, Sung J, Luther G, Gu FX, Levy-Nissenbaum E, Radovic-Moreno AF, Langer R, Farokhzad OC (2007) Formulation of functionalized PLGA–PEG nanoparticles for in vivo targeted drug delivery. Biomaterials 28:869–876

Chiannilkulchai N, Driouich Z, Benoit JP, Parodi AL, Couvreur P (1989) Doxorubicin-loaded nanoparticles: increased efficiency in murine hepatic metastases. Sel Cancer Ther 5:1–11

Chiannilkulchai N, Ammoury N, Caillou B, Devissaguet JP, Couvreur P (1990) Hepatic tissue distribution of doxorubicin-loaded nanoparticles after i.v. administration in reticulosarcoma M 5076 metastasis-bearing mice. Cancer Chemother Pharmacol 26:122–126

Cho K, Wang X, Nie S, Chen Z, Shin D (2008) Therapeutic nanoparticles for drug delivery in cancer. Clin Cancer Res 14:1310–1316

Citri A, Alroy I, Lavi S, Rubin C, Xu W, Grammatikakis N, Patterson C, Neckers L, Fry D, Yarden Y (2002) Drug-induced ubiquitylation and degradation of ErbB receptor tyrosine kinases: implications for cancer therapy. EMBO J 21:2407–2417

Couvreur P, Vauthier C (2006) Nanotechnology: intelligent design to treat complex disease. Pharm Res 23:1417–1450

Cunningham D, Humblet Y, Siena S, Khayat D, Bleiberg H, Santoro A, Bets D, Mueser M, Harstrick A, Verslype C, Chau I, Van Cutsem E (2004) Cetuximab monotherapy and cetuximab plus irinotecan in irinotecan-refractory metastatic colorectal cancer. New Engl J Med 351: 337–345

Damascelli B, Cantù G, Mattavelli F, Tamplenizza P, Bidoli P, Leo E, Dosio F, Cerrotta AM, Di Tolla G, Frigerio LF, Garbagnati F, Lanocita R, Marchianò A, Patelli G, Spreafico C, Tichà V, Vespro V, Zunino F (2001) Intraarterial chemotherapy with polyoxyethylated castor oil free paclitaxel, incorporated in albumin nanoparticles (ABI-007): phase II study of patients with squamous cell carcinoma of the head and neck and anal canal: preliminary evidence of clinical activity. Cancer 92:2592–2602

de Jonge MJA, Slingerland M, Loos WJ, Wiemer EAC, Burger H, Mathijssen RHJ, Kroep JR, den Hollander MAG, van der Biessen D, Lam M-H, Verweij J, Gelderblom H (2010) Early cessation of the clinical development of liplacis, a liposomal cisplatin formulation. Eur J Cancer 46:3016–3021

Eatock MM, Schätzlein A, Kaye SB (2000) Tumour vasculature as a target for anticancer therapy. Cancer Treat Rev 26:191–204

Ekblom P, Thesleff I, Lehto VP, Virtanen I (1983) Distribution of the transferrin receptor in normal human fibroblasts and fibrosarcoma cells. Int J Cancer 31:111–117

El Samaligy MS, Rohdewald P (1983) Reconstituted collagen nanoparticles, a novel drug carrier delivery system. J Pharm Pharmacol 35:537–539

Eliaz RE, Szoka FC (2001) Liposome-encapsulated doxorubicin targeted to CD44: a strategy to kill CD44-overexpressing tumor cells. Cancer Res 61:2592–2601

Farokhzad OC, Langer R (2006) Nanomedicine: developing smarter therapeutic and diagnostic modalities. Adv Drug Deliv Rev 58:1456–1459

Farokhzad O, Jon S, Khademhosseini A, Tran T-N, Lavan D, Langer R (2004) Nanoparticle-aptamer bioconjugates: a new approach for targeting prostate cancer cells. Cancer Res 64: 7668–7672

Feinberg AP, Ohlsson R, Henikoff S (2006) The epigenetic progenitor origin of human cancer. Nat Rev Genet 7:21–33

Fenske DB, Cullis PR (2008) Liposomal nanomedicines. Expert Opin Drug Deliv 5:25–44

Fenske DB, Maclachlan I, Cullis PR (2001) Long-circulating vectors for the systemic delivery of genes. Curr Opin Mol Ther 3:153–158

Ferrara N (2004) Vascular endothelial growth factor: basic science and clinical progress. Endocr Rev 25:581–611

Ferrara N, Hillan KJ, Novotny W (2005) Bevacizumab (Avastin), a humanized anti-VEGF monoclonal antibody for cancer therapy. Biochem Biophys Res Commun 333:328–335

Ferrari M (2005) Cancer nanotechnology: opportunities and challenges. Nat Rev Cancer 5:161–171

Folkman J (1996) Fighting cancer by attacking its blood supply. Sci Am 275:150–154

Gabizon AA (2001) Pegylated liposomal doxorubicin: metamorphosis of an old drug into a new form of chemotherapy. Cancer Invest 19:424–436

Gabizon A, Martin F (1997) Polyethylene glycol-coated (pegylated) liposomal doxorubicin. Drugs 54:15–21

Gatter KC, Brown G, Trowbridge IS, Woolston RE, Mason DY (1983) Transferrin receptors in human tissues: their distribution and possible clinical relevance. J Clin Pathol 36:539–545

Gibril F, Reynolds JC, Doppman JL, Chen CC, Venzon DJ, Termanini B, Weber HC, Stewart CA, Jensen RT (1996) Somatostatin receptor scintigraphy: its sensitivity compared with that of other imaging methods in detecting primary and metastatic gastrinomas. a prospective study. Ann Intern Med 125:26–34

Gillies ER, Fréchet JMJ (2005) Dendrimers and dendritic polymers in drug delivery. Drug Discov Today 10:35–43

Glennie MJ, Johnson PWM (2000) Clinical trials of antibody therapy. Immunol Today 21: 403–410

Goldenberg MM (1999) Trastuzumab, a recombinant DNA-derived humanized monoclonal antibody, a novel agent for the treatment of metastatic breast cancer. Clin Ther 21:309–318

Goldenberg DM, Deland F, Kim E, Bennett S, Primus FJ, van Nagell JR, Estes N, Desimone P, Rayburn P (1978) Use of radiolabeled antibodies to carcinoembryonic antigen for the detection and localization of diverse cancers by external photoscanning. N Engl J Med 298:1384–1386

Gref R, Lück M, Quellec P, Marchand M, Dellacherie E, Harnisch S, Blunk T, Müller RH (2000) 'Stealth' corona-core nanoparticles surface modified by polyethylene glycol (PEG): influences of the corona (PEG chain length and surface density) and of the core composition on phagocytic uptake and plasma protein adsorption. Colloids Surf B Biointerfaces 18:301–313

Grislain L, Courvreur P, Lenaerts V (1983) Pharmacokinetics and distribution of a biodegradable drug-carrier. Int J Pharm 15:335–345

Haldemann AR, Rösler H, Barth A, Waser B, Geiger L, Godoy N, Markwalder RV, Seiler RW, Sulzer M, Reubi JC (1995) Somatostatin receptor scintigraphy in central nervous system tumors: role of blood–brain barrier permeability. J Nucl Med 36:403–410

Haley B, Frenkel E (2008) Nanoparticles for drug delivery in cancer treatment. Urol Oncol 26:57–64

Harries M, Smith I (2002) The Development and clinical use of trastuzumab (Herceptin). Endocr Relat Cancer 9:75–85

Hawley AE, Illum L, Davis SS (1997) Lymph node localisation of biodegradable nanospheres surface modified with poloxamer and poloxamine block co-polymers. FEBS Lett 400: 319–323

Hensley ML, Schuchter L, Lindley C, Meropol N, Cohen G, Broder G, Gradishar W, Green D, Langdon R, Mitchell RB, Negrin R, Szatrowski T, Thigpen JT, Vonhoff D, Wasserman T, Winer E, Pfister D (1999) American Society of Clinical Oncology clinical practice guidelines for the use of chemotherapy and radiotherapy protectants. J Clin Oncol 17:3333–3355

Hermann T, Patel DJ (2000) Adaptive recognition by nucleic acid aptamers. Science 287: 820–825

Hicklin DJ, Ellis LM (2005) Role of the vascular endothelial growth factor pathway in tumor growth and angiogenesis. J Clin Oncol 23:1011–1027

Hirsch LR, Stafford RJ, Bankson JA, Sershen SR, Rivera B, Price RE, Hazle JD, Halas NJ, West JL (2003) Nanoshell-mediated near-infrared thermal therapy of tumors under magnetic resonance guidance. Proc Natl Acad Sci U S A 100:13549–13554

Hobbs SK, Monsky WL, Yuan F, Roberts WG, Griffith L, Torchilin VP, Jain RK (1998) Regulation of transport pathways in tumor vessels: role of tumor type and microenvironment. Proc Natl Acad Sci U S A 95:4607–4612

Hogemann-Savellano D, Bos E, Blondet C, Sato F, Abe T, Josephson L, Weissleder R, Gaudet J, Sgroi D, Peters P, Basilion J (2003) The transferrin receptor: a potential molecular imaging marker for human cancer. Neoplasia 5:495–506

Hong S, Leroueil P, Majoros IN, Orr B, Baker J, Banaszak Holl M (2007) The binding avidity of a nanoparticle-based multivalent targeted drug delivery platform. Chem Biol 14:107–115

Hosoda J, Unezaki S, Maruyama K, Tsuchiya S, Iwatsuru M (1995) Antitumor activity of doxorubicin encapsulated in poly(ethylene glycol)-coated liposomes. Biol Pharm Bull 18:1234–1237

Hrkach JS, Peracchia MT, Bomb A, Lotan N, Langer R (1997) Nanotechnology for biomaterials engineering: structural characterization of amphiphilic polymeric nanoparticles by 1H NMR spectroscopy. Biomaterials 18:27–30

Hyodo K, Yamamoto E, Suzuki T, Kikuchi H, Asano M, Ishihara H (2013) Development of liposomal anticancer drugs. Biol Pharm Bull 36:703–707

Ishida O, Maruyama K, Tanahashi H, Iwatsuru M, Sasaki K, Eriguchi M, Yanagie H (2001) Liposomes bearing polyethyleneglycol-coupled transferrin with intracellular targeting property to the solid tumors in vivo. Pharm Res 18:1042–1048

Jabir NR, Tabrez S, Ashraf GM, Shakil S, Damanhouri GA, Kamal MA (2012) Nanotechnology-based approaches in anticancer research. Int J Nanomedicine 7:4391–4408

Jain RK (1987) Transport of molecules in the tumor interstitium: a review. Cancer Res 47: 3039–3051

Jain RK (1994) Barriers to drug delivery in solid tumors. Sci Am 271:58–65

Jain RK, Baxter LT (1988) Mechanisms of heterogeneous distribution of monoclonal antibodies and other macromolecules in tumors: significance of elevated interstitial pressure. Cancer Res 48:7022–7032

Jain A, Jain S (2008) Pegylation: an approach for drug delivery. a review. Crit Rev Ther Drug Carrier Syst 25:403–447
James JS, Dubs G (1997) FDA approves new kind of lymphoma treatment. Food And Drug Administration. AIDS Treat News (284):2–3
Jayasena SD (1999) Aptamers: an emerging class of molecules that rival antibodies in diagnostics. Clin Chem 45:1628–1650
Jemal A, Bray F, Center MM, Ferlay J, Ward E, Forman D (2011) Global cancer statistics. CA Cancer J Clin 61:69–90
Jensen RT (2000) Editorial: somatostatin receptor-based scintigraphy and antitumor treatment–an expanding vista? J Clin Endocrinol Metab 85:3507–3508
Jiang W, Kim BYS, Rutka J, Chan WCW (2008) Nanoparticle-mediated cellular response is size-dependent. Nat Nanotechnol 3:145–150
Kang KW, Chun M-K, Kim O, Subedi RK, Ahn S-G, Yoon J-H, Choi H-K (2010) Doxorubicin-loaded solid lipid nanoparticles to overcome multidrug resistance in cancer therapy. Nanomed Nanotechnol Biol Med 6:210–213
Kersten S, Desvergne B, Wahli W (2000) Roles of PPARs in health and disease. Nature 405: 421–424
Khazaeli MB, Conry RM, Lobuglio AF (1994) Human immune response to monoclonal antibodies. J Immunother Emphasis Tumor Immunol 15:42–52
Kim ST, Saha K, Kim C, Rotello VM (2013) The role of surface functionality in determining nanoparticle cytotoxicity. Acc Chem Res 46:681–691
Koenig JA, Edwardson JM (1997) Endocytosis and recycling of G protein-coupled receptors. Trends Pharmacol Sci 18:276–287
Koutsopoulos S (2012) Molecular fabrications of smart nanobiomaterials and applications in personalized medicine. Adv Drug Deliv Rev 64:1459–1476
Krasnici S, Werner A, Eichhorn M, Schmitt Sody M, Pahernik S, Sauer B, Schulze B, Teifel M, Michaelis U, Naujoks K, Dellian M (2003) Effect of the surface charge of liposomes on their uptake by angiogenic tumor vessels. Int J Cancer 105:561–567
Krenning EP, Bakker WH, Breeman WA, Koper JW, Kooij PP, Ausema L, Lameris JS, Reubi JC, Lamberts SW (1989) Localisation of endocrine-related tumours with radioiodinated analogue of somatostatin. Lancet 1:242–244
Krenning EP, Kwekkeboom DJ, Bakker WH, Breeman WA, Kooij PP, Oei HY, Van Hagen M, Postema PT, De Jong M, Reubi JC (1993) Somatostatin receptor scintigraphy with [111In-DTPA-D-Phe1]- and [123I-Tyr3]-octreotide: The Rotterdam experience with more than 1000 patients. Eur J Nucl Med 20:716–731
Krohn KA (2001) The physical chemistry of ligand-receptor binding identifies some limitations to the analysis of receptor images. Nucl Med Biol 28:477–483
Kukowska-Latallo JF, Candido KA, Cao Z, Nigavekar SS, Majoros IJ, Thomas TP, Balogh LP, Khan MK, Baker JR (2005) Nanoparticle targeting of anticancer drug improves therapeutic response in animal model of human epithelial cancer. Cancer Res 65:5317–5324
Kumar S, Li C (2001) Targeting of vasculature in cancer and other angiogenic diseases. Trends Immunol 22:129
Kwekkeboom DJ, Bakker WH, Kam BL, Teunissen JJM, Kooij PPM, de Herder WW, Feelders RA, van Eijck CHJ, de Jong M, Srinivasan A, Erion JL, Krenning EP (2003) Treatment of patients with gastro-entero-pancreatic (GEP) tumours with the novel radiolabelled somatostatin analogue [177Lu-DOTA(0), Tyr3]octreotate. Eur J Nucl Med Mol Imaging 30:417–422
Laskin J, Sandler A (2004) Epidermal growth factor receptor inhibitors in lung cancer therapy. Semin Respir Crit Care Med 25(suppl 1):17–27
Lavan DA, Mcguire T, Langer R (2003) Small-scale systems for in vivo drug delivery. Nat Biotechnol 21:1184–1191
Li KCP, Pandit S, Guccione S, Bednarski M (2004) Molecular imaging applications in nanomedicine. Biomed Microdevices 6:113–116
Lobo ED, Hansen RJ, Balthasar JP (2004) Antibody pharmacokinetics and pharmacodynamics. J Pharm Sci 93:2645–2668

Lopes De Menezes DE, Pilarski LM, Allen TM (1998) In vitro and in vivo targeting of immunoliposomal doxorubicin to human B-cell lymphoma. Cancer Res 58:3320–3330
Lu Y, Sega E, Leamon CP, Low PS (2004) Folate receptor-targeted immunotherapy of cancer: mechanism and therapeutic potential. Adv Drug Deliv Rev 56:1161–1176
Lynch I, Cedervall T, Lundqvist M, Cabaleiro-Lago C, Linse S, Dawson KA (2007) The nanoparticle-protein complex as a biological entity; a complex fluids and surface science challenge for the 21st century. Adv Colloid Interface Sci 134–135:167–174
Maeda H, Wu J, Sawa T, Matsumura Y, Hori K (2000) Tumor vascular permeability and the EPR effect in macromolecular therapeutics: a review. J Control Release 65:271–284
Mantyh PW, Demaster E, Malhotra A, Ghilardi JR, Rogers SD, Mantyh CR, Liu H, Basbaum AI, Vigna SR, Maggio JE (1995) Receptor endocytosis and dendrite reshaping in spinal neurons after somatosensory stimulation. Science 268:1629–1632
Marks J (2004) Selection of internalizing antibodies for drug delivery. Methods Mol Biol 248: 201–208
Maruyama K, Takahashi N, Tagawa T, Nagaike K, Iwatsuru M (1997) Immunoliposomes bearing polyethyleneglycol-coupled Fab′ fragment show prolonged circulation time and high extravasation into targeted solid tumors in vivo. FEBS Lett 413:177–180
Matsumura Y, Maeda H (1986) A new concept for macromolecular therapeutics in cancer chemotherapy: mechanism of tumoritropic accumulation of proteins and the antitumor agent smancs. Cancer Res 46:6387–6392
May D, Allen J, Ferrara K (2002) Dynamics and fragmentation of thick-shelled microbubbles. IEEE Trans Ultrason Ferroelectr Freq Control 49:1400–1410
Mehren MV, Adams GP, Weiner LM (2003) Monoclonal antibody therapy for cancer. Annu Rev Med 54:343–369
Mendelsohn J, Baselga J (2000) The EGF receptor family as targets for cancer therapy. Oncogene 19:6550–6565
Miller RA, Maloney DG, Warnke R, Levy R (1982) Treatment of B-cell lymphoma with monoclonal anti-idiotype antibody. N Engl J Med 306:517–522
Minelli C, Lowe SB, Stevens MM (2010) Engineering nanocomposite materials for cancer therapy. Small 6:2336–2357
Moses MA, Brem H, Langer R (2003) Advancing the field of drug delivery: taking aim at cancer. Cancer Cell 4:337–341
Mylonakis N, Athanasiou A, Ziras N, Angel J, Rapti A, Lampaki S, Politis N, Karanikas C, Kosmas C (2010) Phase II study of liposomal cisplatin (Lipoplatin™) plus gemcitabine versus cisplatin plus gemcitabine as first line treatment in inoperable (stage IIIB/IV) non-small cell lung cancer. Lung Cancer 68:240–247
Netti PA, Baxter LT, Boucher Y, Skalak R, Jain RK (1995) Time-dependent behavior of interstitial fluid pressure in solid tumors: implications for drug delivery. Cancer Res 55:5451–5458
Nicholson RI, Gee JMW, Harper ME (2001) EGFR and cancer prognosis. Eur J Cancer 37(suppl 4):9–15
Niethammer AG, Xiang R, Becker JC, Wodrich H, Pertl U, Karsten G, Eliceiri BP, Reisfeld RA (2002) A DNA vaccine against VEGF receptor 2 prevents effective angiogenesis and inhibits tumor growth. Nat Med 8:1369–1375
Nishioka Y, Yoshino H (2001) Lymphatic targeting with nanoparticulate system. Adv Drug Deliv Rev 47:55–64
Nitta S, Numata K (2013) Biopolymer-based nanoparticles for drug/gene delivery and tissue engineering. Int J Mol Sci 14:1629–1654
Otsuka H, Nagasaki Y, Kataoka K (2003) Pegylated nanoparticles for biological and pharmaceutical applications. Adv Drug Deliv Rev 55:403–419
Ottaway CA (1992) Insertion and internalization of vasoactive intestinal peptide (VIP) receptors in murine CD4 T lymphocytes. Regul Pept 41:49–59
Otte A, Mueller Brand J, Dellas S, Nitzsche EU, Herrmann R, Maecke HR (1998) Yttrium-90-labelled somatostatin-analogue for cancer treatment. Lancet 351:417–418
Pan J, Feng S-S (2008) Targeted delivery of paclitaxel using folate-decorated poly(lactide)–vitamin E TPGS nanoparticles. Biomaterials 29:2663–2672

Pan X, Wu G, Yang W, Barth R, Tjarks W, Lee R (2007) Synthesis of cetuximab-immunoliposomes via a cholesterol-based membrane anchor for targeting of EGFR. Bioconjug Chem 18: 101–108
Park JW, Hong K, Kirpotin DB, Colbern G, Shalaby R, Baselga J, Shao Y, Nielsen UB, Marks JD, Moore D, Papahadjopoulos D, Benz CC (2002) Anti-HER2 immunoliposomes: enhanced efficacy attributable to targeted delivery. Clin Cancer Res 8:1172–1181
Pastan I, Hassan R, Fitzgerald DJ, Kreitman RJ (2007) Immunotoxin treatment of cancer. Annu Rev Med 58:221–237
Peer D, Margalit R (2004) Tumor-targeted hyaluronan nanoliposomes increase the antitumor activity of liposomal doxorubicin in syngeneic and human xenograft mouse tumor models. Neoplasia 6:343–353
Peer D, Margalit R (2006) Fluoxetine and reversal of multidrug resistance. Cancer Lett 237: 180–187
Peer D, Karp JM, Hong S, Farokhzad OC, Margalit R, Langer R (2007) Nanocarriers as an emerging platform for cancer therapy. Nat Nanotechnol 2:751–760
Plummer R, Wilson RH, Calvert H, Boddy AV, Griffin M, Sludden J, Tilby MJ, Eatock M, Pearson DG, Ottley CJ, Matsumura Y, Kataoka K, Nishiya T (2011) A phase I clinical study of cisplatin-incorporated polymeric micelles (NC-6004) in patients with solid tumours. Br J Cancer 104:593–598
Potineni A, Lynn DM, Langer R, Amiji MM (2003) Poly(ethylene oxide)-modified poly(B-amino ester) nanoparticles as a pH-sensitive biodegradable system for paclitaxel delivery. J Control Release 86:223–234
Prost AC, Ménégaux F, Langlois P, Vidal JM, Koulibaly M, Jost JL, Duron JJ, Chigot JP, Vayre P, Aurengo A, Legrand JC, Rosselin G, Gespach C (1998) Differential transferrin receptor density in human colorectal cancer: a potential probe for diagnosis and therapy. Int J Oncol 13:871–875
Qaddoumi M, Gukasyan H, Davda J, Labhasetwar V, Kim K-J, Lee VHL (2003) Clathrin and caveolin-1 expression in primary pigmented rabbit conjunctival epithelial cells: role in PLGA nanoparticle endocytosis. Mol Vis 9:559–568
Qian Z, Li H, Sun H, Ho K (2002) Targeted drug delivery via the transferrin receptor-mediated endocytosis pathway. Pharmacol Rev 54:561–587
Ratnam M, Hao H, Zheng X, Wang H, Qi H, Lee R, Pan X (2003) Receptor induction and targeted drug delivery: a new antileukaemia strategy. Expert Opin Biol Ther 3:563–574
Ray S, Reddy PJ, Choudhary S, Raghu D, Srivastava S (2011) Emerging nanoproteomics approaches for disease biomarker detection: a current perspective. J Proteomics 74: 2660–2681
Reddy JA, Low PS (1998) Folate-mediated targeting of therapeutic and imaging agents to cancers. Crit Rev Ther Drug Carrier Syst 15:587–627
Reubi JC (1995) Neuropeptide receptors in health and disease: the molecular basis for in vivo imaging. J Nucl Med 36:1825–1835
Reubi JC (2003) Peptide receptors as molecular targets for cancer diagnosis and therapy. Endocr Rev 24:389–427
Reynolds AR, Moein Moghimi S, Hodivala-Dilke K (2003) Nanoparticle-mediated gene delivery to tumour neovasculature. Trends Mol Med 9:2–4
Richardson DR, Ponka P (1997) The molecular mechanisms of the metabolism and transport of iron in normal and neoplastic cells. Biochim Biophys Acta 1331:1–40
Roettger BF, Rentsch RU, Pinon D, Holicky E, Hadac E, Larkin JM, Miller LJ (1995) Dual pathways of internalization of the cholecystokinin receptor. J Cell Biol 128:1029–1041
Ross JS, Fletcher JA (1998) The HER-2/neu oncogene in breast cancer: prognostic factor, predictive factor, and target for therapy. Oncologist 3:237–252
Roy I, Ohulchanskyy T, Pudavar H, Bergey E, Oseroff A, Morgan J, Dougherty T, Prasad P (2003) Ceramic-based nanoparticles entrapping water-insoluble photosensitizing anticancer drugs: a novel drug-carrier system for photodynamic therapy. J Am Chem Soc 125:7860–7865
Ruoslahti E (2012) Peptides as targeting elements and tissue penetration devices for nanoparticles. Adv Mater 24:3747–3756

Santhakumaran LM, Thomas T, Thomas TJ (2004) Enhanced cellular uptake of a triplex-forming oligonucleotide by nanoparticle formation in the presence of polypropylenimine dendrimers. Nucleic Acids Res 32:2102–2112

Sapra P, Moase EH, Ma J, Allen TM (2004) Improved therapeutic responses in a xenograft model of human B lymphoma (Namalwa) for liposomal vincristine versus liposomal doxorubicin targeted via anti-CD19 IgG2a Or Fab′ fragments. Clin Cancer Res 10:1100–1111

Sapra P, Tyagi P, Allen T (2005) Ligand-targeted liposomes for cancer treatment. Curr Drug Deliv 2:369–381

Sarntinoranont M, Rooney F, Ferrari M (2003) Interstitial stress and fluid pressure within a growing tumor. Ann Biomed Eng 31:327–335

Schally AV (1988) Oncological applications of somatostatin analogues. Cancer Res 48:6977–6985

Schally A, Nagy A (1999) Cancer chemotherapy based on targeting of cytotoxic peptide conjugates to their receptors on tumors. Eur J Endocrinol 141:1–14

Schütz CA, Juillerat-Jeanneret L, Mueller H, Lynch I, Riediker M (2013) Therapeutic nanoparticles in clinics and under clinical evaluation. Nanomedicine 8:449–467

Shiau AK, Barstad D, Loria PM, Cheng L, Kushner PJ, Agard DA, Greene GL (1998) The structural basis of estrogen receptor/coactivator recognition and the antagonism of this interaction by tamoxifen. Cell 95:927–937

Shmeeda H, Mak L, Tzemach D, Astrahan P, Tarshish M, Gabizon A (2006) Intracellular uptake and intracavitary targeting of folate-conjugated liposomes in a mouse lymphoma model with up-regulated folate receptors. Mol Cancer Ther 5:818–824

Sikora K (2002) The impact of future technology on cancer care. Clin Med 2:560–568

Singh M (1999) Transferrin as a targeting ligand for liposomes and anticancer drugs. Curr Pharm Des 5:443–451

Slamon DJ, Leyland-Jones B, Shak S, Fuchs H, Paton V, Bajamonde A, Fleming T, Eiermann W, Wolter J, Pegram M, Baselga J, Norton L (2001) Use of chemotherapy plus a monoclonal antibody against HER2 for metastatic breast cancer that overexpresses HER2. New Engl J Med 344:783–792

Speth PAJ, Hoesel QGCM, Haanen C (1988) Clinical pharmacokinetics of doxorubicin. Clin Pharmacokinet 15:15–31

Steinhauser I, Spänkuch B, Strebhardt K, Langer K (2006) Trastuzumab-modified nanoparticles: optimisation of preparation and uptake in cancer cells. Biomaterials 27:4975–4983

Steinman RM, Mellman IS, Muller WA, Cohn ZA (1983) Endocytosis and the recycling of plasma membrane. J Cell Biol 96:1–27

Storm G, Belliot SO, Daemen T, Lasic DD (1995) Surface modification of nanoparticles to oppose uptake by the mononuclear phagocyte system. Adv Drug Deliv Rev 17:31–48

Sudimack J, Lee RJ (2000) Targeted drug delivery via the folate receptor. Adv Drug Deliv Rev 41:147–162

Suri S, Fenniri H, Singh B (2007) Nanotechnology-based drug delivery systems. J Occup Med Toxicol 2:16

Tannock IF (1998) Conventional cancer therapy: promise broken or promise delayed? Lancet 351(suppl 2):9–16

Torchilin VP (2005) Recent advances with liposomes as pharmaceutical carriers. Nat Rev Drug Discov 4:145–160

Turek JJ, Leamon CP, Low PS (1993) Endocytosis of folate-protein conjugates: ultrastructural localization in KB cells. J Cell Sci 106:423–430

Uner M, Yener G (2007) Importance of solid lipid nanoparticles (SLN) in various administration routes and future perspectives. Int J Nanomedicine 2:289–300

Verdun C, Brasseur F, Vranckx H, Couvreur P, Roland M (1990) Tissue distribution of doxorubicin associated with polyisohexylcyanoacrylate nanoparticles. Cancer Chemother Pharmacol 26:13–18

Veronese FM, Pasut G (2005) Pegylation, successful approach to drug delivery. Drug Discov Today 10:1451–1458

Waldherr C, Pless M, Maecke HR, Haldemann A, Mueller Brand J (2001) The clinical value of [90Y-DOTA]-D-Phe1-Tyr3-octreotide (90Y-DOTATOC) in the treatment of neuroendocrine tumours: a clinical phase II study. Ann Oncol 12:941–945

Waldherr C, Pless M, Maecke H, Schumacher T, Crazzolara A, Nitzsche E, Haldemann A, Mueller Brand J (2002) Tumor response and clinical benefit in neuroendocrine tumors after 7.4 GBq (90)Y-DOTATOC. J Nucl Med 43:610–616

Wang YM, Sato H, Adachi I, Horikoshi I (1996) Preparation and characterization of poly(lactic-co-glycolic acid) microspheres for targeted delivery of a novel anticancer agent, taxol. Chem Pharm Bull 44:1935–1940

Wang AZ, Langer R, Farokhzad OC (2012) Nanoparticle delivery of cancer drugs. Annu Rev Med 63:185–198

Warenius HM, Galfre G, Bleehen NM, Milstein C (1981) Attempted targeting of a monoclonal antibody in a human tumour xenograft system. Eur J Cancer Clin Oncol 17:1009–1015

Weiner LM, Adams GP (2000) New approaches to antibody therapy. Oncogene 19:6144–6151

Whitesides GM (2003) The 'right' size in nanobiotechnology. Nat Biotechnol 21:1161–1165

Wu J, Liu Q, Lee RJ (2006) A folate receptor-targeted liposomal formulation for paclitaxel. Int J Pharm 316:148–153

Wu J, Lu Y, Lee A, Pan X, Yang X, Zhao X, Lee R (2007) Reversal of multidrug resistance by transferrin-conjugated liposomes co-encapsulating doxorubicin and verapamil. J Pharm Pharm Sci 10:350–357

Xie Y, Bagby TR, Cohen M, Forrest ML (2009) Drug delivery to the lymphatic system: importance in future cancer diagnosis and therapies. Expert Opin Drug Deliv 6:785–792

Yan F, Kopelman R (2003) The embedding of meta-tetra(hydroxyphenyl)-chlorin into silica nanoparticle platforms for photodynamic therapy and their singlet oxygen production and pH-dependent optical properties. Photochem Photobiol 78:587–591

Zhang Y, Zhang J (2005) Surface modification of monodisperse magnetite nanoparticles for improved intracellular uptake to breast cancer cells. J Colloid Interface Sci 283:352–357

Zhang L, Gu FX, Chan JM, Wang AZ, Langer RS, Farokhzad OC (2007) Nanoparticles in medicine: therapeutic applications and developments. Clin Pharmacol Ther 83:761–769

Index

N. Maulik and T. Karagiannis (eds.), *Molecular Mechanisms and Physiology of Disease: Implications for Epigenetics and Health*, DOI 10.1007/978-1-4939-0706-9,

MIX
Papier aus verantwortungsvollen Quellen
Paper from responsible sources
FSC® C105338

If you have any concerns about our products,
you can contact us on
ProductSafety@springernature.com

In case Publisher is established outside the EU,
the EU authorized representative is:
Springer Nature Customer Service Center GmbH
Europaplatz 3, 69115 Heidelberg, Germany

Printed by Libri Plureos GmbH
in Hamburg, Germany